Lecture Notes of the Institute for Computer Sciences, Social Informatics and Telecommunications Engineering 410

More information about this series at https://link.springer.com/bookseries/8197

Qing Guo · Weixiao Meng ·
Min Jia · Xue Wang (Eds.)

Wireless and Satellite Systems

12th EAI International Conference, WiSATS 2021
Virtual Event, China, July 31 – August 2, 2021
Proceedings

Editors
Qing Guo
Harbin Institute of Technology
Harbin, China

Min Jia
Harbin Institute of Technology
Harbin, China

Weixiao Meng
Harbin Institute of Technology
Harbin, China

Xue Wang
Harbin University of Science
and Technology
Harbin, China

ISSN 1867-8211 ISSN 1867-822X (electronic)
Lecture Notes of the Institute for Computer Sciences, Social Informatics
and Telecommunications Engineering
ISBN 978-3-030-93397-5 ISBN 978-3-030-93398-2 (eBook)
https://doi.org/10.1007/978-3-030-93398-2

This Springer imprint is published by the registered company Springer Nature Switzerland AG
The registered company address is: Gewerbestrasse 11, 6330 Cham, Switzerland

Preface

We are delighted to introduce the proceedings of the 12th edition of the European Alliance for Innovation (EAI) International Conference on International Conference on Wireless and Satellite Systems (WiSATS 2021). This conference brought together researchers, developers, and practitioners around the world who are leveraging and developing wireless and satellite technology for a smarter global communication architecture. The theme of WiSATS 2021 was "Wireless and Satellite Services for Personal Communications, Multimedia, and Location Identification".

The technical program of WiSATS 2021 consisted of 79 full papers. The conference main track was organized into three sessions. Aside from the high-quality technical paper presentations, the technical program also featured a keynote speech and a technical workshop. The keynote speaker was Yue Gao from University of Surrey, UK. The workshop was on Integrated Space and Onboard Networks (ISON), which aimed to gain insights into current research and the future development of integrated onboard and space networks.

Coordination with the steering chairs, in particular Imrich Chlamtac, was essential for the success of the conference. We sincerely appreciate their constant support and guidance. It was also a great pleasure to work with such an excellent organizing committee team for their hard work in organizing and supporting the conference. We are grateful to the Technical Program Committee who have completed the peer-review process for technical papers and helped to put together a high-quality technical program. We are also grateful to Conference Manager, Aleksandra Sledziejowska for her support and all the authors who submitted their papers to the WiSATS 2021 conference and workshops.

We strongly believe that the WiSATS conference provides a good forum for all researchers, developers, and practitioners to discuss all science and technology aspects that are relevant to wireless and satellite services. We also expect that the future WiSATS conferences will be as successful and stimulating as this year's, as indicated by the contributions presented in this volume.

Qing Guo
Weixiao Meng
Min Jia
Youping Zhao
Feifei Gao
Chunxiao Jiang
Nan Cheng
Xin Liu
Lidong Zhu
Min Lin
Shushi Gu

Organization

Steering Committee

Imrich Chlamtac	University of Trento, Italy
Weixiao Meng	Harbin Institute of Technology, China
Min Jia	Harbin Institute of Technology, China

Organizing Committee

International Advisory Committee

Tomaso De Cola	German Aerospace Center (DLR), Germany
Hsiao-Hwa Chen	National Cheng Kung University, Taiwan
Chenhua Sun	The 54th Research Institute of China Electronics Technology Group Corporation, China
Qihui Wu	Nanjing University Aeronautics and Astronautics, China
Linling Kuang	Tsinghua University, China
Min Sheng	Xidian University, China
Qinyu Zhang	Harbin Institute of Technology, China
Gengxin Zhang	Nanjing University of Posts and Telecommunications, China
Zan Li	Xidian University, China
Mugen Peng	Beijing University of Posts and Telecommunications, China
Shi Jin	Southeast University, China
Xuemai Gu	Harbin Institute of Technology, China

General Co-chairs

Qing Guo	Harbin Institute of Technology, China
Weixiao Meng	Harbin Institute of Technology, China
Min Jia	Harbin Institute of Technology, China

Technical Program Committee Chairs

Youping Zhao	Beijing Jiaotong University, China
Feifei Gao	Tsinghua University, China
Chunxiao Jiang	Tsinghua University, China
Nan Cheng	Xidian University, China
Xin Liu	Dalian University of Technology, China
Lidong Zhu	University of Electronic Science and Technology of China, China

Min Lin	Nanjing University of Posts and Telecommunications, China
Shushi Gu	Harbin Institute of Technology, China

Sponsorship and Exhibit Chair

Dongbo Li	Harbin Institute of Technology, China

Local Chair

Shaochuan Wu	Harbin Institute of Technology, China

Workshops Chair

Hongbin Chen	Guilin University of Electronic Technology, China

Publicity and Social Media Chair

Wei Wu	Harbin Institute of Technology, China

Publications Co-chairs

Haibo Zhou	Nanjing University, China
Xue Wang	Harbin University of Science and Technology, China

Web Chair

Xuanli Wu	Harbin Institute of Technology, China

Industry Panels Co-chairs

Zhisong Hao	The 54th Research Institute of China Electronics Technology Group Corporation, China
Xiongwen He	Beijing Institute of Spacecraft System Engineering, China

Tutorials Chair

Xuejun Sha	Harbin Institute of Technology, China

Technical Program Committee

Youping Zhao	Beijing Jiaotong University, China
Feifei Gao	Tsinghua University, China
Chunxiao Jiang	Tsinghua University, China
Nan Cheng	Xidian University, China
Xin Liu	Dalian University of Technology, China
Lidong Zhu	University of Electronic Science and Technology of China, China

Min Lin	Nanjing University of Posts and Telecommunications, China
Shushi Gu	Harbin Institute of Technology, China
Liu Chungang	Hebei Normal University, China
Tong Liu	Harbin Engineering University, China
Zhang Jiayan	Harbin Institute of Technology, China
Xinlin Huang	Tongji University, China
Bo Li	Harbin Institute of Technology at Weihai, China
Shaochuan Wu	Harbin Institute of Technology, China
Xiaojin Ding	Nanjing University of Posts and Telecommunications, China
Yong Wang	Chongqing University of Posts and Telecommunications, China
Xujie Li	Hohai University, China

Contents

Space Information Networks and Spacecraft System

Ad Hoc Networks; Optical Communications and Networks

Satellite Network Transmission

Inter Satellite Link Interference Detection and Analysis of NGSO Satellite System

Shiyao Meng(✉), Min Jia, Qing Guo, and Xuemai Gu

School of Electronics and Information Engineering, Harbin Institute of Technology, Harbin, Heilongjiang, China
jiamin@hit.edu.cn

Abstract. In view of the fact that non geostationary satellite constellation is becoming more and more complex and it is difficult to calculate and observe directly, this paper mainly studies the inter satellite interference of NGSO constellation. Firstly, the distribution of NGSO and various interference scenarios are comprehensively analyzed to enrich and improve the existing inter constellation interference scenarios of satellite system, and the direct interference between constellations in the same system is analyzed. Then, the data of Leo electromagnetic satellite are analyzed, and the inter satellite link interference of NGSO constellation system is calculated. Using the concept of relative interference time, the time distribution characteristics of interference are visualized, and the time characteristics of interference in different scenes are analyzed. The inter satellite link interference of large-scale NGSO satellite constellation system is analyzed from multi-dimensional perspective, and the simulation is carried out. The implementation results of the strategy prove that there is interference in the inter satellite links of large NGSO satellite constellation system. The corresponding signal analysis strategy is designed, and the data processing method of interference data is proposed.

Keywords: NGSO · Inter satellite link · *EIRP* · Electromagnetic situation

1 Introduction

1.1 Development of NGSO Constellation Status of NGSO Constellation

The Internet of things is the key direction and an important part of the current development of information technology. The current ground Internet of things is facing a variety of difficulties, including the impact of weather, capacity, network resources and so on. Therefore, people begin to expand the current ground Internet of things, and open up its re air field, in which the largest main battlefield is the space information network. The space information network integrates geosynchronous orbit satellites, medium orbit satellites and low orbit satellites to process the network information in space [1, 2].

Among them, NGSO satellite communication system has small time delay, strong anti attenuation ability of signal, and can achieve global seamless satellite coverage by improving the coverage ability of the system.When the electromagnetic data of NGSO satellite is modeled, it is necessary to mine the electromagnetic data of NGSO satellite.

Q. Guo et al. (Eds.): WiSATS 2021, LNICST 410, pp. 3–13, 2022.
https://doi.org/10.1007/978-3-030-93398-2_1

So far, the number of NGSO satellite systems has been increasing, and the latest SpaceX project will eventually reach more than 4000 NGSO constellations [3–5].

1.2 Research of Interference Analysis for NGSO Constellation

There are all kinds of interference in satellite communication. The construction of electromagnetic interference model for different scenarios has always been a key research direction of the majority of scholars.

Satellite communication links are divided into uplink and downlink. Some scholars have studied the coordination method of uplink and downlink interference in multi beam satellite communication system and the co channel interference cancellation technology [6].

However, there is still a lack of inter satellite link interference for NGSO satellites, which can not accurately reflect the different gain of different angles and the continuous interference changes at different times [7, 8].

So the main purpose of this paper is to complete the inter satellite link interference of NGSO constellation. According to ITU report, if the inter satellite links of two NGSO constellations use the same fixed service frequency band of satellites and the inter satellite links increase continuously, it may cause the interference of inter satellite links to affect the system. At present, the main NGSO constellation is basically used Ka and Ku band, and the service frequency bands used by each system may produce coincidence. In addition, NGSO constellation, such as Starlink, has launched two groups of more than 1000 satellites, and there are still a third batch of launch plans. Moreover, there are inter satellite links in Starlink system, which will lead to the future low orbit satellite system more complex and need to consider inter satellite interference [9–11].

2 Construction of Inter Satellite Link Interference Model

2.1 NGSO Constellation Construction

Several common NGSO satellite constellations are simulated: Oneweb, Starlink, to obtain satellite orbit related parameters, the electromagnetic interference model is used to calculate several electromagnetic parameters including antenna gain and *EIRP*, and the *EIRP* of uplink ground station transmitter and downlink satellite transmitter in inter satellite interference is calculated [8, 11]. *EIRP* is the effective isotropic radiated power. In satellite communication, the radiation power of satellite or ground station in a specified direction is equal to the transmitting power of power amplifier multiplied by the gain of antenna in ideal state, which is used to measure the intensity of interference and the ability of transmitter to transmit strong signal [9, 12].

The position of the ground station (72.826 ° e, 0) is set. In the constellation of simulation, a period of time when the satellite passes the top is selected for simulation, and the step is set to 1s. The azimuth and distance data of the satellite to the ground station and the ground station to another satellite are obtained. The constellation satellite transmission power, ground station transmitter transmission power, satellite communication frequency and other data are obtained by consulting literature.

2.2 System Model

Interference Scene Analysis

In NGSO satellite communication, an important situation that may produce interference is the interference scene of inter satellite link. A satellite in NGSO constellation will be subject to a variety of potential interference scenarios. It is mainly divided into uplink and downlink scenarios. For the communication between constellation A and earth station A in Fig. 1, the interference in the uplink is mainly the interference from satellites of other constellations and other possible earth stations. The main interference in the downlink is divided into the interference of other constellation to the earth station and the possible interference of GSO constellation to the earth station.

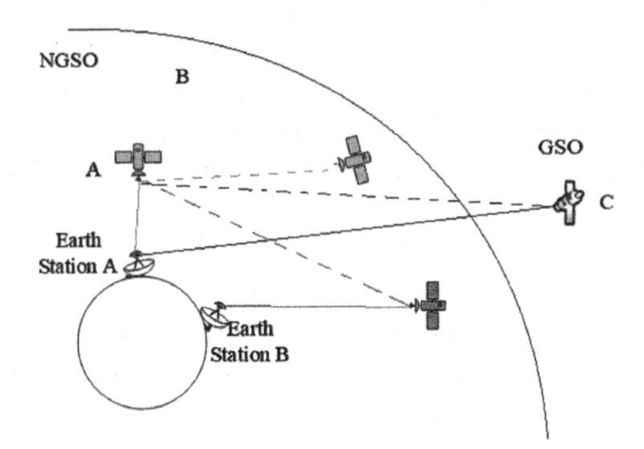

Fig. 1. Schematic diagram of electromagnetic spatial distribution and interference complexity

In the discussion of inter satellite interference scenarios, the adjacent satellite interference scenarios are generally considered. In this problem, we mainly consider the interference of the uplink channel in the satellite adjacent system. In other words, the actual useful signal in the process of satellite communication is the useful signal strength sent by the ground station, while the interference is caused by the information sent by other constellations.

However, in some scenarios, when the local satellite a exchanges information with the local earth station, the output information of the neighboring satellite B may also be transmitted to the receiver of the local satellite a, causing direct interference between the two stars. It can be seen that this situation will occur in the uplink state of the main link.

In analyzing the interference in the uplink of non-geostationary orbit constellations, the overall model is shown in Fig. 1. In complex interference analysis, geostationary orbit satellites and non-geostationary orbit satellites need to be considered. When examining the uplink of non-geostationary satellites, The main link of communication is the link transmitted to the constellation by the Earth station of a non-geostationary orbit satellite, Among them, the most important interference scenario is the useless signals emitted by other earth stations. In addition, it is necessary to consider the interference of signals emitted by other non-geostationary satellites and the interference caused by some geostationary satellites.

Interference Calculation

No matter what kind of inter satellite interference is caused, the same data processing method can be used to quantify the interference and know whether it is necessary to adjust the system, such as adaptive power control. Then we need to calculate the best sampling time of the system, which can reduce the amount of data calculation as much as possible on the premise of correctly judging the interference between the systems. The following formula is needed to calculate the optimal sampling time:

$$\Delta t_{step-down} = \frac{\Delta t_{down}}{N_{step-down}} \tag{1}$$

In the calculation, we need to obtain the selected constellation orbit height and satellite angular velocity data.

In the above formula, $\Delta t_{step-down}$ means the step size of the downlink time of the system, that is the best sampling time interval of the system, Δt_{down} is the time required to interfere with the main lobe radiation range of the antenna of the NGSO satellite passing through the earth station, which will be affected by the antenna type and amplification capacity of the system, as well as the orbit height of the satellite. $N_{step-down}$ indicates the sampling times of the main lobe radiation area of the receiving antenna in the ground station of the disturbed system. It is affected by the main lobe width and other factors. It is the ratio of the 3 dB width to the off-axis angle sampling interval of the main lobe of the receiving antenna in the earth station of the disturbed NGSO system. Its value can also be replaced by a value greater than or equal to $\sqrt{\frac{24}{\Delta R}}$, where ΔR is the resolution dI of the interference signal power I received by the NGSO disturbed earth station. According to experience, its value can be set as 0.5dB.

The value of can be determined by the ratio of the angle of the main lobe radiation area of the satellite receiving antenna through the earth station to the angular velocity of the satellite in the circular orbit. In this scene, the final time step is 6 s according to the above method.

Generally, according to the regulations of the International Telecommunication Union, as long as the influence of the interference signal from the adjacent satellite on the equivalent noise of the receiving system of the satellite reaches 6%, it is generally recognized that there is interference in the adjacent satellite system and needs to be adjusted. Therefore, 6% is regarded as the threshold. If the relative increment of equivalent noise temperature can be expressed as $\frac{\Delta T}{T}$, it can be expressed as:

$$\frac{\Delta T}{T} = 6\% \tag{2}$$

The conversion to DB is 12.2 db.

When considering inter satellite interference, power is chosen to judge. In the calculation process, the data processing of load-to-dry ratio is needed. So the relationship between the load-to-dry ratio of the final transmission and the carrier to dry ratio transmitted by the inter satellite link itself is as follows:

$$\frac{C}{I_{th}} = \frac{C}{N_{th}} + 12.2\ \text{dB} \tag{3}$$

The downlink carrier to interference ratio is:

$$\frac{C}{I_{dn}} = C_{dn} - I_{dn} = \frac{C}{T_{dn}} - \frac{I}{T_{dn}} \tag{4}$$

When considering the power spectral density to analyze the above, the above formula can be changed into:

$$\frac{C}{I_{dn}} = (\text{EIRPD} + G_{er} - L_d) - (\text{EIRPD}' + G'_{er} - L'_d) \tag{5}$$

When the power spectral density is considered to analyze the above, in the calculation of the uplink load to dry ratio, the carrier power of the antenna output end of the earth station at the receiving end can be changed into:

$$C_{dn} = EIRP_s - L_d + G_{er} \tag{6}$$

$EIRP_s$ is $EIRP$ value of carrier satellite, L_d is loss value of downlink, G_{er} is the receiving gain of ground station.

When considering the system margin, M can get the formula according to the threshold calculated before:

$$M = \frac{C}{I} - (\frac{C}{N_{th}} + 15.2) \tag{7}$$

According to the radio rules, when judging whether there is interference, the system margin can be used. When the system margin is greater than 0, the interference of the system is not serious and will not affect the channel of the main link. When it is less than 0, the interference will affect the communication of the main link, so the system needs to be operated, such as adaptive power control, to improve the channel environment.

For the general system, the frequency band overlap between uplink and downlink, so in order to meet the above requirements, the threshold will be increased by 3 dB. Therefore, the downlink carrier to interference ratio threshold will be 15.2 db higher than the system's own carrier to noise ratio, and the carrier to noise ratio of the system can be obtained as follows:

$$\frac{C}{N_{th}} = \frac{E_b}{N_0} + R_s - BW_0 + M(dB) \tag{8}$$

Table 1. Inter satellite link interference analysis algorithm.

Computational Procedure
1. Obtain the inter satellite data and calculate the data acquisition step size $\Delta t_{step-down}$.
2. Calculated the inter satellite link angle to obtain the antenna amplification G_{er}.
3. Calculate the signal-to-noise ratio of the originator $\frac{C}{I_{th}}$.
4. Calculate the interference received by the system L_d.
5. Calculate the carrier to noise ratio received by the receiver C_{dn}.
6. Comparison with the system decision value M specified by ITU.

The specific algorithm is shown in Table 1. In this way, the relative increment of equivalent noise temperature can be calculated according to the satellite constellation parameters, and it can be judged according to ITU-R S.324 [13] whether the two systems need to coordinate.

2.3 Analysis of Interference Problem

There are many kinds of interference and environmental factors in satellite communication system. When considering the interference between satellites, the free space transmission model is generally considered, and the atmospheric loss is also considered.

Free Space Loss

Free space is an ideal model for long-distance transmission of radio waves. Suppose that the transmitting power of a directional antenna is P_T, the power gain of the transmitting end is G_T, the opening area of the receiving end is A_R, and the power gain of the receiving end is G_R. The free space transmission loss is defined as L_f as follows:

$$Lf = \left(\frac{4\pi d}{\lambda}\right)^2 = \left(\frac{4\pi df}{c}\right)^2 \tag{9}$$

When the free space transmission loss is converted to dB, it is:

$$[Lf] = 92.44 + 20\lg d(km) + 20\lg f(GHz) \tag{10}$$

Then the power P_R to the receiving end is as follows:

$$PR = \frac{P_T G_T A_R \eta}{4\pi d^2} = P_T G_T G_R \left(\frac{\lambda}{4\pi d}\right)^2 = \frac{P_T G_T G_R}{L_f} \tag{11}$$

Atmospheric Loss

In addition, the attenuation of electromagnetic wave transmission in clear weather is mainly due to the absorption of water vapor and oxygen in the lower atmosphere, and the density of water vapor and oxygen is uneven, which decreases with the increase of altitude. In addition, along the ray path, the temperature of atmospheric environment decreases with the increase of height, so some environmental models require temperature correction.

Rain Loss

The influence of rain on communication system is considered. According to the calculation method of rain attenuation specified by ITU-R P.681-1 [14], it can be concluded that the rain attenuation of space is related to transmission distance and communication frequency. In the actual communication system, the communication spectrum may not be fixed, but the fixed transmission distance should be. At the same transmission distance, the larger the communication frequency, the greater the free space loss and atmospheric attenuation loss.

3 Simulation Setup133

Through the real-time data of NGSO satellite system obtained by communication system simulation, the angle relationship, distance relationship and current time of two independent satellites in the system at a certain time can be obtained. Because at some time, there may be no signal transmission between the two satellites, there is no interconnection, so the data will not be obtained at this time. In the final data processing, the interference value is shown as Nan, that is, it does not exist.

After that, we need to classify each group of data, extract the data according to the interference situation, then select the appropriate system parameters for various interference situations, and design and calculate the relationship between inter satellite interference and its influencing factors.

Considering different interference scenarios, cosine theorem is used to calculate various angles to obtain antenna gain of different angles, and finally interference situation is obtained, as shown in Fig. 2 and Fig. 3.

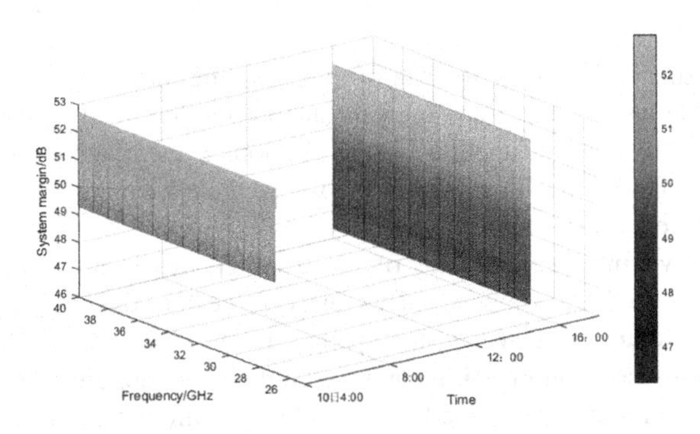

Fig. 2. Inter satellite interference of Starlink system to OneWeb system.

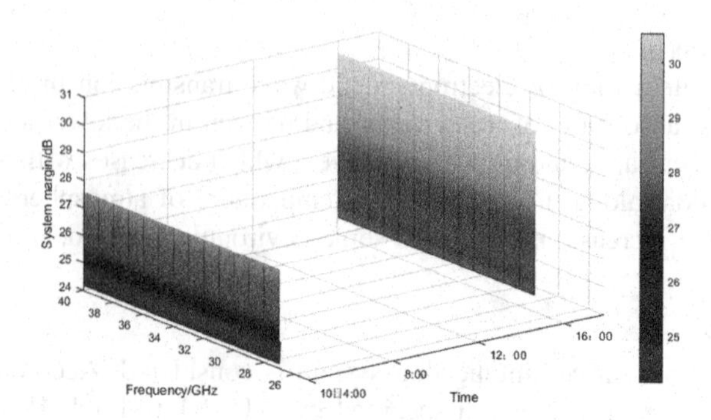

Fig. 3. Inter satellite interference of OneWeb system to system Starlink.

According to the simulation, the judgment value of interference between Starlink, and Oneweb on the same layer is always greater than 0, so it can be seen that it will not interfere with the communication of the current link. In the case of Starlink on the same layer, it is within 3 km, so the interference situation is basically unchanged, and the decision value is about 6 dB. The analysis of the judgment value shows that the interference of Oneweb to Starlink system is slightly larger than that of Starlink to Oneweb system. When the communication frequency is about 20 GHz, the rainfall attenuation is about 5 dB under the condition of light rain with daily rainfall of about 5 to 10 mm. In Starlink system, the inter satellite interference may be caused at some time under the condition of rainfall, which may cause interference to NGSO communication system, but it is relatively small.

The multi-dimensional analysis of interference and time and frequency shows that in the frequency range from 26 GHz to 40 GHz, in another word, in the range of Ku, with the increase of frequency and free space loss, the final SINR of the system will decrease and the possibility of interference will increase. As for the relationship between interference and time, with the change of time, the angle and distance between satellites will also change, and the final interference decision will be as follows.As is shown in the previous image, this is the interference feature of inter satellite data. When considering the time distribution characteristics of inter satellite interference, a new reference standard is introduced, which is defined as the relative interference time, that is, the ratio of the time that a system is affected by interference in one day to the total operation time of the system.

When analyzing and mining the data of inter satellite interference between NGSO satellite constellations, an important analysis standard is the relative interference time of the system. In the adjacent satellite system, the longer the relative interference time is, the longer the interference time is, and the worse the environment is, so the system needs to be adjusted.

For the relative interference time of electromagnetic data processing, Starlink constellation 1007 to 1027 are selected to analyze the interference of Oneweb constellation 0007 in one day. In the test frequency range of 1 to 24 MHz, the relative interference time is shown in Fig. 4. It can be seen that there are great differences in the relative time of communication interference between different satellites on this day.

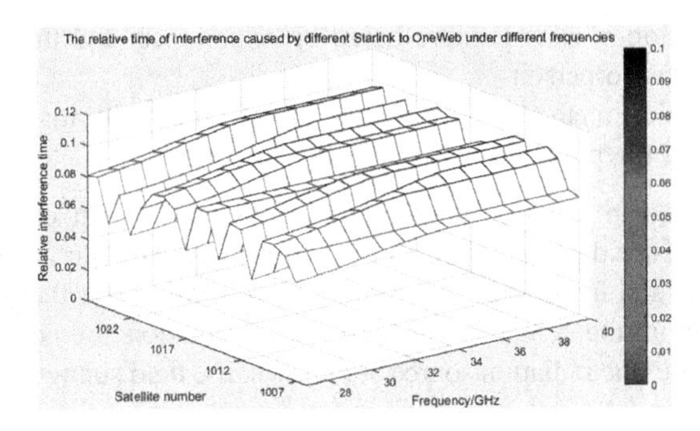

Fig. 4. Relative interference time of Starlink to OneWeb in one day under different frequencies.

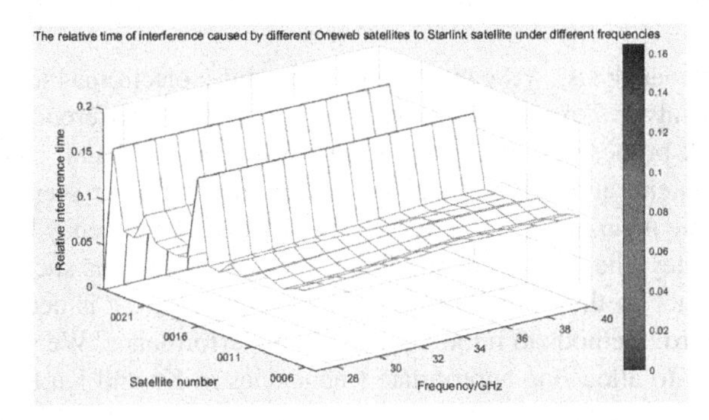

Fig. 5. Relative interference time of OneWeb to Starlink in one day under different frequencies.

Among these 20 satellites, the average time of communication interference between different satellites is 1 The time of communication interference between 1016 and 1017 satellites is about 2 h, and the time of interference between 1012, 1018, 1022 and 1024 satellites is the smallest, about 50 min. In addition, it can be seen that the greater the frequency of communication, the more time of system interference.

Accordingly, the interference of OneWeb on Starlink is shown in Fig. 5.

For the interference analysis and data processing of the system, in order to obtain the characteristics of the interference data, it is necessary to process the data according to the following steps:

(1) Obtain the real jamming situation of the jammed area, including its time, location, respective emitter parameters and other data.
(2) The jamming data are classified according to different jamming sources.
(3) According to the interference data, an appropriate interference data analysis model is established, and the prediction data are obtained by using the model, and then the model parameters are optimized by comparing with the actual data.

When building prediction model based on actual data, Markov prediction method is usually used to analyze and predict situation information according to probability transfer matrix. However, Markov model has the following disadvantages:

(1) The state partition of data is more based on experience, and the partition results vary from person to person.
(2) Markov prediction method uses the principle of probability maximization to calculate, ignoring the possible low probability information.

In addition, we can use the interference analysis algorithm designed before to analyze the interference data in the region by combining the emitter location algorithm, the source division and the desired emission data. Assuming that there are k radiation sources in the region, the center frequency of each radiation source and the distance from the grid point to the radiation source are known, the field strength of the grid point can be obtained.

4 Conclusion

According to the analysis, we can see that the space electromagnetic interference phenomenon has always existed and is very complex. The interference caused by the inter satellite link proposed in this paper exists in Ka and Ku band, and will have an impact on the system, and will be more serious in rain and other scenarios.

In addition, the main trend of interference in Ka and Ku frequency bands is that the higher the frequency, the greater the interference. Under conditions such as rain, it may have a great impact on the communication system. Therefore, it is necessary to adopt some power control methods to improve the system performance. We should also give full consideration to allocating appropriate frequencies in Ka and Ku frequency bands to reduce system interference.

References

1. Han, H.: Analysis of LEO communication constellation inter satellite links. Satellites Netw. (08), 40–42 (2018)
2. Dong, S., Yao, X., Gao, X., Han, Z., Yan, Y., Sun, Y.: Evaluation method of communication interference in GSO satellite system deployment. J. Beijing Univ. Aeronaut. Astronaut. **46** (11), 2184–2194 (2020)
3. Liu, J., Shi, Y., Fadlullah, Z.M., Kato, N.: Space-air-ground integrated network: a survey. IEEE Commun. Surv. Tutor. **20**(4), 2714–2741 (2018)
4. Wang, Y., Ding, X., Zhang, G.: A novel dynamic spectrum-sharing method for GEO and LEO satellite networks. IEEE Access **8**, 147895–147906 (2020)
5. Han, Q., Nie, J., Liu, W., Wang, F.: Interference intensity and feasibility analysis of Ka band inter satellite link. J. Cent. South Univ. (Nat. Sci. Edn.) **45**(03), 769–773 (2014)
6. Xie, X., Yan, S., Chen, J.: Anti jamming performance analysis of inter satellite link. Aerosp. Meas. Technol. **31**(06), 53–56 (2011)
7. Author, F.: Contribution title. In: 9th International Proceedings on Proceedings, pp. 1–2. Publisher, Location (2010)
8. Dai, J., Feng, X.: Simulation and analysis of spectrum interference and its avoidance in LEO satellite system. Inf. Technol. (02), 79–84 + 91 (2021)

9. Xi, C., Deng, Y., Yang, B., Wen, X., Cai, J., Yuan, Y.: Research on electromagnetic emission visualization technology of LEO constellation system. Electron. Test (09), 49–51 (20200)
10. del Portillo, I., Cameron, B.G., Crawley, E.F.: A technical comparison of three low earth orbit satellite constellation systems to provide global broadband. Acta Astronaut. **159**, 123–135 (2019)
11. Chen, Z., Jin, J., Hao, Z., Ting, L.: Spectral coexistence between LEO and GEO satellites by optimizing direction normal of phased array antennas. China Commun. **15**(06), 18–27 (2018)
12. Xiao, Q., Geng, X., Chen, J., Meng, L., Li, N., Zhang, Y.: Study on the calibration method of interference magnetic field of Leo magnetic survey satellite. Acta Geophys. Sinica **61**(08), 3134–3138 (2018)
13. ITU-R. Analytical method for estimating interference between non-geo stationary mobile-satellite feeder links and geostationary fixed-satellite networks operating co-frequency and codirectionally: ITU-R S.1324 [S]. ITU, Geneva (1997)
14. ITU-R P.681-11 Propagation data and prediction methods required for the design of telecommunication systems. ITU-R P Serise Recommendations Radio wave Propagation, Gevena (2007)

A Fast Acquisition Algorithm Based on High Sampling Rate FFT for LEO Satellite Signals

Deyang Kong(✉), Zhisong Hao, and Yongfei Hou

The 54th Research Institute of China Electronics Technology Group Corporation, 589 Zhongshan West Road, Shijiazhuang, China

Abstract. In the communication of LEO satellite system, there is a high relative velocity and acceleration between the receiver and the transmitter due to the high-speed motion of the platform. This leads to doppler frequency offset in the communication link. In large frequency offset and high dynamic communication environment, signal frame capture is the key technology to realize signal demodulation and link establishment. The acquisition bandwidth of low symbol rate signal is very small, which leads to the long acquisition time of frequency sweeping method. In this paper, the fast acquisition problem of low-speed FDMA signal with large frequency offset and high dynamic is solved by using high sampling rate partially matched filter FFT algorithm. The algorithm can realize the frequency estimation and fast acquisition of the signal when the frequency offset is 4 times the symbol rate. Compared with the traditional method, the algorithm greatly improves the acquisition bandwidth and speed.

Keywords: LEO satellite communication · Fast signal acquisition · Frequency offset estimation

1 Introduction

In recent years, with the application of LEO satellite constellation and various micro satellites, LEO satellite communication has been widely used [1]. However, compared with geostationary satellites, the high-speed motion of LEO satellite brings large Doppler frequency offset [2]. For example, when the symbol rate is more than 10 kbps, the Doppler frequency offset is as high as tens of kHz, which is far beyond the symbol rate, so it is difficult to capture the signal. This leads to a very bad communication environment for low-speed satellite signals. The low-speed FDMA signal of the geostationary orbit (GEO) is usually captured in a frequency sweep mode, and then the frequency offset is estimated through the pilot frequency [3]. However, the frequency offset estimation range of this method is half of the pilot rate, and the sweep step is related to the symbol rate and the length of the capture frame header, which results in a longer sweep time when the symbol rate is low [4]. This paper proposes a fast capture

Q. Guo et al. (Eds.): WiSATS 2021, LNICST 410, pp. 14–22, 2022.
https://doi.org/10.1007/978-3-030-93398-2_2

algorithm, which realizes the fast capture of the signal by means of frequency estimation and search on the frame header at different sampling rates. MATLAB simulation verifies that the algorithm can work normally when the signal noise ratio (SNR) is 3dB and the frequency offset range is 4 times the symbol rate.

2 Principle

This algorithm is mainly used for the fast acquisition of low symbol rate FDMA signals under large frequency offset. This algorithm estimates the frequency offset exceeding the symbol rate by performing frequency when the signal sampling rate is 8 times, then estimates the remaining frequency offset when the signal sampling rate is 1 times, and outputs the frame header position to achieve frame capture. The main process of this algorithm is shown in Fig. 1.

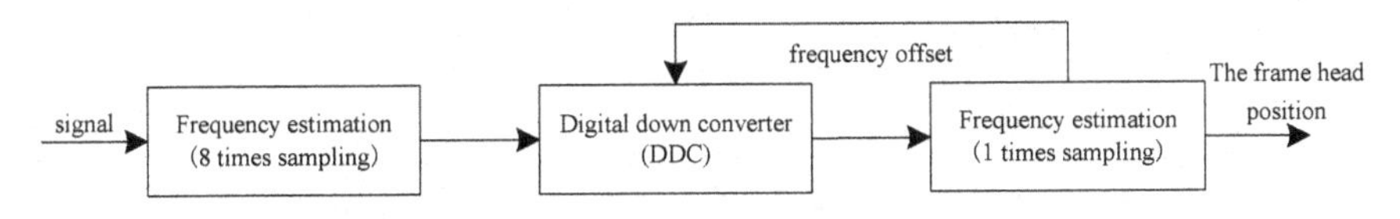

Fig. 1. The algorithm flow chart

The basic principles of frequency estimation are as follows.

Assuming that S(t) is the signal to enter Analog Devices (AD), then the S(t) can be expressed as

$$S(\mathrm{t}) = \mathrm{d(t)}e^{j(\omega_0 + f_0)t + \theta_0} + n(t) \tag{1}$$

In Eq. (1), ω_0 is the carrier frequency; f_0 is the frequency offset; θ_0 is the Signal phase; $n(t)$ is Gaussian white noise; d_t represents the transmitted data; Within one frame

$$\mathrm{d(t)} = \begin{cases} Pn_i(t) + j \cdot Pn_q(t) & 0 \leq \mathrm{t} \leq 64\mathrm{T}_s \\ d_i(t) + j \cdot d_q(t) & 64\mathrm{T}_s < \mathrm{t} \leq m\mathrm{T}_s \end{cases} \tag{2}$$

Ts is the symbol period, m is the number of symbols per frame. The format of the transmitted data frame is that the first 64 data is a fixed pseudo-random sequence, and the following is the data to transmit.

The signal S(n) sampled by AD can be expressed as

$$
\begin{aligned}
S(\mathrm{n}) &= \sum_{n=0}^{n=N} \delta(nT_c)\cdot\left[\mathrm{d(t)}\cdot \mathrm{e}^{j(\omega_0+f_0)t+\theta_0}+n(t)\right] \\
&= \begin{cases} \displaystyle\sum_{n=0}^{n=64*T_s/T_c} \delta(nT_c)\cdot\left[(Pn_i(t)+j\cdot Pn_q(t)) * \mathrm{e}^{j(\omega_0+f_0)t+\theta_0}+n(t)\right] \\ \displaystyle\sum_{n=64*T_s/T_c+1}^{n=mT_s/T_c} \delta(nT_c)\cdot\left[(d_i(t)+j\cdot d_q(t)) * \mathrm{e}^{j(\omega_0+f_0)t+\theta_0}+n(t)\right] \end{cases}
\end{aligned}
\tag{3}
$$

In Eq. (2), T_c is the sampling period and T_s/T_c is an integer; N is the number of sampling points. Assuming that the signal entering ad is a baseband signal, then the sampled signal is S(n).

$$
S(\mathrm{n})=\begin{cases} \displaystyle\sum_{n=0}^{n=64*T_s/T_c} \delta(nT_c)\cdot\left[(Pn_i(t)+j\cdot Pn_q(t)) * \mathrm{e}^{jf_0t+\theta_0}+n(t)\right] \\ \displaystyle\sum_{n=64*T_s/T_c+1}^{n=mT_s/T_c} \delta(nT_c)\cdot\left[(d_i(t)+j\cdot d_q(t)) * \mathrm{e}^{jf_0t+\theta_0}+n(t)\right] \end{cases}
\tag{4}
$$

$PN(\mathrm{n})$ is the local sequence generated by the PN sequence generator.

$$
PN(\mathrm{n})=\sum_{n=0}^{n=64T_s/T_c} \delta(nT_c)\cdot\left[Pn_i(t)+j\cdot Pn_q(t)\right]
\tag{5}
$$

$S_1(p)$ is the result of the conjugate multiplication of $S(\mathrm{n})$ and $PN(\mathrm{n})$.

$$
S_1(p)=\sum_{p=0}^{p=(m-64)\cdot T_s/T_c}\left(\sum_{n=p}^{n=(64T_s/T_c)+p} S(n)\cdot PN(n)\right)
\tag{6}
$$

When p is 0, that is, when the local code and the random sequence in the selected signal are exactly aligned, the S1 can be expressed as

$$
\begin{aligned}
S_1 &= \sum_{n=0}^{64*T_s/T_c} \delta(nT_c)\cdot\left\{\left[Pn_i(t)+j\cdot Pn_q(t)\right]\cdot e^{jf_0t+\theta_0}+n(t)\right\}\cdot PN(n) \\
&= \sum_{n=0}^{64*T_s/T_c} \delta(nT_c)\cdot\left\{\left[Pn_i(t)+j\cdot Pn_q(t)\right]\cdot e^{jf_0t+\theta_0}+n(t)\right\}\cdot\left[Pn_i(t)-j\cdot Pn_q(t)\right] \\
&= \sum_{n=0}^{64*T_s/T_c} 2\delta(nT_c)\left\{\left[1-j\cdot Pn_i(t)Pn_q(t)\right]\right\}\cdot e^{jf_0t+\theta_0}+n(t)\cdot\left[Pn_i(t)-j\cdot Pn_q(t)\right] \\
&= \sum_{n=0}^{64*T_s/T_c} 2\left\{\left[1-j\cdot Pn_i(nT_c)Pn_q(nT_c)\right]\right\}\cdot e^{jnf_0T_c+\theta_0}+n(nT_c)
\end{aligned}
\tag{7}
$$

The $64T_s/T_c$ points fast Fourier transform is S_1 [5]. Then S_1 can be expressed as

$$
\begin{aligned}
S_1(k) &= \sum_{n=0}^{N-1} S_1 e^{-i2\pi kn/N} \\
&= \sum_{n=0}^{N-1}\sum_{m=0}^{64*T_s/T_c} 2\left\{\left[1-j\cdot Pn_i(mT_c)Pn_q(mT_c)\right]\right\}\cdot e^{jmf_0T_c+\theta_0}\cdot e^{-i2\pi kn/N}
\end{aligned}
\tag{8}
$$

When $\sum_{m=0}^{64*T_s/T_c} Pn_i(mT_c)Pn_q(mT_c)\approx 0$, we can get

$$
S_1(k)=\sum_{m=0}^{64*T_s/T_c} 4\pi\cdot e^{\theta}\cdot\delta(\mathrm{k}-mf_0T_c) \tag{9}
$$

When $k=mf_0T_c$, we can get $\max(S_1(k))=4\pi e^{\theta}$. The frequency offset value in the signal can be calculated according to the position of the maximum value. Because of the high-speed movement of the platform where the transmitter is located, there is a Doppler shift in the received signal. Assuming that the range of frequency offset is $[-f_{\max}, f_{\max}]$; Doppler rate of change is a; Doppler model is cosine model; we can get

$$
f(t)=f_{\max}\cdot\cos(\omega t) \tag{10}
$$

Derivative of $f(t)$ is $f'(t)=f_{\max}\cdot\omega\sin(\omega t)$. When $t=0$, the change rate reaches the maximum.

$$
a=f_{\max}\cdot\omega \tag{11}
$$

So we can get

$$f(t) = f_{\max} \cdot \cos(\frac{at}{f_{\max}}) \tag{12}$$

The frequency offset of $64T_s/T_c$ signal sampling points is

$$\begin{aligned} f(n) &= \sum_{n=0}^{64*T_s/T_c} f(t) \cdot \delta(nT_c) \\ &= \sum_{n=0}^{64*T_s/T_c} f_{\max} \cos(\frac{at}{f_{\max}}) \cdot \delta(nT_c) \end{aligned} \tag{13}$$

When $a = 0$, the frequency offset of the intercepted signal is a fixed value. There is only one larger value in $S_1(k)$, and the remaining values are very small. When $a \neq 0$, the frequency component of the intercepted signal is not single, and there are many larger values of $S_1(k)$. Define λ

$$\lambda(m) = \begin{cases} 10 * \log_{10}(\frac{S_m(p) + S_m(p+1) + S_m(64 * T_s/T_c)}{\left[\sum_{n=1}^{64*T_s/T_c} S_m(n)\right] - S_m(1) - S_m(2) - S_m(64 * T_s/T_c)}) & p = 1 \\ 10 * \log_{10}(\frac{S_m(p) + S_m(p-1) + S_m(p+1)}{\left[\sum_{n=0}^{64*T_s/T_c} S_m(n)\right] - S_m(p) - S_m(p-1) - S_m(p+1)}) & 1 < p < 64T_s/T_c \\ 10 * \log_{10}(\frac{S_m(1) + S_m(p-1) + S_m(p)}{\left[\sum_{n=1}^{64*T_s/T_c} S_m(n)\right] - S_m(1) - S_m(p-1) - S_m(p)}) & p = 64T_s/T_c \end{cases} \tag{14}$$

In Eq. (13), m is the initial position of the intercepted signal; $S_m(n)$ is the result of $S_1(k)$ when the initial position of the intercepted signal is m; p is the position of the maximum value in $S_m(n)$. Define the detection is successful when $\lambda \geq 15$. When $T_c = T_s$, the frequency resolution of the Fourier transform is f_s. Representative frequency range is $[-f_s/2, f_s/2]$. We can get $f_0 = (p - 32)f_s/128$. The error of frequency offset estimation is $f_s/128$. When $T_c = 8T_s$, the frequency resolution of the Fourier transform is $8f_s$. Representative frequency range is $[-4f_s, 4f_s]$. We can get $f_0 = (n - 256)f_s/128$.

3 Simulation

In this simulation, the rate of symbols is 3.29Ksys and the frequency offset is 5 kHz. The framer in the signal is 64 bit PN sequence and the length of FFT used in the simulation is 512.

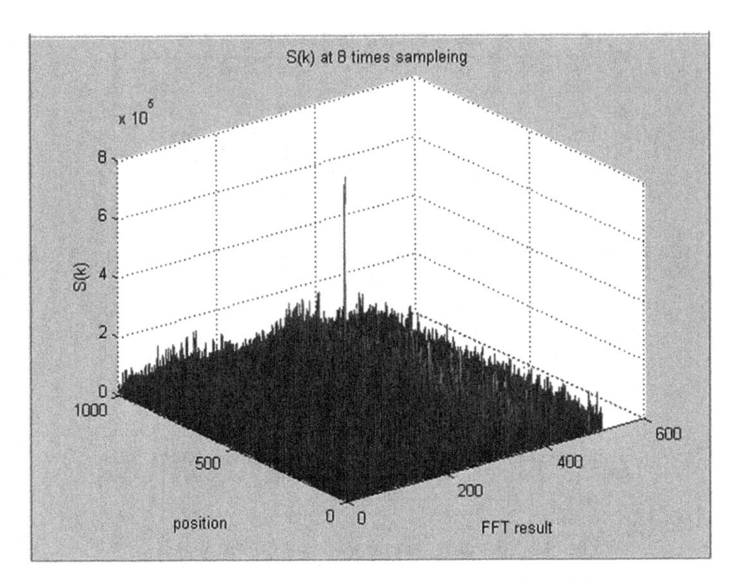

Fig. 2. S(k) simulation result at 8 times sampling

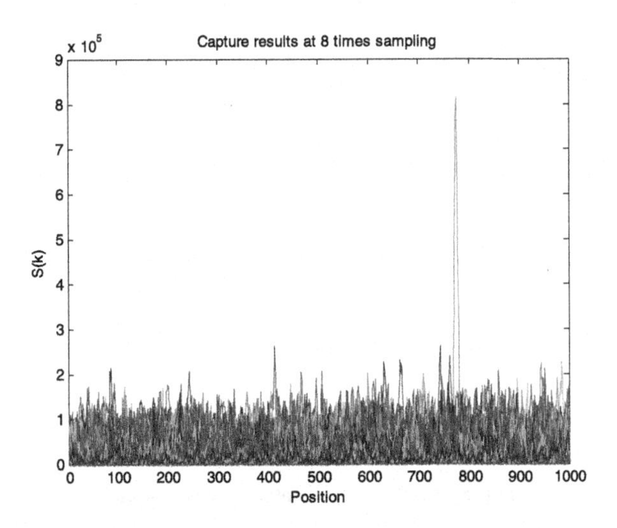

Fig. 3. S(k) cross-sectional view on the x-axis

Figure 2 is the result of S(k) at 8 times sampling. Through the Fig. 2 we can see that S(k) is very big at some points. Thus points' position represent the framer's position and the frequency offset.

Figure 3 and Fig. 4 are cross-sectional views of the capture results on the x-axis and y-axis at 8 times sampling. As shown in Fig. 3, the peak is the capture position, that is, the synchronization position of the signal. Figure 4 shows the position of the peak in the result of an FFT, which is p in Eq. 14. The position of this point represents

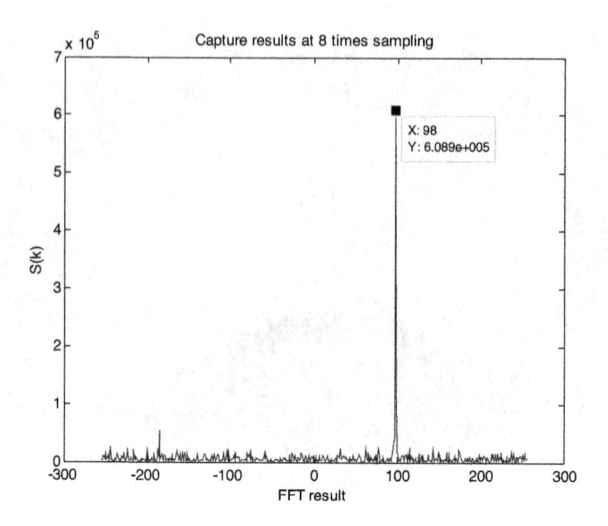

Fig. 4. S(k) cross-sectional view on the y-axis

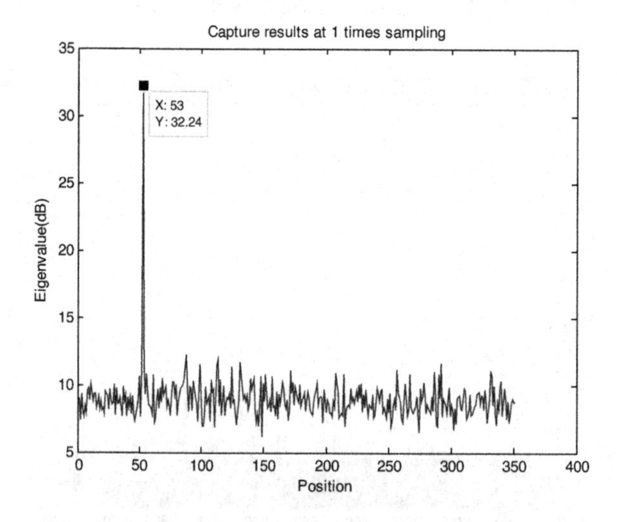

Fig. 5. Capture results at 1 time sampling

the frequency offset. From the figure, it can be seen that the peak position is 98. So we can get the estimated value of the frequency offset is 5.037 kHz.

Figure 5 shows the capture result of 1x sampling, and the captured position is the frame synchronization position. As shown in the figure, the calculation result at the 53rd sample point is significantly larger than other positions, which means that this point is a synchronous position.

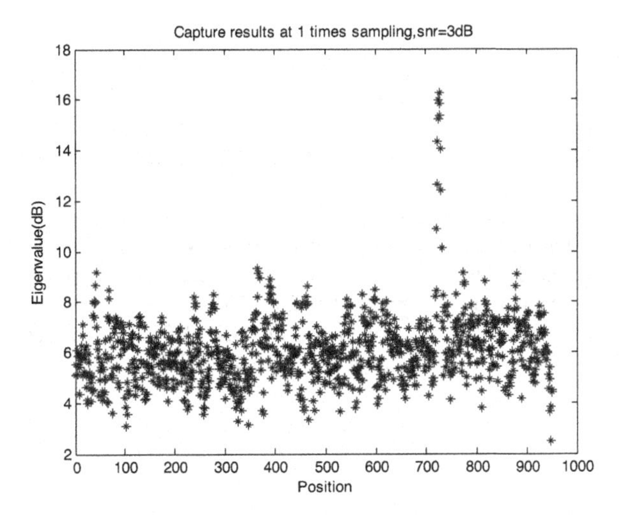

Fig. 6. Capture results when the SNR is 3 dB

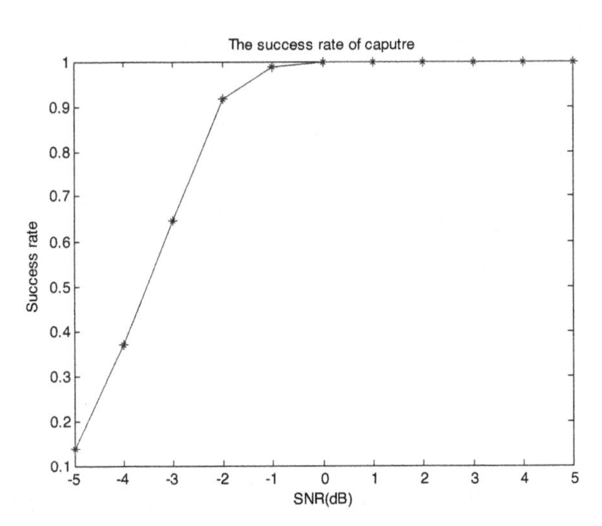

Fig. 7. Acquisition success rate at different SNR

Figure 6 shows the capture result when the signal-to-noise ratio is 3dB. The frequency offset can be estimated from the 8x sampling capture result, and the precise synchronization position can be captured when 1x sampling. Figure 7 shows the capture success rate at different SNR. At each SNR, 1000 acquisition simulations are carried out, and finally calculate the acquisition success rate.

References

1. Tawk, Y., Jovanovic, A., Leclere, J., et al.: A new FFT-based algorithm for secondary code acquisition for galileo signals. In: Vehicular Technology Conference. IEEE (2011)
2. Sun, G., Huang, Q., Zhu, L.: A fast acquisition algorithm based on FFT for DSSS signal and FPGA realization. IEEE Computer Society (2009)
3. You, Y.H., Song, H.K.: Efficient sequential detection of carrier frequency offset and primary synchronization signal for 5G NR systems. IEEE Trans. Veh. Technol. **PP**(99), 1 (2020)
4. Venkataramanan, R., Prabhu, K.: Estimation of frequency offset using warped discrete-Fourier transform. Signal Process. **86**(2), 250–256 (2006)
5. Wang, Y., Zhang, X., Zheng, K.: An improved PMF-FFT algorithm based on all phase preprocessing. J. Telemet. Track. Command **39**, 16–20 (2018)

Satellite Staring Beam Scheduling Strategy Based on Multi-agent Reinforcement Learning

Hongtao Zhu(✉), Zhenyong Wang, Dezhi Li, and Qing Guo

School of Electronics and Information Engineering, Harbin Institute of Technology, Harbin 150001, China
{zhuhongtao,ZYWang,lidezhi,QGuo}@hit.edu.cn

Abstract. Low Earth Orbit (LEO) satellites are an important part of Space-Air-Ground Integrated Networks (SAGIN), which play an irreplaceable role in providing global communication and emergency communication. With the development of phased array technology, many satellites begin to try to use staring beam technology, which can make the beam serve a hot spot on the ground as long as possible by adjusting its phased array parameters, so as to reduce the impact of fast switching on the service performance of LEO satellites. In the satellite service time, how to balance the load of each satellite and meet the communication needs of hot spots is an important problem to be considered. Excellent beam allocation strategy can reduce the network handover rate and signaling overhead. In this paper, the satellite staring beam scheduling problem is transformed into a two-dimensional model, and we propose a novel satellite beam scheduling strategy based on multi-agent reinforcement learning that aims to maximize system performance. Each satellite is regarded as an individual agent, and the decision is to provide communication beam for the current hot spot area. Compared with the beam allocation algorithm based on KM, simulation results show that the proposed strategy can effectively reduce the handoff rate of hot spots when the coverage is satisfied.

Keywords: Low orbit satellite · Multi-agent reinforcement learning · Staring beam scheduling

1 Introduction

Satellite communication can provide seamless wireless signal coverage to support and expand ground communication, which has become an important research direction of 5G and future 6G [1–3]. With the large-scale deployment of OneWeb, StarLink, TeleSat and other mega constellations, LEO satellite shows its advantages in reducing communication delay, providing wide area coverage, and not affected by the ground environment. However, due to the high mobility of LEO satellite, terminals need to switch frequently to maintain the connection of communication links. This will increase the signaling overhead and drop call rate

Q. Guo et al. (Eds.): WiSATS 2021, LNICST 410, pp. 23–34, 2022.
https://doi.org/10.1007/978-3-030-93398-2_3

of the system, and seriously affect the system throughput. Therefore, more and more scholars have studied the handoff problem of LEO satellite, hoping to reduce the impact of high-speed mobility.

According to whether the direction angle of satellite beam in LEO system is adjustable, LEO system is usually divided into two types: one is satellite fixed cell system (SFCS), the other is earth fixed cell system (EFCS) [4]. EFCS is also called staring beam satellite system. In [5], a new satellite handoff strategy based on the potential game of mobile terminal in LEO satellite communication network is proposed. In the software defined satellite network (SDSN) architecture, the author regards satellite handoff as a bipartite graph and proposes a terminal random-access algorithm based on the target of user space maximization. In [6], a fixed beam LEO satellite model is introduced, and the author proposes a method to analyze the throughput of fixed beam according to its coverage time. In [7], the authors propose a performance comparison of fixed and dynamic channel allocation techniques in a LEO satellite system, and they study the case of earth-fixed cell systems with different kinds of fixed and mobile users. In [8], the author presented a comprehensive literature review on applications of deep reinforcement learning (DRL) in communications and networking. In [9], a dynamic channel reservation (DCR) strategy based on deep Q network is proposed for multi-service LEO satellite communication system, which can improve the overall quality of service (QoS) of the system. Inspired by this, we will try to solve the staring beam scheduling problem by reinforcement learning.

The rest of the paper is organized as follows. In Sect. 2, we introduce the LEO satellite beam scheduling system model and an optimal problem is proposed under the constraints of the number of satellite beams and satellite capacity. In Sect. 3, we solve the problem by using a multi-agent DQN learning algorithm. Simulation results are analyzed in Sect. 4 and conclusions are drawn in Sect. 5.

2 System Model

We consider the problem of LEO satellite beam scheduling during satellite operation, as shown in Fig. 1. The set of satellites is denoted by $\mathcal{M} = \{1, 2, ..., M\}$. In this paper, the satellite is equipped with a multi beam phased array antenna system, so it can gaze at one or more hot spots by adjusting the parameters of the antenna array. Each satellite can form up to K beams, which is denoted by $\mathcal{K} = \{1, 2, ..., K\}$. The earth's surface is divided into a fixed number of hot spots according to the degree of user service demand, and the set of communication hot spots is denoted by $\mathcal{N} = \{1, 2, ..., N\}$. Figure 1 shows the coverage of two orbiting satellites to the hot spot area on the ground. The yellow dotted line is the satellite orbit. From t1 to t2, area A is within the coverage of Leo1, so Leo1 can adjust the antenna to maintain the connection to area A. At t3, leo1 exceeds the visible range of area A and starts to serve area F. At t2, because both leo1 and leo2 are within the visible range of area A, that is, a region may have multiple satellites covering at the same time. Also when the service satellite of a region leaves, it is necessary for this region to access another satellite to

Fig. 1. LEO satellite system model of staring beam.

ensure that the communication will not be interrupted. Therefore, staring beam satellite system mainly involves global satellite beam scheduling problem.

In order to simplify the model, this paper uses a two-dimensional plane model to model the staring beam scheduling problem, as shown in Fig. 2. A series of hot spots (red spots) are evenly distributed in the plane area. The satellite (blue dot) moves along the fixed orbit (yellow dotted line) at the same speed. When the satellite moves beyond the plane area, it will appear from the other side of the map and continue to move along the track. Each satellite can provide staring services for one or more regions at the same time, which is constrained by the maximum number of beams and the elevation angle between the satellite and the staring region. And each region can also accept the service of multiple satellites.

When the satellite serves a hot spot area, this paper assumes that the satellite can completely cover this area to avoid the discussion of incomplete beam coverage. When dealing with staring beam scheduling, there are three main factors considered in this paper:

1) Satellites should provide services to the nearest hot spots as far as possible to improve the service quality;
2) When the satellite moves out of a hot spot area, other satellites should continue to cover the area to reduce the drop rate of the area;
3) The satellite load capacity and the capacity of hot spot area should be considered in beam scheduling. If the satellite capacity is not enough to fully serve the current region, other nearby satellites should participate in the service to ensure the service quality.

Assume that the capacity of the satellite is $U = \{u_1, u_2, ..., u_N\}$, the remaining beam of each satellite is $B = \{b_1, b_2, ..., b_M\}$, and the capacity requirement

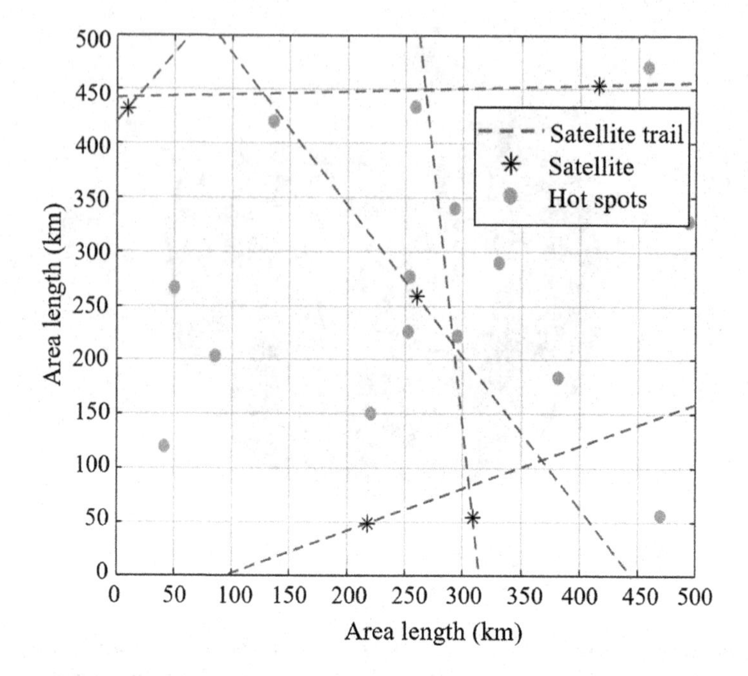

Fig. 2. Two dimensional model of staring beam satellite system.

of hot spot area is $L = \{l_1, l_2, \ldots, l_N\}$. When satellite i is connected with hot spot j, satellite i will provide services for hot spot j as much as possible. At this time, the remaining capacity of satellite i is:

$$s_i' = s_i - \min(u_i, l_j) \tag{1}$$

the remaining capacity requirement of hot spot j is:

$$u_j' = u_j - \min(u_i, l_j) \tag{2}$$

and the number of remaining beams of satellite i is:

$$b_i' = b_i - 1 \tag{3}$$

Considering the two-dimensional plane model we built, the elevation relationship between the hot spot area and the satellite will be transformed into the Euclidean distance between them. If the satellite i coordinate is $(x_{i,1}, x_{i,2})$ and the hot spot area j coordinate is $(y_{j,1}, y_{j,2})$, then the Euclidean distance between the satellite and the hot spot area can generate the weight matrix of M rows and N columns:

$$\mathbf{W} = \begin{bmatrix} w_{11} & w_{12} & \cdots & w_{1N} \\ w_{21} & w_{22} & \cdots & w_{2N} \\ \vdots & \vdots & \ddots & \vdots \\ w_{M1} & w_{M2} & \cdots & w_{MN} \end{bmatrix} \tag{4}$$

where $w_{ij} = \sqrt{\sum_{k=1}^{2} (x_{ik} - y_{jk})^2}, x_i \in X, y_j \in Y$. Then we introduce the beam allocation matrix F of satellite and hot spot area as follows:

$$\boldsymbol{F} = \begin{bmatrix} f_{11} & f_{12} & \cdots & f_{1N} \\ f_{21} & f_{22} & \cdots & f_{2N} \\ \vdots & \vdots & \ddots & \vdots \\ f_{M1} & f_{M2} & \cdots & f_{MN} \end{bmatrix} \tag{5}$$

where $f_{m,n} \in \{0, 1\}$, $f_{m,n} = 1$ means that satellite m allocates a beam to hot spot area n. The service radius of the satellite is R. And the matrix F can be obtained by:

$$f_{m,n} = \begin{cases} 0, & w_{m,n} > R, or\ u_m = 0, or\ l_j = 0 \\ 1, & others \end{cases} \tag{6}$$

We introduce $c_{i,j}$ to represent the capacity of satellite i allocated to hot spot area j, then the capacity allocation matrix C can be expressed as:

$$\boldsymbol{C} = \begin{bmatrix} c_{11} & c_{12} & \cdots & c_{1N} \\ c_{21} & c_{22} & \cdots & c_{2N} \\ \vdots & \vdots & \ddots & \vdots \\ c_{M1} & c_{M2} & \cdots & c_{MN} \end{bmatrix} \tag{7}$$

In this paper, incomplete service rate, handover rate and insufficient capacity rate are used to measure the performance of beam allocation. Incomplete service rate P_b refers to the proportion of hot spot area whose business requirements can't be met. It can be given by:

$$P_b = \frac{\sum\limits_{j \in \mathcal{N}} g_j}{N} \tag{8}$$

where g_j is:

$$g_j = \begin{cases} 0, & \sum\limits_{i \in \mathcal{M}} C_{ij} < l_j \\ 1, & \text{others} \end{cases} \tag{9}$$

Handover rate P_h is used to measure the frequency of satellite switching hot spots in the service process, which is given by:

$$P_h = \frac{\sum\limits_{n \in \mathcal{N}} \sum\limits_{m \in \mathcal{M}} h_{m,n}^t}{\sum\limits_{n \in \mathcal{N}} \sum\limits_{m \in \mathcal{M}} f_{m,n}^t} \tag{10}$$

where $h_{m,n}^t$ is:

$$h_{m,n}^t = \begin{cases} 0, & f_{m,n}^t = f_{m,n}^{t+1} \\ 1, & \text{others} \end{cases} \tag{11}$$

P_c is the insufficient capacity rate of hot spot area demand, which is a supplement to P_b. It can be expressed as:

$$P_c = \frac{\sum\limits_{i \in \mathcal{M}} \sum\limits_{j \in \mathcal{N}} c_{i,j}}{\sum\limits_{j \in \mathcal{N}} l_j} \tag{12}$$

Then the staring beam scheduling problem is formulated as follows:

$$\min_{c_{m,n}} \quad P = \alpha_1 P_b + \alpha_2 P_h + \alpha_3 P_c \tag{13}$$

$$\text{s.t.} \quad \sum_{n \in \mathcal{N}} f_{m,n} \leq K, \quad \forall m \in \mathcal{M} \tag{13a}$$

$$\sum_{n \in \mathcal{N}} c_{m,n} \leq u_m, \quad \forall m \in \mathcal{M} \tag{13b}$$

$$f_{m,n} \in \{0,1\}, \quad \forall m \in \mathcal{M}, \forall n \in \mathcal{N} \tag{13c}$$

$$c_{m,n} \in [0, u_m], \quad \forall m \in \mathcal{M}, \forall n \in \mathcal{N} \tag{13d}$$

where α_1, α_2, and α_3 are positive parameters.

3 Multi-agent Deep Q-Learning Algorithm

Deep Q-Learning (DQN) algorithm is a classic and effective reinforcement learning algorithm, which can solve complex problems in many communication scenarios. We extend the deep Q-learning to the multi-agent cases to solve the problem of staring beam scheduling. In the multi-agent DQN model, the learning and decision of each agent are realized by DQN algorithm. $s_{e,t}^{m,k}$, $a_{e,t}^{m,k}$, $r_{e,t}^{m,k}$, and $s_{e,t+1}^{m,k}$ represent the state, action, reward and the next state of agent (satellite) i at time t of the e-th training round. The online Q function fitted by neural network and the objective Q function are randomly initialized. With the continuous interaction between the agent and the environment, the generated action sequence is stored in the experience pool. In order to minimize the error function $L(\theta^m)$, a batch of sequences are randomly selected from the experience playback pool every certain interval. $L(\theta^m)$ is given by

$$L(\theta^m) = \left(r_j^m + \gamma \max_{A^i} Q^m \left(s_{j+1}^m, a^m; \theta^{m-} \right) - Q^m \left(s_j^m, a_j^m; \theta^m \right) \right)^2 \tag{14}$$

where γ is the discount factor, θ^{m-} is the parameters of target value network, and θ^m is the parameters of online value network.

The multi-agent deep Q-learning algorithm is shown in Algorithm 1. State $s_{e,t}^m = \left[W_{e,t}^i, a_{e,t-1}^m, a_{e,t-2}^m, C_{e,t}^m \right]$ represents the allocation state of the k-th beam of satellite m of the e-th round during training at the time t. The reward for satellite m performing action $a_{e,t}^{m,k}$ under the k-th beam at the time t in the e-th round is $r_{e,t}^{m,k}$. and it can be expressed as

$$r_{e,t}^{m,k} = -r\_l_{e,t}^{m,k} - a \times r\_b_{e,t}^{m,k} + b \times r\_c_{e,t}^{m,k} \tag{15}$$

Algorithm 1. Multi-Agent Deep Q-Learning Algorithm

1: **Initilization:**

- Satellite capacity set S, hot spot area capacity demand set U, satellite beam number set B, satellite coverage radius R, satellite geographic coordinates $(x_{i,1}, x_{i,2})$, $i \in \mathcal{M}$,hot spots area geographic coordinate $(y_{j,1}, y_{j,2})$, $j \in \mathcal{N}$;
- Initialize the status of each satellite $s_{1,1}^{m,k}, m \in \mathcal{M}, k \in \mathcal{K}$;
- Initialize the action value function of each agent with random parameters $Q^i\left(s^i, a^i; \theta^i\right), i \in \mathcal{M}$;
- $epoch = \{1, 2, ..., E\}$, $time = \{1, 2, ..., T\}$;

2: **for** $e \in epoch$ **do**
3:   **for** $t \in time$ **do**
4:     **for** $k \in \mathcal{K}$ **do**
5:       **for** $m \in \mathcal{M}$ **do**
6:         Using ε-greedy based exploration strategy $\pi^{\varepsilon}\left(s_{e,t}^{m,k}\right)$ to get action $a_{e,t}^{m,k}$, the reward $r_{e,t}^{m,k}$, and the transition state $s_{e,t+1}^{m,k}$, store sequence $\left(s_{e,t}^{m,k}, a_{e,t}^{m,k}, r_{e,t}^{m,k}, s_{e,t+1}^{m,k}\right)$ in D^m, update U, L, F, and C;
7:       **end for**
8:       **if** $t > 200$ and $t\%5 == 0$ **then**
9:         **for** $m \in \mathcal{M}$ **do**
10:           Select a batch of sequence $\left(s_j^{m,k}, a_j^{m,k}, r_j^{m,k}, s_{j+1}^{m,k}\right)$ from D^m randomly;
11:           Set $y_j^m = \begin{cases} r_j^m \\ r_j^m + \gamma \max_{A^m} Q^m\left(s_{j+1}^m, a^m; \theta^{m-}\right) \end{cases}$;
12:           $L\left(\theta^m\right) = \left(\mathrm{y}_j - Q^m\left(s_j^m, a_j^m; \theta^m\right)\right)^2$;
13:           Gradient descent update θ^m from $L\left(\theta^m\right)$;
14:           update $\theta^{m-} = \theta^m$;
15:         **end for**
16:       **end if**
17:     **end for**
18:   **end for**
19: **end for**

where a is the penalty coefficient of disconnection, and b is the reward coefficient of connection.

Here $r\_l_{e,t}^{m,k}$ is the distance penalty, which aims to make the satellite serve the nearest area as far as possible. $r\_b_{e,t}^{m,k}$ is the disconnection penalty and $r\_c_{e,t}^{m,k}$ is the connection reward. In this way, the handover rate can be reduced. And the three parameters is given by

$$r\_l_{e,t}^{m,k} = \sqrt{\sum_{k=1}^{2}\left(x_{ik} - y_{jk}\right)^2}, x_i \in X, y_j \in Y \tag{16}$$

$$r\_b_{e,t}^{m,k} = \begin{cases} -La & c_{m,n}^{t-1} = 1 \; and \; c_{m,n}^{t} \neq 1 \\ 0 & \text{else} \end{cases} \tag{17}$$

$$r\_c_{e,t}^{m,k} = \begin{cases} \sqrt{\sum_{j=1}^{2} \left(x_{m,j} - y_{n,k}\right)^2}, x_m \in X, y_{a_n} \in Y & c_{m,n}^{t-1} = c_{m,n}^{t} = 1 \\ 0 & \text{else} \end{cases} \tag{18}$$

By synthesizing the three kinds of rewards and adjusting their coefficients, the agent is expected to learn the corresponding strategies to meet the performance requirements.

4 Simulation Results

In this section, we present the simulation results of proposed strategy what we called multi-agent DQN and compare it with Beam scheduling algorithm based on KM. In this paper, we consider that a reasonable beam scheduling scheme should minimize the sum of the connection distances between the satellite and the hot spot region, that is, to find the minimum value of the edge weight of the bipartite graph $G = (V, E)$. So the problem is transformed into the optimal matching of bipartite graph, which can be solved by the classical KM algorithm.

The simulation parameters are summarized in Table 1. We consider a square area with 500 km side length, where hot spot areas are randomly distributed within the area. When the number and capacity of satellites are fixed, we focus on the impact of the maximum number of beams and the number of hot spots on the algorithm performance. In terms of satellite capacity, refer to StarLink single satellite capacity (17 Gbps) and Iridium system single satellite capacity (7.5 Gbps), the satellite capacity is set to 3 Gbps and the maximum capacity demand of hot spot area is 1 Gbps.

Table 1. Simulation parameters

Parameter	Value
Number of satellites M	5
Number of satellite beams K	3
Maximum service range of satellite R	250 km
Maximum capacity of satellite s_{max}	3 Gbps
Number of hot spots N	15
Maximum capacity demand of hot spots u_{max}	1 Gbps
Area size La^2	$250 \times 250\,\text{km}^2$
Satellite velocity v	10 km/s
Simulation duration T	300 s

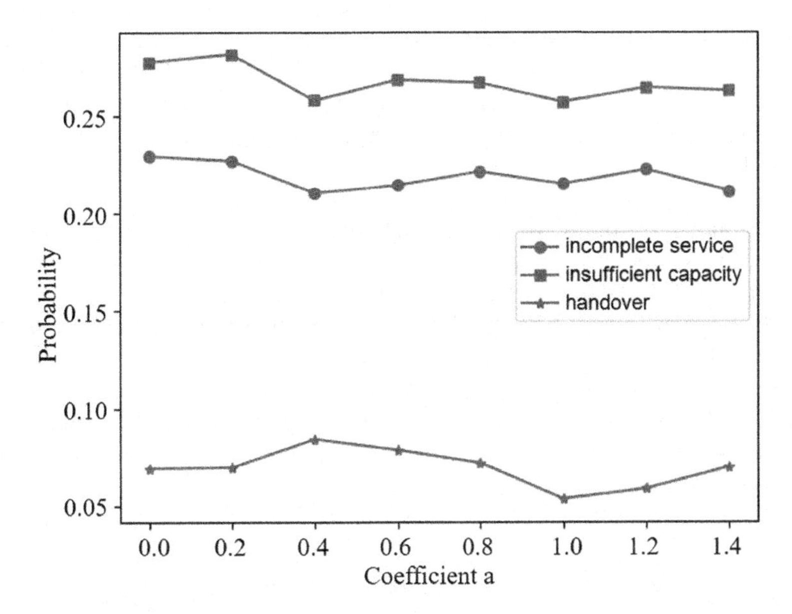

Fig. 3. The performance of the system varies with the penalty coefficient a of disconnection.

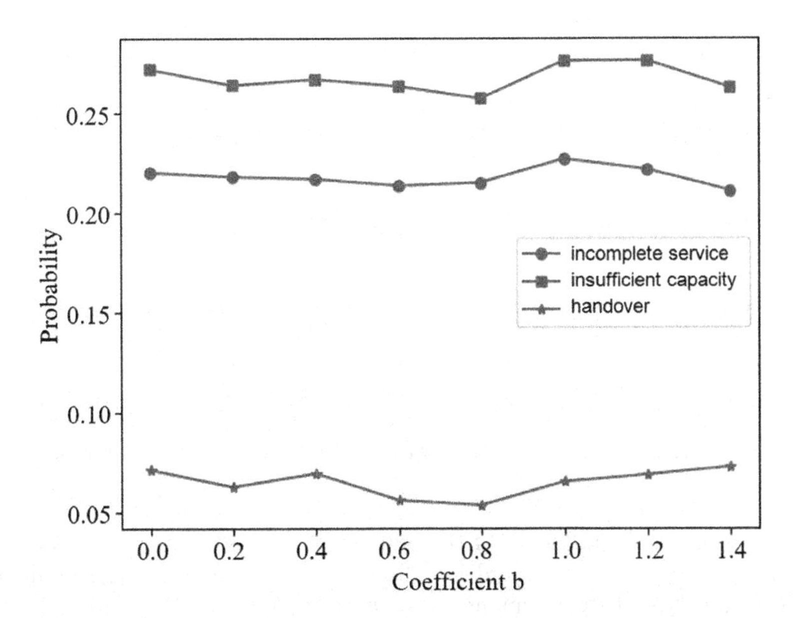

Fig. 4. The performance of the system varies with the reward coefficient b of connection.

In Fig. 3 and Fig. 4, we compared the effect of the penalty coefficient a of disconnection and the reward coefficient b of connection on the beam scheduling performance. It can be found that when $a = 1.0$ and $b = 0.75$, the handover rate of the system reaches the lowest, and the rate of incomplete service and insufficient capacity rate are also at a low value. So we set $a = 1.0$ and $b = 0.75$ in the following simulation.

Figure 5 shows the influence of the number of beams of satellite on the handover rate, incomplete service rate and insufficient capacity rate. With the increase of the maximum number of beams, the performance of KM algorithm and multi-agent DQN algorithm is improved. When the number of beams is small, the performance of KM algorithm is better than that of multi-agent DQN algorithm. However, with the increase of the number of beams, their coverage performance is almost the same, but the multi-agent DQN is significantly better than KM algorithm in the handover rate. This is because the punishment of satellite switching is strengthened in the training, so the satellite connection strategy is adjusted.

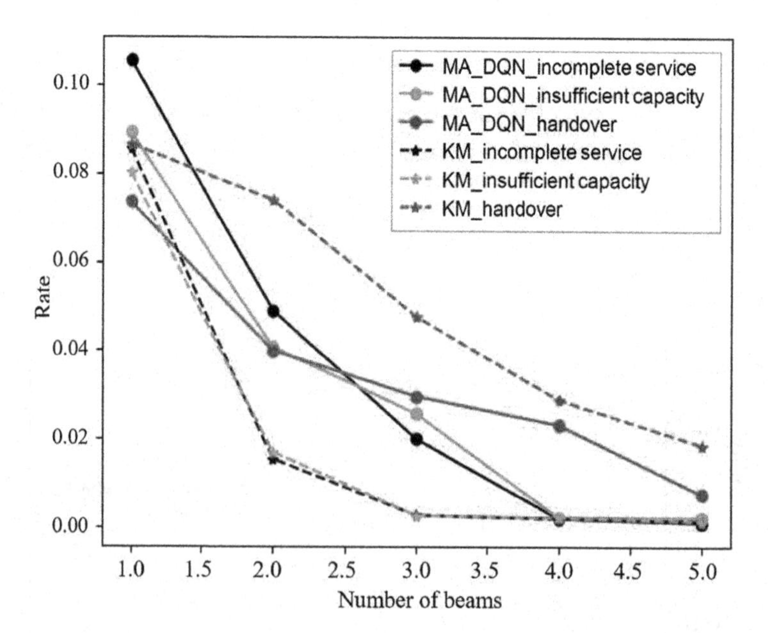

Fig. 5. The performance of the system varies with the maximum number of beams.

Figure 6 shows the influence of the number of hot spot area on the handover rate, incomplete service rate and insufficient capacity rate. As the number of hot spots increases, the performance of both algorithms decreases. The handover rate of multi-agent DQN algorithm is always better than that of KM algorithm. But its coverage performance is not as good as KM. In general, by adjusting the reward setting, the reinforcement learning algorithm achieves better performance

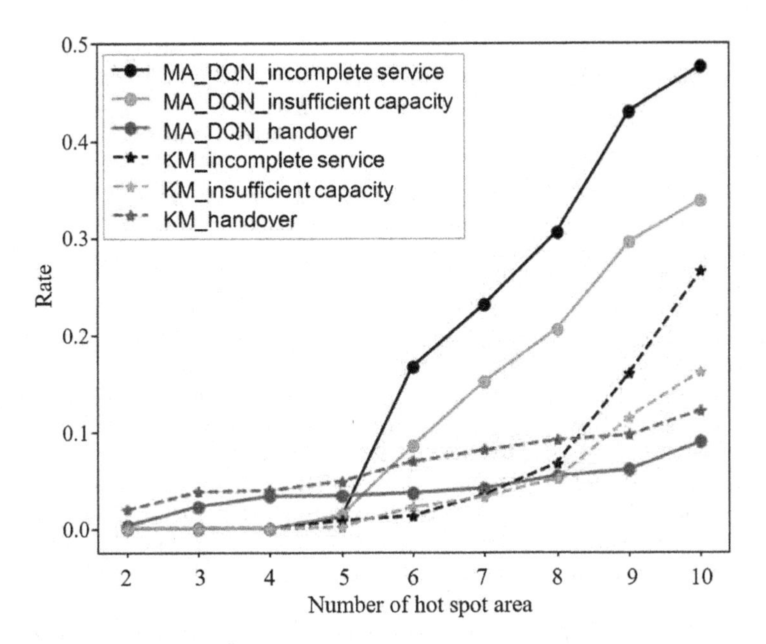

Fig. 6. The performance of the system varies with the number of hot spot area.

in the handover rate. In the future, when the number of satellites is increasing, the coverage performance will not be the main consideration, while reducing the handover rate can greatly reduce the signaling overhead of the system, which indicates that multi-agent DQN algorithm is a desirable staring beam scheduling algorithm.

5 Conclusion

In this paper, we have investigated the staring beam scheduling problem in LEO satellite network. By establishing a two-dimensional model, an optimal problem is proposed under the constraints of the number of satellite beams and satellite capacity. We have solved the problem by multi-agent DQN algorithm. Compared with the beam scheduling algorithm based on KM, multi-agent DQN algorithm can adjust the weight to change the scheduling strategy to meet the optimization requirements. Simulation results have shown that the proposed algorithm can reduce the handover rate, which is of great significance to reduce the network signaling overhead.

References

1. Chen, S., Sun, S., Kang, S.: System integration of terrestrial mobile communication and satellite communication - the trends, challenges and key technologies in B5G and 6G. China Commun. **17**(12), 156–171 (2020). https://doi.org/10.23919/JCC.2020.12.011

2. Rinaldi, F., et al.: Non-terrestrial networks in 5G beyond: a survey. IEEE Access **8**, 165178–165200 (2020). https://doi.org/10.1109/ACCESS.2020.3022981
3. Giordani, M., Zorzi, M.: Non-terrestrial networks in the 6G era: challenges and opportunities. IEEE Netw. **35**(2), 244–251 (2021). https://doi.org/10.1109/MNET.011.2000493
4. Restrepo, J., Maral, G.: Cellular geometry for world-wide coverage by non-geo satellites using "earth-fixed cell" technique. In: Proceedings of GLOBECOM 1996. 1996 IEEE Global Telecommunications Conference, vol. 3, pp. 2133–2137 (1996). https://doi.org/10.1109/GLOCOM.1996.592010
5. Wu, Y., Hu, G., Jin, F., Zu, J.: A satellite handover strategy based on the potential game in LEO satellite networks. IEEE Access **7**, 133641–133652 (2019). https://doi.org/10.1109/ACCESS.2019.2941217
6. Baik, J.S., Kim, J.H.: Analysis of the earth fixed beam duration in the LEO. In: 2021 International Conference on Information Networking (ICOIN), pp. 477–479 (2021). https://doi.org/10.1109/ICOIN50884.2021.9333981
7. Boukhatem, L., Beylot, A., Gaiti, D., Pujolle, G.: Performance analysis of dynamic and fixed channel allocation techniques in a LEO constellation with an "earth-fixed cell" system. In: Globecom 2000 - IEEE. Global Telecommunications Conference. Conference Record (Cat. No.00CH37137), vol. 2, pp. 1145–1149 (2000). https://doi.org/10.1109/GLOCOM.2000.891316
8. Luong, N.C., et al.: Applications of deep reinforcement learning in communications and networking: a survey. IEEE Commun. Surv. Tutor. **21**(4), 3133–3174 (2019). https://doi.org/10.1109/COMST.2019.2916583
9. Li, Z., Xie, Z., Liang, X.: Dynamic channel reservation strategy based on DQN algorithm for multi-service LEO satellite communication system. IEEE Wirel. Commun. Lett. **10**(4), 770–774 (2021). https://doi.org/10.1109/LWC.2020.3043073

Optimization of Joint Power and Bandwidth Allocation for Multiple Users in a Multi-spot-Beam Satellite Communication

Heng Wang, Shijun Xie, Ganhua Ye(✉), and Bin Zhou

The 63rd Research Institute of National University of Defense Technology, Nanjing, China

Abstract. Multi-spot-beam techniques have been widely applied in modern satellite communication systems, due to the advantages of reusing the frequency of different spot beams and constructing flexible service networks. As the on-board resources of bandwidth and power in a multi-spot-system are scarce, it is important to enhance the resource utilization efficiency. To this end, this paper initially presents the formulation of the problem of joint power and bandwidth allocation for multi-users, and demonstrates that the problem is one of convex minimization. An algorithm based on the Karush-Kuhn-Tucker (KKT) conditions is then proposed to obtain an optimal solution of the problem. Compared with existing separate power or bandwidth algorithms, the proposed joint allocation algorithm improves the total system capacity and the fairness between users. A suboptimal algorithm is also proposed, to further reduce computational complexity, with a performance level much closer to that of the optimal allocation algorithm.

Keywords: Multi-spot-beam satellite communication system · Joint bandwidth and power allocation · Convex optimization · Optimal allocation algorithm · Low computational complexity allocation algorithm

1 Introduction

In recent years, as an important component of internet, satellite communication has played a key role in seamless internet access. In a modern satellite communication system, the satellite has multiple-spot-beams, each one of which covers different areas of the earth. Thus the multi-spot-beam system can reuse the frequencies of the different spot beams, to significantly increase the total system capacity. In addition, the system can provide high power density to a particular spot beam, by allocating more resources to it, thereby supporting high traffic rates to small antenna terminals [1].

However, the on-board resources of bandwidth and power are scarce and expensive in multi-spot-beam satellite systems. As a result, it is crucial for us to improve the resource utilization efficiency. To this end, dynamically allocating these resources to each user according to their traffic demands is a viable solution.

In previous works, separate optimal power or bandwidth allocation for spot beams have been investigated by [1–4] and [5]. In these works, the metric to evaluate the

Q. Guo et al. (Eds.): WiSATS 2021, LNICST 410, pp. 35–49, 2022.
https://doi.org/10.1007/978-3-030-93398-2_4

system performance is to minimize the deficit between the traffic demand and the capacity allocated, taking into account a compromise between the total system capacity and the proportional fairness among spot beams. It was proved that it is need to allocate more resource to the spot beam with higher traffic demand to get fairness between among spot beam, thus the total system capacity decreased, due to concavity of the capacity function with a fixed power or bandwidth allocation. To overcome this drawback, in this paper we propose a joint bandwidth and power allocation algorithm. Moreover, we propose solving the problem of joint bandwidth and power allocation for multiple users in each spot beam. As a result, the constraints and complexity are greater than for those problems mentioned in the above referenced works.

In [6], the joint bandwidth and power allocation of downlink transmissions were investigated. The object of the optimization problem was to maximize the system capacity and fairness between each link. Fairness was achieved by assigning different weights to different links, which were the reciprocals of the average long term rates. However, the author only solved for a maximum of two-user allocation simultaneously, based on the Concave Envelope Theorem, and ignored larger simultaneous multi-user allocations. In [7], the optimal joint bandwidth and power allocation in wireless, multi-user networks, both with and without relays, was proposed. The author focused on a scenario in which a source served multi-users with different channel conditions, simultaneously. The optimization objective was to maximize the total system capacity. However, the author failed to consider the traffic demands of each user. The results showed that for a set of users served by one source, all the power from that source was allocated only to the user having the highest channel gain. It is obvious that the conclusion is questionable, when the traffic demands of the user with the highest channel gain does not exceed the source capacity. In [8], a joint power and bandwidth allocation algorithm with Quality of Service (QoS) support in heterogeneous wireless networks was proposed, using convex optimization methodology. The terminal was supported to access different wireless networks simultaneously, and the objective of the convex optimization was to maximize the system capacity without regard for the fairness amongst the users. However, the conclusions obtained in this work cannot be applied to our system, because the users cannot access different spot beams in parallel, in multi-spot-beam satellite systems.

In this paper, the objective is to solve for the optimal joint bandwidths and power allocation for users in a multi-spot-beam Satellite System. We initially formulate the problem of joint power and bandwidth allocation for users as a nonlinear optimization problem, and demonstrate that the optimization problem is a convex optimization problem. The object of our optimization is to match the capacity allocated to each user, as closely as possible to the traffic demand, taking into account a compromise between the total system capacity and the proportional fairness between the users. Then we propose an algorithm based on the Karush-Kuhn-Tucker (KKT) conditions to achieve an optimal solution. Compared with the individual power or bandwidth optimal allocation algorithms, the proposed joint bandwidth and power allocation algorithm improves the total system capacity and the fairness between users. A suboptimal algorithm is also proposed, to reduce the computational complexity, the performance of which is much closer to the optimal algorithm.

The remainder of this paper is organized as follows: Sect. 2 formulates the optimization problem of joint bandwidth and power allocation, utilizing a compromise between system capacity maximization and proportional fairness between the users; demonstrating that the optimization problem is a convex minimization type problem. Section 3 proposes an optimal joint bandwidth and power allocation algorithm based on KKT conditions, and a suboptimal algorithm to reduce the computational complexity. Section 4 presents the simulation results and compares the performance of the low computational complexity algorithm with that of the optimal algorithm, and finally, in Sect. 5 the conclusions are presented.

2 System Model

2.1 Downlink Capacity of Users

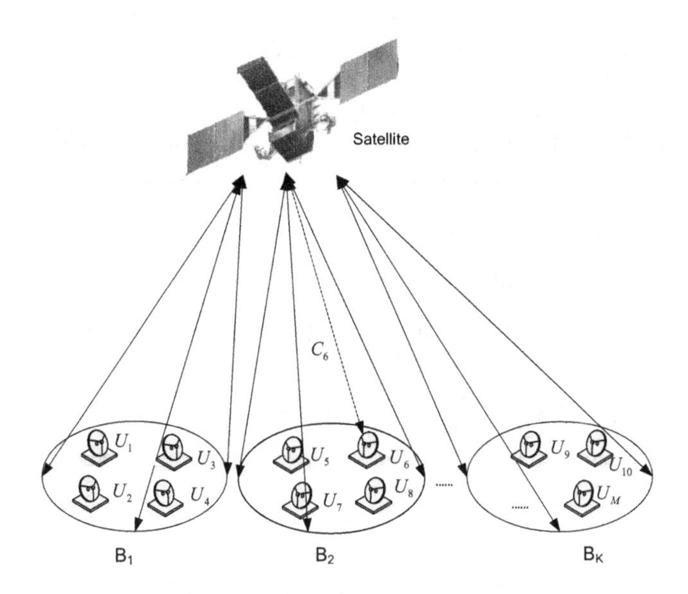

Fig. 1. System configuration of a multi-spot-beam satellite system with multiple users.

The configuration of a multi-beam satellite system with multiple users is shown in Fig. 1. The system consists of K beams B_i, $I \in \{1,\ldots, K\}$, and M users U_i, $I \in \{1,\ldots, M\}$. The set of users which are served by the beam B_i is denoted by $\mathcal{N}_{B_i}$. The traffic demand of the user U_i is T_i, the power and bandwidth allocated to the user U_i are P_i and W_i, and the signal attenuation factor of the user U_i is α_i^2. It is noted that α_i^2 consists mainly of the effects of weather conditions, free space loss and antenna gain. The total power and bandwidths of the system are P_{total} and W_{total}.

Using time sharing for Gaussian broadcast channels [9], we obtain the Shannon bounded capacity C_i for the user U_i as:

$$C_i = W_i \log_2\left(1 + \frac{\alpha_i^2 P_i}{W_i N_0}\right) \quad (1)$$

where N_0 is the noise power density of each user. It is noted that interbeam interference from the sidelobes of adjacent spot beams to degenerate the Shannon capacity. However, in this paper, we ignore interbeam interference, because we consider very narrow spot beams over a large number of spot beams [1]. It is observed from (1) that the user capacity C_i is increased as the bandwidth or power allocated to the user increases. However, the total bandwidth and power of the satellite is fixed, so the capacity of the system is limited.

In the multi-beam satellite system, there is always a power or bandwidth pre-allocation for each beam. Therefore, in this paper we analyze the following four situations: (a) No pre-allocation for any beam. (b) There is only power pre-allocation for each beam. (c) There is only bandwidth pre-allocation for each beam. (d) There are both power and bandwidth pre-allocations for each beam. Let P_{B_i} and W_{B_i} denote the pre-allocated power and bandwidth of the i-th beam.

If the total system resources of power and bandwidth are sufficient to support the traffic demand generated by all the users, it seems meaningless for us to make efforts to improve the resource utilization efficiency. Therefore, we only focus on the resource allocations for scenarios where the total traffic demand exceeded the total available system capacity.

2.2 Optimization Problem Formulation

There are many metrics to evaluate the system performance, and different metrics may lead to different allocation results. Therefore, it is very important to choose an appropriate metric. Motivated by J. P. Choi and V. W. S. Chan [1], in this paper the metric is designed to minimize the deficit between the traffic demand and the capacity allocated, taking into account a compromise between the total system capacity and the proportional fairness between the users. The problem is formulated as follows:

$$\min_{\{P_i\},\{W_i\}} \sum_{i=1}^{M} (T_i - C_i)^2 \quad (2)$$

s.t.

$$C_i = W_i \log_2\left(1 + \frac{\alpha_i^2 P_i}{W_i N_0}\right) \le T_i,\ \forall i \quad (3)$$

$$\sum_{i=1}^{M} P_i \le P_{total} \quad (4)$$

$$\sum_{i=1}^{M} W_i \le W_{total} \quad (5)$$

$$\sum_{i \in \mathcal{N}_{B_j}} P_i \leq P_{B_j} \tag{6}$$

$$\sum_{i \in \mathcal{N}_{B_j}} W_i \leq W_{B_j} \tag{7}$$

The constraint (3) indicates that the allocated resources should not exceed the traffic demands of each spot beam. Conditions (4, 5, 6 and 7) imply the constraints for the total system power, total system bandwidth, and the power and bandwidth for the j-th spot beam, respectively.

As mentioned in Subsect. 2.1, in this paper we analyze four cases. Different cases result in different constraints for the optimization problem. Case (a) does not use constraint numbers (6 and 7). Case (b) does not use constraint number (7). Case (c) does not use constraint number (6). Case (d) uses all four constraints.

Without a loss of generality, we first solve the optimization problem with all four constraints, numbers (4–7). Introducing the non-negative Lagrangian multipliers μ, λ, $\boldsymbol{\rho} = [\rho_1, \rho_2, \ldots, \rho_K]$, and $\boldsymbol{\sigma} = [\sigma_1, \sigma_2, \ldots, \sigma_K]$, yielded the Lagrange function, given as:

$$\begin{aligned} L(\mathbf{P}, \mathbf{W}, \boldsymbol{\rho}, \boldsymbol{\sigma}, \lambda, \mu) &= \sum_{i=1}^{M} (T_i - C_i)^2 - \mu\left(W_{total} - \sum_{i=1}^{M} W_i\right) \\ - \lambda\left(P_{total} - \sum_{i=1}^{M} P_i\right) &- \sum_{i=1}^{K} \rho_i \left(P_{B_i} - \sum_{j \in \mathcal{N}_{B_i}} P_j\right) - \sum_{i=1}^{K} \sigma_i \left(W_{B_i} - \sum_{j \in \mathcal{N}_{B_i}} W_j\right) \end{aligned} \tag{8}$$

where $\mathbf{P} = [P_1, P_2, \ldots, P_M]$ and $\mathbf{W} = [W_1, W_2, \ldots, W_M]$.

According to the KKT conditions, we obtain the following equations:

$$\frac{\partial L}{\partial P_i} = \frac{2\alpha_i^2 W_i}{(W_i N_0 + \alpha_i^2 P_i) \ln 2}(T_i - C_i) - \lambda - \rho_j = 0, \; i \in \mathcal{N}_{B_j} \tag{9}$$

$$\begin{aligned} \frac{\partial L}{\partial W_i} &= 2(T_i - C_i)\left[\frac{C_i}{W_i} - \frac{W_i P_i}{\ln 2(N_0 W_i^2/\alpha_i^2 + P_i W_i)}\right] \\ &- \mu - \sigma_i = 0 \end{aligned} \tag{10}$$

It is clear from (9) that the non-negative λ and ρ_i means that $T_i \geq C_i$. As a result, constraint number (3) is satisfied.

It is known that when the optimization problem is convex, and a feasible solution satisfies the KKT conditions, then the solution is a global optimal solution to the optimization problem [10]. Fortunately, the optimization problem mentioned above is a convex type, the proof for which is shown in the appendix. Therefore, in the next section we propose an iterative algorithm based on the KKT conditions. Although the optimization problems are different for different cases, the proposed algorithm solves them well within the same architecture.

3 Proposed Joint Bandwidth and Power Allocation Algorithm

3.1 Optimal Allocation Algorithm

When the Lagrangian multiplier variables are given, the optimal P_i is obtained from (9) by numerical calculation methods, e.g., the Golden Section Method. If the optimal $P_i < 0$, then P_i is set to zero.

Substituting the optimal P_i into (8), we obtain the optimal W_i from (10) by using the Golden Section method. Similarly, if the optimal $W_i < 0$, then W_i is set to zero.

Here, we only have one problem to solve, which is how to search the Lagrangian multipliers. Motivated by W. Yu and G. Ding [11, 12], we use the sub-gradient method to update the Lagrangian multipliers, which are obtained according to the following equations:

$$\mu^{n+1} = \left[\mu^n - \Delta_\mu^n\left(W_{total} - \sum_{i=1}^{M} W_i\right)\right]^+ \tag{11}$$

$$\lambda^{n+1} = \left[\lambda^n - \Delta_\lambda^n\left(P_{total} - \sum_{i=1}^{M} P_i\right)\right]^+ \tag{12}$$

$$\rho_i^{n+1} = \left[\rho_i^n - \Delta_\rho^n\left(P_{B_i} - \sum_{j\in\mathcal{N}_{B_i}} P_j\right)\right]^+ \tag{13}$$

$$\sigma_i^{n+1} = \left[\sigma_i^n - \Delta_\sigma^n\left(W_{B_i} - \sum_{j\in\mathcal{N}_{B_i}} W_j\right)\right]^+ \tag{14}$$

where $[x]^+ = \max\{0, x\}$, n is the iteration number and Δ is the iteration step size.

The above sub-gradient update method is guaranteed to converge to the optimal as long as the iteration step chosen is sufficiently small [11–14].

The whole process of the proposed optimal joint bandwidth and power allocation algorithm is summarized as follows:

Step 1. Set appropriate initial values for the Lagrangian multipliers and the bandwidth of each user.
Step 2. Substitute the values of the bandwidth of each user and the Lagrangian multipliers into (9), and then calculate the optimal power allocated to each user.
Step 3. Substitute into (10), both the power values for each user obtained from Step 2 and the Lagrangian multipliers, and then calculate the optimal bandwidth allocated to each user.
Step 4. Substitute the values of the power and the bandwidth of each user, which are separately obtained from Steps 2 and 3, into (11)–(14), and then update the Lagrangian multipliers.

Step 5. If the conditions of $|\mu^{n+1}(W_{total} - \sum_{i=1}^{M} W_i)| < \varepsilon$, $|\lambda^{n+1}(P_{total} - \sum_{i=1}^{M} P_i)| < \varepsilon$, $|\rho_i^{n+1}(P_{B_i} - \sum_{j \in \mathcal{N}_{B_i}} P_j)| < \varepsilon$, $\forall i \in \{1, \ldots, K\}$, and $|\sigma_i^{n+1}(W_{B_i} - \sum_{j \in \mathcal{N}_{B_i}} W_j)| < \varepsilon$, $\forall i \in \{1, \ldots, K\}$ are simultaneously satisfied, terminate the algorithm; otherwise go to Step 2.

According to the above process, it is shown that the computational complexity is $O(4SK + 2SMT)$, where M is the number of the users and K is the number of the spot beams, S is the number of iterations, and T is the computational complexity of the Golden Section method. It is noted that either the S or T are independent of K and N. Therefore, the computational complexity of the proposed algorithm is linear in the number of the spot beams and users.

As mentioned above, different cases result in different optimization problems. For different optimization problems, we only need to remove the corresponding Lagrangian multipliers in (8), for the optimal solution to be obtained by the same algorithm.

3.2 Low Computational Complexity Allocation Algorithm

In this subsection we present a suboptimal algorithm to further reduce the computational complexity. The performance of this algorithm is much closer to that of the optimal algorithm. As the coverage of each spot beam is limited, the channel conditions of the users in the same spot beam are always the same. In such circumstances, the performance is equal to that of the optimal algorithm. The low computational complexity algorithm is based on spreading the spot beam power evenly over the whole spot beam bandwidth for all the users in the same spot beam. Let P_{ai} and W_{ai} denote the powers and bandwidths allocated to the i-th spot beam. The bandwidths and powers allocated to the users in the same spot beam will thus have the following relationship:

$$\frac{P_j}{W_j} = \frac{P_k}{W_k} = \frac{P_{ai}}{W_{ai}}, \quad \forall j, k \in \mathcal{N}_{B_i} \tag{15}$$

As a result, the capacity allocated to the user is given as follows:

$$C_j = W_j \log_2\left(1 + \frac{\alpha_j^2 P_{ai}}{N_0 W_{ai}}\right), \quad \forall j \in \mathcal{N}_{B_i} \tag{16}$$

According to (9), we obtain the following equations for the users in the same spot beam.

$$\frac{T_j - C_j}{T_k - C_k} = \frac{N_0/\alpha_k^2 + P_{ai}/W_{ai}}{N_0/\alpha_j^2 + P_{ai}/W_{ai}}, \forall j, k \in \mathcal{N}_{B_i}, \ j \neq k \tag{17}$$

$$W_k = A_{jk} W_j + B_{jk}, \forall j, k \in \mathcal{N}_{B_i}, j \neq k \tag{18}$$

where: $A_{jk} = \frac{P_{ai}M_j/W_{ai} + N_0 M_j/\alpha_k^2}{P_{ai}M_k/W_{ai} + N_0 M_k/\alpha_j^2}$,
$B_{jk} = \frac{P_{ai}T_k/W_{ai} + N_0 T_k/\alpha_j^2 - P_{ai}T_j/W_{ai} - N_0 T_j/\alpha_k^2}{P_{ai}M_k/W_{ai} + N_0 M_k/\alpha_j^2}$,
and $M_k = \log_2\left(1 + \frac{\alpha_k^2 P_{ai}}{N_0 W_{ai}}\right)$, $M_j = \log_2\left(1 + \frac{\alpha_j^2 P_{ai}}{N_0 W_{ai}}\right)$.

Since the sum of the bandwidth allocated to users in the same spot beam is equal to the bandwidth allocated to the i-th spot beam, we obtain the following equation:

$$\sum_{j \in \mathcal{N}_{B_i}} W_j = W_{ai} \tag{19}$$

According to (18) and (19), we obtain the bandwidth allocated to the users in the same spot beam:

$$W_j = \frac{W_{ai} - \sum_k B_{jk}}{\sum_k A_{jk} + 1}, \quad \forall j, k \in \mathcal{N}_{B_i}, k \neq j \tag{20}$$

As a result, the power allocated to the users is given as:

$$P_j = \frac{W_j}{W_{ai}} P_{ai}, \quad \forall j \in \mathcal{N}_{B_i} \tag{21}$$

When there is a power or bandwidth pre-allocation in the i-th spot beam, $P_{ai} = P_{B_i}$ or $W_{ai} = W_{B_i}$. Otherwise, the power or bandwidth allocated to the i-th spot beam can be calculated according to J. P. Choi and U. Park [1, 5], where the traffic demand of the spot beam is the sum of the traffic demand of the users in it, and the signal attenuation factor of the spot beam is the mean value of the signal attenuation factor of the users in it. In summary, the whole process of the low computational complexity algorithm is given as follows:

Step 1. Calculate the bandwidth and power allocated to each spot beam according to Choi and Chan and Unhee Park [1, 5].
Step 2. For the users in the same spot beam, calculate the bandwidth and power allocated to each user according to (18) and (19).
Step 3. If the bandwidth and power allocated to a user is smaller than zero, then set the bandwidth and power allocated to the user equal to zero, and go to step 2 to recalculate the bandwidth and power allocated to the remaining users; otherwise terminate the algorithm.

According to the above process, it is seen that the computational complexity of step 1 is $O(K)$ [1, 5]. The computational complexity of step 2 is $O(M)$. Therefore, the total computational complexity of the suboptimal algorithm is much less than the optimal algorithm.

4 Performance Analysis and Simulation Results

For the simulation, we set up a Ka band multi-spot-beam satellite communication system. The system has four spot beams and 20 users, and each spot beam has five users. The total power of the satellite is 200 W, and the total bandwidth of the satellite is 500 MHz. The noise power spectral density parameter N_0 is e^{-6}. The traffic demand of a user in each spot beam increases from 30 Mbps to 70 Mbps, by steps of 10 Mbps.

4.1 Efficiency of the Proposed Joint Allocation Algorithm

To verify the efficiency of the proposed optimal joint allocation algorithm, we compare the algorithm with the following 3 algorithms in case (d). The signal attenuation factors α_i^2 of all the users are set to be 5.

a. Uniform bandwidth allocation and uniform power allocation (UBUP).
b. Uniform bandwidth allocation and optimal power allocation (UBOP).
c. Uniform power allocation and optimal bandwidth allocation (OBUP).

Since channel conditions of users in different beams are the same, the allocation results of users in different spot beam are the same. Therefore, we only need to plot the allocation results for the first beam.

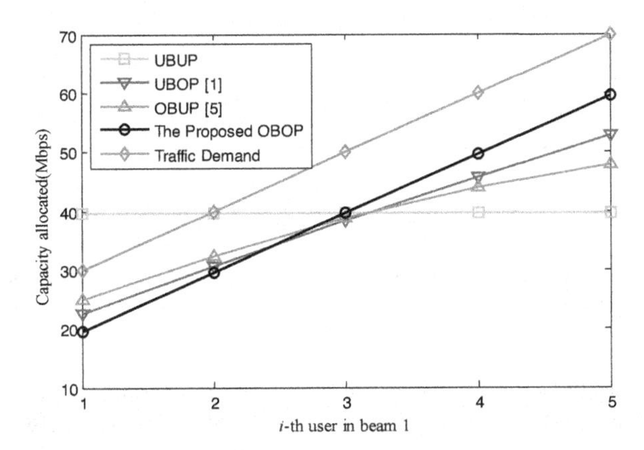

Fig. 2. Comparison of capacity allocated to the i-th use in beam 1 for the four algorithms.

Figure 2 shows the comparison of the capacity allocated to the i-th use in beam 1 for the four algorithms. Table 1 shows the comparison of the total system capacity for the four algorithms. It is known that the traffic demand of the users increases linearly, thus to obtain fairness between the beams, the separate optimal allocation algorithms (UBOP, OBUP) will provide more power or bandwidth resources to higher traffic demand users. However, due to the concavity of the capacity function with a fixed bandwidth or power allocation, the capacity allocated to each user is not linearly increased. It is clearly seen from Fig. 2 that the capacity curve is concave. The UBUP

Table 1. Comparison of total capacity for the four algorithms

Algorithm	$\sum Ci$
UBUP	753.985 Mbps
UBOP [1]	761.272 Mbps
OBUP [5]	751.822 Mbps
The Proposed OBOP	792.455 Mbps

algorithm allocates resources to each user regardless of the traffic demand, resulting in user one being allocated more resources than are needed, causing resource waste. The OBOP algorithm dynamically allocates bandwidth and power resource to each user, thus the capacity curve is almost linear, and the total system capacity is improved. This conclusion is also demonstrated by the data in Table 1.

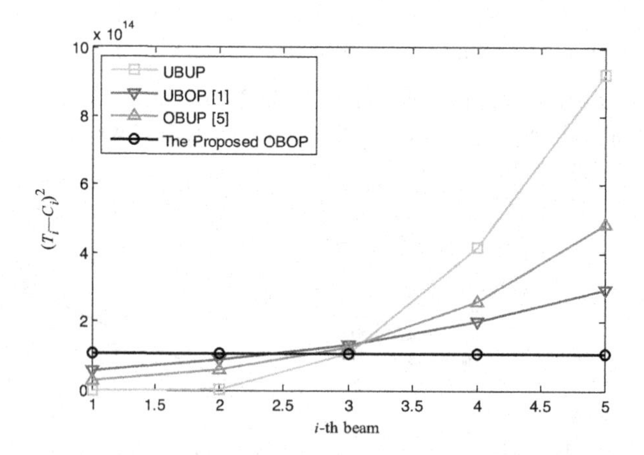

Fig. 3. Comparison of the deficit between the traffic demand and the capacity allocated to beam 1 for the four algorithms.

Figure 3 shows the deficit between the traffic demand and the capacity allocated to the *i*-th user in beam 1. Table 2 presents the sum of the deficit between the traffic demand and the capacity allocated to each user. It is seen from Fig. 3 that in the optimal joint allocation algorithm (OBOP), the deficit between traffic demand and capacity allocated is almost the same, so the fairness between the beams is much better than for that of the separate optimal allocations (UBOP and OBUP). This conclusion is also shown in Table 2. Together with the conclusion above regarding total system capacity, we can conclude that the performance of the optimal joint allocation algorithm (OBOP) is much improved compared with the individual optimal algorithms (UBOP and OBUP).

Table 2. The total sum of $(T_i - C_i)^2$ of the four algorithms

Algorithm	$\sum(Ti - Ci)2$
UBUP	6.0606E15
UBOP [1]	3.0894E15
OBUP [5]	3.8173E15
The proposed OBOP	2.1537E15

4.2 Performance of the Low Computational Complexity Allocation Algorithm

To show the performance of the low computational complexity algorithm (LBLP), we compare it with the optimal allocation algorithm (OBOP) in the following three scenarios.

Scenario 1: The channel conditions of all users are the same. The signal attenuation factors α_i^2 of all the users are set to 5 (Table 4).

Table 3. The total system capacity of two algorithms in the four cases in scenario 1.

	Case (a)	Case (b)	Case (c)	Case (d)
OBOP	792.48 Mbps	775.51 Mbps	790.38 Mbps	789.48 Mbps
LBLP	792.48 Mbps	775.51 Mbps	790.38 Mbps	789.48 Mbps

Scenario 2: The channel conditions of each user in the same beam are the same, while the channel conditions of users in different beams are different. The signal attenuation factors α_i^2 of the users in four beams are set to be 10/2, 10/2.5, 10/3, and 10/3.5, respectively, and the signal attenuation factors α_i^2 for users in the same beam are set to be the same (Tables 5 and 6).

Scenario 3: We compare the performance of the two algorithms when the channel condition of each user is different. The signal attenuation factor α_i^2 of each user conforms to uniform distribution between 5 and 3.5 (Table 7).

Table 4. The objective function value of two algorithms in the four cases in scenario 1.

	Case (a)	Case (b)	Case (c)	Case (d)
OBOP	2.15E15	2.83E15	2.21E15	3.14E16
LBLP	2.15E15	2.83E15	2.21E15	3.14E16

Table 5. The total system capacity of two algorithms in the four cases in the scenario 2.

	Case (a)	Case (b)	Case (c)	Case (d)
OBOP	655.81 Mbps	645.76 Mbps	652.52 Mbps	645.25 Mbps
LBLP	652.40 Mbps	645.76 Mbps	652.52 Mbps	645.25 Mbps

Table 6. The objective function value of two algorithms in the four cases in scenario 2.

	Case (a)	Case (b)	Case (c)	Case (d)
OBOP	6.03E15	6.28E15	6.08E15	6.33E16
LBLP	6.08E15	6.28E15	6.08E15	6.33E16

Table 7. The total system capacity of two algorithms in the four cases in scenario 3.

	Case (a)	Case (b)	Case (c)	Case (d)
OBOP	652.84 Mbps	648.89 Mbps	653.41 Mbps	652.70 Mbps
LBLP	650.84 Mbps	646.56 Mbps	650.68 Mbps	651.48 Mbps

Table 8. The objective function value of two algorithms in the four cases in scenario 3.

	Case (a)	Case (b)	Case (c)	Case (d)
OBOP	6.09E15	6.86E15	6.15E15	6.92E16
LBLP	6.12E15	6.87E15	6.18E15	6.94E16

From the above tables, it is seen that the performance of the LBLP algorithm is the same as that of the OBOP algorithm, for the four cases when the channel conditions of each user are the same. When the channel conditions of each user in the same beam are the same, the performance of the LBLP algorithm is the same as that of the OBOP algorithm for cases (b)–(d). When the channel condition of each user is different, the value of the objective function of the LBLP algorithm is little more than that of the OBOP algorithm, thus the fairness between each user of the LBLP algorithm is lower, however, the total system capacity of the LBLP algorithm is improved over that of the OBOP algorithm. Therefore, the performance of LBLP algorithm is much closer to the OBOP algorithm, especially when the channel conditions of each user in the same beam are the same.

4.3 The Impact of Pre-allocations for Each Beam

From Tables 3 through 8, it is shown that in the same scenario, the value of the objective function of case (a) is lower than the other three cases. In other words, the power or bandwidth pre-allocations to each beam deteriorate the total system performance. Because when there is no power or bandwidth pre-allocation to each of the beams, the power and bandwidth allocated to each user can be more flexible.

5 Conclusion

In this paper, we sought to solve a problem of the joint power and bandwidth allocations for multiple users in a multi-beam satellite communication system. To this end, we first formulated the problem as a convex optimization problem. Then we proposed an optimal joint allocation algorithm and a low computational complexity algorithm. The optimal joint allocation algorithm was more efficient than the separate bandwidth or power allocation algorithm. The performance of the low computational complexity algorithm was very close to that of the optimal joint allocation algorithm.

Appendix

From the analysis in Sect. 3, it is shown that the constraints can indeed be ignored.

Taken together with the fact that the constraints (4, 5, 6 and 7) are linear, to prove the optimization problem is convex, we only need to prove that $\sum_{i=1}^{M}(T_i - C_i)^2$ is convex [10].

It is known that the sum of convex functions is also convex. Therefore, to prove that $\sum_{i=1}^{M}(T_i - C_i)^2$ is convex, we just need to prove the following function is convex:

$$f(P_i, W_i) = (T_i - C_i)^2 \tag{22}$$

where $C_i = W_i \log_2\left(1 + \frac{\alpha_i^2 P_i}{W_i N_0}\right)$.

It is known that is the Hessian of one function is semi-definite, thus the function is convex [10]. The Hessian of $f(P_i, W_i)$ is given as follows:

$$H_f = \begin{bmatrix} \frac{\partial^2 f(P_i, W_i)}{\partial P_i^2} & \frac{\partial^2 f(P_i, W_i)}{\partial P_i \partial W_i} \\ \frac{\partial^2 f(P_i, W_i)}{\partial P_i \partial W_i} & \frac{\partial^2 f(P_i, W_i)}{\partial W_i^2} \end{bmatrix} \tag{23}$$

To prove that H_f is positive semi-definite, we obtain the following equations:

$$\begin{aligned}\frac{\partial^2 f(P_i, W_i)}{\partial P_i^2} &= 2\left(\frac{\partial C_i}{\partial P_i}\right)^2 - 2(T_i - C_i)\frac{\partial^2 C_i}{\partial P_i^2} \\ &= 2\left(\frac{\partial C_i}{\partial P_i}\right)^2 + 2(T_i - C_i)\frac{W_i}{\ln 2(N_0 W_i/\alpha_i^2 + P_i)^2}\end{aligned} \tag{24}$$

$$\begin{aligned}|H_f| &= \frac{\partial^2 f(P_i, W_i)}{\partial P_i^2}\frac{\partial^2 f(P_i, W_i)}{\partial W_i^2} - \frac{\partial^2 f(P_i, W_i)}{\partial P_i \partial W_i}\frac{\partial^2 f(P_i, W_i)}{\partial P_i \partial W_i} \\ &= 4(T_i - C_i)\frac{C_i^2}{W_i \ln 2(N_0 W_i/\alpha_i^2 + P_i)^2}\end{aligned} \tag{25}$$

When $T_i \geq C_i$, it is obvious that (24) and (25) are non-negative. Therefore, H_f is positive semi-definite, and $\sum_{i=1}^{M} (T_i - C_i)^2$ is convex, thus the optimization problem is convex. As a result, the solution obtained from the joint bandwidth and power algorithm based on KKT conditions is the global optimal solution of the optimization problem.

References

1. Choi, J.P., Chan, V.W.S.: Optimum power and beam allocation based on traffic demands and channel conditions over satellite downlinks. IEEE Trans. Wirel. Commun. **4**(6), 2983–2993 (2005)
2. Yang, H., Srinivasan, A., Cheng, B., et al.: Optimal power allocation for multiple beam satellite systems. In: Proceedings of IEEE Radio and Wireless Symposium, pp. 823–826, January 2008
3. Feng, Q., Li, G., Feng, S., et al.: Optimum power allocation based on traffic demand for multi-beam satellite communication systems. In: International Conference on Communication Technology (ICCT), pp. 873–876, September 2011
4. Park, U., Kim, H.W., Ku, B., et al.: Optimum selective beam allocation scheme for satellite network with multi-spot beams. In: SPACOMM 2012: The Fourth International Conference on Advances in Satellite and Space Communications, pp. 78–81, April 2012
5. Park, U., Kim, H.W., Ku, B., et al.: A dynamic bandwidth allocation scheme for a multi-spot-beam satellite system. ETRI J. **34**(4), 613–616 (2012)
6. Kumaran, K., Viswanathan, H.: Joint power and bandwidth allocation in downlink transmission. IEEE Trans. Wirel. Commun. **4**(3), 1008–1016 (2005)
7. Gong, X., Vorobyov, S.A., Tellambura, C.: Joint Bandwidth and power allocation with admission control in wireless multi-user networks with and without relaying. IEEE Trans. Signal Process. **59**(4), 1801–1813 (2011)
8. Miao, J., Hu, Z., Yang, K., et al.: Joint power and bandwidth allocation algorithm with Qos support in heterogeneous wireless networks. IEEE Commun. Lett. **16**(4), 479–481 (2012)
9. Cover, T.M., Thomas, J.A.: Elements of Information Theory. Wiley, New York (1991)
10. Boyd, S., Vandenberghe, L.: Convex Optimization. Cambridge University Press, Cambridge (2004)

11. Yu, W., Lui, R.: Dual methods for nonconvex spectrum optimization of multicarrier systems. IEEE Trans. Commun. **54**(7), 1310–1322 (2006)
12. Ding, G., Wu, Q., Wang, J.: Sensing confidence level-based joint spectrum and power allocation in cognitive radio networks (unpublished)
13. Wang, R., Vincent, K.N.L., Lv, L., et al.: Joint cross-layer scheduling and spectrum sensing for OFDMA cognitive radio systems. IEEE Trans. Wirel. Commun. **8**(5), 2410–2416 (2009)
14. Antonio, G.M., Wang, X., Georgios, B.G.: Dynamic resource management for cognitive radios using limited-rate feedback. IEEE Trans. Signal Process. **57**(9), 3651–3666 (2009)

A Networking Transmission Method in the Case of Limited Satellite Transceiver

Yaoxu He, Hongyan Li(✉), Peng Wang, Hang Liu, and Fan Qi

State Key Laboratory of Intergrated Service Networks, Xidian University, Xi'an 710071, China
hyli@xidian.edu.cn

Abstract. Due to the cost and hardware limitations of the satellite, it is difficult for a single satellite to establish multiple data communication links at the same time. When the satellite network carries a large amount of data transmission, efficient network transmission in this resource-constrained network has become a challenge. Considering the constraints of a single transceiver, we propose a topology planning method based on connectivity fairness, and formulate the rules for satellite transceiver rotation; on this basis, we design a low-delay routing method under resource constraints, which is characterized by time-varying graphs, and uses traffic to calculate single or multiple paths that meet transmision requirements, which not only ensures equal connectivity opportunities, but also reduces the delay of end-to-end paths. Finally, relying on the iridium constellation, under the connectivity fairness topology configuration method, we simulate and analyze the algorithm, and prove that the network transmission has a lower path delay than the satellite over-the-top transmission.

Keywords: Resource constraints · Connectivity fairness · Low-delay routing · Networking transmission

1 Introduction

With the development of science and technology, satellite communication has become the main force of information transmission and communication, integrated into people's life and become an inseparable part of human daily life. The importance of satellite communication is not only reflected in ordinary communication transmission, but also has a profound impact on military national defense, production safety and economic development [1]. Now there is another bright spot in satellite communication-small satellite network. Since the concept of satellite communication was put forward, it has made great progress. In recent years, with the endless emergence of affordable and innovative commercial spot technology solutions, and the continuous progress of microelectronics and microsystem technology, the size of satellite components has been continuously reduced. so that people can design small satellites. Because of its low development cost and low energy consumption, small satellite network is playing a more and more important role in today's satellite communication technology. Its emergence has greatly bridged the gap of data shortage in

Q. Guo et al. (Eds.): WiSATS 2021, LNICST 410, pp. 50–61, 2022.
https://doi.org/10.1007/978-3-030-93398-2_5

many vertical industries, and with the arrival of the era of the Internet of things, satellites also play an important role in M2M communications.

However, due to the limitations of the cost and hardware conditions of the small satellite network, for example, a small satellite can only carry one transceiver, so that a single satellite can only establish a single data communication link at the same time, and it is impossible to build a complete end-to-end transmission path. in this context, how to quickly return a large amount of information from the satellite to the ground station has become a thorny problem. At the same time, the satellite nodes in the satellite network will operate periodically according to the orbit, and the communication links between the nodes will be disconnected frequently, so the satellite network topology is time-varying. In view of the above problems, this paper proposes a "step-by-step" strategy to meet the communication requirements of conventional tasks, by considering the time evolution characteristics of network topology and link connections in each period of time, to coordinate multi-satellite nodes to achieve joint scheduling of multiple transceivers, in order to ensure a fair connectivity opportunity for each satellite node, and then to ensure the fairness of task transmission among different satellite nodes. Enable all nodes in the satellite network to access the available links between them as well as possible. The goal of the strategy is to provide equal opportunities for all network nodes to exchange data traffic by establishing appropriate plans, and on this basis, an efficient resource scheduling method with low delay is established.

Reference [2] discusses an efficient broadcast and multicast tree construction algorithm for all-wireless multi-hop networks under the joint constraints of a limited number of transceivers and a limited number of available frequencies on network nodes. By using the method of formula derivation and simulation, this paper deduces the performance of energy-saving multicast of session traffic in wireless network under the condition of limited number of transceivers. The resource-constrained satellite network of intermittent Unicom is analyzed in reference [2], and a connection plan design (Contact Plan Design, CPD) is proposed to efficiently schedule the link according to the optimization objective in this environment. An analysis framework for resource-constrained small satellite networks is proposed in reference [3–6]. The extended traditional time spread graph is used as a tool to optimize the delay and throughput. These literatures have analyzed the resource-constrained satellite networks, but the common disadvantage is that the analysis process is based on mathematical formulas, which leads to the complexity and obscurity of the analysis process. And the resource scheduling problem of satellite networks with limited transceivers is still being studied, so we propose a new strategy to solve this problem.

In view of the limited satellite transceiver, this paper first proposes a "step-by-step" strategy based on connectivity fairness to ensure that each satellite node has the same connectivity opportunity; on this basis, a low-delay routing method based on "step-by-step" strategy is proposed, and one or more end-to-end low-delay paths satisfying service transmission are constructed according to the relationship between service resources and link available resources. Finally, we simulate and analyze the proposed strategy and method based on the iridium constellation, and the results show that compared with the satellite over-top transmission mode, single-path transmission has lower path delay. Multi-path transmission is suitable for link resource constraints to obtain lower delay.

2 Topology Planning Based on Fairness

In order to meet the fairness of inter-satellite transmission, a method of rotation of transceiver angle between satellites in constellation is designed, so that all satellites can get fair communication opportunities in a long time range. As shown in Fig. 1 below, a simplified satellite constellation is constructed, consisting of 16 satellite nodes, numbered 1 × 16. In the satellite constellation, the relative position between satellites does not change, in a given topology, each satellite can only have the opportunity to communicate with neighboring nodes. Because each satellite carries only a single transceiver. Therefore, if two satellites need to communicate between satellites, they need to point each other's transceivers to each other's location. As shown in Fig. 1, if satellite 1 needs to communicate with satellite 4, satellite 1 needs to point its transceiver to satellite 4, while satellite 4 points its transceiver to satellite 1.

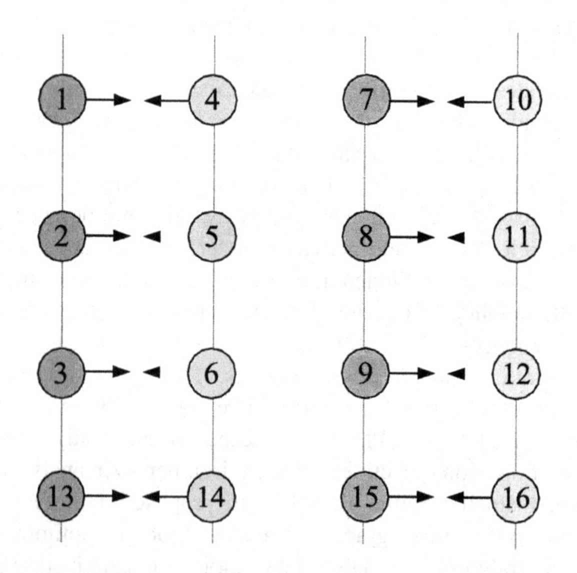

Fig. 1. Satellite constellation A possible transceiver orientation.

Based on the above description, we specify the orientation of the transceivers of all satellites in the satellite constellation at the initial time, and design the periodic rotation method of the satellite in a certain time range T, so that each satellite has a fair communication opportunity with all its neighboring nodes.

For a given constellation, we propose a satellite transceiver rotation method for fair communication. First of all, it is stipulated that during the initial period of time, all satellites communicate with satellites in adjacent orbits. This is shown in the topology of the first period of time in Fig. 2. Then when any satellite in the topology determines the rotation direction of its transceiver at a uniform speed, its neighbor node satellite transceiver rotates in the opposite direction. Therefore, when the rotation direction of any one satellite is determined, the rotation direction of other satellites in the

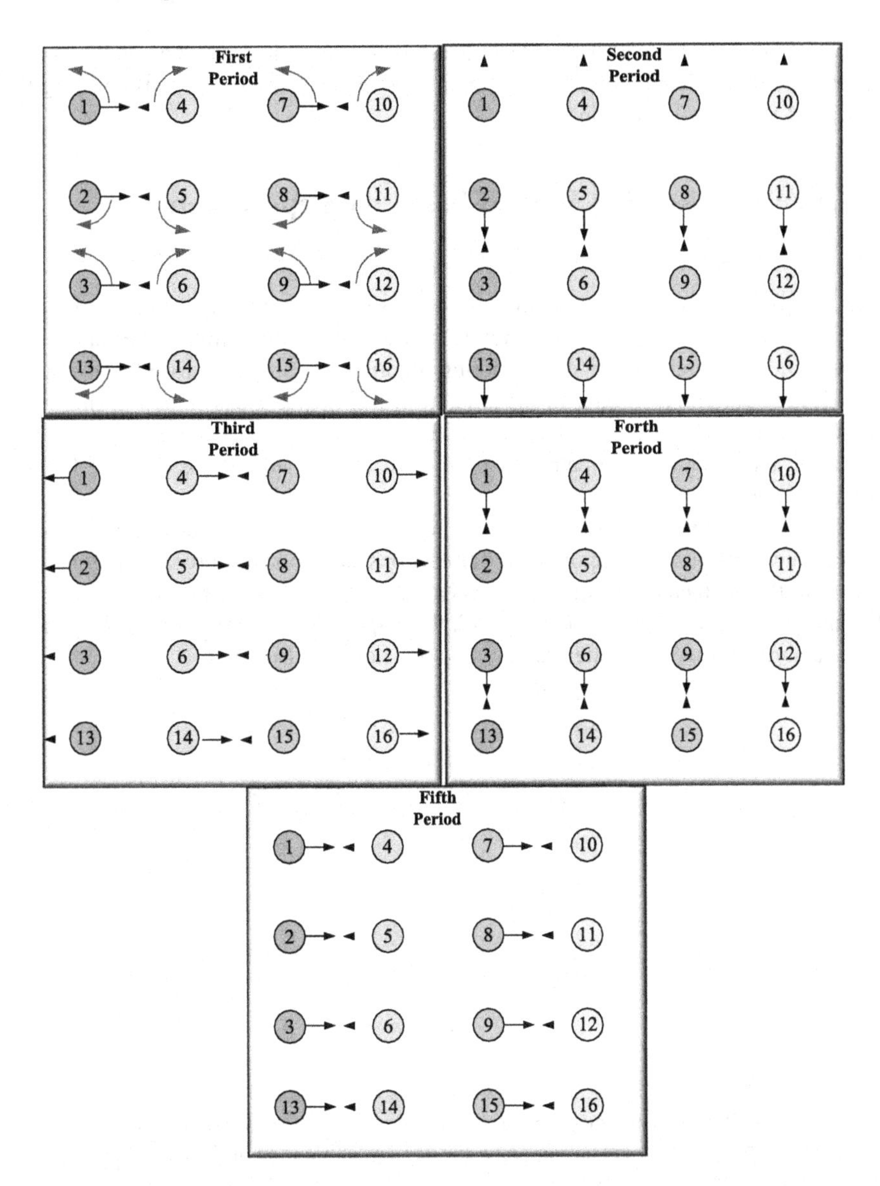

Fig. 2. Rotation method of each satellite transceiver in satellite constellation.

constellation can be determined. Based on this rotation method, the communication opportunity between the satellite and its neighbor nodes is the same in a rotation period. As shown in Fig. 2, satellite 5 rotates counterclockwise at a uniform speed, and its neighbor nodes 2, 4, 6 and 8 all rotate at a uniform speed clockwise, and in the next four periods, they can communicate with 6, 8, 4 and 2 respectively.

Therefore, based on the satellite transceiver rotation method, the satellite transceiver can communicate with each neighbor node in a rotation cycle, thus ensuring the fairness of the communication.

3 Low Delay Routing Method Based on Step-by-Step Strategy

In this paper, the topology planning method based on connectivity fairness is summarized, and the steering rules of satellite nodes with a single transceiver are given. based on this research, a low-delay routing algorithm is proposed to transfer data as quickly as possible. According to the amount of traffic that needs to be transmitted in different scenarios, we jointly consider the available satellite node resources and link resources, and obtain single-path and multi-path routing algorithms suitable for different scenarios.

The satellite with a single transceiver rotates periodically according to the configured rules, resulting in topological connections and link resources changing with time, with the characteristics of a time-varying network. Considering the accuracy and efficiency of the time-varying graph model, this paper uses the time expansion graph to model a single transceiver satellite network, as shown in Fig. 3.

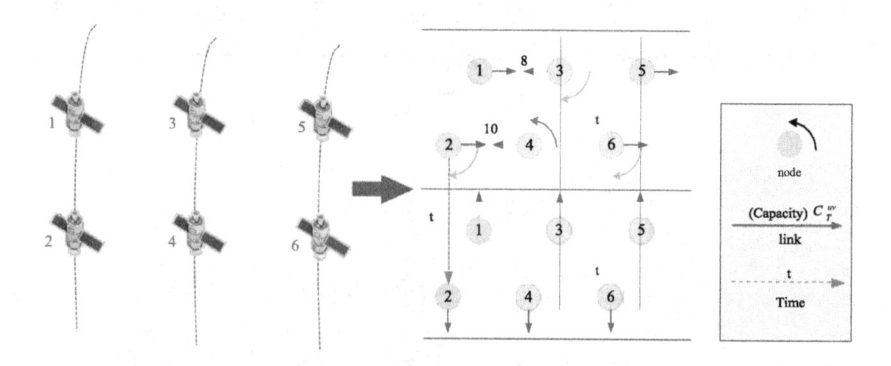

Fig. 3. The satellite network model of a single transceiver satisfies the steering rules of the satellite transceiver, and the green link represents the switching time of the satellite transceiver. (Color figure online)

The rotation direction of each satellite node is opposite to that of the neighboring satellite nodes, and the rotation time is t, so the period is Tunable 4t. Because the rotation handoff time between satellite nodes is much longer than the propagation delay of satellite links, link resources only consider the amount of data that can be transmitted by the link in a period of time. The links between different periods represent the longest time it takes for satellite nodes to switch neighbors, that is, rotation time t.

The core idea of low-delay routing algorithm based on connectivity fairness is to find one or more paths that satisfy low-delay transmission at one time. First of all, an end-to-end path with the lowest cost is calculated according to the transmission time of the data request, and the routing rule is similar to the CGR algorithm (reference). Then it is compared with the amount of data to be transmitted, and the above steps are repeated until enough end-to-end paths are found that all the data can be transmitted. The algorithm flow of solving single path and multi-path is given in Algorithm 1 and 2 respectively (Table 1).

Table 1. The description of single path algorithm

Algorithm1 single path algorithm	
1.	Input: Topology G, Source node s, Destination node d, Amount of data M, Start time t_{start}.
2.	Output: The shortest delay of the $s - d$ is min_delay.
3.	Initialize the distance of each node is the graph G, $A[v] = \infty, \forall v \neq s; A[s] = t_{start}$, add s to the priority queue Q.
4.	While Q is not empty do {
5.	$u = \min(Q)$;
6.	If u is d, then
7.	break;
8.	For each node v adjacent to u do $A[v] = A[u] + uv_{delay}$; $parent[v] = u$;}
9.	$C_{min}^{uv} = \min\_capacity(path)$
10.	If $C_{min}^{uv} \geq M$ then
11.	$\min\_delay = A[u]$;
12.	Output min_delay with the shortest delay of $s - d$.

In the single path algorithm, the links in different periods are associated together by using the switching time of the satellite transceiver to form a shortest delay path to ensure the reliability of data transmission. It is worth noting that when the volume of traffic is graeter than the data that can be carried by the path the formula for calculat-ing the delay is as follws:

$$\min\_delay = ceil\left(\frac{M}{C_{min}^{uv}}\right) * A[u] \quad (1)$$

Due to the constraints of satellite node resources in a single cycle, the number of end-to-end paths is limited, so the delay is composed of the sum of path delays in multiple periods. Unlike a single path, the flow of the multipath algorithm is as follows. The algorithm calculates multiple paths at one time according to the traffic, but needs to update the topology resources before calculating the next path to prevent the use of duplicate edges in multiple paths and does not meet the constraints of a single satellite transceiver. It is precisely because of the constraints of satellite resources that it is impossible for a satellite node to have more than 1 neighbors in a period of time, that is, there are no duplicate edges in multiple paths. The satellite rotates according to a certain period, and the data to be transmitted can be transmitted along the calculated different paths, making full use of the link resources of the network. The delay calculation formula of multi-path is as follows (Table 2):

$$\mathrm{min\_delay} = \mathrm{ceil}\left(\frac{\mathrm{M}}{C_{min}^{uv}}\right) * A[u] \tag{2}$$

Table 2. The description of multipath algorithm

Algorithm 2 multipath algorithm	
1.	Input: Topology G, Source node s, Destination node d, Amount of data M, Start time t_{start}.
2.	Output: The shortest delay of the s – d is min_delay .
3.	Initialize $capacity_{record} = 0$, $\mathrm{temp_{delay}} = 0$.
4.	While $capacity_{record} < M$ {
5.	Run Algorithm 1;
6.	$capacity_{record} = C_{min}^{uv} + capacity_{record}$, $\mathrm{temp_{delay}} = \max(\mathrm{temp_{delay}}, A[u])$;
7.	update(G);}
8.	Output min_delay with shortest delay of s – d.

In order to clearly explain the two routing algorithms, a simplified satellite constellation with six nodes is constructed, and the nodes are marked with the number 1/6 respectively. The nodes represent the satellite nodes at different times, and the link indicates that there is a communication opportunity between the two satellites. The transceiver can be directed to each other to build a communication link. As shown in Fig. 4 below, each satellite has only one set of transceivers, and each satellite has communication opportunities only with neighboring nodes. For example, satellite 1 may communicate with satellite 2 or satellite 3. Satellite 3 may communicate with satellite 1 or satellite 4 or satellite 5. Assuming that the amount of data to be transmitted is 10 m, the link capacity on each link is known, and the rotation period of the satellite is 4, now the satellite node 1 needs to communicate with the satellite node 6 and transmit the data.

Figure 4 shows the schematic diagram of single-path and multi-path solution. First of all, a time spread graph model is constructed, which correlates the topologies of different periods and characterizes the distributable relationship of transceiver resources of constellations in different periods. The model is shown in Fig. 5 below.

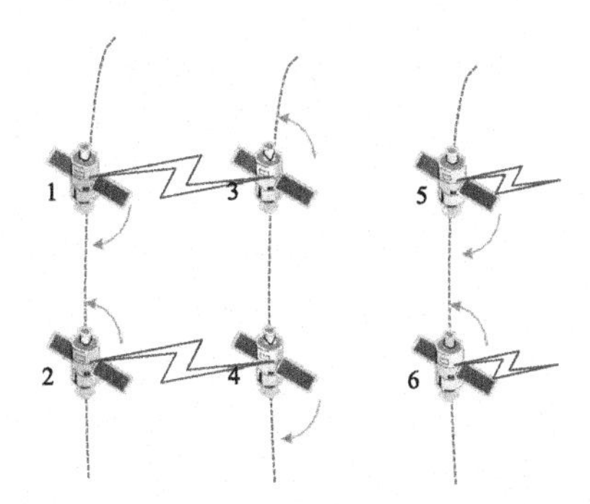

Fig. 4. A simplified 6-star network constellation in which each satellite node turns to the opposite of its neighbor satellite node.

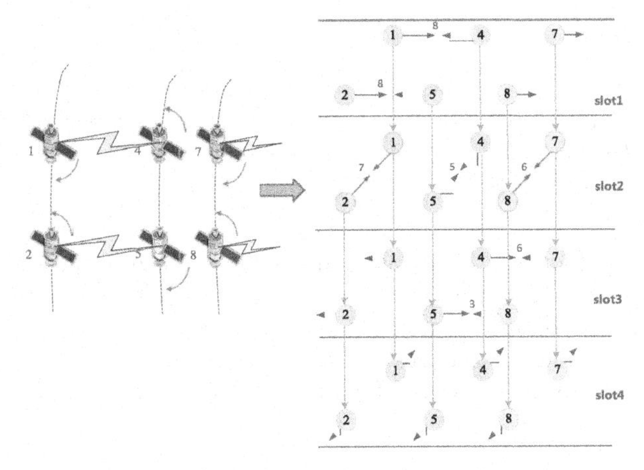

Fig. 5. The time expansion chart of the satellite network shows the topological connection of a cycle.

As shown in Fig. 6, a constellation consisting of six satellites uses a step-by-step strategy to run an one-cycle topology. Node 1 is selected as the source node and satellite 8 is selected as the destination node. If the traffic to be transmitted is 3 m, then follow the purple path 1-3-4-6 in the figure on the left, and it will take 2 periods of time to transmit. However, at this time, the transmission traffic is 10 m, and because the link capacity is only 3 m, if the single-path algorithm is used, the path will follow the purple path for 3 times, and it will take 14 periods of time; if the multi-path algorithm is used, a total of 3 paths will be calculated, and marked red in the figure on the right, the maximum delay of the three paths is 6 periods. Compared with the singlepath algorithm, it is 8 times faster (Fig. 6).

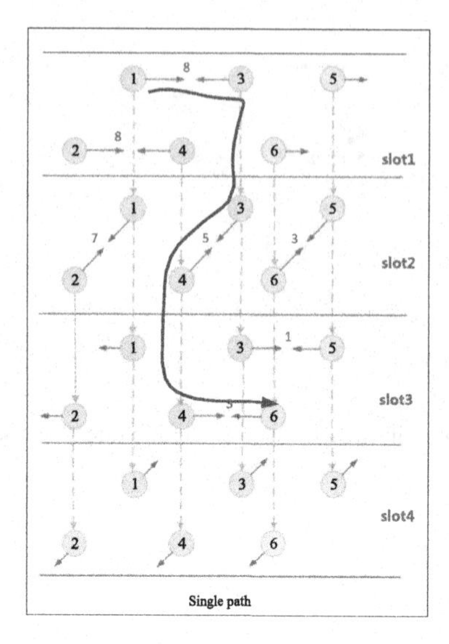

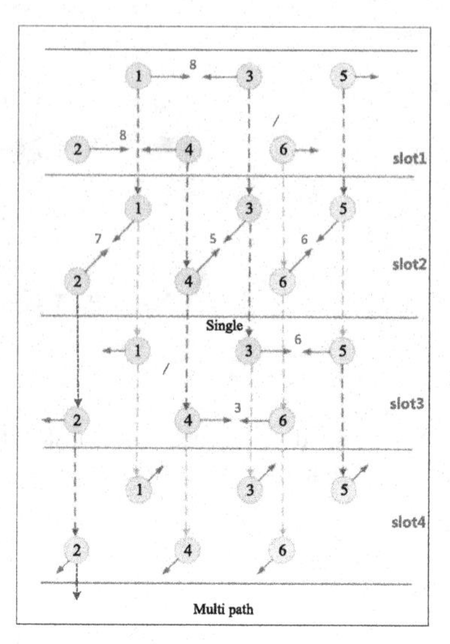

Fig. 6. The schematic diagram of the algorithm shows that the red path on the left is the result of the single-path algorithm, and the red link on the right is the three paths obtained by the multi-path algorithm (except for handoff delay).

Under the condition that the resources of satellite transceivers are limited, the low-delay routing method based on synchronous walking strategy aims at low delay, and the algorithm can be adjusted according to the amount of data to be transmitted, so as to meet the needs of different scenarios and improve the resource utilization of the network. it provides a reliable end-to-end transmission path for services.

4 Performance Evaluation and Simulations

In this paper, the communication connection between satellite networks is established by networking, and the information is sent from the originating node to the node that establishes a communication connection with the ground station, which greatly shortens the information transmission waiting time for sending information from the originating node to the ground station.

In order to evaluate the proposed algorithm, we select the iridium constellation for simulation. We configure each node in the constellation as the synchronous walking strategy proposed above, each satellite has four neighbors, rotates according to certain rules, 100 s switches, 400 s is a cycle. The Iridium constellation has six orbits, and the number of satellites in each orbit is 12. Here, the first satellite in each orbit and the sixth satellite are selected to transmit data to the ground station.

In the case of sufficient network link resources, the constructed single path can meet the data transmission. We simulate the 2G transmission service on each selected satellite node and compare the delay required by single path and over-top transmission. The simulation results are shown in Fig. 7.

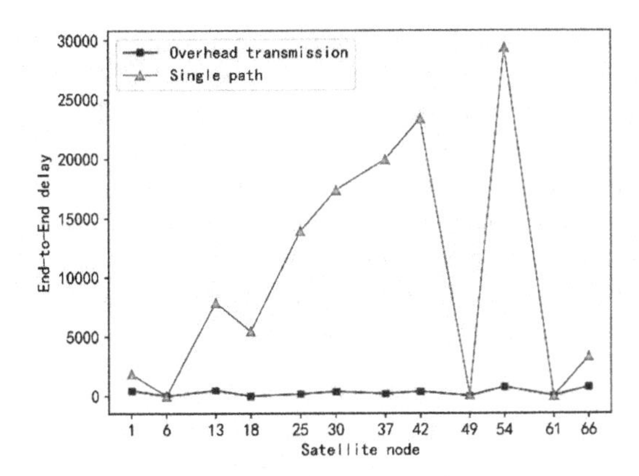

Fig. 7. The schematic diagram of the algorithm shows that the red path on the left is the result of the single-path algorithm, and the red link on the right is the three paths obtained by the multi-path algorithm (except for handoff delay). (Color figure online)

The simulation results show that under the condition of a single transceiver, the delay required for satellite over-the-top transmission is much larger than that for single-path transmission. By correlating and utilizing the link resources of different periods, the delay of the selected path can be greatly reduced. If the network link resources are tight, it can not guarantee that it can be transmitted all at once, so it is necessary to build multiple paths for transmission at the same time. By reducing the link capacity resources, we simulate the multipath algorithm and compare it with the single path. The simulation results are shown in Fig. 8.

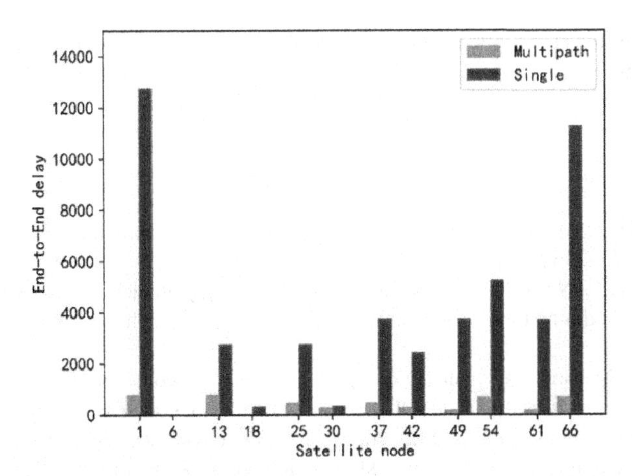

Fig. 8. Comparison of single-path and multi-path (ordinate is end-to-end delay, unit is second, Abscissa is the label of satellite node).

We can see intuitively that the delay required for multi-path transmission is much lower than that for single-path transmission of the same traffic. The reason is that the previously proposed topology planning method is periodic. The algorithm looks for multiple end-to-end reliable paths at one time, makes full use of link resources in different periods, and reduces the delay of the path.

Under the condition of ensuring the connectivity fairness of a single transceiver satellite, the low-delay routing algorithm constructs a low-delay end-to-end path to ensure the transmission of traffic. Under the condition of resource first, multi-path transmission can often achieve better performance.

5 Conclusion

This paper studies the problem of network transmission in satellite network under the condition of limited transceiver. Considering the problem of fair transmission in satellite networks, a "step-by-step" strategy is proposed to establish a fair satellite network transmission mechanism, and on this basis, the relative size of service bandwidth and link bandwidth is considered. two routing algorithms, single-path and multi-path, are proposed, which correlate and utilize the topology resources in different periods to ensure the end-to-end delay and improve the utilization of network resources. The algorithm has analysis and consideration for different situations and has good practicability. Finally, based on the specific satellite network scenario, the performance of single-path and multi-path routing algorithms is simulated and compared with that of traditional non-networking satellite network topology. verify the advantages of satellite network transmission under the constraint of a single transceiver, and reveal that the "step-by-step" strategy and network transmission have important practical significance for the communication between satellite and ground station.

Acknowledgments. This work is supported by the National Natural Science Foundation of China (61871456).

References

1. Zhang, J.: The development prospect and prospect of satellite communication. Digit. Technol. Appl. (005), 23–23 (2017)
2. Wieselthier, J.E., Nguyen, G.D., Ephremides, A.: Energy-efficient multicasting of session traffic in bandwidth-and transceiver-limited wireless networks. Clust. Comput. **5**(2), 179–192 (2002)
3. Huang, T.: Research on connection plan in resource-constrained satellite networks. Chongqing University of posts and Telecommunications, Chongqing (2019)
4. Madoery, P.G., Finochietto, J.M., Fraire, J.A.: Traffic-aware contact plan design for disruption-tolerant space sensor networks. Ad Hoc Netw. **47**(Sep.), 41–52 (2016)

5. Fraire, J.A., Finochietto, J.M.. Design challenges in contact plans for disruption-tolerant satellite networks. IEEE Commun. Mag.: Articles News Events Interest Commun. Eng. **53**(5), 163–169 (2015)
6. Fraire, J., Finochietto, J.M.: Routing-aware fair contact plan design for predictable delay tolerant networks. Ad Hoc Netw. **25**(Feb. Pt.B), 303–313 (2015)

Design and Implementation of a High Efficiency Space Packet Routing Algorithm on a Spacecraft

Sheng Yu(✉), Bo Zhou, Jiaxiang Niu, Jianbing Zhu, Jian Guo, and Duo Wang

Beijing Institute of Spacecraft System Engineering, CAST, Beijing, China

Abstract. Space packet protocol is a widely used network layer protocol on a spacecraft. A modern spacecraft, such as the Tianhe core module of Chinese space station, has a very complex on-board network. Routing space packets in time is a challenge for such a spacecraft. To solve this problem, we design and implement a high efficiency space packet routing algorithm based on hash routing table. First, we introduce the typical structure of a space packet. Second, we introduce the concept of logical address and hardware address of a space packet. Third, we introduce the construction of the routing table based on the pairs of logical address and hardware address. At last, we implement a routing search algorithm based on the hash table and evaluate the performance the routing algorithm through experiments. We successfully applied the algorithm on the Tianhe core module and achieved good performance.

Keywords: Space packet protocol · Spacecraft · Routing algorithm · Hash table

1 Introduction

A spacecraft onboard network is referring to the local network that connects all devices on a spacecraft and exchanges telecommand (TC), telemetry (TM) and other data. State-of-the-art spacecraft onboard networks [1, 2] use the CCSDS (Consultative Committee for Space Data Systems) AOS (Advanced Orbiting System) protocol [3] as its data link layer protocol and SpaceWire [4], Ethernet, or MIL-STD-1553B bus as its physical layer implementation.

Space packet protocol is a widely used network layer protocol on spacecraft. Space packets are used in both ground-to-space networks and spacecraft onboard networks. A typical example using space packets is the TC application: A ground device encapsulates satellite commands in space packets, and uploads these packets to the satellite via ground-space data link. A router on the satellite receives these commands and send them to other devices via the spacecraft onboard network.

In a complex spacecraft, how to efficiently route these space packets is a challenging problem. The core module of Chinese space station, i.e., Tianhe, has a very complex on-board network, with more than 200 different devices connected together, and the on-board network uses the space packet protocol. A central unit is in charge of

Q. Guo et al. (Eds.): WiSATS 2021, LNICST 410, pp. 62–68, 2022.
https://doi.org/10.1007/978-3-030-93398-2_6

routing all space packets generated by both ground devices and on-board devices. Packet routing is a very time critical mission. In most case, the central unit needs to find a route for a space packet as soon as possible, thus, a high efficiency routing algorithm is desired.

In this paper, we proposed and implemented a high efficiency space packet routing algorithm based on hash table, and we used the algorithm on the Tianhe core module. First, we introduce the typical structure of a space packet. Second, we introduce the concept of logical address and hardware address of a space packet. Third, we introduce the construction of the routing table based on the pairs of logical address and hardware address. At last, we implement a routing search algorithm based on the hash table and evaluate the performance the routing algorithm through experiments.

2 Space Packet Protocol

The space packet protocol is defined in the CCSDS standard CCSDS 133.0-R-1 [5]. The space packet protocol is located in the network layer. A typical structure of a space packet used on the Tianhe core module is illustrated in the Fig. 1.

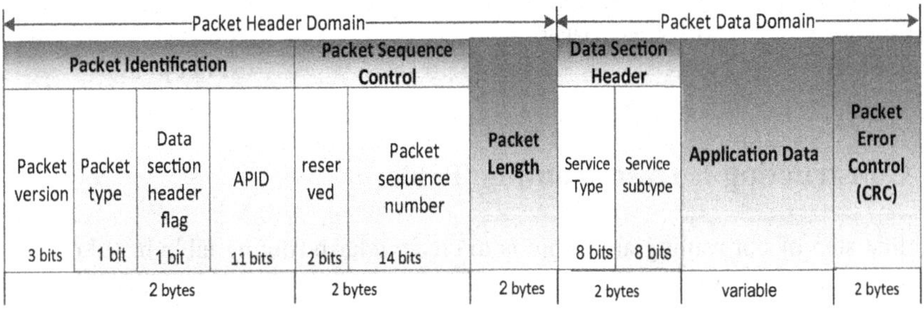

Fig. 1. A typical structure of a space packet used on the Tianhe core module

On the Tianhe core module, the meaning and possible value of each field of the space packet is listed below (Table 1).

Table 1. The space packet structure used on the Tianhe core module

Section name	Meaning	Possible value	Length
Packet version	The version of the space packet protocol	[0, 3]	3 bits
Packet type	The type of the packet	0: a telemetry packet 1: a telecommand packet	1 bit

(*continued*)

Table 1. (*continued*)

Section name	Meaning	Possible value	Length
Data section header flag	Indicates whether the packet has a data section header, which concludes service type and service subtype	0: no 1: yes	1 bit
Application Process Identifier (APID)	ID of the destination application which accepts and processes the packet	[0x000, 0x7FF]	11 bits
Packet sequence number	The sequence number of the packet	[0x0000, 0x3FFF]	14 bits
Packet length	The length of the packet (in bytes)	[0, 836]	16 bits
Service type	An application usually has many service types and this value defines the service type of the packet	[0x00, 0xFF]	8 bits
Service subtype	Each service type may have many subtypes and this value defines the service subtype of the packet	[0x00, 0xFF]	8 bits
Application data	Payload of the packet	/	Variable
Packet error control field	CRC of the packet	[0x0000, 0xFFFF]	16 bits

3 Constructing a Hash Routing Table

The first step of our routing algorithm is to create a hash routing table based on all the routing information we have. Each entry of the hash routing table is a linked list node, which contains a piece of the routing information. Each node has three elements, which are a pointer to the next node, a logical address and a hardware address. The logical address consists of the type of the packet, the APID of the packet, the service type and the service subtype of the packet, and the length of logical address is 4 bytes.

The hardware address defines the destination address of a packet. The format of the hardware address is variable depends on the type of the network used on a spacecraft. In our case, we use three types of format of the hardware address. We use many 1553B buses and an Ethernet bus on the Tianhe core module. If the destination device of a packet is a Remote Terminal (RT) of one of the 1553B buses, the hardware address of the routing entry consists of a bus id, a RT address and a RT sub address of the destination device. If the destination device of a packet is on the Ethernet bus, the hardware address we use is the IP address of the destination device. If the destination of a packet is another process on the central unit itself, the hardware address is the process id of the destination. The length of hardware address is 16 bytes.

We use E_i to represent an entry of the routing table, so $E_i = \langle pNode_i, LA_i, HA_i \rangle$. Here $pNode_i$ is the pointer to the next node of the routing table. LA_i and HA_i are a logical address and a hardware address respectively. To evenly distribute routing entries within the routing table, we select a large hash key $K_D = 429677$. The size of the hash table is represented by S_T. The index of E_i in the hash table Idx_i is calculated as follows:

$$Idx_i = (LA_i \, mod \, K_D) \, mod \, S_T \tag{1}$$

Hash collisions happen when multiple entries have a same index. We use linked list to avoid hash collisions. If E_i and E_j have a same index, then $pNode_i$ equals the start address of E_j.

4 Forwarding a Space Packet

4.1 Finding a Route

A central unit, also known as the Kernel Process Unit (KPU) on the Tianhe core module, is in charge of collecting and forwarding space packets to other devices. These devices may connect to one of the MTD-STL-1553B buses or the Ethernet bus. Whenever KPU receives a space packet, it finds a route for the packet first.

KPU parses the packet and constructs the logical address of the packet, then it uses the Eq. (1) to calculate the index corresponding to the logical address. KPU uses the index to find the entry that contains the logical address in the routing table. This process is called hash search process. If the search succeeds, KPU get the hardware address for the packet, otherwise, KPU cannot find a route for the packet and it will drop the packet. Then, KPU forwards the packet to its destination based on the type of the hardware address (Fig. 2).

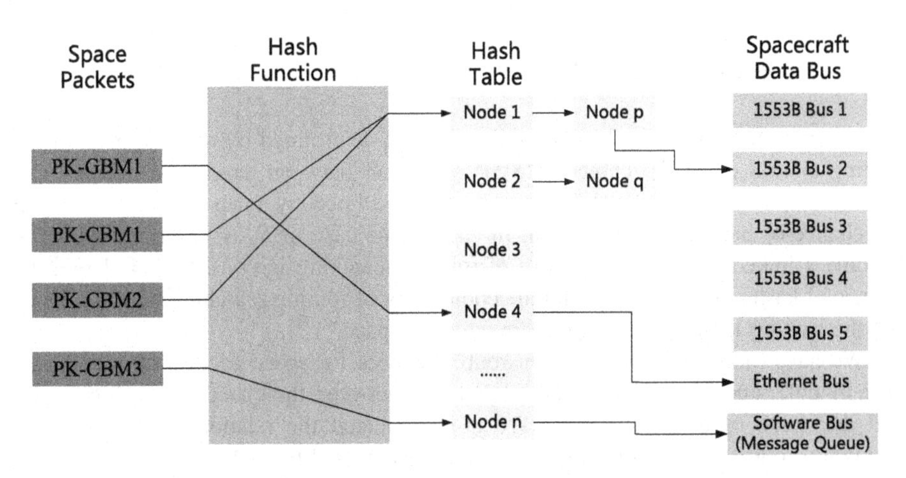

Fig. 2. The process of finding routes for space packets using hash search algorithm

4.2 Forwarding a Space Packet on 1553B Buses

If the hardware address of the packet is a 1553B address, KPU will forward the packet on one of the 1553B buses. First, KPU gets the bus_id from the hardware address and put the packet into a data buffer for the bus. Second, KPU generates a control word and a command word for the packet and put them into the buffer of a 1553B chip along with the packet. Third, the 1553B chip sends the packet to its destination based on the control and command word.

4.3 Forwarding a Space Packet on Ethernet Bus

If the hardware address of the packet is an IP address, KPU will forward the packet on the Ethernet bus using TCP/IP protocols. The TCP/IP protocol stack we use on the Tianhe core module is a version of the LWIP protocol [6]. In our case, the packet is always wrapped into an IPv4 data packet and is transmitted to the destination using UDP protocol. The packets which are send on Ethernet bus are telemetry data packets and are tolerant to packet loss.

4.4 Forwarding a Space Packet to Other Process

If the hardware address of the packet is a process id, KPU will forward the packet to another process on the KPU using message queue. In this case, the service type and service subtype in the packet are useful. The destination process receives the packet and deal with it based on the service type and service subtype of the packet. For example, the destination process can send telemetry packets to the ground via ground-space data link.

5 Performance Evaluation

The performance of the hash routing algorithm is determined by the size of the hash routing table S_T, and the choice of S_T is a tradeoff between temporal efficiency and spatial efficiency. If S_T is small, the routing table will occupy small memory space, but the chance of hash collision is high. In an extreme case, if S_T is 1, the routing table become a linked list and complexity of routing searching is $O(n)$. If S_T is large, the chance of hash collision is low and the complexity of routing searching becomes $O(1)$, but the routing table will occupy large memory space.

We have more than 500 different routing entries for space packets on the Tianhe core module, and we evaluated the relationship between the chance of hash collision and the value of S_T (see Fig. 3). We also evaluated the relationship between the maximum routing search time among all the packets and the value of S_T (see Fig. 4). The value of S_T we use in the experimental is increased exponentially.

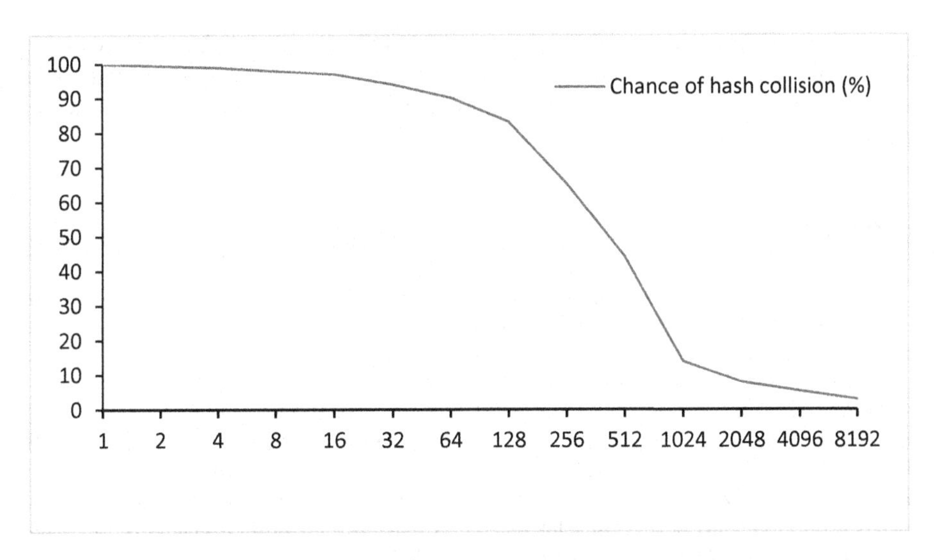

Fig. 3. Chance of hash collision versus the size of the routing table

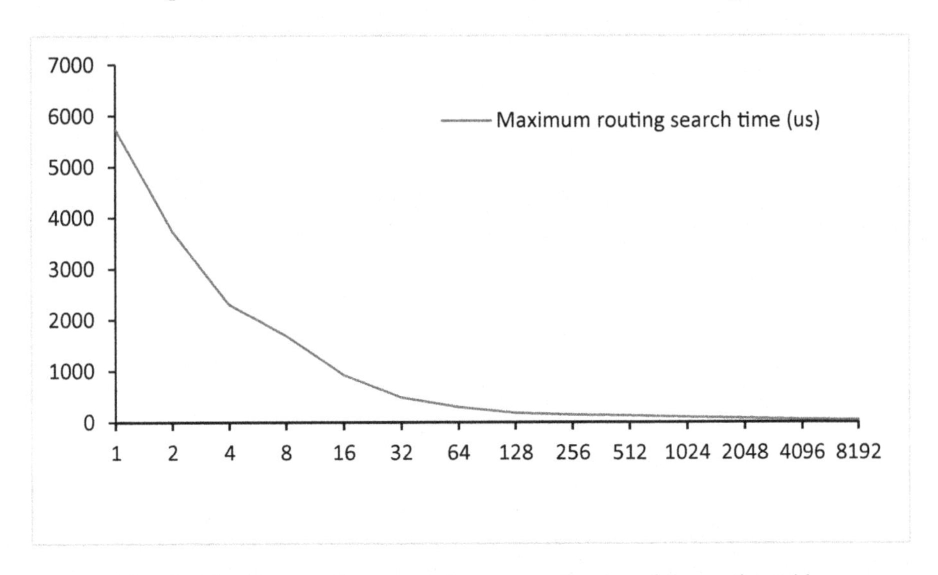

Fig. 4. Maximum routing search time versus the size of the routing table

Based on the experiment results, we choose $S_T = 1024$. The maximum routing search time is below 100 μs and the memory usage is acceptable.

6 Summary

In this paper, we design and implement a high efficiency space packet routing algorithm based on hash routing table. The algorithm is used on the Tianhe core module of the Chinese space station. It solves the difficult problem of routing different space packets in the complex ground-to-space network and spacecraft onboard network on the Tianhe core module. We introduce the method of creating the hash routing table and forwarding a space packet. Finally, we present the experimental result of how the size of the hash routing table effects the performance of the routing algorithm.

References

1. Niu, Y., Zhao, W.: Design and analysis of a strong fault-tolerant on-board spacewire bus network. J. Comput. Appl. **34**(9), 2497–2500 (2014)
2. Mao, C., Guan, Y., David, J.: Research and design of on-board dynamic SpaceWire router. J. Electron. Inf. Technol. **32**(8), 1904–1909 (2010)
3. CCSDS 732.0-B-2: AOS Space Data Link Protocol. CCSDS blue book, July 2006
4. ECSS-E-ST-50-12C: SpaceWare - Links, Nodes, Routers and Networks. ESA-ESTEC (2012)
5. CCSDS 133.0-B-1: Space Packet Protocol. CCSDS blue book, September 2012
6. Yu, S., Qiu, Q., Zhou, B., et al.: Design and implementation of spacecraft onboard network based on LWIP. In: 2018 Eighth International Conference on Instrumentation & Measurement, Computer, Communication and Control (IMCCC) (2018)

A Design of GNSS Based Time Management System for Telecommunication Satellite

Nuo Xu(✉), Jia Guo, Tao Wang, Weiwei Gao, Rui Wang, and Weiyu An

Institute of Telecommunication and Navigation Satellite, China Academy of Space Technology, Beijing 100094, China

Abstract. Telecommunication satellites have shown the trend of network development, and the onboard payload processing and inter satellite link communication put forward higher requirements for satellite time synchronization. Taking telecommunication satellite as the application object, this paper carries out the design of GNSS (Global Navigation Satellite System) based time management system of telecommunication satellite, establishes the time management model, designs the time management system architecture, and analyzes the design strategy of time management applicable to telecommunication satellite. Furthermore, an instance of time synchronization error analysis is introduced.

Keywords: Telecommunication satellite · Time management · GNSS

1 Introduction

With the increasing demand of users for networked communication, the field of telecommunication satellite tends to develop in the direction of space-based networking, in order to improve the service quality of the system, expand the service scope of the system, and expand the business needs of users. For example, the onboard payload processing of telecommunication satellite, the establishment and maintenance of inter satellite links all have high requirements for time management. If the time synchronization error exceeds the specified range, it will cause link interruption, service discontinuity and other problems [1, 2].

Taking telecommunication satellite as the application object, this paper proposes a design method of time management system based on GNSS. Through GNSS timing, satellite bus network broadcasting, high-precision time-frequency source timing, the 1PPS (one pulse per second) time reference signal generation and so on, the satellite time management system is constructed to meet various requirements of telecommunication satellite equipment.

The remainder of this paper is structured as follows. Section 2 presents the working principle, model and system architecture of telecommunication satellite time management. In Sect. 3, we describe the operation mode and strategy design of time management. The analysis of the time synchronization error is presented in Sect. 4. We conclude in Sect. 5.

Q. Guo et al. (Eds.): WiSATS 2021, LNICST 410, pp. 69–75, 2022.
https://doi.org/10.1007/978-3-030-93398-2_7

2 Architecture of Satellite Time Management System

2.1 Time Management Working Principle

The time management function is the way to realize the time unification among constellation, whole satellite, inter satellite and satellite-ground. It realizes the time synchronization of satellite-ground and inter satellite communication of telecommunication system, and meets the requirements of time calibration and punctuality for on orbit autonomous operation.

Time synchronization is divided into absolute time synchronization and relative time synchronization. Relative time synchronization refers to the time synchronization between all nodes in the system. Absolute time synchronization refers to not only completing time synchronization in the system, but also synchronizing with coordinated universal time (UTC). The GNSS receiver provides the system time reference [3], calculates the UTC time, provides it to the whole satellite platform and payload time management equipment, and then distributes it to the satellite time users.

Satellite communication system realizes absolute time synchronization and relative time synchronization through ground time service, GNSS time service, satellite ground link, inter satellite link. Among them, ground time service and GNSS time service are absolute time synchronization means, while satellite ground link and inter satellite link are relative time synchronization means.

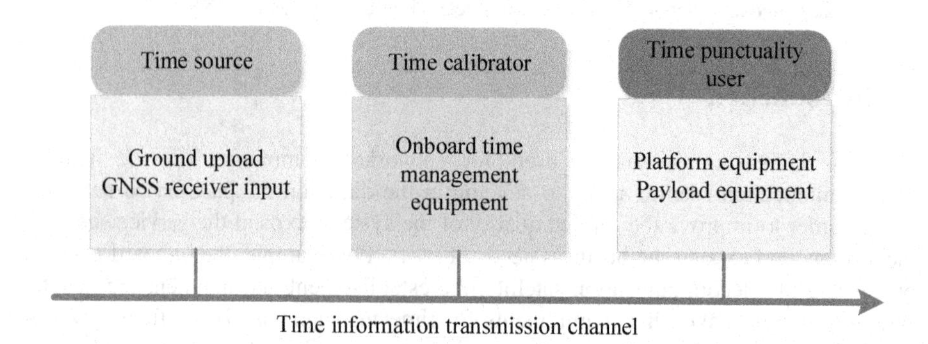

Fig. 1. Time management model

As shown in Fig. 1, the time management model of telecommunication satellite consists of four parts: time source, time calibrator, time punctuality user and time information transmission channel.

- The time source is used to initialize the time on the satellite and provide the time reference. Its input sources include the ground absolute time service telecommand upload and GNSS receiver input.
- Time calibrators provide time calibration means and correction compensation methods, including uniform time calibration, incremental time calibration and absolute time calibration. Time distribution is realized by broadcasting, point-to-point transmission with onboard time management equipment.

- The time punctuality user is the terminal application object of time information, including satellite platform equipment such as attitude and orbit control unit, payload equipment such as communication processing unit, etc.
- The time information transmission channel is used for the distribution and transmission of PPS and absolute time information, including satellite bus network such as 1553B bus and point-to-point interface such as RS422 serial data interface.

2.2 Time Management System Architecture

Figure 2 shows an architecture of time management system of telecommunication satellite. Under the normal operation mode of the satellite, the UTC time calculated by GNSS receiver is used as the time source. GNSS receiver uses high stable clock to generate 1PPS signals, and outputs two channels of 1PPS to satellite management unit and payload time unit respectively through RS422 serial data interface. The satellite management unit and the payload time unit rely on the input time information and 1PPS to trigger their own counters for time calibration and maintenance. The satellite management unit releases the time of the platform and payload time punctuality users through 1553B bus, and the payload time unit releases the time of the payload time punctuality users through RS422 serial data interface.

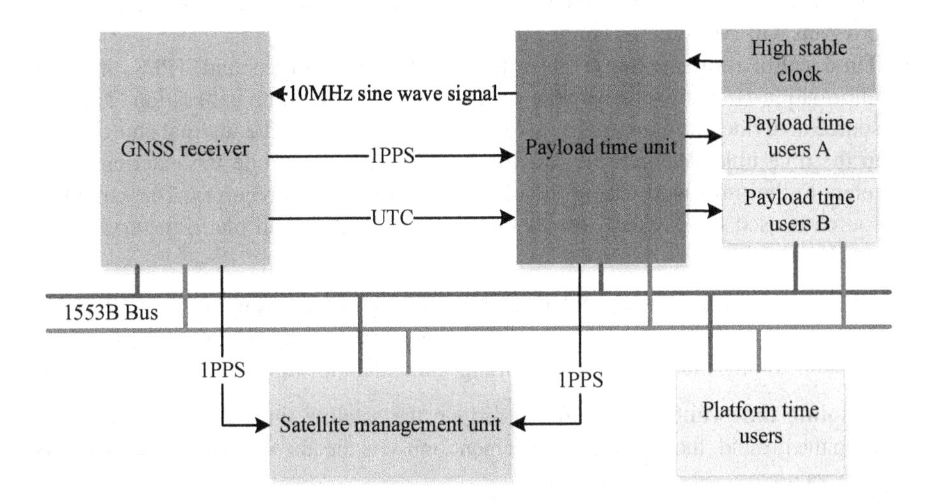

Fig. 2. Time management system architecture

3 Time Management Strategy Design

3.1 Operation Mode of Time Management

In this paper, onboard time management mainly includes two operation modes: satellite payload service operation mode based on high precision time synchronization, and satellite safe operation mode based on common precision time synchronization.

- **Satellite payload service operation mode:** GNSS receiver generates 1PPS for each time user. Within 100 ms after the 1PPS is generated, the GNSS receiver sends the time data corresponding to the 1PPS to the satellite management unit and the payload time unit through the bus and RS422 serial port respectively. When the time user receives the 1PPS, its own timer starts to subdivide the time within one second, and generates local time system according to the received time data.
- **Satellite safe operation mode:** Satellite management unit regularly distributes time information through the bus [4]. The logic of this time synchronization mechanism is simple, but it depends on the software to operate the time, which will lead to large uncertainty of time delay. If the time data is not sent successfully and needs to be resent twice, or the lower computer software is interrupted by other terminals during the processing of time, the time delay will be longer [5]. Therefore, this kind of operation mechanism is generally suitable for the application scenarios which require low accuracy of time synchronization.

3.2 Time Management Specific Design

- **Time initialization strategy:** the default initial time value of power on of satellite management unit is 0, and the ground sets the onboard time benchmark through time service telecommand. In case of software reset or shutdown, the time recovery program will be executed immediately.
- **Time calibration strategy:** according to the time source and 1PPS, the time information is sent to the satellite management unit for time calibration. The time source is divided into internal source and external source. The internal source refers to the time uploaded into the satellite by telecommand, while the external source refers to the time acquired by the satellite through GNSS receiver. The same time source is used for payload service time synchronization and platform service time synchronization.

The whole satellite timing includes GNSS timing management and ground timing. The deviation of 1PPS can be corrected by sending telecommand to adjust the phase. The ground timing includes absolute timing, incremental timing and uniform timing.

- **Absolute time calibration:** after receiving the satellite time setting telecommand from the ground, the satellite management unit sets the absolute time to the received time value.
- **Incremental timing:** after receiving the telecommand from the ground, the satellite management unit adds the received incremental value to its own time.
- **Uniform time calibration:** Each time the satellite management unit passes a time calibration interval, the onboard software will increase the absolute time of the satellite management unit by a time increment (positive or negative) that can be set by the ground telecommand.

- **Time distribution strategy:** GNSS receiver and payload time unit generate 1PPS at each UTC integer second time. The satellite management unit counts the frequency signal output by the local clock source to generate the local time, compares it with the external time reference, outputs the reference time according to a certain time coding format, and sends the time information to the lower computer through the bus.

- **Punctuality strategy:** the punctuality part is usually composed of time information receiving part, local time code generator. It mainly completes the receiving of time information sent by time service part, local time correction and time application. At present, the punctuality part of satellite design often exists in all terminal users in need.
- **Fault strategy:** the satellite management unit receives the 1PPS from GNSS receiver and payload time unit, and uses local clock as backup time service means to support the autonomous selection of clock source and ensure the high reliable operation of satellite time management system.

The 1PPS channel A is defined as the main signal of GNSS receiver while the channel B is defined as the backup signal. The 1PPS channel C is defined as the main signal of payload time unit while the channel D is defined as the backup signal. And the channel E is the local clock of satellite management unit. The switching strategy is as follows:

The satellite management unit identifies the time-frequency characteristics of the 1PPS. If the 1PPS is interrupted in orbit, it switches according to the priority order of channel A $\rightarrow$ B $\rightarrow$ C $\rightarrow$ D $\rightarrow$ E. Channel A, B, C and D can ensure the normal operation of satellite payload service. When the GNSS receiver is not turned on or fails to output time information and 1PPS, the satellite will adopt the common precision time synchronization mechanism. The satellite management unit will switch to rely entirely on the local clock of channel E as the time-frequency source to realize the system time synchronization and maintenance function, and ensure the safe and stable operation of the satellite platform, and in a certain period of time to maintain the operation of the satellite load. The ground can use this period of time for abnormal disposal.

In addition, if the satellite management unit is reset or cut off due to failure, the time information and important data of the satellite management unit are recovered from the platform service unit, payload service unit and other lower computers according to the preset priority, which can automatically restore the clock source use status before the failure and complete the autonomous time synchronization. If the lower computer fails to recover, the ground needs to send time service telecommand for time recovery.

4 Accuracy Analysis of Time Management

The error of time management system of telecommunication satellite is divided into hardware delay error (t_h), time transmission error (t_t), time locking error (t_l) and user timing error (t_u) [6]. As shown in Fig. 3, a time synchronization error model of telecommunication satellite should be built according to the time information transmission order of error sources. The time synchronization error of satellite could be estimated for specific time users through the whole process transmission analysis, so as to verify whether the time management accuracy meets the requirements.

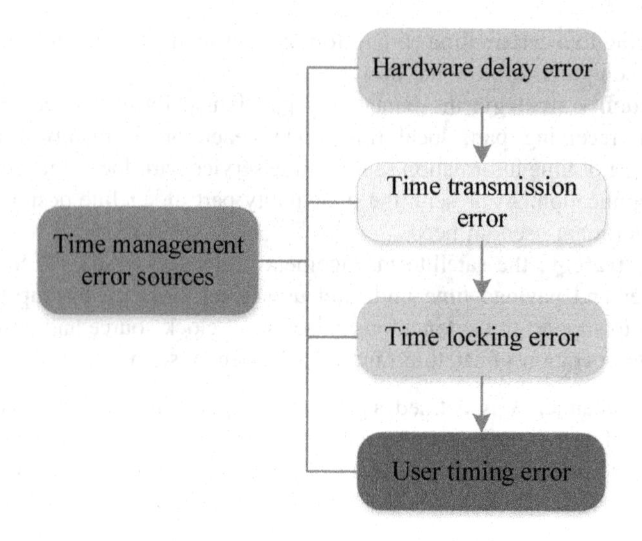

Fig. 3. Time synchronization error model

Combining the devices design and the chips user manuals, the error analysis of the operation mode of the satellite payload service based on the high-precision time synchronization is carried out. The PPS error generated by GNSS receiver is 0.5 μs. The maximum inherent transmission delay of the PPS receiving chip of the satellite management unit and the payload time unit is 40 ns, and the maximum FPGA delay of the satellite time latch is about 1μs. The maximum relative time error of satellite management unit and payload time unit is 35 μs.

The time synchronization delay characteristics are defined as follows, and the total delay T is calculated by Eq. 1).

$$T = t_h + t_t + t_l + t_u \tag{1}$$

Finally, the high-precision timing error with PPS is less than 37 μs, which meets the requirements of a telecommunication satellite payload application for time synchronization accuracy better than 40 μs.

5 Conclusion

This paper analyzes the working principle of time management based on GNSS, constructs the model and system architecture, and carries out the operating mode research and strategy design. Combined with the error theory analysis, it provides new ideas and design reference for the design of telecommunication satellite system. In the future, the experimental platform will be built to verify the design. In addition, with the development of space-based networking of telecommunication satellite system, the connotation and extension of satellite time management will be further expanded.

References

1. Kodheli, O., Lagunas, E., et al.: Satellite communications in the new space era: a survey and future challenge. IEEE Commun. Surv. Tutor. **23**(1), 1–45 (2020)
2. Perez, A.I., Vazquez, M.A., et al.: Signal processing for high-throughput satellites: challenges in new interference-limited scenarios. IEEE Signal Process. Mag. **36**(4), 112–131 (2019)
3. Hou, J.H., et al.: Research of synchronization for satellite time and frequency based on GPS. Mod. Electron. Tech. **33**(15), 8–10 (2010)
4. Zhao, H.P., He, X.W., et al.: Space data system. Beijing Institute of Technology Press, Beijing (2018)
5. Gao, J.J., et al.: A high reliable method for time management of onboard software. Comput. Meas. Control **21**(3), 806–808 (2013)
6. Gardner, F.M.: Phaselock Technique. Wiley, New York (2005)

Attitude Determination Method Based on GNSS Signals Null Steering Technology

Jianlou Zhuang[1,2(✉)], Chengbin Kang[1,2], and Jianjun Zhang[1,2]

[1] China Academy of Space Technology, Beijing, China
[2] Institute of Telecommunication and Navigation Satellites, Beijing, China

Abstract. GNSS signals are usually used for positioning, but some applications also use them to determinate direction, such as geography north. The attitude of a vehicle could be determined by using multiple short baselines. However, the measurement accuracy is limited by the length of the baseline, so it also limits its application. This paper demonstrates an attitude determination method using GNSS Signals Null Steering (GSNS) technology by a miniaturized phase controlled antenna array. The error of DOA estimation is simulated and analyzed under the condition of received signal's amplitude quantization error and the radiation patterns of antenna cells are inconsistent. The results show that the method has certain engineering feasibility.

Keywords: DOA (Direction of Arrival) · GNSS antenna array · ADM (Attitude Determination Module) · Null steering

1 Introduction

In recent years, with the deployment and operation of giant constellations such as OneWeb and Starlink, as well as the development of Xingwang constellation, telecommunication satellite constellations have become one of the current research hotspots in the space and communications field [1, 2], and present characteristics such as broadband interconnection, and integration of communication and navigation [3].

According to ITU, satellite telecommunication services are divided into fixed satellite service (FSS) and mobile satellite service (MSS). With the development of satellite communication technology, Earth station In Motion (ESIM) [4] communication technology between users and geostationary orbit (GEO) satellites has emerged. Generally speaking, the target satellite of ESIM is GEO satellite, and beacons can be used. It is relatively simple for ESIM users to capture and track the target satellite with narrow beam antenna (hereinafter referred to as pointing access) [5]. For non-GEO communication satellites, due to the low orbit and the fast motion speed relative to the user, the user's pointing access needs to be completed by pointing calculation instead of beacon, and there are some problems such as the low updating frequency of almanac data used in calculation, and the inconsistent coordinates of body attitude and almanac data, which are more difficult than ESIM. Navigation satellite systems (including BD-3 and other global and LEO enhanced navigation satellite systems) can provide body attitude by GNSS signals [6], which solves the problem of inconsistent coordinates to a

Q. Guo et al. (Eds.): WiSATS 2021, LNICST 410, pp. 76–85, 2022.
https://doi.org/10.1007/978-3-030-93398-2_8

certain extent. However, there are constraints such as the long baseline required and unfavorable miniaturization application.

By the popularity of GNSS receiver, if can the user via the navigation satellite broadcast nearly real-time update of non-GEO satellite almanac data, and users can be implemented through miniaturization of equipment GNSS on board, the user to obtain the goal of the almanac data and body attitude with the same space-based coordinate system (such as BDCS) as a reference, will be able to solve the above problems. For this reason, this paper proposes a method of connecting non-GEO satellite with user pointing assisted by navigation satellite. The core of this method is a miniaturized GNSS multi-DOA (Direction of Arrival) attitude determination method using phase-controlled null steering technology. Preliminary simulation results show that this method has certain engineering feasibility.

2 Beam Access by GNSS Attitude Determination Module

To enable users' high gain antenna beam to access the non-GEO satellite through the auxiliary of the GNSS signals, as shown in Fig. 1, it mainly includes 3 steps: first, the GNSS Attitude Determination Module (ADM) receive GNSS signals from GNSS Antenna array, and second, calculates the body attitude through the measurement of GNSS signals, and finally calculates the beam direction and drives the antenna beam to point to the target non-GEO satellite.

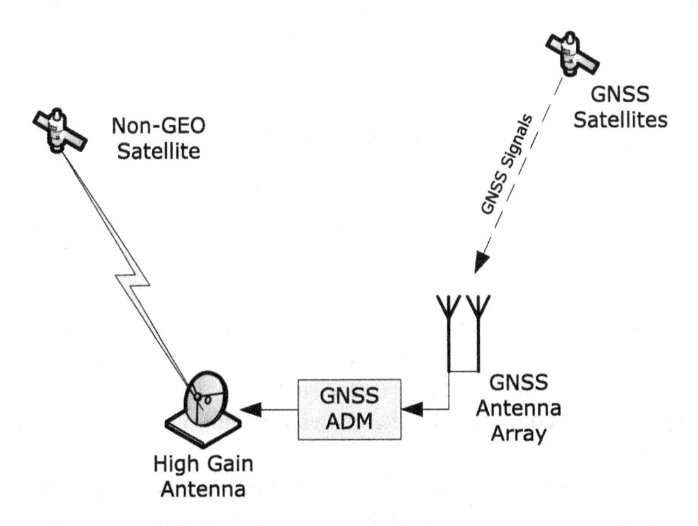

Fig. 1. User beam access to non-GEO satellite by GNSS ADM

After obtaining the almanac data of the target satellite, the direction of the target satellite can be calculated as long as the body attitude is determined (i.e. the attitude determination). Most users adopt inertial equipment (gyro) or radio direction measurement technology [6, 7] to determine the body attitude.

Unmanned aerial vehicle (UAV), for example, is widely used in combined attitude sensor [8], such as MEMS gyroscope (angular velocity), MEMS accelerometer (gravity direction), and magnetometer (geomagnetic measurement direction). The method of attitude determination uses accelerometer to determine horizontal plane and uses magnetometers to determine magnetic north. It is enough for uses of the stability control of flight in a short time, but is insufficient for satellite communications as its inconsistence to space-based coordinates system.

GNSS attitude determination mainly uses the short baseline method to calculate the geographic true north [9]. Combined with gravity measurement, the accuracy of the shot baseline method could be about 0.1°–0.2°/1 m [10, 11], which basically meet the application requirements of pointing access for medium UAV and other platforms. However, the baseline length is still the main bottleneck for minimization application.

Therefore, we need to research and development of miniaturization GNSS attitude determination technology, in order to solve the current problem of the long baseline. It could also be achieved by improving navigation signal carrier frequency, but in this paper the L band is concerned to carry out the analysis in view of the current use.

3 GNSS Attitude Determination Technologies

Assuming that the user body coordinate system is $\mathbf{A} = [\hat{x} \quad \hat{y} \quad \hat{z}]$, and the space-based coordinate system is $\mathbf{G} = [\hat{X} \quad \hat{Y} \quad \hat{Z}]$, and the coordinate transformation relationship between them is: $\mathbf{A}^{\mathrm{T}} = \mathbf{T}_{\mathrm{G2A}}\mathbf{G}^{\mathrm{T}}$. Attitude determination is to determine the transformation matrix $\mathbf{T}_{\mathrm{G2A}}$. Where, the hat ^ means a unit vector, and the superscript ^{T} means the transpose of a matrix.

3.1 Multi-baseline and Multi-DOA Attitude Determination Methods

As shown in Fig. 2, suppose the direction of the satellite i is $\hat{r}_i$, and the vector of the baseline j is $\vec{L}_j$. In the body coordinate system and space-based coordinate system, there are respectively: $\hat{r}_i = \boldsymbol{\rho}_i\mathbf{A}^{\mathrm{T}} = \mathbf{r}_i\mathbf{G}^{\mathrm{T}}, \vec{L}_j = \boldsymbol{\lambda}_j\mathbf{A}^{\mathrm{T}} = \mathbf{L}_j\mathbf{G}^{\mathrm{T}}$, i.e. $\boldsymbol{\rho}_i\mathbf{T}_{\mathrm{G2A}} = \mathbf{r}_i, \lambda_j\mathbf{T}_{\mathrm{G2A}} = \mathbf{L}_j$.

It can be seen that in order to determine the attitude of the body, two methods can be used: one is to measure the baseline vector of the body in the space-based coordinate system, and compare it with the known baseline vector in the body coordinate system; the other is to measure the direction vector of the navigation satellite in the body coordinate system, and compare with the known direction vector of the satellite in the space-based coordinate system.

The first method is the common called multi-baseline method, when there are multiple baselines, we have $[\boldsymbol{\lambda}_j]\mathbf{T}_{G2A} = [\mathbf{L}_j]$. In the calculation, $\boldsymbol{\lambda}$ is the local measurable quantity and $\mathbf{L}$ is the real measurement. In particular, for the case of three baselines, such as the three antenna method and the four antenna method [9], it can be solved directly:

$$\mathbf{T}_{\mathrm{G2A}} = [\lambda_j]^{-1}[\mathbf{L}_j] \tag{1}$$

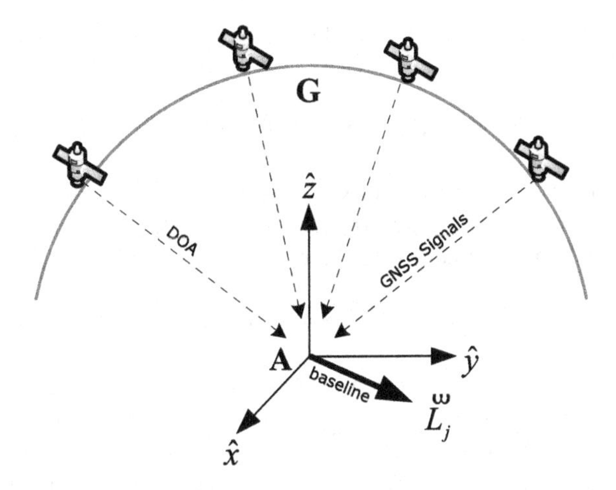

Fig. 2. GNSS attitude determination

The second method is the so-called multi-DOA attitude determination method, when several satellites are observed we have $[\boldsymbol{\rho}_i]\mathbf{T}_{\mathrm{G2A}} = [\mathbf{r}_i]$. In the calculation, $\boldsymbol{\rho}$ is the local observation and $\mathbf{r}$ can be calculated according to the almanac data. It can be solved by the least square method:

$$\mathbf{T}_{\mathrm{G2A}} = \left([\boldsymbol{\rho}_i]^{\mathrm{T}}[\boldsymbol{\rho}_i]\right)^{-1}[\boldsymbol{\rho}_i]^{\mathrm{T}}[\mathbf{r}_i] \tag{2}$$

3.2 Limitation of Carrier Phase Measurement

The carrier phase measurement (CPM) of conventional GNSS receiver can be used for both short-baseline method and multi-DOA method, but there is a limitation of the required baseline length.

In the space-based coordinate system, the carrier phase of multiple satellites (≥ 4) is observed to measure the j-th baseline:

$$\mathbf{P} + 2\pi\mathbf{N} = k\mathbf{R}\mathbf{S}^{\mathrm{T}} + \boldsymbol{\varepsilon} \tag{3}$$

Where $\mathbf{P}$ is the carrier phase vector, $\mathbf{N}$ is the integer ambiguity resolution, $\mathbf{R} = [\mathbf{r}\ 1]$ is the direction matrix, $\boldsymbol{\varepsilon}$ is the error vector, $\mathbf{S} = [\mathbf{L}\ c\tau]$ is the baseline vector with clock difference, c is the velocity of light in free space, τ is the clock offset of GNSS receiver, and k is the carrier wave number. $\mathbf{P}$, $\mathbf{N}$, $\mathbf{R}$ and $\boldsymbol{\varepsilon}$ are expanded according to the number of satellites in the column direction. It can be obtained from formula (3) that:

$$\mathbf{S}^{\mathrm{T}} = \left(\mathbf{R}^{\mathrm{T}}\mathbf{R}\right)^{-1}\mathbf{R}^{\mathrm{T}}(\mathbf{P} + 2\pi\mathbf{N} - \boldsymbol{\varepsilon})/k \tag{4}$$

Then the baseline vector $\mathbf{L}$ is taken out from $\mathbf{S}$ and solved $\mathbf{T}_{\mathrm{G2A}}$ by formula (1).

In the body coordinate system, the direction of the *i*-th satellite is measured by multi-baseline method:

$$\mathbf{P} + 2\pi\mathbf{N} = k\mathbf{V}\mathbf{W}^{\mathrm{T}} + \boldsymbol{\varepsilon} \tag{5}$$

Where $\mathbf{V} = [\boldsymbol{\lambda}\ 1]$ is the baseline matrix and $\mathbf{W} = [\boldsymbol{\rho}\ c\tau]$ is the direction vector with clock difference. **P**, **N**, **V**, and $\boldsymbol{\varepsilon}$ are expanded according to the baseline number in the column direction. When the number of baselines is 3, formula (5) shows that:

$$\mathbf{W}^{\mathrm{T}} = \mathbf{V}^{-1}(\mathbf{P} + 2\pi\mathbf{N} - \boldsymbol{\varepsilon})/k \tag{6}$$

Then the direction vector $\boldsymbol{\rho}$ is taken out from **W** and $\mathbf{T}_{\mathrm{G2A}}$ solved by formula (2).

It should be pointed out that the above two methods are equivalent. In general, the baseline vector and the satellite vector are related by the attitude. In the GNSS carrier phase observation Eq. (3) or (5), the carrier phase just relates the baseline vector and the satellite vector, too. Therefore, the measurement of the carrier phase is equivalent to the attitude. Whether the baseline vector or the satellite vector is regarded as a known quantity, the effect is equivalent.

Further more, to obtain the carrier phase a phase-locked loop (PLL) is usually necessary and the phase precision is difficult to increase. High-precision measurements both rely on long baselines, and there is also the problem of solving the integer ambiguity resolution, which is not conducive to miniaturization application.

In view of the current development level of high-precision phase shifting technology, such as a high-precision phase shifter equivalent to 24 bits implemented in 1 GHz [12], this paper further proposed the attitude determination method based on GNSS signals null steering (GSNS) technology uses multi-DOA method for miniaturization applications.

3.3 GSNS Technology for DOA Estimation

A simple idea of miniaturization is to use a half wavelength spaced GNSS antenna array to scan the DOA of GNSS signals by phased differential beam. For L-band the size of the 4 cells array antenna can be controlled at about 200 mm*200 mm. The 4 cells array can be simplified to 2 cells array for DOA estimation. The unit spacing *d* of cells is half wavelength, has gain pattern of $f(\theta)$, and cell2 has a *B*-bit digital phase shifter, as shown in Fig. 3 below.

For a certain DOA θ_0, when the phase shifter is searched step by step, the received signal amplitude is a function of the phase shift quantity ϕ_v:

$$p(\phi_v) = f(\theta_0) + f(\theta_0)\exp(\mathrm{j}(kd\sin\theta_0 + \phi_v)) \tag{7}$$

When the null point appears in the above formula, the DOA can be obtained:

$$\theta_0 = \arcsin((\pi - \phi_v)/k/d) \tag{8}$$

so it is called GNSS Signals Null Steering (GSNS) technology. The relationship between the angular resolution and the phase-shifting resolution can be obtained as:

$$\mathrm{d}\phi_v/\mathrm{d}\theta_0 = -kd\cos\theta_0 \tag{9}$$

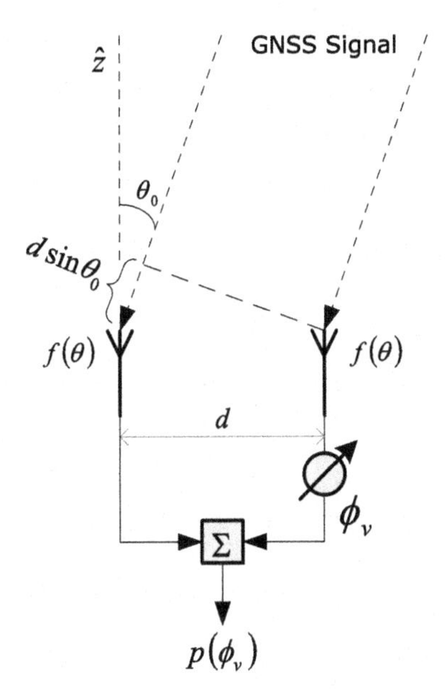

Fig. 3. 2 cells array for DOA estimation

It can be deduced that in order to achieve 0.2° angle DOA estimation accuracy, when the incoming wave is approximate to the vertical incidence, it needs at least 0.6° stepped phase shifting; when the incoming wave is approximately 70° incident, it needs at least 0.2° stepped phase shifting. Though the required phase shifter bits number B could be calculated by:

$$B \geq \log_2\left(180^\circ/0.2^\circ\right) \approx 10 \tag{10}$$

For this bits number it is feasible in engineering.

For an actual example, the cell is a micro-strip fed patch antenna and both the amplitude and phase radiation pattern is simulated by HFSS®. The cell space is 1/3 wavelength, and the received signal has an amplitude error(pp) of 0.5 dB and phase error(pp) ±1°, and the receiver gets the signal power by C/N0 estimation which has a quantization error of 0.5 dB. Assumed the DOA is 1.2°, we could get the phase scanning curve as shown in Fig. 4, in which the x-axis is marked as varied phase shifter (ϕ_v) and y-axis is marked as received signal amplitude calculated by Eq. (7), thus the null point indicates the estimation of θ_0 which could be calculated by Eq. (8). Though as shown in Fig. 4, the phase scanning curve has a random but zigzag noise and strongly shakes near the null point, it makes the DOA estimation difficult, but we could use a template curve to eliminate the noise and get a more accurate DOA estimation.

The template curve $T(\phi_v)$ is drawn in Fig. 4 as a red or a more smooth curve, which just comes from Eq. (7) by applying ideal cell pattern so

$$T(\phi_v) = A\cos(\theta_0')\left[1 + \exp(\mathrm{j}(kd\sin\theta_0' + \phi_v))\right] \tag{11}$$

Where, A and θ_0' are variables to be optimized aimed to make the template curve most reach the curve under test, and θ_0' is just the DOA estimation we pursuing when optimization is obtained.

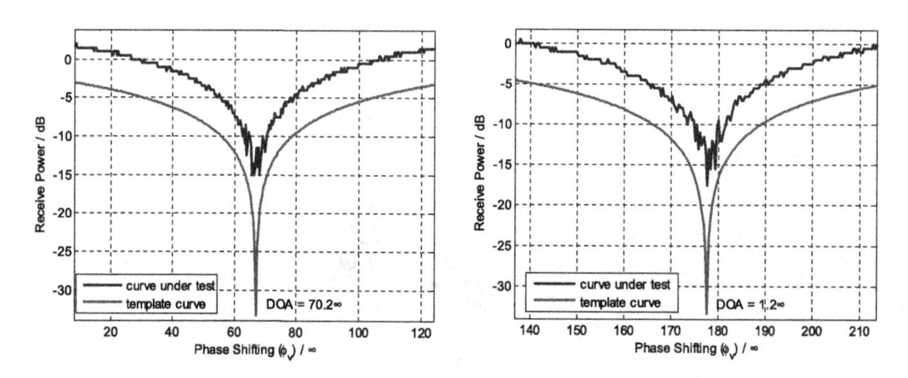

Fig. 4. Phase scanning curve, left: DOA 71.2°, right: DOA 1.2°

For common phase shifters the range of phase variety is 0° to 180°, but it should make the phase scanning curve asymmetric when DOA is near 0° because the null point appears as phase shifter touches its limit 180°. Thus, a wide range phase shifter is used which has a phase variety of 0° to 250° and only ±60° phase scanning around null point is reserved to get a symmetric phase scanning curve. For example as shown in Fig. 4 right, when DOA is 1.2°, the phase scanning curve null point is near 180°, but the actual phase scanning range is up to 120° to 240°.

Except for finding the minimum value of the GSNS curve to determinate θ_0', we could use least-square method move the template curve by searching A and θ_0' in two dimensions grade calculation, and this is called template curve optimization method, which could significantly improve the accuracy of DOA estimation.

To illustrate this, for an example of the same configuration of Fig. 4, using a phase shifter of 9 bits and 100 times statistics per DOA, we could get the DOA estimation error (1σ) shown as Fig. 5.

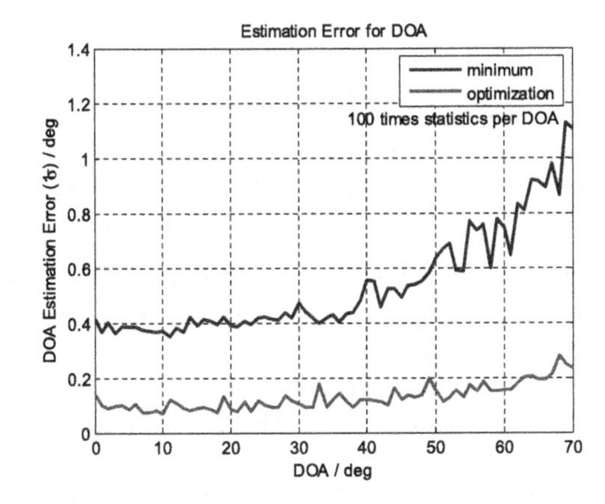

Fig. 5. Estimation error for DOA (Color figure online)

In Fig. 5, the blue curve is the DOA estimation error using minimum method to find null point, and the red curve using the template curve optimization method mentioned above. It could be seen that the estimation error is decrease to about 1/5 of the minimum method after using the template curve optimization method. So, we could get about 0.2° accuracy in 0° to 70° DOA range only by a 1/3 wavelength cell space and 9 bits phase shifter, and it has an advantage of miniature and engineering.

Furthermore, the direction vector ρ of the target satellite can be obtained by using two orthogonal 2 cells arrays, and then $\mathbf{T}_{G2A}$ can be solved by using formula (2).

4 Typical Applications

The miniaturized GNSS multi-DOA ADM enables users to realize positioning and attitude determination without relying on inertial navigation, accelerometer, magnetometer and other means. For example, the miniaturized GNSS multi-DOA ADM can be widely used for attitude measurement and attitude control of mobile platforms such as automobiles, UAVs, micro/nano satellites, etc., and can also be used for attitude measurement independent of gravity and geomagnetism, and play a role in special geological conditions.

Typical application scenarios such as a high gain flat antenna integrated with four elements GNSS multi-DOA ADM can independently sense its body attitude and adjust its beam direction, so as to realize pointing access with target non-GEO satellite.

With the completion of BD-3 global navigation satellite system and the development of non-GEO constellations, the planning of the next generation navigation satellite system is put on the agenda under the background of telecommunication and navigation integration. Taking full advantage of the popularity of navigation signal and GNSS receiver, the miniaturized GNSS multi-DOA ADM technology can provide systematic solutions for the user's antenna beam access of non-GEO constellation, which will further promote the development of telecommunication and navigation integration.

5 Conclusion

In this paper, the method of attitude determination using GNSS signal is researched and analyzed, in which the short baseline method has some disadvantages in miniaturization, and the GNSS Signals Null Steering (GSNS) technology is proposed in this paper which is suitable for miniaturization. Now we could sum up the following conclusions:

1) The attitude determination method based on GNSS signal could be divided into two forms: multi-baseline attitude determination and multi-DOA attitude determination, and both of them are equivalent;
2) The common short baseline method belongs to multi-baseline attitude determination, and the GSNS method proposed in this paper belongs to multi-DOA attitude determination, and thus they are equivalent, too. But because of the difference of the observation means, the GSNS method has better performance;
3) According to the characteristics of GSNS curve, this paper proposes a universal GSNS template curve optimization method which greatly improves the accuracy of DOA estimation.

In this paper, the DOA estimation accuracy of a 2 cells GNSS antenna array is simulated and analyzed. The simulation results show that the DOA estimation error will decrease to 1/5 of the original after using the template curve optimization method, even if the received signal has an amplitude quantization error and the radiation patterns of antenna cells are inconsistent. So it is possible to reduce the number of phase shifters and the cell space, which makes the miniaturization of GNSS ADM more practical.

However, it should be recognized that there are still some key technologies, such as more powerful algorithm to decrease the DOS estimation error, which still needs further research.

References

1. Liu, Y.: Research on the development of foreign LEO high-throughput communication satellite constellation. Int. Space (461) (2017)
2. Guan, Z., He, M.: Hongyan constellation communication system opens a new era of mass satellite communication consumption. In: 2016 Aerospace International Development Forum, Beijing (2016)
3. Wang, Y., Li, X., Wang, X.: Preliminary design of ground communication application system of LEO constellation. Commun. Technol. **52**(3) (2019)
4. Tian, W., Han, Z., Mu, X.: Analysis of frequency rules of Ka band mobile to China Communication. In: The 12th Annual Conference of satellite communication, Beijing (2016)
5. Cui, X., Xi, X., Zhang, Y.: High dynamic Ku/ka dual antenna civil aviation communication system design. Comput. Meas. Control **27**(5) (2019)
6. Li, S., Liu, L., Yan, G.: Determination of spacecraft attitude using GPS carrier coherent measurement technology. Aerosp. Control (4) (1998)
7. Liu, J., Zhu, X.: Comparative study on multiple methods of radio direction finding. Inf. Commun. (11) (2017)
8. Xue, Y.: Attitude fusion and navigation method of small rotor UAV. University of Electronic Science and Technology (2016)

9. Wu, M., Hu, X., Fu, L.: Satellite orientation technology. National defense science and Technology Industry Press, Beijing (2013)
10. Um482 Product Manual of Multi Frequency High Precision Positioning and Orientation Module of The Whole System. https://www.unicorecomm.com
11. He, X., Wu, M.: Low cost GNSS receiver chip carrier phase measurement technology. Navigat. Control **17**(5) (2018)
12. Wang, X., et al.: Design of high precision numerical control microwave phase shifter. J. Microwave **25**(3) (2009)

Construction of Knowledge Graph Based on Conventional Jamming Patterns

Shaoqin Kou(✉), Yintao Niu, and Hui Zhang

National University of Defense Technology, Nanjing 210007, China

Abstract. Construct a knowledge graph of anti-conventional jamming patterns based on the prior knowledge of conventional jamming patter, Firstly, adopt a manual extraction method to extract the prior information of the conventional jamming patterns and form a structured RDF data set; Then construct the anti-jamming knowledge graph pattern layer ontology from the bottom up and form the classification of anti-jamming decisions: Finally, use Java language to input the constructed data set and ontology classification into Neo4j to form an anti-jamming knowledge graph based on conventional jamming patterns. The result of the expansion of anti-jamming knowledge graph shows that the relationship between the entities of anti-jamming decision-making can be displayed directly by using the knowledge graph, which provides a basis for further implementation of low-complexity anti-jamming decision-making.

Keywords: Conventional jamming pattern · Anti-jamming · Knowledge graph

1 Introduction

With the rapid development of fifth-generation communication, massive wireless connectivity is in great demand. However, due to the openness and broadcast nature of wireless channels, wireless communications are increasingly vulnerable to malicious jamming attacks, where a third-party node injects jamming power into the legitimate users, hindering the legitimate transmission by decreasing the signal-to-interference-plus-noise ratio (SINR) [1]. Therefore, anti-jamming is an essential ability for wireless communications.

The type of jamming attacks can be divided into two categories: natural interference and human jamming [2]. Human jamming can also be divided into two categories: unintentional jamming and malicious jamming. Unintentional jamming generally refers to the jamming of wireless communication system caused by communication, radar, navigation, broadcasting and other wireless equipment itself and its radiation outside the band. In order to interrupt the transmission of specific information, some organizations or individuals release a malicious wireless signal for a specific wireless communication signal, so that the target communication signal could not be received normally, which called intentional jamming. Common human jamming patterns mainly include single tone, multi-tone, narrow band, wide band, linear sweep, periodic pulse, tracking jamming, etc. All the jamming has a serious impact on wireless communication. Therefore, it is necessary to take strong anti-jamming technical measures.

For a long time, the main technical measures of communication anti-jamming include direct spread, frequency hopping, time hopping, UWB and so on. These

Q. Guo et al. (Eds.): WiSATS 2021, LNICST 410, pp. 86–97, 2022.
https://doi.org/10.1007/978-3-030-93398-2_9

technologies have also been combined with jamming detection/sensing, adaptive anti-jamming, anti-jamming decision-making and other technologies to effectively improve the anti-jamming performance of wireless communication systems. For example, in Literature [3], the frequency adaptive anti-jamming technology based on jamming detection technology was proposed, which effectively improved the ability of wireless communication to avoid fixed jamming. Literature [4] proposes an adaptive frequency-hopping anti-jamming technology based on sweep jamming perception, which can avoid sweep jamming in real time according to sweep frequency law. Literature [5] proposes a method to design the optimal receiver against periodic pulse jamming, which simulates the signal of ranging equipment through the Gaussian mixture model, and drives the parameters of the Gaussian mixture model according to the characteristic properties of the signal to eliminate the periodic pulse jamming caused by the ranging equipment by designing the optimal receiver. Literature [6] proposes an anti-jamming method that uses reinforcement learning to learn the sweep jamming behavior and then obtains the optimal anti-jamming strategy. Literature [7] presents a tracking jamming detection method based on multi-channel detection. All the above put forward corresponding anti-jamming methods for different jamming patterns respectively, but two problems are still unsolved. First, most of the existing literature proposes algorithms for a specific jamming pattern, and it is difficult to provide effective anti-jamming strategies for a variety of different jamming patterns. Second, the lack of common sense reasoning ability, it is difficult to effectively use the prior knowledge of conventional jamming patterns to achieve low complexity real-time anti-jamming.

Knowledge Graph is a branch of artificial intelligence that uses graph model to describe the relationship between knowledge and modeling the relationship between everything in the world, which is of great significance for explaining artificial intelligence [8]. A graph is made up of nodes, which can be entities or abstractions, and edges, which can be properties of entities or relationships between entities. The early concept of knowledge graph comes from Semantic Web, which is the result of the interaction and inheritance of Semantic Web, knowledge representation, ontology, Semantic Web, natural language processing and other related technologies. Since 2012, when Google first used the Knowledge Graph to improve the search experience and improve the quality of search, it has attracted a lot of attention from academia and industry. As a new method of knowledge representation and a new idea of knowledge management, knowledge graph has been widely used in knowledge search, intelligent question-answering, big data analysis, natural language understanding and auxiliary device interconnection.

Therefore, we introduce the knowledge graph into the field of communication anti-jamming to improve the intelligence level of communication anti-jamming from the perspective of knowledge utilization. The main innovation of this paper is that the anti-jamming prior knowledge is represented in the form of knowledge graph, which has the advantages of strong expansibility and low complexity. The structure of this paper is as follows: in the second part, the system model is proposed and eight kinds of conventional jamming patterns and their anti-jamming strategies are briefly introduced. In the third part, the establishment method of anti-jamming knowledge graph of conventional jamming patterns is given. In the fourth part, the general jamming pattern knowledge graph library based on Neo4j and its application are given.

2 System Model

2.1 Anti-jamming Knowledge Graph Architecture

The main objective of the anti-jamming knowledge graph is to store the prior knowledge of anti-jamming so as to realize the low-complexity anti-jamming decision under the conventional jamming pattern. The anti-jamming prior knowledge is mainly stored in the graph database in the form of triples. The construction technical process of anti-jamming knowledge graph is shown in Fig. 1.

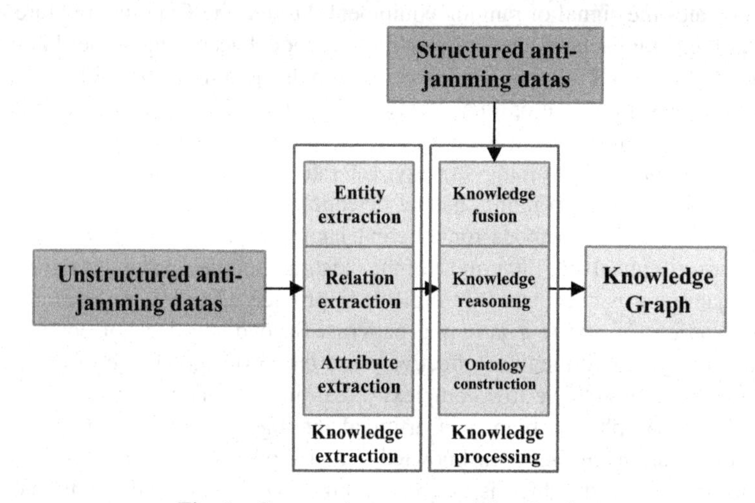

Fig. 1. Construction process of the Graph

The architecture of knowledge graph includes the logical structure of knowledge graph itself and the technology (system) architecture used to construct knowledge graph.

Logical Structure. In terms of logic structure, anti-jamming knowledge graph can be divided into data layer and pattern layer. Knowledge is stored in the form of "entity-relations-entity" triples or "property-value" pairs at these two layers.

The data layer stores facts and instances. The entities of the data layer are specific jamming patterns and information, such as single-tone aiming jamming, jamming attributes, etc. The pattern layer stores concepts, rules, categories, etc., and the entities of the pattern layer are generally abstracts, also known as ontologies, such as the category of jamming patterns and the classification of anti-jamming modes.

Construction Method. This paper adopts the bottom-up approach to construct the atlas: with the help of certain technical means, entities are extracted from some open linked data, resource patterns are extracted from publicly collected data, and new patterns with high confidence are selected, which are manually reviewed and added to the knowledge base, and then the top-level ontology pattern is constructed.

The advantage of this method is that it can generalize ontology according to data pattern type and determine framework classification upward in the absence of authoritative ontology data support, which is in line with the cognitive process of knowledge.

2.2 Anti-jamming Strategies of Conventional Jamming Patterns

Conventional jamming patterns mainly refer to eight common man-made intentional jamming, such as single-tone aiming jamming, multi-tone (comb)jamming, wide-band blocking jamming, partial band jamming, sweep jamming, periodic pulse jamming, tracking jamming, high-speed collision jamming, etc. The definitions and expressions of these jamming are shown in Literature [9]. The main attributes and anti-jamming strategies of conventional jamming are as follows:

Table 1. Main attributes of different jamming and their anti-jamming strategies

Number	Jamming pattern	Main attributes of jamming	Anti-jamming strategies
1	Single-tone aiming jamming	Start-end frequency, bandwidth	Adaptive frequency change, power adaptation
2	Partial band jamming	Center frequency, bandwidth, power, start frequency, end frequency	Adaptive frequency change, power adaptation
3	Multi-tone jamming	Frequency point or channel	Adaptive frequency change, power adaptation
4	High-speed collision jamming	Total bandwidth, instantaneous power, start frequency, end frequency	Adaptive frequency change, power adaptation
5	Wide-band blocking jamming	Total bandwidth, total power, center frequency	Jamming bandwidth is not beyond the working bandwidth of wireless communication system: adaptive frequency change and power adaptation
			Jamming bandwidth exceeds the operating bandwidth but with less power: modulation, channel coding, and power adaptation
			Jamming bandwidth exceeds the working bandwidth and with larger power: blind source signal processing or replacing the working frequency band
6	Sweep jamming	Start frequency, end frequency, instantaneous power, instantaneous bandwidth, sweep cycle	Adaptive frequency hopping, power adaptive
7	Periodic pulse jamming	Center frequency, bandwidth, cycle, duty cycle, instantaneous power	Adaptive time hopping, power adaptive
8	Tracking jamming	Bandwidth, instantaneous power, tracking rate	Accelerating frequency hopping or adaptively changing the working frequency band

3 Establishment of Communication Anti-jamming Knowledge Graph of Conventional Jamming Patterns

3.1 The Construction Process of Anti-jamming Knowledge Graph of Conventional Jamming Patterns

Referencing literature [10], according to the logical framework of the anti-jamming knowledge graph, the construction process of the anti-jamming knowledge graph proposed in this paper mainly consists of the following steps: (1) Extract the entity knowledge of the jamming pattern from the existing conventional jamming pattern data; (2) The extracted knowledge is processed and transformed into structured RDF data form through knowledge reasoning; (3) The ontology is constructed upward according to the extracted structured triad data; (4) Entity and ontology are respectively corresponding to data layer and pattern layer to form anti-jamming knowledge graph.

3.2 Construction of Anti-jamming Knowledge Graph

Knowledge Extraction. The main purpose of anti-jamming knowledge graph construction is to reduce the complexity of decision-making, so the key information is anti-jamming decision-making and the information or characteristics of jamming patterns related to it. The data layer knowledge is extracted from the above table on this basis. Since the scale of the data is small: there are only 8 kinds of conventional jamming patterns, the entity extraction can be manually carried out directly based on experience.

Entity Extraction. Entity extraction is to identify the named entity from the data and indicate the category. In this paper, entity extraction is carried out in Table 1.

Entities can be extracted by jamming types: single-tone aiming jamming, multi-tone (comb) jamming, wide-band blocking jamming, partial frequency band jamming, linear sweep jamming, periodic pulse jamming, tracking jamming and high-speed collision jamming.

Entities can be extracted by anti-jamming strategy: adaptive frequency change, power adaptation, adaptive frequency hopping, adaptive time hopping and accelerating frequency hopping rate.

"Main attributes" in Table 1 belong to the scope of attributes and will be researched in attribute extraction.

Relation Extraction. Relation extraction mainly obtains semantic relations between entities through modeling text data.

The entity relationship of the data layer as shown in Table 1 mainly reflect in the relationship between the entity of the interference pattern and the entity of the anti-interference strategy. "Entity-relationship-entity" can be extracted as: "single-tone aiming jamming—anti-jamming strategy—adaptive frequency change + power adaptation", and the remaining entities can be extracted in the same way.

Attribute Extraction. The purpose of attribute extraction is to construct attribute list and enrich entity connotation.

The corresponding properties of the jamming pattern entities can be extracted from the data in Table 1. "Entity—Attribute—Value" can be extracted as: "Single tone aiming jamming—main attribute—starting and stopping frequency of interference + bandwidth", and the remaining attributes are as above.

Knowledge Reasoning. Knowledge reasoning is based on the existing ontology relations and establishes new relations through reasoning, which can expand and enrich the knowledge stored in the graph, and excavate new knowledge from the existing knowledge, which is the key link in the construction of anti-jamming knowledge graph.

Taking the entity relationship as an example, "single-tone aiming jamming—anti-jamming strategy—adaptive frequency change + power adaptation", this relationship of data layer is relatively single, from which we need to reason and dig out new knowledge and relationship:

First of all, consider the relationship between the two entities of "single-tone aiming interference" and "adaptive frequency modulation". According to the classification of basic anti-jamming modes in literature [11], communication jamming patterns can be divided into aiming jamming, arresting jamming and multi-target jamming. Among them, aiming jamming is used to interfere with a particular channel communication. Then single-tone aiming jamming belongs to the category of aiming jamming, whose principle is that the characteristics of the jamming spectrum are similar to the signal spectrum, and the position and occurrence time on the frequency axis are completely or approximately coincident. Its anti-jamming strategy "adaptive frequency change" is to change the communication frequency to no jamming or weak jamming frequency.

The process of inferring and excavating the deep relationship between the two entities: single-tone aiming interference refers to targeting the jamming signal at the frequency of the communication signal; Adaptive frequency modification is to make the communication signal avoid the jamming frequency -> if still targeted by jamming signal, the signal is still jamming -> Adaptive frequency modification can be successful only if there is a frequency that is not targeted -> Single-tone aiming jamming targets only a specific frequency, there are other frequencies that are not targeted -> Adaptive frequency change can resist single-tone aiming interference.

Finally, a new attribute is excavated. Since the goal of anti-jamming knowledge graph is to make anti-jamming decisions, this attribute is described from the perspective of anti-jamming parties. It is summarized as "frequency gap exists", and the corresponding opposite attribute is "no frequency gap" due to the opposite nature of the attribute. Then the relation group corresponding to the two entities is obtained: adaptive frequency change - application situation - frequency gap exists; Adaptive frequency change - not applicable - no frequency gap; Single-tone aiming jamming - pattern features - frequency gap exists.

By inferring other deep relationships, a set of such relationships is obtained: adaptive frequency hopping - applicability - frequency gap exists; Adaptive frequency hopping - not applicable - no frequency gap; Sweep jamming - pattern feature - no frequency gaps, which conflicts with the data in Table 1, so further study of the attribute "frequency gap exists" is required. Comparing the frequency characteristics of single-tone aiming jamming and sweep jamming, it can be seen that the frequency of single-tone aiming jamming does not change in the whole time period, while the frequency of sweep jamming changes constantly in the whole time period, so the frequency gap of single-tone aiming jamming is fixed, and the frequency gap of sweep jamming does not exist for the whole time period. The basic principle of adaptive frequency hopping is basically similar to that of adaptive frequency changing. The difference is that frequency changing is completed at a time point, and the frequency will not change after completion. Frequency hopping changes the frequency at each frequency hopping instantaneous. Therefore, attributes can be further divided into full time and transient cases. Under instantaneous observation, sweep jamming can be regarded as a narrow-band jamming with the same pattern characteristics as single-tone aiming jamming, so the attribute of "frequency gap exists" would be changed to "instantaneous frequency gap exists".

The relation group corresponding to sweep jamming is changed to: adaptive frequency hopping - application situation - instantaneous frequency gap exists; Adaptive frequency hopping - not applicable - instantaneous no frequency gap; Sweep jamming - pattern characteristics - instantaneous frequency gap exists.

The single-tone aiming jamming relation group was changed to: adaptive frequency change - applicable situation - full time frequency gap exists; Adaptive frequency change - not applicable - no full time frequency gap; Single-tone aiming jamming - pattern features - full time frequency gap exists. Both of these relationship groups were established.

Finally, the entity set of the anti-jamming knowledge graph of the conventional jamming pattern is obtained.

Ontology Construction. Ontology construction abstracts entities to construct ontology. Ontology construction can adopt either manual construction or data-driven construction. Since the scale of conventional jamming pattern data is small, manual construction is adopted in this paper.

The construction process of anti-jamming knowledge graph ontology is to abstract the entity network of the data layer and construct the pattern layer from the data layer.

Construction Principles. Completeness: In order to make full use of knowledge graph to express knowledge, the classification of schema graph should be able to point to all entities in the data layer according to the path of corresponding classification.

Uniqueness: In order to avoid ambiguity and unclear relationship representation, the classification path of the schema diagram should point to the unique corresponding entity, so as to avoid the problem of multiple paths with the same result.

Selection of Classification Mode. The first step classification of anti-jamming knowledge graph data layer can be classified according to the classification mode of jamming system [11], which can be divided into aiming jamming, arresting jamming and multi-target jamming. However, this classification method cannot cover all jamming patterns. For example, sweep jamming can be classified as instantaneous aiming jamming, and can also be classified as blocking jamming within a sweep period, which cannot meet the uniqueness requirement.

The jamming pattern can also be divided into frequency domain jamming, power domain jamming and time domain jamming according to the composition elements of the jamming volume according to the classical optimal jamming theory [2]. Because the influence of a particular jamming mode on communication is not unique, it cannot meet the uniqueness requirement.

According to the conventional optimal communication jamming theory, when the volume of the communication signal is larger than the volume of the jamming signal, anti-jamming can be realized. From this perspective, we can try to classify anti-jamming strategies. Different from the perspective of jamming pattern, the anti-jamming strategy usually only deals with one domain of the signal, so it can meet the unique requirement when it is classified by domain. Meanwhile, according to the optimal jamming theory, the anti-jamming decision is a reverse operation to realize the optimal jamming condition, and its action space is all contained in the action space of changing the volume element of communication signal, so the domain division also meets the requirement of completeness. Therefore, this paper will divide anti-jamming strategies into frequency domain, power domain, time domain and so on according to anti-jamming elements.

Construction Results. Adaptive frequency changing, adaptive frequency hopping and accelerating frequency hopping are all anti-jamming strategies in frequency domain. It is obvious that the adaptive frequency modification only changes the working frequency band of communication from one frequency to another, and the bandwidth does not change. Adaptive frequency hopping and accelerating frequency hopping are the anti-jamming methods of frequency hopping, and frequency hopping itself belongs to the anti-jamming method of spread spectrum communication. Therefore, these anti-jamming methods in the frequency domain can be further classified into 'non-spread spectrum mode' and 'spread spectrum mode'.

Power adaptation is a power domain anti-jamming strategy. The basic principle is to adjust the power of communication signal to meet the requirement of bit error rate or transmission rate according to the change of interference power. Therefore, power adaptation can be classified as 'adjusting the power of communication signal'.

Adaptive time hopping is an anti-jamming strategy in time domain. Time division multiplexing technology is to divide the time of transmission information of the whole channel into several time slices (time slots for short), and allocate these time slots to each information source for use. Analogous to the concept of time division multiplexing, adaptive time hopping can also be regarded as dividing the whole communication time period into segments. Communication signals and jamming signals are regarded as two information sources or users, and all time slots are allocated to these two users. When there is jamming in the time slot, it is equivalent to allocating these time slots to users of the jamming source. When there is no jamming in the time slot, it is regarded as assigned to the communication user. Therefore, this anti-jamming mode similar to time division multiplexing can be summed up as 'time division multiplexing like mode'.

In the practical application, the commonly used anti-jamming strategies also include modulation coding. Different from the above three anti-jamming strategies, modulation coding belongs to the anti-jamming strategy in the modulation coding domain, so modulation coding mode should be added to the classification of anti-jamming strategies, which indicates that the ontology scale will be expanded according to the actual application in the process of ontology construction.

4 Results and Application

4.1 Overview of Anti-jamming Knowledge Graph

This paper uses the graph database Neo4j to store and construct the anti-jamming knowledge graph. Neo4j is a graph database based on the attribute graph model. Its storage management designs special storage schemes for nodes, node attributes, edges, and side attributes in the attribute graph structure, which can access graph data more efficiently than relational databases in the storage layer.

According to the construction method in Sect. 3, the processed RDF data were input into Neo4j software through the Java program written to form the anti-jamming knowledge graph of conventional interference style as shown in Fig. 2.

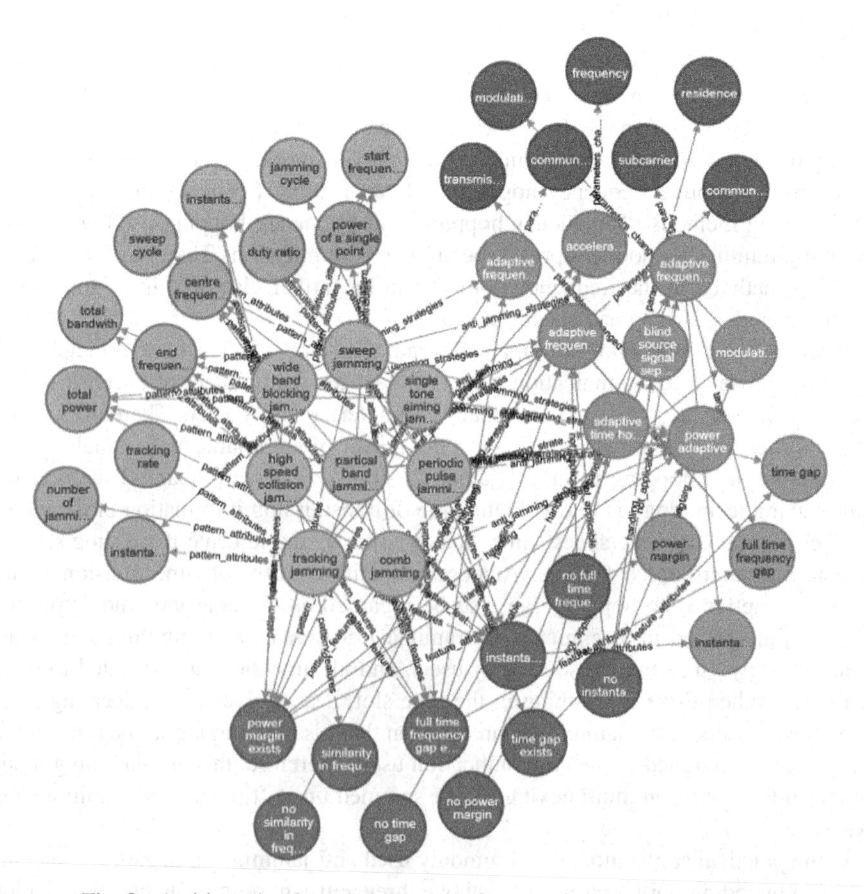

Fig. 2. Anti-jamming knowledge graph of conventional jamming patterns

4.2 Relationship Expansion of Anti-jamming Knowledge Graph

Jamming Patterns. As shown in Fig. 3, the anti-jamming knowledge graph is expanded from the perspective of jamming pattern, so that the relationship directly linked to different jamming pattern can be searched. For example, selecting wide-band blocking jamming is from eight kinds of conventional jamming in the anti-jamming graph, its triplet relationship can be searched to obtain the style characteristics, jamming attributes and anti-jamming methods. These relations are also related to some other jamming styles, indicating that there is a certain relationship between several jamming styles.

Fig. 3. Relationship expansion from perspective of jamming patterns

Pattern Features. From the perspective of pattern features, the anti-jamming knowledge graph can search for specific patterns and anti-jamming methods pointed by pattern features. In practical application, the steps of perceiving specific pattern can be eliminated, and the corresponding processing methods can be found directly through the search of pattern features. For example, by searching the feature "full time frequency gap exists" among several pattern features, the jamming pattern, processing mode and important parameters of this feature can be searched, and the corresponding processing means of different features can be quickly found (Fig. 4).

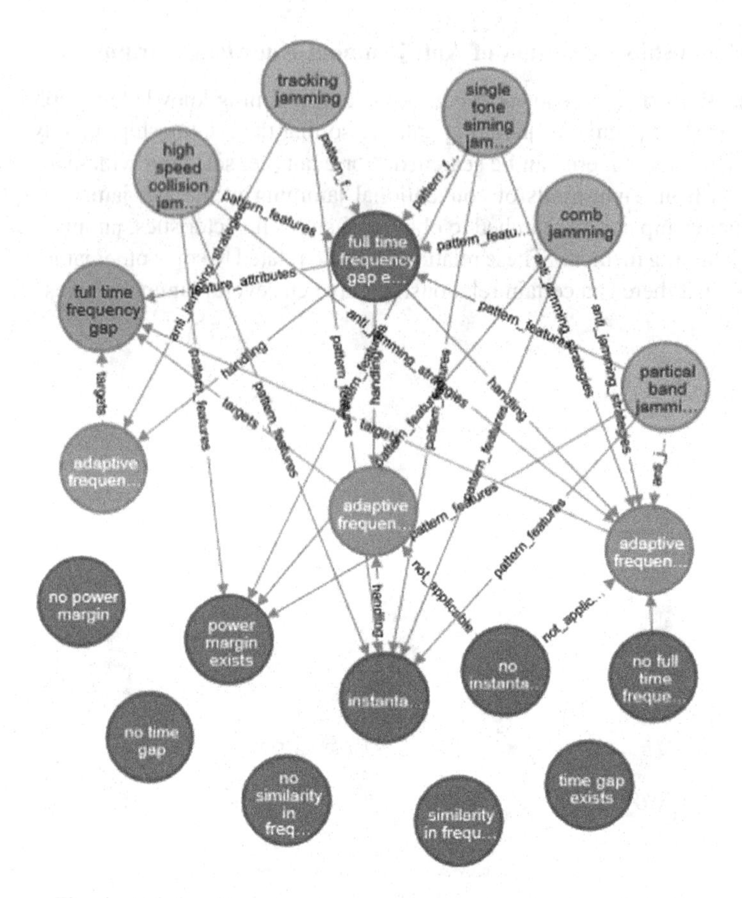

Fig. 4. Relationship expansion from perspective of pattern features

5 Conclusion

In this paper, an anti-jamming knowledge graph construction process based on conventional jamming patterns is proposed. Using this process, the conventional jamming pattern information is extracted and processed to form the data layer entity of anti-jamming knowledge graph. The pattern layer ontology of anti-jamming knowledge graph is constructed by sorting the data layer upward and classifying it. Based on the graph database software, the anti-jamming knowledge graph relationship was expanded from different perspectives. Next step, to deal with the different jamming pattern information input of sensing modules, propose an efficient knowledge graph searching and decision-making method, and form an effective decision-making system based on anti-jamming knowledge graph.

References

1. Wang, X., Wang, J., Xu, Y.: Dynamic spectrum anti-jamming communications: challenges and opportunities. IEEE Commun. Mag. **58**, 79–85 (2020). https://doi.org/10.1109/mcom.001.1900530
2. Yao, F.: Communication Anti-jamming Engineering and Practice. Pubilshing House of Electronics Industry, Beijing (2012). (in Chinese)
3. Chung, S.T., Goldsmith, A.J.: Degrees of freedom in adaptive modulation: a unified view. IEEE Trans. Commun. **49**(9), 1561–1571 (2001)
4. Niu, Y., Yao, F., Wang, M., Chen, J.: Anti-chirp-jamming communication based on the cognitive cycle. Int. J. Electron. Commun. (AEÜ) **66**, 547–560 (2012)
5. Saaifan, K.A., Elshahed, A., Henkel, W.: Cancellation of distance measuring equipment interference for aeronautical communications. IEEE Trans. Aerosp. Electron. Syst. **53**, 3104–3114 (2017). https://doi.org/10.1109/TAES.2017.2728958
6. Han, C., Niu, Y., Xia, Z.: Detection algorithm and anti-jamming method for linear sweeping jamming with low complexity. Appl. Res. Comput. **37**(01), 267–270+274 (2020). (in Chinese)
7. Xia, Z., Chen, J., Niu, Y.: Performance analysis of a detection method for follower jamming in wireless communication. J. Terahertz Sci. Electron. Inf. Technol. **17**(5) (2019). (in Chinese)
8. Wang, H., Qi, G., Chen, H.: Knowledge Graph. Publishing House of Electronic Industry, Beijing (2019)
9. Niu, Y., Yao, F., Chen, J.: Fuzzy jamming pattern recognition based on statistic parameters of signal's PSD. J. China Ordnance **7**(1), 15–23 (2011)
10. Ruan, T., Sun, C., Wang, H.: Construction of traditional Chinese medicine knowledge graph and its application. J. Med. Inf. **37**(4) (2016). (in Chinese)
11. Adamy, D.L.: EW 104: EW Against a New Generation of Threats. Song Zhu. Publishing House of Electronic Industry, Beijing (2017). (in Chinese)

Effects of Link Disruption on Licklider Transmission Protocol for Mars Communications

Jie Liang[1], Ruhai Wang[1(✉)], Xingya Liu[2], Lei Yang[3], Yu Zhou[3], Bin Cao[3], and Kanglian Zhao[4]

[1] Phillip M. Drayer Department of Electrical Engineering, Lamar University, Beaumont, TX, USA
{jliang, rwang}@lamar.edu

[2] Department of Computer Science, Lamar University, Beaumont, TX, USA
xliu@lamar.edu

[3] School of Electronics and Information Engineering, Soochow University, Jiangsu, People's Republic of China
{20195228043,20195228016}@stu.suda.edu.cn

[4] School of Electronic Science and Engineering, Nanjing University, Jiangsu, People's Republic of China
zhaokanglian@nju.edu.cn

Abstract. Some studies on Licklider transmission protocol (LTP) in cislunar and Mars communications have been done. However, little work is currently available in studying how link disruption affects LTP for Mars communications. This study presents a discussion of link disruption on LTP for Mars communications based on experimental data block transfer. The study focuses on how link disruption duration and starting time affect the number of transmission attempts taken, round-trip time (RTT) and goodput performance for Mars communications.

Keywords: Space networks · Network protocols · Wireless networks · DTN

1 Introduction

Delay/disruption tolerant networking (DTN) [1] has been recognized as the primary space networking technology for interplanetary Internet [2]. Since the introduction of DTN technology, different data transport protocols are expected to be operable on the DTN protocol stack under its core protocol, bundle protocol (BP) [3], to provide specific services depending on individual application need [4]. Developed as DTN's main transport protocol, the Licklider transmission protocol (LTP) [5, 6] is developed to operate over unreliable space communication channels for efficient data delivery in space [7].

Link disruption events in interplanetary Internet are generally caused by a lack of limited relay transit capability between a data node on the remote planet and the Earth. They can occur over both the direct planet-Earth link and the broadly adopted indirect, relay-based data links. Taking a typical Mars communications architecture presented in

Q. Guo et al. (Eds.): WiSATS 2021, LNICST 410, pp. 98–108, 2022.
https://doi.org/10.1007/978-3-030-93398-2_10

[7] for discussion, the Mars rover serving as a data source node periodically goes to the opposite side to Earth because of the periodic rotation of the Mars. As a result, the data links connecting the source node on the Mars surface and the destination node at Earth ground station are frequently disrupted. This leads to frequent interruptions to the communication channels. The consequence is that transmission performance and efficiency of data delivery are severely degraded, especially in Mars communications having an extremely long link propagation delay.

Some studies have been done for reliable data transport protocols [8–11] in space, DTN's BP [12–22] and LTP in space [23–35]. Most of these studies focus on design, analysis and implementation of space networking and data transport protocols with emphases on BP and LTP. Little work is currently available in studying how link disruption (either periodic or random) affects LTP for Mars communications. This paper presents a discussion of link disruption on LTP for Mars communications based on experimental data block transfer. The study focuses on how link disruption duration and starting time affect the number of transmission attempts taken, round-trip time (RTT) and goodput performance for Mars communications.

2 System Model

In [30], analytical models are built in estimating the RTT and data block delivery time of LTP in cislunar communication. Although they are built for LTP in Moon communication, they are applicable in Mars communications, and thus are revisited. In [30], the transmission time for a block, T_{Block}, is approximated as

$$T_{Block} = \frac{\left(L_{Seg} + L_{Frame\_Head}\right)}{R_{Data}} \times \frac{N_{Bundle} \times \left(L_{Bundle} + L_{Bundle\_Head}\right)}{L_{Seg}} \tag{1}$$

in which the overhead length L_{Frame_Head} is formulated as

$$L_{Frame\_Head} = L_{LTP-Head} + L_{UDP-Head} + L_{IP-Head} + L_{Link-Head} \tag{2}$$

For all the terms in (1) and (2), they are self-explanatory. Refer to [3] for the details. With the data block transmission derived, the RTT can be estimated as

$$\begin{aligned} RTT &= 2 \times T_{OWLT} + T_{Block} + T_{RS} + \boldsymbol{E}\left(T_{Disrupt}\right) \\ &= 2 \times T_{OWLT} + \frac{\left(L_{Seg} + L_{Frame-Head}\right)}{R_{Data}} \\ &\times \left\lceil \frac{N_{Bundle} \times \left(L_{Bundle} + L_{Bundle-Head}\right)}{L_{Seg}} \right\rceil + \frac{L_{RS}}{R_{RS}} \\ &+ \boldsymbol{E}\left(T_{Disrupt}\right) \end{aligned} \tag{3}$$

in which the terms are self-explanatory, and $\boldsymbol{E}\left(T_{Disrupt}\right)$ denotes the time effect to data block delivery caused by the link disruption event.

3 Experimental Setup and Configurations

The same testbed (SCNT) used in [30] was adopted as the infrastructure for the data block delivery experiments in this paper. Refer to [23] for a detailed description of the testbed. The Interplanetary Overlay Network (ION) distribution v3.6.2 [36] was adopted as the LTP protocol implementation.

To evaluate the sole reliable data delivery service of LTP, the BP custody transfer option was disabled so that the effect of BP reliability service on LTP transmission is removed. To achieve reliable transmission service of LTP, the data of entire LTP block is configured red. LTP data block contains 40 bundles with each bundle of 5-Kbytes.

For the setting and parameters of link disruption events, they are presented under each subsection of Sect. 4.

4 Numerical Experimental Results and Discussion

The experiments are divided into two cases:

Case 1—Reliable data block transmission in presence of a single link disruption event with different durations;
Case 2—Reliable data block transmission in presence of more than one link disruption event with different starting times.

Table 1. Settings of link disruption events for two sets of experiments in case 1.

	Link disrupt. starting time (min)	Durations of link disruption event (min)
Experiment-1	7	5, 10, 20, 30, 45, and 60
Experiment-2	13	5, 10, 20, 30, 45, and 60

4.1 Single Link Disruption Event with Different Durations

The setting of the link starting time and various link disruption durations for two sets of experiments in Case 1, named as Experiment-1 and Experiment-2, are listed in Table 1. Experiment-1 and Experiment-2 are different mainly on their link disruption starting time, 7 min and 13 min from the beginning of the transmission, respectively. However, for each set of the experiments, various link durations are configured for the single link break event, ranging from 5 min to 60 min.

Figure 1 presents the extra number of transmission attempts because of the break for successful block delivery for two sets of experiments in Case 1 with a propagation delay of 10 min (600 s). It is observed that for both Experiment-1 and Experiment-2, the extra transmission attempts increase with respect to an increase in the link disruption. This is reasonable. A longer link disruption results in a long waiting time at the sender, leading to an increase of the checkpoint (CP) timer expirations at the sender and thus an increase of the attempts.

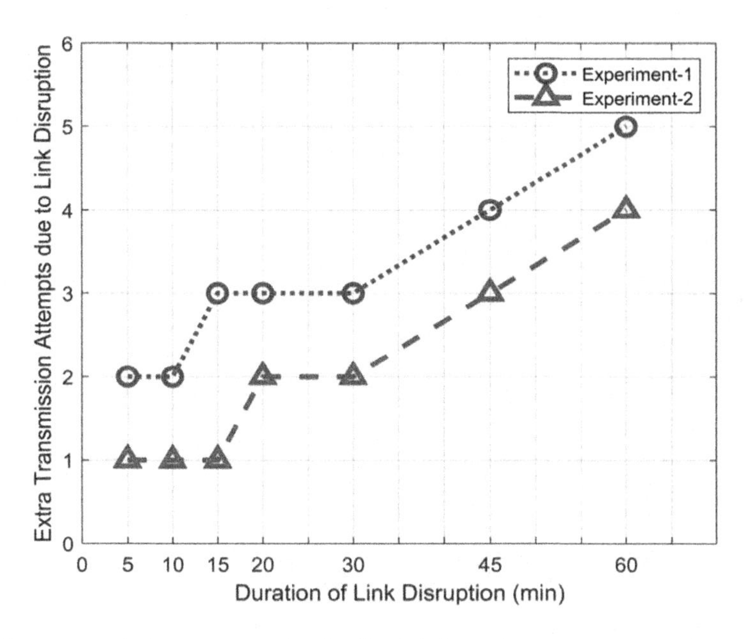

Fig. 1. Extra transmission attempts taken for block delivery for two sets of experiments in Case 1 with respect to various durations of link break.

In comparison, the data delivery runs in Experiment-1 consistently have one extra transmission attempt for all the experimented durations of a link disruption except for 15 min for which Experiment-1 has two extra transmission attempts. The consistent difference of the extra transmission attempt is because of the difference of the starting times. The link disruption in Experiment-2 starts six minutes later than the one in Experiment-1. Regardless of the duration of link disruption, the disruption event in Experiment-1 starts at 7 min from the start of the transmission. This means the disruption event surely causes the CP and a part of the data lost. Therefore, statistically, an extra retransmission attempt is needed.

In contrast, the disruption event in Experiment-2 starts at 13 min from the beginning of the transmission. At the time the disruption starts, the segments already arrive, and therefore, the link disruption may only affect the transmission of the RS over the opposite ACK link. This leads consistently to one or two fewer retransmission attempts required for successful block delivery than Experiment-1.

The effect of the link break can also be observed from the RTT and goodput performance variation trends. Figure 2 presents a comparison of the corresponding RTT between Experiment-1 and Experiment-2. Obviously, the RTT shows a similar increasing pattern as in Fig. 1. The main reason is that the duration of the link disruption event experienced during block delivery is also counted as a part of the RTT.

In comparison, the measured RTT lengths are the same for all the runs in both sets of the experiments even though their starting times are different. This is because the link disruption is introduced before the block delivery ends and therefore, its duration always contributes to the length of RTT.

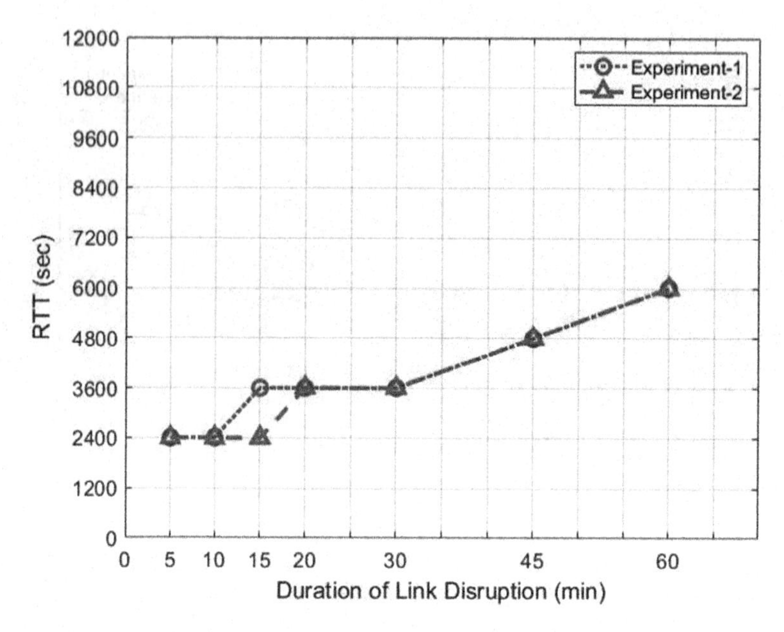

Fig. 2. A comparison of RTT for block delivery between two sets of experiments in Case 1 with respect to various durations of link break.

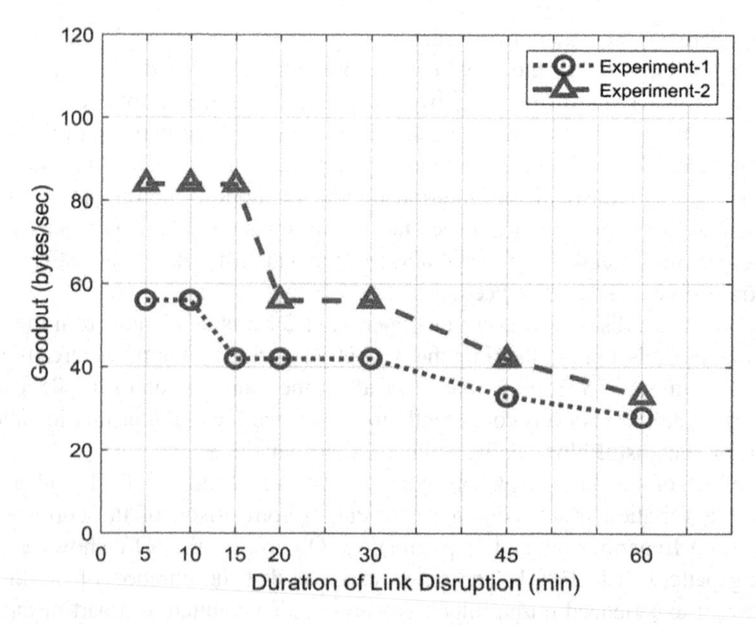

Fig. 3. Goodput performance for block delivery for two sets of experiments in Case 1 with respect to various durations of link break.

Figure 3 presents the corresponding goodput performance. The increase of the RTT in Fig. 2 surely results in a longer block delivery time. As the block size is the same for all the runs, the goodput consistently shows degradation. It is also observed that the difference in goodput between Experiment-1 and Experiment-2 is declining, and it tends to get merged for a very long disruption. This is because the lengthy link disruption has much more significant effect to the block delivery time than their difference in starting time.

4.2 More than One Link Disruption Event with Different Starting Times

The experiments in Case 2 are configured to run with two link disruption events that start at different times. Table 2 lists the settings of the link disruption starting times for the first and the second link disruption and the durations of both events for Experiment-1 and Experiment-2. For each set of Experiment-1 and Experiment-2, while the starting time for the first link disruption event is fixed, various starting times are configured for the second disruption event. Both disruption events have the same duration, 10-min. Figure 4 and Fig. 5 present the extra transmission attempts and the RTT with respect to the time interval between two link disruption events for successful block delivery, respectively, between Experiment-1 and Experiment-2. For the comparison of the extra transmission attempts, the runs in Experiment-1 consistently have three extra transmission attempts for the link disruption durations of 5–20 min and increase to four starting from 25 min. In comparison, the runs in Experiment-2 show opposite variation trend for which they consistently have two extra transmission attempts for the link disruption durations of 5–15 min and drop to one starting from 25 min.

Table 2. Settings of link disruption events for two sets of experiments in case 2.

	1st link disrupt. starting time (min)	2nd link disrupt. starting time (min)	Durations of link disruption event (min)
Experiment-1	7	22, 27, 32, 37, 42 and 47	10
Experiment-2	13	28, 33, 38, 43, 48, and 53	10

Experiment-1 and Experiment-2 have a difference of one extra transmission attempt for the link disruption durations of 5–15 min. It is reasonable because the link disruption event in Experiment-2 starts much later while the duration of the disruption is equal. This is discussed in in Sect. 4.1. The significant difference in the extra transmission attempts between Experiment-1 and Experiment-2 is on their variation trend. Their significant difference is related to the time at which the second disruption starts. For the runs in Experiment-1, the effect of the second link disruption still exists after the transmission starts for twenty-three (23) minutes, which continue to affect the subsequent data transmission. In comparison, for the runs in Experiment-2, there is no data flow existing after the transmission starts for seventeen (17) minutes. This means

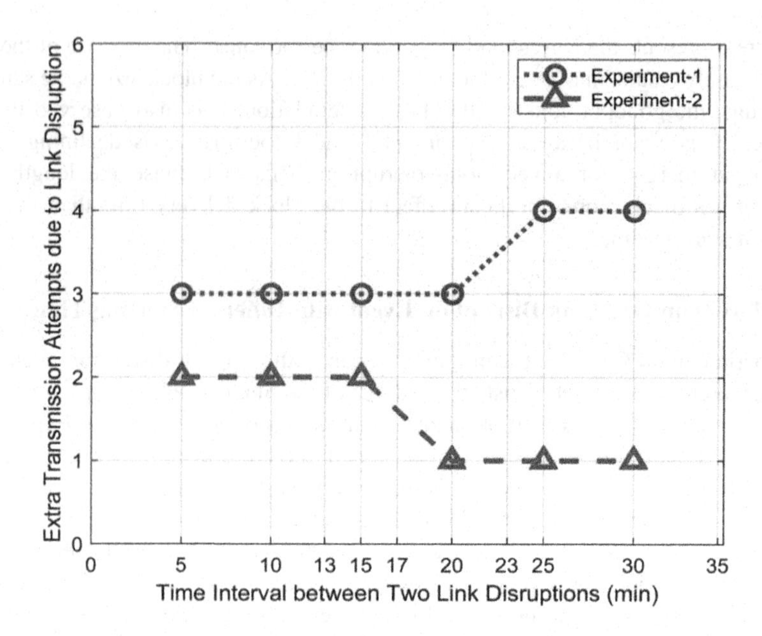

Fig. 4. A comparison of extra transmission attempts taken for block delivery between two sets of experiments in Case 2.

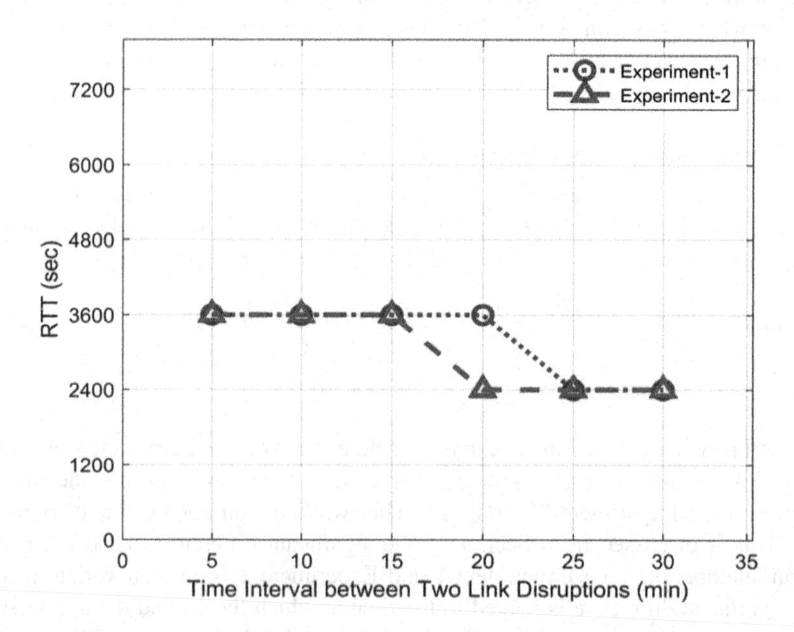

Fig. 5. RTT for block delivery for two sets of experiments in Case 2 with respect to time interval between two link disruption events.

that the second disruption is actually ineffective to the data delivery and therefore, the extra transmission attempts drops by one. This also explains the declining trend of RTT and why the numerical values of the RTT are either 3600 s or 2400 s.

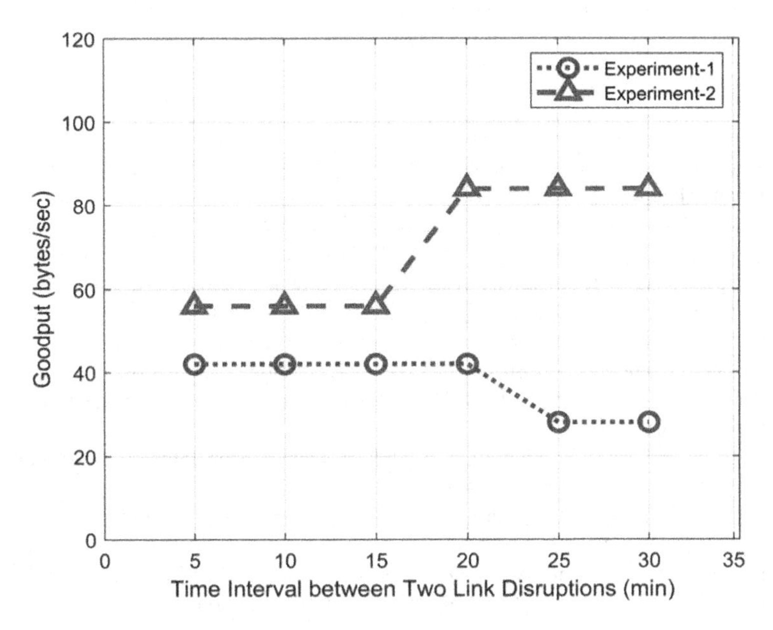

Fig. 6. A comparison of goodput for successful block delivery between two sets of experiments in Case 2 with respect to time interval between two link disruption events.

Figure 6 illustrates the corresponding goodput between two sets. While the data block size is consistently the same, the block delivery time is mainly contributed by the transmission attempts. In other words, more extra transmission attempts are taken, the longer the block delivery time is. This implies that the goodput performance in Experiment-1 and Experiment-2 is expected to be inversely proportional to the transmission attempts in Fig. 4. This is exactly what the goodput performance comparison in Fig. 6 illustrates for with respect to the goodput for each of Experiment-1 and Experiment-2 and their relationship.

5 Conclusions

It is found that the link disruption leads to extra data attempts and thus total number of transmission attempts. The experiment results shows that the extra transmission attempts and thus total attempts increase with respect to an increment of the duration of link disruption. As a result, the RTT and data block delivery time also increase, leading to goodput performance degradation.

For a given link disruption event, earlier it starts, more likely the number of transmission attempts increases and thus longer data block delivery time and lower the goodput performance are. For a transmission involving more than one link disruption event, their effect to all three aspects of transmission performance can't be linearly added. It is related to the connections among the link disruption events.

References

1. Burleigh, S., et al.: Delay-tolerant networking: an approach to inter-planetary Internet. IEEE Commun. Mag. **41**(6), 128–136 (2003)
2. The Space Internetworking Strategy Group (SISG). "Recommendations on a strategy for space internetworking," IOAG.T.RC.002.V1, Report of the Interagency Operations Advisory Group, NASA Headquarters, Washington, DC 20546-0001, USA, 1 August 2010
3. Scott, K., Burleigh, S.: Bundle protocol specification. IETF Request for Comments RFC 5050, November 2007. http://www.ietf.org/rfc/rfc5050.txt
4. Consultative Committee for Space Data Systems. "Rationale, scenarios, and requirements for DTN in space," CCSDS 734.0-G-1. Green Book. Issue 1. Washington, DC, USA: CCSDS, August 2010
5. Burleigh, S., Ramadas, M., Farrell, S.: Licklider Transmission Protocol-Motivation. IRTF Internet Draft, October 2007
6. Ramadas, M., Burleigh, S., Farrell, S.: Licklider Transmission Protocol Specification. Internet RFC 5326, September 2008
7. Consultative Committee for Space Data Systems. "Solar system internetwork (SSI) architecture," CCSDS 730.1-G-1. Green Book. Issue 1. Washington, DC, USA: CCSDS, July 2014
8. Wang, R., Gutha, B., Horan, S., Xiao, Y., Sun, B.: Which transmission mechanism is best for space Internet: window-based, rate-based, or a hybrid of the two? IEEE Wirel. Commun. **12**(6), 42–49 (2005)
9. Wang, R., Horan, S.: Protocol testing of SCPS-TP over NASA's ACTS asymmetric links. IEEE Trans. Aerosp. Electron. Syst. **45**(2), 790–798 (2009)
10. Wang, R., Horan, S.: The impact of van Jacobson header compression on TCP/IP throughput performance over lossy space channels. IEEE Trans. Aerosp. Electron. Syst. **41**(2), 681–692 (2005)
11. Wang, R., Gutha, B., Rapet, P.K.: Window-based and rate-based transmission control mechanisms over space-Internet links. IEEE Trans. Aerosp. Electron. Syst. **44**(1), 157–170 (2008)
12. Wang, R., Qiu, M., Zhao, K., Qian, Y.: Optimal RTO timer for best transmission efficiency of DTN protocol in deep-space vehicle communications. IEEE Trans. Veh. Technol. **66**(3), 2536–2550 (2017)
13. Sabbagh, A., Wang, R., Burleigh, S.C., Zhao, K.: Analytical framework for effect of link disruption on bundle protocol in deep-space communications. IEEE J. Sel. Areas Commun. **36**(5), 1086–1096 (2018)
14. Sabbagh, A., Wang, R., Zhao, K., Bian, D.: Bundle protocol over highly asymmetric deep-space channels. IEEE Trans. Wirel. Commun. **16**(4), 2478–2489 (2017)
15. Wang, R., et al.: Modeling disruption tolerance mechanisms for a heterogeneous 5G network. IEEE Access **6**, 25836–25848 (2018)

16. Feng, C., Wang, R., Bian, Z., Doiron, T., Hu, J.: Memory dynamics and transmission performance of bundle protocol (BP) in deep-space communications. IEEE Trans. Wirel. Commun. **14**(5), 2802–2813 (2015)
17. Zhao, K., Wang, R., Burleigh, S., Sabbagh, A., Wu, W., Sanctis, M.D.: Performance of bundle protocol for deep-space communications. IEEE Trans. Aerosp. Electron. Syst. **52**(5), 2347–2361 (2016)
18. Wang, R., Sabbagh, A., Burleigh, S., Zhao, K., Qian, Y.: Proactive retransmission in delay-/disruption-tolerant networking for reliable deep-space vehicle communications. IEEE Trans. Veh. Technol. **67**(10), 9983–9994 (2018)
19. Yang, L., Wang, R., Zhou, Y., Liu, X., Zhao, K., Burleigh, S.C.: Hybrid retransmissions of bp for reliable deep-space vehicle communications in presence of link disruption. IEEE Trans. Veh. Technol. **70**(5), 4968–4983 (2021)
20. Wang, G., Burleigh, S., Wang, R., Shi, L., Qian, Y.: Scoping contact graph routing scalability. IEEE Veh. Technol. Mag. **11**(4), 46–52 (2016)
21. Yu, Q., Sun, X., Wang, R., Zhang, Q., Hu, J., Wei, Z.: The effect of DTN custody transfer in deep-space communications. IEEE Wirel. Commun. **20**(5), 169–176 (2013)
22. Yang, G., Wang, R., Sabbagh, A., Zhao, K., Zhang, X.: Modeling optimal retransmission timeout interval for bundle protocol. IEEE Trans. Aerosp. Electron. Syst. **54**(5), 2493–2508 (2018)
23. Wang, R., Burleigh, S., Parik, P., Lin, C.-J., Sun, B.: Licklider transmission protocol (LTP)-based DTN for cislunar communications. IEEE/ACM Trans. Netw. **19**(2), 359–368 (2011)
24. Wang, R., et al.: Which DTN CLP is best for long-delay cislunar communications with channel-rate asymmetry? IEEE Wirel. Commun. **18**(6), 10–16 (2011)
25. Sun, X., et al.: Performance of DTN protocols in space communications. ACM/Springer Wirel. Netw. (WINET) **19**(8), 2029–2047 (2013)
26. Hu, J., Wang, R., Sun, X., Yu, Q., Yang, Z., Zhang, Q.: Memory dynamics for DTN protocol in deep-space communications. IEEE Aerosp. Electron. Syst. Mag. **29**(2), 22–30 (2014)
27. Zhao, K., Wang, R., Burleigh, S., Qiu, M., Sabbagh, A., Hu, J.: Modeling memory variation dynamics for the licklider transmission protocol in deep-space communications. IEEE Trans. Aerosp. Electron. Syst. **51**(4), 2510–2524 (2015)
28. Yu, Q., Burleigh, S., Wang, R., Zhao, K.: Performance modeling of LTP in deep-space communications. IEEE Trans. Aerosp. Electron. Syst. **51**(3), 1609–1620 (2015)
29. Wang, R., Wu, X., Wang, T., Liu, X., Zhou, L.: TCP convergence layer-based operation of DTN for long-delay cislunar communications. IEEE Syst. J. **4**(3), 385–395 (2010)
30. Yu, Q., et al.: Modeling RTT for DTN protocols over asymmetric cislunar space channels. IEEE Syst. J. **10**(2), 556–567 (2016)
31. Shi, L., et al.: Integration of Reed-Solomon codes to licklider transmission protocol (LTP) for space DTN. IEEE Aerosp. Electron. Syst. Mag. **32**(4), 48–55 (2017)
32. Yang, G., Wang, R., Burleigh, S., Zhao, K.: Analysis of licklider transmission protocol (LTP) for reliable file delivery in space vehicle communications with random link interruptions. IEEE Trans. Veh. Technol. **68**(4), 3919–3932 (2019)
33. Wang, R., Wei, Z., Zhang, Q., Hou, J.: LTP aggregation of DTN bundles in space communications. IEEE Trans. Aerosp. Electron. Syst. **49**(3), 1677–1691 (2013)
34. Yang, Z., et al.: Analytical characterization of licklider transmission protocol (LTP) in cislunar communications. IEEE Trans. Aerosp. Electron. Syst. **50**(3), 2019–2031 (2014)

35. Zhou, Y., Wang, R., Zhao, K., Burleigh, S.: A study of cross-layer BP/LTP data block size in space vehicle communications over lossy and highly asymmetric channels. IEEE Trans. Veh. Technol. **69**(12), 16126–16141 (2020)
36. Burleigh, S.: Interplanetary overlay network design and operation v3.6.2. JPL D-48259, Jet Propulsion Laboratory, California Institute of Technology, CA, March 2020. http://sourceforge.net/projects/ion-dtn/files/latest/download

Space Information Networks and Spacecraft System

A Cloud Service Architecture for SDN-Based Space-Terrestrial Integrated Network

Xin Ning[1,2], Chengzu Huang[3(✉)], Chenlu Wang[3], and Yuhuai Peng[3]

[1] Innovation Center of Satellite Communication System (CNSA), Beijing 100048, China
[2] Institute of Telecommunication and Navigation Satellites (CAST), Beijing 100081, China
[3] Northeastern University, Shenyang 110819, China
2071862@stu.neu.edu.cn

Abstract. With the development of communication technologies such as 5G, terrestrial networks have been able to provide efficient network services. However, the terrestrial network construction relies on base stations, which leads to the inability to achieve coverage in many areas of the world. To solve this problem, more and more researchers put forward the integrated network of terrestrial network and satellite network to achieve global coverage of the network. In this paper, we considered multiple factors and proposed a SDN-based cloud service architecture for space-terrestrial integrated network, which realizes the unified network management and provisioning through SDN. For the cloud service architecture, we introduce the components of the whole cloud service architecture in this paper, which are divided according to different levels and the functions of each part are explained in detail. Using the proposed cloud service architecture, different user requirements will be satisfied, and the quality of service is greatly improved.

Keywords: Space-terrestrial integrated networks · Cloud service · SDN · Architecture

1 Introduction

With the rapid economic development, the scope of human activities has spread all over the world. However, in remote areas such as mountainous areas and the sea, traditional terrestrial networks cannot achieve full coverage. Therefore, the communication requirements on a global scale cannot be met. In recent years, due to the rapid development of satellite technology, more and more satellites have been launched into space. The resulting space network provides many communication services for ground users, greatly facilitating people's production and life. But the existing satellite systems are isolated from each other. Without perfect space networks, the user demand of high speed, low latency and large connection cannot be met.

At present, many recently proposed studies combine space networks with terrestrial networks. With the global coverage capability of space networks, even in areas that cannot be covered by traditional terrestrial networks, users will obtain communication

Q. Guo et al. (Eds.): WiSATS 2021, LNICST 410, pp. 111–117, 2022.
https://doi.org/10.1007/978-3-030-93398-2_11

services by directly accessing space networks. However, it is not easy to construct and manage a space-terrestrial integrated network formed by space networks and terrestrial networks. The rapid topology changes, long transmission delay, and high heterogeneity between satellite network and terrestrial network all make the existing network architecture unable to achieve networks cooperation.

To solve this problem, a cloud service architecture for SDN-based space-terrestrial integrated network is proposed, which realizes global communication services. In this architecture, the SDN controller performs unified scheduling of space network and terrestrial network, and collects information from them in real time. Thus, it realizes networks cooperation and provides better quality of experience.

2 Related Work

The concept of an integrated space-terrestrial network is based on terrestrial network and integrates the space-based network into the existing terrestrial network to form a unified and efficient interconnection network [1]. From the development point of view, early integrated network was constructed by GEO satellites [2]. GEO satellites have a large coverage area and only need fewer satellites to achieve global coverage, but their transmission delays are large, which cannot provide real-time and efficient services for delay-sensitive applications. As a result, the current satellite network mainly relies on the satellite constellation composed of LEO satellites as the access network to provide services [3]. From the perspective of public demand, the literature [4] elaborated on the development of satellite constellations and proposed network slicing technologies for different service characteristics to achieve flexible message forwarding. The literature [5] analyzed the current situation of integrated network construction. From the composition of space-based networks and terrestrial networks, the research summarized the current development of domestic and international networks and pointed out the direction for future development. To achieve the integration of various heterogeneous networks, the literature [6] pointed out the problems of dynamic routing, link exposure, and time-varying topological information faced by the integrated space-terrestrial network and proposed future research priorities. Now, most of research consider combining SDN technology to achieve deep network integration. For example, in [7], the network architecture consisting of SDN/NFV was proposed, and the dynamic management of the network and the efficient use of resources were analyzed in detail. The literature [8], to improve the efficiency of the current network management, a new STIN architecture was proposed to satisfy the rapid retrieval of information by different users. The literature [9] combined with railway operation examples, in order to solve the practical problems such as poor railway communication signals and slow data transmission, an air-sky network fusion scheme for railway systems was proposed. To address the problem of efficient routing in networks, the literature [10] provided a detailed analysis of routing policies and described the various challenges of achieving efficient routing in networks. The literature [11] presented a comprehensive overview of different satellite communication networks, focusing the research on the physical layer for the problem of exposed satellite communication links that are vulnerable to eavesdropping, and detailing the security techniques currently used in the physical layer. The literature [12] proposed a H-STIN network architecture based on the current development of the Internet of Things. It combined mobile and satellite networks, and a comparative analysis of the different communication protocols that are still available. The literature [13] pointed out the development trend of the future network, combined with different application scenarios, explained the key technical need to be solved, and put forward corresponding suggestions. To improve the utilization of space-based information, a space-based information application service system based on cloud architecture was proposed by [14]. It introduced the overall composition of the system and explained the specific role of each part. However, the above paper only considers certain aspects and lacks an explanation of the overall structure. In this paper, we analyze the composition of this network from a holistic perspective, and then introduce the cloud service architecture, explain the functions of each layer.

3 The Space-Terrestrial Integrated Network Architecture

The space-terrestrial integrated network architecture is composed of three parts: the space-based backbone network, the space-based access network, and the terrestrial network. The structure is as shown in Fig. 1.

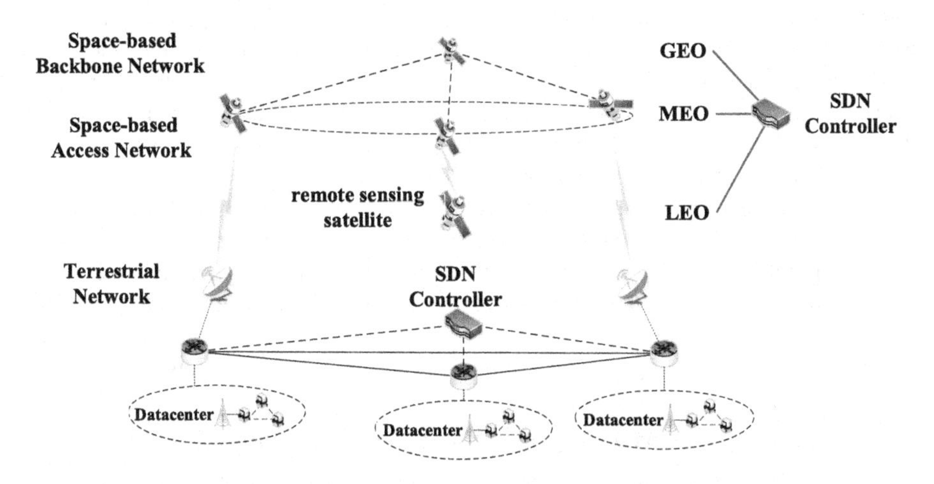

Fig. 1. The architecture of the space-terrestrial integrated network

3.1 SDN-Based Heterogeneous Network Convergence Architecture

This section focuses on the role of each heterogeneous network in the overall network architecture.

The Space-Based Backbone Network. The space-based backbone network consists of GEO satellites, which interconnected by a laser link. It is responsible for connecting the satellite nodes in the network and can be used as the SDN controller, which analyzes and forwards the commands issued by the ground data center and transmits commands to the MEO/LEO orbiting satellites. In addition, with the large storage and computing capabilities of GEO satellites, it maintains the flow tables in the network and manages all flow tables. When a data packet comes in from the ground data center or space-based access network, the data packet can be routed to the designated location according to the stored flow table information. For different QoS requirements, the space-based backbone network can also choose to forward the incoming data after preliminary processing, which effectively reduces the load on the link and improves the quality of service.

The Space-Based Access Network. The space-based access network is mainly composed of MEO/LEO satellites. Compared with GEO satellites, medium-low orbit satellites are closer to the ground, have lower transmission delay, and less information loss during transmission. Therefore, it is a suitable choice for the space-based access network. The LEO satellites are composed of communication, navigation, and remote sensing satellites. The geographic location, weather conditions and other information of the access user collected through the satellite provide the access user with appropriate channels and network resources to maximize the use of resources rate. At the same time, the MEO/LEO satellites can act as switches for the entire SDN architecture. After a user accesses it, different data packets are distributed according to the command of the SDN controller.

Terrestrial Network. The terrestrial network consists of the existing cellular network, mobile Wi-Fi, which together with the satellite network forms a communication network with global coverage. It is interconnected by optical fiber links to form a unified data center in which data can be forwarded and stored. To facilitate the management of the integrated network, the SDN is used to separate the control plane and the data plane, and operations such as the distribution of the flow table are realized through the SDN controller and the OpenFlow protocol. NFV technology is also used to map the network resources into a virtualized resource pool, and the network is managed uniformly through the SDN controller to realize real-time monitoring and control of the network status. Based on the information monitored at different moments, the SDN controller can dynamically adjust the network resource allocation policy, routing path selection and realize the dynamic configuration management of the network.

3.2 The Space-Terrestrial Integrated Cloud Service Architecture

To fully utilize and share the information resources of the above-mentioned networks, with the help of virtualization, cloud computing and other technologies, the terrestrial network and space network resources are integrated, with a view to providing various users with personalized service content and improve the overall service quality. This section introduces the components of the Space-Terrestrial Integrated Network Cloud Service Architecture and describes their corresponding functions. The structure is as shown in Fig. 2.

Infrastructure Layer. The infrastructure layer contains a variety of bare metal, which mainly includes the real physical devices such as computing, storage, and network of each space-based node and ground-based node. Relying on the cloud service

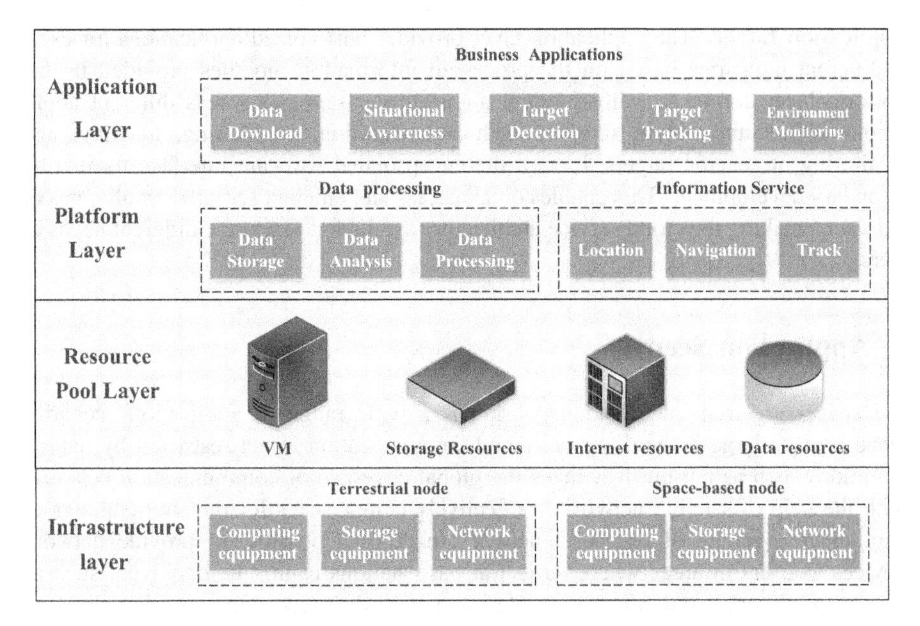

Fig. 2. The architecture of the space-terrestrial integrated cloud service

architecture to form a unified hardware device, it achieves efficient management of each device. As the lowest part of the cloud service architecture, the infrastructure layer provides basic hardware support for the various functions of the entire cloud service, ensuring that the cloud service platform can run stably and efficiently.

Resource Pool Layer. The resource pool layer contains virtual machines and containers. It adopts virtualization technology to virtualize the hardware devices of the infrastructure layer and abstract it into a virtualized resource pool, forming a resource pool layer where each real physical resource can be configured and managed to achieve efficient, stable, and flexible device configuration. It provides the appropriate device resources for different user needs, while providing support for data processing, storage, analysis, and other operations at the platform layer.

Platform Layer. The platform layer manages virtualized resources and provides users with basic communication services. With the help of the powerful computing and storage capabilities of the ground data center, it can also realize the storage of multi-source data of all terrestrial networks and space-based networks, and comprehensively process these data to provide users with intelligent services. According to the status, intelligent decision-making is adopted to improve the resource utilization rate of the whole system and reduce the network congestion rate. The processing of data at the platform layer includes the processing of various data such as radar images, navigation data, geographic location information, and weather conditions to form effective basic information. It is classified and managed in a reasonable manner to provide effective information services to the users.

Application Layer. The application layer provides customized applications for users in different industries based on the processed information modules provided by the platform layer combined with different algorithms. Users can access different applications online and obtain services such as navigation, positioning, tracking, and monitoring of specific targets through the encapsulated software interface formed by secondary development. This enables quick access and efficient services for all types of end-users, making the cloud service architecture available to meet the different needs of users.

4 Application Scenarios

The space-terrestrial integrated network has a wide range of application scenarios because of its large network capacity and communication is not restricted by natural conditions such as terrain. It realizes the global coverage of communication network. With the help of satellite network, it effectively compensates for the shortcomings of insufficient coverage of terrestrial network, making it possible to provide network services to users in areas where terrestrial base stations cannot be deployed, such as oceans and deserts. And when encountering natural disasters such as earthquakes and mudslides, even if the ground network is destroyed, the satellite networks are still available to realize the communication, navigation, and other services to the disaster area. In addition, with the increasing popularity of mobile communication devices, people nowadays rely more and more on mobile communication devices such as cell phones. However, currently on high-speed trains, airplanes, and other means of transportation, due to various conditions, it cannot provide efficient Internet services. With the help of the integrated network, this problem can be effectively solved, and a better quality of experience will be provided on the journey.

5 Conclusions

Based on the existing satellite network architecture and terrestrial network infrastructure with SDN technology, this paper presents rational integration of cloud services architecture to achieve seamless coverage of satellite-terrestrial network. At the same time, virtualization technology is used to map the calculation, storage, and network resources in the network into a unified resource pool, which realizes the comprehensive management of heterogeneous network resources. On this basis, a platform layer and an application layer are built to process multi-source information and provide convenient application services for different access users.

Acknowledgement. This work was supported in part by the National Key Research and Development Program of China under Grant 2020YFB1807900, in part by the Aeronautical Science Foundation of China under Grant 2020Z066050001, and in part by the Fundamental Research Funds for the Central Universities under Grant N2116013.

References

1. Wu, W.: Reflections on the development and construction of space-ground integration information network. Telecommun. Sci. **33**(12), 3–9 (2017)
2. Sun, C., Zhang, J., Zhao, W., Xiao, Y.: Comparative analysis of GEO and LEO broadband satellite communication system. Radio Commun. Technol. **46**(05), 505–510 (2020)
3. Liu, J., Pan, J., Cao, S.: Analysis of the development trend of low-orbit Internet constellation service. Aerosp. China (07), 17–21 (2015)
4. Li, D., Zhu, D., Shen, J.: Networking architecture and slicing technology of space-ground cooperative network based on full-dimension definability. Strateg. Study CAE **23**(2) (2021)
5. Wu, W.: Survey on the development of space-integrated-ground information network. Space-Integrated-Ground Inf. Netw. **1**(01), 1–16 (2020)
6. Yao, H., Wang, L., Wang, X., Lu, Z., Liu, Y.: The space-terrestrial integrated network: an overview. IEEE Commun. Mag. **56**(9), 178–185 (2018)
7. Zhang, H., Nie, J.: Architecture of space-terrestrial integrated network based on SDN/NFV. J. Mil. Commun. Technol. **38**(02), 33–38 (2017)
8. Li, J., Xue, K., Liu, J., Zhang, Y., Fang, Y.: An ICN/SDN-based network architecture and efficient content retrieval for future satellite-terrestrial integrated networks. IEEE Netw. **34** (1), 188–195 (2020)
9. Teng, Y., Li, X., Song, M.: SDN enabled space-terrestrial integrated network architecture of railway system. Chin. J. Internet of Things **4**(03), 30–41 (2020)
10. Jiang, L.: Research on control architecture and routing technology of satellite communication port in space-terrestrial integrated network. Telecom Power Technol. **35**(09), 189–190 (2018)
11. Li, B., Fei, Z., Zhou, C., Zhang, Y.: Physical-layer security in space information networks: a survey. IEEE Internet of Things J. **7**(1), 33–52 (2020)
12. Chien, W., Lai, C., Hossain, M.S., Muhammad, G.: Heterogeneous space and terrestrial integrated networks for IoT: architecture and challenges. IEEE Netw. **33**(01), 15–21 (2019)
13. Liu, Y., Huang, T., Liu, J.: Challenges and opportunities of future network. Radio Commun. Technol. **46**(01), 1–5 (2020)
14. Liu, G., Lu, Z., Qiu, L., Zhou, B.: Study on the spatial information service system based on cloud architecture. J. China Acad. Electron. Inf. Technol. **13**(05), 526–531 (2018)

Research on Beam Interference Optimization Strategy of LEO Constellation

Huajian Zhang(✉), Bo Yang, Chao Xi, Xiao Yang, Jiangyan Hu, Jirong Wang, Xueyuan Qiao, and Chuanguang Fu

Space Star Technology Co., Ltd., Beijing, China
zhanghj@spacestar.com.cn

Abstract. The LEO constellation is a near earth polar orbit satellite constellation, which will cause a large number of beam overlap in the high latitude during the motion, and frequency multiplexing is usually used in LEO constellation to improve the efficiency of system resource use. All these factors will cause the interference between beams. The frequency interference of the whole constellation can be reduced to a large extent by closing the beam strategy of partial overlapping points, but due to the high speed movement of the satellite beams, it is still unable to guarantee the full period of the whole constellation without interference or interference below the threshold value. The dynamic adjustment strategy of the beam resource designed in this paper is based on the periodic closing strategy, using the sub-band division and frequency dynamic allocation technology, so as to better realize the global beam frequency resource allocation and interference optimization.

Keywords: LEO constellation · Interference · Beam on/off · Optimization strategy

1 Introduction

With the diversification and rapid increase of LEO satellite constellations, there are many kinds of interference between LEO satellite constellations and GSO satellite constellations. Potter, Bob researched how traditional carrier monitoring, ground system monitor and control (M&C) and data analytic products can be utilized by LEO operators to monitor the performance of their complete satellite network, drilling down to ground systems, satellite performance, beam pointing and power usage to minimize interference with GEO satellites [1]. Braun, Christophe and others have shown that many satellite operators are planning to deploy NGSO systems in Ku, Ka, and V bands. The coexistence of satellites between systems will face new challenges, because the heterogeneity of the constellation leads to increased interference levels and complex interactions. The existing spectrum adjustment may not be enough to ensure that GSO is protected from NGSO interference [2]. As the satellite advances from the equator to the polar regions, certain coverage areas will overlap to a certain extent, which may cause severe inter-beam interference (IBI) [3]. The implementation of any LEO satellite system's high-throughput satellite system will cause a conflict between the radio

Q. Guo et al. (Eds.): WiSATS 2021, LNICST 410, pp. 118–134, 2022.
https://doi.org/10.1007/978-3-030-93398-2_12

spectrum of the fixed-satellite service and the satellite broadcasting service (Ku, Ka, Q/V band) [4].

Aiming at the interference problem: for the interference mitigation technology between GEO and NGEO systems, Sharma et al. proposed an adaptive power control technology suitable for uplink and downlink to reduce interference [5]. Mendoza et al. proposed a RF interference analysis method for dynamic satellite constellation simulation based on the inter-satellite link (ISL) of LEO satellites, analyzed and designed reasonable ISL parameters, and reached the conclusion that the spectrum of LEO and GSO network coexisted [6]. Su, Yongtao adopted a shutdown strategy when LEO satellite caused interference to GSO satellite communication. If the closed LEO satellite provided service for covered users, the progressive pitching method and coverage expansion method were adopted [7]. Garcia et al. proposed a method of measuring interference on the large-scale LEO satellite communication constellation system by SINR [8]. Leyva-Mayorga et al. proposed a dynamic ISL resource allocation method based on the non-interference environment [9]. Based on interference analysis, Li, R. et al. proposed an adaptive beam power control method, which attempted to maximize the throughput of LEO satellite [10]. Saiko, V. et al. proposed a multipath signal processing method to improve the interference immunity of inter-satellite communication link [11].

When the satellites in the LEO constellation motion from low latitude to high latitude, the distance between the satellites in adjacent orbits decreases gradually, causing a considerable number of beams to fall into the coverage of other satellites completely or partially, leading to the waste of some beam resources and the problem of co-frequency interference.

Because of the LEO satellite constellation system available frequency resources are limited, even in the case of static allocation without interference, with the movement of the satellites, the beams of different satellites will appear adjacent, overlapping, and covering interference. Therefore, the problem of beam interference suppression under the whole constellation is a prominent problem that must be solved in constellation system.

Beam frequency resource allocation result quality in the whole constellation is closely related with the beams interference, the beam of resource dynamic adjustment strategy designed in this paper is distributed by track, reduce the overlapping interference with beam in contiguous or overlapping area of the beam basis on closing overlapping beam, using beam resource dynamic adjustment strategy, the global users of frequency assignment problem attribute to a sub-band distribution problem which greatly simplifies the whole constellation frequency assignment problem and the analysis of the whole constellation interference, by optimizing the number of sub-bands and adjusting the period to achieve a certain balance between beam capacity and interference, at the same time, it takes into account the problems of less sub-band division, less interference, and less adjustment frequency. While suppressing beam interference, it also achieves better constellation beam frequency resource allocation, which not only supports the dynamic allocation of local beam frequency resource, but also reduces the interference factors between beams in the satellite constellation to the greatest extent.

2 Periodic Beam Closing Strategy

2.1 Beam on/off Policy

In order to solve the interference problem caused by the overlapping of satellite beams at high latitudes, the beam on/off strategy in this paper realizes the global single repeating cover on the premise of seamless global beam coverage, and reduces the number and frequency of satellite beam on/off control in the whole constellation.

The simulated LEO constellation has 6 orbital planes, each of which has 9 satellites, and each satellite has 52 spot beams (as shown in Fig. 2), Fig. 1 shows the beam coverage of the LEO constellation, their relative position to the satellite is fixed. The satellites moving relative to each other on both sides of the seam area are interlaced about every 5 min 40 s, and the satellites in the non-reversing slit area are interlaced about every 53 min. For all beams, the on/off state will have the same period as the satellite's orbit periodicity (about 107 min).

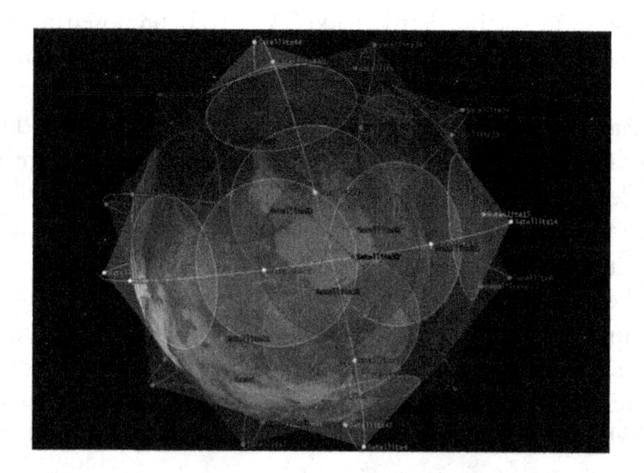

Fig. 1. Satellite beam coverage as seen from above the North Pole (10° elevation)

Periodic beam on/off processing is considered for the main lobe based on the optimized beam pattern. The strategy (flowchart shown in Fig. 3) gives priority to the beam on/off control of odd orbital plane (1, 3, 5 orbital plane). The first step is to complete the beam single cover decision for all beams on the first orbital plane. When the satellite in the third orbital plane overlaps with the satellite in the first orbital plane near the poles, turn off the satellite beam in the third orbital plane which falls completely into the first orbital plane. Similarly, when the satellite in the fifth orbital plane overlaps with the satellite in the first or third orbital plane near the poles, turn off the satellite beam in the fifth orbital plane which falls completely into the first or third orbital plane. After completing the beam assignment to the odd orbital planes (as shown in Fig. 4), the even orbital planes (2, 4, 6 orbital planes) are used as supplementary orbital planes for beam allocation, and the allocation method is the same as that of odd orbital planes.

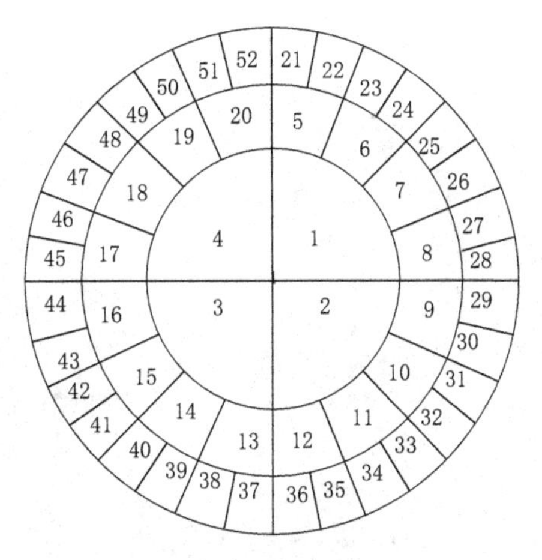

Fig. 2. Schematic diagram of point beam

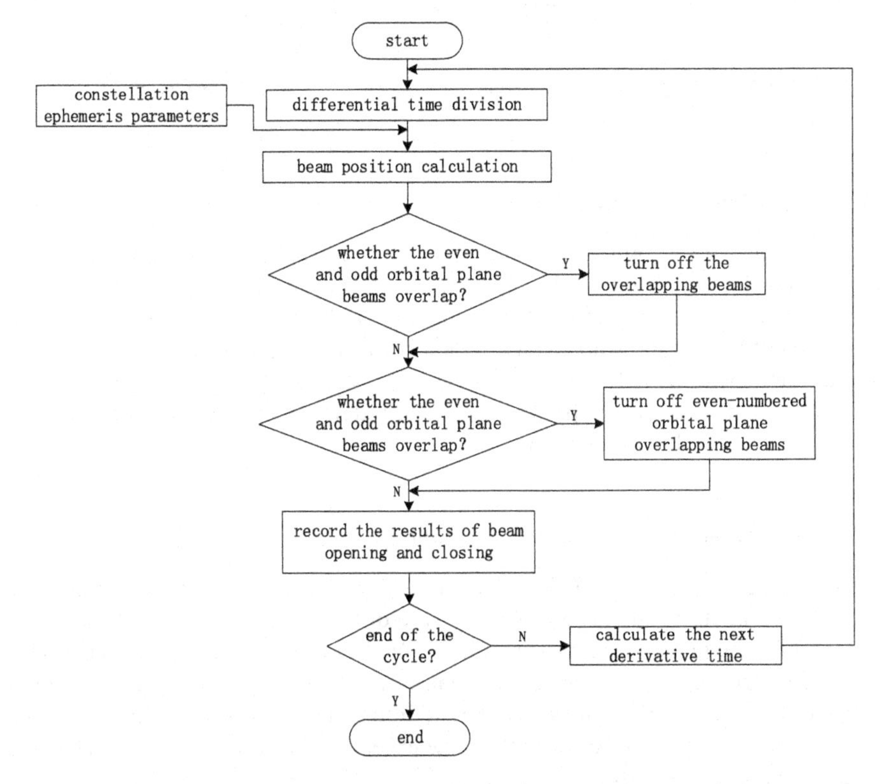

Fig. 3. The planning process of periodic beam-opening and closing

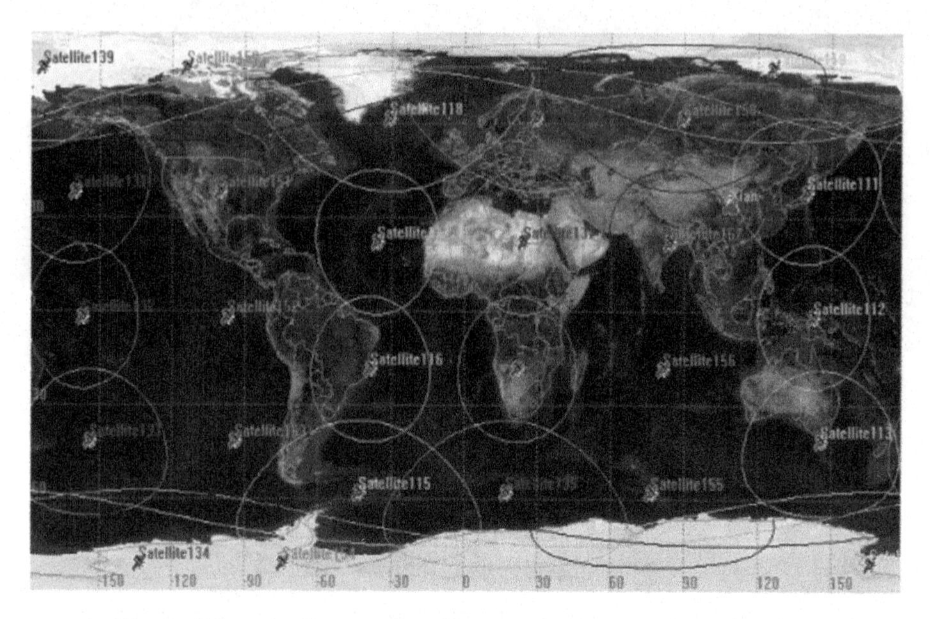

Fig. 4. Schematic diagram of satellite coverage on the main orbital plane

2.2 Simulation Results and Analysis

Since the beams of odd orbiting satellite has been able to realize the complete coverage of the region above 60° north and south latitude, when the even orbiting satellite moves into that area, it can completely close all the beams of the corresponding satellite.

The simulation of the whole constellation by using the periodic on/off strategy of the beam, and the results of the change of the on/off state of all beams with time in 12 h are as follows. The abscissa represents time, the ordinate represents the beam number of the satellite, and the red line represents the beam opening state during this period.

The following is the time sequence diagram of the first satellite in each orbital plane (Figs. 5, 7, 8, 9 and 10):

The simulation results are analyzed as follows:

1. The beam state has a periodicity of about 107 min, satellites in the same orbit have similar beam states and have a certain phase difference.
2. Except for the first and sixth orbital planes, the beams of all odd orbital planes also have a certain similarity (there are differences in individual beams). The beam states of the corresponding satellites in different orbits (such as satellite 1 in 1 orbit and satellite 1 in 5 orbit) are basically the same except for individual satellites. This rule also applies to even orbital.
3. Because the beam is preferentially allocated to the first orbital surface, the beam remains open throughout the entire process, but the 6th orbit satellite beam turns on/off more frequently than other even orbital planes due to the seam.

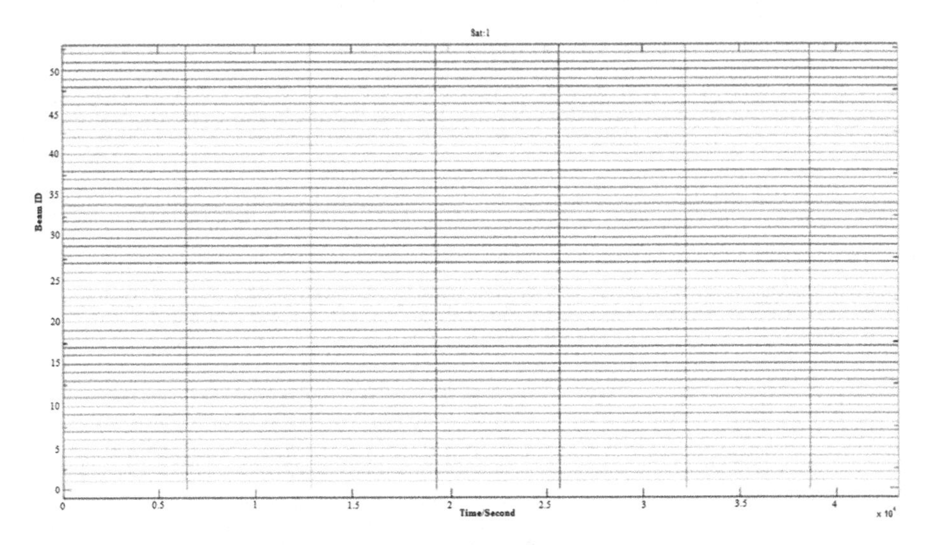

Fig. 5. Beam status of satellite 1 in 1 orbit

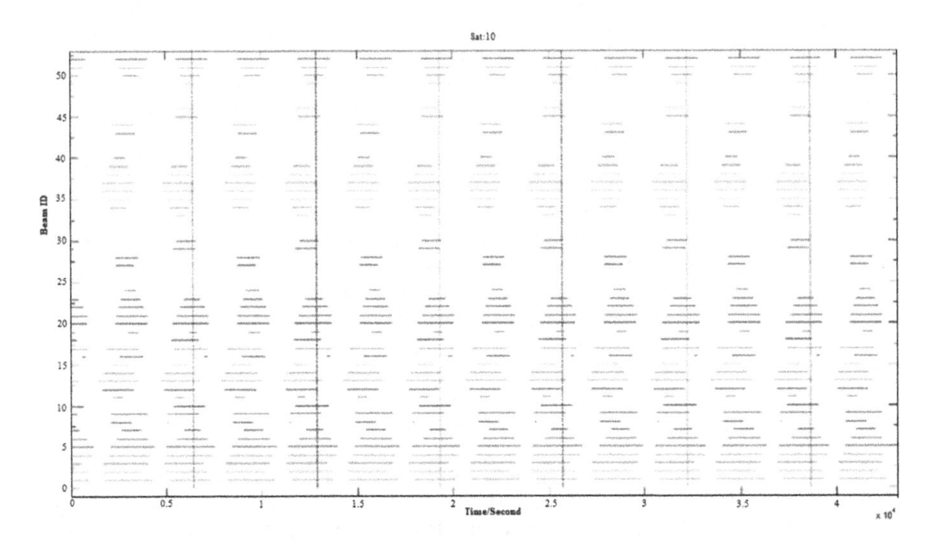

Fig. 6. Beam status of satellite 10 in 2 orbit

4. Satellite beams on odd orbital take more than 50% longer to turn on than even orbital, and many of which never turn on during the entire cycle.
5. The number of open beams varies steadily with time and remains between 1530 and 1590.
6. The first orbital remains open during the full cycle, and the beam opening time of the other odd orbital is also significantly longer than that of the even orbital. In order to maintain the normal operation of the satellites in the whole constellation

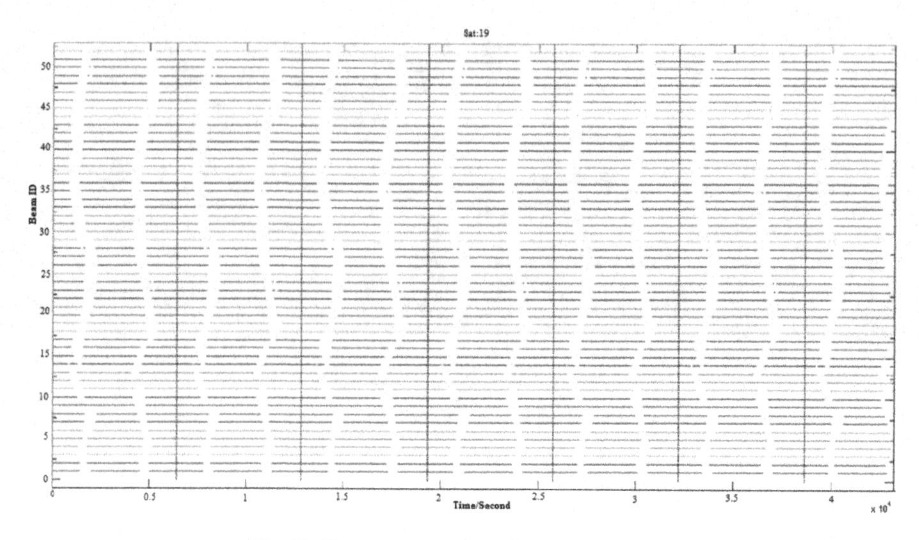

Fig. 7. Beam status of satellite 19 in 3 orbit

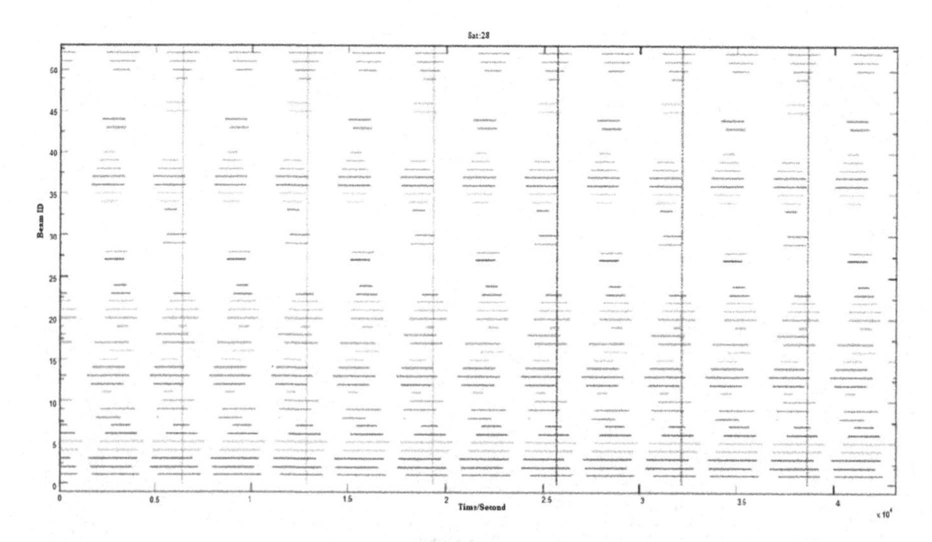

Fig. 8. Beam status of satellite 28 in 4 orbit

and avoid the long-term operation of the satellites in a certain orbital plane, we can rotate in each orbital plane in a certain period to balance the burden.

7. Dynamic frequency allocation prevents the beams on both sides of the seam from interference, thereby solving the problem of frequent opening and closing of the sixth orbit beam.

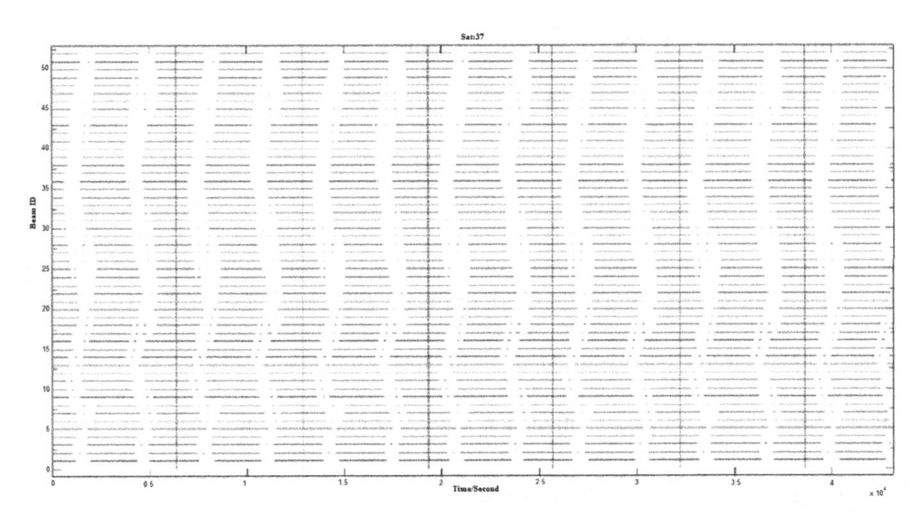

Fig. 9. Beam status of satellite 37 in 5 orbit

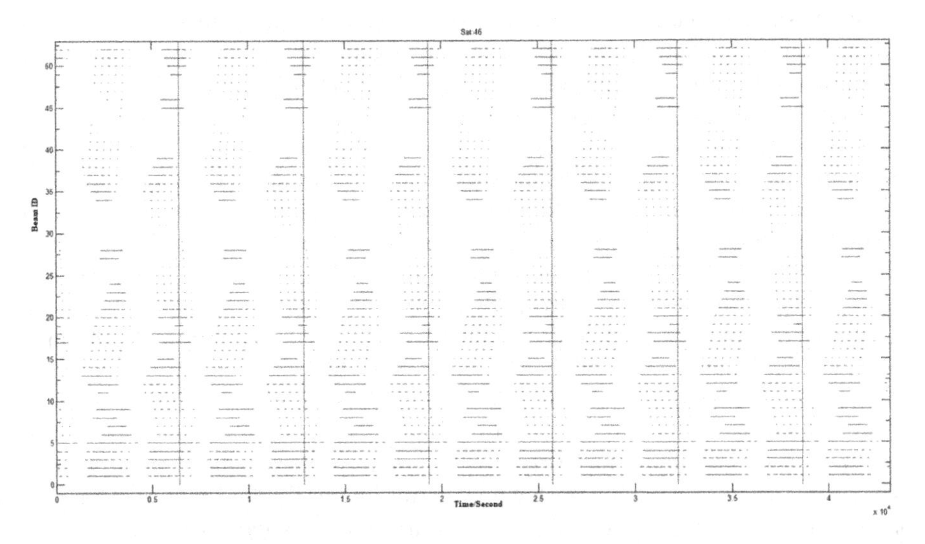

Fig. 10. Beam status of satellite 46 in 6 orbit

3 Dynamic Adjustment Strategy of Beam Resources

The periodic on/off strategy of spot beam greatly reduces the interference problem of the whole constellation, but it does not consider the interference problem of the cross area. In order to improve resource utilization, frequency multiplexing is usually adopted to divide the user's bandwidth into several sub-bands and allocate them to different beams. The more sub-bands are divided, the greater the choice of frequency reusability and the smaller the interference, but the less the number of carriers available to users under a single beam; the less the sub-bands are divided, the more serious the inter beam interference in the constellation. Due to the relative motion of satellites, it is difficult to ensure that the initial allocated non-interference sub-bands remain non-interference in the whole process. If the allocated sub-bands of satellite point beam are not adjusted dynamically during the operation cycle of the constellation, the difficulty of initial sub-band allocation and the probability of interference will be increased.

In order to take into account the number of sub-band divisions and interference, the constellation operating period can be divided into several equally spaced time slices. In each time slice, the sub-band allocation of each point beam in the constellation remains unchanged. Prioritize the allocation result of the previous time slice as the initial value of the next time slice and perform interference detection. If there is interference, re-adjust and allocate the beam sub-bands in the interference area. In the case of a fixed time slice, by repeatedly iterating the beam frequency allocation scheme, which can obtain a better beam frequency allocation scheme within the time slice.

3.1 Frequency Dynamic Allocation Strategy

Initial Allocation

In order to simplify the process, first simulate an initial frequency allocation plan, and use the result of the initial frequency allocation as input for iterative calculations. The allocation result is the same for all 54 satellites. The initial frequency allocation generation process is as follows (Fig. 11):

First, assign a frequency point to the i-th beam, and calculate whether the beam will produce interference exceeding the threshold with other beams. In order not to affect the communication quality, the beam interference threshold is set to -12dB, if it exceeds threshold, judge whether other available frequency points also produce interference exceeding the threshold; if it still exceeds, it indicates that there is a problem in the frequency point allocation of the first i-1 beam, firstly, reallocate the frequency point of the i-1 beam, and so on the interference generated by the beam is within the threshold.

Assign a frequency point to the i-th beam, and calculate whether this beam will cause interference with other beams that exceeds the threshold; if it exceeds, then determine in turn whether other frequency points available for allocation also cause interference that exceeds the threshold; if it still exceeds, it means that there is a problem in the frequency allocation of the i-1 beam. Then the frequency points of the i-1th beam are re-allocated, until the interference generated by all beams is within the threshold.

Interference calculation is to calculate the interference value of the i-th beam to other beams. The calculation process is as follows (Fig. 12):

By simulating the influence factors of the interference pattern of each point beam of the satellite, the interference values of other beams under the same satellite for each beam are calculated. Through the above iterative method, a sub-band and carrier allocation strategy lower than the interference threshold value (−12dB) is assigned to each beam (Figs. 13, 14 and 15).

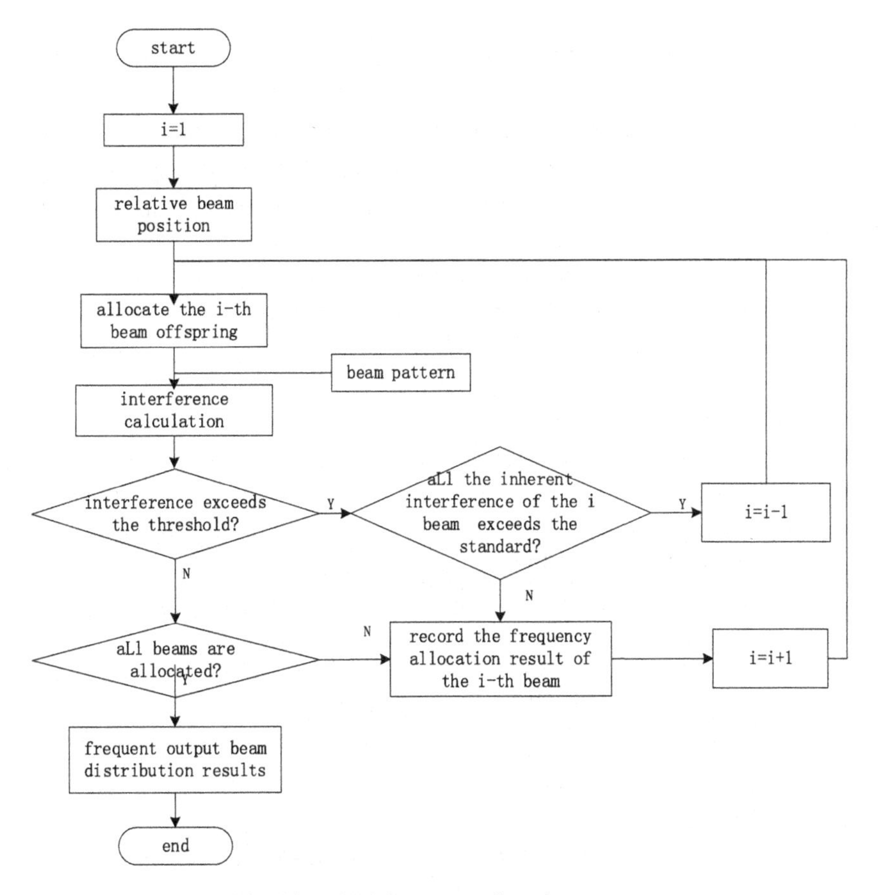

Fig. 11. Initial frequency allocation process

If the number of access terminals under a certain beam exceeds the number of users that can be carried by the initial sub-band during the actual system operation, the operation control system will then separately increase the available sub-bands or carriers for the satellite beam. The premise of allocating sub-bands or carriers is that the allocated carriers will not cause interference to other users and will not be interfered by the same carriers in other beams.

We randomly simulate 780 user terminals under a single satellite, and the terminals are randomly distributed in each spot beam; the L-band 7M user bandwidth is divided into 12 sub-bands, and each sub-band is divided into 15 carriers. The simulation results are shown in Fig. 6. In the figure, the black number represents the beam number, the red number represents the sub-band number obtained through allocation of the beam, and the two red numbers indicate that the beam has obtained additional sub-bands allocated by the system on the basis of the existing initial sub-bands (Fig. 16).

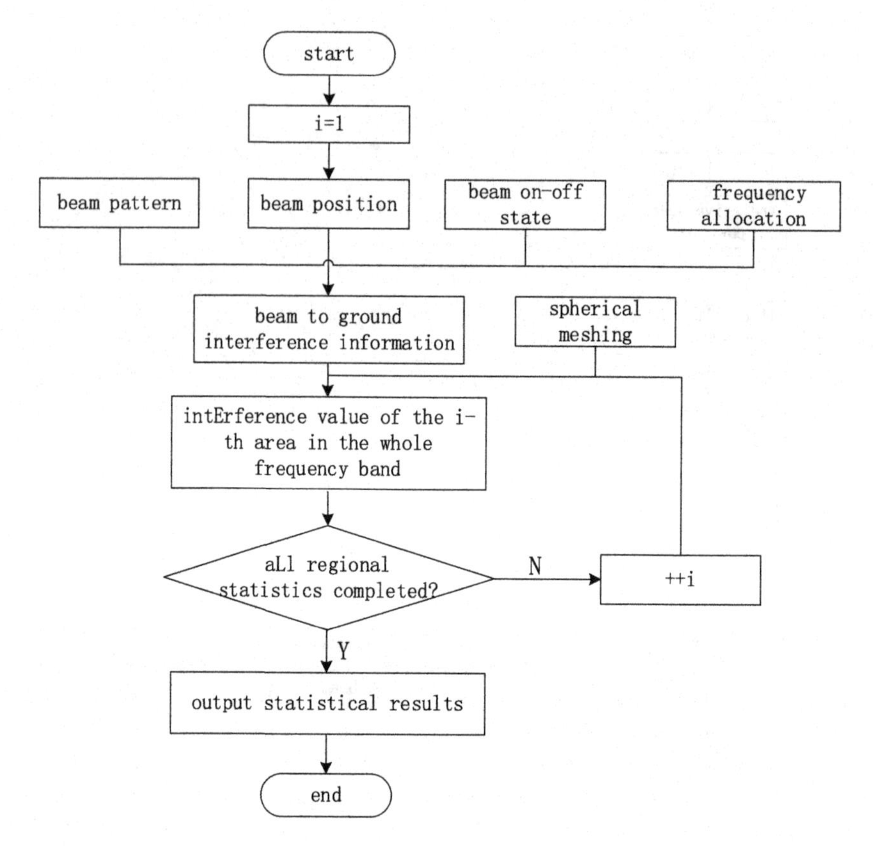

Fig. 12. Interference calculation process

Dynamic Allocation

According to the beam periodic closing strategy, the beam opening and closing have the same periodicity (about 107 min) as the constellation operation, and the beam frequency allocation scheme only needs to consider one cycle, which can greatly reduce the amount of calculation.

Divide a satellite's operating cycle into a number of differential times with a length of ΔT. The satellite state, the beam on/off state within a differential time and the frequency allocated to it are regarded as fixed. The earth's surface is evenly divided into 10560 sampling points, and a constellation operating cycle is divided according to the time slice length ΔT, assuming that M is the number of global subbands, and N is the number of time slices.

According to the current time and ephemeris parameters, the real-time position of the constellation satellites can be obtained. Since the beam position is fixed relative to the satellite, the position of all beams can be calculated from the position of the satellite. Beam pattern is obtained from satellite antenna data, and the on/off state of the beam can be obtained from the periodic closing strategy. In this paper, the method of gridding the earth's surface is adopted, and the spherical surface is divided into several

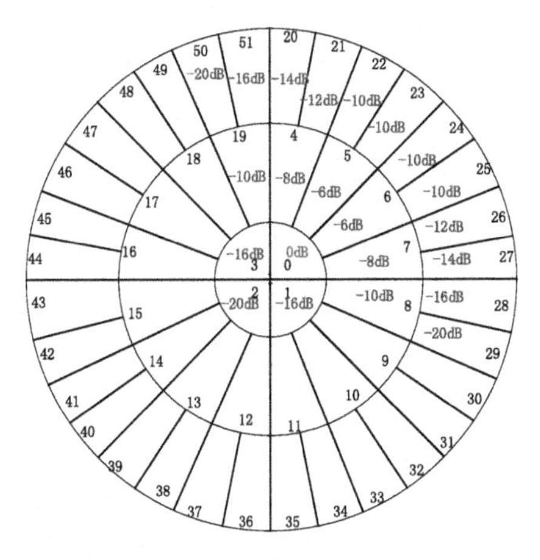

Fig. 13. Pattern interference influence factor of inner spot beam

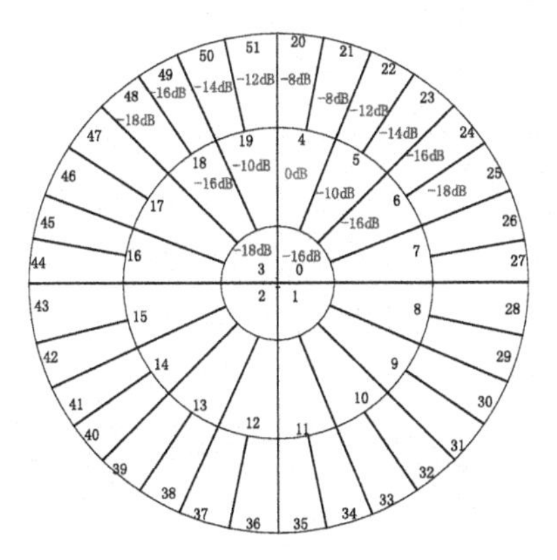

Fig. 14. Pattern interference influence factor of mid-level spot beam

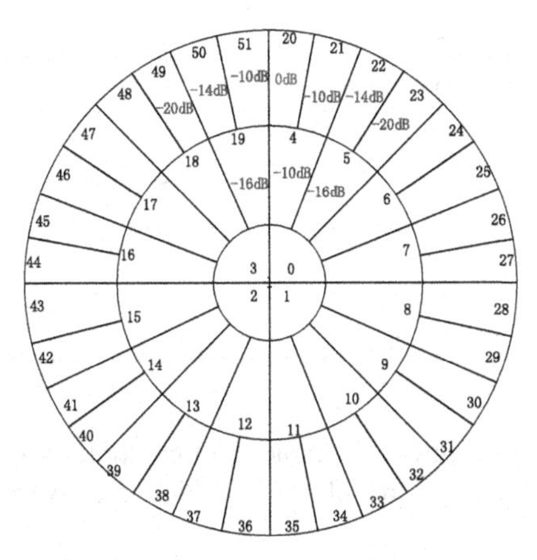

Fig. 15. Pattern interference influence factor of outer spot beam

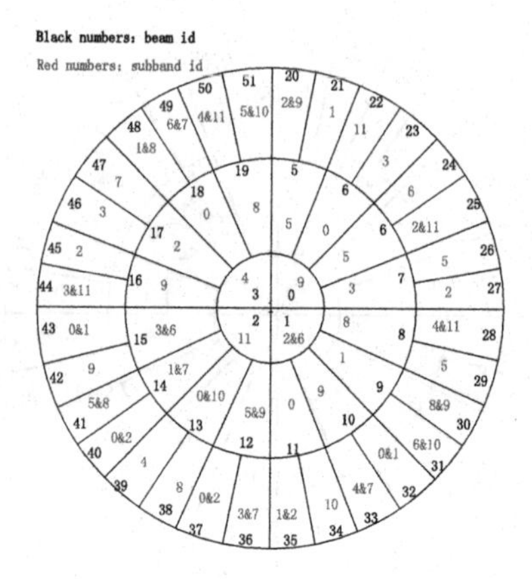

Fig. 16. Schematic diagram of sub-band allocation

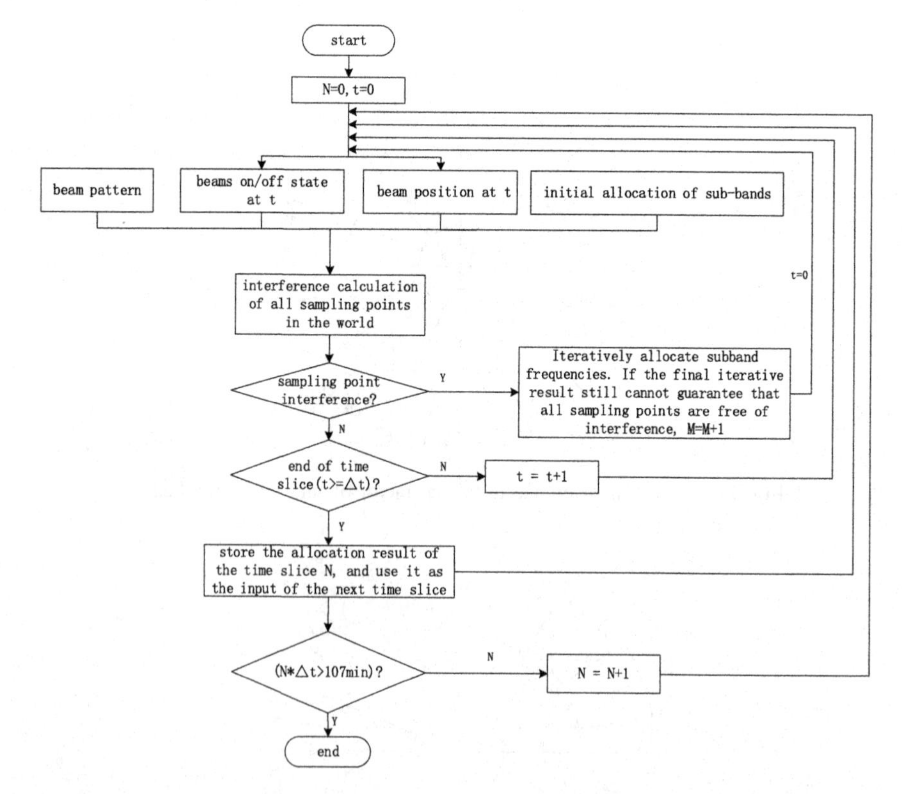

Fig. 17. Constellation frequency resource optimization allocation processing flow

cells. This way, while ensuring considerable calculation accuracy, numerical solutions can be used instead of analytical solutions, which greatly reduces the amount of calculation.

In the process of frequency allocation, the interference values of all regions in the world can be calculated by combining the beam position, the beam on/off state and the beam pattern. If the interference in a location region exceeds the threshold value, the beam allocation state is reprogrammed. After many iterations, the frequency allocation of the whole constellation beam in [T0, T0 + ΔT] can be obtained, and the results are used as the input for the iterative calculation of the beam sub-band of [T0 + ΔT, T0 + 2ΔT], until a broadcast frequency allocation scheme can be obtained without interference in the whole constellation period. There is no situation of being in two or more beams and allocating the same beam sub-bands at any place at any time in the world. During the iteration process, the beams of the same frequency band do not overlap, and the iteration period is the entire constellation period, the flowchart is as follows (Fig. 17):

Set the time slice length ΔT, and iteratively change the frequency of the sub-bands through the algorithm processing flow. If after all the frequencies are iterated, all sampling points cannot be satisfied without interference, the number of sub-bands needs to be re-divided, and then the iterative calculation is performed again until meet the algorithm termination condition for global beam interference-free.

3.2 Simulation Results and Analysis

Set different time slice lengths ΔT and the number of sub-bands M, and perform simulation according to the above procedure. Under the constraint of full constellation without interference, the relationship between the selection of time slice length and the required number of sub-bands is shown in the following table (Table 1).

Table 1. The relationship between the selection of time slice length and the number of interference-free sub-bands

Time slice length (ΔT)	The number of sub-bands that satisfy the interference-free (M)
6 min	21
5 min	16
4 min	12

The simulation results show that when the set time slice length is longer, more sub-bands need to be set to meet global interference-free. When the number of sub-bands is set less, beam frequency adjustments are more frequent, after comprehensively considering the beam capacity and the dynamic adjustment period, this scheme selects that the beam frequency dynamic adjustment duration in the constellation is 4 min, and the frequency allocation of the global interference-free beam can be satisfied by the division of 12 sub-bands.

Based on the above-mentioned dynamic adjustment strategy, this paper performs beam control and dynamic allocation of frequency resources for 54 LEO satellites. The simulation results before and after the optimization of the beam and resource show that the optimized beam can satisfy the global single repeat cover. At the same time, the overlapping coverage is greatly reduced, and the frequency of adjacent and intersecting beams is not repeated, which not only reduces the waste of beam resources, but also satisfies the interference suppression results of the beams (Figs. 18 and 19).

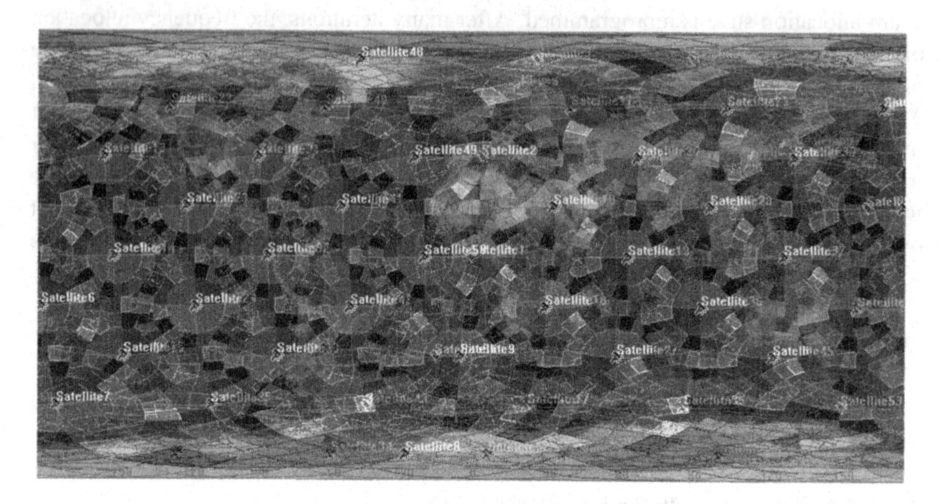

Fig. 18. Global beam coverage map before dynamic adjustment of beam resources.

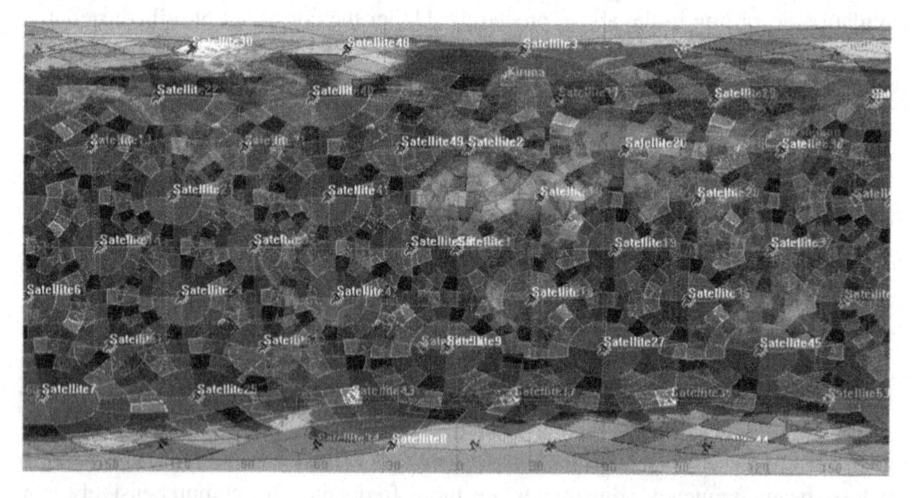

Fig. 19. Global beam coverage map after dynamic adjustment of beam resources.

The simulation results are analyzed as follows:

1. Due to the relative movement of the satellite and the ground, the coverage time of a single beam can be up to 3–4 min. When performing global beam frequency allocation (including broadcast and user carriers), it is not necessary to use a fixed frequency allocation scheme for each satellite beam, otherwise will greatly increase the difficulty of frequency allocation and the occurrence of interference.

2. Through reasonable sub-band division and the setting of the dynamic beam frequency adjustment duration, a better allocation of global beam frequency resources can be achieved and no interference or below the interference threshold can be achieved.
3. There are two types of constellation operation: large period and small period. The large period of constellation operation is 6397 s (106 min and 37 s), and the small period of constellation operation is 6397/9 s (about 11 min and 50 s). Regardless of the rotation of the earth and the number of satellites, only from the constellation's own beam sub-band allocation, if it can ensure that the allocated beam sub-bands within a small period of 11 min and 50 s (approximately 3 4-min time slices) are free of global interference. Then it can be guaranteed that there will be no interference in the entire constellation operating cycle.
4. Implementation process of non-interference frequency resource allocation: first calculate the sub-band allocation of each beam in the time slice (which can be converted into the corresponding latitude interval) of each satellite in the world within a small period of time. When a satellite arriving from the same orbital plane moves to the corresponding latitude interval, it uses the results of the beam sub-band or frequency assigned by the satellite in front of it under the condition that there is no interference in the interval.

4 Conclusion

The periodic beam closing strategy designed in this paper ensure the global coverage and suppress interference between beams by closing overlapping beams. A beam sub-band division and frequency dynamic allocation is proposed on the basis of periodic beam closing strategy, in this way, the interference between beams is avoided and the constraint condition of global coverage without interference is met. On the basis of the research in this article, we will continue to optimize the interference for the movable beam and beam hopping of the mega-constellation in the future.

References

1. Potter, B.: The growing LEO/GEO interference challenge. In:AIAA/USU Conference on Small Satellites, 09 August 2018
2. Braun, C., Voicu, A.M., Simic, L., Mahonen, P.: Should we worry about interference in emerging dense NGSO satellite constellations? In: 2019 IEEE International Symposium on Dynamic Spectrum Access Networks (DySPAN) (2019)
3. Liu, S., Lin, J., Xu, L., et al.: A dynamic beam shut off algorithm for LEO multibeam satellite constellation network. IEEE Wirel. Commun. Lett. **9**(10), 1730–1733 (2020)
4. Anpilogov, V.R., Gritsenko, A.A., Chekushkin, Y.N., Zimin, I.V.: A conflict in the radio frequency spectrum of LEO-HTS and HEO-HTS systems. In: 2018 Engineering and Telecommunication (EnT-MIPT) (2018). https://doi.org/10.1109/ent-mipt.2018.00034
5. Sharma, S.K., Chatzinotas, S., Ottersten, B.: Inline interference mitigation techniques for spectral coexistence of GEO and NGEO satellites (2014). info:eu-repo/semantics/article

6. Mendoza, H.A., Corral-Briones, G., Ayarde, J.M., Riva, G.G.: Spectrum coexistence of LEO and GSO networks: an interference-based design criteria for LEO inter-satellite links. In: 2017 XLIII Latin American Computer Conference (CLEI) (2017)
7. Su, Y., Liu, Y., Zhou, Y., Yuan, J., Cao, H., Shi, J.: Broadband LEO satellite communications: architectures and key technologies. IEEE Wirel. Commun. **26**(2), 55–61 (2019). https://doi.org/10.1109/mwc.2019.1800299
8. Garcia, A.: Low Earth orbit satellite communication networks. Computer Science, Information General Works (2018)
9. Leyva-Mayorga, I., Soret, B., Popovski, P.: Inter-plane inter-satellite connectivity in dense LEO constellations http://arxiv.org/abs/2005.07965 (2020)
10. Li, R., Gu, P., Hua, C.: Optimal beam power control for co-existing multibeam GEO and LEO satellite system. In: 2019 11th International Conference on Wireless Communications and Signal Processing (WCSP) (2019)
11. Saiko, V., Domrachev, V., Gololobov, D.: Improving the noise immunity of the inter-satellite communication line of the LEO-system with the architecture of the "distributed satellite." In: 2019 IEEE International Conference on Advanced Trends in Information Theory (ATIT) (2019). https://doi.org/10.1109/atit49449.2019.9030501

A Full Lifecycle-Oriented Flexible Method for Satellite Design

Yaohua Zhou[1,2](✉), Cuina Liang[1,2], and Jingquan Ji[1,2]

[1] China Academy of Space Technology, Beijing, China
[2] Institute of Telecommunication and Navigation Satellites, Beijing, China

Abstract. In the process of satellite development, user requirements may continue to change with the deepening of understanding. Due to the difficulty of satellite in-orbit maintenance, while considering the cost and time of satellite production, under the pressure of satellite batch production, this paper proposes a full life cycle-oriented navigation satellite flexible design, which can shorten the development cycle of the satellite when the requirements change, upgrade the software in the ground and in orbit flight stages, reduce the development cost of satellite and enhance the value of satellite.

Keywords: Flexible design · Satellites · Lifecycle

1 Introduction

The satellite design is carried out after the user's task requirements are specified. Mission requirements include mission definition, technical indicators and development cycle. The content of satellite design is to carry out mission analysis, put forward the whole satellite scheme and carry out the whole satellite design [1]. With the development of new technology and the extension of satellite life, users' requirements may change. The design of the satellite must be able to quickly meet the changes of user requirements, whether in the ground development stage or in orbit stage. The design of the satellite should reduce the development cycle and the increase in funds brought by the change as far as possible. Keeping the flexibility of design is always the goal of satellite designers.

In recent years, flexible, modular, plug and play, software redefinition and other concepts are gradually introduced into satellite design, such as flexible thermal control system [3], plug and play satellite [3–5], software-defined satellite [6, 7], etc. The above research only involves some functions of the satellite, and can not be carried out from the perspective of the satellite system. The design of the whole satellite shall be able to adapt to the changes of hardware and software caused by user requirements in the whole life cycle. This paper presents a full life cycle-oriented flexible method for satellite design, which solves the problems of long periods and high costs to realize the change.

Q. Guo et al. (Eds.): WiSATS 2021, LNICST 410, pp. 135–141, 2022.
https://doi.org/10.1007/978-3-030-93398-2_13

2 Definition of Satellite Flexibility

Flexibility is defined as the ability to effectively handle environmental changes or uncertainty caused by the environment. Unlike other products, satellites have to experience complex and harsh space environments during launch and in orbit. The cost and risk of satellite in orbit maintenance are very high, even not repairable.

The flexibility of satellite should have the ability of in orbit maintenance and be able to upgrade its functions. Maintenance refers to the functional aspect of the software. A full lifecycle refers to two phases of satellite development on the ground and flight in orbit. In the ground development stage, the flexibility of the satellite means that it can be compatible or easy to complete modification when hardware products and software functions are changed. In orbit stage, the flexibility of the satellite means that it can be upgraded when the software function is changed.

The overall goal of satellite flexible design for the whole life cycle is to reduce the cost. It has two meanings:

- Shorten the increase in development cycle caused by the change, and reduce the cost of test and test caused by the change of equipment and the whole satellite;
- Improve the ability of in orbit software upgrade, enhance the value of satellites, and reduce the total cost of launching satellite again.

3 Methods Supporting the Design of Flexible for Full Lifecycle

A method is here defined as a tool that supports the designer in designing satellites in a systematic way, taking relevant parameters into account. Systematic in this context means the theory of system engineering. The requirements are decomposed into different parts of the satellite, and then integrated into the whole satellite, through system research and problem solving. The flexible design process, as well as the relevant parameters should be truly structured in such a method (Fig. 1).

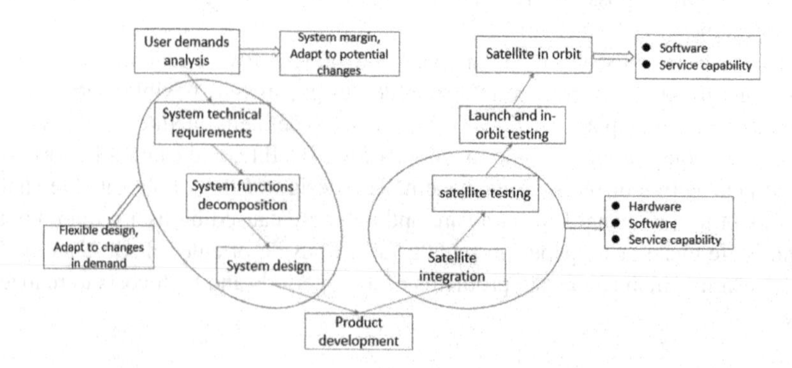

Fig. 1. Flexible design and implementation of satellites

How to develop a design that adapts to change? The core of the flexible design process is to adapt to changes. As shown in Fig. 2. When the design requirements are determined in phase 1, it is necessary to consider the impact of the external interface changes and the internal interface changes of the satellite on the whole satellite from the perspective of the system. In phase 2, the flexible design requirements should be considered from the satellite architecture, hardware and software. Architecture is the key to satellite flexible design, which ensures that changes will not have a disruptive impact on satellite design under the premise of given mission requirements. Hardware and software are the means to realize the flexible design.

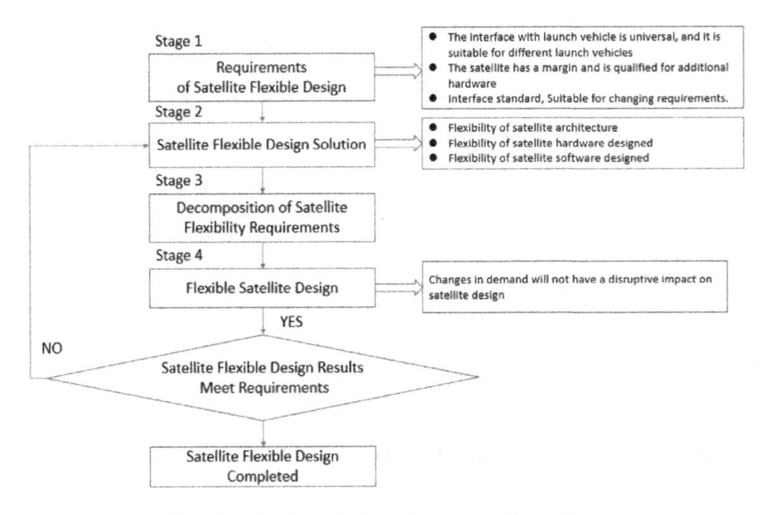

Fig. 2. Flexible designed process of satellites

4 Flexible Design of the Satellite

The flexible design of satellite starts with architecture design. For the efficient and reliable management and control requirements of on-board tasks such as attitude and orbit control, energy, thermal control, and load, there are centralized and distributed architectures. After comprehensive analysis and evaluation, the hierarchical distributed architecture is adopted, the onboard computer is the center, the interface unit is the middle layer, and the general module and standard interface are adopted. When the demand changes, the system flexibility is realized through the middle layer buffer.

- Hardware design such as electric heating interface and mechanical performance matching, which can adapt to the changes of external new equipment requirements, mainly in the ground development stage.
- The whole satellite operation software design, such as plug and play of equipment, independent health management, and independent operation ability of the whole satellite, adapts to the changes of internal and external software functions, mainly in the ground development stage and in orbit flight stage.

- The whole satellite application software design, such as in orbit reprogrammable digital signal generator [10, 12], realizes flexible navigation load and adapts to satellite value promotion, mainly in the ground development stage and in orbit flight stage (Fig. 3).

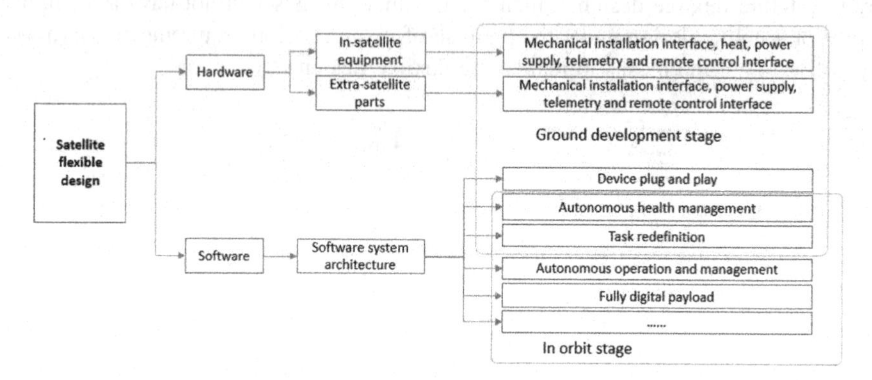

Fig. 3. Satellite flexible architecture

4.1 Flexibility of Satellite Hardware Designed

The hardware flexible design of the satellite realizes the installation of a single machine and the integration of the whole satellite. The flexible design is carried out from the following three aspects.

1. The flexible design is realized through the standard interface of the local cabin board. By adopting compatible, envelope and transfer designs on the cabin board, the installation requirements of new stand-alone products can be met. The compatibility and envelope design is carried out in the whole satellite design stage, and can be realized in the whole satellite production stage. The switching design has no impact on the whole satellite production, and can be carried out in the whole satellite design stage and development stage (Fig. 4).

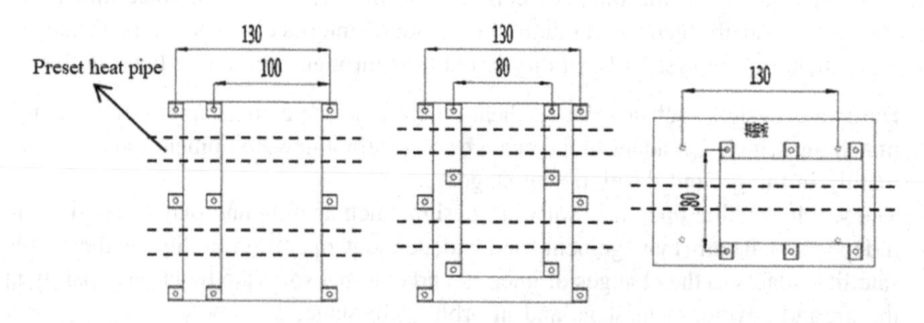

Fig. 4. Schematic diagram of compatibility, envelope and transfer design

2. The flexible design is realized through the standard interface of a modular independent cabin. The hatch adopts the standard interface with the hatch section. Different hatch plates can be carried out in the whole satellite design stage and development stage. The hatch plates shall be uniformly designed at the design stage, and two or more different hatch plates of the same type shall be produced when they are put into production. In the whole satellite development stage, the cabin plate is directly installed according to the needs of different products to realize the flexible configuration of products. Compared with the flexible design of partial bulkhead, the independent bulkhead is suitable for the replacement of a large number of equipment or the configuration of new functional equipment in the development stage (Fig. 5).

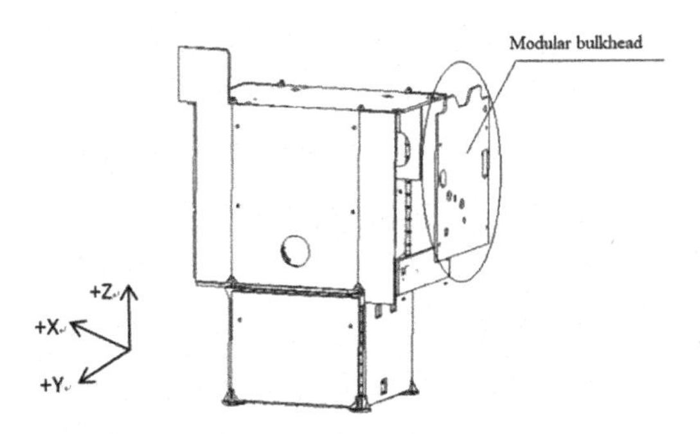

Fig. 5. Modular bulkheads, the interface of the bulkhead is universal

3. The hardware flexibility design is realized through the transfer of second intermediate structure. This design can realize the decoupling between the whole satellite and the load terminal design, and meet the development requirements of the whole satellite. The secondary structure is fixed on the main structure of the whole satellite, does not transfer the main load, and does not significantly affect the structure of the fundamental frequency and vibration mode of the whole satellite (Fig. 6).

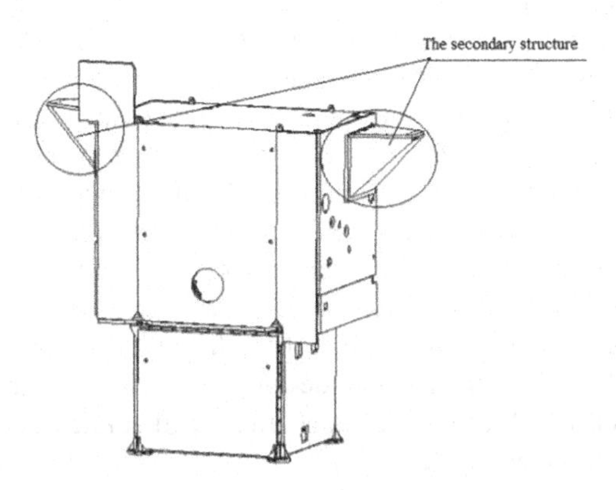

Fig. 6. Add secondary intermediate structure

4.2 Flexibility of Satellite Software Designed

The flexible design of satellite software realizes the rapid upgrading of software functions. Software flexible design is based on the software hierarchical design of the integrated electronic system. Under the unified bottom hardware resources and operating system configuration, it supports the plug and play of equipment and the realization of various satellite application services (Fig. 7).

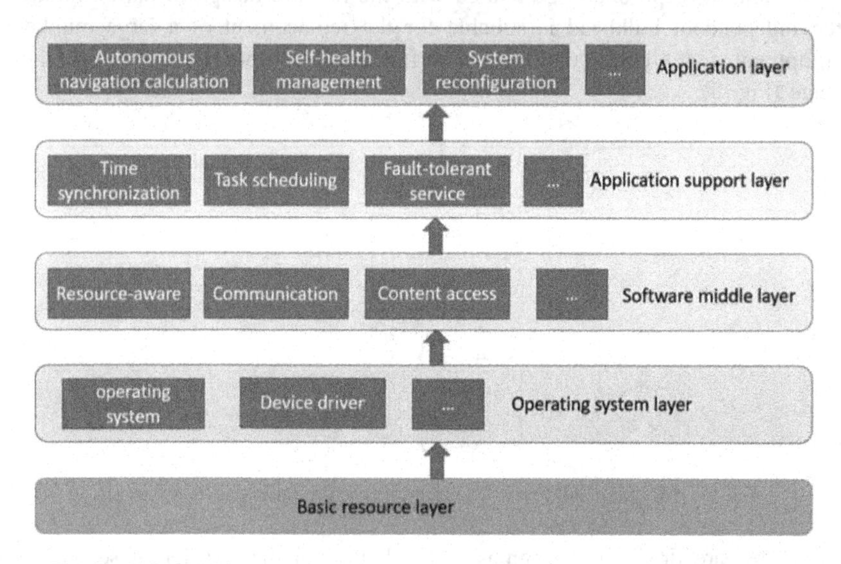

Fig. 7. Software hierarchical design of integrated electronic information system

In the ground development stage, plug-and-play technology is applied to realize the flexible design support of hardware. Plug and play technology can automatically identify access devices and conduct dynamic monitoring and configuration. After the equipment is installed, it is connected to the standard power supply and distribution, information, and other interfaces on the satellite, and can be found, identified, and configured independently on the satellite to meet the needs of task changes.

In orbit stage, digital load technology is used to realize the flexible design of service-oriented software. Taking the navigation satellite as an example, the digital load is to amplify and frequency transform the RF signal, digitize the broadband IF signal to its digital signal processing unit, and process the data onto the signal processing software to complete the corresponding different task functions.

According to the traditional whole satellite design, when the requirements change in the development stage, it is necessary to restart the interface coordination and satellite design, with a long iteration cycle. If the flexible design is adopted, interface coordination, overall design, and satellite development will be changed from serial to collaborative parallel, which can realize new or changed configurations at a small cost and shorten the development cycle.

5 Conclusion

With the improvement in user requirements, the satellite has a shorter and shorter period from contract signing to delivery, and the satellite life is longer and longer. Using flexible design can shorten the coordination iteration cycle at the initial stage of design and improve the value of in orbit satellites.

In the future, as the top-level guidance of design, flexible design theory and method need to further strengthen research, develop a new hardware flexible design system based on the modular cabin, modular cabin board and standardized single machine interface, and coordinate the relationship between modularity and system optimization;

The development cycle of new technologies is getting shorter and shorter, and the satellite life is getting longer and longer. Under the constraints of the limited resources of the whole satellite, improving the value of the whole satellite is the goal that users always pursue. Based on the uncertainty of future new technology prediction and the in-depth mining of satellite value, it will be an important development direction from the traditional fixed resources and fixed functions to the direction of flexibility and redefinition.

References

1. Xu, F.: Satellite Engineering. China Aerospace Publishing House, Beijing (2002)
2. Ning, X., Wang, Y., Song, X., Zhang, B., Zhang, J.: Modular and flexible thermal control architecture of satellite platform. Spacecraft Eng. **21**(02), 50–55 (2012)
3. Wang, X.: Research progress of plug-and-play satellite system technology. Satell. Netw. (05), 70–74 (2014)
4. Wang, X., Dong, Y.: Research on plug and play modular satellite architecture. Spacecraft Eng. **21**(05), 124–129 (2012)
5. Zhang, S., Wang, B., Sun, Z.: The design method of micro-satellite flexible platform based on standardized function modules. Aerosp. Stand. (02), 1–6 (2009)
6. Chen, J., Wang, C., Liang, X.: Progress in foreign software-defined satellite technology. Satell. Netw. (04), 50–53 (2018)
7. Huang, J., Chen, X., Li, Z.: Software-defined integrated electronic design and implementation of micro-nano satellites. Mod. Electron. Technol. **42**(02), 30–32 (2019)
8. Li, C.: Theory and Practice of Project Group Management. Publishing House of Electronics Industry, Beijing (2014)
9. Beyond GPS III, NTS-3: the next chapter, February 2019. www.gpsworld.com
10. Harris all-digital GPS III payload, March 2016. www.gpsworld.com

Spacecraft Information Flow Collaborative Design Model and Application

Ruijun Li(✉), Yanfang Fan, Jian Guo, Hongjun Zhang, and Nan Pei

Beijing Institute of Spacecraft System Engineering, Beijing, China

Abstract. To depict information flow structure of spacecraft system, the paper introduced information flow analysis method and defined tow information flow factors, the entity and the relationship. The relationships among entities were analyzed and the information mapping rules of activities and roles were given with information flow as main line. Based on these information mapping rules which were adapted to the spacecraft development process and engineering practice, models of collaborative design activity, role and information alternation were built. Finally, the application of the Electronic Data Sheet Manage System (EDS) showed. The results show that the method and models proposed in the paper are feasible and effective, which satisfy the collaborative design requirements of the spacecraft information flow.

Keywords: Information flow · Collaborative design · Mapping rule · Model

1 Introduction

The information flow design of spacecraft system is a part of the overall design of spacecraft. With the continuous development of spacecraft technology and the accumulation of experience, the means of spacecraft development are constantly enriched, and the functions, logic, information flow, software protocol and interface of spacecraft system are becoming more and more complex. Traditional mode of implementation can't meet the needs of complex system design, development and integration testing.

In view of the fact that spacecraft development is a system engineering involving multi-disciplinary integration, and its development process has high requirements for the control of the overall state of the product, in order to solve the above problems, all parties propose to adopt the information flow collaborative design method to improve the efficiency of satellite product design and reduce the cost of product development. At present, information transmission has become a key research field in information management, as research showing in [2–4]. However, these studies are from the perspective of macro and management, and stay at the conceptual level, lack of theoretical and practical. As a result, they can't support real system design. Based on the information flow analysis method, combined with the development process of spacecraft in China and its application in engineering practice, this paper defines the information flow elements, studies the information mapping rules, establishes the collaborative design model, role model and information interaction model of spacecraft information flow, and describes the complex and diverse information processing and interaction relationship in spacecraft information flow design. Thus, it provides a new idea and

Q. Guo et al. (Eds.): WiSATS 2021, LNICST 410, pp. 142–154, 2022.
https://doi.org/10.1007/978-3-030-93398-2_14

method for spacecraft information flow collaborative design modeling and information demand analysis. Finally, this paper gives the operation effect of the information flow collaborative design platform based on the model.

2 Information Flow Elements

According to the basic principle of information flow analysis method, the characteristics and change rules of information in the whole interaction process are the key and difficult points of the research [5]. Therefore, the primary task is to clarify the information flow elements of spacecraft development and their relationships. According to the general method of architecture design theory, spacecraft information flow development elements can be divided into entity and relationship.

2.1 Entity

Definition 1: Entity is all the objects related to the design of information flow system under the traction of specific task requirements, including collaborative design information, collaborative design role and collaborative design activities. It is represented as:

$$CE = \{CE_i \mid i \in [1, m]\} \tag{1}$$

In formula (1), m is the entity category [6]. The level and quantity of similar entities are determined by the form and view granularity of the collaborative design system of information flow.

Collaborative design information is a general term for all data generated and processed in the process of spacecraft development. It is a direct or indirect expression of the existing mode, operation state, change situation, load execution task and the result of the mission, etc., and it is expressed as:

$$CE_{info} = \left\{ I_{i,j}^{(\omega)} / i \in [1, X_0], j \in [1, I_0], \right\} \tag{2}$$

In formula (2), ω is the system structure granularity, I_0 is the information type, and x0 is the form type divided by granularity.

Collaborative design role refers to the collaborative design entity with the functions of model creation, personnel allocation, form initiation, filling in, submission, approval, consulting, calling back, simulation verification and auxiliary design [7], and it is represented as:

$$CE_{Role} = \left\{ R_{i,j}^{r} / i \in [1, J_0], j \in [1, R_0], \right\} \tag{3}$$

In formula (3), r is the system structure granularity, R_0 is the total number of collaborative design roles, and J_0 is the form type divided by granularity.

Collaborative design activities are a collection of ordered behaviors that can accomplish certain research tasks for the purpose of collaborative design of spacecraft information flow, and it is expressed as:

$$CE_{action} = \left\{A_{i,j}^{a}/i \in [1, H_0], j \in [1, A_0],\right\} \tag{4}$$

In formula (4), a is the system structure granularity, A_0 is the total number of collaborative design activities, and H_0 is the form type divided by granularity.

Collaborative design task is an ordered set of activities organized to achieve specific collaborative design intent, and it can be expressed as:

$$CE_{task} = \{T_i/i \in [1, T_0], j \in [1, A_0],\} \tag{5}$$

In formula (5), T_0 is the total number of tasks, and task T_i is expressed as:

$$T_i = A_{k+1} \cup A_{k+2} \cup \ldots \cup A_{k+j} \tag{6}$$

In formula (6), $k \in [1, H_0]$, $j \in [1, A_0]$.

2.2 Relationship

Definition 2: Relationship refers to the information flow between different entities, which is identified by the directed information demand line, expressed as:

$$CR = \{CR_i/i \in [1, n]\} \tag{7}$$

In formula (7), n is the number of association relationships. According to the function, the relationship can be divided into task allocation relationship, approval relationship and task dependency relationship.

The above two information flow elements are the necessary conditions for modeling the spacecraft information flow collaborative design architecture, and also the basis for ensuring that the architecture meets the modeling syntax normalization and information completeness. The internal relationship is shown in Fig. 1.

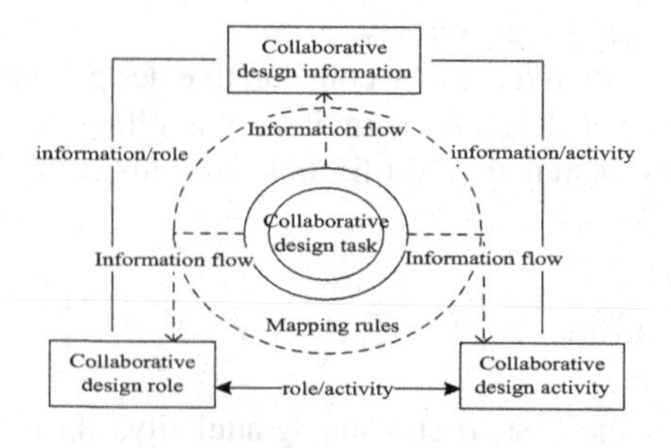

Fig. 1. Relationship between entities

A ternary relationship is formed around collaborative design tasks, collaborative design activities, collaborative design roles and information, which is recorded as formula (8), and this is the key of information flow collaborative design system modeling.

$$CR = \gamma\left(CE_{action}, CE_{node}, CE_{info}\right) \tag{8}$$

3 Information Mapping Rules

Definition 3: Information mapping rules refer to the basic rules followed by the association information relationship between different entities, including activity → activity information mapping, activity → role information mapping, role → activity information mapping and role → role information mapping [5], and it can be expressed as:

$$CF = \{CF_i / i \in [1, p]\} \tag{9}$$

In formula (5), p is the number of rules. According to the basic modeling principle of IDEF0 diagram [6], the specific process of information mapping is as follows.

Rule 1: Activity - activity information mapping

Activity - activity information mapping is a way of association between collaborative design activities, which defines the information interaction between activities and reflects the input, output and control information of collaborative design activities, as shown in Fig. 2.

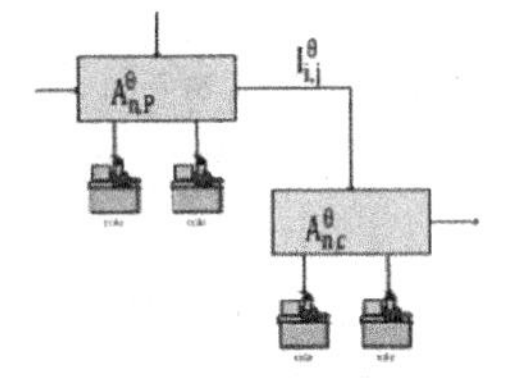

Fig. 2. Activity - activity information mapping

The structure granularity θ of the information flow collaborative design system should be firstly determined; And then for any collaborative design information $I_{i,j}^{\theta}$, we can determine its producing collaborative design activities and ending collaborative design activities; Information $I_{i,j}^{\theta}$ is treated as activity $A_{n,P}^{\theta}$ output data constraints, and as activity $A_{n,c}^{\theta}$ input (or control) data constraints; So we can build activities-activity information mapping between $A_{n,P}^{\theta}$ and $A_{n,c}^{\theta}$ as formula below:

$$CF_1 = A^{\theta}_{n,P} \xrightarrow{I^{\theta}_{i,j}} A^{\theta}_{n,c} \tag{10}$$

Rule 2: Activity - role information mapping

Activity-role information mapping is the way to associate collaborative design activities with collaborative design roles, indicating all roles involved in collaborative design activities, and defining the information interaction between activities and roles, as shown in Fig. 3 below:

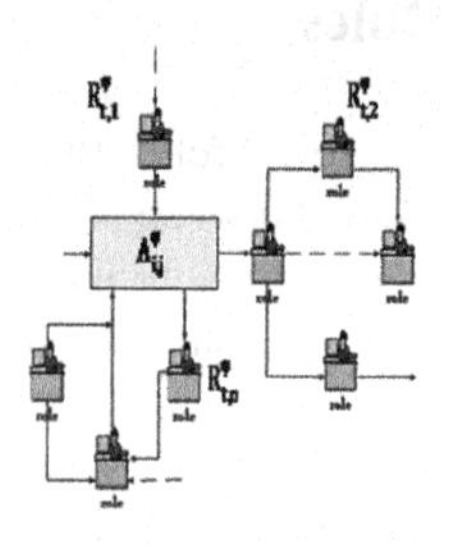

Fig. 3. Activity - role information mapping

The structure granularity φ of the information flow collaborative design system should be firstly determined; Based on the development process of various activities in spacecraft information flow collaborative design practice, for any the collaborative design activity $A^{\varphi}_{i,j}$, the collaborative design role that triggers the activity is determined as $R^{\varphi}_{t,1}$, $R^{\varphi}_{t,2}$,…, $R^{\varphi}_{t,n}$, and it can be expressed as set $R^{\varphi}_{t,r}$:

$$R^{\varphi}_{t,r} = R^{\varphi}_{t,1} \cup R^{\varphi}_{t,2} \cup \ldots \cup R^{\varphi}_{t,n} \tag{11}$$

Similarly, we can determine control roles set of $A^{\varphi}_{i,j}$ as $R^{\varphi}_{t,z}$, and output roles set as $R^{\varphi}_{t,c}$. So we can build activity - role information mapping of activity $A^{\varphi}_{i,j}$ as:

$$CF_2 = A^{\varphi}_{i,j} \rightarrow R^{\varphi}_{t,r} \cup R^{\varphi}_{t,z} \cup R^{\varphi}_{t,c} \tag{12}$$

Rule 3: Role > activity information mapping

The role-activity information mapping indicates all the activities the role participates in, and defines the information interaction between roles and activities, as shown in Fig. 4.

The structure granularity ϑ of the information flow collaborative design system should be firstly determined; According to the assignment of responsibilities, rights and responsibilities of each participant in the collaborative design of spacecraft information flow, for any the collaborative design role R, we can determine its participating activities $A^{\vartheta}_{h,1}$, $A^{\vartheta}_{h,2}$, …, $A^{\vartheta}_{h,n}$, and these can be expressed as set A^{ϑ}_{h}:

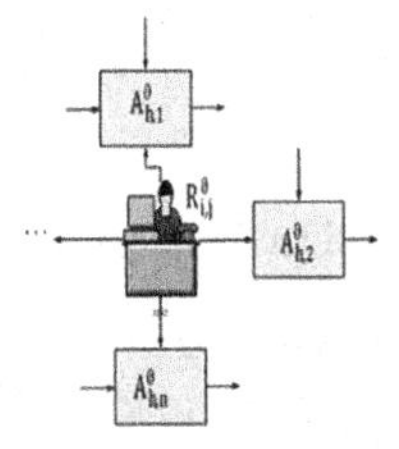

Fig. 4. Role activity information mapping

$$A_h^{\vartheta} = A_{h,1}^{\vartheta} \cup A_{h,2}^{\vartheta} \cup \ldots \cup A_{h,n}^{\vartheta} \tag{13}$$

By establishing connections from $R_{i,j}^{\vartheta}$ to all elements of activity set A_h^{ϑ}, we can build role-activity information mapping of $R_{i,j}^{\vartheta}$ as:

$$CF_3 = R_{i,j}^{\vartheta} \rightarrow A_{h,1}^{\vartheta} \cup A_{h,2}^{\vartheta} \cup \ldots \cup A_{h,n}^{\vartheta} \tag{14}$$

Rule 4: Role-role information mapping.

Role-role information mapping is a way to associate roles in collaborative design, which clarifies the association relationship and interaction information between roles, as shown in Fig. 5.

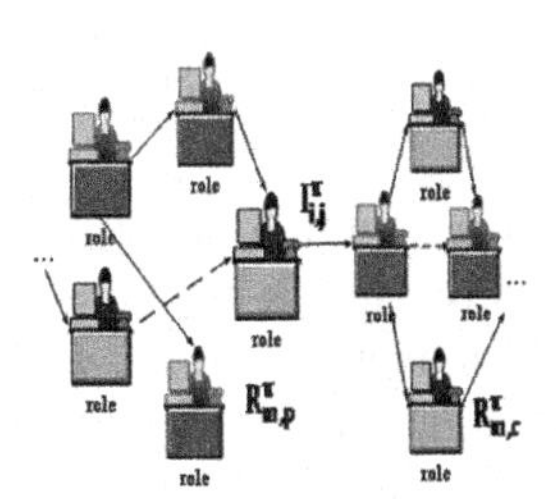

Fig. 5. Role-role information mapping

The structure granularity π of the information flow collaborative design system should be firstly determined; And then for any collaborative design information $I_{i,j}^{\pi}$, we can determine its producing role $R_{m,p}^{\pi}$ and its ending role $R_{m,c}^{\pi}$. By establishing connection from role $R_{m,p}^{\pi}$ to role $R_{m,c}^{\pi}$, we can build role-role information mapping of information $I_{i,j}^{\pi}$ as:

$$CF_4 = R_{m,p}^{\pi} \xrightarrow{I_{i,j}^{\pi}} R_{m,c}^{\pi} \tag{15}$$

4 Modeling of Spacecraft Collaborative Design System Based on Information Flow

4.1 Spacecraft Collaborative Design Process

Firstly, the collaborative design activities and roles are decomposed hierarchically, according to the structure granularity of spacecraft information flow collaborative design system. And then based on the requirements of spacecraft development process and information mapping rules, the activity model, role model and information interaction model are established respectively. The specific steps are shown in Fig. 6.

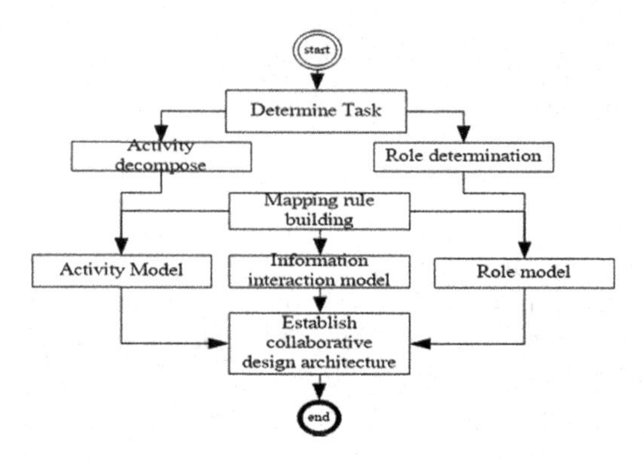

Fig. 6. Process of spacecraft collaborative design system based on information flow

(1) According to the characteristics of the spacecraft model, collaborative design activities, roles and collaborative design hierarchy are determined;
(2) According to the information mapping rules 1–3, the interaction information among activities is determined, all the roles involved in the activities are clarified, and the activities involved in each role are defined, and the collaborative design activity model is constructed;
(3) According to information mapping rule 4, the interaction information between roles is defined, and the role model of collaborative design is constructed;
(4) According to the activity and role model of collaborative design, the information relationship between different collaborative design or top nodes is determined, and the information interaction model is constructed;
(5) The collaborative design architecture based on information flow is designed by collaborative design activities, roles and information interaction model.

4.2 Activity Model of Spacecraft Collaborative Design

The activity model of spacecraft collaborative design reflects the hierarchical information flow interaction among collaborative design activities, which can be expressed by special ternary functions among activities, roles and information:

$$OAM = \{CEaction, CEinfo, CERole\} \quad (16)$$

The collaborative design of spacecraft information flow is divided into two levels: subsystem and single machine; At the same time, collaborative design activities are divided into nine categories: project creation, personnel allocation, form initiation, form design, form submission, form approval, form consulting, form rollback, simulation verification and aided design, and the corresponding activity set is:

$$CE_{action} = \{A^2_{n,t} | n\epsilon[1,6], t\epsilon[1,A_0]\} \quad (17)$$

The information set between activities is:

$$CE_{info} = \{I^2_{p,q} | p\epsilon[1,X_0], q\epsilon[1,I_0]\} \quad (18)$$

And the role set participating in the activity is:

$$CE_{Role} = \{R^2_{f,t} | f\epsilon[1,H_0], l\epsilon[1,R_0]\} \quad (19)$$

4.3 Role Model of Spacecraft Collaborative Design

The role model of spacecraft collaborative design reflects the complex and diverse information interaction relationship and hierarchical information requirements among collaborative design roles, which can be expressed by special functions between roles and information:

$$ONM = \{CERole, CEinfo\} \quad (20)$$

The roles of cooperative design for spacecraft information flow are divided into six categories: project manager, project administrator, information master, subsystem master, device manager and software manager. Then the corresponding role set is:

$$CE_{Role} = \{R^2_{m,k} | m\epsilon[1,6], k\epsilon[1,R_0]\} \quad (21)$$

And the information set between roles is:

$$CE_{info} = \{I^2_{i,j} | i\epsilon[1,X_0], j\epsilon[1,I_0]\} \quad (22)$$

4.4 Information Interaction Model of Spacecraft Collaborative Design

The information interaction model of spacecraft collaborative design reflects the diversified information interaction relationship among the nodes in the collaborative design activities in the process of collaborative design implementation. It can be expressed by the special ternary function of collaborative design information, roles and activities, which is recorded as:

$$OIM = \{CEinfo, CERole, CEaction\} \tag{23}$$

The collaborative design of spacecraft information flow is divided into two levels: subsystem and device; And then for any collaborative design information $I^2_{i,j}$, we can determine its producing role $R^2_{m,p}$ and its ending role $R^2_{m,c}$. Thus, the corresponding information interaction model can be expressed as:

$$\{R^2_{m,p}|A^2_{m,p}\} \xrightarrow{I^2_{i,j}} \{R^2_{m,c}|A^2_{m,c}\} \tag{24}$$

5 Practice of Spacecraft Information Flow Collaborative Design

As mentioned above, based on the general principles of collaborative design, combined with the development process of spacecraft in China and its application scenarios in engineering practice, a set of spacecraft information flow collaborative design system model is given, and the model is applied to the implementation of EDS (electronic data sheet) integrated management system.

5.1 Division of Information Flow Elements

In the design practice of spacecraft information flow system, we should not only avoid the long coordination chain and heavy management workload, but also facilitate the full communication and cooperation among the system, subsystem and device (software). The structure granularity of collaborative design system is defined as two-level working mechanism. On this basis, the information flow elements are determined as follows: the collaborative design information is determined as telemetry parameters, remote control instructions, data protocols (intra satellite bus protocol, inter satellite link protocol, satellite ground telemetry, remote control protocol, etc.) and application data expression (instruction group, instruction sequence, thermal control table, etc.); The collaborative design roles are determined as project administrator, information master, subsystem master, device manager and software manager; Collaborative design activities include form initiation, form design, form submission, form approval, form rollback, simulation verification, aided design, etc.; The relationship between entities can be divided into three types: task allocation relationship, approval relationship and task dependency relationship.

5.2 Collaborative Design Activity Model

According to the rule of activity information mapping, the internal information mapping of spacecraft information flow collaborative design activity is established, as shown in Table 1.

The activity diagram of establishing the collaborative design model of information flow is shown in Fig. 7.

Table 1. Mapping relationship between information flow collaborative design activities

Source activity	Destination activity	Information type
Project creation	Personnel allocation	I_{CC}
Personnel allocation	Form initiation	I_{CC}
Form initiation	Form design	I_{IS}
Form design	Form submission	I_{IS}
Form submission	Form approval	I_{CC}
Form consulting	Form rollback	I_{CC}
Form rollback	Form design	I_{CC}
Form consulting	Form approval	I_{CC}
Aided design	Simulation verification	I_{IS}
Simulation verification	Form design	I_{IS}

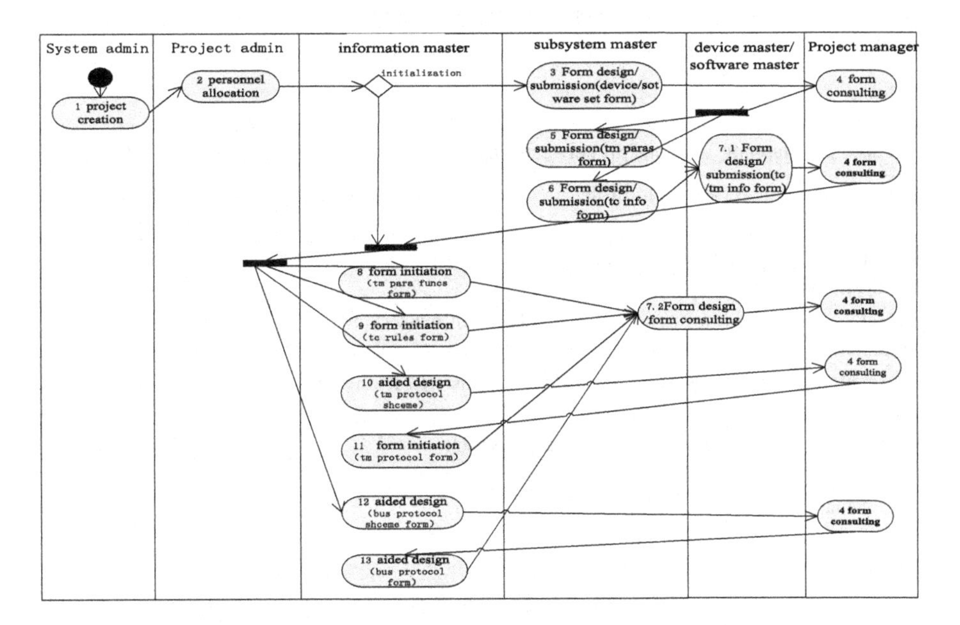

Fig. 7. Activity diagram of information flow collaborative design model

According to the mapping rules of activity-role and role-activity information, 9 activity-role information mappings and 24 role-activity information mappings are established, as shown in Table 2.

5.3 Role Model of Collaborative Design

According to the rules of role-role information mapping, the following 11 role-role information mappings are established, as shown in Table 3.

Table 2. Activity-role mapping information flow collaborative design model

Activity	Participate role
Project creation	System administrator
Personnel allocation	Project administrator
Form initiation	Information master/subsystem master
Form design	Information master/subsystem master/device master/software master
Form submission	Information master/subsystem master/device master/software master
Form consulting	Information master/subsystem master/device master/software master
Form rollback	Information master/subsystem master/device master/software master
Aided design	Information master/subsystem master
Simulation verification	Information master/subsystem master

Table 3. Mapping relationship between information flow collaborative design activities

Source role	Destination role	Information type
System administrator	Project administrator	I_{CC}
Project administrator	Information master/subsystem master/device master/software master	I_{CC}
Information master	Subsystem master/device master/software master	$I_{IS}/I_{CC}/I_{wc}$
Subsystem master	Device master/software master	$I_{IS}/I_{CC}/I_{wc}$
Device master	Device master	I_{IS}
Software master	Software master	I_{IS}

5.4 Information Interaction Model

According to the given four types of information mapping rules, based on the collaborative design activity model and collaborative design role model, combined with the current spacecraft verification process and engineering practice experience, this paper analyzes the relationship between the source point, destination point, production activity and consumption activity of different types of information flow layer by layer, and clarifies the information transfer process and interaction between collaborative design activities and roles, The model of spacecraft collaborative design system based on information flow is established, as shown in Fig. 8 below.

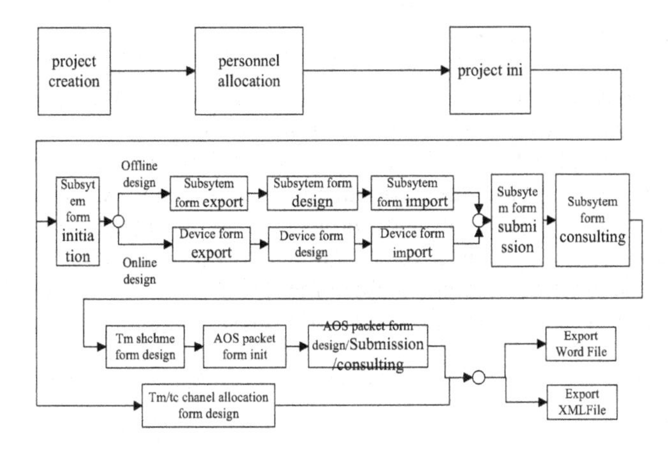

Fig. 8. Model of spacecraft information flow collaborative design system

Based on this model, the EDS integrated management system platform is established. Up to now, EDS has been involved more than fifty projects, and has been put into trial in several research models. The experimental results show that, compared with the traditional spacecraft development model, the information flow collaborative design model can more clearly describe the complex information interaction between the spacecraft generation and development parties, more intuitively display the dynamic whole process of the whole spacecraft development activities, and more systematically assist the development parties to coordinate and agree The design and implementation of complex large system are realized efficiently.

Through the analysis of the trial situation of the model in progress, it can be seen that the use process of EDS integrated management system conforms to the actual situation and habits of each model development, the modeling method meets the requirements of the information circulation and utilization design of spacecraft, and the collaborative design modeling elements can cover all links of the whole life cycle of satellite development, the baseline release, technical status change, greatly reduce the cost of point-to-point manual communication, significantly improve the error risk caused by single point design and implementation, greatly improve the production and development efficiency of spacecraft, with an increase rate of more than 40%.

6 Conclusion

Based on the contents of digital development of spacecraft information flow and the new requirements and problems faced by traditional development mode, combined with the general theory of collaborative design, this paper presents a collaborative design mode of information flow suitable for the development of spacecraft in China, and applies it to the construction of EDS integrated management system of our Institute. Through the statistical analysis of the practical application results, it is proved that the information flow collaborative design model proposed in this paper comprehensively considers the responsibilities, rights and responsibilities of all parties in the collaborative design, takes the information flow as the main line, comprehensively considers the factors such as roles, activities and information, realizes the architecture modeling, refines the process analysis of collaborative design activities, and highlights the design characteristics of different types of information. The rules of collaborative design [9] are standardized, and the general idea, objectives and implementation means are feasible, which are in line with the actual operation scenarios and requirements. The related information flow protocol modeling method and parameter format model are universal and flexible, which can cover most of the daily production needs, and provide rich theoretical support and a large number of practical verification data for spacecraft development and design and management mode transformation.

References

1. Yu, H., Hao, W., Yuan, J., et al.: Development schemes of spacecraft system engineering techniques. Spacecraft Eng. **18**(1), 1–7 (2009)
2. Ahlswede, R., Cai, N.: Network information flow. Inf. Theory **46**(4), 1204–1216 (2000)
3. Klyubin, A.S., Polani, D.: Organization of the information flow in the perception-action loop of evolved agents. In: Proceedings of the 2004 NASA/DOD Conference on Evolvable Hardware, 24–26, pp. 177–180 (2004)
4. Alghathbar, K., Wijesekera, D.: Analyzing information flow control policies in requirements engineering. In: Proceedings of the Policies for Distributed Systems and Networks, POUCY 2004, pp. 193–196 (2004)
5. Xu, B., Zhang, L.: Muti-dimensional architecture modeling for cyber physical systems. In: Jeong, H., S. Obaidat, M., Yen, N., Park, J. (eds.) Advances in Computer Science and its Applications. LNEE, vol. 279, pp. 101–105. Springer, Heidelberg (2014). https://doi.org/10.1007/978-3-642-41674-3_16
6. Waters, J., Ceruti, M.G.: Modeling and simulation of information flow: a study of infodynamic quantities. In: The 15th International Command and Control Research and Technology Symposium (ICCRTS 2010), pp. 178–183 (2010)
7. Guo, J., Li, R., Fan, Y., et al.: Digital collaborative design of spacecraft information flow. Spacecraft Eng. **29**(4), 59–65 (2020)

Design of a Spacecraft Network System Based on Space-Ground Integrated Network

Guangri Li(✉), Rong Dang, Kai Ding, and Lintao Wang

Beijing Institute of Spacecraft System Engineering, Beijing 100094, China

Abstract. Since the rapid development of the internet technology, the space-ground integrated network is becoming an inevitable trend of development in communication between space and ground. At present, the internet technology to realize the integrated communication has been applied to ground stations. Similarly, the internet technology can be used in the communication system of spacecraft. But comparing to the network used on the ground, the space-link between the spacecraft and ground stations has different characteristics, such as the high propagation delay, the narrow bandwidth, the asymmetry transmission speed of the bidirectional space-link and the complicated Electromagnetic Compatibility (EMC) environment in the spacecraft. These can make the space-ground integrated network to be inefficient and unreliable. To resolve these problems, an efficient and reliable space network system for the spacecraft is designed based on the space Ethernet switch and the space-ground gateway. And the error-control mechanism of application layer, the IP package partition strategy of data link layer and the cable layout method of physical layer has been proposed to optimize the TCP/IP protocols used in the ground. The spacecraft network system can be used in the space-ground integrated network of GEO, MEO or LEO spacecraft and the deep space exploration.

Keywords: Space-ground integrated network · Spacecraft · Efficient and reliable · Network system

1 Introduction

With the development of network technology, the integrated network between space and ground becomes the inevitable development trend. At present, the ground TT & C communication system has realized the integrated communication based on network, while the communication inside the spacecraft is still based on the bus or point-to-point communication. If the information exchange between the internal equipment of the spacecraft adopts network to realize the integrated communication with the ground network, the communication protocol can be unified, and the communication efficiency can be improved. At present, the integrated communication based on network is mainly in the theoretical stage [1–5], and its engineering application is less. The international space station has used the network for the high-speed transmission of experiment data, and carried out on orbit tests, but it has not been applied to the data transmission task of spacecraft platform.

Q. Guo et al. (Eds.): WiSATS 2021, LNICST 410, pp. 155–165, 2022.
https://doi.org/10.1007/978-3-030-93398-2_15

In order to realize the integrated network communication between spacecraft and ground, it is necessary to apply ground network technology to the spacecraft TT & C communication system. Compared with the ground network environment, the space-link of spacecraft has the characteristics of long-time delay, limited bandwidth, unbalanced bandwidth of bidirectional space-link and complex electromagnetic environment inside the spacecraft. The direct application of ground network technology will lead to problems of low communication efficiency and poor reliability for the integrated space-ground network. In view of the above problems, this paper puts forward a solution of space network system to achieve efficient and reliable integration of space-ground network communication.

2 Architecture of Spacecraft Network

With the rapid development and wide application of Ethernet technology, the integration technology of space communication and ground network has become the main development trend in the future. The proposed draft of red book [6] provides guidance for the implementation of IP packet transmission over CCSDS space data link protocol between spacecraft and ground system, so that the communication between space and ground can start from the network layer to adopt consistent protocol to realize integrated network communication, and establish an integrated network communication system between space and ground.

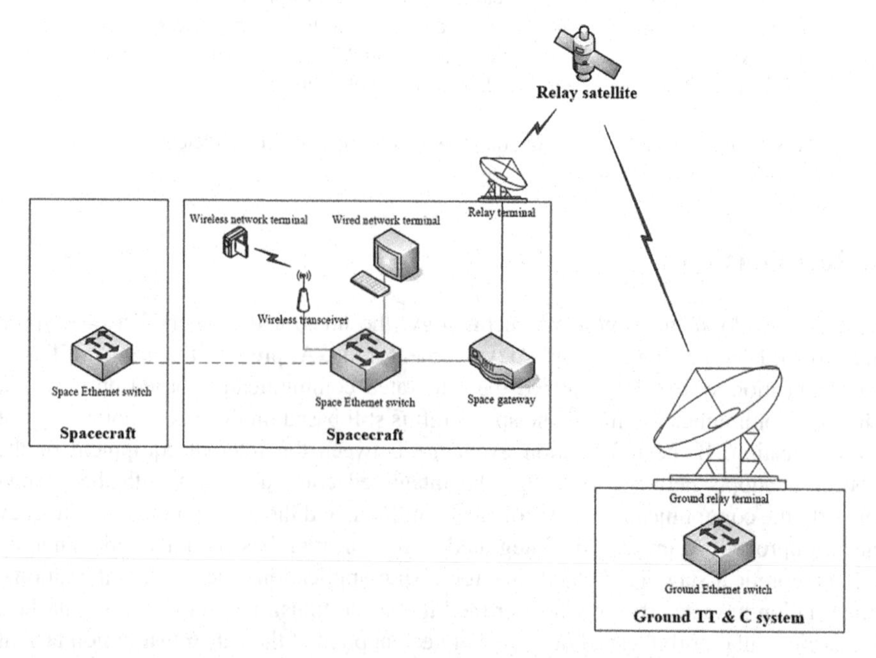

Fig. 1. Architecture of space-ground integrated network system

Through the analysis of Ethernet technology and IP over CCSDS protocol in space application to construct the space-ground integrated network system architecture based on space Ethernet switch, wireless transceiver equipment and space-ground gateway, it can support the wired or wireless access of self-developed or commercial network terminal, the communication between different spacecraft, and the space-ground integrated network communication through relay satellite system. The system architecture is shown in Fig. 1.

1. The space Ethernet switch is responsible for the access of wired network terminals and networking between different spacecraft. It supports the standard Ethernet protocol and interface, and exchanges the data of network terminals inside the spacecraft with other spacecraft network terminals and the space gateway.
2. The wireless transceiver is responsible for the access of the wireless network terminal. It supports WIFI protocol and connects the data of the wireless network terminal to the space Ethernet switch to realize the conversion between the wireless network protocol and the Ethernet protocol. It also can exchange data with other spacecraft network terminals and the space gateway through the space Ethernet switch.
3. The space gateway is responsible for the bidirectional space-link data transmission between the spacecraft network and the ground network. It transforms the IP protocol of the space network [7] and the CCSDS AOS space link transmission protocol [8], completes the mapping between the spacecraft network and the ground network protocol, and realizes the gateway function of the integrated space network [9].

3 Protocol Stack of Spacecraft Network

Aiming at the limitation of long-time delay, limited bandwidth and inconsistent rate of bidirectional space-link, the multi-layer network protocol is optimized, which effectively reduces the conversion link of space protocol, and improves the efficiency and reliability of space-ground integrated network.

The network protocol stack of spacecraft adopts five-layer structure, including application layer, transmission layer, network layer, data link layer and physical layer. Seeing Fig. 2 for details.

3.1 Application Layer

According to the requirements of spacecraft data, it can be divided into three types:

1. Low speed, real-time and high reliability data, such as control and telemetry data;
2. High speed, real-time and fluent data, such as image and voice data;
3. High speed, reliability, but no real-time requirement data, such as high-speed experiment data.

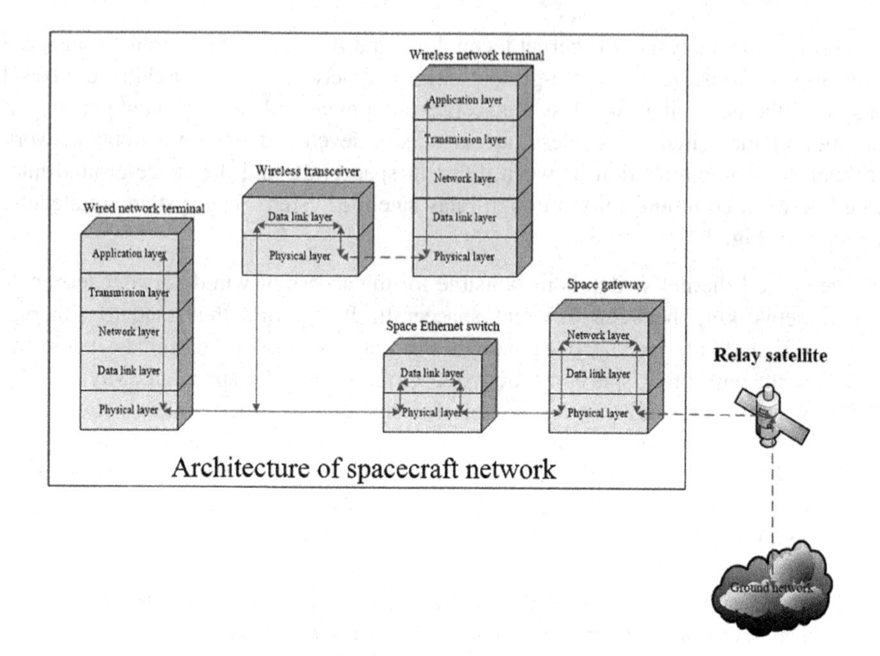

Fig. 2. Spacecraft network protocol stack

Due to the burst transmission mode of network data, once a variety of data are sent at the same time, it will lead to instantaneous exceeding of the link bandwidth and the loss of data. Therefore, a reasonable control strategy for the network flow must be adopted to avoid network congestion.

According to the characteristics of spacecraft data, three kinds of control strategies for the network flow are adopted, and their corresponding relationships are shown in Table 1.

Table 1. Flow control strategy for spacecraft data

Control strategy	Data
Control strategy 1	Telemetry and control data
Control strategy 2	Image and voice data
Control strategy 3	Experimental data

1. Control strategy 1: for low-speed data with periodicity or randomness and real-time requirements, no flow control is carried out;
2. Control strategy 2: for high-speed data with real-time and fluency requirements, it is required to control the average rate of statistics in a certain period of time. It needs to meet the Eq. (1):

$$V_t \leqslant V_{avg} \tag{1}$$

V_t: the average rate of statistics in t time
V_{avg}: average rate requirement.

3. Control strategy 3: for the high-speed data without real-time requirement, in order to ensure the smooth output data rate of the terminal, the time interval between adjacent data packets is specified, and the output data is required to be buffered and smoothed in the application layer. The time interval of application layer data packets should meet the following Eq. (2):

$$t \geq \frac{L}{v_{avg}} - \frac{L}{v_{\max}} \tag{2}$$

t: The time interval between two consecutive frames of data
L: The length of the data in this frame, bits
V_{avg}: average rate requirement
V_{max}: port transmission rate, 100 Mbps.

3.2 Transport Layer

Transport layer adopts connection-oriented transport protocol (TCP) and non-connection-oriented transport protocol (UDP).

TCP Transport Service. TCP adopts acknowledgement mechanism to ensure the reliability of data transmission. Because the format of acknowledgement packet is consistent with that of sending datagram, the overhead of network transmission is increased. The bidirectional delay of the direct to ground link is about 3 ms, and that of the relay satellite link is usually more than 500 ms, which is less than the default TCP retransmission time of 3s. However, due to the imbalance of the bidirectional bandwidth of the transmission link, the downlink bandwidth is usually more than 20 times of the uplink bandwidth, which leads to the low transmission efficiency when using TCP protocol, and the data rate is basically constrained at about 30 kbps. Therefore, for data with high reliability and integrity requirements and low speed (less than 30 kbps), TCP protocol can be used; for data with high speed (more than 30 kbps), TCP protocol is not allowed to be used for data transmission between space and ground. For data with reliability and integrity requirements, it is recommended to adopt corresponding mechanism in application layer to ensure the reliability of data transmission.

UDP Transport Service. UDP protocol adopts connectionless mechanism to ensure the real-time data transmission. Because there is no need to establish a connection, it is not constrained by the imbalance of the bidirectional transmission bandwidth, and it can make full use of the downlink bandwidth to transmit data. Therefore, UDP protocol can be used for data with high real-time requirement and high rate (more than 2 Mbps) to ensure the transmission delay as small as possible.

3.3 Network Layer

Network layer uses IP protocol to realize the data exchange of the network layer. The IP datagram is used to mark the source address, destination address and other information when the network forwards the datagram, so as to realize the routing of the IP datagram.

ICMP Protocol is used in network layer to feedback Internet control messages, which is used to detect and report various errors in IP packet transmission. Terminal equipment and platform equipment select the protocol according to the demand.

ARP protocol is used in network layer to realize the mapping between IP address and physical address. The network of spacecraft only allows ARP broadcast inside the spacecraft, and does not transmit ARP broadcast packets between space and ground.

3.4 Data Link Layer

Data link layer of spacecraft network mainly includes Ethernet data link layer, wireless WIFI data link layer and space-ground transmission data link layer.

Ethernet Data Link Layer. The data link layer of wired network terminal adopts the MAC frame format of IEEE Std 802.3TM-2005, and does not use LLC sublayer. In order to avoid the MAC address conflict of the spacecraft network terminal equipment, the terminal developer needs to purchase the commercial network card and configure the spacecraft terminal MAC address as the commercial network card MAC address. Make sure that the MAC address is not used by other terminals.

Wireless WIFI Data Link Layer. The wireless terminal is connected to the Ethernet of spacecraft through the wireless transceiver device, which can communicate with other Ethernet terminals of spacecraft and the ground in two directions. The wireless transceiver device adopts the two-layer forwarding data mode, stores the MAC forwarding table internally, extracts the data link layer information of the incoming data, forwards the data according to the destination MAC address, and compares the source MAC address with the destination MAC address. The corresponding relation of the port is updated to the MAC forwarding table. The wired network interface of wireless transceiver meets the 10Base-T/100base-TX standard of IEEE 802.3 and 802.3u specifications, and the radio frequency interface meets the IEEE 802.11g/n transmission specification.

Data Link Layer of Space Ground Transmission. The data link layer adopts IP over CCSDS protocol. The gateway device encapsulates the Ethernet MAC frame in the AOS frame data area through CCSDS package protocol, which realizes the conversion between the Ethernet transmission protocol and the ground link transmission protocol. Due to the mismatch between the length of data frame (generally defined as 1024 bytes) and the length of MAC frame (generally defined as 1500 bytes), invalid data will be filled at the end of data frame of the link if the IP packet is directly divided according to the length of MAC frame for encapsulation. It will result in the waste of link bandwidth. In order to solve this problem, this paper proposes a network packet segmentation strategy based on the data frame format of the link. The IP packet length is divided into MAC frames according to the link data frame length, which ensures that the IP packet length is consistent with the link data frame length, and avoids invalid filling packets in the link. It can greatly improve the utilization of the link. Compared with 74% in MAC frame length division, the utilization is increased to 86% and the improvement is more than 10%.

3.5 Physical Layer

The network terminals of spacecraft are distributed inside and outside the cabin. The vacuum environment outside the cabin is alternating high and low temperature, and inside the cabin is the constant temperature environment. The network signal transmission inside and outside the cabin must use the through cabin connector.

There is signal crosstalk problem in the transmission of 100 MHz Ethernet signal by through cabin connector [10]. In order to solve this problem, by analyzing the electric field distribution of the connector, a low electromagnetic interference arrangement method of cross arrangement and cable group shielding is proposed for the multi-channel high-density connector, which optimizes the electric field distribution between the wires in the connector, reduces the mutual influence of the transceiver line, and improves the performance of the near end. The crosstalk is optimized by 8dB, which solves the problem of crosstalk between high-speed transmission lines and multi-channel high-density connectors, and realizes the stable transmission of 100 MHz Ethernet signal for network terminals inside and outside the cabin in the space environment.

4 Network Communication Mode

4.1 Autonomous Addressing Mechanism for Spacecraft Network

ARP protocol can be used to establish and maintain the corresponding relationship between MAC address and IP address in the internal network terminal of spacecraft. The space switch maintains the storage table of the switch through the corresponding relationship between MAC address and port of the received data. When the switch receives a frame of data packet, it looks up whether there is a storage item of the destination MAC address in the storage table. If there is no such item, it will send a broadcast frame to all terminals. When the other party feeds back the data information, it can store and update the relationship between the MAC address and the port.

The table stored in the switch and terminal is provided with the latest update time information. The data exchanged by the terminal and the switch is required to maintain and update the storage table. When the update is not carried out for a long time, the corresponding items in the table should be deleted to prevent the normal communication of the network affected by the network topology and terminal status changes. ARP query is performed again before the ARP cache table of terminal is aged, and the storage content is updated with feedback information. The autonomous addressing mechanism of spacecraft network is shown in Fig. 3:

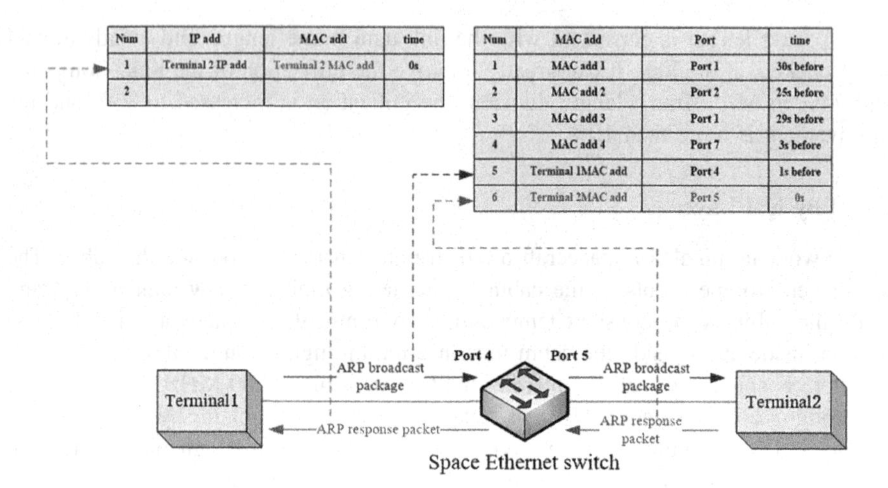

Fig. 3. Autonomous addressing mechanism for spacecraft network

4.2 Dynamic Routing Mechanism in Space-Ground Network

Three-layer data forwarding mode is adopted between the spacecraft and the ground network. The communication mode between the spacecraft network terminal and the ground network terminal is different from that of the internal network of the spacecraft. The protocol processing function of the space-ground link needs to be completed by the space-ground gateway.

When the spacecraft network terminal needs to send the data to the ground terminal, the network terminal judges whether the destination IP address is an internal address before transmitting the data. If it is an internal address, layer 2 exchange is carried out directly. If it is an external address, the default gateway MAC address is queried through ARP protocol, and the IP packet is sent to the gateway through the switch. After receiving the packet, the gateway encapsulates the IP packet and transmits it to the ground through CCSDS AOS protocol. When the ground receives the data, it forwards the IP packet to the corresponding ground network terminal. The data processing process of up-link is consistent with that of down-link. Seeing Fig. 4 for the dynamic routing mechanism of the space-ground network.

5 Experiment

In order to verify the design of the spacecraft network architecture and protocol stack proposed in this paper, the experiment is carried out through the space-ground test. The network equipment of spacecraft, space-ground link and ground station network are all real equipments. In the test, the transmission delay test, image and voice transmission test are carried out.

The transmission delay of space-ground IP packets is about 526 ms, as shown in Fig. 5.

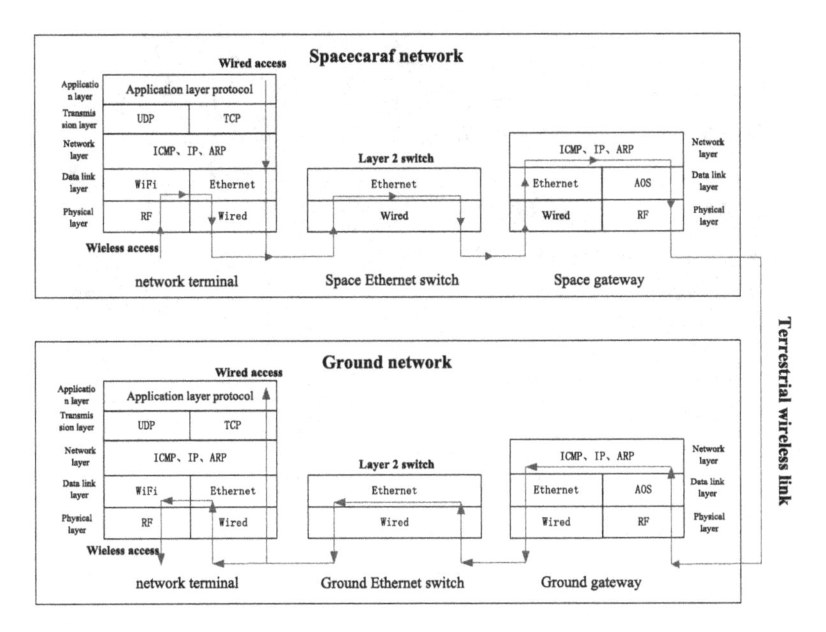

Fig. 4. Dynamic routing mechanism in space-ground network

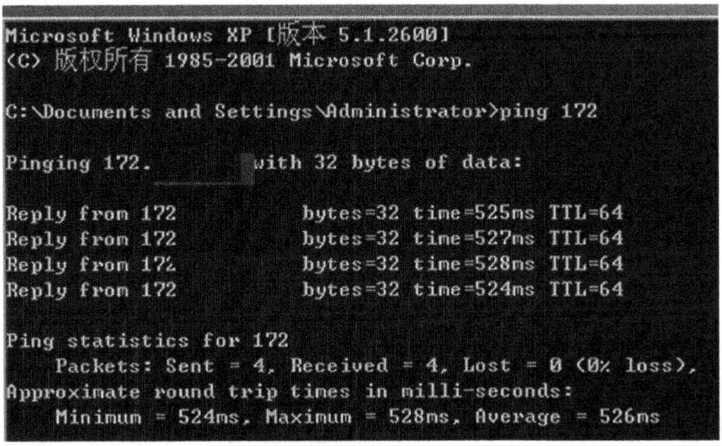

```
Microsoft Windows XP [版本 5.1.2600]
(C) 版权所有 1985-2001 Microsoft Corp.

C:\Documents and Settings\Administrator>ping 172

Pinging 172.          with 32 bytes of data:

Reply from 172            bytes=32 time=525ms TTL=64
Reply from 172            bytes=32 time=527ms TTL=64
Reply from 172            bytes=32 time=528ms TTL=64
Reply from 172            bytes=32 time=524ms TTL=64

Ping statistics for 172
    Packets: Sent = 4, Received = 4, Lost = 0 (0% loss),
Approximate round trip times in milli-seconds:
    Minimum = 524ms, Maximum = 528ms, Average = 526ms
```

Fig. 5. Transmission delay test

The HD image down-link is normal, and the image is clear and smooth, as shown in Fig. 6.

The voice transmission is normal, the voice rate is controlled evenly with an average of 60 packets per second (see Fig. 7 for details), the voice is smooth, and there is no packet loss.

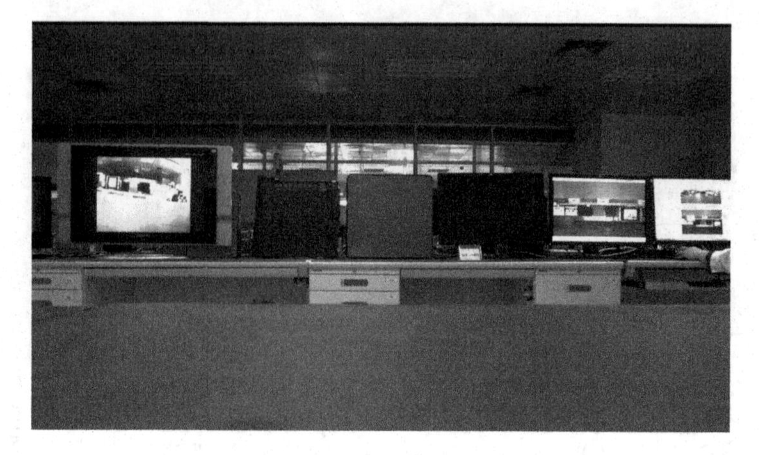

Fig. 6. Image transmission test

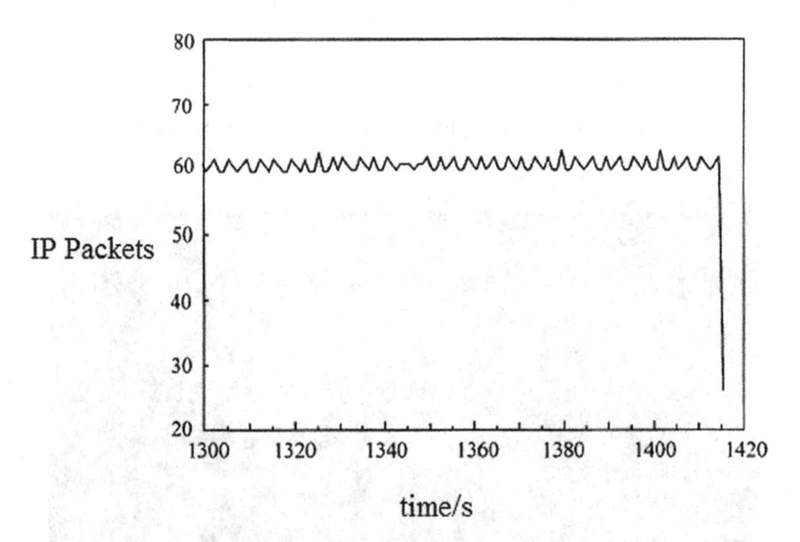

Fig. 7. Voice transmission test

6 Conclusion

In this paper, a network architecture and protocol stack for spacecraft are proposed and verified, which can realize efficient and reliable space-ground integrated network communication.

The spacecraft network system can be applied to the space-ground integrated network of high, medium and low earth orbit spacecraft represented by manned spacecraft, and can also be applied to deep space exploration missions. At the same time, it provides a technical way for commercial network products to access the spacecraft network, promotes the on-orbit application of commercial products, and has great social benefits and application prospects.

References

1. Ning, L., Chi, T.W.: Space Internet and space measurement and control. Space Electron. Technol. **6**(3), 116–121 (2009)
2. Shen, R.: China aerospace integration Internet concept. Chin. Eng. Sci. **8**(10), 19–30 (2006)
3. Jiang, L.: IP over CCSDS Spatial Network Communication Key Technologies. Graduate University of Chinese Academy of Sciences, Beijing (2009)
4. Kai, D., Song, C., Ke, Z., Yin, L.: Design of spacecraft IP network in integrated space network. Spacecraft Eng. **26**(4), 67–73 (2017)
5. Hu, X.Y.: IP over CCSDS analyzes. Satell. Netw. **9**(9), 34–40 (2010)
6. CCSDS. CCSDS 702. 1-B-1. IP over CCSDS space links. CCSDS, Washington D.C. (2012)
7. Xie, X.R.: Computer Network, 5th edn. Publishing House of Electronics Industry, Beijing (2008)
8. CCSDS. CCSDS 701. 0-B-3. Advanced orbiting systems, networks and data links: architectural specification. CCSDS, Washington D.C. (2001)
9. Yin, L., Ke, Z., Kai, D.: Design and implementation of spacecraft gateway in the integrated network of heaven and earth. Spacecraft Eng. **25**(1), 77–83 (2016)
10. Chen, Y., Li, W.Y., Yang, F., Zhao, R., Xia, Z.J., Zhang, Z.: Influence of shielding method on cable transmission performance. **3**(5), 55–58 (2018)

A Dynamic Hashing Method for Storage Optimization of Spacecraft Verification Database

Hongjing Cheng(✉) and Yanfang Fan

Beijing Institute of Spacecraft System Engineering, Beijing 100094, China

Abstract. In order to facilitate the repetition, deduction and playback of the problems in the test process, the corresponding requirements for the retention of test data are put forward in the current test process of spacecraft data management software. In order to deal with the operation based on data itself and the higher level of extraction and analysis based on data, the use of database to store data has become the highest performance and the most convenient way. Due to the large amount of test data, using a single table storage method can cause the time of data query to increase rapidly until it is unacceptable. However, several commonly used table partitioning methods also have their own defects which make it difficult to apply in this scenario. In this paper, a dynamic hashing method (DSH algorithm) is proposed to partition the data according to the characteristics of generating and using the test data of spacecraft data management. The data is hashed through partitioning process feedback to ensure the bias balance of queue hashing. The dynamic partitioning method ensures that the partition fragmentation is controllable and the number of partitions is reduced. It improves the speed of data storage and query, and improves the efficiency of storage space. So that in the same hardware platform, running a larger and more complex tube software testing work provides a stable and reliable solution.

Keywords: Data management test · Database partition · Dynamic hash

1 The Introduction

As the brain of the spacecraft, the spacecraft data management subsystem is responsible for the data organization, processing, transfer and transmission within and between the spacecraft and the satellite-ground link. The quality and life of the spacecraft are closely related to the stability and reliability of the tube subsystem. Because of its importance and complexity, the development and testing of the subsystem software are paid more attention to. Spacecraft data management software testing database provides basic data for software testing application tools to ensure the accuracy of data time sequence, content and correlation.

The amount of data produced in the testing process of several tubes of software is very large, up to tens of megabytes per second according to different models. The insertion and search speed of these data are greatly affected by the huge amount of data. In the process of using data, the application software directly faces the user, so the

Q. Guo et al. (Eds.): WiSATS 2021, LNICST 410, pp. 166–176, 2022.
https://doi.org/10.1007/978-3-030-93398-2_16

speed of reading, organizing and displaying data is required to be high. In order to improve the reading speed of data, it is necessary to set up an effective index and partition the data. The index design of the database will not be further explored in this article, but the index itself should be considered in the establishment of the index on the impact of insert speed.

It can be seen from the test data that adding a reasonable number of indexes can obviously improve the speed of data retrieval (see Fig. 1). However, if the index is not added properly, the data insertion speed may be slowed down (see Fig. 2). When the index design is complete, the partition is created with the primary index. The significance of data partitioning lies in that all the data that need to be written is stored in the specified storage area according to certain rules, which makes different data be written into different disk pre-divided space in the physical storage of the database, which can further improve the speed of data reading.

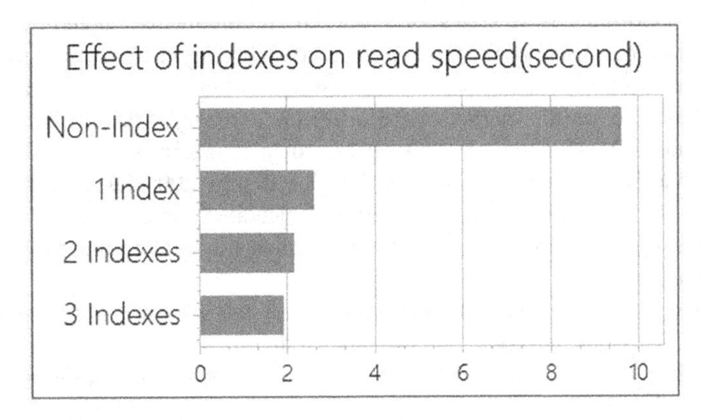

Fig. 1. Effect of indexes on read speed (mysql 5.7, 1000 items were retrieved from the table with 1,200,000 records)

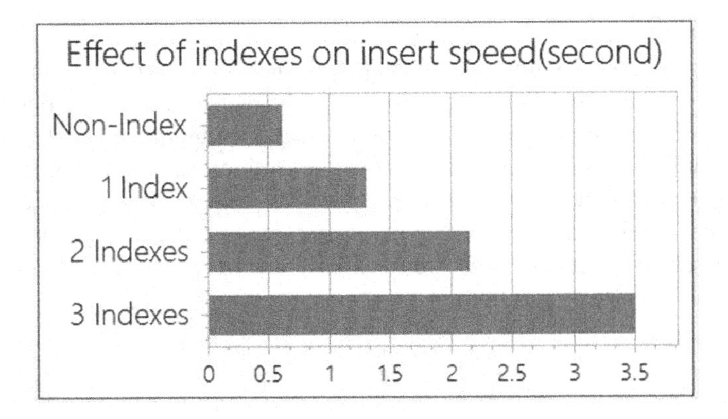

Fig. 2. Effect of indexes on insert speed (mysql 5.7, 1000 items were inserted to the table with 1,200,000 records)

There are obvious criteria for partition creation: the more partitions there are, the more expensive it is to manage and maintain. The greater the difference in the amount of data stored within a partition, the less meaningful the data partition is. Therefore, in the design of the test database of the spacecraft data management software, it is necessary to choose an algorithm that can reduce the partition fragmentation as far as possible and divide the storage evenly to complete this task.

2 Selection of Common Algorithms

Hash Algorithm, can map binary data of any length to a shorter binary string, and it is difficult for different data to map to the same Hash value. It can be understood as a spatial mapping function, mapping from a large value space to a very small value space. This mapping method from large to small ensures that the mapping process is not reversible. Therefore, hash algorithm is often used in encryption algorithm.

Hashing algorithm is a series of algorithm requirements, including the basic requirements of input sensitivity, irreversible, conflict avoidance. Any algorithm that satisfies these requirements can be called a hash algorithm. The existing MD5, SHA1, SHA2, Siphash etc. on the market can all meet the demand of data hashing.

For example, MD5 code can convert any data into 256-bit source code value. Assuming that 50,000 e-books need to be placed into 100 classified bookshelves according to their contents, MD5 code can be calculated for the binary files of each book, and the value obtained is the bookshelf number by dividing the result by 100 to get the residual value.

In the design of spacecraft data management test database, the first level is divided according to the parameter code, and the input source is used as the partition basis. In different models, the number of parameters is on the order of thousands to tens of thousands, and it is not possible to create separate partitions for each parameter. Therefore, after the use of the hashing algorithm, it is still necessary to take the remainder again to restrict all data classification within a certain range.

Among the current hash algorithms, the algorithms that can convert arbitrary data into fixed-length source code are MD5, SHA and other algorithms. These algorithms can fulfill the requirements of hashing algorithm well, and meet the requirements of input sensitivity, irreversibility and conflict avoidance. However, in database design, the need for conflict avoidance is not strong, and the requirement for partition balance is high. After the traditional hash algorithm is used to merge the remainder, under the condition that the input is completely random, the partition equilibrium of itself can be almost evenly distributed. However, the spacecraft test database storage does not simply depend on the number of categories stored, but also depends on the amount of data in the category. At this point, you can only attempt to redistribute the data by increasing the number of partitions.

3 Dynamic Hash DSH Algorithm

Spacecrafts test data has such a feature that some parameters are updated very frequently, while some parameters are updated at a low speed, and the speed difference may be more than one hundred times. If you do not weight parameter data volumes and hash only by parameter names, you may end up with several large data volumes of parameters being partitioned in the same partition.

Therefore, the first step of DSH algorithm is to ensure the uniformity of the hash algorithm under the condition of permitting collisions.

3.1 The First Round of Uniform Hashing of DSH Algorithm

Create a hash map space in memory, the length of which is determined by the computer based on the hardware environment, and can be assumed to be 100 and the number of mapped Spaces is 5.

After receiving the new parameter, judge whether there is already a mapping. If there is, enter the original mapping, and add one to the mapping space data count. If not, select the queue with the least data count in the mapped space (see Fig. 3).

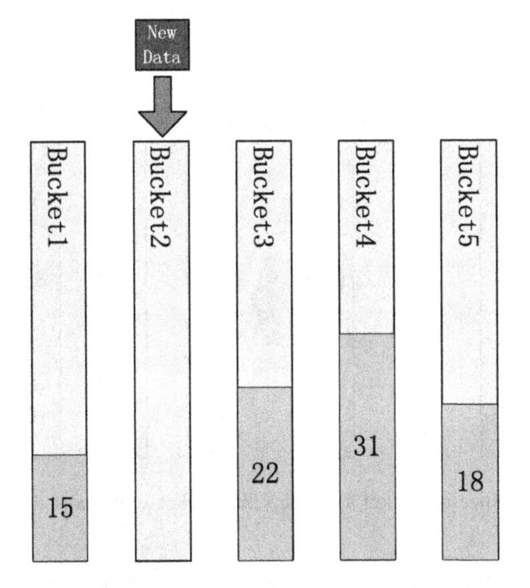

Fig. 3. Insert new data into a Bucket with least data

When all mapped Spaces are non-zero, such as a minimum of 15, the data is written and the memory space is cleared (see Fig. 4).

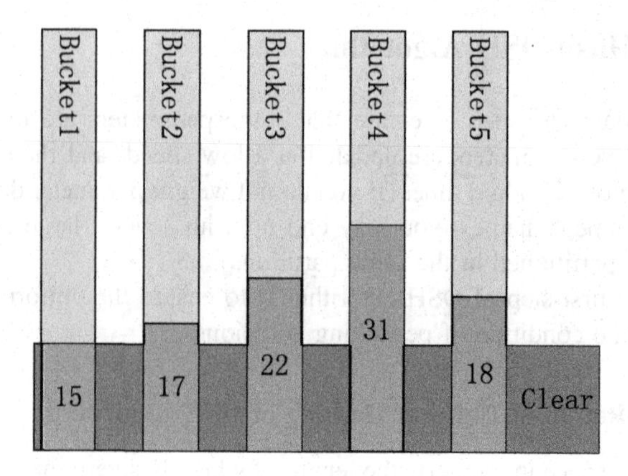

Fig. 4. Remove Buckets' fixed data when all Buckets have a certain amount of data.

In some special cases, the single-mapped space becomes full before the overall memory cleanup threshold is reached due to the explosion of data volume in the single-mapped space. At this time, the memory space will be cleaned up to the maximum value of the current whole space, and the cleaned data will be written into the database (see Fig. 5).

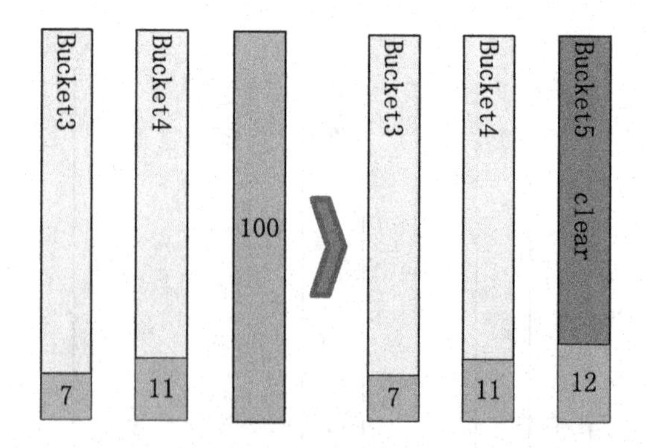

Fig. 5. Clear the special bucket and keep the bucket with most data count remain.

With the use of DSH, the stacked data moves from a single partition to each partition, ensuring partition balance.

On the basis of evenly divided data, the storage mode still has the possibility of optimization. Due to the uncertainty of the number of partition Settings, individual parameters with small data volume may occupy individual partitions.

Using DSH algorithm, the statistical data in the process of data generation is used to merge and store the data with small amount and low frequency, while the data with large amount is stored in a proportional exclusive partition. The number of partitions is small and the uniformity of data partition can be guaranteed.

3.2 DSH Algorithm Sub-round Dynamic Partitioning

The system calculates the partitioning results of the existing data once according to the default partition statistical unit time (90s). A total of L rounds of statistics were conducted. until

The system calculates the partitioning results of the existing data once according to the default partition statistical unit time (90s). A total of L rounds of statistics were conducted. until

$$\sum_{r=0} f(r) - f(l) < k \tag{1}$$

This indicates that statistics have converged. On this basis, partition calculation is carried out. Let the partition number of the current test model design be s, where I, j, k $\in$ S. If any I, j, k exists, the following relation can be satisfied:

$$n_i + n_j < n_k \tag{2}$$

Then partition I and partition j can be combined into a partition, creating a mapping relationship such that all

$$x \in n_j \rightarrow x \in n_i \tag{3}$$

And $s = s - 1$; Delete n_j partition and retraverse s so that it does not exist

$$n_i + n_j < n_k \tag{4}$$

At this point, the number of partitions is compressed to a minimum, and the partition uniformity is guaranteed. According to the measured data, when the initial S is set as m, the value of S after multiple cycles is at

$$\overline{4} \leq s \leq \overline{2} \tag{5}$$

Therefore, in the initial setting of the algorithm, when the user's expected partition value is set to s, the algorithm will use 2*s partition number to start iteration in the first round of partitioning, and the final partition result may be greater than or less than s. Specific data values are affected by the largest number of parameter values.

After the real-time test of more than 17,000 parameters with a total of 400MB data, the partitioning results of DSH algorithm and MD5 residual algorithm are shown in the figure below:

It can be seen that when the number of partitions is 32 and 64, the result of partitioning is that some partitions are too large due to insufficient hashing (see Fig. 6 and Fig. 7). When the number of partitions increases to 128, the largest data block is partitioned, which also leads to a large number of empty partitions, wasting system resources and increasing management costs (see Fig. 8).

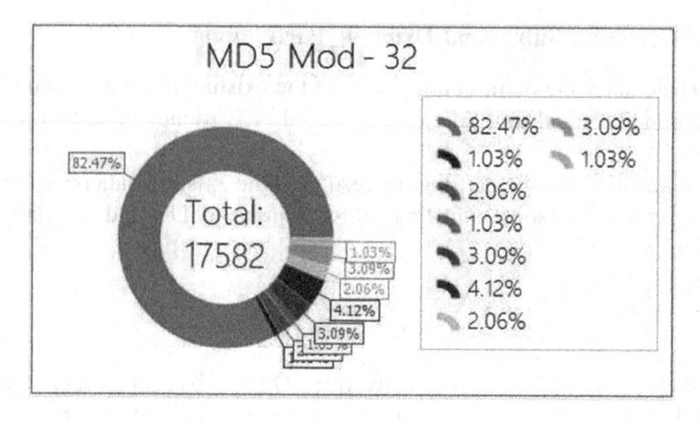

Fig. 6. 32 partitions for MD5 mod algorithm. The biggest partition own 82% data size in all 17582 records.

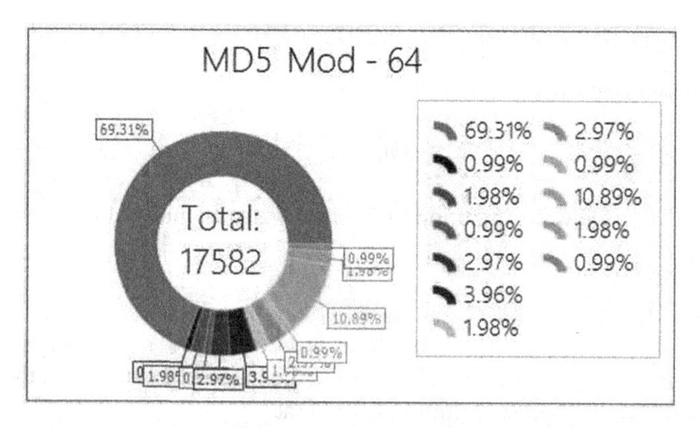

Fig. 7. 64 partitions for MD5 mod algorithm. The biggest partition own 69% data size in all 17582 records.

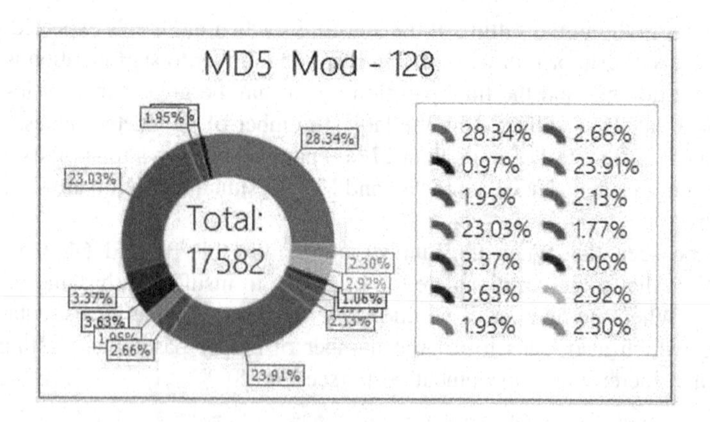

Fig. 8. 128 partitions for MD5 mod algorithm. The biggest partition own 28% data size in all 17582 records. And many partitions are not assigned data.

After the use of DSH algorithm, the data partition has a good result:

When DSH algorithm is used, it can be seen that when the number of partitions is 4, 9 and 14, good partition results can be obtained. The appropriate number of partitions can be specified according to the number of parameters of the specific model without considering the influence of the selection of the number of partitions on the final partition effect (see Fig. 9, Fig. 10 and Fig. 11).

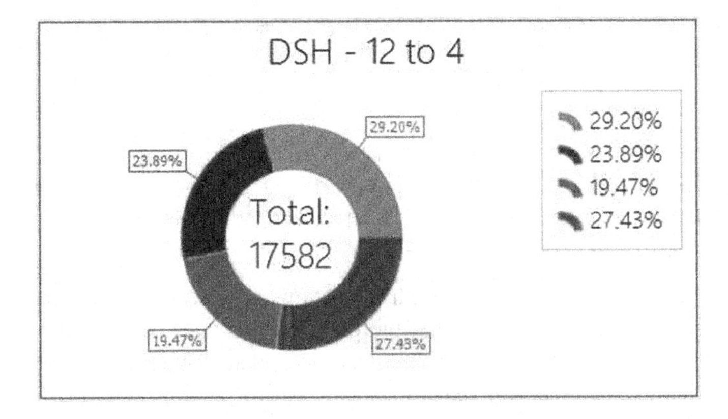

Fig. 9. The initial designation of 12 partitions is reduced to 4 partitions after calculation by DSH algorithm, and the largest partition is 10% larger than the smallest partition.

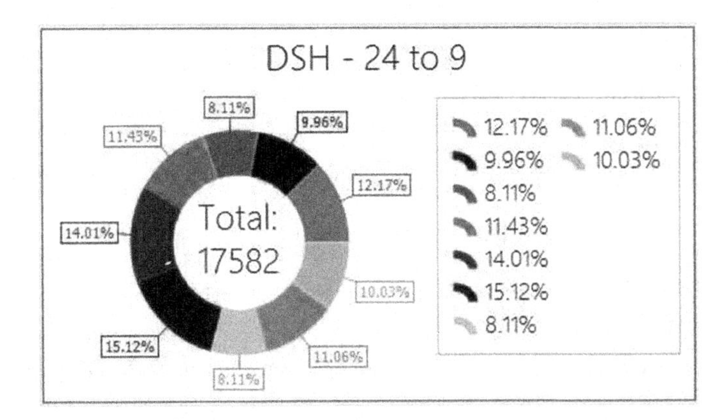

Fig. 10. The initial designation of 24 partitions is reduced to 9 partitions after calculation by DSH algorithm, and the largest partition is 7% larger than the smallest partition.

After the use of DSH algorithm, additional computing resources are needed due to the addition of dynamic partition algorithm, so the time spent in the data insertion process is recorded. It can be seen:

Inserts are fastest when the data is not partitioned or indexed at all, because there is no sorting of the database data.

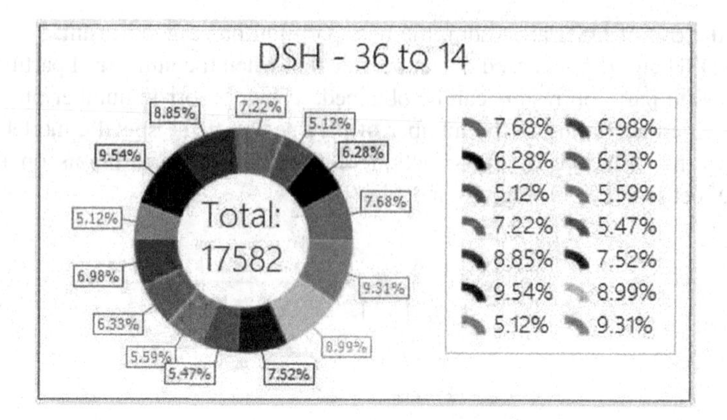

Fig. 11. The initial designation of 36 partitions is reduced to 14 partitions after calculation by DSH algorithm, and the largest partition is 7% larger than the smallest partition.

When DSH algorithm and other hash algorithms are used, the difference in insertion speed is negligible, indicating that the computing power consumed by the algorithm itself is not a decisive factor in the database update scenario and does not affect the insertion efficiency (see Fig. 12).

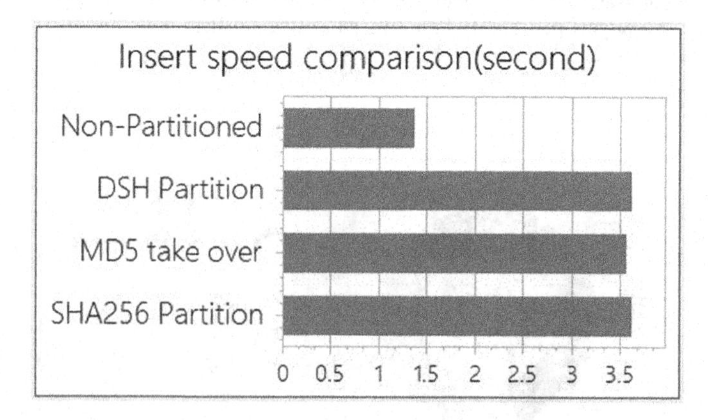

Fig. 12. In the algorithm with partition, the insertion speed difference is less than 1%, which is negligible.

In order to verify the data stored by DSH algorithm, read the 15-h test data after querying the data, and read the 1000 records before the specified parameter code:

It can be seen that the DSH algorithm is close to the MD5 residual algorithm with better partition in time, but it still has an advantage of about 15% (see Fig. 14). The speed of DSH algorithm is about one times higher than that of MD5 redundancy algorithm with poor partition. Data without adding any index partitions consumes a lot of time resources (see Fig. 13).

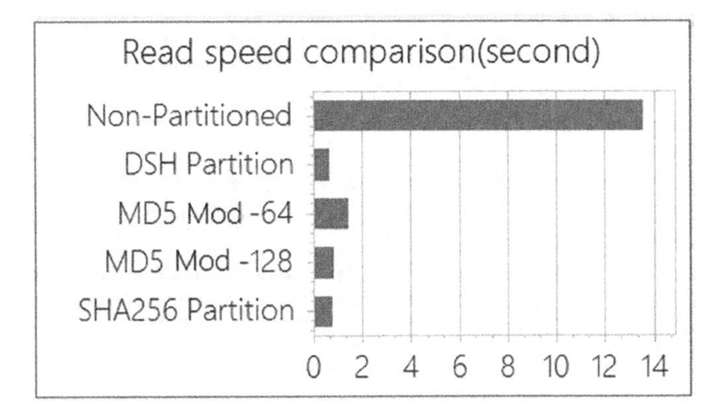

Fig. 13. Data reads are unacceptably slow without any partitioning.

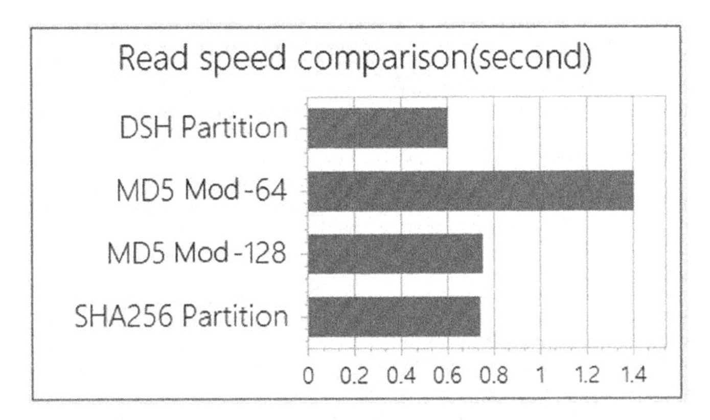

Fig. 14. DSH reads 15% faster than the already over-partitioned MD5 mod algorithm, and reduces a lot of administrative overhead.

4 Conclusion

In this paper, a dynamic hashing method is proposed to partition the test data before warehousing for spacecraft data management software, which has many test data, high requirement of storage and reading speed, and poor performance of the current storage algorithm. DSH algorithm can dynamically count the data of the same parameters in the stored procedure and make uniform hashing and partition adjustment based on it, which can effectively improve the rationality of data partition, reduce the cost of system management and avoid partition failure. The partition algorithm studied in this paper can not only be used in the storage and processing of spacecraft test data, but also be extended to all data processing scenarios requiring uniform data partition. In the future, this algorithm should be verified and promoted in more environments to ensure its effectiveness.

References

1. Aumasson, J.-P., Bernstein, D.J.: SipHash: a fast short-input PRF (2012)
2. Wang, Z., Zhang, H., Qin, Z., Meng, Q.: Fast attack algorithm on the MD5 hash function (2006)
3. Liang, J., Lai, X.: Improved collision attack on hash function MD5 (2007)
4. Teh, J.S., Samsudin, A., Akhavan, A.: Parallel chaotic hash function based on the shuffle-exchange network. Nonlinear Dynamics (2015)
5. Kim, D.-C., Hong, D., Lee, J.-K., Kim, W.-H., Kwon, D.: LSH: a new fast secure hash function family. In: Lee, J., Kim, J. (eds.) ICISC 2014. LNCS, vol. 8949, pp. 286–313. Springer, Cham (2015). https://doi.org/10.1007/978-3-319-15943-0_18

A Method of On-Board Dynamic Health Management of TT&C Transponder of Spacecraft

Hongjun Zhang(✉), Yulan Shi, and Shuai Wang

Beijing Institute of Spacecraft System Engineering, Beijing 100094, China
zhanghongjunbuaa@sina.com

Abstract. A method of on-board dynamic health management for spacecraft TT&C transponder is proposed. The on-board working state of TT&C transponder is a process of dynamic change in a certain and limited range. The CTU belonging to the OBDH subsystem collects the working state of the TT&C transponder in real time, and judges whether the operation of TT&C transponder is normal by judging whether the current state is consistent with the expected state. The CTU software designed the state machine of health state detection of TT&C transponder, established a health management data model, completed the autonomous health detection of TT&C transponder and the autonomous reset operation, and realized the on-board abnormal removal of TT&C transponder.

Keywords: TT&C transponder · Central Terminal Unit (CTU) · Dynamic detection · On-board health management

1 Introduction

With the development of China's space industry, the types and quantity of spacecraft in orbit are increasing day by day, and their life span is getting longer and longer. On-orbit management has entered the stage of multi-spacecraft parallel management [1]. Spacecraft in orbit is an important part of life cycle of spacecraft. Compared with the spacecraft products themselves, users are more concerned about the operation service of spacecraft. Spacecraft in orbit fault diagnosis and processing and in orbit data analysis and reuse technology is particularly important.

The ability of data processing and management of each subsystem by spacecraft onboard computer has been greatly improved, which provides a good foundation of spacecraft autonomous management [2]. The research on spacecraft autonomous health management technology has been continuously deepened, and a relatively systematic design method of autonomous health management system has been formed [3–5]. The TT&C transponder is an important equipment of the TT&C subsystem of the spacecraft. If there is an on-orbit fault, the telemetry and telecommand channels of the spacecraft will be disconnected, thus affecting the normal work of the spacecraft. Statistical data show that the on-orbit failure rate of spacecraft TT&C subsystem reaches 4% [6].

Q. Guo et al. (Eds.): WiSATS 2021, LNICST 410, pp. 177–185, 2022.
https://doi.org/10.1007/978-3-030-93398-2_17

A common fault removing method is to periodically close and open the TT&C transponder, used to avoid the TT&C transponder working abnormal. This method does not judge whether the TT&C is working abnormal in real time, that is, no matter whether the TT&C transponder is working abnormal, it will be forced to close and then open the transponder. In this way, when the TT&C transponder is working normally, it will also be closed, so there is a risk that the earth station injection instruction will not be properly received during the period from the TT&C transponder is closed to the open.

This paper presents a method for dynamic health detection and recovery of the TT&C transponder. Spacecraft health management system can obtain the health status of the TT&C transponder in real time, and judge whether the TT&C transponder is working abnormally. Once the TT&C transponder is abnormal, it can send fault relief and recovery instructions to make the transponder return to normal.

2 Principle of Fault Detection

2.1 Health State

Finite state machine, which represents a finite number of states and the mathematical model of their transitions and actions, is widely used in computer, communication, digital operation design, software engineering and other fields [7, 8]. Spacecraft and its components will be in a certain state at any time [9], and monitoring and judging the state is an effective fault detection method.

The state changes of a typical TT&C transponder during normal operation include state1, state2 and state3. When the TT&C transponder is working normally, it changes in the order of "State1 → State2 → State3", and the three state parameters change in turn once for a state cycle (called the state cycle of the TT&C transponder), as shown in Fig. 1.

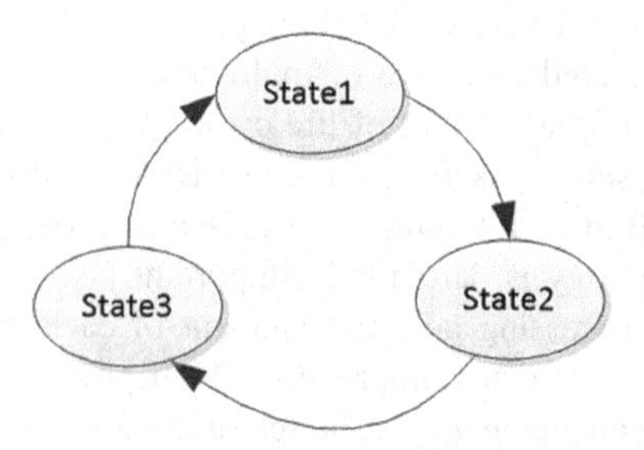

Fig. 1. TT&C transponder state cycle

If the TT&C transponder is always working normally, it will change in the order of "State1 → State2 → State3". If the working state of the TT&C transponder is no longer changing or the order of state change is not "State1 → State2 → State3", it indicates that it is working abnormally.

2.2 Fault Decision

The central terminal unit (CTU) collects the working states of the TT&C transponder periodically (this period is called CTU collection period) and transmits them down through real-time telemetry [10]. The working state of the TT&C transponder changes after CTU collection. Since there are three state parameters in each state cycle of the TT&C transponder, CTU continuous collection for three times (i.e., three collection periods) will complete the collection of one state cycle of the TT&C transponder. Corresponding relationship between CTU collection period and state cycle of the TT&C transponder as shown in Fig. 2.

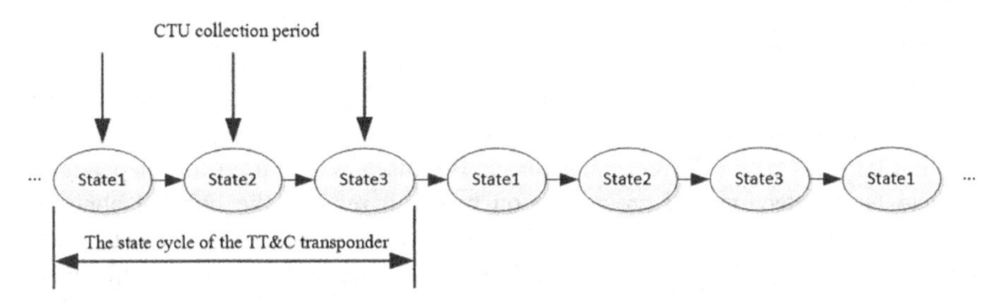

Fig. 2. Relationship between CTU collection period and state cycle of the TT&C transponder

CTU gets its state parameters from the TT&C transponder in each collection period, and then retrieves the working state relation table of the TT&C transponder, as show in Table 1, to determine whether it is the expected state parameters. If not, it indicates that the current state cycle of the TT&C transponder is abnormal. If the TT&C transponder is abnormal for three consecutive state cycles, it is determined that the TT&C transponder is abnormal.

Table 1. Expected state of the TT&C transponder

No.	The current state of the TT&C transponder	The next expected state of the TT&C transponder
1	State1	State2
2	State2	State3
3	State3	State1

If CTU determines that one of the state parameters in the state cycle of the TT&C transponder is not the expected state parameter, indicating that the state cycle is abnormal, then the remaining state parameters in the state cycle will not need to be judged, and will directly wait for the arrival of the next sate cycle to start the judgement again. Assuming that the first expected state parameter in the current state cycle is State1, but the actual state parameter collected by CTU is not state1, it indicates that this state cycle is abnormal. The remaining two state parameters in this state cycle do not need to be judged, and the judgement will be started again after the arrival of the next state cycle, as shown in Fig. 3.

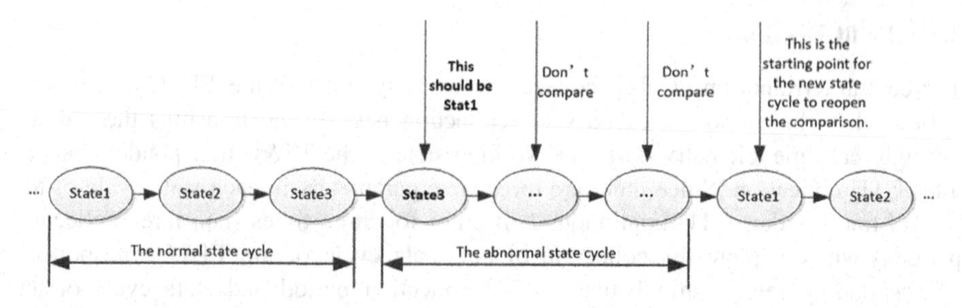

Fig. 3. Example of normal status period and fault status period of T&C transponder

2.3 Fault Removal

After determining that the TT&C transponder is abnormal, CTU will automatically send the reset instruction of the TT&C transponder to reset it, so as to realize the abnormal removal of the TT&C transponder. Because CTU is used to independently complete fault detection and processing on board, it reduces the links of abnormal discovery, fault determination, operation and processing on the ground, so as to realize quick fault removal and improve the reliability of the system quickly and effectively.

3 Design and Application of Health Management System

CTU is the center of spacecraft data interaction and has the advantage of unique data resources. As spacecraft system-level software, CTU software is responsible for spacecraft system-level health management. Based on this, CTU shall automatically remove the abnormality of the TT&C transponder in orbit [11].

3.1 System Composition

The health management system for the TT&C transponder is composed of the TT&C transponder and CTU. The TT&C transponder is the tested object, while CTU is the health detection and execution unit. In CTU, hardware and software jointly complete the health state detection and recovery, as shown in Fig. 4.

The health detection module drives the state acquisition module to complete the collection of the working state of the TT&C transponder within each collection period, and to complete the health judgment of the working state of the TT&C transponder. The health recovery module obtains the health results generated by the health detection module. If abnormal operation of the TT&C transponder is found, the instruction data will be transmitted to the instruction execution module, and the instruction execution module will generate instruction signals according to the instruction data and output them to the TT&C transponder, so as to realize the state recovery of the TT&C transponder.

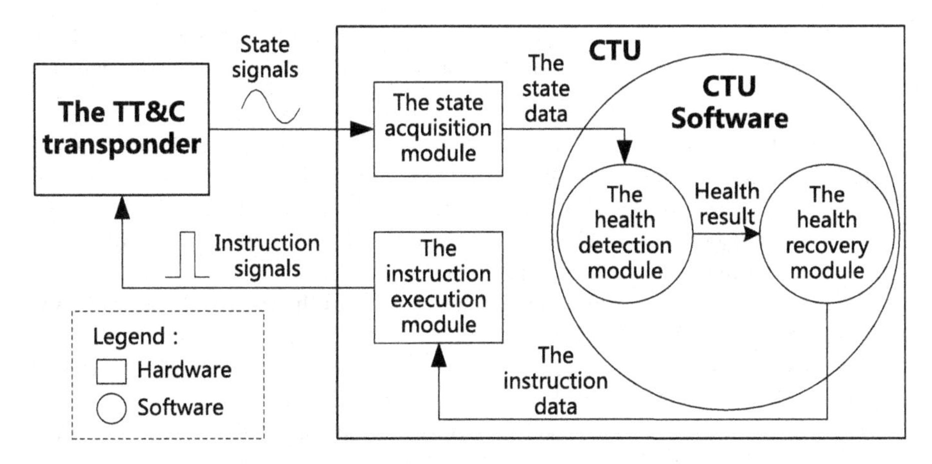

Fig. 4. Health management system composition

3.2 Design of Fault Detection and Recovery Process

With the increasing of software complexity, it is a very practical and effective way to solve software development problems by using abstract means to raise the level of abstraction [12].

The health state detection of the TT&C transponder is a dynamic detection process, that is, CTU needs to detect whether it is abnormal in the process of dynamic state changes of the TT&C transponder, which is different from other health detection process that only need to determine one or more static parameters. Based on this, CTU software designs a state machine to dynamically detect the working state of the TT&C transponder, as shown in Fig. 5.

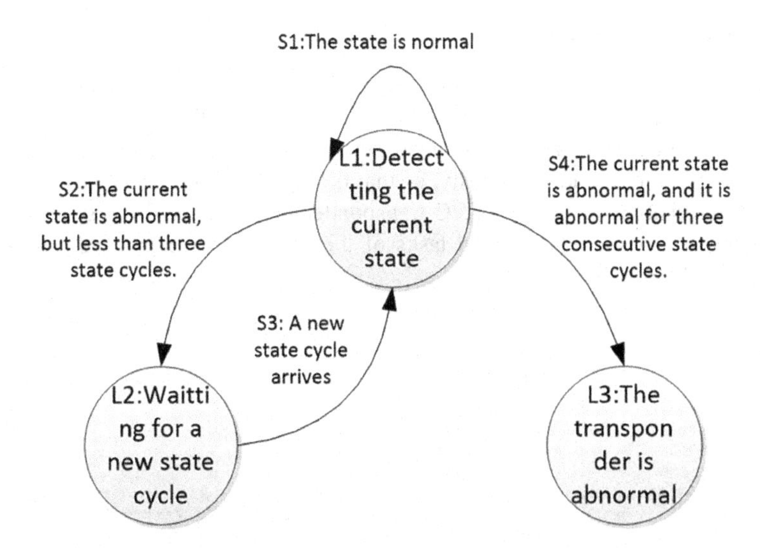

Fig. 5. State transition diagram of health detection

In order to facilitate CTU software to accurately detect the working state of the TT&C transponder, a health detection data model is established. The model elements include:

1) The current states of the TT&C transponder. This state is the state collected by CTU from the TT&C transponder in each collection period.
2) The expected states of the TT&C transponder. CTU collects the current state from the TT&C transponder in each collection period, and then looks up the table to get the expected state of the TT&C transponder.
3) State count. It is used to record the number of states in the state cycle of the TT&C transponder judged by CTU. The state count is 0, 1 and 2, which correspond to one state parameter in the state cycle respectively. When the state count changes to 2, it indicates that the judgment of the current state cycle is completed.
4) Abnormal state cycle count. Used to record the number of continuous abnormal state cycle of the TT&C transponder.
5) Flag of waiting for the new state cycle. Since the TT&C transponder has three state parameters in one state cycle, the remaining state parameters do not need to be judged if the current state cycle is abnormal. When the flag is set to true, wait for the next state cycle to come and start the judgment again.
6) Abnormal flag of the TT&C transponder. If the TT&C transponder is abnormal for three consecutive state cycles, the TT&C transponder should be determined to be abnormal, and the flag should be set as true. CTU will remove the transponder abnormal after detecting this flag (Fig. 6).

Based on the above state transfer and data model, CTU software completes the collection of the state parameters, determination and autonomous removal of abnormal conditions of the TT&C transponder. The specific process is shown in Fig. 7.

3.3 On-Orbit Applications

The health management method proposed in this paper has been applied in many resource survey satellites in China. There was an anomaly in the TT&C transponder of a satellite in an invisible TT&C arc outside China. CTU software detected the anomaly in time and recovered it autonomously, so that the satellite could work normally when it entered the visible arc of the TT&C transponder in China. And the ground station could inject the subsequent payload tasks of the satellite normally, and ensured the continuity of the satellite's work.

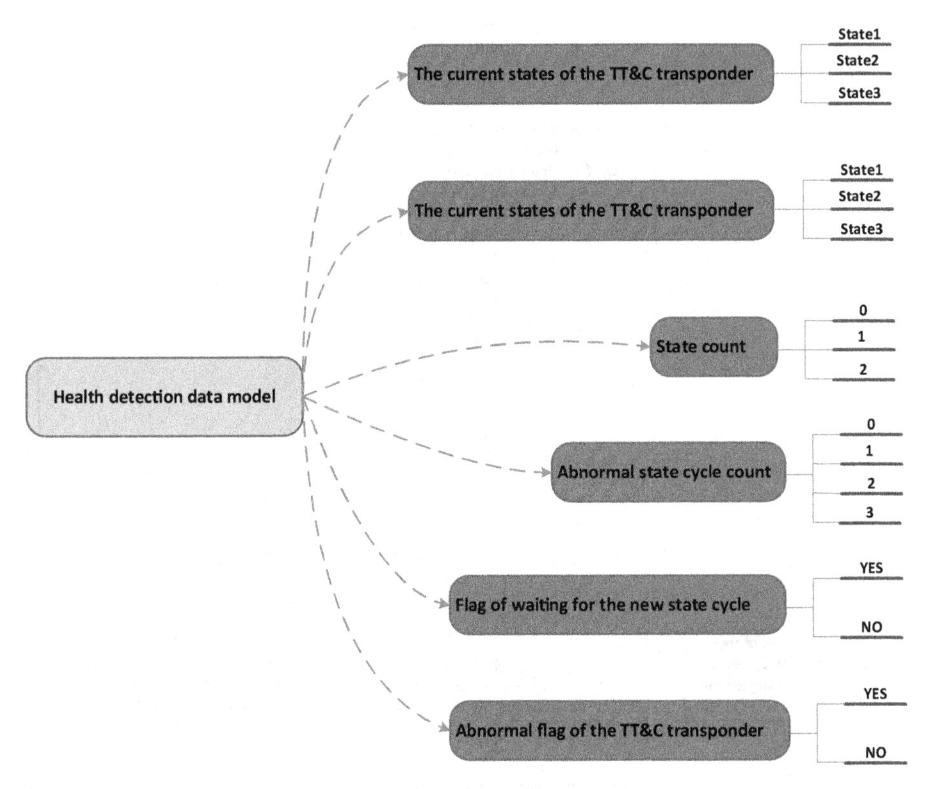

Fig. 6. Data model of health detection

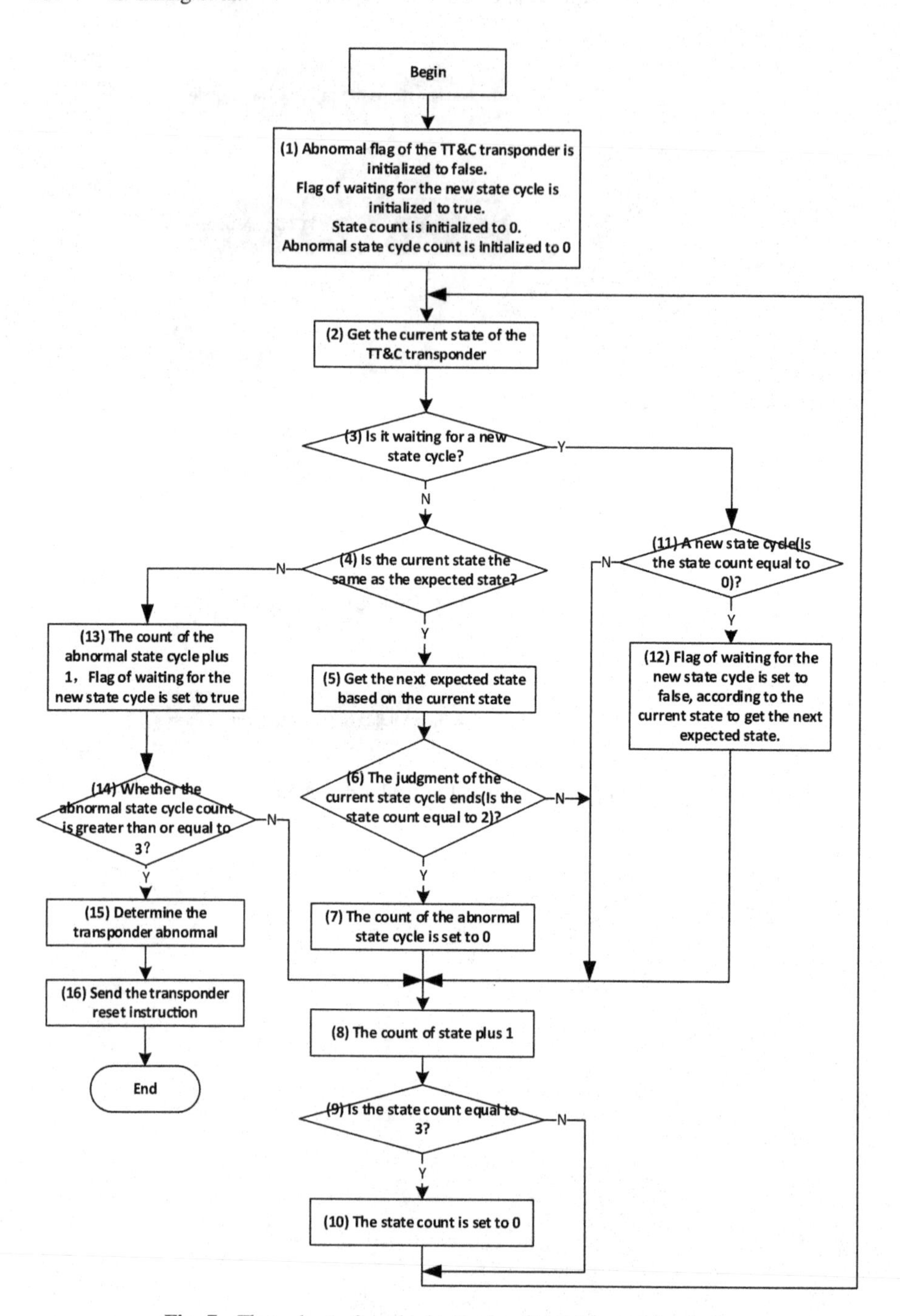

Fig. 7. Flow chart of method relieving T&C transponder fault

4 Conclusion

The on-orbit health management method of the TT&C transponder proposed in this paper is accomplished by CTU. CTU software designed the detection state machine and the health management data model, then CTU software collects the states of the TT&C transponder in real time, so that CTU can accurately and timely determine whether the TT&C transponder is abnormal in orbit, and autonomously reset the TT&C transponder, so that the transponder can be timely restored to the normal working state, to ensure the safety of spacecraft in orbit operation.

References

1. Wang, H., Guo, Y., Qin, W.: Application of expert knowledge of on-orbit management. Spacecraft Eng. **17**(4), 67–71 (2008)
2. Shan, C., Li, Y., Wang, L.: Study on methods of abnormal alarm and fault diagnosis for in-orbit satellites. J. Spacecraft TT&C Technol. **30**(3), 7–10 (2011)
3. Wang, W., Wang, X., Xu, H., et al.: Design and implementation of autonomous health management system for GF-3 satellite. Spacecraft Eng. **26**(6), 40–46 (2017)
4. Liang, K., Deng, K., Ding, R., et al.: Autonomous on-orbit health management architecture and key technologies for manned spacecrafts. Manned Spaceflight **20**(2), 116–121 (2014)
5. He, X., Guo, J., Li, Y., et al.: Autonomous health management requirements and software architecture for deep space probe. Control Theory Appl. **36**(12), 2066–2073 (2019)
6. Tan, C., Hu, T., Wang, D., et al.: Analysis on foreign spacecraft in-orbit failures. Spacecraft Eng. **20**(4), 130–136 (2011)
7. Yang, S., Wang, X., Li, M.: RVD flight mission planning and scheduling method based on finite state machine. J. Beijing Univ. Aeronaut. Astronaut. **45**(9), 1741–1746 (2019)
8. Jia, J.: Research on Fault Diagnostics for Multiple State Machine Model. Harbin Institute of Technology, Harbin (2016)
9. Yan, Q., Luo, Y.: Establishment of expert knowledge models for spacecraft faults and its application. J. Spacecraft TT & C Technol. **28**(4), 44–47 (2009)
10. Tan, W., Hu, J.: Spacecraft Systems Engineering. China Scientific and Technology Press, Beijing (2009)
11. Zhang, Q., Guo, J., Dong, G.: Space Data System. China Scientific and Technology Press, Beijing (2016)
12. Liu, X.: Research on Model Driven Safety Verification for Embedded System Design. Nanjing University of Aeronautics and Astronautics, Nanjing (2015)

Structured Design Method of Spacecraft Telecommand Information Flow Based on Electronic Data Sheet

Nan Pei(✉), Shuo Yang, Yanfang Fan, Yong Xu, Qiang Mu, and Hongjing Cheng

Beijing Institute of Spacecraft System Engineering, Beijing 10094, China

Abstract. This paper proposes a structured description method of spacecraft telecommand based on electronic data sheet. The collaborative design is realized by constructing command tree. The command tree model is composed of multi-level data sheet, which contain component combination and interface. A basic component is the smallest unit of command that describes data fields by standard types and attributes. The EDS system and general command generation tool are developed to implement the application of the model. With the characteristics of clear hierarchy, flexibility and high expansibility, it solves the problems of asynchronous information and understanding differences, reduces the repeated generation and check work in telecommand application, and improves the work efficiency and correctness.

Keywords: Electronic data sheet · Structured design · Component

1 Introduction

With the development of space missions, the system functions, information flow, protocols and interfaces have become more and more complex, and the scale of commands has also been increasing. Generally, format requirements and design results of telecommand are described through documentation. There are the following problems. Data is inconsistent in different documents, and error is occurred in unstructured data transmission, and a lot of understanding and input work is repeated. These problems lead to decreased efficiency and increased errors [1].

According to the requirements of standardized and structured information design and transmission, relevant research have been carried out. The SOIS Electronic Data Sheets (SEDS) [2, 3] and XML-based Telemetric and Command Exchange (XTCE) protocol which proposed by Consultative Committee for Space Data System (CCSDS) [4] are more authoritative.

SEDS focus on the definition of data interaction interface for satellite-borne information flow, while the XTCE focuses on the data interaction interface between the satellite and the ground station.

This paper proposes a structured representation of command based on model. In the process of information flow modeling, the specification defined by SEDS and XTCE

Q. Guo et al. (Eds.): WiSATS 2021, LNICST 410, pp. 186–193, 2022.
https://doi.org/10.1007/978-3-030-93398-2_18

standards is greatly borrowed. The command information is divided into multiple fields and abstracted to form a model. A single-layer command format is formed through the combination of basic models, and then a tree structure is formed through the relationships among levels to achieve data structuring. To complete the structured model, a visual data sheet is provided to designers in a collaborative design tools. Designers of different systems can use data sheets to realize data interaction. Finally, the results can be transformed into standard interactive interfaces to realize structured transmission.

2 The Process of Spacecraft Information Flow Design

The process of information flow design is a V-shaped structure which transmit requirements from top to bottom and collect data from bottom to top. It is also the realization of information interface between various protocol layers according to SOIS [5, 6].

In system-level, the telecommand protocols of data link layer and transport layer are designed and the interface is identified. Requirements for the services and entities are passed through the interface, such as space packet service in transfer layer and PUS in application support layer.

Requirements are passed layer by layer to the final design entity. The design entity may be a device or software. After the completion of the design, the telecommand information from the entity will be transmitted upward, including the format, coding, using conditions and criteria. After the step-by-step transmission, a complete command information flow is formed.

According to the above procedures, a structured and visualization model is provided by electronic data sheet for command protocol design. The design results are eventually converted into XML files which comply with XTCE and SEDS standards. The output files can be read directly by the computer system of user, which can greatly improve work efficiency and effectively standardize the design.

3 Design of Electronic Data Sheet

3.1 Design Principles

The frequently used spacecraft remote control standards including PCM telecommand, the packet telecommand and AOS telecommand. Hierarchical design is used in all of three, and packet is defined and used in packet telecommand and AOS telecommand. Although the content of the command is not defined, the interface between each other is standardized because all users work with packets [7].

The European Cooperation for Space Standardization (ECSS) has developed a standardized telemetry and command format, ECSS-E-70-41A Packet Utilization Standard (PUS) [8], which defines specific types and contents through service types and sub service types.

The electronic data sheet should fully considers the above standard protocols, and follows the principles below:

Flexible. The existing protocol design should be covered.
Extensible. Future extensions of new protocols should be considered.
Structured. The interface data is machine-readable to avoid ambiguity.

3.2 Design of Basic Component

Component is defined as the smallest unit that makes up the command format. The basic element is shown in Fig. 1. The attributes of component include name, identifier, length, and type. The name attribute describe the component meaning, such as block head, command type, etc. The identifier attribute is the universally unique identity, which can be referenced. The length attribute is the length of the Component in bytes. The type attribute is used to distinguish between different basic component. Different types of component have different extended attributes, including algorithm, value, range and extended information.

According to the analysis of the commonly used command structure, the types of components are divided into seven categories: enumeration, constant, check, padding, count, data field and user input.

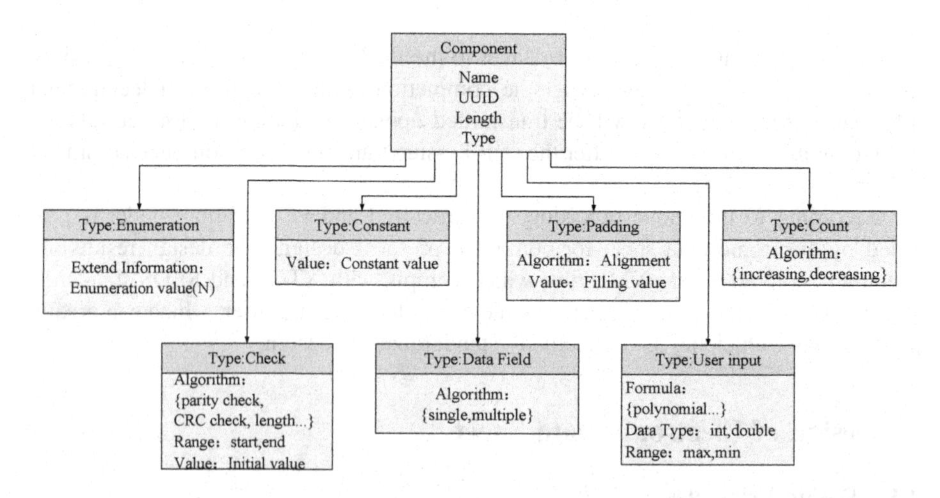

Fig. 1. Types and attributes of basic component

Enumerated types are used to describe units with a finite number of classes which can be extended as interfaces to lower-layer, such as the APID in the packet. The extended attributes of enumeration type include enumeration value and meaning. The enumeration value and meaning correspond one to one, and there can have multiple enumeration descriptions.

Constant type is used to describe a fixed data unit, such as a fixed header in a command. The extended attribute of constant type only contains the value, which is the fixed value of the data unit.

Check types are used to describe a unit computed by some algorithm from other units. Such as checksum or data field length and so on. The extended attribute of check types include algorithms, ranges and values. The algorithm includes parity check, CRC, ISO, length by byte, length by word, etc. It can be expanded according to the need. The range is used to identify the calculated range, such as calculate length from start to end. The value is optional and is used as the initial value of the algorithm.

Padding type is used to describe the filling data unit to reach the required length. The extended attribute include algorithms and values. The algorithm describes the filling method, such as filling by even bytes, filling by fixed length, etc. The field value is the data value of padding.

Count type is used to describe the unit of data that generates the count, such as the sequence number. The extended attributes of the counting type only includes algorithms, such as increasing and decreasing, which can be expanded according to needs. When an command code is generated, the count type fields can be counted and changed as needed.

Data field type is used to describe a unknown data unit, such as the packet data field of an packet. The detail of data field type will be described at the lower-level. The extended properties of the data field type include algorithms, used to select whether multiple data unit can be included in this data field.

User input type is used to describe a data unit which value is input by user instantly when the command is generated. The extended attribute of user input type includes formula, data type and valid range. The specific value will be obtained by real-time calculation when the command is generated according to these attributes.

3.3 Design of Container Data Sheet

An electronic data sheet for a physical device or a software is defined as a container that consists of an ordered series of components.

The container can be composed of name, ID, affiliated device entity, component combination and interface. ID can be uniquely referenced. Affiliated device entity is used to describe the specific position in the system. Components are combined into an ordered set that describes the format of the command. Interfaces include key-value to specify branches in the telecommand tree.

3.4 Design of Interface Connection Between Containers

Considering the attributes and design flow of telecommand information, a tree structure is formed through multiple nodes.Each node corresponds to a container. The root node reflects the top-level information. The intermediate node can have one father node and several children node. The branching relationship between container is defined by the key-value. Leaf nodes describe the command code.

Containers are connected through interfaces shown in Fig. 2. The upper layer container contains a data field type unit as a data association and a set of values for the enumeration type unit as all key values. The number of branches is the same as the number of key values. The child container selects one key value to form a branching relationship. If the parent container does not include enumeration type, only one branch can be connected.

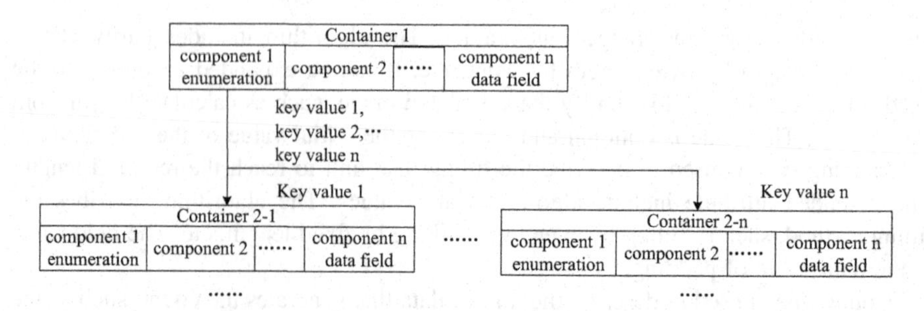

Fig. 2. Branch connection relationships between containers

When there is no data field type component in a container, it means that this is the lowest level. Generally, fixed code are described by enumerating fixed command, and variable parameter are described by using user input component. The formula, data type and other extended information are used for calculation when generating command.

4 Application of Electronic Data Sheet in Telecommand Information Flow Design

4.1 Design and Application Based on Layered Model

The telecommand design process is completed by constructing the command tree. The top-level user establishes the root sheet, design the rules into the container structure, and pass it to the lower-level user. The lower-level user, refines the command design in this level. Due to the restriction of structure, the lower-level users can only design in strict accordance with the requirements of the higher level. Finally, the bottom-level user completes the leaf sheet design to close the branch of the tree. After all branches are closed, the complete command tree is constructed, including all command formats, code, and rules.

The layered tree structure has strong extensibility, and conforms to the design method that SEDS describes the device information and service interface information in SOIS architecture.

The interface is connected through the format of command and key values between device entities, and the data interaction between protocol layer is realized. It is suitable for collaborative design among multiple users and can realize the design requirements of "Plug and Play" for spaceborne equipment. New business models and applications of spacecraft information flow design can be easily and flexibly expanded.

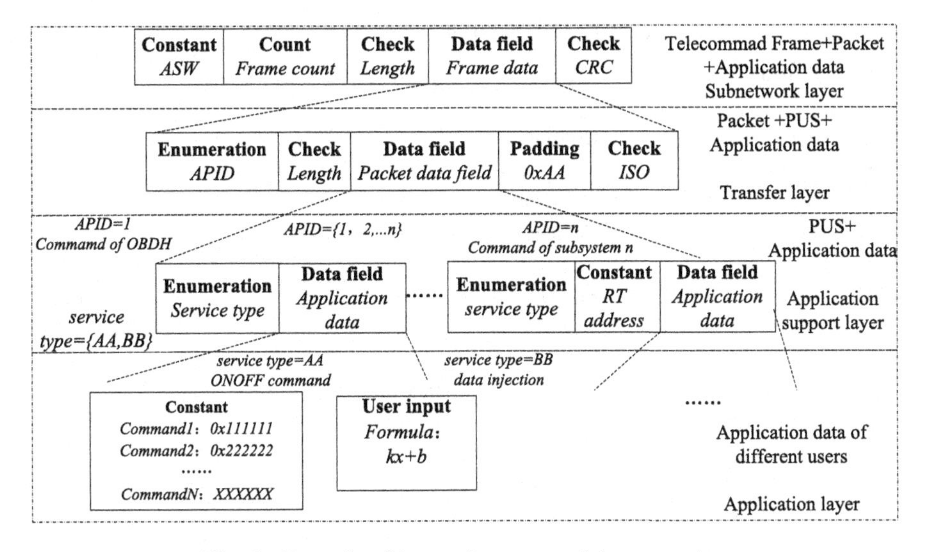

Fig. 3. Example of instruction tree model construction

The application example is shown in the Fig. 3. Different terminals in the application layer have their own application data format. The application data is packaged in application support layer using the PUS protocol. Telecommand packet of different APIDs if formed in network layer and transferred to data link layer. This process represents the interface and relationships of different levels.

Collaborative Information Flow Digital Design Platform which called EDS (Electronic Data Sheet Manage System) has been developed by CAST based on this model. The EDS system provide information collection, formatting, sharing, change management and other functions for spacecraft information flow design and verification [9]. At the same time, it provides third-party data interfaces for applications, including spacecraft software design [10], system test and in-orbit operation.

The formatted data sheet is designed in EDS system. Designers record their design result in the database through distributed sheet filling. Meanwhile, the system provides designers with a unified data source to ensure that the design data are uniform, accurate and clear for all users.

The electronic data sheet provides the interface for users, which is shown in Table 1. Each row is a component, and multiple rows form a container. The user can select the component type through the pull-down menu, and fill in the attribute information. The system can automatically verify the information, to ensure the design is correct. After the completion of the filling, the system will automatically generate the model and pass it through the interface.

Multiple data sheet are connected to form a command tree. Each node in the tree is an electronic data sheet. The system converts the complete command tree to form a standard interface file output.

Table 1. Data sheet of command format.

Component name	Length (Byte)	Type	Algorithm	Range	Value	Enumeration details
Packet version number	0.375	Constant	/	/	0	/
Packet type	0.125	Constant	/	/	1	/
Secondary head flag	0.125	Constant	/	/	0	/
APID	1.375	Enumeration	/	/	/	0x401:real-time 0x402:delay
Sequence flags	0.25	User input	/	/	/	/
Sequence count	1.75	Count	Increasing	/	/	/
Packet data length	2	Check	Length by byte	[8, 8]	/	/
Packet data field	0	Data field	/	/	/	/

4.2 Output and Application of Structured File

In the system design, integration test, the research of telemetry information design and processing based on XTCE has been carried out and applied [11, 12].The electronic data sheet of command format generated in the different layer is gathered into an command tree, which is converted by software tools to generate XML (Extensible Markup Language) interface files for transmission. The structured information is transmitted to software design platform, sub-system test platform, integrated test system, on-orbit data management system and other application scenarios.

The structured output file is in good agreement with the XTCE definition. By defining reference type, set and container, XTCE standard describes all attribute information of container, parameter and argument used in telemetry and telecommand. The interaction interface of spacecraft information flow data is completely defined.

First, standard and custom types are defined. And then, the description of the set defines all the parameters and arguments used by the spacecraft and associated then to the types. Finally, the parameters or arguments are arranged in order to form the command format.

The component model defined in this article corresponds to the type sets including ParameterTypeSet and ArgumentTypeSet in XTCE. The instantiated components that are filled into the data sheet with specific parameters corresponding to the data sets including ParameterSet and ArgumentSet in XTCE. The data sheet corresponds to the Container in XTCE, and the command tree corresponds to the ContainerSet. Finally, command code corresponds to MetaCommand in XTCE.

Standardized command generation tool is developed, which can display the structure of telecommand after read and transfer the XML file. The tool can generate commands automatically and quickly. The command code can be reverse solved according to structured format using to be checked. In this way, a general tool can adapt to multiple risks without repeated development and configuration. It will play an important role in testing and on-orbit operating.

5 Conclusion

In this paper, a telecommand information model is proposed based on layered assembly model. Different types of basic components are combined into data sheet and connected through interfaces to form a tree structure. The electronic data sheet is filled through multi-level collaborative design. With the characteristics of clear hierarchy, flexibility and strong expansibility, it solves the problems of data asynchronization and understanding of differences, reduces the repeated generation and check work, and improves the work efficiency and correctness.

Later, different levels of interface parsing methods can be developed to achieve plug-and-play and automatic code generation.

References

1. Jonathan, W.: Using CCSDS Standards to Reduce Mission Costs. NASA Goddard Space Flight Center, Greenbelt (2017)
2. CCSDS. 876.0-R-3 Spacecraft onboard interface services - XML specification for electronic data sheets: Washington, CCSDS (2018)
3. CCSDS. 660.0-B-1 XML telemetric and command exchange (XTCE), Washington: CCSDS (2017)
4. CCSDS. 876.1-R-2 spacecraft onboard interface services—specification for dictionary of terms for electronic data sheets, Washington: CCSDS (2016)
5. Yang, L., Chen, B., Zhang, R.: Application of SEDS in service and protocol system. Comput. Measur. Control **26**(11), 248–251 (2018)
6. Zhang, X., Lyu, L., An, J.: Instance design of EDS based on SOIS. Comput. Eng. Des. **41** (9), 2670–2677 (2020)
7. He, X.: Design and application of a common spacecraft telecommand data format. Spacecraft Eng. **17**(1), 94–99 (2018)
8. European Cooperation for Space Standardization: ECSS-E-70-41-A Space engineering: ground systems and operations telemetry and telecommand packet utilization. Noordwijk, ECSS (2003)
9. Guo, J., Li, R., Fan, Y., et al.: Network coordinated digital design and practice of spacecraft information flow. Spacecraft Eng. **29**(4), 59–65 (2020)
10. Zhang, H., Pan, L., Yu, M.: A method of automatic code generation for spacecraft OBDH software. In: Jia, M., Guo, Q., Meng, W. (eds.) WiSATS 2019. LNICSSITE, vol. 280, pp. 275–282. Springer, Cham (2019). https://doi.org/10.1007/978-3-030-19153-5_28
11. Liu, Y., Li, Z., Ding, X., et al.: Satellite telemetry data processing method based on XTCE. J. Telemetry Tracking Command **38**(2), 27–31 (2017)
12. Zhang, H., Guo, J., Kuang, D., et al.: Design of spacecraft telemetry transfer universal interface. Spacecraft Eng. **28**(6), 46–51 (2019)

A Telecommand Scheme for Spacecraft Based on High-Speed and Low-Speed Hybrid TT&C System

Siyang Zhao(✉), Yong Xu, Weiwei Liu, Minfang Yu, and Yuehua Niu

Beijing Institute of Spacecraft System Engineering, Beijing 100094, China

Abstract. In order to meet the increasing demand for high-speed data uplink of spacecraft, more and more spacecraft have added high-speed TT&C (telemetry, track and command) channels besides retaining the traditional low-speed TT&C channels. This paper proposes a telecommand scheme under the existing spacecraft electronic equipment architecture and uplink data protocol framework, which combines the data processing of high-speed and low-speed TT&C channels, optimizes the data flow design, and gives an efficient and reasonable uplink Space data link protocol, the data fusion of high-speed and low-speed TT&C links is realized, which improves the reliability of the spacecraft and the flexibility of uplink data transmission.

Keywords: Uplink · Telecommand · Space data link · Protocol · TT&C system

1 Introduction

With the continuous growth of the size of spacecraft software and the increasing demand for software on-orbit dynamic loading and reconfiguration, it is more and more urgent to improve the high-speed uplink information transmission capability of spacecraft. The high-speed and low-speed hybrid TT&C (telemetry, track and command) system is a new TT&C system developed in recent years to meet the above needs. On the basis of conventional TT&C system, the new system expands the capability of uplink high-speed data transmission, supplements the design of the uplink channel to accommodate high-speed data transmission and gives consideration to the modulation mode and data structure of the telecommand information.

The existing spacecraft has formed mature data flow design, protocol selection, interface design and stand-alone electrical device design under the conventional low-speed TT&C system. The low-speed uplink channel and high-speed uplink channel of most existing spacecraft are completely independent, and the data protocol chosen is also different. This brings a lot of inconvenience to the uplink data processing on the ground and on-board. In this paper, an uplink telecommand scheme is proposed based on the existing spacecraft electronic architecture and uplink data protocol framework. By combining the data of high-speed and low-speed channel, the existing design of data flow is optimized, the data protocol is carefully selected, and the cost of

Q. Guo et al. (Eds.): WiSATS 2021, LNICST 410, pp. 194–200, 2022.
https://doi.org/10.1007/978-3-030-93398-2_19

modification of the existing software and electronic equipment is comprehensively considered, this new telecommand scheme can support the high-speed and low-speed hybrid TT&C system, so as to adapt to the current and future 5–10 years of intra-satellite communication and inter-satellite communication requirements.

2 Requirement of Spacecraft Telecommand

There are two main types of uplink telecommand data sent to the spacecraft via TT&C system: Direct Command and Upload Data. In recent years, Direct Command has been gradually reserved with only Direct ON/OFF (DOO) Command. In addition to the traditional Routed ON/OFF (ROO) Command, Memory Load (ML) Command, OBDH (On-board Data Handling Subsystem) House-keeping instructions, bus instructions, on-orbit maintenance data and inter-satellite data have been added. With the increasing demand for on-orbit maintenance and reconfiguration of large-scale/super-large-scale software for spacecraft, updates of on-orbit intelligent processing related algorithms/applications, increasing demand for autonomous constellation operation and large data interaction, the amount of upload data increases in series, and the amount of single upload data increases from KB level to tens of MB level. The high-speed and low-speed hybrid TT&C system can make up for the shortcoming of low speed in traditional system and better meet the needs of spacecraft for high-speed ascending. The following Fig. 1 shows the classification of spacecraft uplink data.

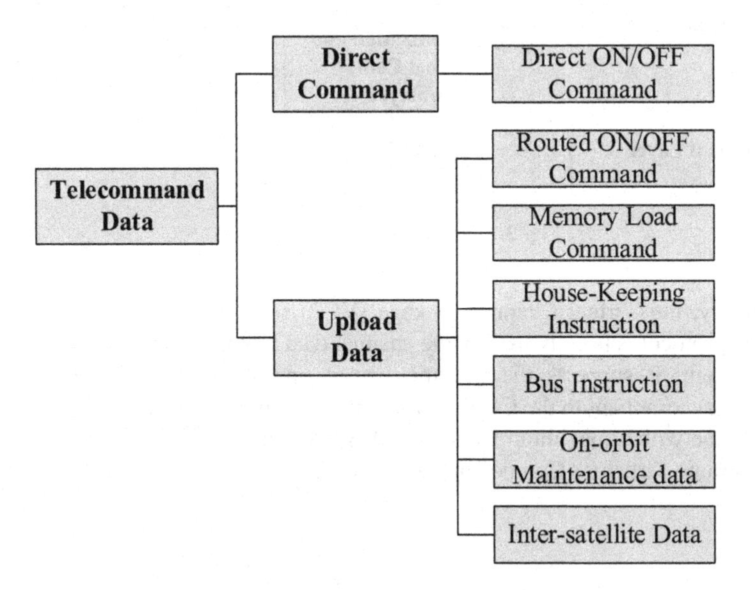

Fig. 1. Classification of spacecraft uplink data

3 Space Data Link Protocols

The space data link protocol used by domestic spacecraft mainly follows the relevant provisions of the Military Standard PCM telecommand protocol and the CCSDS (The Consultative Committee for Space Data Systems) protocol system [1–5] (corresponding Military Standard will be published soon). In recent years, newly developed spacecraft have gradually moved to use CCSDS protocol system.

CCSDS protocols currently widely used in domestic TT&C systems and spacecrafts include TC Synchronization and Channel Coding [4], TC Space Data Link Protocol [1], AOS Space Data Link Protocol [2] and Space Package Protocol [3]. Unified Space Data Link Protocol [5] is the latest space data link protocol introduced by CCSDS. As an improvement of AOS Space Data Protocol, it improves the support of spacecraft constellation and inter-satellite data transmission services in the original space link protocol. It can be used as one of the next generation space data link protocol. The following Fig. 2 shows Relationship of CCSDS Layers with OSI Layers.

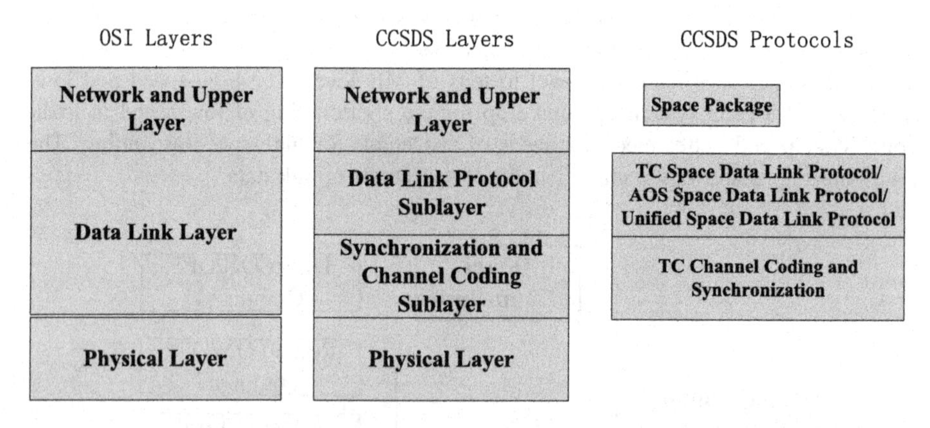

Fig. 2. Relationship with OSI layers

Currently, the Virtual Channel Package (VCP) service in the CCSDS TC Space Data Link protocol is usually used in the channel data link layer under the conventional TT&C system. It supports the transmission of one or more telecommand source packages that compliant to the CCSDS Space Package protocol in a variable-length TC transfer frame with a maximum length of 1024 bytes. The following Fig. 3 shows TC Transfer Frame Structure Component of conventional TT&C Channel.

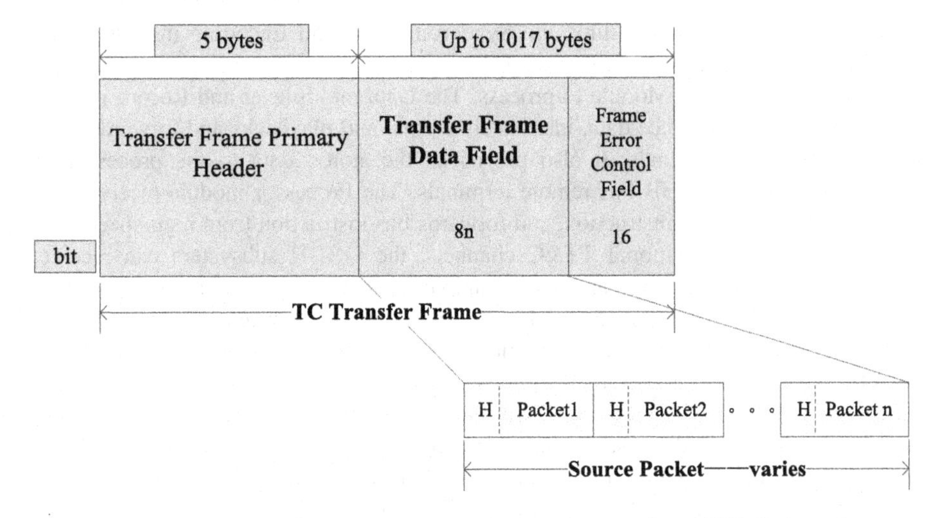

Fig. 3. TC transfer frame structure component of conventional TT&C channel

4 Design of Telecommand Scheme

4.1 Main Ideas

The main aspects to be considered in the telecommand scheme of the hybrid TT&C system are as follows:

1) The on-board data flow design should be adapted to both the high-speed and low-speed TT&C link. Considering the requirements of data encryption and decryption, data storage, on-board data distribution and inter-satellite data forwarding, barrel effect should be avoided to reduce the system application efficiency.
2) Considering the high-speed TT&C link can completely replace the conventional low-speed TT&C link to increase the redundance of TT&C link.
3) The design should consider the mature scheme of the existing system and reduce the design overhead.

4.2 Design of Data Flow

In the conventional low-speed TT&C link, the Synchronization and Encoding module of OBDH subsystem receives the telecommand PCM signal through the on-board transponders, and after channel decoding and decryption, the uplink TC transfer frame is sent to the Instruction module and the Processor module. The Instruction modules receive and decode direct ON/OFF command and output. The processor module recognizes and receives upload data. The Processor module handles upload data or forwards it to other 1553B bus remote terminals, or sends it to other mid-speed bus terminals, such as LVDS (Low-Voltage Differential Signaling) and the SpaceWire (SPW) bus terminals, via the Data multiplexer and Record module.

In the high-speed TT&C link, the Synchronization and Encoding module of the OBDH subsystem receives high-speed TC transfer frames and sends them to the Data multiplexer and Record Module to process. The Data multiplexer and Record module stores on-orbit maintenance data, inter-satellite data, and playbacks and forwards them to mid-speed bus terminals. It also playbacks the stored data to the processor or forwarding them to 1553B bus remote terminals. The Processor module receives and executes house-keeping instruction, and forwards bus instruction from high-speed link.

Through the conventional TT&C channels, the OBDH subsystem can receive, process and store direct command and all upload data. Through the high-speed TT&C channel, the OBDH subsystem can handle all upload data except direct command. Therefore, all direct ON/OFF instructions of the OBDH subsystem have routed ON/OFF instructions backup, the OBDH subsystem can also handle all telecommand functions in high-speed TT&C link. With this design, upload data can be connected to all 1553B bus and medium-speed bus terminals on the whole satellite through high-speed or low-speed TT&C channels by the OBDH computer, which improves the reliability of the system and the flexibility of data transmission. The following Fig. 4 shows the design of data flow on hybrid TT&C link.

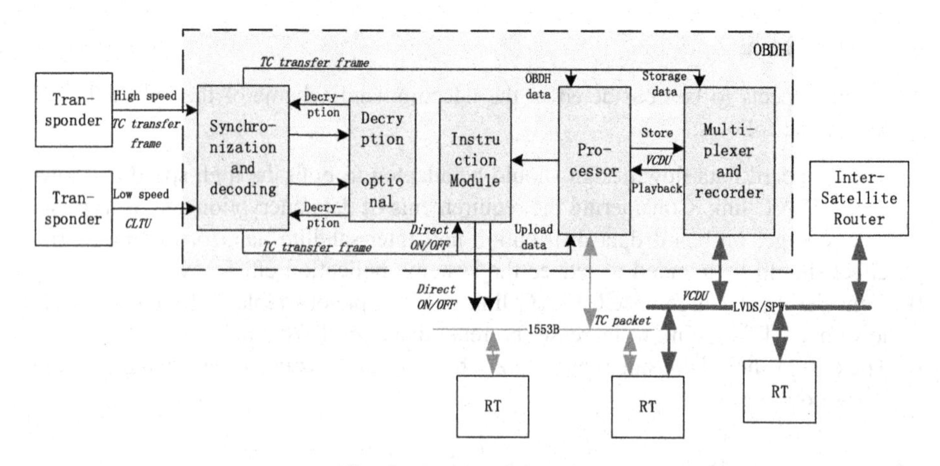

Fig. 4. Design of data flow

4.3 Design of Protocol

Considering the integration of ground data processing and on-board data processing with conventional TT&C uplink data, the TC transfer frame format of high-speed TT&C channel can be selected as CCSDS TC Space data link protocol compliant with conventional TT&C channel. Considering the length of data field required by the proprietary frame format of the high-speed uplink channel and the additional data bit overhead reserved for channel encryption and decryption, the effective data field of the TC transfer frame is set to variable length and maximum 892 bytes. The following Fig. 5 shows the TC Transfer Frame Structure Component of High-speed TT&C Channel.

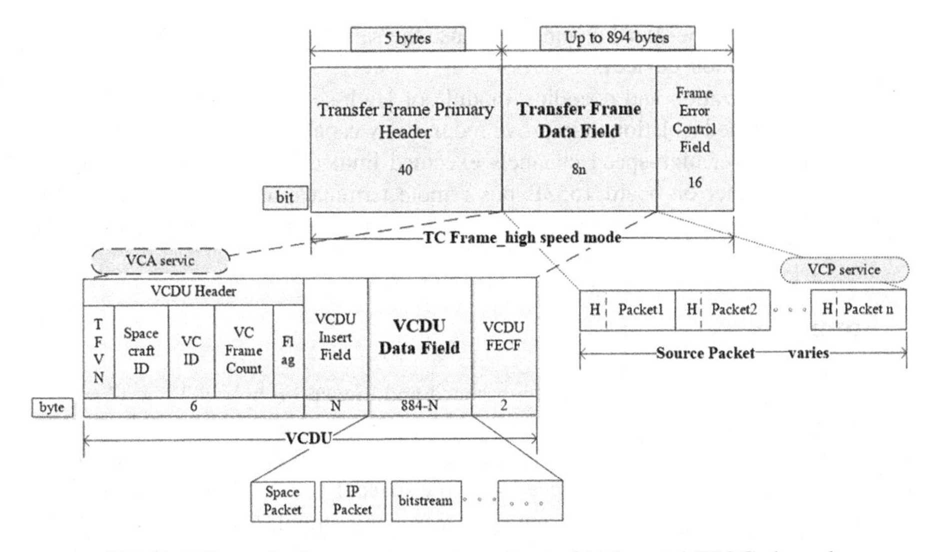

Fig. 5. TC transfer frame structure component of high-speed TT&C channel

When data is sent to the OBDH processor, the VCP service is used to transfer the TC space packet, and the effective data field of the TC transfer frame becomes no longer than 892 bytes.

When data is sent to Data multiplexer and Record Module of the OBDH, a Virtual Channel Access (VCA) service is used, and the data field of TC transfer frame is a fixed length of 892 bytes, including a Virtual Channel Data Unit (VCDU) that compliant with the AOS Space data link protocol. The services provided by this VCDU internally under the AOS Space Data Link Protocol include TC space packages, IP packages, bitstream data, user-defined data, and so on. These VCDUs can be stored and played back/forward to 1553B bus and medium speed bus such as SpaceWire and LVDS.

4.4 Further Recommendation

Further protocol recommendations and other design recommendations include:

1) The spacecraft identification of the TC Transfer Frame and the VCDU should be the same, which represents the spacecraft receiving the upload data.
2) The virtual channel identification of the TC transfer frame is used to indicate the direction of the data: Direct Command, Upload Data sent to the OBDH computer software, and upload data sent to the Multiplexer and record module. The virtual channel identification is used to indicate different data users. The value of this identification should be uniformly specified with the VCDUs of the Data Transmission channel.
3) For storing data that requires a sequence of data receipts, a Zero-start sequential count is required at the VCDU counter. The Data multiplexer and Record Module of OBDH computer will store or forward the data frame only after it receives a frame that meets the expected count.

4) High-speed and low-speed channels can use the same data frame structure to share channel decryption devices.
5) The Synchronization and decoding module of High-speed and low-speed channel are recommended isolation to improve redundancy capability and system reliability.
6) Instructions over high-speed channels executed immediately by OBDH computer software or other on-board 1553B bus remote terminals, and the data rate of the instructions should take into account the ability of the user to receive instructions.

5 Summary

In view of the rapid development of high-speed and low-speed hybrid TT & C system in recent years, this paper proposes a telecommand scheme under the existing spacecraft electronic system architecture and space data protocol framework, which combines the data processing of high-speed and low-speed TT&C channels, optimizes the data flow design, and gives an efficient and reasonable uplink data link protocol, the data fusion of high and low speed TT&C links is realized, which improves the reliability of the system and the flexibility of uplink data transmission.

References

1. TC Space Data Link Protocol. Recommendation for Space Data System Standards, CCSDS 232.0-B-2. Blue Book. Issue 2. CCSDS, Washington, D.C., September 2010
2. AOS Space Data Link Protocol. Recommendation for Space Data System Standards, CCSDS 732.0-B-2. Blue Book. Issue 2. CCSDS, Washington, D.C., July 2006
3. Space Packet Protocol. Recommendation for Space Data System Standards, CCSDS133.0-B-1. Blue Book. Issue 1. CCSDS, Washington, D.C., September 2003
4. TC Synchronization and Channel Coding. Issue 3. Recommendation for Space DataSystem Standards (Blue Book), CCSDS 231.0-B-3. CCSDS, Washington, D.C., September 2017
5. Unified Space Data Link Protocol. Recommendation for Space Data System Standards, CCSDS 132.1-B-1. Blue Book. CCSDS, Washington, D.C., October 2018

Spacecraft Electronic Datasheet Based Onboard Spacecraft Test System Design

Yanfang Fan(✉), Hongjing Cheng, Ruijun Li, and Zhang Tao

Institute of Spacecraft System Engineering, Beijing, China

Abstract. To reduce spacecraft development costs and risks, improve quality of building and test equipment, increase spacecraft software reuse and test system reuse, the paper presents how to develop a SEDS based test system that can automatically extract SOIF and protocol information from SEDS to build an efficient and organized testing environment. The test system established by this method is consistent with the concept of model-driven software organization. Practice has proved that this test system architecture can effectively use the information transmitted by SEDS to establish a spacecraft difference free test environment, which can not only save labor cost, but also boosts manufacture efficiency, and improves test quality by improve the quality of test basis input.

Keywords: Spacecraft · Spacecraft electronic datasheet · Test system · Model

1 Introduction

In the past, each satellite has its own principles to design hardware/software interface, so that, during the spacecraft design and implementation process, the interface function and compatibility between different devices and/or subsystems cannot be verified until most of integration work complete, which results in identifying potential problems. However, nowadays, the payload and platform equipped by spacecraft is becoming more and more complex, and the manufacture period requirements are getting shorter and shorter, the satellite manufacturing industry must transform from the classical type to productization.

Spacecraft manufactures tried multiple methods to achieve productization. One of the most effective tries is known as Spacecraft Onboard Interface (SOIF) [1]. SOIF is a serial of standards for the interchange of information, and the interconnection of subsystems and devices onboard of a spacecraft. The standardization will allow for the enhanced reuse of spacecraft equipment and software, which leads to the productization [2]. CCSDS has been branching out to provide new standards on SOIF for many years [3]. Recently, Chinese satellite manufacturers began to engage in this research too, and has made a lot of meaningful attempts [4]. For example, in order to mandate manufactures following and apply the SOIF, Chinese Academy of Space Technology (CAST) has developed an information system called spacecraft electronic datasheet system (SEDS) that documented the SOIF and standard protocol specifications in terms of an electronic datasheet, and provide XML file for the design data exchange and utilization.

Q. Guo et al. (Eds.): WiSATS 2021, LNICST 410, pp. 201–206, 2022.
https://doi.org/10.1007/978-3-030-93398-2_20

To maximize the productization advantage, our team developed a modular layered test system, which will firstly extract structure information from the Spacecraft electronic datasheets and manual freely create appropriate processing components, and secondly connect each component together according to the SOIF. The whole testing environment creating procedure is not only a procedure of instantiation of spacecraft datasheet model, but also a digital simulation process, which will provide verification environment for the correctness of the information exchange protocol and simulate the compatibility of each real device interface.

The paper proposed a SEDS based test system, which design processing components' interface according the stipulation of the SOIF. Simultaneously, it can simulate and verify the consistency between the realization of the onboard hardware/software and with their user requirements. The following sections will explain how the component models of the test system have been created through the SEDS, and how the test system will be established through these components. The evaluation of the trial effect of the system will be given at the end of the paper.

2 SEDS Based Modeling

In order to effectively model the information provided by the SEDS, we observe the SEDS through three orthogonal views. The first view is the Protocol view, which describes the protocols and services that are to be implemented in the onboard spacecraft system in order to provide the users with the advantages of the SOIF architecture. The second view is the Services View, which describes the data communications services that are provided to the users. And finally, the Interoperability view, which describes how the data has been exchanged between different spacecraft data busses.

From the perspective of Protocol view, the process of Onboard software development can be described as a nested process. On the contrary, the process of data processing by the test system can be modeled as a de-nesting process. The correspondence between hierarchical transmission protocol of the spacecraft and data process component model shows in Fig. 1.

The Protocol view model abstracted information flow as a list of data package, which was composed with data segments and sub-data packages. Each layer can be modeled as a data process component. The function of this component is to decompose the current package to data segments and sub-packages, and assign corresponding attributes to them.

From the perspective of Service view, each application programming interface must bond to a set of procedure and function calls to access services offered by SOIF. From this view, the underlying hierarchy is not only not visible, but not of interest. Users see only a set of APIs that are uniformly accessible from each application. Hence the test system framework needs to automatically create a runtime testing instance which is composed by a set of service components or data processing components, according to the information provided by the SEDS. Therefore, the Service view model abstracted onboard services as a set of components which correspond to the service access points exposed by the SOIF stack. This services view of SEDS Model is shown in Fig. 2.

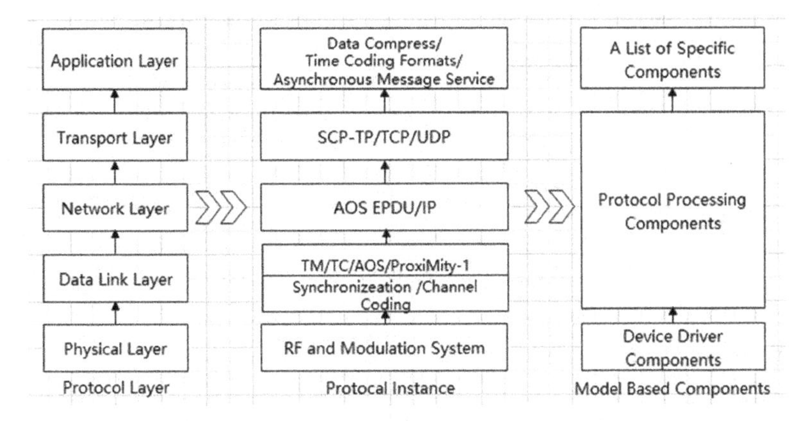

Fig. 1. The correspondence between protocol and component model

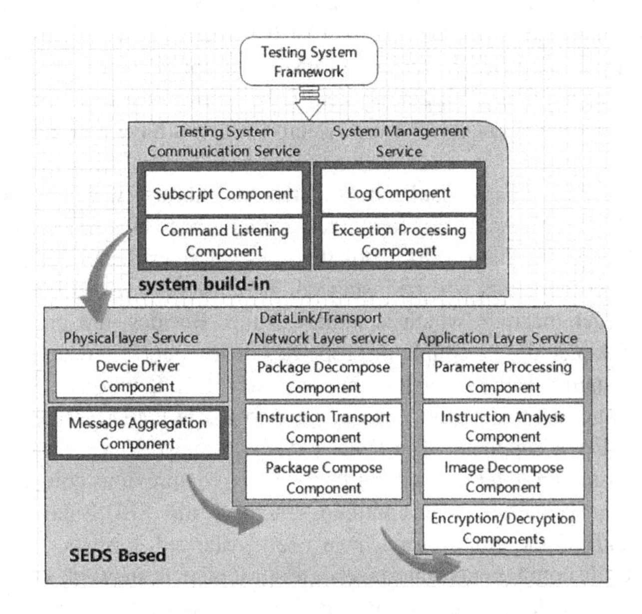

Fig. 2. The services view of SEDS model

The Interoperability view is different from the other two views, it doesn't consider what the test system can do, but how to design a SOIF compliant hardware-driver components, which will be instanced to connect any real hardware device or virtual hardware to a runtime instance. Therefore, the Interoperability view model abstract spacecraft hardware interfaces as a couple of standard hardware-driver components, and the test system framework knows how to extract information provided by the SEDS to instance an appropriate hardware-driver component to meet the needs of a particular testing request. The Interoperability view of SEDS Model is shown in Fig. 3.

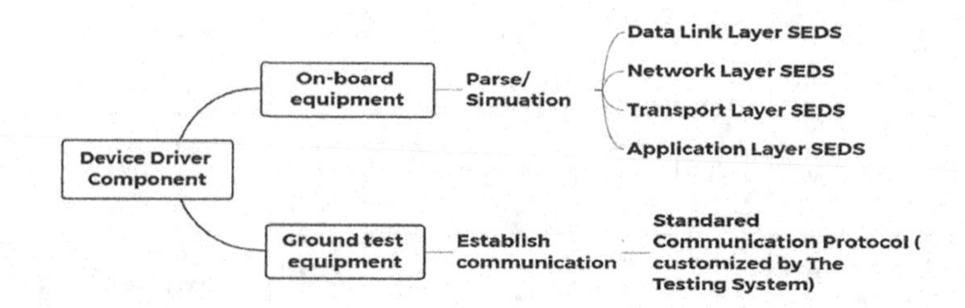

Fig. 3. The Interoperability view of SEDS model

3 System Architecture

The test system is composed of the system framework and a set of components that with different interfaces and functions.

The System framework is the core part of the test system, by which the whole system is controlled, managed, and monitored. It is mainly used to complete the tasks which including loading SEDS, managing instanced components, rendering UI, handling system exceptions and logging system logs, storing test data, etc. All the tasks that the framework is responsible for are system-level and have nothing to do with the testing business work.

Each component was encapsulated as a dynamic library, which possess different interfaces and functions. As described in Sect. 2, there are twelve different categories of components. Firstly, the framework loads the physical layer SEDS that specifies what kinds of device components will be instanced. The interface would correspond to the SOIF of the target machine which is under testing. Besides, the interfaces of each component are depicted by a couple of properties called interface descriptor, including interface type, name, and especially, the successor SEDS unique identify number. This SEDS unique identify number will be used to locate the successor component of the current device component. Then the system framework will instance the next component whose input interface will perfectly match the previous component's output interface. Similar processes will continue, until all the SEDS have been loaded sequentially, and all the components have been instanced correspondingly. Finally, according to each components' interface descriptor, the framework connects all the components together. In other words, as long as the components are instanced based on the SEDS, the test system would run exactly as the reverse sequence of the onboard system runs. The process of establishing a test instance is shown in the Fig. 4.

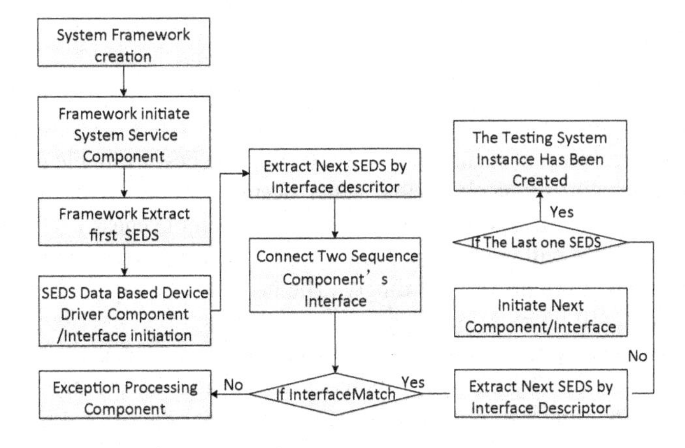

Fig. 4. Schematic diagram of establishing a test instance

4 Experiment and Effects

The SEDS based test system has some distinct properties as follows:

1. No operator is needed in front of the test system establishment.
2. Subjectivity aroused by human data reading is avoided.
3. Realize a high efficiency, one-key automatic test system establishment.

Typical application tests are carried out in integrated electronic subsystems functional tests of more than 10 spacecrafts of five different fields. For application layer, currently SEDS includes five types of parameters, which are analog signal collection, Bi-lever electronic signal collection, temperature variable collection, digital signal collection, etc. Each parameter has properties in SEDS, such as name, Code identity, data type, unit, length, byte order, processing function, calibration, etc. For transport Layer, currently SEDS includes 1553B bus transmission protocol. For Network layer, currently SEDS includes AOS EPDU protocol. For Data link layer, currently SEDS includes telemetry, telecontrol, AOS, BCH coding, etc. With the guidance of SEDS, the test system can correctly instance all the service component in order. Each component's interfaces can correctly reflect design requests of the target machine's interfaces. The results of experiments show that the test instance composed by these component instances can satisfy the functional test requests, and can provide all kinds of test data products that, from all levels, prove the system implementation correctness.

In addition, as described above, the functions and interfaces of each component are designed in accordance with the common requirements of spacecraft development and the standard interface specifications. The test instance established based on SEDS will work correctly with no interface mismatch faults, as long as there are not any design faults in the SEDS. However, exception warnings are often thrown out during the instantiation process, which happens because the instanced components with interface design faults cannot be connected as expected. Through the way, the test system is helpful in early verification of the spacecraft system interface design.

5 Summary

By performing of SEDS, not only the spacecraft design procedure, but also its testing procedure benefits from the SOIF.

Firstly, this capability of identifying and selecting the best available data source(s) for an application will ultimately enable better design support. With the continuous progress of the spacecraft development industry, the completeness of the spacecraft development standards will make great progress too. Such, the test system components design will continue to improve accordingly. The benefits of early verification feature of the test system will be more obvious.

Secondly, the test system can automatically create a test instance just according to the information provided by the SEDS, without and manual intervention.

Finally, by extracting appropriate interface or protocol data from SEDS, the test system framework can instance matchable device components that can effectively exchange test data between the target machine and the test system, without and additional hardware interface adaption or adjustment.

All these features mentioned above will reduce the costs associated with integration and test, and potentially all the testing case are reusable in further by a similar spacecraft.

References

1. Smith, J.F.: The spacecraft onboard interface standardization activity. In: IEEE Space OPS 2003 Conference, pp.1–7 (2003)
2. Hansen, L.J., Lanza, D., Pasko, S.: Developing an ontology for standardizing space systems data exchange. In: Big Sky Aerospace Conference, IEEE and AIAA, Big Sky, MT (2012)
3. Drenkow, G.: Future test system architectures. In: IEEE A&E Systems Magazine, pp. 27–32 (2005)
4. Jian, G., Ruijun, L., Yanfang, F., Xiangdong, S.: Digital collaborative design method for spacecraft information flow. Spacecr. Eng. **2**, 59–65 (2020)

Design of Universal Software Architecture for Spacecraft Autonomous Thermal Control

Tian Lan(✉), Zhenhui Dong, Ma Zhu, Hongjun Zhang, and Wenjuan Li

Beijing Institute of Spacecraft System Engineering, Beijing 100094, China

Abstract. The traditional design method of spacecraft autonomous thermal control software is difficult to reuse because of the tight coupling between software logic and corresponding user requirements. Based on the present and foreseeable autonomous thermal control requirements, a universal software architecture is proposed. The process of thermal control is divided into five steps that include temperature acquisition, temperature preprocessing, duty cycle calculation, energy optimization and on-off switching of heater. The generalized software components are realized. Different requirements of different spacecraft can be met by setting different parameters and assembling software components. The engineering practice proves that proposed universal autonomous thermal control software architecture can significantly reduce the time cost of software development, and has excellent adaptability to different requirements. Proposed software architecture has been applied to some agile remote sensing satellites and deep space probes, and it can provide a reference for future spacecraft software design.

Keywords: Spacecraft · Avionics · Autonomous thermal control · Software architecture · Universal design

1 Introduction

As a basic function of the spacecraft, effective thermal control is essential to ensure proper functioning of the devices in spacecraft. Thermal control can be divided into two categories: passive thermal control and active thermal control. Passive thermal control takes coating, heat pipe, multi-layer thermal insulation material and phase change material as control means, and active thermal control takes shutter, electric heater, fan and fluid loop as control means [1]. Under the guidance of agile remote sensing satellites, manned spaceflight, deep space exploration and other missions, China's spacecraft thermal control technology has made great progress [2, 3]. Based on passive thermal control, active thermal control has become a key mean to meet the requirements of complex tasks in various fields [4–6].

With the evolution of intelligent spacecraft, it is not necessary to set a specific device for thermal control. The common choice of current spacecraft is to integrate the thermal control function into the onboard data handling (OBDH) system [7]. With the construction of space-based information network, the performance improvement of spacecraft has generated an urgent need for precise thermal control; traditional OBDH

Q. Guo et al. (Eds.): WiSATS 2021, LNICST 410, pp. 207–216, 2022.
https://doi.org/10.1007/978-3-030-93398-2_21

system is insufficient to accomplish more and more heavy workload [8]. On the other hand, the new missions require that the information system of spacecraft be promoted to the direction of information-processing-centered and multi-spacecraft collaborative work [9].

To satisfy the urgent need of intelligent and networked spacecraft, the avionics system is rapidly evolved. After decades of development, China's avionics system has established a business and protocol framework that selectively applied CCSDS and ECSS standard [10–12]. Under the guidance of demand analysis and overall design, a centralized avionics system [13] was designed for micro and small spacecraft and a distributed avionics system [14] was designed for medium and large spacecraft. On the basis of intelligent avionics system and according to the principles of universalization, intelligentialize and networking, the development of avionics software based on software components is promoted [15, 16].

Electric heater is widely used in active thermal control. Based on intelligent avionics system, thermal control of spacecraft is evolved towards the direction of precision, intelligence, and joint optimization design with other fields. The joint optimization design can be further divided into two directions: one is to ensure energy balance; another is to optimize the impact of thermal control on energy. By adjusting the number of heater channels involved in thermal control, the peak power can meet the energy constraints. Zhang H B et al. proposed that when predicting spacecraft energy imbalance, the thermal control should be transferred to the minimum mode to reduce the number of active heaters [17]. Wei Y Q et al. proposed to reduce the peak power by time-sharing batch switching of heaters when judging the total thermal control power is at risk of exceeding the energy constraint [18]. In active thermal control, the on-off of heaters cause the fluctuation of the load power of battery. Lan T et al. propose using duty cycle mode to control heater's on-off, and rationally arrange the heaters on-off combination in each time slot within a control period to suppress the fluctuation of power, so as to relieve the negative influence of thermal control on voltage and current [19]. Those two design directions are not in conflict with each other and can be used in thermal control simultaneously.

Though there are various joint optimization designs, the thermal control software architecture of current spacecraft has not been unified yet. Causing by the strong coupling of the code, traditional thermal control software needs a lot of changes when the functions are changed, upgraded or expanded. The amount of Chinese spacecraft in orbit will grow significantly in the foreseeable future. Generalized design of thermal control can effectively help the design and implement of onboard software, promote flight support effect and increase the on-board maintenance capability of avionics system. To improve the scalability and portability of thermal control software, an universal architecture based on software components is proposed. According to the existing thermal control requirements and the "energy, thermal control and avionics" joint optimization needs, the influence of control elements on thermal control performance is analyzed. Based on avionics system, an universal software architecture compatible with existing and foreseeable spacecraft optimization needs is implemented and testified.

2 Analysis of Control Elements of Active Thermal Control

To accomplish precise thermal control, the heat capacity of control object and thermal environment should be planned. To ensure the effective implementation of the thermal control strategy, it is necessary to analysis and optimizes the performance of control elements, such as temperature accuracy, thermal control cycle, thermal control time slot, heater switching mode, energy constraints and software control process.

2.1 Temperature Accuracy

The temperature accuracy depends on both the acquisition accuracy and the processing of raw data. To improve acquisition accuracy, AD converter with better acquisition accuracy and stability should be selected. Traditional OBDH system saves the control threshold directly in the form of raw data, compares the raw data with the threshold, and decides the on-off state of heater according to comparing results. In the solution, the temperature accuracy is greatly affected by the voltage stability of the acquisition circuit. If the tension voltage of the acquisition circuit is changed, the stratification value corresponding to the same temperature would be changed accordingly. As the corresponding threshold remains unchanged, the voltage fluctuation will lead to the change of the actual temperature threshold used for control, and affecting the control effect. As an effective way to solve the problem, the raw data can be converted into temperature according to the homologous calibration voltage of the acquisition circuit.

2.2 Control Period and Time Slot of Thermal Control

Thermal control period refers to the time interval between two adjacent executions of thermal control logic to update the heater on-off state and duty cycle. Thermal control time slot refers to the time interval of switching the heater on-off state according to the duty ratio within a control period. The thermal control period should be an integer multiple of the thermal control time slots. A thermal control period should contain at least one thermal control time slot.

The control period influences the responding speed of temperature change, and the amount of time slot determines the granularity of the duty cycle. More precise thermal control effect requires shorter control period and more time slots. The foundation to realize shorter control period is higher computing power, and the foundation to realize more time slots is heater batch switching capability. Those two aspects demands promote the transformation from traditional OBDH system to avionics system.

2.3 Switch Mode of Heater

When the heater needs precision control in a control period, the duty cycle control should be applied. Traditional OBDH system uses the pulse command to control the on and off of the heater. Subject to the instruction sending ability, it can only realize duty ratio temperature control in a continuous way, and the control effect is limited by the command sending ability. To reduce the temperature fluctuation of the controlled object, the batch switching of heater on-off state is a good choice. In current avionics system, the pulse command is replaced by electronic switch, which has the ability to batch switch the heater on-off state rapidly and effectively.

2.4 Energy Constraints

Electricity is the main energy source of spacecraft. The thermal control power consumption influences the energy balance of spacecraft. In current designs, the constraints of energy on thermal control are mainly reflected in the energy balance and peak power consumption. Most of those designs need to forecast the energy imbalance by testifying whether the thermal control actions can meet energy constraints. When the spacecraft energy imbalance is predicted, thermal control should be transferred to the minimum working mode to decrease the number of active heaters and reduce power consumption, or adopted to time-sharing switch mode to reduce the peak power consumption.

3 Universal Architecture for Active Thermal Control

3.1 Universal Software Architecture Based on Software Components

Based on the commonness and characteristics of the existing solutions of spacecraft active thermal control, the key steps of thermal control are summarized.

Temperature Raw Data Acquisition. Temperature acquisition is generally realized by thermistor measurement circuit. The acquisition circuit that supports homologous calibration can meet the control requirements of all existing spacecraft.

Temperature Raw Data Preprocessing. Different spacecraft uses temperature in different ways: using raw data directly, using converted data based on thermistor coefficients, or using statistics of multiple thermistors, such as maximum, minimum or average. The software architecture needs to support various patterns of temperature processing needs.

Thermal Control Logic. Thermal control logic includes generalized logic and personalized logic. Through related control logic, software determines how to perform on-off control or duty cycle control of the heater based on temperature. Generalized control logic includes On-Off control algorithm, proportional control algorithm, proportional integral (PI) control algorithm and proportional integral differential (PID) control algorithm. Personalized control logic involves collaborative control between multiple heaters, such as specialized control of fuel tanks and thrusters.

Energy Optimization Constraints. From the percept of energy, the influence of the constraints would be reflected in heater's on-off. When the predicted peak power exceeds the constraints, the power used by thermal control can be reduced by switching off some heaters or reducing the heating time.

Heater Switch Control. The control of heater can be classified into two categories, on-off control and duty cycle control. The on-off control can be implemented as a special case of duty cycle control, so the duty control mode can be a universal solution.

Based on the analysis of the above five elements, the universal software components are implemented in temperature raw data acquisition and heater switch control, and the configurable software components are introduced in the other three elements: temperature raw data preprocessing, heater control logic and energy optimization design. Spacecraft with different backgrounds can insert, modify or expand the software components to achieve specific intention.

3.2 Hardware Composition of Avionics System

As a basis of proposed software architecture, the avionics system consists of System Management Unit (SMU) and Data Interface Units (DIU), as shown in Fig. 1. The devices adopt universal module design; DIU has the same computing power as SMU. Data exchange among modules is realized through the internal bus inside device, and communication between devices is implemented by intra-satellite data bus.

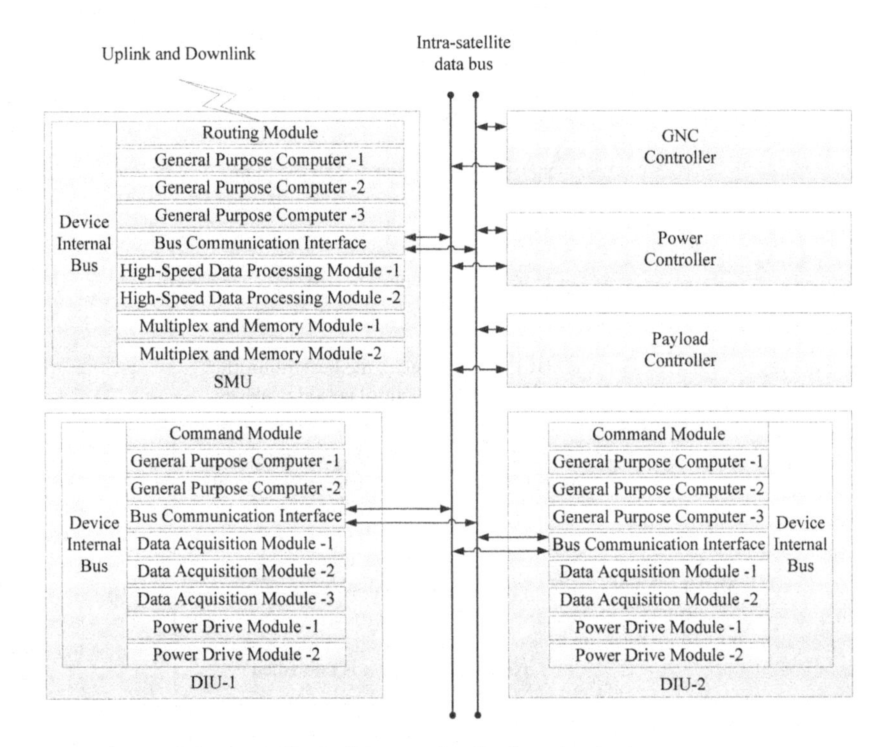

Fig. 1. A block diagram of a distributed avionics system

The applications of high performance General Purpose Computer (GPC) module achieve the balance of distributed computing ability available system-wide. The task division between SMU and DIU is optimized, more complex control tasks are assigned to the DIU within computing power support range. For example, the autonomous thermal control is realized by DIU, and the control logic is running on the first GPC module of each DIU. The raw temperature and homologous calibration voltage are collected by the Data Acquisition Module (DAM), the heater switch is controlled by the Power Drive Module (PDM), and the safety switch is controlled by the Command Module (CM).

The independent PDM ensures that DIU can batch switch the heater's on-off state in a short time slice. With its support, the action of heaters in thermal control can be described as a matrix indexed by heater and time slot. Each element in the matrix represents the on-off state of heater in a certain thermal control time slot. The action of thermal control can be abstracted as updating the matrix according to the temperature periodically, and switching the heater's state according to the matrix.

3.3 Control Process of Software Architecture

The control process of proposed architecture is shown in Fig. 2.

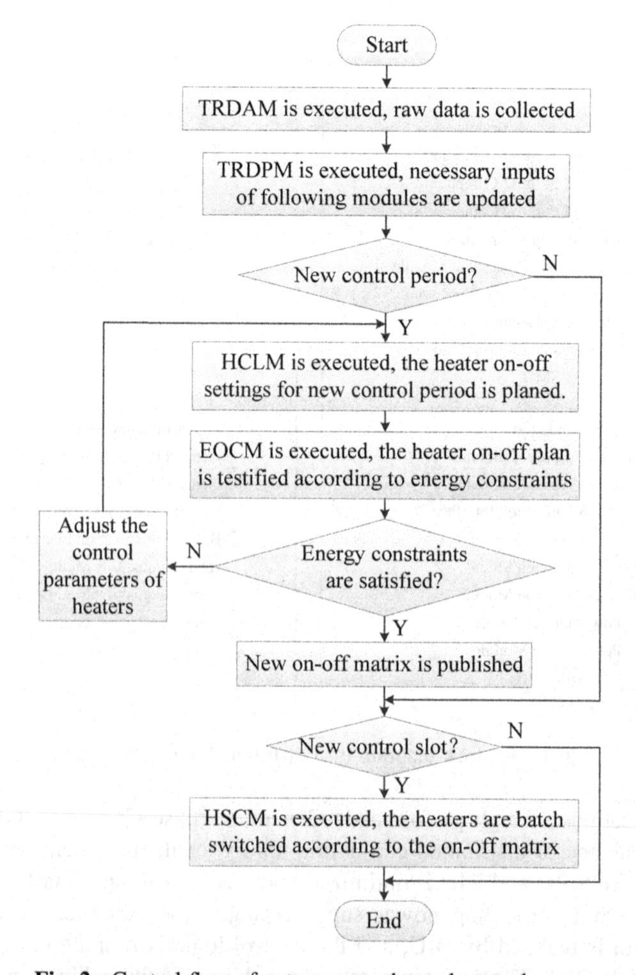

Fig. 2. Control flow of autonomous thermal control process

As shown in Fig. 2, the autonomous thermal control process is called periodically by external logic as a process, and the internal process is controlled according to the schedule cycle. According to the control object, the process can be further divided into the control of the thermistor and the control of the heater. The control of thermistor is divided into two modules: the Temperature Raw Data Acquisition Module (TRDAM) and the Temperature Raw Data Preprocessing Module (TRDPM). The control of heater is divided into three modules: heater control logic module (HCLM), energy optimization constraint module (EOCM) and heater switch control module (HSCM).

Every time the autonomous thermal control process is called, the TRDAM is first executed. In the TRDAM, the original data needed to be used in thermal control is updated. To eliminate the coupling, an independent telemetry acquisition process is designed to complete the data acquisition, and the collected data is stored in the global data pool. When the thermal control process runs, it directly obtains the latest raw data from the global data pool and refreshes it to the thermistor data structure.

After the TRDAM, the TRDPM is executed. In TRDPM, configurable software components are provided. Users can use different plugins to achieve raw data format conversion, data statistics, thermistor health monitoring and other functions according to personalized needs. Virtual temperature can be generated according to the statistical results for subsequent heater control. Through those two modules, the processing requirements of thermistor in existing spacecraft autonomous thermal control can be met, and the input data needed for heater control can be obtained.

The input of HCLM is the control state of the heater and the output temperature of the thermistor, and the output of HCLM is the switching matrix of the heater in the next control cycle. As an example shown in Fig. 3, the matrix illustrates the distribution of switching states of 10 heaters in a control period divided into 10 time slots. According to the switching states of each time slot and the power consumption of corresponding heater, the total power consumption of every time slot can be obtained.

Heater No.	Power consumption /W	Time slots in a control period									
		1	2	3	4	5	6	7	8	9	10
1	10	ON	ON	ON	ON	ON	ON	ON	ON	ON	ON
2	10	ON	ON	ON	ON	ON	ON	ON	OFF	ON	ON
3	10	ON	ON	ON	ON	ON	ON	ON	ON	OFF	OFF
4	10	OFF	ON	ON	ON	OFF	ON	OFF	ON	ON	ON
5	5	ON	OFF	OFF	OFF	ON	OFF	ON	ON	ON	ON
6	5	OFF	OFF	OFF	OFF	ON	OFF	ON	ON	ON	ON
7	13.1	ON	OFF	OFF	OFF	OFF	OFF	OFF	ON	ON	ON
8	10.6	OFF	ON	ON	ON	OFF	OFF	OFF	OFF	OFF	OFF
9	11.9	OFF	OFF	OFF	OFF	ON	ON	OFF	OFF	OFF	OFF
10	12.1	OFF	OFF	OFF	OFF	OFF	OFF	ON	OFF	OFF	OFF
Total power consumption/W		48.1	50.6	50.6	50.6	51.9	51.9	52.1	53.1	53.1	53.1

Fig. 3. Heater switching state matrix compatible with duty cycle control

The EOCM takes the total power consumption of each time slot and the energy constraints as inputs, judge whether the power consumption can meet energy constraints or not, and outputs the switching choice of the heaters when current power consumption could not meet the energy constraints. When the power consumption is at risk of exceeding the energy constraints, the control state of the heater is switched according to the corresponding strategy, and the HCLM needs to execute again.

If the power consumption of the switch matrix output by the HCLM can meet the energy constraints, the HSCM is executed. In HSCM, the heaters are batch switched according to the on-off states in switch matrix through the interface with the power drive module.

In the three modules, HCLM and EOCM provide configurable software components support and run once each control cycle. The HSCM runs once each time slot. Through cooperation of those three modules, the control needs of the heater in spacecraft autonomous thermal control can be met.

4 Analysis and Evaluation of Application Effect

The development of thermal control software in four satellites is analyzed and evaluated. Satellite 1 adopts the traditional thermal control software architecture; Satellite 2 inherits the universal thermal control logic of Satellite 1 and upgrades the personalized control logic. Satellite 3 adopts the design of proposed universal thermal control software architecture, and Satellite 4 is developed base on the software architecture of satellite 3. In Satellite 4, the special thermal control logic is changed, and a new energy constraint strategy is applied. The utilization data of software development of the four satellites are shown in Table 1.

Table 1. Utilization data of spacecraft thermal control software architecture

Satellite No.	Lines of software components reused	Lines of application layer software	The number of global variables and arrays	The number of structures	Development cost/(man-day)
1	0	5167	150	18	60
2	0	5696	126	19	25
3	7591	1288	23	0	36
4	7591	1430	29	0	5

Compared with the traditional architecture of spacecraft thermal control software, proposed software architecture has achieved significant improvement in the development efficiency, and the software complexity is effectively controlled. The application effect of proposed software architecture in spacecraft software development process is shown in Fig. 4.

As shown in Fig. 4 (a), the software development efficiency of Satellite 3, which adopted the universal thermal control software architecture for the first time, showed no significant change compared with previous satellites. Compared with Satellite 2, which inherited the traditional thermal control software development mode of Satellite 1, the development efficiency of Satellite 4, which continued to apply the software architecture of Satellite 3, was greatly improved. The development cost is reduced by 66.5% and the efficiency is increased about 3 times.

As shown in Fig. 4 (b), the number of global variables and arrays required by the thermal control software of satellite 3 and satellite 4 based on software components is significantly reduced compared with that of the traditional thermal control software. The number of global variables and arrays of satellite 4 is only about 23% of the traditional thermal control software in satellite 2. The thermal control related structures of satellite 3 and satellite 4 are encapsulated in the software components. The software components contain a total of 10 structures, nearly half of the structures used by traditional thermal control software. After the adoption of proposed universal thermal control software architecture, the development efficiency of satellite 3 and 4 is greatly improved. At the same time, the robustness of thermal control is also improved.

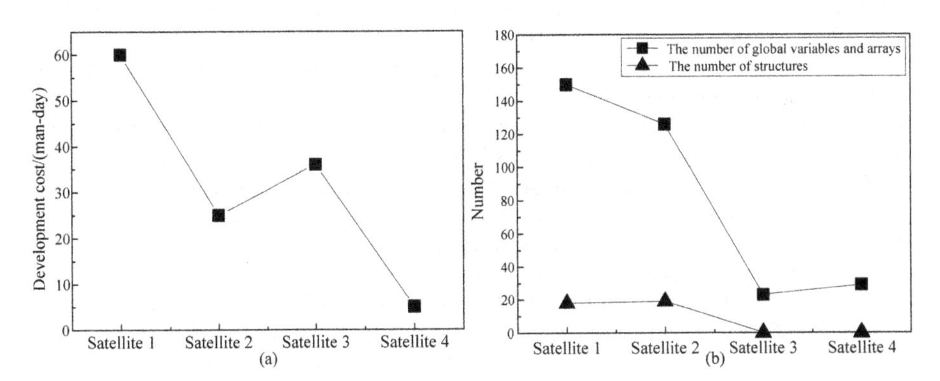

Fig. 4. Comparison of software development efficiency and complexity of thermal control in different satellites

5 Conclusions

Based on the envelopment analysis of existing spacecraft thermal control requirements, a universal autonomous thermal control software architecture is proposed. The application effect is statistically analyzed after the engineering practice. Research shows that the configurable software components in the preprocessing of raw temperature, the control logic and the energy optimization, can realize common control logic of thermal control and can effectively support the personalized thermal control needs, energy constraint strategies of different spacecraft.

The application of proposed software architecture does not increase the occupancy rate of system resources. In the process of software development, the generalizable and configurable software architecture can significantly improve the efficiency of software development, reduce the complexity of code design and improve the robustness of code. Proposed software architecture has been applied to several agile remote sensing satellites and deep space probes, and has obtained positive effect, it can be a reference for future spacecraft software design.

References

1. Hou, Z.Q., Hu, J.G.: Spacecraft Thermal Control Technology-Theory and Application, 1st edn. China Science Technology Press, Beijing (2007)
2. Ning, X.W., Li, J.D., Wang, Y.Y., et al.: Review on construction of new spacecraft thermal control system in China. Acta Aeronautica et Astronautica Sinica **40**(7), 022874 (2019)
3. Jia, Y., Liu, Q., Xiang, Y.C., et al.: The role of deep space exploration in promoting spacecraft thermal control technologies. Spacecr. Environ. Eng. **33**(2), 115–120 (2016)
4. Li, C.L.: Research on space optical remote sensor thermal control technique. J. Astronaut. **35** (8), 863–870 (2014)
5. Man, G.L., Cao, J.F., Meng, F.K.: Research on a thermal management system for docking spacecraft combination. Spacecr. Eng. **20**(6), 32–37 (2011)
6. Xiang, Y.C., Chen, J.X., Zhang, B.Q.: Thermal control for jade rabbit rover of Chang'E-3. J. Astronaut. **36**(10), 1203–1209 (2015)
7. Guo, J., Chen, Y., Shao, X.G.: Intelligent control technology for spacecraft thermal autonomous management. Spacecr. Eng. **21**(6), 49–53 (2012)
8. Tong, Y.L., Li, G.Q., Geng, L.Y.: A review on precise temperature control technology for spacecraft. Spacecr. Recov. Remote Sens. **37**(2), 1–8 (2016)
9. Zhao, H.P.: To build a highway to spacecraft intelligentization with avionics technology. Spacecr. Eng. **24**(6), 1–6 (2015)
10. He, X.W.: Service and protocol architecture design of spacecraft avionics system. Spacecr. Eng. **26**(1), 71–78 (2017)
11. He, X.W., Li, N., Xu, Y., et al.: Requirements analysis of intelligent spacecraft avionics system and discussion of its architecture. Spacecr. Eng. **27**(4), 82–89 (2018)
12. Liu, S., Wang, H.M.: Overall design technology of integrated electronic system configuration for next generation satellites. Space Electron. Technol. **12**(6), 90–94 (2015)
13. Zhan, P.P., Cao, Y.T., Zhang, C.T., et al.: High functional density avionics system for satellites. Chin. Space Sci. Technol. **40**(1), 87–93 (2020)
14. Liu, W.W., Cheng, B.W., Wang, L.Y., et al.: Design of distributed spacecraft avionics system. Spacecr. Eng. **25**(6), 86–93 (2016)
15. Zhang, Y.H., Yuan, J., Guo, J.: Design and reconfigurable general tm based on software components. Spacecr. Eng. **22**(4), 62–67 (2013)
16. Zhang, Y.H., Yuan, J., Yu, J.H., et al.: Design of spacecraft OBDH software framework based on onboard route. Spacecr. Eng. **24**(6), 70–74 (2015)
17. Zhang, H.B., Pan, Y.Q., Feng, W.J., et al.: An intelligent control strategy of spacecraft electric heating. Spacecr. Eng. **25**(4), 48–53 (2016)
18. Wei, Y.Q., Guo, J., Zhang, H.J., et al.: An autonomous thermal control method for spacecraft based on power balance. PRC Patent CN202011068753.8 (2020)
19. Lan, T., Mu, Q., Jiang, L.F., et al.: Power fluctuation suppression method for spacecraft autonomous thermal control. Spacecr. Eng. **30**(1), 72–78 (2021)

A Network Traffic Collection System for Space Information Networks Emulation Platform

Chengpeng Kuang, Dongxu Hou, Qi Zhang, Kanglian Zhao(✉), and Wenfeng Li

School of Electronic Science and Engineering, Nanjing University, Nanjing 210023, People's Republic of China
{mf1923032,mg1923075}@smail.nju.edu.cn, zhaokanglian@nju.edu.cn

Abstract. This paper presents a network traffic collection system for space networks emulation platform. The system uses virtualization technology and Elasticsearch database technology to collect the network traffic of gigabits per second. On this basis, a parallel processing method is proposed to further utilize the resources and improve the collection performance. Corresponding experiments are carried out to evaluate the collection performance of the system. The experimental results show that the system can collect massive and high-speed network traffic in real time and improve the collection rate.

Keywords: Space information networks emulation · Network traffic collection · Elasticsearch · Parallel processing

1 Introduction

The space information networks system is a network system which takes the space platform as the carrier to acquire, transmit and process spatial information. It is an indispensable guarantee to serve the national economic construction and national security. In order to verify the relevant space technologies and networking communication protocols, it is necessary to develop an emulation platform to emulate the space networks application scenarios.

The Nanjing University team has constructed a laboratory scale network emulation platform which applies virtualization technologies to emulate space information networks [1]. As shown in Fig. 1, the space networks emulation platform is divided into four planes, including the logic plane, the control plane, the data plane and the analysis plane. The logical plane is responsible for the actual network scenario. The control plane receives the parameters of the logic plane and creates the emulation scene. The data plane generates real data flow between network nodes to ensure the authenticity of emulation. The analysis plane defines the specific parameters of the network according to the actual situation of the network. And it monitors, statistics and feeds back the emulation results.

Q. Guo et al. (Eds.): WiSATS 2021, LNICST 410, pp. 217–225, 2022.
https://doi.org/10.1007/978-3-030-93398-2_22

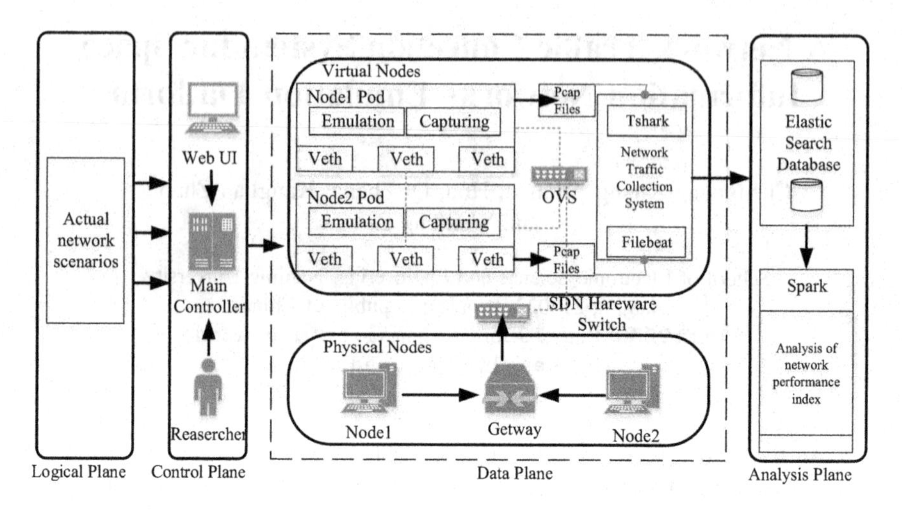

Fig. 1. Architecture of space information networks emulation platform.

In order to evaluate the performance of space networks scenarios, it is necessary to conduct in-depth analysis of traffic data in the emulation network environment. Network traffic collection system is an important part of the emulation platform. Under the limited platform resources, the network traffic collection system must be able to meet the collection, storage and analysis requirements of massive high-speed traffic, such as the acquisition of gigabit traffic per second. The system can analyze the specific protocol to facilitate the subsequent analysis. At the same time, the network traffic collection system should also have the functions of high stability, expandability, easy maintenance and easy recovery after failure.

Network collection is actually network measurement. Network measurement includes active measurement and passive measurement. Researchers have proposed a variety of network data collection methods and developed the corresponding system for different networks. The mainstream network collection system is roughly divided into three categories. The first is the network collection system based on SNMP protocol [2]. The basic principle is to exchange MIB table information between the management terminal and the managed terminal through the request answer mode of SNMP protocol [3]. However, SNMP cannot distinguish the proportion and distribution of different types of network services in the total network traffic, and the collected network traffic information is very rough. The second network collection system is based on flow technology [4]. In the network device that starts the flow mode, the data packets will be recorded and counted according to their own characteristics [5]. But this way of working is to use hardware technology to achieve fast forwarding of network data, which is not suitable for network emulation. The third collection system is based on the probe capture network [6], mainly through the use of data collection tools as a probe to monitor the specified network card [7], directly capture network messages for analysis, which is widely used to analyze the low-speed traffic between servers or workstations in simple network. The above methods cannot meet our needs. For the massive high-speed traffic collection of the emulation platform, container technology and big data technology are used to collect and store all data packets in real time, as well as parse and filter the traffic at the same time.

This paper uses virtualization container technology to share network namespace, and uses Tshark to capture and parse network traffic [8]. Using big data technology and Elasticsearch database to realize data storage function, a method of parallel collection is proposed. At the same time, reasonable allocation of system resources, distributed deployment, load balancing through file sharing, network traffic collection performance of the emulation platform is further studied.

The rest of this paper is organized as follows. In Sect. 2, we will introduce the realization of network traffic collection system architecture. In Sect. 3, we introduce the optimization of system parameters and the parallel processing method. In Sect. 4, we do experiments with the parallel processing method and analyze the experimental results. At last, the conclusion is given in Sect. 5.

2 Architecture of Network Traffic Collection System

The network traffic collection system is integrated in the data plane of the space networks emulation platform. First of all, container technology and big data technology are used. Based on the ELK architecture [9], the collection system can capture, parse and store network traffic data. Secondly, in order to further improve the collection performance, the system parameters are optimized and a parallel processing method is proposed.

The architecture of our designed system is illustrated in Fig. 2, which can be divided into network traffic capturing module, network traffic parsing module, network traffic filtering module, network traffic storage module and visualization module.

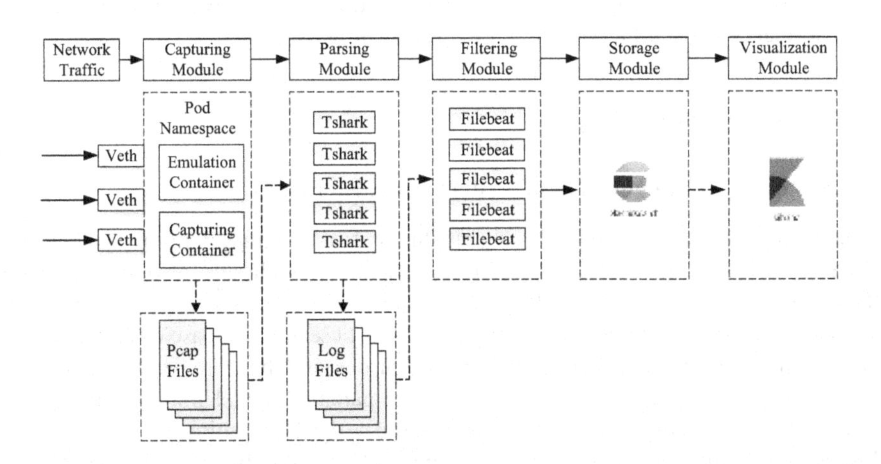

Fig. 2. Architecture of network traffic collection system.

Network Traffic Capturing Module: This module is to ensure that the network traffic data is completely captured without loss of packet, so as to achieve flexible and configurable network capture. Each virtual node of the emulation platform includes an emulation node and a capture node. The emulation container is responsible for the emulation of network protocols. The capture container is responsible for the capture of network traffic. Docker containers share network resources [10], and the network traffic processed by the emulation container is visible to the capture container. To provide extensibility, the container is managed by orchestration, for example, Kubernetes [11].

Network Traffic Parsing Module: All captured packets have a lot of redundant information. This module implements the parsing function. It parses out the fields we need, such as source IP, destination IP, packet length, protocol type, and other useful information, and writes them to the specified log file. For example, Tshark is primarily used for packet capture and analysis in a command-line environment, where we listen on container ports and implement parsing.

Network Traffic Filtering Module: This module is mainly responsible for filtering information according to rules. The information is written into the database in the form of fields to facilitate the direct use of queries after taking out. Filebeat is used to collect log data [12]. It realizes the function of preprocessing and can continuously collect new contents of log files and write them into the database.

Network Traffic Storage Module: This module stores useful information for subsequent analysis. It requires full utilization of storage resources. Data is easy to deploy, manage, extend, and query, such as Elasticsearch databases [13].

Visualization Module: Data visualization helps effectively understand and analyze massive amounts of data, which is at the end of the system. Kibana [14], with the help of the powerful search engine Elasticsearch, provides columns that can be used for retrieval and can realize analysis and positioning of massive data.

The above modules work together to finally store traffic data in the database in JSON format. Data can be written and displayed in real time.

3 Optimization Schemes

3.1 Optimized Configuration of System Parameters

We hope that the network traffic filtering and the write rate of the enclosure can keep up with the conversion resolution rate of traffic capture module and resolution module Tshark. However, for the default Filebeat and Elasticsearch cluster configuration parameters, the index rate of data writing to the Elasticsearch database cannot meet the requirements of emulation. While Filebeat and Elasticsearch have no fixed optimization method, so we need to study how to configure parameters and improve the speed. Emulation experiments were carried out to discuss the best settings of Filebeat and Elasticsearch in the system. Performance is judged by the number of packets processed per second.

The experimental scenario is shown in Fig. 3. Two emulation nodes are used. Each node is composed of an emulation container and a traffic collection container. They share a network namespace. The two nodes send 1 Gbps packet traffic. The collection container collects, captures and filters the traffic, and writes it to Elasticsearch database. By recording the rate of each operation step, we observe the performance under different parameter configurations.

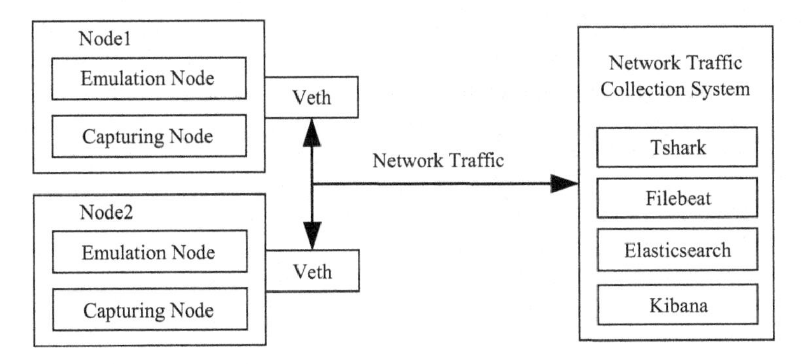

Fig. 3. Topology of test experiment.

Elasticsearch is a distributed database, so the indexing of Elasticsearch is usually broken into different parts. The data distributed on different nodes is called shards. The number of primary shards affects write performance. There are two ways to refresh when writing data. Synchronization means that memory data is brushed to disk at fixed intervals.

Figure 4 shows the impact of different primary shards and different synchronization modes on performance. Performance improves as the number of primary shards increases. Compared with synchronous writing, asynchronous writing has a great improvement in write performance.

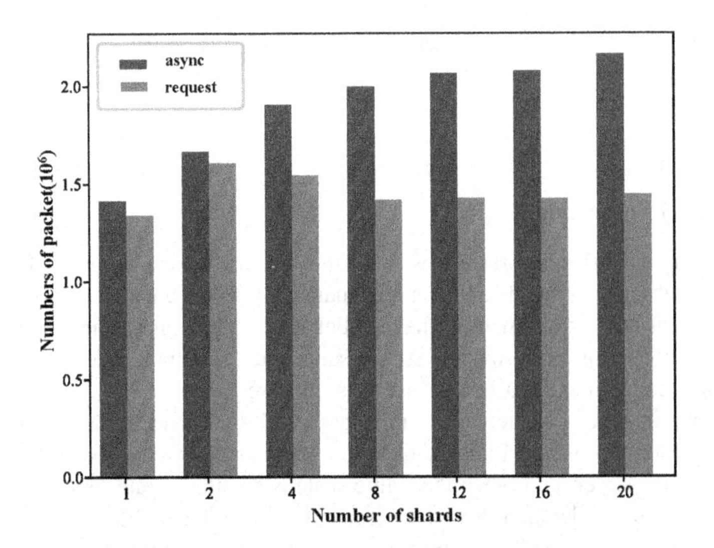

Fig. 4. Number of shards and writing method.

Table 1 shows the performance impact of different replica shards. Replica shards are mainly used to backup and restore data to improve the security and reliability of the system. We can temporarily set the replica shards to 0 when a large amount of data is written, and then restore it later. This not only improves the writing performance, but also ensures the security of the system.

Table 1. The effect of the number of replicas on performance

Numbers of replicas	Elasticsearch database writing speed (packet/s)
0	36123
1	32023
2	30379
3	30175

The results show that when the number of primary shards is 8 and the number of replica shards is 0, the write rate reaches a reasonable value.

3.2 Real-Time Tracking Performance

Based on the above optimized configuration, it can be seen from the results listed in Table 2 that the data rate written to Elasticsearch database by the traffic collection system can keep up with the capture and resolution rate of Tshark.

Table 2. Real-time tracking performance.

Time (s)	Capturing speed (packets/s)	Elasticsearch database writing speed (packet/s)	Difference
30	33902	33792	0.62%
60	33211	33343	0.39%
90	33413	33112	0.91%
120	33905	34091	0.55%

3.3 Parallel Processing

Under the limited hardware resources, with the increase of the number of emulation nodes and network traffic data, the performance of serial data collection has been unable to meet our requirements. The parallel data collection method is adopted to improve the collection performance. At the same time, distributed deployment is carried out in order to make reasonable use of system resources.

As shown in Fig. 5, all network traffic captured by a Tshark is serially stored in a log file and monitored by a Filebeat file. This method does not make full use of the resources of the server, and when the traffic scale is small, the rate of serial collection method can meet the demand. So we collect traffic in segments based on a fixed time period. Every once in a while, we turn on a Tshark capture and parse the traffic in the current time period, and turn on a Filebeat to monitor and filter log files to achieve a parallel effect. It saves the parsed data in JSON format to log files and databases. In this way, in subsequent analysis, the queried data can be used directly without additional processing, which saves processing time.

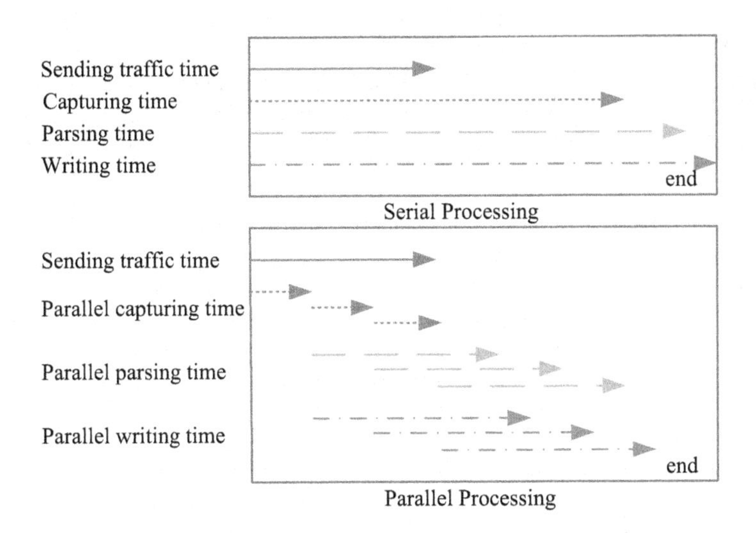

Fig. 5. Serial processing and parallel processing.

At the same time, as shown in Fig. 6, Tsharks and Filebeats are deployed in a distributed way. Tshark is deployed on a server. It shares the log files on another server via NFS files. Files are monitored and filtered by Filebeat. This kind of distributed deployment makes reasonable use of server resources and reduces the contention of CPU and other resources between traffic collection system and emulation system.

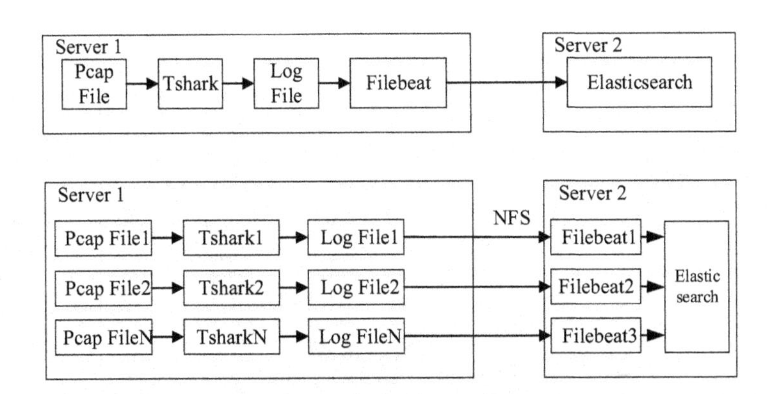

Fig. 6. Distributed deployment of Tshark and Filebeat.

4 Experiment Results

4.1 The Effect of the Number of Tsharks on Performance

The amount of parallel processing is one metric. The number of different Tsharks has an impact on write performance. Table 3 shows that with the increase of the number of Tsharks, the write rate of a single thread will be reduced. But the overall write rate will increase greatly due to the increase of the total number. It can be observed that limited by hardware resources, performance cannot be infinitely linear improved. When the number of Tsharks is 20, the system resource bottleneck is probably reached.

Table 3. The effect of the number of Tsharks on performance.

Numbers of Tsharks	Total speed (10^3*packets/s)	Single speed (10^3*packets/s)
1	31	31
2	60	30
4	115	28.75
6	165	27.5
10	232	23.2
20	302	15.1
30	324	10.8

4.2 Comparsion of Writing Time Between Different Methods

Figure 7 shows the time comparison required for single serial and multi-channel parallel processing when processing the same traffic data. The results show that the parallel mode has improved performance and nearly doubled.

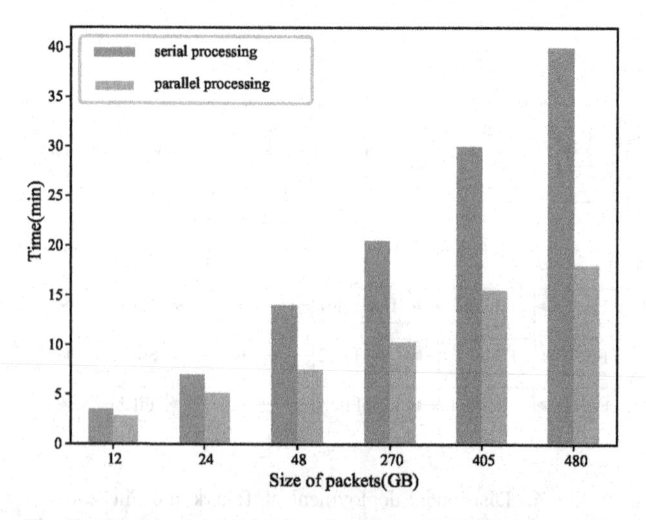

Fig. 7. Comparsion of writing time between different methods.

5 Conclusion

This paper introduces a network traffic collection system based on container technology and Elasticsearch database. Based on the space networks emulation platform, we evaluate the performance of the system to collect massive high-speed network traffic. The system can collect gigabits network traffic per second. Moreover, the system parameters are optimized and a parallel processing method is proposed. The experimental results show that the system can collect massive high-speed network traffic in the space networks emulation platform in real time under the condition of limited hardware resources.

Acknowledgements. This work is supported by the 13th Five-Year Civil Aerospace Technology Pre Research Project, the Fundamental Research Funds for the Central Univesities under Grant 021014380187 and the National Natural Sciences Foundation of China under Grant 62131012.

References

1. Lu, T., Zhang, W., Ni, X., et al.: A scalable network emulation architecture for space internetworking. In: 2016 IEEE International Conference on Communication Systems (ICCS). IEEE (2016)
2. Case, J.D.: Simple Network Management Protocol (SNMP). RFC (1990)
3. Stallings, W.: SNMP and SNMPv2: the infrastructure for network management. IEEE Commun. Mag. **36**(3), 37–43 (1998). https://doi.org/10.1109/35.663326
4. Kapoor, R.D., Calabrese, A.D., Dubey, R.K., et al.: Sample netflow for network traffic data collection. US (2007)
5. Zhao, X.F., Yi-Dong, X.U.: Monitoring system on campus network based on NETFLOW and SNMP. Comput. Technol. Dev. (2008)
6. Harvey, J.P.: Network sniffer for monitoring and reporting network information that is not privileged beyond a user's privilege level. US (2000)
7. Wang, Y., Zhang, N.: Principles and implementations of network sniffer. Appl. Res. Comput. (2003)
8. Tshark. http://www.wireshark.org/docs/man-pages/Tshark.html
9. Bajer, M.: Building an IoT data hub with Elasticsearch, Logstash and Kibana. In: 2017 5th International Conference on Future Internet of Things and Cloud Workshops (FiCloudW). IEEE Computer Society (2017)
10. Boettiger, C.: An introduction to Docker for reproducible research. Oper. Syst. Rev. **49**(1), 71–79 (2015)
11. Bernstein, D.: Containers and cloud: from LXC to Docker to Kubernetes. Cloud Comput. **1** (3), 81–84 (2014)
12. Ya-Rong, Z., Jin-Gang, Y.U.: Analysis system based on filebeat automated collection of kubernetes log. Comput. Syst. Appl. (2018)
13. Dhulavvagol, P.M., Bhajantri, V.H., Totad, S.G.: Performance analysis of distributed processing system using shard selection techniques on Elasticsearch. Procedia Comput. Sci. **167**, 1626–1635 (2020)
14. Andreassen, O., Alicia, D., Charrondière, C.: Monitoring mixed-language applications with elastic search, Logstash and Kibana (ELK) (2015)

Design and Implementation of SEDS in Spacecraft Software

Lijun Yang[1], Bohan Chen[1], and Xiongwen He[1,2](✉)

[1] Beijing Institute of Spacecraft System Engineering, Beijing 10094, China
[2] Tsinghua University, Beijing 10084, China

Abstract. In order to meet the requirements of rapid integration and test of onboard software, the research status of SEDS in the field of CCSDS standard interface service (SIOs) of Space Data System Advisory Committee is analyzed, and this paper discusses how to apply and implement SEDS standard in Chinese spacecraft software. Based on the software layered architecture, the top-level architecture of SEDS reference is designed during the development of the integrated electronic system. Taking the telemetry acquisition function of spacecraft as a use case, this paper designs and implements the application of SEDS and the extension method of SEDS subsequent application. It is verified with SEDS as the input in each stage of software development and testing. With SEDS as the input, the design verification is carried out in each stage of spacecraft software development. The results show that the application of SEDS can realize the standardization of aerospace software data interface and shorten the software development cycle through the reuse of electronic data forms and the automatic generation of relevant codes in each stage of development.

Keywords: Spacecraft software · SEDS · SOIS

1 Introduction

SEDS (SOIS Electronic Data Sheet) is an Electronic Data Sheet used to describe device information and service interface information in the Spacecraft Onboard Interface Services (SIOS) architecture of the Consultative Committee for Space Data Systems (CCSDS).Data description includes: the interface of two-way data exchange between SOIS layers; Service implementation of mapping between two groups of interfaces; The command and parameters that make up the interface; State machines, variables, and behaviors of components; The types, variables, encodings, and terms used as references above.

Based on the requirements of current spacecraft software integration and test cycle shortening, SEDS can describe the equipment information and business interface information, automatically generate onboard software code, test cases and related documents through tools, so as to reduce the time of onboard software integration, test and maintenance, and ensure the consistency of data in each development stage. The implementation and application of this standard can shorten the development cycle, reduce the risk and cost of spacecraft software. This paper first analyzes the current

Q. Guo et al. (Eds.): WiSATS 2021, LNICST 410, pp. 226–236, 2022.
https://doi.org/10.1007/978-3-030-93398-2_23

situation of SEDS research and application abroad, and then designs and implements SEDS application based on hierarchical software architecture.

2 Research Status

At present, SEDS is first applied in NASA and ESA. NASA designed and implemented SEDS in CFE core flight software architecture; ESA is currently in progress, including the development of support related tools.

NASA core software system CFE has been used in many models, based on the software bus as the core, the software components communicate with each other, not only for spacecraft, but also for other aircraft. In the core flight software architecture of NASA's CFE, SEDS electronic data form is used to automatically generate configuration code of onboard software for software assembly. SEDS can be used to define the interface information of devices and software construction services. SEDS can be used to generate software code. On the contrary, SEDS can be generated by scanning the parameter definition or header file of the code. After that, the remote control command code needed for test can be generated by SEDS, which can also be used for telemetry data analysis. NASA has developed tools that use software component SEDS and task configuration files as header files for code generation. The header file contains the message definition and engineering unit transformation. Currently, the tool is integrated into the CFS (core flight system) creation system and used by several NASA centers.

Scisys company in Europe has been researching SEDS and is developing tools to support SEDS. The tool is developed in Java language, and the configuration code can be generated automatically according to SEDS. Its application process is as follows: in the early stage of software development, SEDS file of equipment information can be generated by tools. SEDS file can generate documents of parameter information, test, etc., and can also generate part of component information of software, such as device driver. Then, in software development, when new devices are added or device information is updated, SEDS can be upgraded directly; SEDS can also be generated from the model or parameter information of the software system. When the software system data and equipment information change, the data can be modified through SEDS, and the modified SEDS file can be automatically updated through the tool. EDS can be directly used as its input in comprehensive test, because SEDS directly contains all its document data.

3 SEDS Design

3.1 Top-Level Design

The software architecture based on SOIS is a layered architecture, each layer contains a variety of county protocol and related services. The business and protocol of SOIS standard are encapsulated as reusable software components, and the software components are divided into application layer and middleware layer, which constitute the whole software together with the operating system layer. In the software system, software architecture, software component forming architecture model and component product model library are used to support software development and SEDS generation; The data dictionary mainly contains the description of semantics, types and structures according to constraints; SEDS library is based on architecture, component library and data dictionary to describe equipment, parameters, interfaces and other information (Fig. 1).

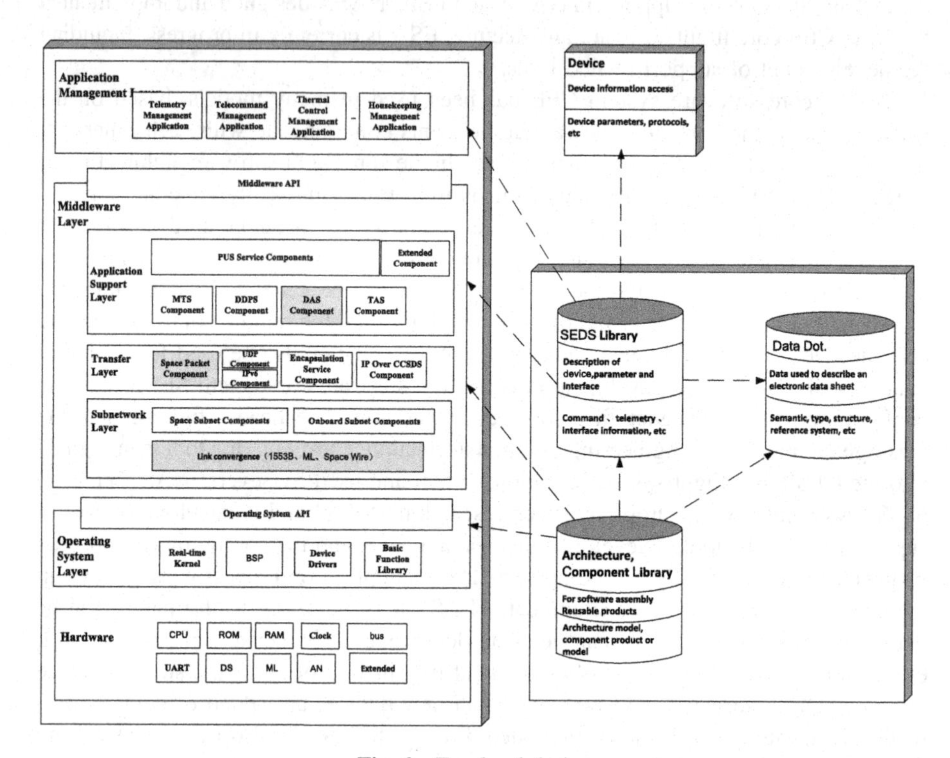

Fig. 1. Top-level design

1) Operating system and hardware layer
   Hardware layer is the bottom layer of software and the basis of software operation. It includes all kinds of hardware of onboard computer, including processor, memory, fixed memory, timer, various interfaces, etc. The interface of the operating system is encapsulated to provide a unified application programming interface for the operating system. SEDS description is not currently involved. SEDS device description mainly refers to the devices connected to CTU and interacting with CTU, not the CTU devices themselves.
2) The middleware layer
   Between the operating system layer and the application layer is the middleware layer, which is mainly a general service system, and its internal is divided into three layers. The levels are as follows:

- Subnetwork layer, a set of service components used to support the upper components, including satellite-borne subnet components, space subnet components, satellite-borne link convergence and so on.
- Transfer layer, transfer layer, mainly used for data transmission, includes the network layer and transport layer of CCSDS. The transport layer includes UDP components and implements the UDP transport protocol.
- Application support layer. The application support layer covers the basic application functions of spacecraft software system, and provides standard services for platform application support.

In the application support layer, transfer layer and subnetwork layer, SEDS is mainly used to describe the interface of components and the description of related configuration parameter tables, such as device table, routing table and so on.

3) The application layer

The application layer is the application software corresponding to the general function, including telecontrol, telemetry, internal management, time management, thermal control management, power management, unlocking and rotation control and so on. Because of the basic service support of the underlying software, the implementation of application layer only needs to combine different basic services according to specific logic. In the application layer, SEDS is mainly used to describe command parameters, telemetry parameters and so on. Once described, these parameters can be used for code generation, developer test, system test and even ground system.

3.2 SEDS Design

Taking the command sending function in spacecraft software as an example, from top to bottom, it corresponds to the software application layer, application support layer, transfer layer, sub-network layer, operating system and hardware layer.D1 and D2 are different devices, D1 is connected to the CTU through the ML channel;D2 is connected to CTU through 422.The ground command is firstly distributed through Telecontrol processing of the application layer, and then data is sent to the space packet protocol of the transfer layer through message sending primitive of the message transmission service of the application support layer. After unpacking, the command are sent to the subnet layer and finally executed by the device.

In this process, SEDS starts from the device. When SEDS describes the device information, since different devices support different data, the SEDS described by the device includes the access interface of the device, the functional interface of the device, the access protocol of the device, the virtual control steps of the device, the use information of the subnet layer and so on, such as SEDS1 and SEDS2. With the hierarchical structure of the system, the higher the upper layer, the higher the aggregation degree of SEDS and the less the number of SEDS. For example, seds5 integrates the relevant information of seds1 and seds2, and then adds the access interface of SPP (Space packet protocol). SEDS mainly describes the data exchange interface between services, including the input or output parameters, commands, business primitives, the relationship among them and the state machine representing the relationship between services (Fig. 2).

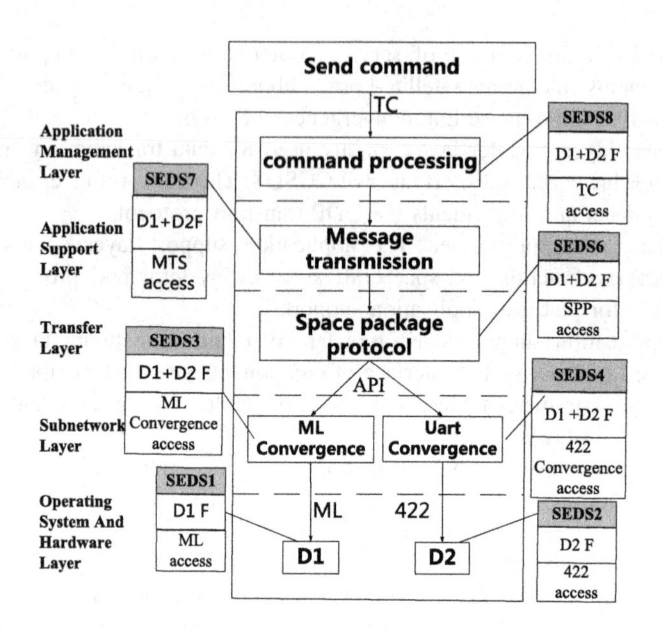

Fig. 2. SEDS describes the device, service interface

4 SEDS Implementation

In the CAST software architecture, there are 27 software components. We chose a typical example of sending a memory load (ML) command, which runs through the various layers of the architecture, as shown in Fig. 3. The application of SEDS is illustrated by this example. Sending a ML command involves the following services:

1) Device Access Service of the application support layer;
2) Space Packet Protocol of the transfer layer;
3) Packet Service of the subnetwork layer;
4) ML Convergence Service (Convergence_ML) of the convergence layer.

4.1 Command Sending Process

The ML command is sent from the application layer to the data link layer. The specific sending process is as follows:

1) Application Management layer: the device access service of the intelligent node accesses the simple intelligent node;
2) Application support layer: configure the device identifier and value identifier of the non-intelligent node 3: device id = 0x8, value id = 0x0. In the device access service, the device access type, which is sending data to the device DAP, is obtained

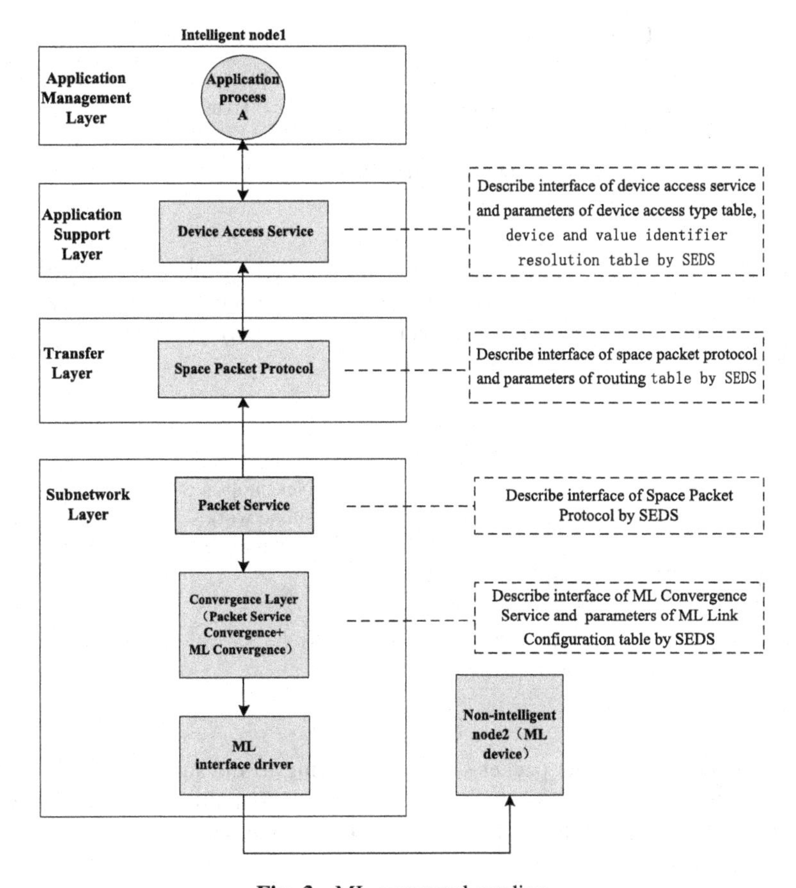

Fig. 3. ML command sending

by searching the device access type table (Table 1) by device id. Then, through searching device and value identifier resolution table (Table 2) by device id and value id, the network address (APID) is found, and then the command and data are encapsulated into a space packet and sent to space packet protocol of the transfer layer.

Table 1. Device access type table

Name	Device id (2Byte)	Corresponding device access type DAP (2Byte)
ML interface 1	0x8	24 (Sending data to the device DAP)
ML interface 2	0x9	24 (Sending data to the device DAP)

Table 2. Device and value identifier resolution table

Device id (2Byte)	Value id (2Byte)	Network address (2Byte)	Start address (4Byte)	Length (2Byte)
0x8	0x0	0x7 (DEVICE_ID_DEV_3)	0	1000

3) Transfer layer: In the space packet protocol, the routing table is searched by the network address (apid = 0x7), the underlying service is identified as the subnetwork packet service, and the subnetwork identifier is LINK_ML (subnetwork id). Packet is then sent to subnetwork packet service.

Table 3. Routing table

Network address (2Byte)	Mask (2Byte)	Next hop subnetwork id (2Byte)	Next hop subnetwork address (2Byte)	Assistant parameters (4Byte)
APID_OBC_A (0x420)	0x7E0	LINK_LOCAL (0x0)	0	0
DEVICE_ID_DEV_3 (0x8)	0x7FF	LINK_ML1 (0x6)	0	0

4) Subnetwork layer: In the packet service, according to the subnetwork id, the link type and the corresponding component instance are found, the externally provided interface is called according to the link type and component instance, and the command is issued.

Table 4. ML link configuration information

Link	Link id (2Byte)	Link type (2Byte)	Driver Master (4Byte)	Driver Slave (4Byte)
ML1	0x6	0	3	1
ML2	0x7	0	3	2

4.2 Interface and Parameter Realization

During the instruction sending process, the parameter configuration and interface are as follows:

1) Parameter configuration
   The parameter configurations to be described are shown in Tables 1, 2, 3 and 4.

```
<ContainerDataType name="device_type_table_1">
  <EntryList>
    <FixedValueEntry fixedValue="8" name="device_id" type="uint16_t"/>
    <FixedValueEntry fixedValue="24" name="cor_dap" type="uint16_t"/>
  </EntryList>
</ContainerDataType>
<ContainerDataType name="device_type_table_2">
  <EntryList>
    <FixedValueEntry fixedValue="9" name="device_id" type="uint16_t"/>
    <FixedValueEntry fixedValue="24" name="cor_dap" type="uint16_t"/>
  </EntryList>
</ContainerDataType>

<ContainerDataType name="device_value_table">
  <EntryList>
    <FixedValueEntry fixedValue="8" name="device_id" type="uint16_t"/>
    <FixedValueEntry fixedValue="0" name="value_id" type="uint16_t"/>
    <FixedValueEntry fixedValue="7" name="dv_apid" type="uint16_t"/>
    <FixedValueEntry fixedValue="0" name="start_adr" type="uint32_t"/>
    <FixedValueEntry fixedValue="1000" name="dv_length" type="uint16_tl"/>
  </EntryList>
</ContainerDataType>

<ContainerDataType name="routing_table">
  <EntryList>
    <FixedValueEntry fixedValue="0x420" name="ro_apid" type="uint16_t"/>
    <FixedValueEntry fixedValue="0x7e0" name="ro_mask" type="uint16_t"/>
    <FixedValueEntry fixedValue="0" name="link_id" type="uint16_t"/>
    <FixedValueEntry fixedValue="0" name="next_link_id" type="uint16_t"/>
    <FixedValueEntry fixedValue="0"  name="s_routing_parameter" type="uint32_t"/>
  </EntryList>
</ContainerDataType>
<ContainerDataType name="routing_table">
  <EntryList>
    <FixedValueEntry fixedValue="0x7" name="ro_apid" type="uint16_t"/>
    <FixedValueEntry fixedValue="0x7ff" name="ro_mask" type="uint16_t"/>
    <FixedValueEntry fixedValue="6" name="link_id" type="uint16_t"/>
    <FixedValueEntry fixedValue="0" name="next_link_id" type="uint16_t"/>
    <FixedValueEntry fixedValue="0" name="s_routing_parameter" type="uint32_t"/>
  </EntryList>
</ContainerDataType>
```

2) interface

a) Device access services of the application support layer

External interfaces required:

status_t (*tpPacketSend_funcp)(uint16_t src_apid, uint16_t dest_apid,
    uint8_t* packet_buffer_p, uint32_t length, uint32_t qos)

b) Spatial packet protocol of the transport layer

Externally provided interfaces:

status_t (*tpPacketSend_funcp)(uint16_t src_apid, uint16_t dest_apid,
    uint8_t* packet_buffer_p, uint32_t length, uint32_t qos)

External interfaces required:

status_t (*snPacketSend_funcp) (uint8_t qos, uint8_t priority, uint8_t
channel, uint8_t
    next_link, uint8_t next_link_address, uint8_t
  *packet_buffer_p,uint32_t length)

c)Package service at the sub-network layer

Externally provided interfaces:

status_t (*snPsSend_funcp)(uint8_t qos, uint8_t priority, uint8_t channel,
uint8_t
    next_link_id, uint8_t next_sn_address, uint8_t
  *packet_buffer_p,uint32_t length)

Other interfaces required:

status_t (*snDclMLInterface_funcp)(dcl_ml_com_t *obj_p, uint8_t
priority, uint32_t length,
    uint8_t *packet_buffer_p)

d) ML aggregation service of the data link layer

Externally provided interfaces:

status_t snDclMLInterface(dcl_ml_com_t * obj_p, uint8_t prority, uint32_t
    length, uint8_t *packet_buffer_p)

```xml
<!-- This is the set of all interface types used by component types in this namespace -->
<DeclaredInterfaceSet>
  <Interface name="tpPacketSend_funcp">
    <ParameterSet>
      <Parameter name="src_apid" readOnly="true" type="uint16_t" mode="async" />
      <Parameter name="dest_apid" readOnly="true" type="uint16_t" mode="async" />
      <Parameter name="packet_buffer_p" readOnly="true" type="uint8_t*" mode="async" />
      <Parameter name="length" readOnly="true" type="uint32_t" mode="async" />
      <Parameter name="config" readOnly="true" type="uint32_t" mode="async" />
      <Parameter name="qos" readOnly="true" type="uint32_t" mode="async" />
    </ParameterSet>
  </Interface>
</DeclaredInterfaceSet>
<DeclaredInterfaceSet>
  <Interface name="tpPacketSend_funcp">
    <ParameterSet>
      <Parameter name="src_apid" readOnly="true" type="uint16_t" mode="async" />
      <Parameter name="dest_apid" readOnly="true" type="uint16_t" mode="async" />
      <Parameter name="packet_buffer_p" readOnly="true" type="uint8_t*" mode="async" />
      <Parameter name="length" readOnly="true" type="uint32_t" mode="async" />
      <Parameter name="config" readOnly="true" type="uint32_t" mode="async" />
      <Parameter name="qos" readOnly="true" type="uint32_t" mode="async" />
    </ParameterSet>
  </Interface>
</DeclaredInterfaceSet>
<DeclaredInterfaceSet>
  <Interface name="snPacketSend_funcp">
    <ParameterSet>
      <Parameter name="qos" readOnly="true" type="uint8_t" mode="async" />
      <Parameter name="priority" readOnly="true" type="uint8_t" mode="async" />
      <Parameter name="channel" readOnly="true" type="uint8_t*" mode="async" />
      <Parameter name="next_link" readOnly="true" type="uint8_t" mode="async" />
      <Parameter name="next_link_address" readOnly="true" type="uint8_t" mode="async" />
      <Parameter name="packet_buffer_p" readOnly="true" type="uint8_t*" mode="async" />
      <Parameter name="length" readOnly="true" type="uint32_t" mode="async" />
    </ParameterSet>
  </Interface>
</DeclaredInterfaceSet>
```

4.3 Verification

Based on 43 software components of layered software architecture, CCSDS spatial link protocol, spatial domain protocol, 9 standard services and protocols of SOIS, 12 services and aggregation protocols of pus protocol of ECSS, and 46 service interface SEDS are developed. Seventeen SEDS are developed for the equipment connected with CTU, which are used as the input of unit test, assembly test and verification test. In the current software development process, saving about 30% of the time, automatic code generation accounted for 46%. At the same time, it solves the inconsistency of interface state caused by most file problems.

5 Concludes

In the layered architecture of spacecraft, SEDS can be used to describe equipment information and service interface information, which is suitable for all stages of spacecraft software development. SEDS can automatically generate software code, remote control instructions, telemetry analysis, test cases and data interface documents through tools, so as to shorten the software development cycle, improve the efficiency of software development, and reduce the uncertainty and inconsistency of documents by automatically generating documents. Using SEDS to describe the data system is helpful to the rapid integration and testing of software, and its reusability and reliability are helpful to improve the development efficiency, so as to shorten the development cycle of the whole spacecraft.

Acknowledgement. This work is supported by the 'The National Key Research and Development Program' of China (No. 2018YFB1800301).

References

1. CCSDS 876.1-R-1 Spacecraft Onboard Interface Service – Specification for dictionary of terms for electronic data sheets for onboard components CCSDS (2016)
2. CCSDS 876.0-R-1 Spacecraft Onboard Interface Services – XML specification for electronic data sheets for onboard devices CCSDS (2016)
3. CCSDS.232.0-B-3 TC Space Data Link Protocol. CCSDS, Washington (2015)
4. CCSDS.732.0-B-3 AOS Space Data Link Protocol. CCSDS, Washington (2015)
5. CCSDS TBD.0-G-0 Electronic Data Sheets and Common Dictionary of Terms for Onboard Devices and Components
6. CCSDS. 871.2-M-1 Spacecraft Onboard Interface Services – Device Virtualization Service. CCSDS, Washington (2014)
7. Kawamura, M., Kawamura, M., Kawamura, M., et al.: Nonspacecraft onboard interface with spacecraft and spacecraft. J. Spacecr. Spacecr. **8**(2), 153–164 (2013)
8. Kawamura, M., Kawamura, M., Kawamura, M., et al.: Spacecraft onboard interface series-message transfer service. J. Spacecr. Spacecr. Interface **8**(1), 1–8 (2012)
9. CCSDS. 871.1-M-1 Spacecraft Onboard Interface Services – Device Data Pooling Service. CCSDS, Washington (20120
10. He, X., He, X.: Design of a spacecraft integrated electronic system service and protocol architecture. Spacecr. Eng. (2017)

A Large-Scale Emulation Method of AOS Protocol Based on Space Network Emulation Platform

Qi Zhang, Dongxu Hou, Chengpeng Kuang, Kanglian Zhao(✉), and Wenfeng Li

School of Electronic Science and Engineering, Nanjing University, Nanjing 210023, People's Republic of China
{mg1923075,mf1923032}@smail.nju.edu.cn, zhaokanglian@nju.edu.cn

Abstract. Advanced Orbiting System (AOS) protocol is a link layer protocol for space network communication specified by CCSDS. On the basis of the proposed semi physical emulation platform in our previous work, this paper implements the AOS gateway and further proposes the design of AOS scale emulation. Corresponding experiment is conducted to validate the scale AOS emulation in the proposed emulation platform. The experimental results show that the large-scale AOS protocol emulation scheme is effective and any two nodes can communicate with each other.

Keywords: AOS · Space network · Scale emulation · Semi physical emulation platform

1 Introduction

Space Information Networks (SIN) are network systems based on space platforms, such as satellites, stratospheric balloons, and so on [1]. In recent years, with the continuous improvement and development of inter satellite communication technology, its application prospects and economic value are also increasing [2–4]. Those technologies used in the terrestrial network may not be compatible with those technologies used in SIN [5]. To solve the problem of the difference between the terrestrial network and the space network, many new technologies have been developed. These technologies need to be verified repeatedly before they can be put into used. We have proposed the semi physical emulation platform for space networks to verify these technologies [6].

The general structure of the semi physical emulation platform for space networks is shown in Fig. 1. According to the space network protocol standard specified by CCSDS [7], the Advanced Orbiting System (AOS) protocol is widely used in the data link layer of various space networks except similar parts of the terrestrial networks [10]. Obviously, the emulation platform should support AOS emulation.

However, AOS frame cannot be transmitted in Ethernet. At present, the emulation method of AOS protocol is to add AOS frame header for data packets and use channel simulator such as the space network channel or cortex (CRT) to realize frame

Q. Guo et al. (Eds.): WiSATS 2021, LNICST 410, pp. 237–243, 2022.
https://doi.org/10.1007/978-3-030-93398-2_24

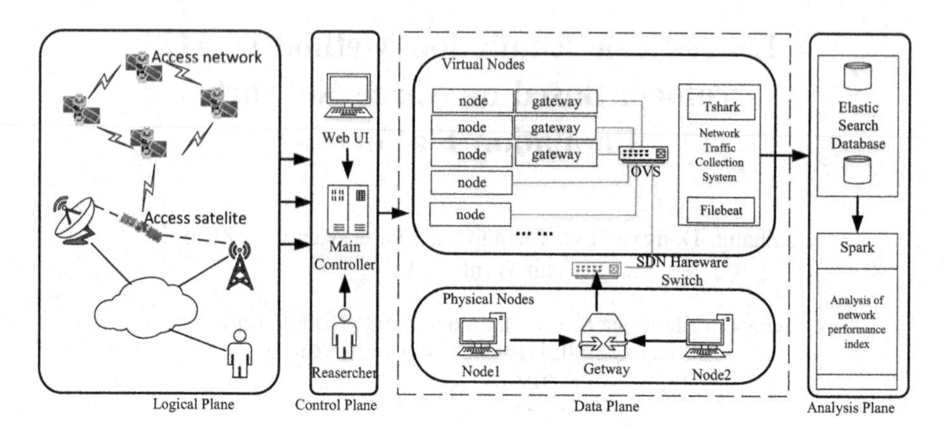

Fig. 1. Architecture of emulation platform

transmission [11–14]. Although these methods can truly restore the communication process of AOS protocol, they are expensive and cannot be used for large-scale emulation. Therefore, an AOS emulation method based on emulation platform is urgently needed to realize AOS scale emulation.

This paper implements the AOS gateway and further proposes the method of AOS scale emulation based on emulation platform, and the effectiveness of this method is verified.

The rest of the paper is organized as follows. In Sect. 2, according to the requirements of emulation platform, the design of AOS large-scale emulation is proposed. In Sect. 3, we implement the design and tests the related functions. Finally, the conclusion is made in Sect. 4.

2 Design for AOS Scale Emulation

2.1 Design of the AOS Gateway

In the emulation platform, AOS protocol emulation is mainly realized by AOS gateway, and the process is shown in Fig. 2. The gateway removes packet's Ethernet header and adds AOS frame header, IP header and Ethernet header. Packets reception is a reverse process.

The structure of AOS gateway is shown in Fig. 3. In a point-to-point emulation scenario, the packet of the sender is captured by the gateway. After removing its Ethernet packet header, it is sent to the next program to encapsulate the AOS frame header. In order to transmit AOS frame in Ethernet, it is necessary to add IP header and Ethernet header to AOS frame. After completing the above steps, the packet can be sent to the network.

Similarly, after receiving a packet with AOS frame, the gateway first removes the external Ethernet header and IP header. The remaining AOS frame structure is sent to the next program. The program sends the IP packets contained in the AOS frame to the next program, encapsulates the Ethernet head and sends it to the host.

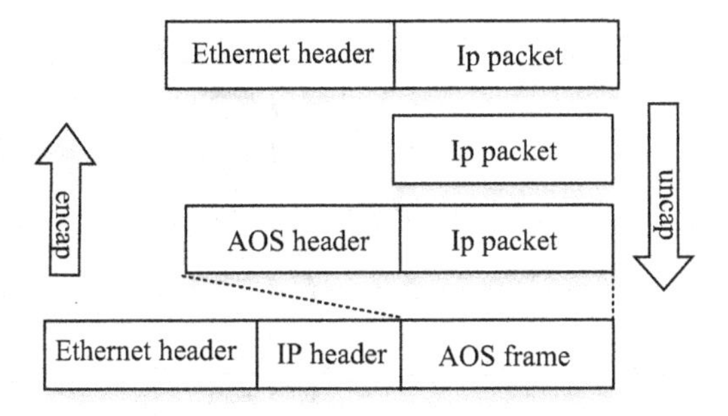

Fig. 2. AOS emulation principle

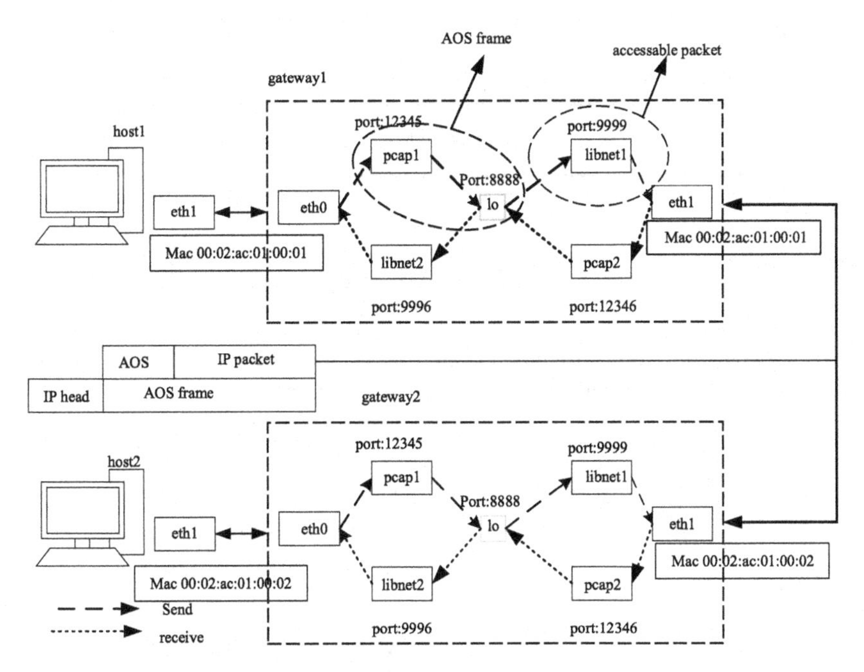

Fig. 3. Architecture of gateway

2.2 Method of AOS Scale Emulation

The implementation of AOS scale emulation is based on the multi links information stored in the database, as shown in Fig. 4. In the figure, each link is determined by two gateways. All links correspond to the data table.

The original Ethernet header was removed during packet processing. In order to transmit packets in the network, we need to construct the Ethernet header again. Therefore, the MAC address of the gateway should also be stored in the data table. The process of automatically uploading MAC address by gateway is shown in Fig. 5. The gateway reads the local MAC address and the corresponding network card name,

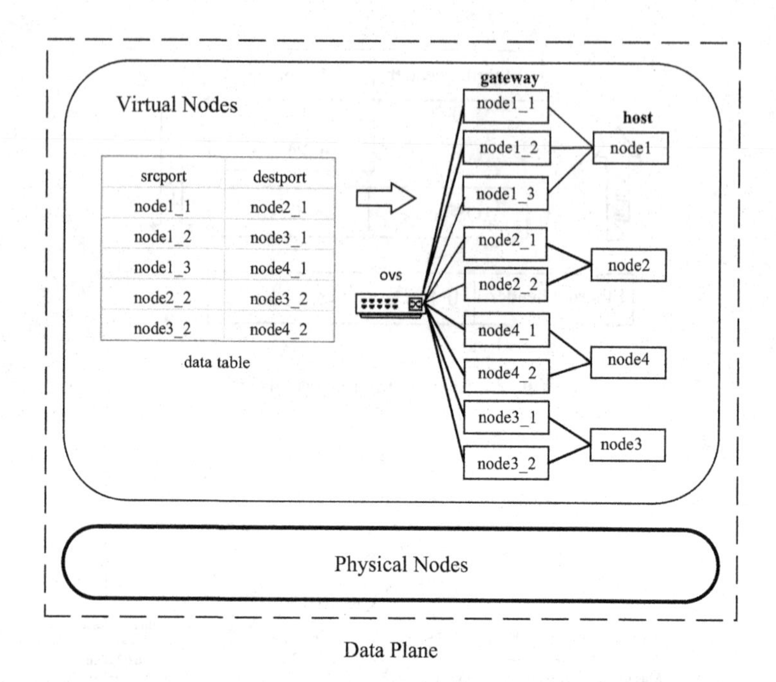

Fig. 4. Structure of AOS scale emulation.

uploads them to the corresponding location in the data table, and reads the MAC address of the destination gateway.

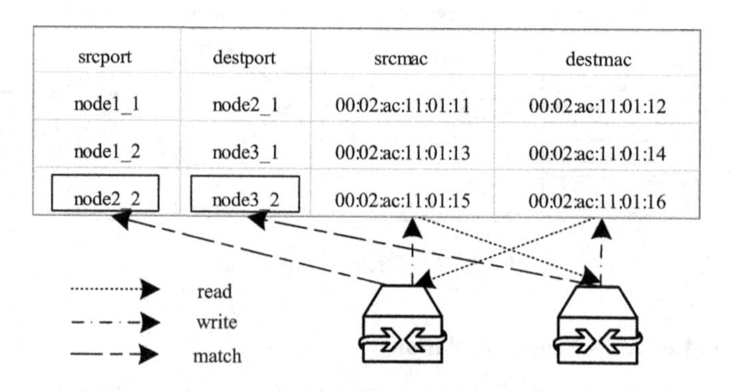

Fig. 5. Format of data table

3 Experiment and Discussion

3.1 Experimental Scenario

The AOS scale emulation design is verified by experiments. In order to be able to test the function of AOS gateway in a comprehensive way, we use the network topology shown in the Fig.6. Six nodes are interconnected, and host1 has 5 gateways corresponding to other hosts. In this structure, host1 can communicate with other hosts.

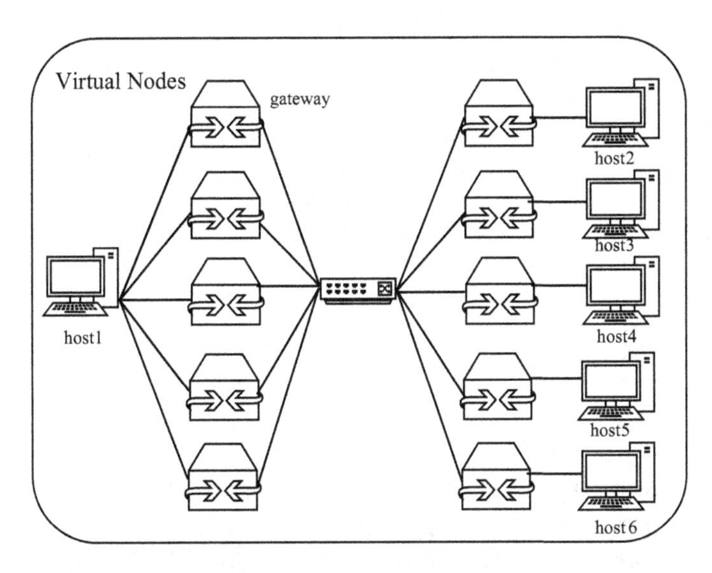

Fig. 6. Experimental link

Each network card of the gateway is listening by a program to capture the packets that need to be processed. The network card of the gateway is monitored by the program. The program captures and processes the packets passing through the network card. Only the packets coming from outside the gateway need to be processed, so the MAC address is used as the basis of whether to capture the packets.

3.2 Results

The communication between nodes can effectively verify the function of the design. To prove this, host1 sends ICMP request messages to other five nodes. The result is shown in Fig. 7(b). Obviously, the communication between two end nodes needs to go through gateway and forwarding node. If ICMP reply message can be received, it indicates that AOS gateway can normally perform encapsulation and uncapsulation services. The packets structure through the gateway is shown in Fig. 7(a).

The experimental results show that two nodes can communicate with each other and the packets contain the structure of AOS frame header.

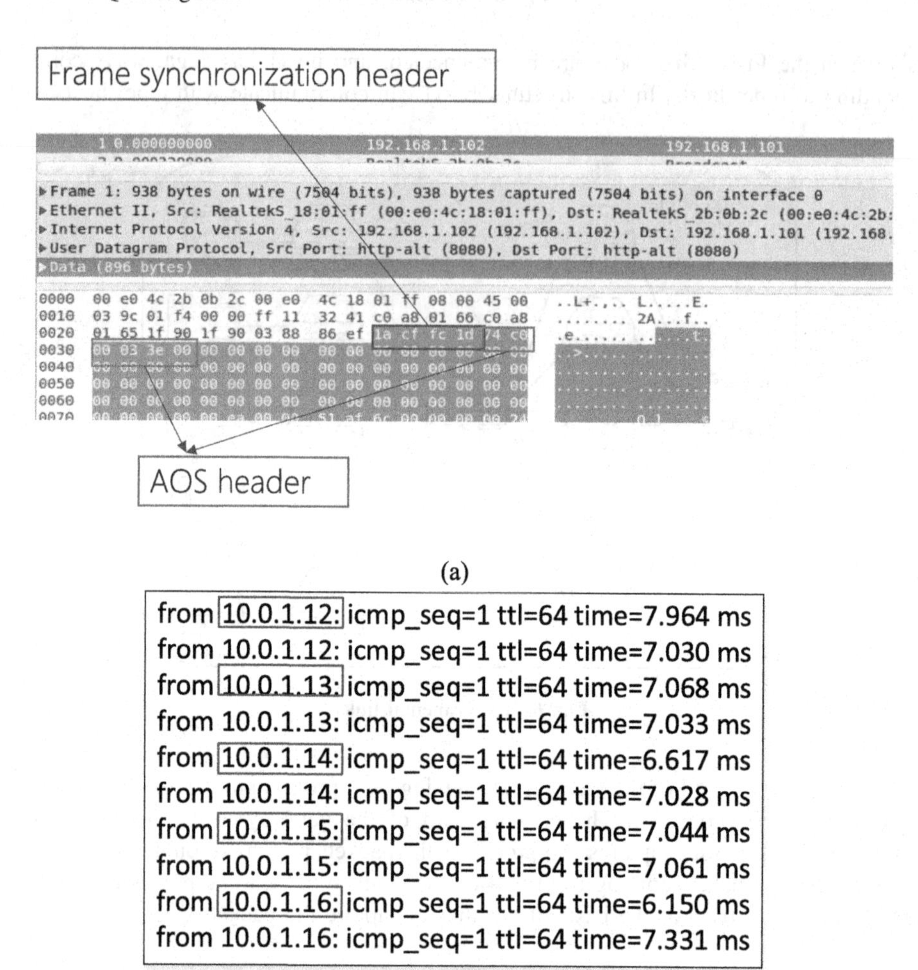
Frame synchronization header
Frame 1: 938 bytes on wire (7504 bits), 938 bytes captured (7504 bits) on interface 0
Ethernet II, Src: RealtekS_18:01:ff (00:e0:4c:18:01:ff), Dst: RealtekS_2b:0b:2c (00:e0:4c:2b:
Internet Protocol Version 4, Src: 192.168.1.102 (192.168.1.102), Dst: 192.168.1.101 (192.168.
User Datagram Protocol, Src Port: http-alt (8080), Dst Port: http-alt (8080)
Data (896 bytes)
AOS header
(a)
from 10.0.1.12: icmp_seq=1 ttl=64 time=7.964 ms
from 10.0.1.12: icmp_seq=1 ttl=64 time=7.030 ms
from 10.0.1.13: icmp_seq=1 ttl=64 time=7.068 ms
from 10.0.1.13: icmp_seq=1 ttl=64 time=7.033 ms
from 10.0.1.14: icmp_seq=1 ttl=64 time=6.617 ms
from 10.0.1.14: icmp_seq=1 ttl=64 time=7.028 ms
from 10.0.1.15: icmp_seq=1 ttl=64 time=7.044 ms
from 10.0.1.15: icmp_seq=1 ttl=64 time=7.061 ms
from 10.0.1.16: icmp_seq=1 ttl=64 time=6.150 ms
from 10.0.1.16: icmp_seq=1 ttl=64 time=7.331 ms

(b)

Fig. 7. Experimental result.

4 Conclusion

In this paper, combined with software emulation and database, the design of AOS scale emulation based on the semi physical emulation platform for space networks is proposed. Corresponding results show that AOS large-scale emulation realizes the normal communication function. It is an ideal way to realize AOS protocol scale emulation.

Acknowledgements. This work is supported by the 13th Five-Year Civil Aerospace Technology Pre Research Project, the Fundamental Research Funds for the Central Univesities under Grant 021014380187 and the National Natural Sciences Foundation of China under Grant 62131012.

References

1. Bai, L., de Cola, T., et al.: Space information networks. Wirel. Commun. **26**(2), 8–9 (2019)
2. Deren, L.I., Shen, X., Gong, J., et al.: On construction of China's space information network. Geomatics and Information Science of Wuhan University (2015)
3. Mukherjee, J., Ramamurthy, B., et al.: Communication technologies and architectures for space network and interplanetary internet. Commun. Surv. Tutor. **15**, 881–897 (2013)
4. Yang, Z., Hewu, L., Qian, W., et al.: Simulation platform for inter-domain protocols validation of integrated space-terrestrial network. J. Commun. **40**, 1 (2019)
5. Yu, Z.A., Bo, W.B., Bg, A., et al.: A research on integrated space-ground information network simulation platform based on SDN. Comput. Netw. **188**, 107821 (2021)
6. Lu, T., et al.: A scalable network emulation architecture for space internetworking. In: 2016 IEEE International Conference on Communication Systems (ICCS), pp. 1–5 (2016). https://doi.org/10.1109/ICCS.2016.7833653
7. what is CCSDS?. https://public.ccsds.org/default.aspx
8. Zhao, J., Zeng, L., Yang, W., Qian, L.: AOS space data link protocol performance analysis and simulation. In: IET International Conference on Communication Technology and Application (ICCTA 2011), Beijing, pp. 287–290 (2011). https://doi.org/10.1049/cp.2011.0676
9. Li, H., Zhou, H., Zhang, H., et al.: EmuStack: an OpenStack-based DTN network simulation platform. In: International Conference on Networking & Network Applications. IEEE (2016)
10. Hooke, A.J.: CCSDS advanced orbiting systems: international data communications standards for the space station freedom. IEEE Netw. **4**(5), 13–16 (1990)
11. Garcia, L.M.: Programming with Libpcap – Sniffing the network from our own application (2008)
12. Li, Q., Wang, R., Ye, T., et al.: Research and simulation on high efficient frame generation model of AOS considering packet extracting time under finite buffer. In: 2014 IEEE 5th International Conference on Software Engineering and Service Science (2014)
13. Turull, D., Hidell, M., Sjö, P.: LibNetVirt: the network virtualization library. In: Workshop on Clouds, Networks and Data Centers (ICC'12 WS - CloudNetsDataCenters). IEEE (2012)
14. Liu, J., Mann, S., Vorst, N.V., et al.: An open and scalable simulation infrastructure for large-scale real-time network simulations. In: INFOCOM 2007, 26th IEEE International Conference on Computer Communications. IEEE (2007)

Towards Self-organized Networking in Large-Scale Nano-satellite Networks

Huan Han, Long Suo(✉), Yuehong Yang, Peng Wang, and Hongyan Li

State Key Laboratory of Integrated Service Networks, Xidian University, Xian 710071, China
lsuo@xidian.edu.cn

Abstract. Nano-satellites have developed rapidly in the fields of communications and remote sensing due to their small size, low cost and flexibility, the number of which has increased dramatically. However, massive amounts of data cannot be effectively transmitted through ground stations or satellite networks due to the growth of transmission demand. To overcome the problem, the nano-satellite network clustering algorithm was designed. Specifically, by clustering nano-satellites, the cluster head can gather remote sensing data and control directives of satellite nodes in the cluster, reduce the collision probability of random transmission, improve the utilization of communication resources and network throughput. We established a random access model of nano-satellites based on a two-dimensional Markov chain, which characterizes the features of the satellite network multiple access process without a carrier sensing mechanism. We also deduce the throughput of the nano-satellite network. We obtain the optimal number of clusters according to the nano-satellite scale, link rate and traffic arrival rate. The performance of non-clustered networks and clustered networks is compared and analyzed by simulation. The result shows that the clustered network reduces the average delay and improves the efficiency of information transmission compared with the non-clustered network.

Keywords: Large-scale Nano-satellite · Multiple access · Markov Chain · Clustering network

1 Introduction

Whether in military communications or civil communications, nano-satellites have great development trends and application prospects [13]. Due to the growth of transmission demand, resource-limited satellite networks face tremendous transmission pressure, since the deployment of ground stations is highly costly. Even worse, the existing clustering algorithm does not consider the optimal number of clusters or give an analysis method for the number of clusters, leading to uneven network clustering [4, 13]. If there are too many clusters, the collision probability between cluster heads will increase and the channel reuse rate will be reduced. Furthermore, efficient access protocols for clustered satellite networks are missing.

Q. Guo et al. (Eds.): WiSATS 2021, LNICST 410, pp. 244–256, 2022.
https://doi.org/10.1007/978-3-030-93398-2_25

In fact, there are many excellent clustering-related works. As in [5], the clustering algorithm assigns unique identifiers to the nodes and selects the node with the smallest ID number as the cluster head node, which is easy to calculate and implement. However, it will cause uneven resources for different clusters. And the selected cluster head will not change, such that the continuous use of cluster head can cause fast energy consumption. In [6], an algorithm based on the geographical location of the cluster head node is proposed, which can artificially control the size of the cluster according to user needs, such that the network load is balanced. However, it does not indicate how to calculate the optimal number of clusters. In [7], an algorithm based on node weights is proposed. The advantage of this algorithm is that it can comprehensively consider a variety of factors, but because the weights are difficult to determine, the selected cluster head may not be the best cluster head in the current network environment. In [8], a LEACH clustering algorithm was proposed. The disadvantage of this algorithm is that the cluster heads are unevenly distributed, because the cluster heads may appear concentrated in a certain area of the network. In [9], the enhanced LEACH is proposed. This algorithm considers the remaining energy of nodes when selecting cluster heads. However, factors such as the optimal number of clusters and node mobility are not considered. Therefore, the existing algorithms can be improved.

To solve the problem, we established a random access model of nano-satellites based on a two-dimensional Markov chain, which characterizes the features of the satellite network multiple access process without a carrier sensing mechanism. We also deduce the throughput of the nano-satellite network under the access mechanism. In order to maximize the throughput, a nano-satellite network clustering algorithm is designed, which considers the arrival rate, the scale of the nano-satellite and the link rate. Finally, we compare and analyze the performance of clustered networks and non-clustered networks through simulation, and the results prove that the clustering algorithm effectively improves the network performance.

2 System Model

In this paper, a random access model of nano-satellites based on two-dimensional Markov chains is established, which characterizes the features of the multiple access process of satellite networks without carrier sensing mechanism. Since the traffic of nano-satellites is not saturated, this paper based on Malone's analysis method [10], calculates the probability τ, which represents a station sending a packet in a slot.

Based on Bianchi's model [11], Malone introduced new states $(0, \mathrm{k_e})$ for $k \in [0, \mathrm{W}_0 - 1]$, representing a station which has just sent a packet without waiting. The Markov chain model is shown in Fig. 1.

According to the probability distribution of Markov chain, it can be known that τ is:

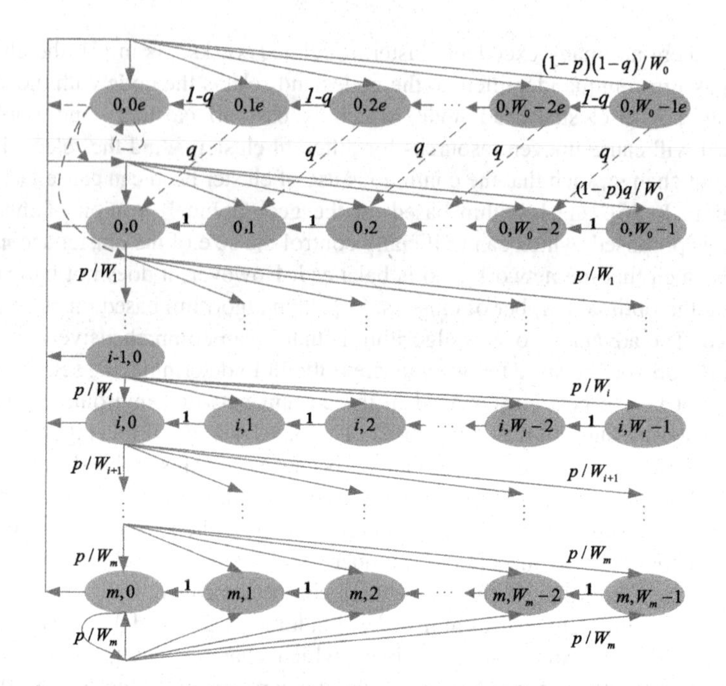

Fig. 1. Unsaturated Markov chain

$$\tau = \left(\frac{q^2 W_0}{(1-p)(1-q)(1-(1-q)W_0)} - \frac{q^2 P_{idle}}{1-q}\right) / \left[\begin{array}{l} (1-q) + \frac{q^2 W_0 (W_0+1)}{2(1-(1-q)W_0)} + \frac{q(W_0+1)}{2(1-q)} \\ \times\left(\frac{q^2 W_0}{1-(1-q)W_0} + (1-P_{idle})(1-q) - qP_{idle}(1-p)\right) \\ + \frac{pq^2}{2(1-p)(1-q)} \times \left(\frac{W_0}{1-(1-q)W_0} - (1-p)P_{idle}\right) \\ \times\left(2W_0 \frac{1-p-p(2p)^{m-1}}{1-2p} + 1\right) \end{array}\right] \quad (1)$$

It can be seen that τ depends on p, q, P_{idle}, W_0 and m. The specific analysis will be introduced in detail in the performance analysis.

Also, we can know the probability q can be written as:

$$q = \min(E_s/\text{mean inter-packet time}, 1) \quad (2)$$

3 Clustering Algorithm

Aiming at the low efficiency of mass information transmission in large-scale nano-satellite communication scenarios, a new and efficient clustering algorithm is designed. The communication scene of a large-scale nano-satellite is shown in Fig. 2, which is composed of a nano-satellite, a low-orbit constellation and a ground station.

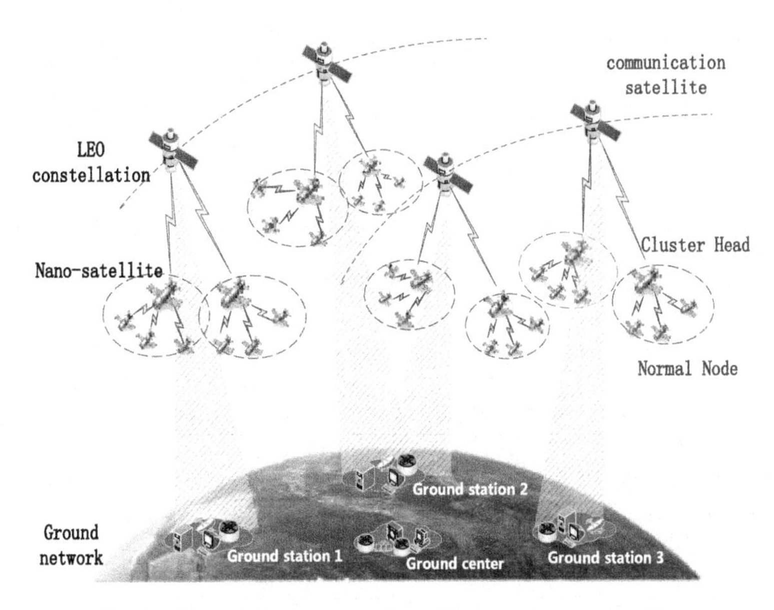

Fig. 2. Communication scenarios of large-scale nano-satellites

Ordinary nodes can use the polling method by cluster head nodes to avoid conflicts; the distributed competition method based on RTS/CTS is used between cluster head nodes to complete the channel resources. The cluster head nodes and ordinary nodes access mode are shown in Fig. 3.

The number N of nano-satellites within the coverage of each satellite can be calculated based on the scale of nano-satellites. If $N < M_{best}$, there is no need to cluster the satellites, the satellites can be directly connected to the communication satellites through competing channel resources. If $N > M_{best}$, the satellites need to be clustered, and M_{best} satellites are considered as the cluster head nodes among N satellites (Fig. 4).

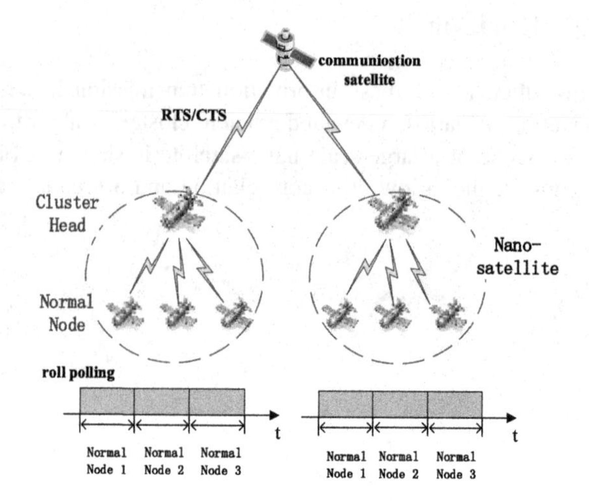

Fig. 3. Cluster head nodes and ordinary nodes access mode

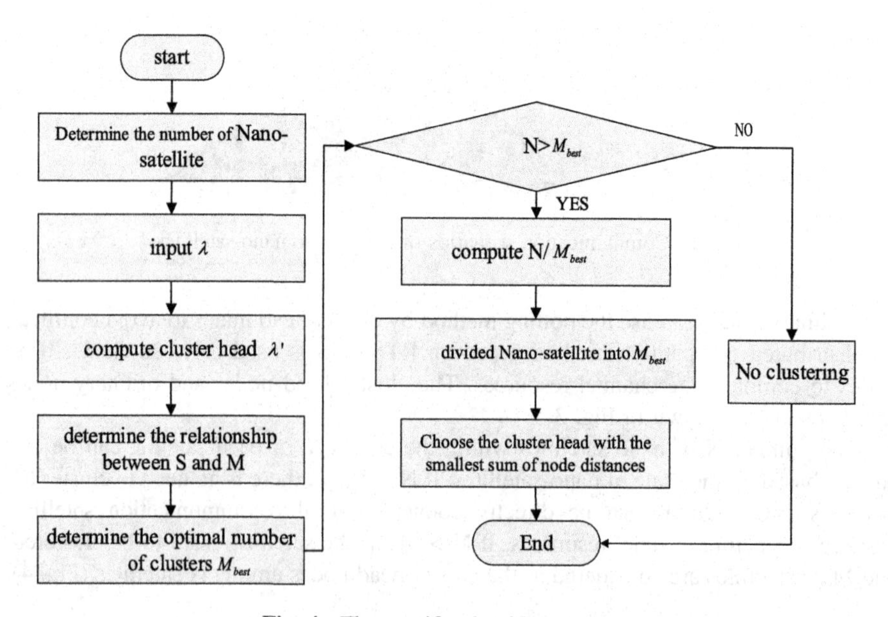

Fig. 4. The specific algorithm process

4 Performance Analysis

4.1 The Arrival Rate

The traffic arrival rate of the cluster head node λ' can be written as:

$$\lambda' = \frac{1}{(\frac{N}{M} - 1) \times (\varepsilon + 2\delta_1) + (\frac{N}{M} - 1)\lambda \times \frac{L}{c}} \tag{3}$$

Among this, λ represents the traffic arrival rate of each satellite, N represents the number of nano-satellites, M represents the number of clusters, ε represents the delay of the cluster head node sending polling frames to the ordinary nodes in the cluster, δ_1 represents the propagation delay from the ordinary node to the cluster head node, and L/c represents data transmission delay.

4.2 The Collision Probability

Since the communication scenario studied in this paper is a space-based information network, when the CSMA/CA protocol is used between clusters, the cluster head nodes cannot listen to each other's state, and can only through RTS/CTS to determine whether the channel is occupied.

As shown in Fig. 5, when station A sends an RTS frame to reserve the channel and waits for the return of the CTS frame, since station B has no listening function, no matter which slot backoff time of station B is completed in the interval between RTS and CTS, station A and station B will collide.

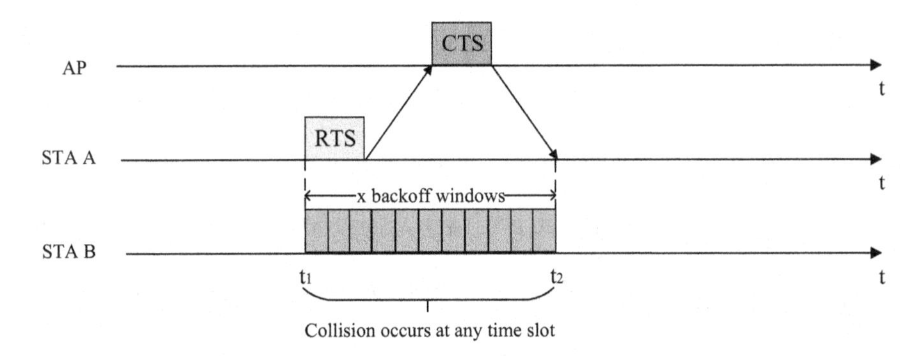

Fig. 5. Data collision slots

The probability of success can be expressed as:

$$1 - p = (1 - \tau)^{x(n-1)} \tag{4}$$

According to the probability of success, the collision probability p is:

$$p = 1 - (1 - \tau)^{x(n-1)} \tag{5}$$

x is the number of time slots included in the interval between RTS and CTS, n is the number of stations, σ is the length of each time slot, δ_2 is the propagation delay, then the value of x is:

$$x = \frac{RTS + CTS + 2\delta_2}{\sigma} \tag{6}$$

4.3 The Time Spent Each State

The probability that at least one packet is being transmitted in a time slot P_{tr} is:

$$P_{tr} = 1 - (1 - \tau)^n \tag{7}$$

Assuming P_s is the probability that is only one packet transmission and the transmission is successful, then P_s is:

$$P_s = \frac{C_n^1 \tau (1 - \tau)^{x(n-1)}}{P_{tr}} = \frac{n\tau(1 - \tau)^{x(n-1)}}{1 - (1 - \tau)^n} \tag{8}$$

Assuming P_{idle} is the probability of channel free, then P_{idle} is:

$$P_{idle} = 1 - P_{tr} = (1 - \tau)^n \tag{9}$$

If the channel is idle, the delay T_{idle} can be written as:

$$T_{idle} = \sigma \tag{10}$$

The probability of successful packet transmission P_{succ} is:

$$P_{succ} = P_{tr} P_s = n\tau(1 - \tau)^{x(n-1)} \tag{11}$$

If there are data packets successfully transmitted in the channel, the delay T_{succ} is represented by:

$$T_{succ} = RTS + SIFS + \delta_2 + CTS + SIFS + \delta_2 + E[P]/c + SIFS + \delta_2 + ACK + \delta_2 \tag{12}$$

The probability of collision of data packets in the channel is:

$$P_{fail} = P_{tr}(1 - P_s) = 1 - (1 - \tau)^n - n\tau(1 - \tau)^{x(n-1)} \tag{13}$$

If the data packets collide in the channel, the time delay T_{fail} is written as:

$$\begin{aligned} T_{fail} &= \frac{1}{2}(RTS + \delta_2) + \frac{1}{2}(RTS + SIFS + CTS + 2\delta_2) \\ &= RTS + \frac{1}{2}(SIFS + CTS + 3\delta_2) \end{aligned} \tag{14}$$

Above all, the expected time spent per state E_s can be written as:

$$\begin{aligned} E_s &= P_{idle}T_{idle} + P_{succ}T_{succ} + P_{fail}T_{fail} \\ &= (1-\tau)^n\sigma + n\tau(1-\tau)^{x(n-1)}(RTS + CTS + E[P] + ACK + 3SIFS + 4\delta_2) \\ &+ \left(1 - (1-\tau)^n - n\tau(1-\tau)^{x(n-1)}\right)\left(RTS + \frac{1}{2}(SIFS + CTS + 3\delta_2)\right) \end{aligned} \tag{15}$$

4.4 System Throughput

Throughput S is the payload transmitted successfully on the channel per time, then S is expressed as the length of the data packet successfully transmitted within the time E_s:

$$S = \frac{P_{succ}E[P]}{E_s} \tag{16}$$

By combining Eqs. (1), (2), (5) and (9), the relationship between the system throughput S and the number of clusters M can be obtained. The number of M when the throughput S is the highest is taken as the optimal cluster number M_{best}.

4.5 The Average Delay

We will compare the performance between non-clustered networks and clustered networks using the average delay (Fig. 6).

Through Eq. (15), we can calculate the average delay T_{ad} of the nano-satellite connecting to the communication satellite and successfully sending the data packet, then the average delay T_{ad} is expressed as:

$$T_{ad} = \frac{L}{\frac{P_{succ}L}{E_s}} = \frac{E_s}{P_{succ}} \tag{17}$$

The average delay of the non-clustered network $T_{not\_cluster}$ is:

$$T_{not\_cluster} = N\lambda \times T_{ad} = N\lambda \times \frac{E_s}{P_{succ}} \tag{18}$$

N represents the number of nano-satellites competing for channel resources, λ represents the traffic arrival rate of nano-satellites.

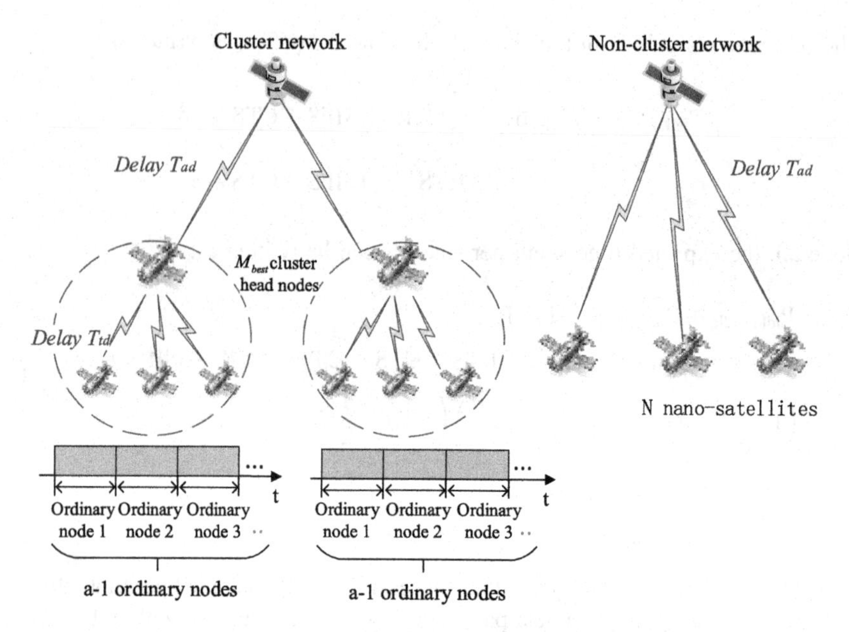

Fig. 6. The average delay of the clustered network and the non-clustered network

The average delay of the clustering network $T_{cluster}$ is:

$$\begin{aligned} T_{cluster} &= M_{best}\lambda' \times (T_{td} + T_{ad}) \\ &= M_{best}\lambda' \times \left(\left(\frac{N}{M_{best}} - 1 \right) \times (\varepsilon + 2\delta_1) + \left(\frac{N}{M_{best}} - 1 \right) \lambda \times \frac{L}{c} + \frac{E_S}{P_{succ}} \right) \end{aligned} \tag{19}$$

5 Simulation and Results Analysis

To verify the performance of the designed nano-satellite network clustering algorithm, this paper with simulations using MATLAB. We compared the average delay of the clustered network and the non-clustered network under different traffic arrival rates. This section first sets up the simulation parameters, then analyzes the performance of the network based on the simulation results (Table 1).

The relationship between the number of clusters and the arrival rate under the conditions of different scales of nano-satellites is shown in Fig. 7. Obviously, the number of clusters increases as the arrival rate or the scale of nano-satellites increases. The number of clusters gradually increases when the arrival rate is low. This is because the load in the network is low, which means the collision probability between clusters is low, and the competition delay is small. The number of clusters increases slowly when the traffic arrival rate is high. This is because the load is high, so the collision probability between clusters is high, and the competition time is prolonged. The selection of the optimal number of clusters aims to balance the polling delay within the

Table 1. Simulation parameter.

Parameter	Value	Parameter	Value
Data packet length	8000 bits	ACK frame length	112 bits
RTS frame length	160 bits	CTS frame length	112 bits
Propagation delay between nano-satellites	0.03 ms	Propagation delay between nano-satellite and communication satellite	1 ms
Back-off window	32	Back-off stage	3
Time slot length	0.8 ms	Traffic arrival rate	(0,1]

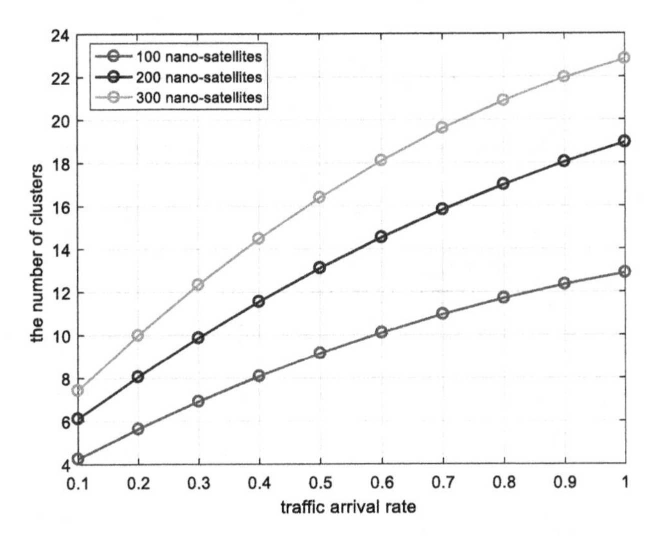

Fig. 7. The relationship between the traffic arrival rate and the number of clusters for different scales of nano-satellites

cluster and the competition delay between clusters, so as to maximize the system throughput and minimize the network delay.

As shown in Fig. 8, it is the relationship between the traffic arrival rate and the number of clusters under the conditions of different link rates.

Figures 9, 10 and 11 show the relationship between the average delay and the arrival rate. As the traffic arrival rate increases, the average delay increases.

Figure 9 shows the relationship between the average delay and the arrival rate of the non-clustered network under the different scales of the nano-satellite. Figure 10 shows the same relationship of the clustered network. It can be seen that as the network scale increases, the average delay also increases due to the probability of node collisions increasing. In general, the larger the scale of the nano-satellite, the more efficient the clustering algorithm is.

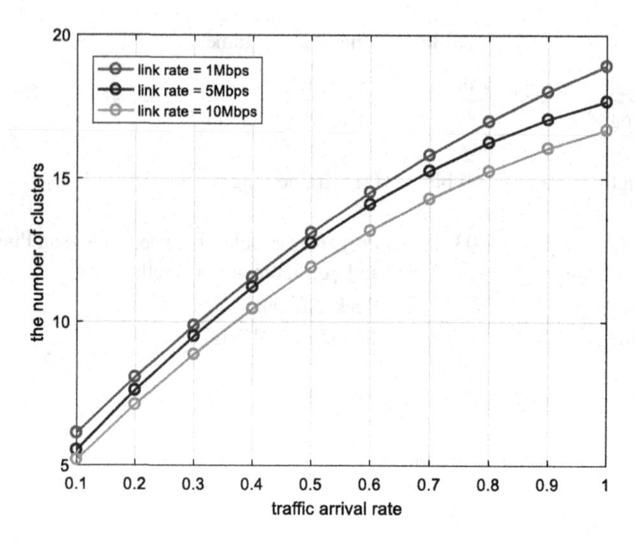

Fig. 8. The relationship between the traffic arrival rate and the number of clusters for different link rate

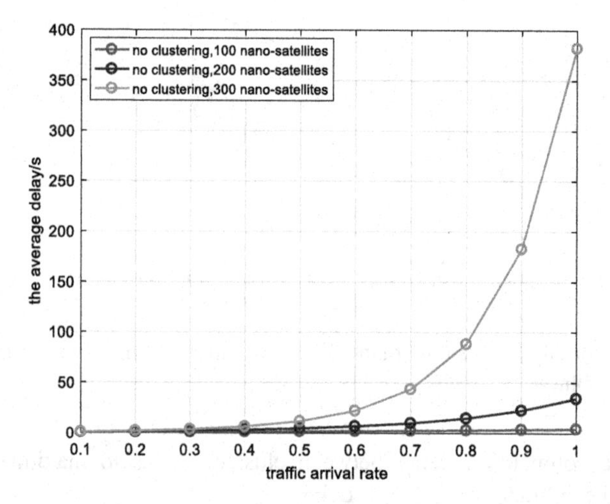

Fig. 9. The relationship between the average delay and the arrival rate for different scales of nano-satellites (non-clustered network)

From Fig. 11 we can see that the delay performance of the clustered network is better than that of the non-clustered network.

In general, the clustering algorithm we designed can effectively reduce the collision probability of nano-satellites and reduce the end-to-end delay.

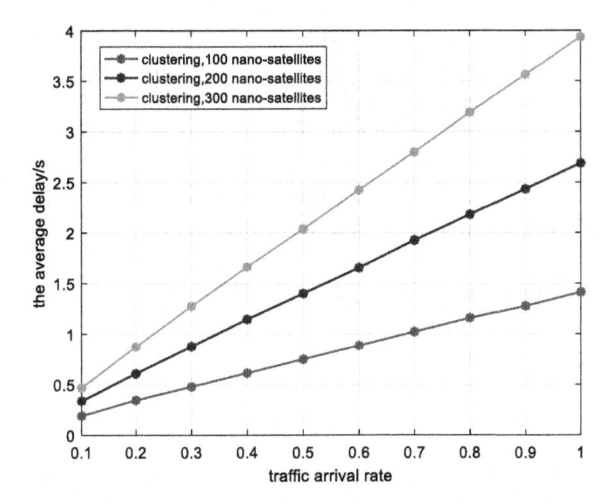

Fig. 10. The relationship between the average delay and the arrival rate for different scales of nano-satellites (clustered network)

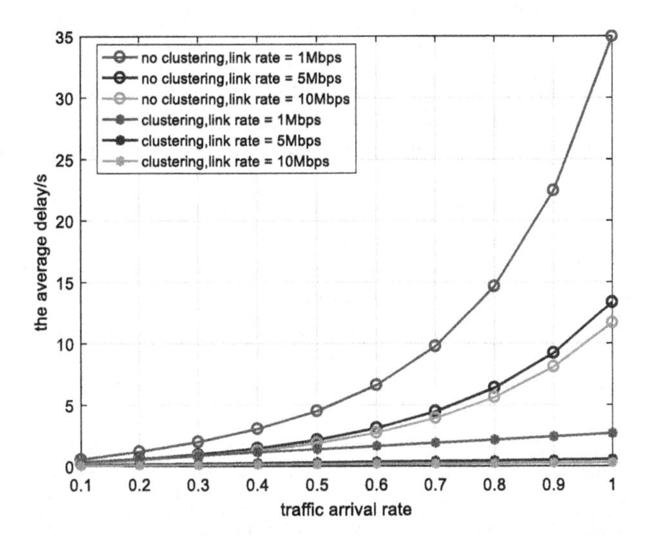

Fig. 11. The relationship between the traffic arrival rate and the average delay for different link rate of the non-clustered and clustered networks

6 Conclusion

A nano-satellite network clustering algorithm based on the optimal number of clusters is designed. We established a random access model of nano-satellites based on a two-dimensional Markov chain, which characterizes the features of satellite network multiple access process without a carrier sensing mechanism. We also deduce the throughput of the nano-satellite network under the access mechanism. To maximize the network throughput, the nano-satellite network clustering algorithm is designed, and the optimal number of clusters is obtained according to the arrival rate, the link rate, and the scale of the nano-satellite. Finally, the performance of non-clustered network and clustered network is compared and analyzed through simulation, and the efficiency of the clustering algorithm is verified. The simulation result shows that the clustered network reduces the average delay and improves the efficiency of information transmission compared with the non-clustered network.

Acknowledgments. This work is supported by the National Natural Science Foundation of China (61871456).

References

1. Evans, D.A., Fu, G.C., Hoveyda, A.H.: Rhodium(I)- and Iridium(I)-catalyzed hydroboration reactions: scope and synthetic applications. ChemInform **114**(17), 5117–5124 (1992)
2. Geng, Y., Liu, J., Ning, Y.: Research on clustering algorithm for space information networks. IEEE (2011)
3. Jiang, C., Yun, R., Zhou, Y., et al.: Low-energy consumption uneven clustering routing protocol for wireless sensor networks. In: 2016 8th International Conference on Intelligent Human-Machine Systems and Cybernetics (IHMSC). IEEE (2016)
4. Rahmani, N., Kousha, H., Darougaran, L., et al.: CAT: the new clustering algorithm based on two-tier network topology for energy balancing in wireless sensor networks. In: 2010 International Conference on Computational Intelligence and Communication Networks. IEEE (2011)
5. Lin, C.R., Gerla, M.: A distributed architecture for multimedia in dynamic wireless networks. In: IEEE Global Telecommunications Conference. IEEE (2002)
6. Shiokawa, S., Chen, D.: Location based clustering scheme considering node mobility in wireless sensor networks. In: International Conference on Ubiquitous and Future Networks. IEEE (2014)
7. Bennani, K., Ghanami, D.E.: Particle swarm optimization based clustering in wireless sensor networks: the effectiveness of distance altering. In: 2012 International Conference on Complex Systems (ICCS). IEEE (2013)
8. Misra, S., Kumar, R.: A literature survey on various clustering approaches in wireless sensor network. In: International Conference on Communication Control and Intelligent Systems. IEEE (2016)
9. Liu, Y., Gao, J., Zhu, L., et al.: A clustering algorithm based on communication facility in WSN. In: WRI International Conference on Communications and Mobile Computing. IEEE Computer Society (2009)
10. Malone, D., Duffy, K., Leith, D.: Modeling the 802.11 distributed coordination function in nonsaturated heterogeneous conditions. IEEE/ACM Trans. Netw. (TON) **15**, 159–172 (2007)
11. Bianchi, G.: Performance analysis of the IEEE 802.11 distributed coordination function. IEEE J. Sel. Areas Commun. **18**(3), 535–547 (2000)

A Dynamic QoS Routing Mechanism for Deep Space Network Based on SDN Architecture

Dong Yan(✉), Ming Gu, Xiongwen He, and Luyuan Wang

Institute of Spacecraft System Engineering CAST, Beijing 100094, People's Republic of China

Abstract. Some characteristics of deep space communication network are well known, such as nodes limited resources, long communication delay and nearly no guarantee for the transmission quality, this paper proposes a dynamic QoS routing mechanism for deep space network based on SDN architecture, which uses heuristic algorithm to discover and update the routing paths dynamically. According to the changing of network topology and performance constantly, the algorithm can select the optimal path according to the current real-time network performance state. In case of the characteristics of deep space network, the heuristic algorithm makes a lot of improvements from the global point of view, so that it can be more suitable for deep space network environment. Finally, the paper verified the performance of the algorithm by simulation experiments.

Keywords: Deep space communication network · Dynamic routing · Heuristic algorithm

1 Introduction

Deep space exploration is an important way for human beings to explore unknown world, along with the implementation of China's lunar exploration project and Mars exploration project. Our country has been in the forefront of the world in the field of deep space exploration, and has gradually established our deep space communication network. Future deep space exploration missions, including follow-up missions of lunar exploration, manned lunar exploration and asteroid exploration, will further expand the scale of China's deep space communication network and the types of space services it carries. It is an important trend of development for the integration of deep space communication network with near-earth satellite network and terrestrial Internet, which is considered by many researchers as an important part of the next generation Internet.

The deep space network is well known for its long delay, asymmetry in uplink and downlink, etc. The network with this characteristic is called DTN (delay tolerant network). The research and construction of the deep space network is still in the preliminary stage. The scale of the network is not very large, with only a small number of nodes, and no higher QoS requirements. At the same time, it is also limited by the computing and storage power of network nodes.

Q. Guo et al. (Eds.): WiSATS 2021, LNICST 410, pp. 257–266, 2022.
https://doi.org/10.1007/978-3-030-93398-2_26

Deep space networks routing often uses static methods or a combination of dynamic and static methods. With the expansion of the deep space network, the types of data increasing sharply. They are not considering the load status of the node, so they may also cause even a sharp drop in the performance of the entire network. [1] In order to alleviate this situation, researchers have tried a variety of solutions. Software-defined network (Software Defined Network, SDN) technology is one of the most promising technologies. It reduces the complexity of the data plane effectively, through the deconsolidation of data plane and control plane, which is easier to realize the centralization of the real-time network status and logic control. In this way, it can achieve more flexible controlling and processing of the network architecture, and applying efficient routing strategies. At present, researchers have carried out a lot of research on routing strategies for satellite networks, and proposed a variety of programs. These schemes use the regularity and periodicity of satellite movement, but most of them only aim at finding shorter paths or maintaining routes by improving flooding methods, without considering the inherent characteristics of deep space network nodes which may lead to routing performance problems. With the enrichment and development of space applications in deep space network, the demand for service quality is also higher and higher.

In order to solve these problems, this paper proposes a dynamic routing mechanism based on SDN Architecture for Deep Space Network (SADR). According to the status of the deep space network QoS guarantee capability, the target node can be found, while avoiding excessive load on nodes and affecting network performance.

2 Related Work

Chen et al. [2, 3] proposed the Satellite Grouping and Routing Protocol (SGRP). SGRP is applied to the LEOMEO network structure of double-layer satellites. It uses the relative position relationship between satellites in different orbits. The upper MEO is collecting the link time of the satellites in its group. The routing table is calculated according to the principle of transmitting data packets along the shortest delay path. When a LEO satellite moves to the position belongs to another group, the grouping member information is changing. At the same time, a new snapshot is generated. The topology in each timeslot can be considered unchanged. This method is mainly applied to near-Earth satellite networks, and does not focus on the application scenarios of deep-space networks.

Lindgren et al. [4] proposed a probabilistic routing protocol using history of encounters and transitivity (PROPHET). Unlike epidemic routing, which blindly forwards messages to all neighboring nodes, each node in probabilistic routing will estimate the probability of reaching the destination node based on historical successful routing information. Each node estimates a predicted distribution probability value for each known identical node, and when two nodes contact, the probability value is updated according to the algorithm. At that same time, the prediction distribution probability value among the node also has transitivity, so that the routing table of the whole network can be calculate. The method designs a routing method based on the probability of contact between nodes, changes the traditional fixed routing method, and can be more suitable for DTN application scenarios. However, the method does not consider the QoS service provision capability of the nodes, and may lead to poor network operation efficiency and stability. Xu et al. [5] proposed a QoS routing algorithm for LEO satellite network, DQA). The algorithm considers states of the satellite node, can be more effective to avoid overloading of nodes. When it performs routing calculations, it combines and optimizes the two performance parameters of satellite network delay and link utilization, which can reduce the probability of network congestion. However, the algorithm is only designed according to specific limited QoS parameters, which limits the applicability of the algorithm, and does a lot of work to the local optimization of the link, which make the algorithm convergence slower.

The deep space network QoS routing algorithm based on SDN architecture proposed in this paper uses the SDN controller to grasp the global information and obtain the real-time status of the network. It adopts a heuristic intelligent optimization method based on the ant algorithm and combines the use of deep space network scenarios. Comprehensive consideration of various QoS parameters to complete the deep-space network QoS intelligent routing calculation, which can prevent a large amount of data from passing through certain high-load nodes, and optimize it from the global perspective.

It further speeds up the collection time of the algorithm and can effectively solve the problems in the above-mentioned research.

3 SDN-Based Deep Space Network Architecture

SDN technology is a new technology in the research of network architecture in recent years. It realizes flexible control and centralized management of the network by separating control from data and combining programmable control [6–8]. For the actual deep space network, the control plane is responsible for collecting network-wide information, and performing complex routing calculations, and then sending the routing rules to the data plane node. The nodes of the data plane no longer need to perform tasks such as complex calculation, state acquisition and heavy storage. They can focus on data forwarding. This feature of SDN technology is very suitable for satellite nodes with severely limited resources. It can effectively reduce node overhead, and ensure more stable and efficient operation of satellite networks. The control plane needs to complete functions such as node search, state collection, routing calculation, rule distribution, etc. It requires higher basic performance such as higher processing power and storage capacity. Therefore, the control plane nodes are always placed on the ground. Although this method can solve the problem of basic capabilities of the control plane, it is not suitable for the condition of relatively limited distribution of ground stations. When satellite nodes are not in the coverage of the ground station or belongs to a long-distance deep-space communication scenario, it is impossible to deliver the control information to the nodes in time. The transmission time may be several hours to more than ten hours, which will be difficult to adapt to the higher real-time and dynamic requirement of future satellite networks. This paper proposes a satellite network model with a master-slave controller structure, called SADR model. Taking the earth-moon communication scenario as an example, the control plane is a two-stage structure. The master controllers are placed on the ground to undertake heavy storage, analysis of the node status of the entire network, complex routing calculations, and mode switching control functions. In addition, this type of node usually has a higher basic performance than the normal satellite node, so it can also ensure a higher work efficiency when implementing the controller function. The slave controller nodes mainly undertake the new nodes discovery, nodes status collection and sent to the master controller. It obtains the routing and forwarding rules from the master controller nodes and sent to the data plane nodes. The schematic diagram of the SADR model is shown in the Fig. 1 below.

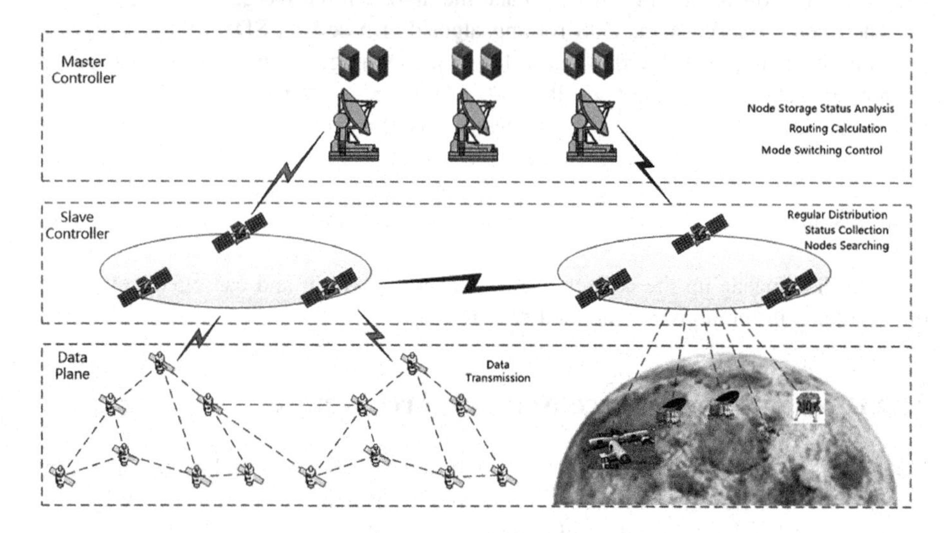

Fig. 1. SADR model figure

4 Dynamic Routing Algorithm Based on Real-Time Status of Deep Space Network Nodes

The current space network routing method mainly improves the path length and topology maintenance, and less considers the limited resources of space nodes and the limited service capacity, which makes it difficult to ensure the efficiency of data transmission and flexible management of the network. Based on the above-designed deep space network model with a two-stage control structure, this paper intends to design routing mechanism from the following aspects: Firstly, it designs a heuristic intelligent dynamic routing algorithm for the characteristics of deep space networks. According to the performance status of deep space network nodes, it finds the optimal path to avoid node overload; then, it uses a global perspective to optimize the algorithm iteratively, which can improve the convergence time of the algorithm.

4.1 Design Goals

When the nodes in the deep space network move in the space, the type of user traffic and the size of application data they serve will change. If these nodes are in one or several important transmission links in the hotspot area, it may cause communication data blocking on these nodes. Given a graph G (V,E) and the connection request from the source node to the destination node, the goal for routing is to find an optimal path with minimum cost from source to the destination, which meets the QoS command of users. In the deep space network, because the node capacity is severely limited, we need to focus on how to make the node load be in a reasonable range, so that the whole network can run stably and efficiently. Therefore, it is particularly important to choose the nodes with strong basic performance to complete routing functions for the realization of QoS routing in satellite networks. The goal of SADR routing algorithm is to find a better path from the source to the destination by combining the short delay path and the basic performance of the node.

(1) Limit of transmission delay: $delay(p(s,d)) \leq D$
(2) Available bandwidth ratio limit: $\varepsilon_{ij} = (C_{ij} - R_{ij})/C_{ij} \geq B$
(3) Link disruption limit: $\omega_{sd} = disruption(p(s,d)) \leq N$

Among them, s and d are the source node and the destination node, i and j are the deep space network nodes, R_{ij} is the data transmission rate on the link (i, j), C_{ij} is the bandwidth of the link (i, j), and R_{ij} can represent the used bandwidth of link (i, j). ε_{ij} can represents the degree of congestion, the smaller the value, the higher the degree of congestion. D, B, and N are set constants, which represent the basic performance requirements for guaranteeing user service QoS.

4.2 SADR Algorithm

M. Dorigo et al. [9–12] proposed the ant algorithm, which is a heuristic algorithm to simulate the behavior of ant colonies. The algorithm introduces a positive feedback parallel mechanism, which has strong robustness, excellent distributed computer capability, and easy integration. It can have a better result for the complex problem of multi-element composition in a reasonable time. The principle of the algorithm is that when ants move, they will release a volatile substance called pheromone on the path they travel.

The probability of other ants in the colony choosing this path is proportional to the size of the pheromone. For deep space network scenario, a path with better link conditions will pass more ants, so that more pheromone will be left on this path. At the same time, more ants will be attracted from the more pheromone on the path through this path, thus forming a positive feedback effect.

SADR designs multi-condition QoS dynamic routing mechanism for deep space network. The probability of ant k selecting the next hop node j at node i is defined as follows:

$$p_{ij}^{k} = \frac{[\mu_{ij}] \times [\eta_{ij}]^{\alpha} \times [\varepsilon_{ij}]^{\beta} \times [\omega_{ij}]^{\gamma}}{\sum_{x \in N(i)} [\mu_{ix}] \times [\eta_{ix}]^{\alpha} \times [\varepsilon_{ix}]^{\beta} \times [\omega_{ix}]^{\gamma}} \tag{1}$$

μ_{ij} is the pheromone on the link (i, j), N(i) is the collection of all neighbor nodes of node i, $\eta_{ij} = 1/d_{ij}$, d_{ij} represents the delay from node i to node j, $\varepsilon_{ij} = (C_{ij} - R_{ij})/C_{ij}$ represents the ratio of available bandwidth from node i to node j, $\omega_{ij} = 1/n_{ij}$, n_{ij} represents the number of link interruptions from node i to node j. α, β, γ indicates the importance of delay, available bandwidth ratio and link interruption to user QoS. SADR will choose the node with shorter delay, larger available bandwidth, smaller number of link interruptions and higher pheromone strength as the basis of next hop routing algorithm. In addition to the above parameters, the basic performance parameters of space network can also include packet loss rate, energy support capability of satellite nodes, etc., which can be flexibly selected and adjusted according to different application environments.

Pheromone plays a very important role in path selection. It changes with the process of searching the destination node. The higher the pheromone strength is, the more ants will be attracted to pass through the node. Pheromone intensity will be updated in several stages.

(1) Pheromone local update
After the ant completes a node selection, the node is placed in the visited node set, and then the pheromone is updated locally.

$$\mu_{ij} \leftarrow (1 - \theta)\mu_{ij} + \Delta\mu_{ij} \tag{2}$$

$$\Delta\mu_{ij} = \mathrm{A} \times [\eta_{ij}]^{\alpha} \times [\varepsilon_{ij}]^{\beta} \times [\omega_{ij}]^{\gamma} \tag{3}$$

$\Delta\mu_{ij}$ means that in the process of searching the next node this time, the pheromone is updated according to the QoS status of the node. θ means that the pheromone volatilization parameter ($0 \leq \theta \leq 1$). A is a constant, which controls the increase of local pheromone update.
In the next hop node selection, it is necessary to make sure that the node is not in the visited node set to avoid the appearance of routing ring, and then determine the selection probability of the next hop node according to the size of pheromone, and select the next hop node according to the probability.

(2) Pheromone Global Update
When the complete path from the source node to the destination node is completed, the pheromone of all nodes on the path is updated globally. The global update method is as follows.
If $\mathrm{T}(P_{sd}) \leq D$ and $(i,j) \in P_{sd}$

$$\mu_{ij} \leftarrow (1 - \varphi)\mu_{ij} + \Delta\mu_{ij}^{'} \tag{4}$$

$$\Delta\mu_{ij}^{'} = B \times [\eta_{ij}]^{\alpha} \times [\varepsilon_{ij}]^{\beta} \times [\omega_{ij}]^{\gamma} \tag{5}$$

If $(i,j) \notin P_{sd}$,

$$\mu_{ij} \leftarrow (1-\varphi)\mu_{ij} \tag{6}$$

For deep space communication scenario, information transmission delay is one of the most important indicators. Only when the delay of the whole communication link is within a reasonable range, communication can be efficient and reliable. D is the end-to-end delay requirement. T(P_{sd}) is the delay for ants to complete the complete path selection. φ is the pheromone global attenuation parameter $(0 \leq \varphi \leq 1)$. B is a constant, and it controls the increase of pheromone global update. For the nodes that are not on the ant path, the updating of pheromone is the attenuation of pheromone.

(3) Extra incentive mechanism
For the ants that complete the whole path selection, SADR algorithm will calculate the QoS state value of the path, which is defined as:

$$\overline{QV(P_{sd})} = (\sum_{ij \in P_{sd}} QV(P_{ij}))/node\_num$$

$$QV(P_{ij}) = [\eta_{ij}]^{\alpha} \times [\varepsilon_{ij}]^{\beta} \times [\omega_{ij}]^{\gamma}$$

QV is the QoS value of a node according to the QoS calculation model. SADR will give extra pheromone rewards to the path that has completed end-to-end communication and the average QoS value of the path is the best.
If $\overline{QV(P_{sd})} \geq \max(\overline{QV(P_{sd})})$, $(i,j) \in P_{sd}$

$$\mu_{ij} \leftarrow \mu_{ij} + \Delta\mu_{ij}^{''} \tag{7}$$

$$\Delta\mu_{ij}^{''} = C \times [\eta_{ij}]^{\alpha} \times [\varepsilon_{ij}]^{\beta} \times [\omega_{ij}]^{\gamma} \tag{8}$$

$QV(P_{sd})$ is the complete path QoS value, C is a constant, which controls the increase of pheromone QoS value.

For deep space network, due to the characteristics of long delay, high dynamic and high packet loss rate, in addition to the nodes and links that meet the delay requirements and have good QoS, the path with less link hops should be selected as far as possible to further ensure the reliability of data transmission on the path, reduce the cost of the whole network system. SADR gives extra reward to the path with less hops.

If $NH(P_{sd}) \leq \min(NH(P_{sd}))$, $(i,j) \in P_{sd}$

$$\mu_{ij} \leftarrow \mu_{ij} + \Delta\mu_{ij}^{'''} \tag{9}$$

$$\Delta\mu_{ij}^{'''} = K/NH(P_{sd}) \tag{10}$$

$NH(P_{sd})$ is the complete hop number of the ant, K is a constant, and controls the increase range of pheromone.

It can be seen from the above that SADR has both local update and global update mechanism. For the complete path from the source node to the destination node with fewer hops and a better QoS state, more pheromone rewards are given. In this way, subsequent traffic will pass this path. When the algorithm described in this paper establishes the QoS state description model, only three of the most typical and commonly used QoS parameters are selected. In actual applications, the required QoS parameters can be adjusted according to requirements.

4.3 SADR Algorithm Description

The goal of SADR is to learn from the ant algorithm and periodically find the optimal path between the source and the destination to meet the user's QoS requirements. Each time a node is selected, the node is added to the set of visited node set. At the same time, a lifetime is set for the path-finding ants. If the destination node is not reached or there is no way to choose during the lifetime, it is considered that the path searching fails. If the destination node is reached during the lifetime, it is considered that the path searching is successful. Each time the ants that completes the path search will compare the corresponding values generated by the previously completed ants in terms of path QoS status, hop count, etc., to continuously obtain a better path, and finally obtain the approximate optimal value under certain conditions.

SADR is mainly divided into two phases: dynamic routing lookup update and routing table distribution. Firstly, in the initialization stage, there is a set of default topological connection relationships to make the entire satellite network maintain data communication in the initial state or emergency state, even if it is not a perfect routing mode. And then, it acquires state information of the deep space network nodes by the slave controller and summarizes the state information to the master controller. The Master controller performs dynamic QoS route calculation according to the SADR algorithm. Because of the constantly moving of satellite nodes, and the changing of space environment, the path found in the previous stage may not be a high-quality path or even a usable path after a certain time interval. Therefore, it is necessary to set an update cycle, and in each update cycle. And finally, in the routing table distribution stage, the master controller distributes the calculated routing table to the slave controller, and then the slave controller forwards the received routing table to nodes covered.

5 Performance Evaluation

This section verifies the performance of end-to-end QoS value and delay with DQA. For different space routing algorithms, the end-to-end delay of satellite nodes is a basic network performance indicator, which can explain the rationality of the network architecture. And the end-to-end QoS status value can reflect the current QoS status of the node and link on the selected path.

The experiment in this section sets the scenario as a combination of lunar surface network and near-Earth space network. The lunar surface network is composed of 3 lunar relay satellites and 30 lunar nodes. The lunar node has 4 communication links. The near-Earth network consists of 3 relay satellites and 6 orbital planes, 11 satellites on each orbital plane, for a total of 66 low-orbit satellites. The nodes also have 4 inter-satellite links (Figs. 2 and 3).

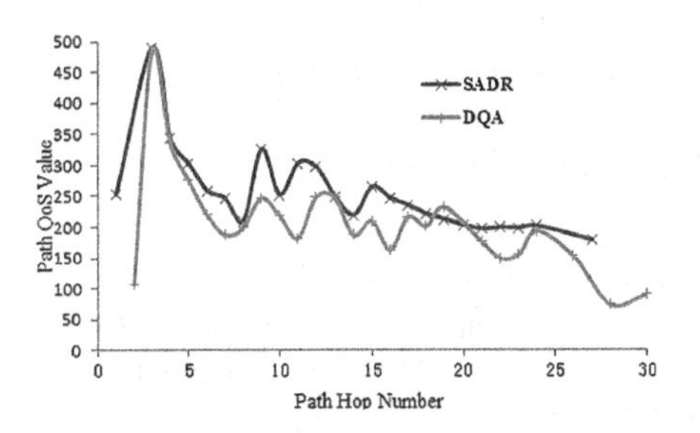

Fig. 2. QoS value for end-to-end

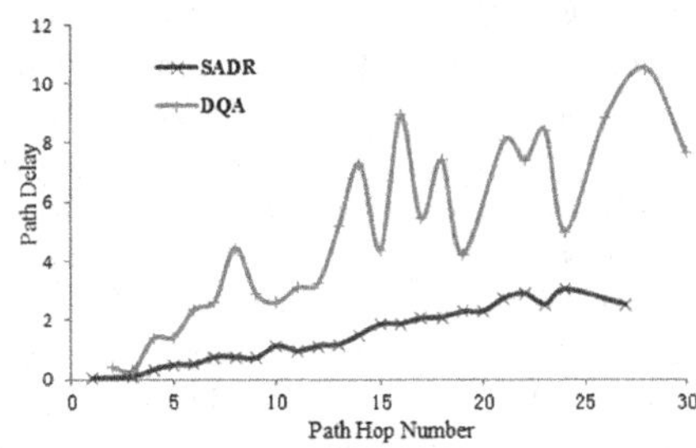

Fig. 3. Delay for end-to-end

6 Conclusion

SADR is a QoS routing algorithm for deep space networks based on SDN architecture. The algorithm is composed of the following parts: 1) SADR dynamic routing lookup process, it acquires state information of the deep space network nodes by the slave controller and summarizes the state information to the master controller. The Master controller performs dynamic QoS route calculation according to the SADR algorithm. 2) SADR routing table distribution stage, the master controller distributes the calculated routing table to the slave controller, and then the slave controller forwards the received routing table to nodes covered. Finally, the simulation results show that SADR has better performance in end-to-end path QOS state value and end-to-end average delay.

Acknowledgement. This work is supported by the 'The National Key Research and Development Program' of China (No. 2018YFB1800301).

References

1. Kwang, M.S., Weng, H.S.: Ant colony optimization for routing and load-balancing: survey and new directions. IEEE Trans. Syst. Man Cybern. **33**(5), 560–572 (2003)
2. Chen, C., Ekici, E., Akyildiz, I.F.: Satellite grouping and routing protocol for LEO/MEO satellite IP networks. In: Proceedings of the 5th ACM International Workshop on Wireless Mobile Multimedia, Rome, Italy, vol. 9, pp. 109–116 (2002)

3. Chen, C., Ekici, E.: A routing protocol for hierarchical LEO/MEO satellite IP networks. ACM Wirel. Netw. J. **11**(4), 507–521 (2005)
4. Lindgren, A., Doria, A., Schelen, O.: Probabilistic routing in intermittently connected networks. SIGMOBILE Mob. Comput. Commun. Rev. **7**, 19–20 (2003)
5. Xu, H., Wu, S.: A distributed QoS routing based on ant algorithm for LEO satellite network. Chin. J. Comput. **30**(3), 361–367 (2007)
6. Fan, Z., Wu, H., Xu, J., et al.: An optimization algorithm for spatial information network self-healing based on software defined network. In: Proceedings of the 12th International Conference on Computer Science and Education (ICCSE), New York, pp. 369–374 (2017)
7. Jiang, Z., Wu, Q., Li, H., et al.: ScMPTCP:SDN cooperated multipath transfer for satellite network with load awareness. IEEE Access **6**, 19823–19832 (2018)
8. Yang, L., Teng, Q., Kong, Z., et al.: Design of spatial information network routing strategy based on SDN architecture. Spacecr. Eng. **28**(5), 54–61 (2019)
9. Caro, G.D., Dorigo, M.: AntNet: distributed stigmergetic control for communications networks. J. Artif. Intell. Res. **9**, 317–365 (1998)
10. Bean, N., Costa, A.: An analytic modelling approach for network routing algorithms that use "ant-like" mobile agents. Comput. Netw. **49**, 243–268 (2005)
11. Ducatelle, F., Di Caro, G., Gambardella, L.M.: Ant agents for hybrid multipath routing in mobile ad hoc networks. In: Proceedings of the 2nd Annual Conference on Wireless on demand Network Systems and Services, Switzerland, pp. 44–53 (2005)
12. Dhillon, S., Van Mieghem, P.: Performance analysis of the AntNet algorithm. Comput. Netw. **51**, 2104–2125 (2007)

Satellite Traffic Forecast Based on Multi-dimensional Periodic Features

Weidong Zhou[1], Yana Qian[2], Kanglian Zhao[1], Wenfeng Li[1(✉)], and Fa Chen[1]

[1] Nanjing University, Nanjing, China
zhaokanglian@nju.edu.cn, leewf_cn@hotmail.com
[2] Shanghai Astronautics Electronic Co., Ltd., Shanghai, China

Abstract. With the development of satellite networks and the increase in business requirements, it is a challenge to better provide high-quality service to users. The accurate prediction of end-to-end traffic can contribute to the realization of congestion control, resource allocation and anomaly detection. Accurate flow forecasts need to consider the short-term and long-term features of the flow. Our paper proposed a model based on deep learning named Multi-Dimensional Temporal Feature Neural Network (MTFNN) to capture both short-term dependencies and long-term dependencies for traffic prediction. MTFNN mainly contains two components: 1) Short-Term Temporal Dependencies which based on Long-Short Term Memory network (LSTM), used to predict the basic trend of traffic. 2) Long-Term Temporal Dynamic Similarity which based on LSTM and Attention mechanisms, used to improve the model's sensitivity to fluctuations and peak prediction. Experiments performed on real-world public traffic datasets show our proposed model has a smaller prediction error and more accurate peak prediction.

Keywords: Traffic flow prediction · LSTM network · Attention mechanism

1 Introduction

The satellite information network is characterized by large coverage and long communication distance, and is an important part of the integration of the sky-ground network. With the construction of satellite information networks, satellite traffic services continue to increase. However, the bandwidth and other resources of satellite nodes are relatively limited. Therefore, further management and allocation methods are needed to avoid the degradation of service quality due to increased services. Traffic distribution can intuitively reflect the current network status, and accurately predict the future network traffic distribution can be used for congestion control, resource allocation, and intelligent routing.

High non-linearity, multi-scalar, self-similarity and long-term dependence are the main characteristics of satellite networks [1]. These statistical characteristics determine the predictability of traffic. Previous work had made traffic predictions by analyzing these characteristics of traffic, and had achieved good research results.

Q. Guo et al. (Eds.): WiSATS 2021, LNICST 410, pp. 267–277, 2022.
https://doi.org/10.1007/978-3-030-93398-2_27

Previous research work has proposed many forecasting models, which can be summarized into two categories: linear algorithms and nonlinear algorithms. The Auto Regressing Moving Average (ARMA) [2], the Auto Regressing Integrated Moving Average (ARIMA) [3, 4] and the Holt-Winter algorithm [5] are most general algorithms applied to network traffic prediction problems. However, these algorithms are based on mathematical theoretical assumptions (such as linear, moving average), and in large-scale networks, these methods have insufficient ability to deal with nonlinear traffic characteristics, which will lead to a decrease in prediction accuracy. Nonlinear prediction methods commonly involve neural network (NN) [6]. Dingde Jiang [7] analyzed the Recurrent Multilayer Perceptron (RMP) to forecast traffic matrices in large scale network. Bermolen and Rossi [8, 9] found that the SVM model has good performance in link load prediction problems. Recurrent Neural Network (RNN) can store the state of each time interval in the network, thus having the characteristics of time memory. RNN is widely used in time series prediction and classification tasks, and has excellent performance in sequence modeling tasks Ramakrishnan [10] compared the RNN and its variants (such as LSTM, Gated Recurrent Unit Network (GRU)) for traffic prediction to linear forecasting models.

However, the above approaches only consider the short-term time dependence of the flow sequence, but ignore the long-term time dependence. Considering short-term time characteristics to predict future flow can achieve good prediction accuracy, but there are deficiencies in the prediction of severe fluctuation sequences and peak flow. Traffic data has strong daily, weekly or even monthly periods, which can be used for future traffic forecasting. Satellites will serve different regions with the orbital period. Switching between regions will produce fluctuations in traffic, and the day-to-night changes in the region itself will also produce fluctuations in traffic. These features have long-term characteristics, and only considering short-term time characteristics will obviously ignore these characteristics which can help the forecasting model to improve the sensitivity to traffic fluctuations. LSTM is a powerful RNN architecture that can be used for time series forecasting. But in traffic prediction, LSTM can only consider the flow sequence of the past several minutes or a few hours (which is determined by time interval), so some long-term temporal features will not be fully used.

In this case, we propose a novel deep-learning mode named Multi-Dimensional Temporal Feature Neural Network (MTFNN), which mainly contains two components: 1) Short-Term Temporal Dependencies employs the Long-Short Term Memory network (LSTM) to capture the short-term dependence, 2) Long-Term Temporal Dynamic Similarity uses attention-based recurrent structures to model the long-term temporal dependencies of each flow, and then predicts flow volume according short-term dependencies and long-term dependencies.

The rest of the paper is organized as follows. Section 2 presents the MTFNN model to make predictions and the dynamic periodicity of traffic. Experiment and results are presented in Sect. 3. The paper is concluded by Sect. 4.

2 Prediction Model

This section describes the details for our proposed model, Multi-Dimensional Temporal Feature Neural Network (MTFNN). Figure 1 shows the architecture of our model, which can be divided into three parts: short-term time dependence, long-term time dependence and final prediction. The short-term time dependence part is based on the historical flow sequence near the forecast time to make the prediction and is used to fit the basic trend of the flow sequence, while the long-term time-dependence part is composed of a traffic input sequence strongly correlated with the forecast time in a longer time interval to capture the periodic characteristics and peak flow and is used to strengthen the model's accuracy of volatility and peak forecasting, and the final forecasting part is mainly to perform the final mapping of the predicted values of the first two parts to obtain The final predicted value can dynamically adjust the impact weight of the short-term and long-term predicted results on the final predicted value, which makes the model more robust and can adapt to different traffic sequences.

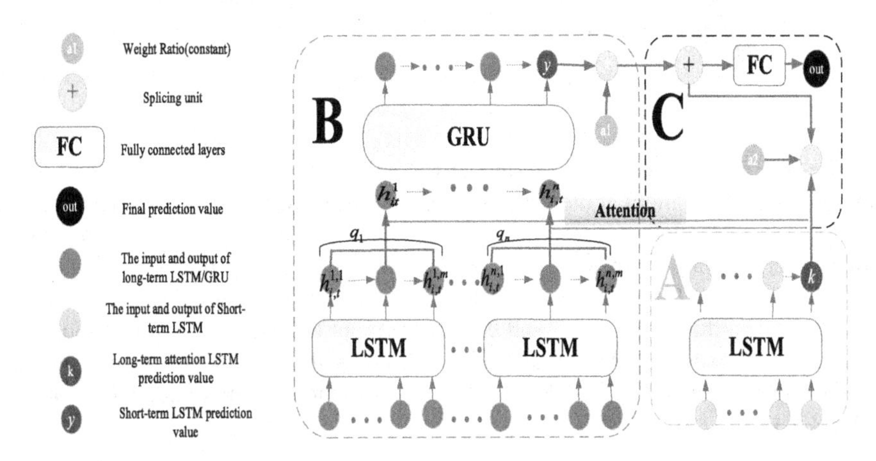

Fig. 1. The Architecture of MTFNN. A: Short-term temporal dependency. B: Long-term temporal network. C: Final prediction.

2.1 Short-Term Temporal Dependency

The short-term dependency part means that we obtain short-term characteristics of traffic and make traffic forecasts based on short-term historical data. Long short-term memory (LSTM) is a special model used to solve the problem of gradient disappearance and explosion during RNN training. The three forget gate structures can fully deal with the time characteristics in the sequence, and show excellent performance in natural language processing and time sequence prediction tasks.

LSTM adds three memory gate structures on the basis of RNN (Fig. 2). The input of the forget gate is the previous unit output h_{t-1} and the current time input data x_t. The output f_t of the forget gate determines what kind of information can pass through the unit. The output i_t of the input gate is a control signal, which determines which new input information can be input to the unit. The output gate determines what information is output [11].

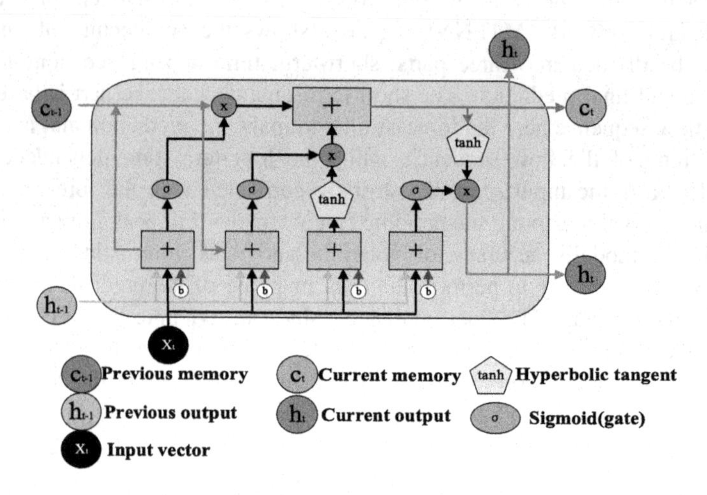

Fig. 2. LSTM architecture.

The specific working process of the LSTM unit is as follows: At time t, the LSTM input x_t includes the input at the current time, the last hidden state h_{t-1} and the last unit state c_{t-1} at the previous time, that is, retain the useful information in the traffic sequence at the last moment. x_t and h_t get the current input unit state c_t through activation function *Tanh*. The current unit state c_t is the sum of the last unit state c_{t-1} passing through the forget gate f_t and the current input unit state c_t passing through the input gate i_t. Finally, LSTM output h is calculated by the current unit state c_t through the activation function and output gate o_t. The calculation formula for each variable is as Eq. (1) to Eq. (5).

$$i_t = \sigma_g(W_{xi} * x_t + W_{hi} * h_{t-1} + W_{ci} * c_{t-1} + b_i) \tag{1}$$

$$f_t = \sigma_g\left(W_{xf} * x_t + W_{hf} * h_{t-1} + W_{ci} * c_{t-1} + b_f\right) \tag{2}$$

$$C_t = f_t \otimes c_{t-1} + i_t \otimes \sigma_c(W_{xc} * x_t + W_{hc} * h_{t-1} + c) \tag{3}$$

$$o_t = \sigma_g(W_{xo} * x_t + W_{ho} * h_{t-1} + W_{co} * c_t + b_0) \tag{4}$$

$$h_t = o_t \otimes \sigma_g(c_t) \tag{5}$$

where: W_{xf}, W_{xi}, W_{xo}, W_{xc} are the training matrices related to the input x_t while W_{hc}, W_{hi}, W_{ho}, W_{hf} are the training matrices related to the last hidden state h_{t-1}, W_{ci}, W_{cf}, W_{co} are diagonal matrices connected to the current unit state a c_t and gate function, b_i, b_f, b_c, b_o are offset factors. σ_g is the activation function and $\otimes$ denotes the Hadamard product.

2.2 Long-Term Temporal Dependency

The long-term dependency part means that we obtain long-term characteristics of traffic and make traffic forecasts based on long-term historical data. Although compared with the traditional time series prediction model, LSTM has a stronger ability to process long-term dependence, but as the time sequence rises, LSTM will face the problem of disappearing gradient and inconspicuous periodicity. In order to avoid the above problems, we need construct a new time series that reflects long-term characteristics, rather than just increase the length of the time series. Usually we use the relative time interval of a clear prediction moment, for example (the same moment yesterday, the same moment last week). A reasonable and clear relative time interval can reflect more time characteristic information at the current moment. Through the analysis of the real data set, we found that the traffic data is not strictly periodic, and there is a certain offset. A week's summary traffic situation of a certain node is selected from the Abilene dataset, as shown in Fig. 3, the interval is one day and the peak traffic appeared between 20:40–23:00 (Although the peak traffic on Thursday appeared at 16:40, the traffic volume at 21:40 is still a local maximum). In satellite network, the relative period interval is set according to the satellite orbit period, which can predict the traffic situation of a certain area in the past period of time, and reduce the abnormal traffic fluctuation caused by the change of satellite coverage area.

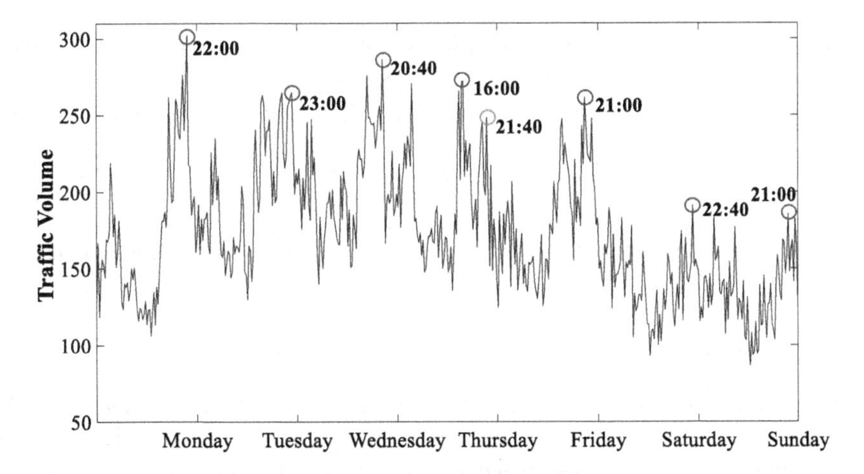

Fig. 3. Long-term periodicity and shifting of traffic

In order to better capture the long-term time characteristics of the time series, we need to set the relative time interval reasonably and consider the deviation of the flow. As shown in Fig. 1, we set the relative time interval to one day, where N represents the past the number of days, and M represents the number of traffic nodes we select in each day, for example, when the next moment we want to predict is 21:00 and M = 5, we select the past 20:50–21:10 a total of 5 moments of traffic matrix. Then we use LSTM to extract the temporal features of the past daily sequence.

Inspired by the Attention paper [12], in the last step, we adopted the idea of attention mechanism to solve the problem of traffic offset, and obtained the weighted result by calculating the correlation between the daily traffic in the past and the traffic at the current forecast moment. First, after LSTM, output the short-term forecasts of each day in the past, and get the weighted summation result $h^n_{i,t}$ of m time interval every day according to the attention. Specifically defined as:

$$h^n_{i,t} = \sum_{q \in Q} \alpha_t h^{n,m}_{i,t} \tag{6}$$

where weight α_t measures the similarity of the time interval m in day $n \in N$. Formally, the weight α_t is defined as:

$$\alpha_t = algin(q,k) = \frac{\exp(score(q,k))}{\sum_{q \in Q} \exp(score(q,k))} \tag{7}$$

where k means the output of short-term LSTM, q means the sequence of m time intervals in day $n \in N$.

In this paper, the alignment function is defined as:

$$\text{score} = \text{sigmoid}\left(v_a^T \tanh(W_a q + Uk + b)\right) \tag{8}$$

where $v_a{}^T$, W_a, U, b are learned parameters. $v_a{}^T$ denotes the transpose of v_a. For each pervious day p, we get the weighted sum presentation $h^p_{i,t}$. Then we use a GRU to preserve the sequential information by using these periodic representations as input.

$$\text{y} = \text{GRU}\left(h^p_{i,t}, h^{p-1}_{i,t}\right) \tag{9}$$

where we regard the output of the last time interval y as the representation of long-term temporal dependency.

2.3 Final Prediction

After getting long-term attention LSTM prediction value y and short-term LSTM prediction value k, we splice y and k through a fully connected layer to achieve the final result. And the out contains spatial and temporal information. However, the long-term characteristics of different traffic sequences have different strengths. Simply splicing y and k give the same weight to y and k (the weight of the influence on the forecast result) and could not get the best prediction result. For different time series, the weight distribution of y and k need to be adjusted flexibly to achieve a better prediction effect.

3 Experiment Settings

3.1 Dataset

Real satellite network traffic data sets are difficult to obtain, and simulated traffic data sets are obtained through simulation formulas and parameters, which lack the variability and complexity of real data. The uncertainty and complexity of real traffic will lead to a decrease in the accuracy of traffic prediction. Therefore, this paper selects ground traffic data sets to conduct preliminary research on satellite network traffic predictions.

According to the topology diagram of the Abilene network (Fig. 4), it can be seen that the nodes in the Abilene network are distributed in various major states in the United States. The distance between the nodes is relatively long and there are direct and indirect links between the nodes, which similar to the satellite topology. We use the ground node as the ground base station, and the node sending traffic is used as the traffic data of the ground base station accessing the satellite, and the traffic received by the node is used as the traffic accessing the ground base station after transmission through the satellite. Therefore, this paper uses the Abilene [13] dataset to evaluate our traffic prediction model. The Abilene network consists of 12 nodes (the two nodes of Atlanta are summarized into one node in the Fig. 4) and 15 links (the connection between the two nodes of Atlanta is also omitted), containing traffic information at 48383 moments. The time interval of the Abilene datasets is 5 min.

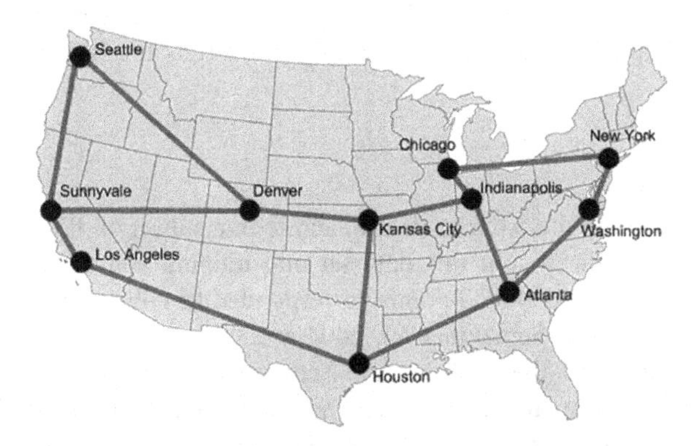

Fig. 4. Topology of Abilene network

We summarized the inflow and outflow of each nodes and spliced them into a matrix. We have performed matrix prediction and node aggregate traffic prediction respectively. Our experiment uses about 30,000 (100 days) of traffic data, of which 80% is used for model training, and the remaining 20% is used to test model performance. Finally, we normalize the data by max-min normalization method.

3.2 Evaluation Metric and Baselines

In order to intuitively reflect the prediction performance of our proposed model, we use the mean square error (MSE) as the evaluation factor

$$MSE = \frac{1}{n}\sum_{i=1}^{n}(\widehat{y}_i - y_i)^2 \tag{10}$$

where y_i is the actual value, $\hat{y}_i$ is the predicted value and n represents the number of predicted sequences.

We compare our model with some widely used time series regression models, including (1) Naive Method (NM) (2) ARIMA (3) Support Vector Machine (SVM) (4) LSTM [14] (the state-of-the-art method in traffic prediction). For model (2), (3), We use the latest implementation in the scikit-learn library.

3.3 Hyper-parameter Settings

We adjust the hyper parameters based on the performance of the validation set. In the temporal part, the length of short-term LSTM sets to 15 (considering the previous 55 min of traffic), for long-term periodic information, P is set to 7 (considering the previous 7 days), and Q is set to 7 (considering 15 min before and after of relative predicted time), the number of hidden elements of LSTM is 64, the entire network adopts the Pytorch architecture and is optimized by Adam. The batch size of the data set is 64, the learning rate is 0.0001. To avoid over-fitting, we introduced the Dropout layer and early-stop mechanism. The training process runs for 500 epochs.

4 Results

We tested our model on the Abilene dataset and took a subset of 30000 traffic volume at each node and similarly split this data set into training and test datasets. For the Abilene data set, these 30000 data points corresponded to volume measurements from May 29, 2004 at 20:00 to September 10, 2004 at 23:55.

This paper compared the performance between our proposed model and other comparison methods in matrix prediction. The matrix is composed of the inflow and outflow of 12 nodes. In addition, ARIMA and SVM output one-dimensional results, we use ARIMA and SVM to forecast each aggregate traffic and the obtained MSE is averaged as the matrix predicted MSE.

As show in Fig. 5, we can see that our proposed model achieves the best performance, and the performance of the methods using the neural network are better than the traditional network prediction method, which shows that the traditional network prediction method has shortcomings in dealing with nonlinearity and complex dynamics.

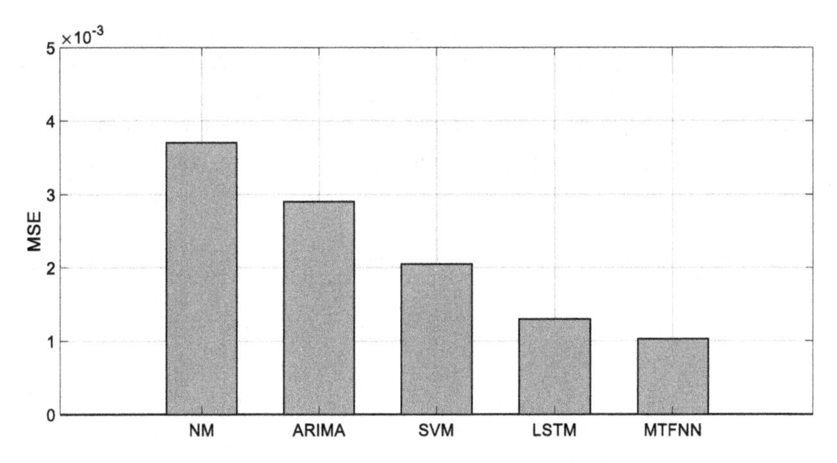

Fig. 5. Comparison of prediction errors for next matrix.

This paper also performed summary traffic prediction for each single node, and in this part, we focus on comparing the performance of LSTM and MTFNN. Compared with traffic matrix forecasting, the cost and overhead of single-node traffic forecasting are smaller. In matrix forecasting, other traffic sequences will have an impact on the current sequence, and these effects may be positive or negative.

The experiment randomly selected four nodes from 12 nodes to predict the inflow and outflow traffic. As shown in Table 1, compared with LSTM, MTFNN achieved a maximum reduction of 25.3% and a minimum of 10.3% in terms of MSE on Abilene. These results well justify the effectiveness of long-term temporal dependence on the prediction quality.

Table 1. Prediction results and comparision of end-to-end traffic

Aggregate traffic	MSE	
	LSTM	MTFNN
C-in	0.000784	0.000659
C-out	0.000244	0.000212
D-in	0.000240	0.000211
D-out	0.001217	0.001087
H-in	0.000554	0.000497
H-out	0.000838	0.000742
L-in	0.000224	0.000200
L-out	0.000677	0.000506

At the same time, the experiment randomly selected a backbone traffic sequence and the prediction situation of the edge traffic sequence. Figure 6 shows the comparison result of the specific predicted value and the true value. In part (a) (we can think of the traffic of satellites passing through low-demand areas), MTFNN can achieve low-value stable fitting and accurate peak calibration (it can be effective by setting the threshold later. Early warning peak traffic). In part (b), the predicted value can better fit the true value. Some of the predicted values are greater than the actual value. This may be due to the large peak in the long-period flow of the predicted node in the past. It is caused by a large time characteristic value, but it does not constitute an obvious misjudgment.

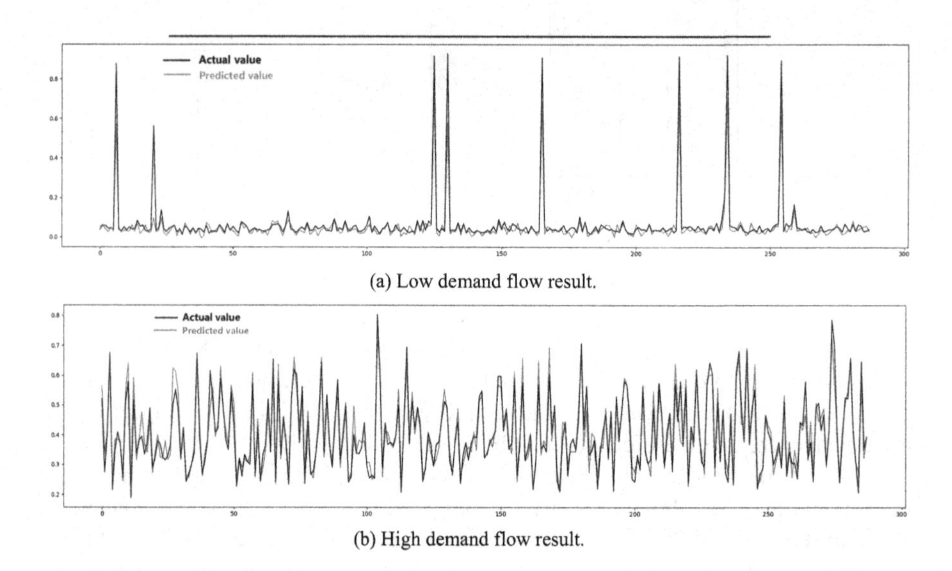

(a) Low demand flow result.

(b) High demand flow result.

Fig. 6. Comparison of predicted value and true value.

5 Conclusion

In this paper, we propose a novel traffic matrix prediction method named Multi-Dimensional Temporal Feature Neural Network (MTFNN). This model considers the multiple dimensions of the flow to predict the future flow. Firstly, we use short-term LSTM and long-term attention LSTM to extract the short-term and long-term temporal dependency of the traffic. The evaluation on the Abilene dataset show that the proposed model has better performance than the previous model. This paper also provides new ideas for traffic forecasting from multiple dimensions. In the space network, in terms of the communication requirements of the satellite orbit coverage area, this paper uses the ground real traffic data set to simulate the satellite traffic, and has achieved good experimental results. Compared with the LSTM model, it is 10.3%–25.3%. The promotion. However, the frequent changes in the topology of the space network and the frequent interruption of the link cannot be simulated by the ground data set. Therefore, the follow-up will use the ground real data set and the simulation data set to make further research work on the satellite traffic prediction.

References

1. Leland, W., Taqqu, M., Willinger, W., Wilson, D.: On the self-similar nature of Ethernet traffic. In: Proceedings of SIGCOMM 1993, pp. 183–193 (1993)
2. Yang, C., Jia, S., Zhang, H., Liu, X.: The prediction model for WSN traffic based on ARMA. Eur. J. Oper. Res. **219**(3), 738–750 (2018)
3. Maequez, P., Pinos, D., Juan, I.O.: Performance comparison in network traffic prediction for polynomial regerssion to P1P versus ARIMA and MWM. In: 2018 IEEE Conference of Russian Young Researchers in Electrical and Electronic Engineering (EUConRus). IEEE (2018)
4. Yu, Y., Wang, J., Song, M., Song, J.: Network traffic prediction and result analysis based on seasonal ARIMA and correlation coefficient. In: Proceedings of International Conference on Intelligent Systems Design and Engineering Applications (ISDEA), vol. 1, pp. 980–983, October 2010
5. Brutlag, J.D.: Aberrant behavior detection in time series for network monitoring. In: LISA, vol. 14, pp. 139–146 (2000)
6. Dharmadhikari, V.B., Gavade, J.D.: An NN approach for MPEG video traffic prediction. In: 2nd International Conference on Software Technology and Engineering, San Juan, USA, pp. V1-57–V1-61 (2010)
7. Jiang, D., Hu, G.: Large-scale ip traffic matrix estimation based on the recurrent multilayer perceptron network. In: IEEE International Conference on Communications, ICC 2008, pp. 366–370. IEEE (2008)
8. Bermolen, P., Rossi, D.: Support vector regression for link load prediction. Comput. Netw. **53**(2), 191–201 (2009)
9. Katris, C., Daskalaki, S.: Prediction of Internet traffic using time series and neural networks. In: Proceedings of International Work-Conference on Time Series Analysis (ITISE), vol. 1, pp. 594–605 (2014)
10. Ramakrishnan, N., Soni, T.: Network traffic prediction using recurrent neural networks. In: 2018 17th IEEE International Conference on Machine Learning and Applications (ICMLA), Orlando, FL, pp. 187–193 (2018). https://doi.org/10.1109/ICMLA.2018.00035
11. Gers, F.A., Schraudolph, N.N., Schmidhuber, J.: Learning precise timing with LSTM recurrent networks. J. Mach. Learn. Res. **3**, 115143 (2003)
12. Vaswani, A., et al.: Attention is all you need. In: Advances in Neural Information Processing Systems, pp. 5998–6008 (2017)
13. Zhang, Y.: Abilene dataset. http://www.cs.utexas.edu/yzhang/research/AbileneTM
14. Azzouni, A., Pujolle, G.: A long short-term memory recurrent neural network framework for network traffic matrix prediction. CoRR, abs/1705.05690 (2017)

Research and Simulation Analysis on the Characteristics of Walker Constellation Network for Global Networking Real-Time Telemetry, Track and Command

Yong Xu(✉), Siyang Zhao, Luyuan Wang, Nan Pei, Hongcheng Yan, Ming Gu, Liang Qiao, Cuilian Wang, Ling Tong, and Wei Wu

Beijing Institute of Spacecraft System Engineering, Beijing 10094, China

Abstract. The resources of ground TT&C (Telemetry, Track and Command) and data transmission station are limited in China, and the visible arc section of satellite is short, which greatly limits the timeliness of satellite on orbit operation control and information return, and affects the full play of satellite efficiency. As a means of response, the relay satellite was born to meet the needs of the full-arc TT&C of key tasks. However, the flexibility of the use of the relay satellite and the number of service objects are still difficult to meet the requirements of ordinary, large number of satellites. Therefore, it is very urgent to realize the real-time telemetry and command of all satellites in the system based on constellation system. In order to meet the requirement of real-time telemetry and command in global network, this paper designs constellation network simulation system, studies the transmission characteristics of ground telemetry and command data and autonomous telemetry and command data in typical constellation configuration, and carries out quantitative simulation analysis. It provides a reference for the optimization of network topology, node cache configuration, routing algorithm and data generation strategy in system design.

Keywords: Walker constellation · TT&C · Global networking

1 Introduction

Due to the advantages of uniform coverage and regular topology, the current satellite constellation system generally adopts Walker Constellation, and constructs inter satellite network in the satellite itself to realize the interconnection between the whole constellation and the ground system [1].

During the operation of satellite system in orbit, the monitoring of working state is basically transmitted to the ground through telemetry parameters. After receiving the data, the monitoring and command station will process and analyze it, and inject the processing parameters and on orbit task arrangement through the remote control channel. Although the autonomous operation technology is adopted, most of the satellite's on orbit operation still depends on the monitoring and task allocation on the ground.

Q. Guo et al. (Eds.): WiSATS 2021, LNICST 410, pp. 278–288, 2022.
https://doi.org/10.1007/978-3-030-93398-2_28

However, for the ground TT&C station and data transmission station, due to the occlusion of the earth, the link and data transmission can only be established in the visible arc or through the relay satellite during the whole orbit period of the satellite. This kind of short arc and intermittent telemetry and command method has many impacts on the use of satellites. From the perspective of satellite safety, it is impossible to obtain the satellite health data outside the visible arc in real time, and the failure cannot be handled in time, resulting in the spread of failure or threatening the safety of the whole satellite. From the perspective of angle of use, it is necessary to wait until the next visible arc comes when temporary demand comes, and receive the downward observation data in next visible arc segment, which seriously affects the use efficiency of the satellite [2, 3].

For this reason, it is an inevitable trend to use the inter satellite link to network the satellite system, to provide telemetry and command services to the whole network satellite through the domestic satellite in the visible arc section, and to realize the global network telemetry and command application. Taking a 24 star constellation system as an example, the global real-time telemetry and command constellation system consists of the space-based constellation network part and the ground-based telemetry and command station, the data transmission station, the mobile station and the data center [4–8].

2 Constellation Network Simulation System Design

Different from the simple communication service mode of traditional ground or relay satellite direct TT&C satellite, it is more difficult to analyze the service capacity, transmission capacity, delay characteristics and node cache of satellite networking transmission. In the design of real-time global TT&C system, in order to facilitate the analysis of the load capacity, transmission delay, node cache setting capacity and other design parameters of the network system. Reasonable, all above design items need to be analyzed and verified by simulation [9, 10] (Fig. 1).

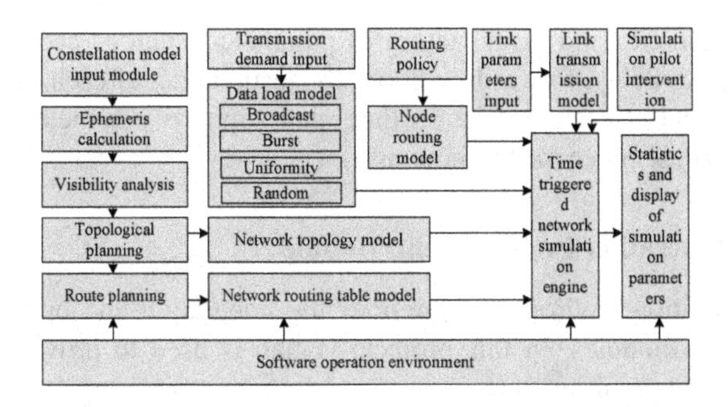

Fig. 1. Framework Design of constellation network simulation system

The figure above shows the structure of constellation networking communication simulation software, mainly including several parts.

2.1 Simulation Input

It includes constellation model input module, transmission demand input module, route processing strategy module, link parameter input module, link parameter input module, which used to set simulation related input parameters and policies.

2.2 Pilot Intervention Assembly

Simulation pilot intervention module is mainly used to adjust simulation start time, end time, simulation step length, simulation process control, fault injection and abnormal event injection.

2.3 Model and Data Generation Module

The model and data generation module generates the data and models needed for simulation, including topology model, routing table model, data load model, link transmission model and node routing processing model, based on track, visibility, chain building planning, routing planning and other domain algorithms.

2.4 Simulation Engine Module

The simulation engine module is the core part of the simulation system, which is responsible for model loading, running, time triggered event deduction, time triggered state transition and update.

2.5 Simulation Parameter Statistics and Display Module

The simulation parameter statistics and display module is used for statistics and visualization of important parameters in the process of network simulation. It supports users to extract statistics and visualization of parameters such as cache queue depth, link load, node load, delay, forwarding hops, congestion, etc. of each node and link in the simulation model at each simulation time, and supports parameter collection and user-defined statistical processing functions.

2.6 Software Operation Environment Module

The software operation environment provides the integration and operation environment of the above modules. In this project, Matlab is used to provide the software integration operation environment.

The simulation process includes input parameter settings such as scene, constellation scale, constraints and transmission load characteristics, statistical parameter extraction and display settings, and finally running network simulation software, which can support network routing rules, algorithms, design correctness, feasibility network load capacity and other simulation evaluation.

3 Modeling of Simulation Model for the Characteristics of Walker Constellation Network

In our study, the 24-star scale constellation network is used as the simulation research scene. The commonly used Walker Constellation, 1500 km altitude orbit and 60° tilt orbit are used. The number of links between stars is no more than 3–4 per satellite, and the network topology map is generated by the strategy of ring chain establishing in orbit plane and dynamic chain establishing across orbit plane (Fig. 2).

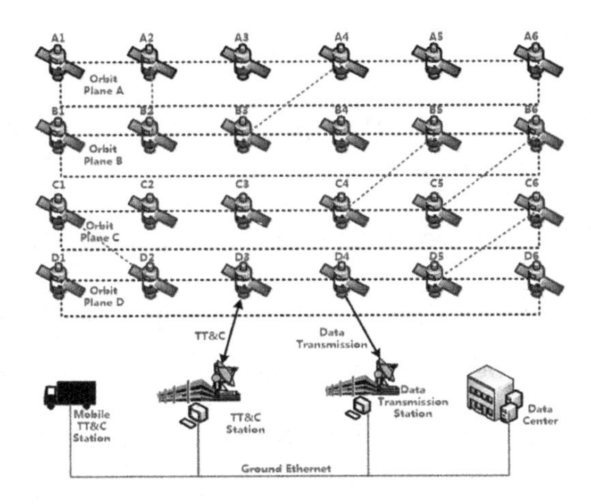

Fig. 2. Global real time telemetry and command constellation system

Generally speaking, Walker Constellation is used in the global network satellite system, which is characterized by circular orbit, uniform distribution of multiple orbital planes, and even arrangement of satellites in the orbital plane. As shown in the figure above [11].

In the aspect of communication demand, as the network load, the telemetry and command data mainly includes the real-time periodic satellite telemetry data, on-demand burst task instruction data, on orbit maintenance data and observation processing task result data transmission. The original load data with a large amount of data is generally not transmitted through the network, but through the data transmission link after using the local star cache (Table 1).

In the above scenarios, the influence of constellation configuration, link bandwidth, number of cross orbit links, transmission requirements, data source characteristics on network performance is studied, which provides basis for the rationality, feasibility and engineering optimization of constellation network design.

Table 1. Characteristics of satellite networking transmission requirements

Type of network load	Transmission requirements	Timeliness	Characteristic	Data volume
Satellite periodic telemetry	Down to the ground	High, average	Periodic	Small
Task instruction	Transmission to satellite	High	On demand burst	Small
On orbit maintenance data	Up star	Commonly	On demand burst	In
Observation processing results	Down to the ground	High	Task burst	In
Multi star state synchronization information	Webcast	High	Sudden task, frequent cycle	Small

4 Influencing Factors and Optimization of Constellation Network Transmission Performance

There are 5 requirements in network performance simulation as follow.

4.1 Support Constellation Configuration Optimization Simulation

Network simulation provides the ability of constellation configuration optimization comparison, taking 24 constellation design as an example, under the fixed link bandwidth situation, for the purpose of network transmission optimization, through the simulation data, analyzes and compares the advantages and disadvantages of 4 orbital plane × 6 satellite and 3 orbital plane × 8 satellite two typical constellation configurations.

4.2 Support Link Bandwidth Verification

In constellation design, there is often a problem, how much is the appropriate inter satellite link rate design, enough, and how much is the margin. Because of the complexity of network topology, it is difficult to analyze and calculate by theoretical means. Therefore, in the design of constellation system, it is necessary to verify the key parameters of inter satellite link, the matching of constellation communication requirements, communication load, and the correctness of link bandwidth design by simulation.

4.3 Support Cross-Orbit Link Optimization

In the optimization of the number of cross-track links, cross-track links is not the more the better, when broadcast data flooding, the more cross-track links, the more copies of broadcast data; However, when used for point-to-point transmission, the more cross-track links, the more conducive to network transmission and data forwarding.

Therefore, after the user determines the data amount requirements of constellation satellite broadcasting and point-to-point data transmission requirements, according to the characteristics of transmission requirements, simulation can be used to optimize the cross-orbit link and design more matching cross-orbit link building rules with transmission sources.

4.4 Analysis of Supporting Transmission Demand Satisfaction Degree

Network load capacity is closely related to the transmission demand of data source, and the data amount of data source is the input of system design. Firstly, the system design needs to comb out the capacity boundary of network transmission service and the transmission demand of users, i.e. Transmission type and transmission packet quantity;

For the demand of broadcast transmission between satellite nodes, the strategy based on flooding will cause a large number of packets replication in the network, each replication will lead to exponential growth in the number of network packets, increase the load of network transmission, lead to node buffer queue growth and path congestion;

The point-to-point transmission demand is linear. After the data source is sent out, the number of data packets will not increase with the transmission process. When the destination node is reached, it will disappear in the network.

Whether the design of constellation network system can meet the transmission requirements of upper applications is also difficult to calculate with theoretical models, so simulation verification and design optimization are needed.

4.5 Support Time Distribution Analysis of Data Source

In addition, the network source node transmission packet generation time characteristics, is also a major factor affecting the network transmission load, network simulation software support for randomness, uniformity, burst high load source node packet generation characteristics of simulation, analysis of its impact on network transmission performance, and gives recommendations to optimization of the source node data output.

5 Simulation and Analysis of Transmission Characteristics of 5 Constellation Network

Different simulation scenarios are constructed according to the network transmission factors, and various design factors are evaluated and optimized in simulation experiments.

5.1 Simulation Analysis of the Influence of Network Topology on Transmission Performance

See Figs. 3, 4 and Table 2.

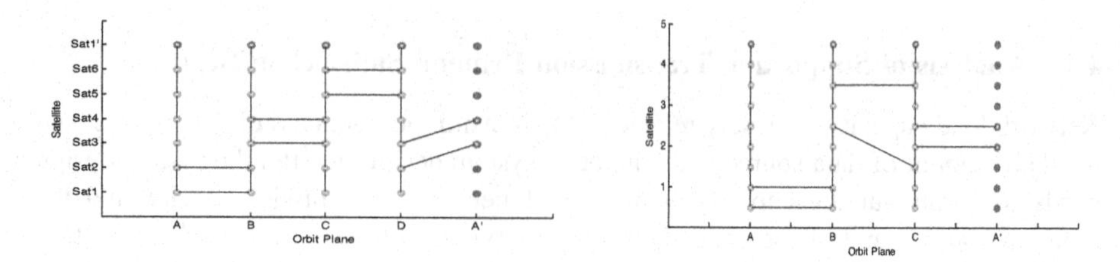

Fig. 3. 4 × 6 and 3 × 8 constellation configurations typical network topology

Table 2. Average queue depth (bandwidth of Inter-satellite-Link is 10 Mbps)

Average queue depth (package)	Single satellite communication requirement (Mbps)			
	3	4	5	6
4 × 6 topology	1.4688	41.1187	281.3026	718.7000
3 × 8 topology	0.0078	17.3990	169.4260	576.2031

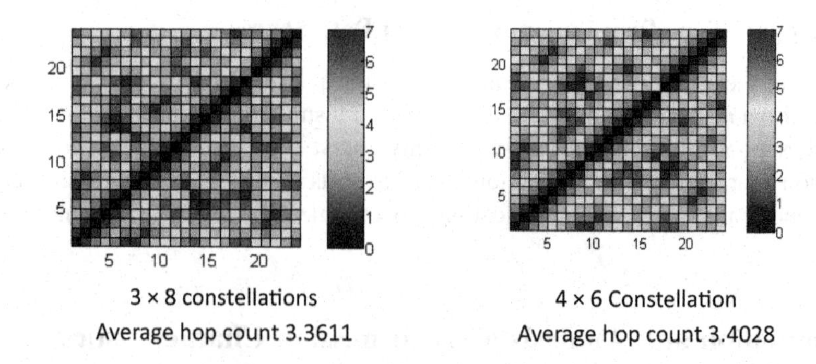

3 × 8 constellations
Average hop count 3.3611
4 × 6 Constellation
Average hop count 3.4028

Fig. 4. End-to-end average routing hops

5.2 Simulation Analysis of the Impact of Link Bandwidth on Network Capacity

The simulation study of the impact of link bandwidth on network capacity can help system designers to design a reasonable transmission bandwidth of inter-satellite links under certain transmission requirements constraints, which not only ensures to meet the transmission requirements, but also optimizes the inter-satellite transmission power and antenna size (Fig. 5 and Table 3).

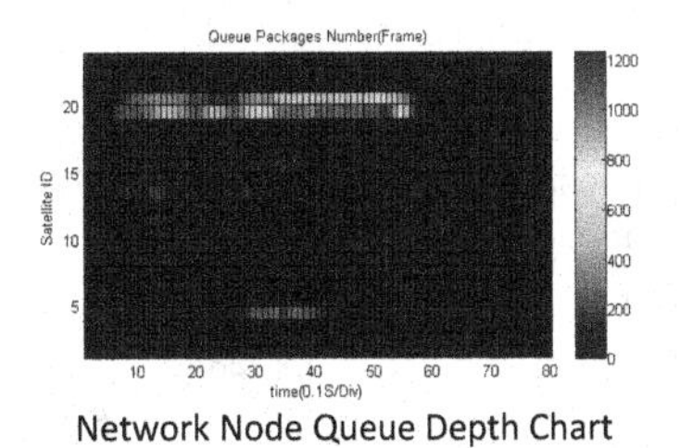

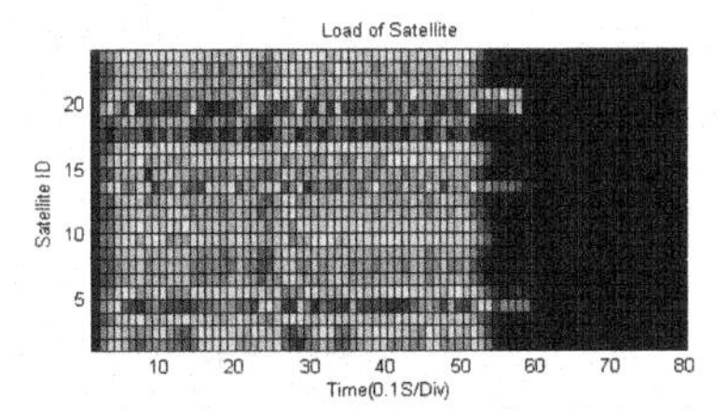

Fig. 5. 4 × 6 Constellation network transmission load (4 Mbps transmission load, bandwidth of inter-satellite-link is 10 Mbps)

Table 3. Average queue depth

Bandwidth of inter-satellite-link	Source node rate (Mbps)				
	1	2	3	4	5
6 Mbps	0	1.68	148.91	570.14	1094.6
8 Mbps	0	0	28.07	238.57	610.27
10 Mbps	0	0	0.23	36.87	239.16
15 Mbps	0	0	0	0	1.87
18 Mbps	0	0	0	0	0.49

The average queue depth represents the congestion situation of the transmission path in the network. In the ideal situation, there should be no queue or the queue depth is very small to ensure the real-time transmission of important data. Therefore, under the current set satellite scale, the transmission demand of 2 Mbps per satellite should be configured with no less than 8 Mbps bandwidth of inter-satellite link. For the demand of 3 Mbps per satellite, the bandwidth of inter-satellite link should be configured at least 10 Mbps. For the requirement of 5 Mbps per satellite, the bandwidth of inter-satellite link should be at least 18 Mbps.

5.3 Simulation Analysis of the Impact of Cross-Track Link Configuration

The more the cross-orbit links are, the more the single-satellite links are required. If the cross-orbit links are too few, it will become the bottleneck of the network communication and interconnection between different tracks. Therefore, it is very important to select the appropriate cross-orbit link for building a constellation system with appropriate scale and adequate performance. Broadcast capacity and routing hops, so that the network system can choose the appropriate link planning strategy according to the demand (Table 4).

Table 4. Number of cross-orbit links

Number of cross-track links	Congestion free upper limit		Route hops	
	Broadcast transmission	End-to-end transmission	Average	Maximal
1	560 kbps	7 Mbps	3.8264	8
2	320 kbps	10 Mbps	3.4028	7
3	600 kbps	16 Mbps	2.7639	5
4	1.12 Mbps	17 Mbps	2.4688	4

As shown in the table above, an increase in the number of cross-rail links helps to reduce the number of route hops and increases the end-to-end transmission capacity. The impact of broadcast data is more complex, because the lifetime of broadcast data caused by the increase of links affects the number of data flooding replication, but the more link branches, the more broadcast packets are replicated in the network, showing a non-monotonic property.

5.4 Simulation Analysis of the Influence of Data Transmission Mode

Table 5. Maximum source node broadcast communication service capability

Link bandwidth	Inter-satellite link bandwidth				
	6 Mbps	8 Mbps	10 Mbps	15 Mbps	18 Mbps
Broadcast transmission	60 Kbps	80 Kbps	110 Kbps	160 Kbps	200 Kbps
End-to-end transmission	1.5 Mbps	2.2 Mbps	2.8 Mbps	4.1 Mbps	4.9 Mbps

The increase of network nodes leads to the increase of the longest hop and the lifetime of broadcast frame, which is not conducive to the improvement of network broadcast performance. The broadcast communication service capacity of source node is about 1/100 of the bandwidth of inter-satellite link, so in the design of application layer, broadcast communication should be reduced as much as possible, and end-to-end transmission mode should be adopted as far as possible, which is more efficient in terms of transmission capacity occupation of inter-satellite link (Table 5).

5.5 Simulation Analysis the Influence of Data Generation Time Distribution

The impact of the time distribution characteristics of satellite communication data sources on the network transmission is mainly reflected in the impact on the depth of buffer queue in the network, using the same bandwidth, topology and average source data rate, the network transmission characteristics of evenly distributed source and randomly distributed source data are simulated, and the simulation results are shown in the following figure (Figs. 6 and 7):

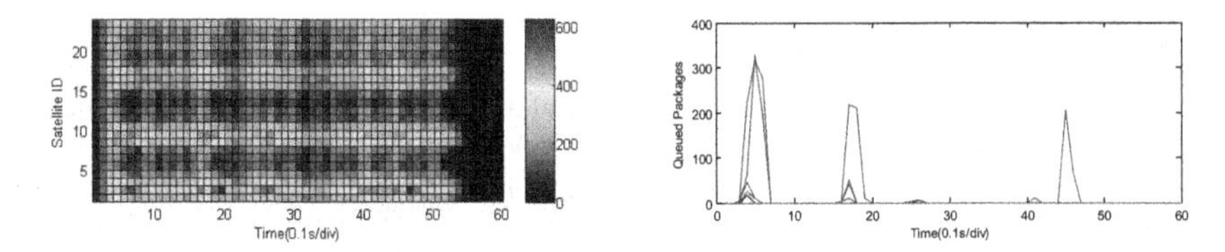

Fig. 6. Influence of randomly distributed sources on the load of network transmission nodes

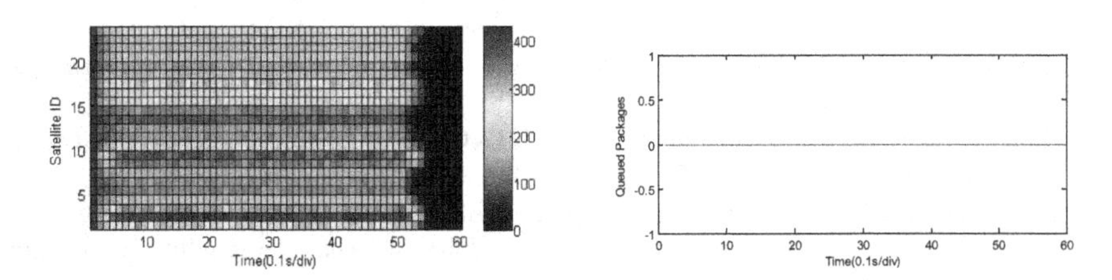

Fig. 7. Influence of uniformly distributed source on network transmission node load

Satellite telemetry, engineering parameters and other state data are uniform periodic communication demand in the time distribution, with less data loads than the network communication load capacity, the network is in a steady state balance, the packet queue will not appear on each satellite node, which means high transmission timeliness.

On the other hand, if the data transmission is on-demand or burst, such as remote control and application event data, it presents the characteristics of random distribution, and in the case of less than the network communication load, the data packet queuing phenomenon will still occur on the high-load nodes because of the uneven data distribution, but because the average communication load is still less than the network capacity, the queuing situation of the intermediate forwarding satellites will still not be continuously accumulated. The network is still smooth, but there is a short-term queuing phenomenon in the burst data concentration period.

6 Conclusion

In constellation network simulation system, the Walker constellation network model, network transmission model, routing and forwarding model, time-driven simulation engine and simulation parameter statistics and display module are designed. The simulation system is built and the simulation scene is run. Through the Walker constellation network simulation system, simulation research is carried out, focusing on the transmission characteristics of ground TT&C data and autonomous TT&C data in the network, which provides reference for network topology, bandwidth capacity configuration, node cache configuration, routing algorithm and data generation strategy design optimization and network transmission capacity analysis and verification in system design.

References

1. She, C., Wang, J., Liu, L., Zhou, M.: Topological dynamics analysis of Walker constellation satellite networks. J. Commun. **27**(8), 45–51 (2006)
2. Liu, B.G., Wu, B.: Application of TDRSS in Chinese space TT&C. J. Spacecr. TT&C Technol. **31**(6), 1–5 (2014). (in Chinese)
3. Li, Y., Sun, H., Zheng, J.: Using tracking and data relay satellite for low and middle earth orbit satellite launch and on-orbit control. Chin. J. Space Sci. **35**(5), 611–617 (2015)
4. Yang, T., Wu, Y., Li, M.: Global satellite mobile communication constellation system space-based inter-satellite-link TC&R scheme. Chin. Space Sci. Technol. **33**(3), 77–82 (2013)
5. Schaire, S., Horne, B., et al.: NASA near earth network (NEN) and space network (SN) CubeSat communications. In: Proceeding of 2016 International Conference on Space Operations. Daejeon, Korea, pp. 1–19. IEEE (2016)
6. Jiang, L.: Development and trends of foreign TT&C and communication networks for small satellites. Telecommun. Eng. **57**(11), 1341–1348 (2017)
7. Sobchak, T., Shinners, D.W., Shaw, H.: NASA space network project operations management: past, present and future for the tracking and data relay satellite constellation. In: 15th International Conference of Space Operations, 28 May–01 June 2018, Marseille, France, pp. 1–12 (2018)
8. Yang, T., Wu, Y.X., Li, M.F.: Global satellite mobile communication constellation system space-based inter-satellite-link TC&R scheme. Chin. Space Sci. Technol. **33**(6), 77–82 (2013). (in Chinese)
9. Li, J., Shao, Q.: Design and optimization on walker constellation based on STK/Matlab. Ord. Ind. Autom. **36**(12), 67–70 (2017)
10. Xiao, X., Shang, X., et al.: Evaluation scheme design for the key performance of inter-satellite link. Space Electron. Technol. **15**(3), 81–86 (2018)
11. Zhang, J., Xi, X., Wang, W.: Sufficient condition and characteristic analysis for a spacecraft rendezvous with walker constellation satellites without orbital. J. Natl. Univ. Def. Technol. **32**(6), 87–92 (2010)

The Development of a Lightweight Network Emulator for Large-Scale Space Network Emulation

Fa Chen, Weidong Zhou, Kanglian Zhao, Wenfeng Li(✉), and Yuan Fang

School of Electronic Science and Engineering, Nanjing University, Nanjing 210023, People's Republic of China
{zhaokanglian,yfang}@nju.edu.cn, leewf_cn@hotmail.com

Abstract. Based on the previously developed network emulator named MininetE [1], this paper proposes an improved scheme of lightweight network emulator for large-scale space information network (SIN) emulation. The improved MininetE still has the ability to emulate network link characteristics such as bandwidth, latency and packet loss, while generating real network traffic. Through the setting of network configuration parameters, not only DTN protocol but also IP-based OSPF protocol can be run in the new MininetE, making it possible to build a SIN emulation environment where multiple network protocols coexist. Based on this, a SIN scenario with the improved MininetE is constructed to evaluate the performance of OSPF and the coexistence of OSPF and DTN respectively. The results show that the emulation platform has the DTN and IP-based routing emulation capability and ensures considerable reliability, which provides a reference for further optimization of MininetE emulation platform in the future.

Keywords: Space information networks · Emulation · MininetE · High fidelity · OSPF

1 Introduction

Space Information Network (SIN) is a backbone communication network composed of several satellites and satellite constellations in orbit, which can provide communication services for various space missions. With the development of SIN, it is very necessary to carry out research on new network technology and evaluate network protocols and algorithms. If we conduct the research in a real space network system, testing software needs to be pre-loaded into satellite communication systems, and then launches the test satellites into space. This approach will not only have a long cycle and high cost, bust also may affect existing satellite network systems.

So we need a emulation platform to conduct the research. The traditional physical emulation will inevitably bring high deployment cost, complicated scene parameter configuration and other problems. On the other hand, if we use the existing network simulator (such as OPNET [2] or ns [3]). They are essentially a discrete event simulator. All simulations are driven by discrete events so they are not as real as mininetE.

Q. Guo et al. (Eds.): WiSATS 2021, LNICST 410, pp. 289–298, 2022.
https://doi.org/10.1007/978-3-030-93398-2_29

Therefore, the Institute of Space-Terrestrial Intelligent Networks (ISTIN) of Nanjing University has developed an extensible space interconnection network emulation platform [4]. This emulation platform can be deployed and switched quickly, and provide accurate spatial simulation results. However, due to the high cost, it needs good hardware resources support, which is not realistic for individuals. In order to satisfy the personal ability to carry out some simple experiments and verification on our own laptop, we developed a lightweight emulation platform which is flexible and reconfigurable for spatial information network protocol [1]. On the basis of keeping lightweight, the platform adds adequate isolation, which makes MininetE simulation platform available for spatial information network protocol simulation.

The goals of a usable emulation platform include not only flexibility and support for quick switching between scenarios, but also authenticity, such as function realism and traffic realism.

Through the setting of network configuration parameters, not only DTN protocol but also IP-based OSPF protocol can be run in the new MininetE. Based on this, the performance of OSPF and the coexistence of OSPF and DTN respectively is evaluated in the scenario which is constructed.

2 Architecture Design

Mininet [5] is a lightweight network simulator based on container virtualization technology. However, like most network simulators, Mininet is widely used in the simulation of ground network, but it lacks the simulation support of spatial information network related protocols. Therefore, MininetE emulation platform is developed based on Mininet terminal host. In addition to Mininet providing isolation of network namespace and mount namespace, MininetE adds isolation support of UTS namespace, PID namespace and IPC namespace to virtual nodes. Figure 1(A) shows If double OSPF processes are started simultaneously in Mininet, the second process will fail. However, Fig. 2(B) shows MininetE can run two or more OSPF processes simultaneously. Because of the adding namespace isolation, processes do not interact with other node's processes. Similarly, Fig. 2(C) shows different nodes can run different protocols without affecting each other. All virtual nodes can use different network protocol systems (such as TCP/IP, CCSDS, DTN) according to different target scenarios, so as to achieve flexible switching of different network protocol simulation scenarios.

3 Improvement to MininetE

In the proposed platform, emulation nodes, links and services are necessary for simulation.

Simulation Nodes: It is composed of MininetE hosts and is the main object of network scale expansion. In order to make the node have a real simulation, the increased

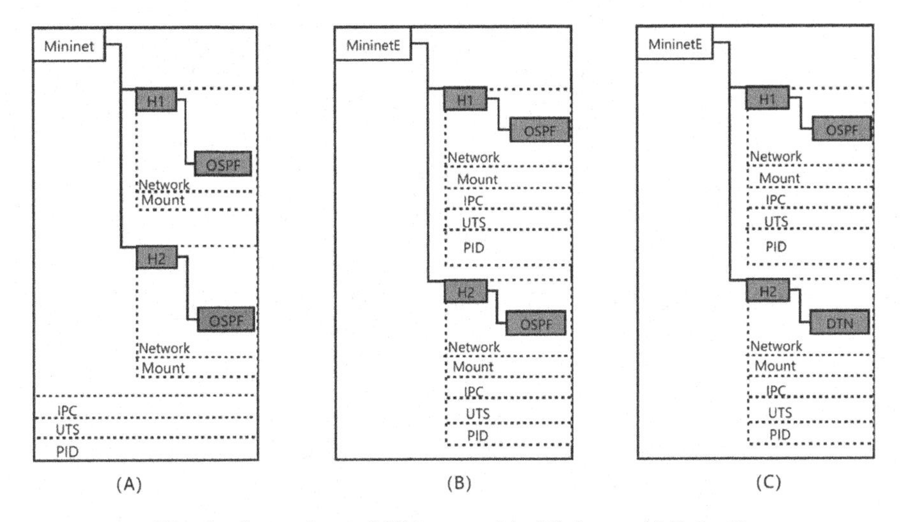

Fig. 1. Comparison of SIN protocol in Mininet and MininetE.

namespace isolation of MininetE can deploy different protocols (such as DTN, TCP/IP) in each simulation node.

Links: Link is necessary for communication between nodes. The on-off of link is determined by the the movement of the satellite. We can use TC NETEM [6] to set the link characteristics.

Switch: We can use ordinary switches or SDN switches, so we can easily control the network topology by the controller.

Controller: The controller is responsible for telling the node and the switch what to do.

In the space information network, due to the movement of satellites, the visibility between satellites may change, and the topology will change accordingly. Switching simulation link is very important. Fortunately, MininetE has an inherent support for software defined (SDN)/Openflow [7]. We develop a link switching method using SDN controller. Our scheme is based on the flow table control of openvswitch by SDN controller. As shown in the Fig. 2, The controller has global topology information. Whenever there is a link state change, it will follow up with the new topology. The controller will send a flow table to the switch to simulate the pass and segment of the link.

As the scale of emulation nodes increases and the nodes in the space information network are in high speed motion, the whole network topology changes dynamically. The connection between nodes is intermittent, and the connection objects of nodes may be switched at some time, and the link parameters will change accordingly. MininetE works by adding different ports, which means different possible neighbor nodes, and switching links by sending flow table.

Figure 3 shows the simulation process. First, We need to determine the simulation scenario. That is to determine the number of simulation nodes, node types, simulation

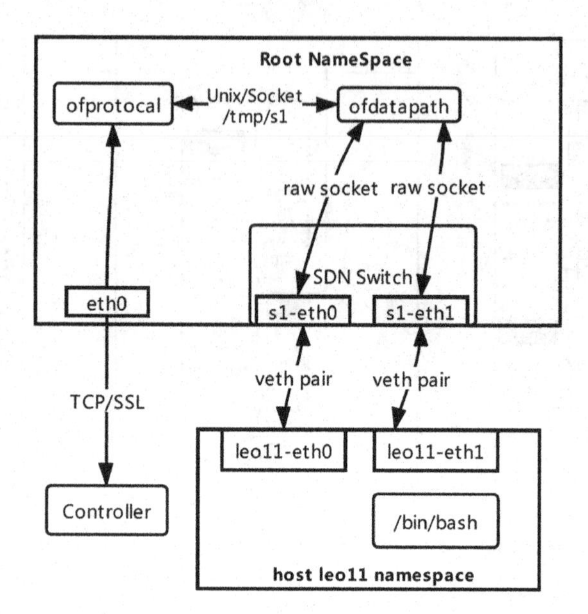

Fig. 2. Controller sending flow table

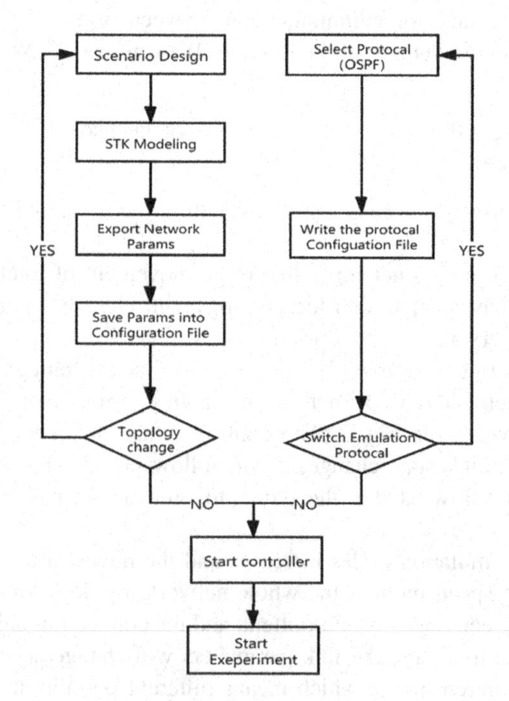

Fig. 3. Simulation process

protocols, etc. After this, we save the network params into configuration file, and then start the controller to begin the simulation.

The simulation node is connected to the SDN switch. Each node runs the program and waits for instructions from the master controller. At the same time, The controller sends a flow table to the switch based on the connection relationship to control the connectivity between the nodes, and tells the nodes what process to run at what time.

4 Experimental Validation

4.1 OSPF Routing Simulation Process

A complete emulation consists of three stages of scene design, experiment operation and data collection.

Scene Design
Firstly, the model is established according to the emulation scene, and detailed simulation parameters can be obtained from such as STK. As shown in the Fig. 4, we build an experiment scenario which contains 14 LEO satellites for space internetworking. The solid line indicates that the links are always connected, and the dashed line indicates that they are intermittently connected. The on-off relationship of the inter orbit link in the experimental period (6000 s) is shown in the Fig. 5. The connection represents the time interval in the connected state. We want to study the performance of OSPF in this satellite network, and show the experimental results on our platform. The 14 simulation nodes are composed of 14 Mininet nodes and connected to SDN switches. After these preparations, the controller is connected to the SDN switch for control.

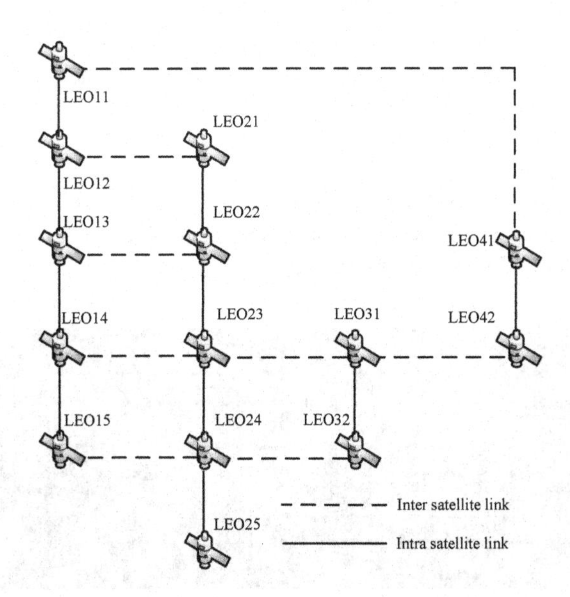

Fig. 4. Satellite network topology in experimental scene.

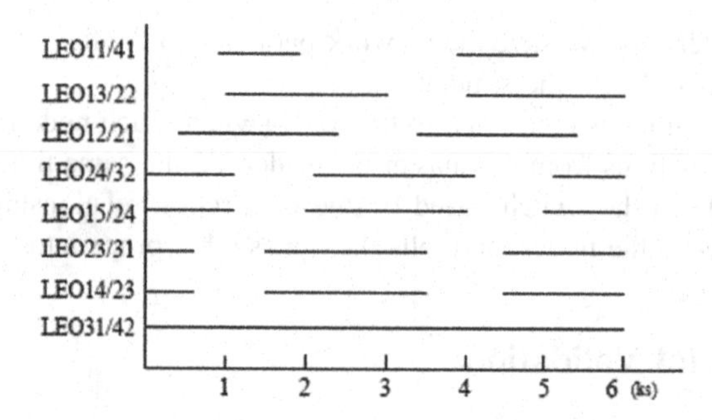

Fig. 5. Inter-satellite link connection relationships

Experimental Operation

The control plane is mainly composed of the main controller. The control receives information and parameters from the logic plane to drive data emulation. Before the implementation of the experiment, the controller should be in operation, and other equipment should be available at the same time. The simulation parameters are written to the configuration file ahead of time. Once the network topology changes, we need to redesign the scenario and repeat the above steps.

In the experimental phase, the controller triggers the simulation node to start. The node then runs the corresponding process, such as OSPF or DTN. The Quagga [8] used to support OSPF routing protocol and installed on the host. At the same time, the software defined network controller will also receive the signal. The controller has global topology information. When a packet arrives at the OpenFlow switch, the flow table will be matched. If no corresponding flow table entry is found, a packet_in message will be sent to the SDN controller. After the controller makes a decision based on a certain routing algorithm, it sends a flow table to the switch.

Data Collection

In the data collection stage, data can be obtained through switches and nodes. Through some commands, we can obtain real-time information such as neighbor state and routing convergence state. we can obtain the real-time neighbor state, routing convergence state and other information.

As shown in Fig. 6, through the Ping between nodes, any two nodes can communicate with each other, and the result of global router information table proves that MininetE supports OSPF routing protocol.

```
mininet> pingall
*** Ping: testing ping reachability
leo11 -> leo12 leo13 leo14 leo15 leo21 leo22 leo23 leo24 leo25 leo31 leo32 leo41 leo42
leo12 -> leo11 leo13 leo14 leo15 leo21 leo22 leo23 leo24 leo25 leo31 leo32 leo41 leo42
leo13 -> leo11 leo12 leo14 leo15 leo21 leo22 leo23 leo24 leo25 leo31 leo32 leo41 leo42
leo14 -> leo11 leo12 leo13 leo15 leo21 leo22 leo23 leo24 leo25 leo31 leo32 leo41 leo42
leo15 -> leo11 leo12 leo13 leo14 leo21 leo22 leo23 leo24 leo25 leo31 leo32 leo41 leo42
leo21 -> leo11 leo12 leo13 leo14 leo15 leo22 leo23 leo24 leo25 leo31 leo32 leo41 leo42
leo22 -> leo11 leo12 leo13 leo14 leo15 leo21 leo23 leo24 leo25 leo31 leo32 leo41 leo42
leo23 -> leo11 leo12 leo13 leo14 leo15 leo21 leo22 leo24 leo25 leo31 leo32 leo41 leo42
leo24 -> leo11 leo12 leo13 leo14 leo15 leo21 leo22 leo23 leo25 leo31 leo32 leo41 leo42
leo25 -> leo11 leo12 leo13 leo14 leo15 leo21 leo22 leo23 leo24 leo31 leo32 leo41 leo42
leo31 -> leo11 leo12 leo13 leo14 leo15 leo21 leo22 leo23 leo24 leo25 leo32 leo41 leo42
leo32 -> leo11 leo12 leo13 leo14 leo15 leo21 leo22 leo23 leo24 leo25 leo31 leo41 leo42
leo41 -> leo11 leo12 leo13 leo14 leo15 leo21 leo22 leo23 leo24 leo25 leo31 leo32 leo42
leo42 -> leo11 leo12 leo13 leo14 leo15 leo21 leo22 leo23 leo24 leo25 leo31 leo32 leo41
*** Results: 0% dropped (182/182 received)
mininet>
```

Fig. 6. Connectivity between nodes

4.2 Hybrid Protocol Emulation

In addition, due to the increasing complexity of space network scenarios, nodes are required to support more than a single network protocol, such as DTN and OSPF routing protocols [9]. It is capable of supporting multiple protocols. As you can see from Fig. 1, MininetE has this ability. In the emulation scenario in Fig. 4, in addition to deploying the OSPF protocol, arbitrarily select several nodes to run the DTN protocol. For example, LEO11 serves as the sending node for data transmission, LEO12 serves as the intermediate node for data forwarding, and LEO13 serves as the receiving node for data. Their communication protocol stack structure is shown in the Fig. 7. The IP layer runs the OSPF routing protocol.

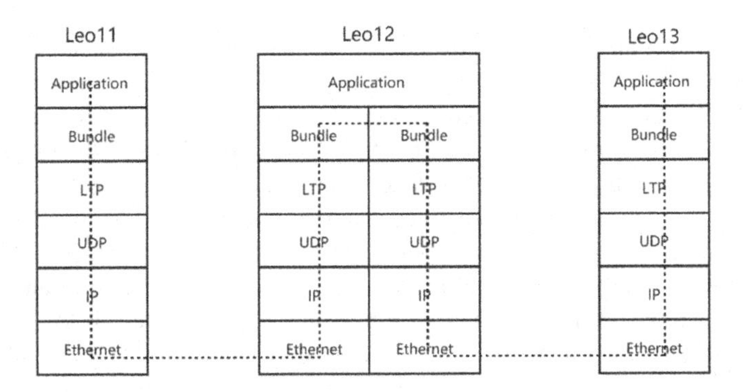

Fig. 7. The node's communication protocol stack

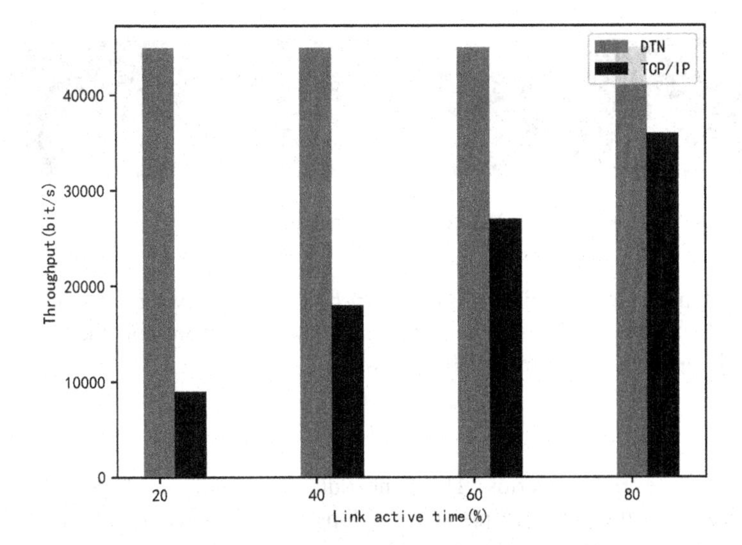

Fig. 8. Data throughput of different link effective time

According to the emulation scenario and experimental configuration, we selected the data throughput of the target node LEO13 as the object of statistics and research, and compared the communication performance of DTN and TCP/IP under different transmission protocols. Figure 8 shows the data throughput rate under the condition that the link time efficiency is 80%, 60%, 40% and 20% respectively [10]. It can be seen that in the deep space communication environment with large delay and high bit error rate, TCP protocol cannot complete the effective transmission of data, and DTN protocol is more suitable to be used in the deep space communication network scenarios.

5 Performance Evaluation

Because the nodes in the space information network are in the state of high speed motion, the whole network topology changes dynamically. The connection between nodes is intermittent, and the connection objects of nodes may be switched at a certain moment. The controller is used to send the flow table to the SDN switch to simulate the link switching and change the network topology. As shown in Fig. 9, The result of traceroute shows that the route converges again. The so-called convergence of routing protocol refers to the process that the routing protocol finds other routers in the network to change and exchange routing entries until all routing entries are exchanged. In this example, as shown in Fig. 3, there is only one hop from leo11 to leo21 at the beginning. When I disconnect the link between leo11 and leo21, OSPF will re select the shortest path for convergence, and the packets will go through leo13, then to leo22, and finally to leo21.

```
root@iZbp1frdqql4a2oq5ctnenZ:/home/admin/mininetE/custom# traceroute 10.0.5.6
traceroute to 10.0.5.6 (10.0.5.6), 64 hops max
  1   10.0.5.6  0.002ms  0.002ms  0.002ms
root@iZbp1frdqql4a2oq5ctnenZ:/home/admin/mininetE/custom# traceroute 10.0.5.6
traceroute to 10.0.5.6 (10.0.5.6), 64 hops max
  1   10.0.2.3  0.002ms  0.002ms  0.002ms
  2   10.0.13.7  0.001ms  0.001ms  0.002ms
  3   10.0.5.6  0.001ms  0.002ms  0.002ms
```

Fig. 9. Route convergence

The judgment interrupt time of OSPF can be calculated by the formula (1), where $t_{linkoff}$ represents time of the link interruption and the $t_{linkoff}$ represents the time nodes to changes the neighbor state from full to down.

$$t_{judge} = t_{Full->Down} - t_{linkoff} \tag{1}$$

We tested several scenarios. The bandwidth and delay are 1000 Mbps/25 ms, 1000Mbps/15ms, 100 Mbps/15 ms and 10 Mbps/15 ms respectively. Measured OSPF interrupt judgment time. The measured OSPF disruption judgment time in MininetE, Docker and NS simulator [11] is shown in the following Table 1.

Table 1. Disruption judgement time.

Bandwidth/Delay	MininetE(s)	Docker(s)	NS(s)
1000 Mbps/25 ms	38.24	37.14	41
1000 Mbps/15 ms	37.13	35.46	41
100 Mbps/15 ms	36.98	34.68	41
10 Mbps/15 ms	36.57	33.64	41

According to Table 1, the judgment interrupt time is basically the same. Considering the possible difference between the software implementation and the introduction of the actual data flow on the emulation platform, this slight difference can be ignored temporarily. By comparing the simulation results of MininetE, Docker and NS, we can see that the improved MininetE can support OSPF and guarantee the reliability.

6 Conclusion

The improved MininetE, which can run emulation scenarios of different protocols on each laptop. With our dynamic topology control method, we can control the spatial network topology. Emulation results show that MininetE has the ability to support DTN network and OSPF protocol. The experimental results are simply compared with docker and NS simulators. The emulation results show that the improved MininetE can support many kinds of spatial network protocols. In the future, more experiments will be carried out to prove that MininetE can meet the requirements of various network protocal and support larger scale emulation.

Acknowledgements. This work is supported by the 13th Five-Year Civil Aerospace Technology Pre Research Project, the Fundamental Research Funds for the Central Univesities under Grant 021014380187 and the National Natural Sciences Foundation of China under Grant 62131012.

References

1. Lin, T., Chen, F., Zhao, K., Fang, Y., Li, W.: MininetE: a lightweight emulator for space information networks. In: Wu, Q., Zhao, K., Ding, X. (eds.) Wireless and Satellite Systems. WiSATS 2020. Lecture Notes of the Institute for Computer Sciences, Social Informatics and Telecommunications Engineering, vol. 357, pp. 48–57. Springer, Cham (2021). https://doi.org/10.1007/978-3-030-69069-4_5
2. Zhao, J., Zhu, X.: Research on a hardware-in-the-loop simulation method for wireless network based on OPNET. In: 2015 IEEE Advanced Information Technology, Electronic and Automation Control Conference (IAEAC), Chongqing, pp. 821–825 (2015)
3. Wang, Z., Cui, G., Li, P., Wang, W., Zhang, Y.: Design and implementation of NS3-based simulation system of LEO satellite constellation for IoTs. In: 2018 IEEE 4th International Conference on Computer and Communications (ICCC), Chengdu, China, pp. 806–810 (2018)

4. Lu, T.: An emulation architecture of routing protocol in space information network. Nanjing University. Master's thesis in Chinese (2018)
5. Lantz, B., Heller, B., McKeown, N.: A network in a laptop: rapid prototyping for software-defined networks. In: ACM Sigcomm Hotnets Workshop (2010)
6. tc-netem. [EB/OL]. https://www.linux.org/docs/man8/tc-netem.html
7. The openflow switch. http://www.openflowswitch.org
8. Chen, Y., He, Y., Zhao, Z., Liang, X., Cui, Q., Tao, X.: DEMO: a Quagga-based OSPF routing protocol with QoS guarantees. In: 2018 24th Asia-Pacific Conference on Communications (APCC), Ningbo, China, pp. 5–6 (2018)
9. Zhao, K., Wang, R., Burleigh, S., Sabbagh, A., Wu, W., Sanctis, M.D.: Performance of bundle protocol for deep-space communications. IEEE Trans. Aerosp. Electron. Syst. **52**(5), 2347–2361 (2016)
10. Sabbagh, A., Wang, R., Burleigh, S.C., Zhao, K.: Analytical framework for effect of link disruption on bundle protocol in deep-space communications. IEEE J. Sel. Areas Commun. **36**(5), 1086–1096 (2018)
11. Xu, M., et al.: Ground-air integrated network intra-domain protocol OSPF+. J. Tsinghua Univ. (Nat. Sci. Edn.) **57**(01), 12–17 (2017). (in Chinese)

Discussion on Application of Model-Based Systems Engineering Method to Human Spaceflight Mission

Zhijie Li(✉), Huan Liu, Yue Qi, and Yaowu Xu

Beijing Institute of Spacecraft System Engineering, Beijing, China

Abstract. Considering the fact that design data exist in different documents dispersedly and are difficult to keep consistent and hard to trace, the concept of MBSE (model—based systems engineering) method is introduced. Firstly MBSE method can graphically support system requirements, design, analysis, verification and validation activities in whole life cycle. Then a general process of MBSE method is summarized, creating a requirement model to guide the construction of functional model and physical architecture model etc. Based on a set of predefined rule the relations among models are set up, by which relevance among design elements and information can be analyzed and searched out. In the end, an example is presented to show how to apply MBSE method to design manned spacecraft docking missions. It is concluded that MBSE method can improve the communication, increase work efficiency, and reduce design risk. The work is a good reference for further application.

Keywords: Systems engineering · Human spaceflight · MBSE · Requirement model · Functional model · Pgysical architecture model

1 Introduction

Manned space engineering is a complex system engineering process, which has the characteristics of large scale, high technology level, high reliability and safety requirements, long development cycle, many participants and huge investment. In the development process, from the overall development requirements of the project to the product realization and on orbit operation, a large amount of design information with complex relationship will be generated, such as system, subsystem, stand-alone design requirements and interface definition. At present, the information is mainly stored and managed in the form of documents. With the task becoming more and more complex, the scale of the system expanding, the subjects involved increasing, and the inconsistent state of system parameters in different documents often occurs, which leads to many security risks. Therefore, a new method and means are urgently needed to change this situation [1].

With the development of computer and information technology, it is more and more easier to use object-oriented, graphical and visual system modeling language to describe the system, and the application proportion of model in system design is also increasing. A method called "model-based system engineering" (MBSE) arises at the

Q. Guo et al. (Eds.): WiSATS 2021, LNICST 410, pp. 299–310, 2022.
https://doi.org/10.1007/978-3-030-93398-2_30

historic moment. In MBSE method, the system requirements, function model and physical architecture model in a graphical and structured way. These models are used as the main way to describe the components of the system and control its evolution process, so as to realize the design and management process of system engineering project. MBSE method can effectively solve the problems of document based method in parameter acquisition and technical status management, and has been applied in the fields of aviation, aerospace, ship and so on. Through decades of development, China has accumulated rich engineering experience in system engineering, but the research on MBSE method is still in its infancy, only in the system modeling language and the development of some small systems. In order to solve this problem, this paper first studies the development of MBSE method, analyzes its advantages, puts forward the general workflow, and then gives an application example of manned spacecraft rendezvous and docking mission design by using this method, explores the transformation from traditional system engineering method to MBSE method, which can provide a reference for the application of MBSE method in manned space mission of our country.

2 Overview of MBSE Method

2.1 The Concept and Advantage of MBSE

The international Association of System Engineering (INCOSE) gave the definition of MBSE method in 《Vision 2020 of system engineering》 published in 2007: MBSE method is the application of formalization and standardization of modeling method in system engineering activities, so that the modeling method can support system requirements, design, analysis, verification and validation activities. These activities start from the conceptual design stage [1]. From the definition, it can be seen that there is no essential difference between MBSE method and traditional document based system engineering method in terms of basic theory and basic process. The difference is mainly in the management mode of design process, work form and display form of design results.

In the document based system engineering method, a few graphics are used to visualize the content that is difficult to describe by text, which is easy for people to understand. At this time, the graphics are only used to assist the text to complete the description of the system. With the development of computer technology, graphic modeling becomes easier, and the proportion of graphic model in the process of system engineering is also increasing, which gradually develops into MBSE method. MBSE method takes the model as the basis of system description, and realizes the whole process from conceptual design, scheme design, and experimental verification to engineering implementation with visualized model [2].

(1) Improve the efficiency of understanding and communication. The first mock exam is more acceptable than the text, and the graphical symbols are written in character. It is intuitive and image, and ensure the integrity of information. It makes the understanding of the same model easier for different personnel to reach agreement, and improves the communication efficiency among different designers.

(2) Data acquisition is easy. The minimum object of document based system engineering method is document. The information users need is scattered in a large number of documents, so it takes a huge amount of work to find. The minimum object of MBSE method is data. Combined with database management method, users can directly obtain the required index parameters, which can greatly reduce the workload of designers.
(3) Good traceability of technical status. MBSE method will continue to establish the relationship between the models in the process of work, through these relationships to achieve the traceability and correlation analysis of the technical status, and complete the comprehensive analysis and control of the technical status.
(4) Integration of design and verification. MBSE emphasizes that design and verification should be considered at the same time in the process of work. By establishing the relationship between verification model and requirement model, function model, and other relevant models used in the design process, the verification coverage analysis is carried out to ensure that all projects meet the verification requirements.

2.2 Typical MBSE Method and Related Tools

Some well-known companies and research institution have gradually formed a method suitable for their own development process through continuous accumulation in the process of using MBSE method for product development. The typical methods are shown in Table 1 [3].

Table 1. Typical MBSE methods

Serial number	Method Name	Organization or Company
1	Harmony system engineering method	IBM Corporation
2	Object oriented system engineering method	INCOSE
3	Unified development process	Rational Corporation
4	Vitech model based system engineering approach	Vitech Corporation
5	State analysis method	Jet Propulsion Laboratory (JPL)
6	Object process method	—

In the early development of MBSE method, system engineers mostly used behavior diagram, function modeling in integrated computer aided manufacturing (IDEF) and other methods to carry out their work [4], but the symbols and semantics used by these methods are different, resulting in information reuse between them. The lack of a unified modeling language and tools severely limits the development of this method.

Unified Modeling Language (UML) is an object-oriented standard modeling language. Since it was approved as a standard by object management group (OMG) in 1997, it has made remarkable achievements in software system design, mechanical design, enterprise business process design, data processing system design and other industries. However, the original UML was proposed to support the development of software system, and the models it provided focused more on the software development process, which had some defects in the process of supporting MBSE method. In order to transform UML into a language suitable for system engineering, a system modeling language (SysML) cooperative organization has defined a system modeling language SysML based on UML 2.0, which is specially extended to support MBSE method [5].

Based on UML and SysML, a variety of MBSE software are developed. At present, there are three kinds of MBSE software: One only supports requirement management, and the other only supports system modeling. The whole system engineering process can be completed only when they are used together; because there are some compatibility problems when they are used together, the third kind of software appears, which integrates the functions of requirement management and system modeling into the same environment.

Table 2 lists several commonly used softwares, among which the corresponding software of serial numbers 1 and 2 only support requirement management, while the corresponding softwares of serial number 3, 4 and 5 only support system modeling, and they should be used together; the corresponding software of serial number 6 is an integrated development environment, under which users can carry out requirements management and system modeling.

Table 2. MBSE tools

Serial number	Software name	Organization or company
1	RequisitePro requirement management tool	Rational
2	DOORS requirements management tool	Telelogic
3	Rational Rose system modeling tool	Rational
4	Telelogic Tau system modeling tool	Telelogic
5	Rhapsody model driven development tool	IBM
6	Cradle system engineering integrated environment	3SL

3 Process of MBSE Method

MBSE method covers the design and management from largescale system, to subsystem and component level. Its basic process is the same as the traditional system engineering method, which starts from the analysis of task requirements are transformed into system function, and various "activities" required by the system to complete the task are defined. Finally, the results of functional analysis are mapped to the equipment and components one by one [6].

The ultimate goal of MBSE method is to build a complete tested and verified system architecture based on the model. In the whole design process, the most basic is to build the system requirement model, function model and physical architecture model [7, 8].

(1) Construction of demand model
Requirement model refers to the set of requirements from the top to the bottom of the system, as well as the logical relationship between them. According to different emphasis, the requirements can be divided into functional requirements, performance requirements, interface requirements, reliability requirements, security requirements, human factors engineering requirements, etc. The requirement model is used to transform the unclear expectations of various stakeholders. These top-level requirements are divided into functional requirements and performance requirements, and are decomposed and allocated in the system. From the system to the subsystem and then to the stand-alone components, this decomposition and allocation process continues until the complete design meeting the requirements is completed. The construction process of requirement model 1 is shown in Fig. 1 [7, 9].

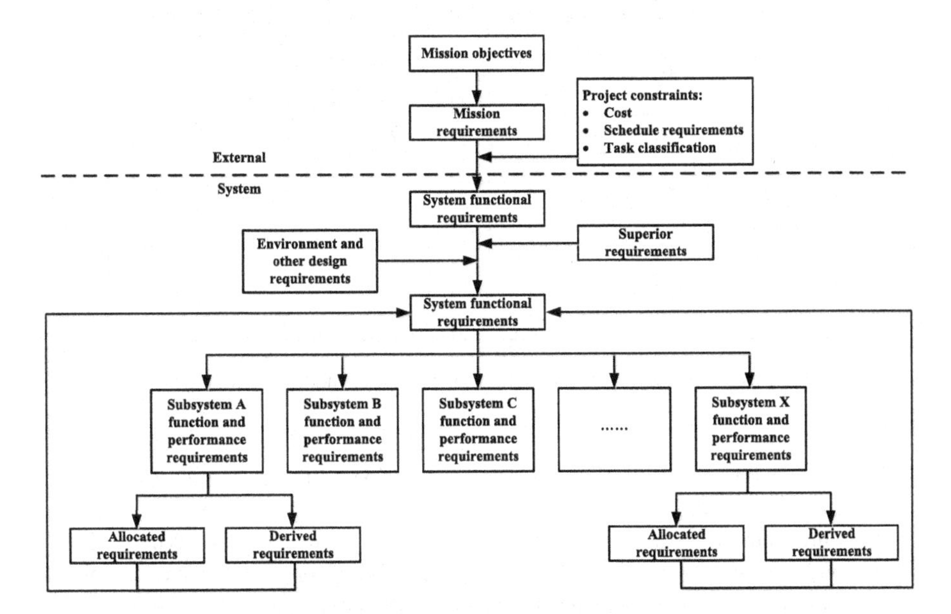

Fig. 1. Construction of requirements model

(2) Function model construction
Function model refers to the set of all functions required by the system to complete the given mission objectives, including the functions corresponding to the system level (such as manned spacecraft), subsystem level (such as measurement and control system), product level (such as sensors), and even smaller units, and the logical relationship between them, which is used to guide the design of system composition. Based on the requirement model, the function model analyzes the system function through logical decomposition. At the same time, based on the

analysis of the mission process, it combs the flight events in the whole process, and then identifies the system function of each level through the flight events, as shown in Fig. 2. After that, the functions are summarized step by step to form the functional module division of the system. In addition, in the process of building functional model, the summarized functions should be matched with the items in the requirement model to ensure that each requirement has corresponding function. For the requirements that are not covered, we should consider whether they are reasonable and whether we need to add corresponding functions to support them. For functions that do not support system requirements, consider deleting them.

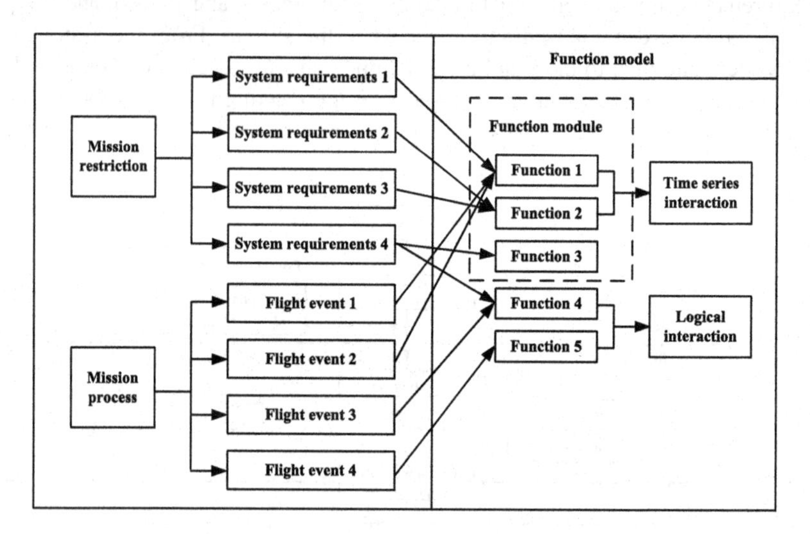

Fig. 2. Construction of functional model

(3) The physical architecture model is used to describe all the elements of the system and the interface relationship between them. It is also composed of the hierarchical structure from the system level to the product, or even smaller units. When building the physical architecture model of the system, based on the demand model and function model, we comprehensively consider the performance index, system efficiency, development cost, system interface, technical risk, etc., carry out multi scheme that can meet the user's needs and better complete the system function. The construction process of physical architecture model is shown in Fig. 3.

In addition to the above three basic models, the system interface model, product structure model, risk analysis and verification model are also needed to complete the whole task design [10]. After building the model for task design, the relationship between different models should be established according to the logic rules made in advance, so as to achieve the comprehensive accessibility of the data in the whole project. The whole process is iterated and refined until the whole design, verification

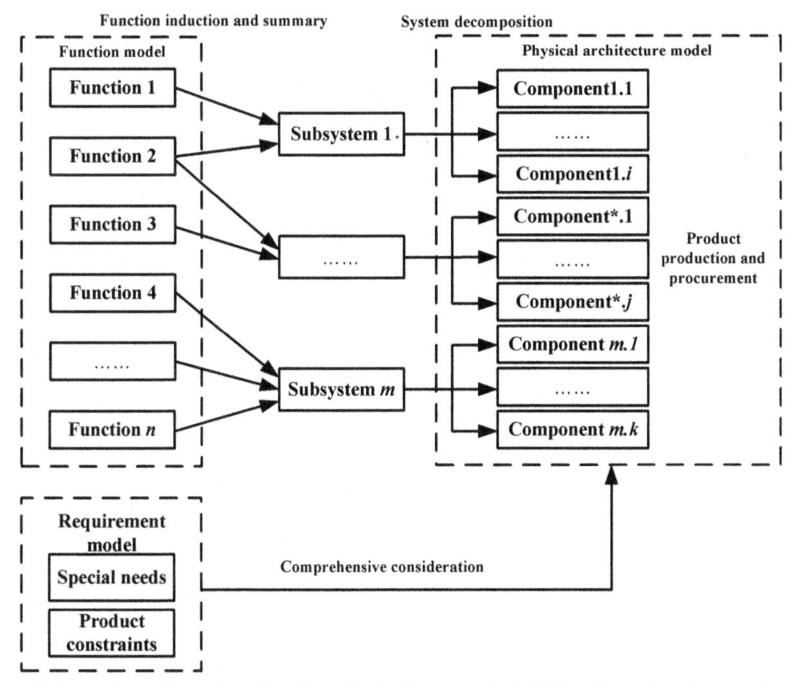

Fig. 3. Construction of physical architecture model

and working process can be clearly described. Finally, a complete, consistent and convenient traceability and query system is established to realize parameter query, coverage analysis and other work, so as to ensure the integration of system design model, avoid design conflicts between various components and reduce risks.

4 Application of MBSE Method in the Mission Design of Manned Spacecraft

This section studies the general process of requirement analysis and system design for manned spacecraft rendezvous and docking mission, including requirement analysis, flight event analysis, function analysis, product structure model creation, and model relationship establishment.

(1) Requirement analysis model
In the analysis of manned spacecraft rendezvous and docking mission requirements, from the measurement and control system, astronaut system and the requirement of the target aircraft for manned spacecraft, a preliminary analysis is carried out. The requirements of manned spacecraft are transformed into the definition of the problem, and they are transformed into a set of statements expressed as "how to complete". Analyze the requirement of manned spacecraft when

completing rendezvous and docking mission, and the specific situation is shown in Fig. 4.

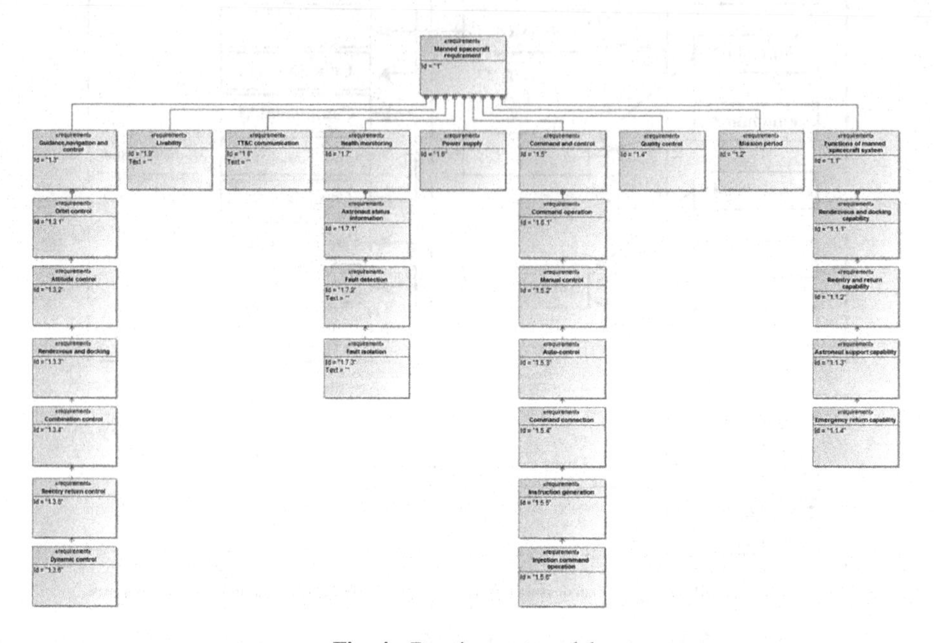

Fig. 4. Requirement model

(2) Flight event analysis model
Flight event analysis is used to convert the high-level requirements in the process of requirement analysis into specific events. Through detailed analysis to ensure the rationality of these events, a preliminary flight plan is obtained, which lays the foundation for functional analysis. When analyzing the flight events of rendezvous and docking of manned spacecraft, it is necessary to comprehensively consider the user's needs, divide the mission process, analyze the stages of completing the mission, and then subdivide each stage to analyze the flight events of the whole process, so as to prepare for the next functional analysis, as shown in Fig. 5. The nodes led out by and in the figure indicate that each event is parallel.

(3) Functional analysis model
Based on the results of flight event analysis, the functional requirement of rendezvous and docking mission for manned spacecraft can be sorted out. In the process of functional analysis, the corresponding functional requirements should be given according to the needs of users. Function model and requirement model feedback information to each other and refine iteratively. The function model obtained from the analysis can further guide the system composition analysis and

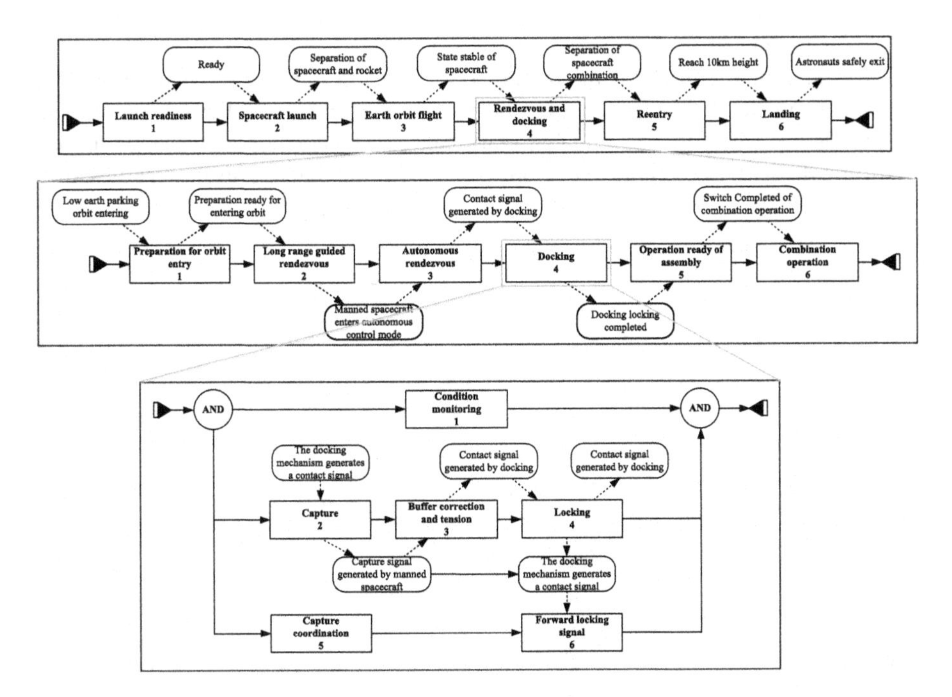

Fig. 5. Flight events analysis

other work. The preliminary results of functional analysis of rendezvous and docking manned spacecraft are shown in Fig. 6.

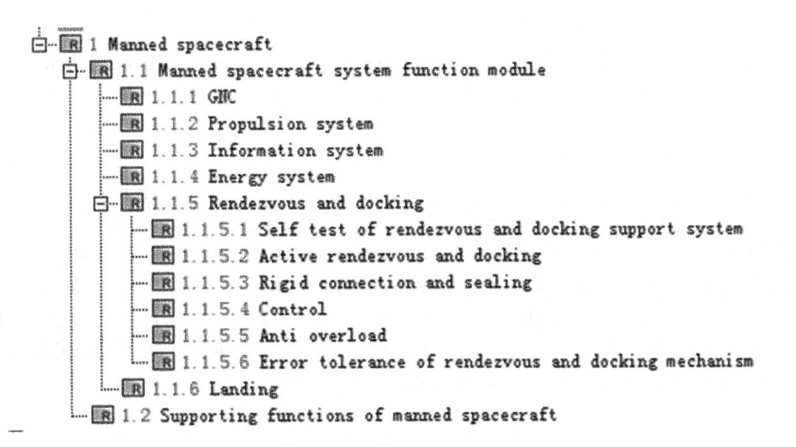

Fig. 6. Function model

(4) Product structure model
Product structure model is the bridge between the design and the real object. The product in the product structure model can be a mature application of existing products (such as the return cabin seat), can also be component (such as the seat lifting device). The product structure model of some manned spaceships is shown in Fig. 7.

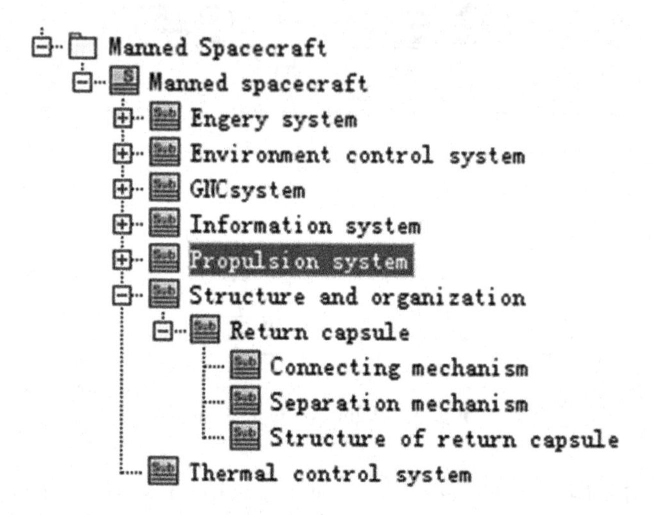

Fig. 7. Product structure model

(5) Each model in MBSE
Each model in MBSE has corresponding logical relations, through which we can use the upper level model to guide the construction of the next level model. The parameters of different models are associated to realize the analysis of the model and the relationship between model parameters. In the whole design process, the tracking of design elements, requirement coverage analysis and simulation verification completeness analysis are realized. The logical relationship between models is mainly composed of relationship orientation and relationship content, as shown in Fig. 8. The content of requirement model is satisfied by function model, and the content of function model is allocated to physical architecture model, which is realized by physical architecture model. Each model itself should be decomposed and refined, and its relationship can be decomposed or related.

In the process of using MBSE method to design the system, designers can design in a unified database, and the parameters in the database can be updated in real time with the modification. Therefore, designers can concentrate on the system design instead of searching for parameters in a large number of documents, which not only improves the work efficiency, but also avoids the errors caused by the untimely update of documents.

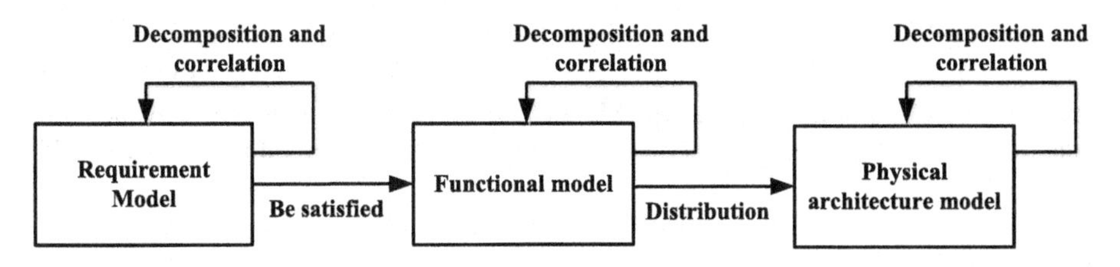

Fig. 8. Relations among primary models

Table 3 shows the statistical results of requirements by using data query, which can easily control the design status and progress. In addition, through the management of the unified database, more data mining and utilization can be realized.

Table 3. Statistic results of requirements

Query name	Number		
All requirements	106		
All requirements	Item status		
	Accepted	11	10.4%
	Adopted	8	7.5%
	Alternative	9	8.5%
	Solidified	0	0%
	Controversial	2	1.9%
	Rejected	3	2.8%
	Under advisement	3	2.8%
	Passed the examination	0	0%
	Suspicious	0	0%
	Pending	0	0%
	Unprocessed	70	66%
	Other	0	0%

5 Conclusion

MBSE method can cover the whole process from conceptual design, system design, and experimental verification to engineering implementation, and describe the system more intuitively. It can easily realize the deterministic and standardized management of system information interface, realize the traceability and coverage analysis of engineering elements such as requirements, design and verification, and ensure the effective control of technical state. In this paper, the MBSE method is applied to rendezvous and docking mission of manned spacecraft and how to use this method to carry out mission design is preliminarily demonstrated. At present, China has accumulated some experience in the system engineering method, but compared with foreign countries, there is still a certain gap in the design standards, top-level specifications and design tools of MBSE method. In the follow-up accumulation and development, China should keep up with the development trend of system engineering method system, further study MBSE method, combined with the mission requirements of China's major aerospace engineering projects, put forward the MBSE method workflow suitable for aerospace, in order to promote the development and application of MBSE method in aerospace field.

References

1. Kunsheng, W., Jianhua, Y.: Research and practice of model–based systems engineering. Aerospace China **11**, 52–57 (2012)
2. Sun, Y., Ma, L.: Model-based systems engineering and systems modeling language. Comput. Knowl. Technol. **7**(31), 7780–7783 (2011)
3. Estefan, J.A.: Survery of model-based systems engineering (MBSE) methodologies. INCOSE, San Diego, CA (2008)
4. Ramos, A., Ferreira, J., Barcelo, J.: Model-based systems engineering: an emerging approach for modem systems. IEEE Trans. Syst. Man Cybern. **42**(1), 101–111 (2012)
5. Caiyun, J., Weiping, W., Li, Q.: SysML: a new systems modeling language. J. Syst. Simul. **18**(6), 1483–1492 (2006)
6. Baozhu, G.: What is systems engineering. Spacecraft Eng. **22**(4), 1–6 (2013)
7. Zhu, Y., Li, Q., Yang, F., et al.: NASA Systems Engineering Handbook. Electronic Industry Press, Beijing (2012). NASA. translated
8. Fuchs, J.: Multi–Disciplinary MBSE Approach in Industrial phases, AIAA 2012–2532. AIAA, Washington D.C. (2012)
9. RussellI, M.: Using MBSE to enhance system design decision making. Procedia Comput. Sci. **8**, 188–193 (2012)
10. Tepper, N.A.: Exploring the use of model-based systems engineering (MBSE) to develop systems architectures in naval ship design, ADA541194. Massachusetts Institute of Technology, Cambridge (2010)

An Agile Platform for Partial Discharge Diagnostic Testing and Monitoring in Air-Insulated Substations

Ziqi Yue[1(✉)] and Zeqiu Liu[2]

[1] Beijing Information Technology College, Beijing, China
yuezq@bitc.edu.cn
[2] Beijing University of Technology, Beijing, China
liuzeqiu@emails.bjut.edu.cn

Abstract. Generally, the high-speed oscilloscopes are selected to acquire the real-time data on site for partial discharge diagnostic testing and monitoring in Air-Insulated Substations (AIS). However, it has some disadvantages, such as, the high cost and the inconvenience to carry and install. In this paper, we proposed an agile software defined radio (SDR) platform, the AD-FMCOMMS5-EBZ mounted on the Zynq®-7000 SoC ZC706 evaluation kit, as a good solution. We will examine the advantages in the platform that allow developers to quickly prototype partial discharge diagnostic testing and monitoring in Air-Insulated Substations. The tools required in this case are MATLAB® and its support packages. As a real-world example of the procedure, we will prototype a platform that determinate partial discharge position directly. The final results verify functionality and performance of the location determination algorithm, and show that the accuracy of the algorithm is acceptable.

Keywords: Partial discharge · Diagnostic testing · Monitoring · Air-insulated substation

1 Introduction

According to IEC-60270, a partial discharge (PD) is defined as a localized electrical discharge that only partially bridges the insulation between conductors and which can or cannot occur adjacent to a conductor [1]. PD can occur at any point within the insulation where the electric field exceeds the local dielectric breakdown strength. PD occurs multiple times and can gradually reduce dielectric breakdown strength.

Partial discharge diagnostic testing and monitoring based on ultra-high-frequency (UHF) sensor array has been applied in air-insulated substations (AIS), especially in location determination of partial discharge sources [2]. Generally, we rely on the high-speed oscilloscopes to acquire the real-time data on

Supported by General Project for Research Plan of Beijing Municipal Education Commission under Grant No. KM201710857002.

Q. Guo et al. (Eds.): WiSATS 2021, LNICST 410, pp. 311–321, 2022.
https://doi.org/10.1007/978-3-030-93398-2_31

site. High-end oscilloscopes have high performance, such as the bandwidth in terms of GHz and the sample rate in terms of GSample/s. However, it has some disadvantages, for example, the high cost and the inconvenience to carry and install.

Some software defined radio (SDR) platform, such as the EVAL-AD-FMCOMMS5-EBZ by Analog Devices, as a daughter board, mounted on the Zynq®-7000 SoC ZC706 evaluation kit by Xilinx, as a motherboard, maybe a good solution. It could support four transceivers at the same time, and the channel bandwidth is up to 56 MHz. In addition, the tuning range (center frequency) is from 1 MHz (with the help of AD-FREQCVT1-EBZ) to 6 GHz covering high frequency (HF), very high frequency (VHF) and ultra high frequency (UHF), even part of medium frequency (MF) and super high frequency (SHF). Last but not least, the cost is low.

The rest of the paper is organized as follows. The base board, the daughter board and the Communications Toolbox Support Package for Zynq®-Based Radio and ADALM-Pluto Radio by MathWorks are described in Sect. 2. The 2-dimension direct position determination algorithm for partial discharge is demonstrated in Sect. 3. Concluding remarks are drawn in Sect. 4.

2 The Agile Platform

2.1 Pre-requisite

The hardware for running PD diagnostic testing and monitoring in air-insulated substations is built with ZC706 evaluation kit, FMCOMMS5 radio hardware and bow-tie antenna array as shown in Fig. 1.

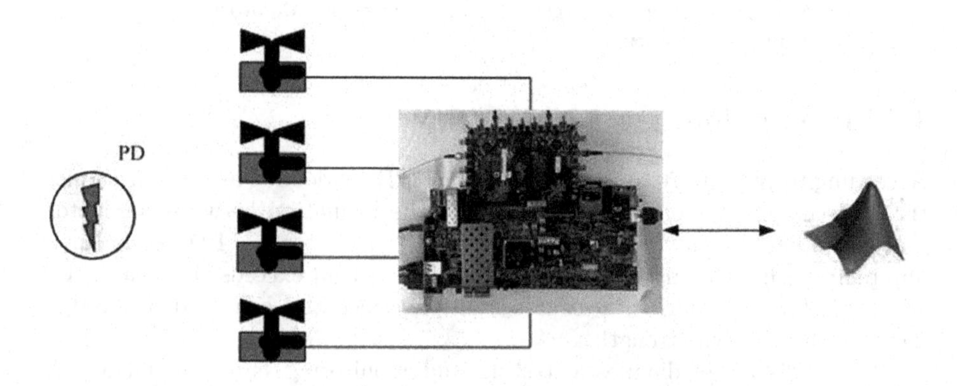

Fig. 1. PD position determination.

The Zynq®-7000 SoC ZC706 evaluation kit, as a base board, includes preverified reference designs and industry-standard FPGA Mezzanine Connectors (FMC) to allow scaling and customization with daughter boards.

The EVAL-AD-FMCOMMS5-EBZ, as a daughter board, is a high-speed analog module designed to showcase dual AD9361 devices in 4×4 multiple-input multiple-output (MIMO) applications. This platform is intended to enable the prototyping and development of many software defined radio applications, such as angle of arrival systems. Up to one thousands user-programmable registers make it an agile platform [3,4]. In many cases, multiple bits or bytes work together to serve a particular function, for example, those used to configure automatic gain control (AGC). Analog Devices provides complete drivers for both bare metal/No-OS (operating systems) and Linux when using the evaluation board.

ADALM-Pluto Radio is a software-defined radio active learning module available from Analog Devices that can generate RF analog signals from 325 MHz to 3800 MHz at up to 61.44 megasamples per second (MSPS) with a 20 MHz bandwidth.

The bow-tie antenna, as shown in Fig. 2, is selected as antenna element [5]. The reflection coefficient (S_{11}) is shown in Fig. 3.

In MATLAB®, Add-ons extend the capabilities by providing additional functionality for specific tasks and applications, for example, connecting to hardware devices and additional algorithms.

By installing the Communications Toolbox™ Support Package for Xilinx® Zynq®-Based Radio (MATLAB R2019b and later versions), we can execute MATLAB® script to prototype and verify PD diagnostic testing and monitoring systems [6].

By installing the Communications Toolbox™ Support Package for Analog Devices® ADALM-Pluto Radio (MATLAB R2018a and later versions), we can run MATLAB® script to calibrate the FMCOMMS5 radio channels and the bow-tie antenna array [7].

2.2 Setup and Procedure

Firstly, we connect the ZC706 evaluation kit and FMCOMMS5 radio hardware by an Ethernet cable. And then, we configure the host computer to work with the Xilinx® Zynq®-Based Radio Support Package. Finally, we open the MATLAB® script for PD diagnostic testing and monitoring.

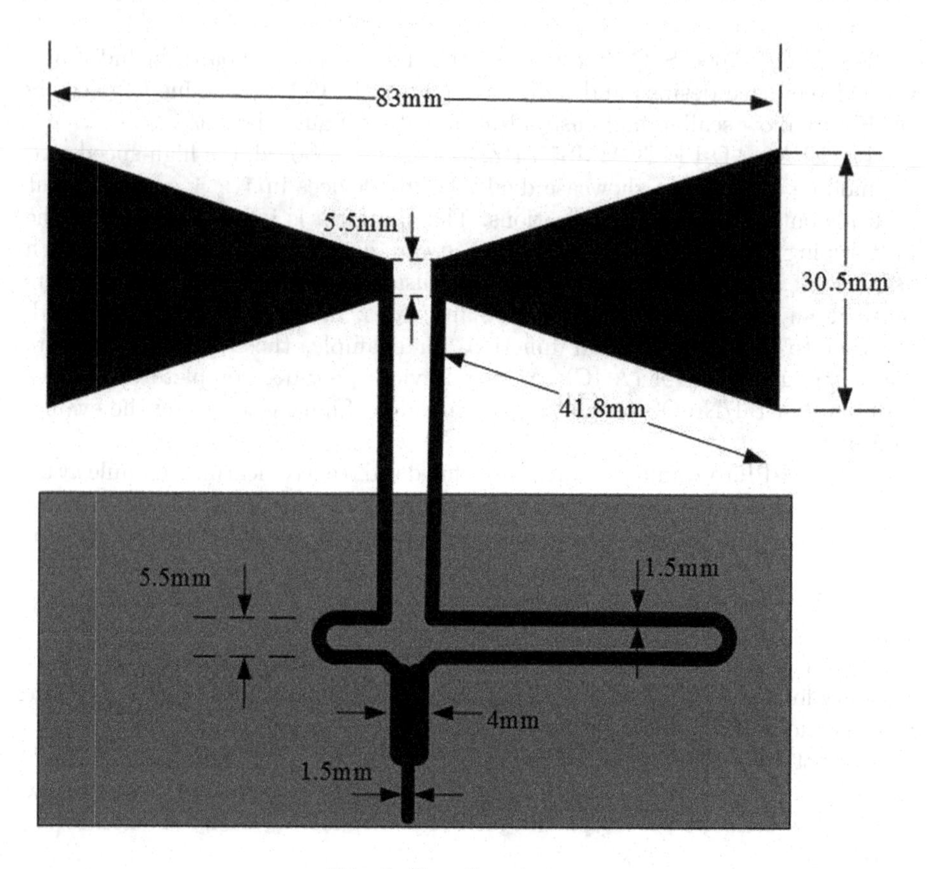

Fig. 2. Bow-tie antenna.

As shown in Fig. 4, for phase calibration among the four FMCOMMS5 radio channels, we connect the power splitter from the Transmitter (TX) port of PlutoSDR to the four receiver channels of FMCOMMS5, and then estimate and correct the phase offset among the channels.

We could display sinusoidal signals in the time domain to check whether all signals are in synchronization after IQ rotation as shown in Fig. 5.

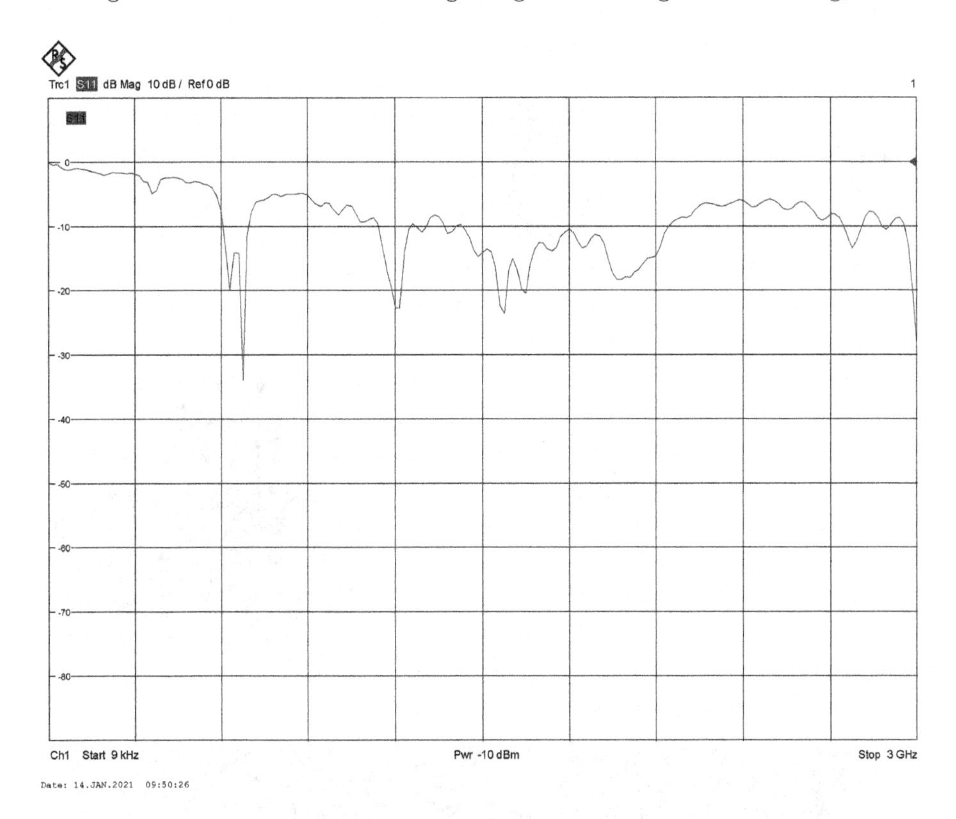

Fig. 3. Reflection coefficient.

As shown in Fig. 6, to calibrate bow-tie antenna array, we connect the power splitter from the four antenna elements to the Receiver (RX) port of the PlutoSDR, and then generate signal at the Transmitter (TX) port of the PlutoSDR, to estimate and correct the phase offset among the antenna elements.

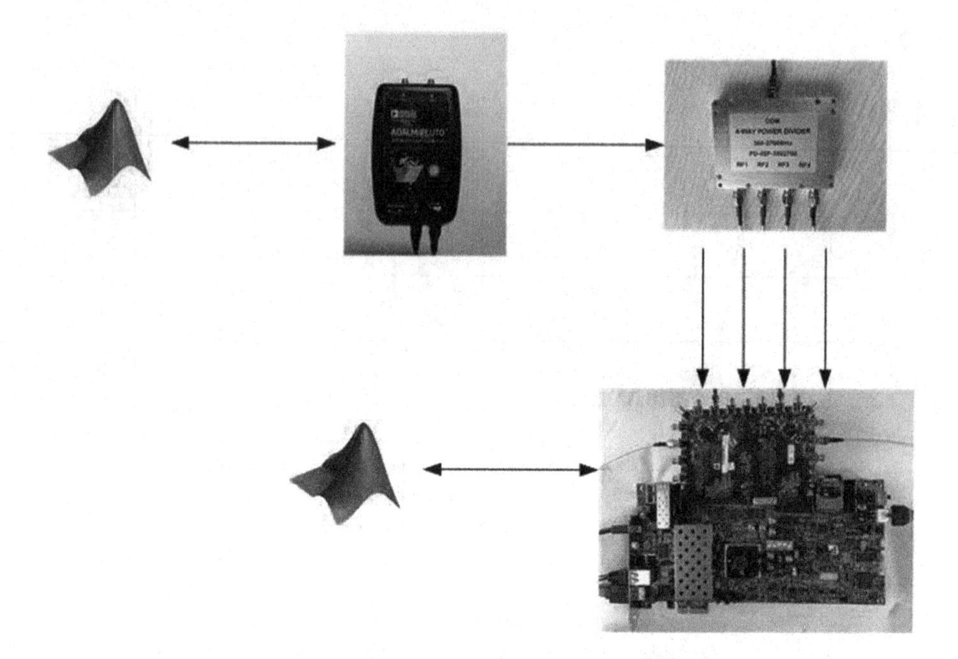

Fig. 4. Phase calibration for EVAL-AD-FMCOMMS5-EBZ.

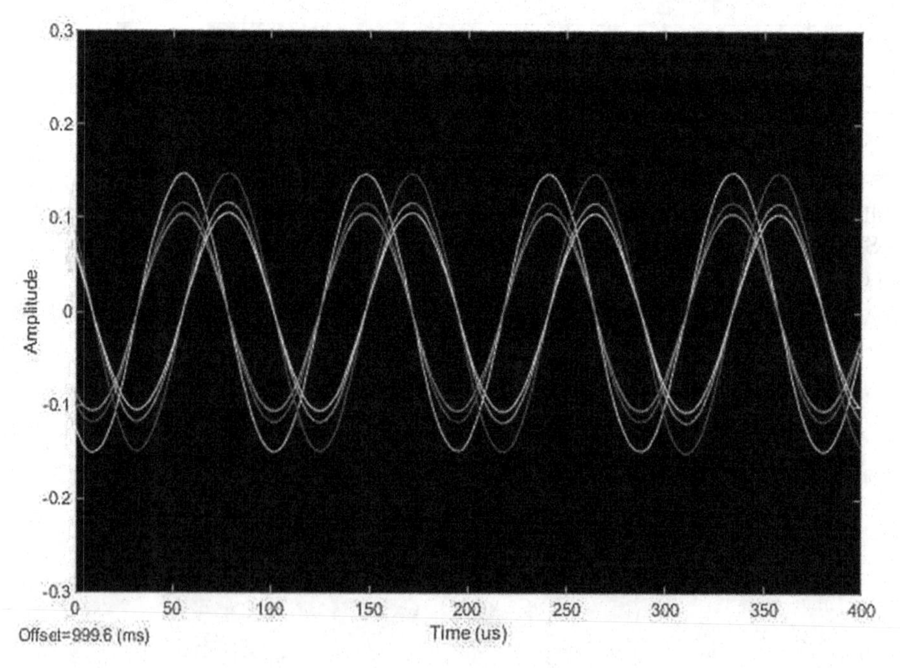

Fig. 5. Four channel signals are in synchronization.

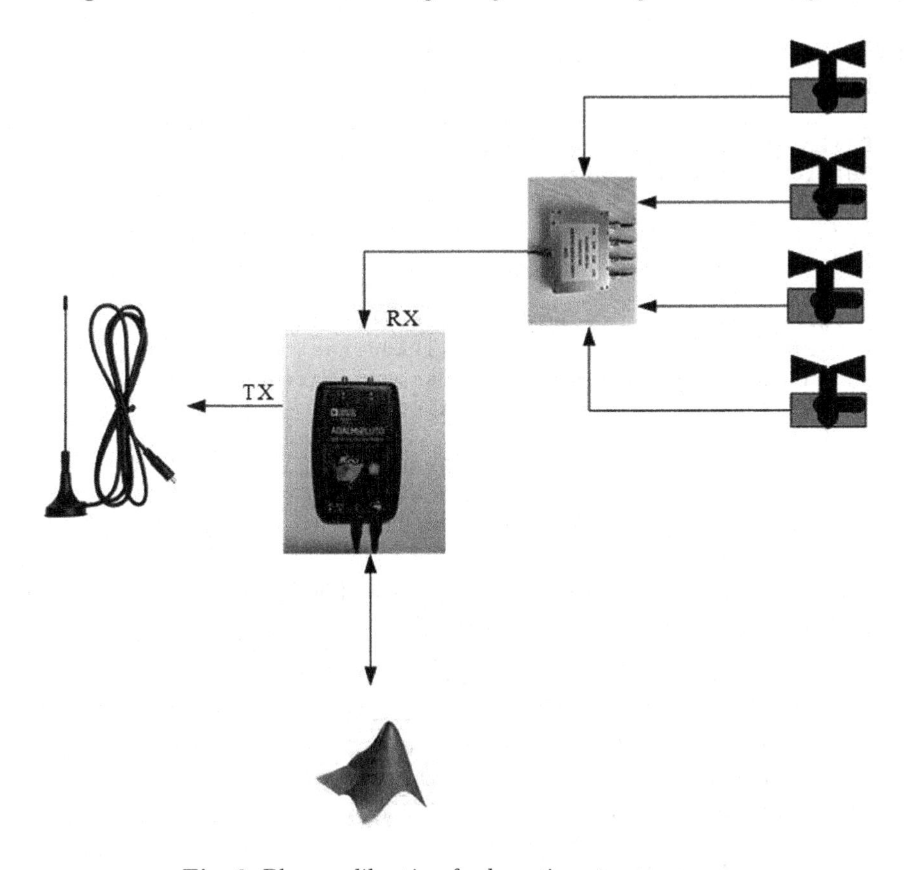

Fig. 6. Phase calibration for bow-tie antenna array.

Carefully disconnect the power splitter from the Tx port of PlutoSDR to the FMCOMMS5 without powering off the ZC706 evaluation kit, and connect the bow-tie antenna array to the four channels of the FMCOMMS5 using equal length SMA cables.

2.3 Verifying Algorithms in MATLAB

The first step, in designing a PD diagnostic testing and monitoring algorithm, is to access some source data. We can use the Zynq® SDR Rapid Prototyping Platform. Analog Devices provides a MATLAB System object™ that is capable of receiving data from the FMCOMMS5 platform over Ethernet.

The System object™ allows us to select a tuning frequency and sampling rate, collect receive samples using the radio hardware, and bring the receive samples directly into the MATLAB workspace as a MATLAB variable. The required code is very short. A few lines of code to set up the FMCOMMS5, a few more to set up the MATLAB System object™ sdrReceiver. A sample of the code is shown as follows

```
DOARx.SDRDeviceName = 'FMCOMMS5';
DOARx.IPAddress = '192.168.3.2';

RadioBasebandRate = 1e6;
RadioFrameLength = 4000;

sdrReceiver = sdrrx( DOARx.SDRDeviceName, ...
  'IPAddress',            DOARx.IPAddress, ...
  'BasebandSampleRate', RadioBasebandRate,...
  'GainSource',           'Manual', ...
  'Gain',                 20, ...
  'SamplesPerFrame',      RadioFrameLength, ...
  'ChannelMapping',       [1 2 3 4],...
  'OutputDataType',       'double');
```

A few lines of code to capture I/Q samples and write them to a MATLAB variable. A sample of the code is shown as follows

```
[data, valid, ~] = sdrReceiver();
```

A few lines of code to phase calibration for EVAL-AD-FMCOMMS5-EBZ. A sample of the code is shown as follows

```
data = zynqRadioDOAEstimation...
AmplitudeNormalization(data);

% Calculating the phase offset
% between 2 channels
[angle1,angle2,angle3] = zynqRadioDOAEstimation...
FindPhaseDifference4Channels(data);

% IQ Rotation to reduce the phase offset
% between all channels to 0
data = zynqRadioDOAEstimation...
RotateIQ4Channels(data,angle1,angle2,angle3);
```

A few lines of code to release the System object™ sdrReceiver. A sample of the code is shown as follows

```
release(sdrReceiver);
```

3 Experiments

As shown in Fig. 7, we consider two stations ($L = 2$) placed at one side of a 10 (m)×10 (m) square and a single PD source placed at the square center. Each station is equipped with a uniform linear array (ULA) of two antenna elements ($M = 2$) whose broadside towards the square center.

The channel response from the PD source to each station is calibrated in advance, and the propagation speed is assumed to be $c = 3 \times 10^8$ [m/s].

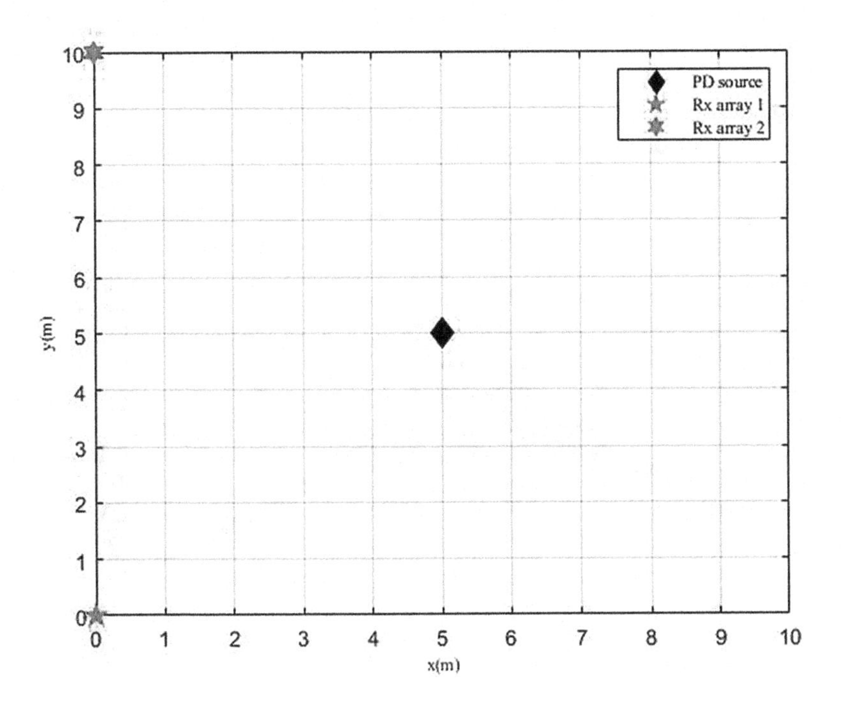

Fig. 7. Laboratory layout.

Following similar coding structures for interacting with the agile platform, we can explore PD position determination as an example of diagnostic testing and monitoring in air-insulated substations [8]. The location determination algorithm was performed in two steps: first, a coarse grid search with 1 (m) resolution over the square generate an initial location estimate; and then, a fine grid search with 0.1 (m) resolution over the neighborhood of the peak detected in the coarse search. The spectrum of the PD source is near 1 GHz, and each location determination is based on single capture of the PD signal for a period of 100 samples ($N_s = 100$).

We performed 500 experiments ($N_e = 500$). The location determination algorithm is executed for each experiment. Then, we evaluate the performance based on the Root Mean Square (RMS) of error given by

$$RMS = \sqrt{\frac{1}{N}\sum_{i=1}^{N_e}(\hat{x}_i - x_0)^2 + \frac{1}{N}\sum_{i=1}^{N_e}(\hat{y}_i - y_0)^2}. \quad (1)$$

where (x_0, y_0) is the PD source location and $(\hat{x}_i, \hat{y}_i)$ is the i-th location estimate.

We plotted one cost function in Fig. 8 for the experiment setup. The cost function has a peak at the location which is close to PD source, the accuracy of the proposed algorithm is acceptable. The RMS of error or miss distance is 0.21 (m).

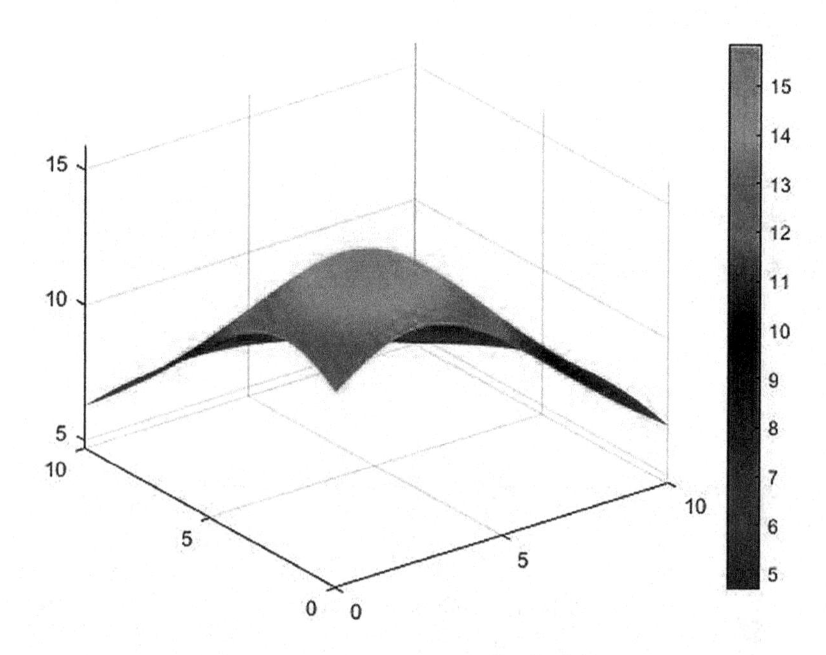

Fig. 8. Cost function for the layout in Fig. 7.

4 Discussion and Conclusions

In this paper, we proposed an agile platform for partial discharge diagnostic testing and monitoring in air-insulated substations. Through experiments in laboratory, the accuracy of the location determination algorithm is shown to be acceptable.

In the future, we will explore other algorithms and implementation of diagnostic testing and monitoring in air-insulated substations by the agile platform.

Acknowledgment. The authors would like to thank the editor and anonymous reviewers for useful comments. This work is supported by General Project for Research Plan of Beijing Municipal Education Commission under Grant No. KM201710857002.

References

1. IEC International Standard 60270: High Voltage Test Techniques-Partial Discharge Measurements; International Electrotechnical Commission, Geneva, Switzerland (2015)
2. Liu, Q., Zhu, M.X., Wang, Y.B., Deng, J.B., Zhang, G.J.: Partial discharge orientation method in substation based on ultra-high frequency phased array theory. Proc. CSEE **37**(20), 6126–6135 (2017)
3. AD9361 Reference Manual, Rev. A. Analog Devices Inc. (2014)
4. AD9361 Register Map Reference Manual, Rev. 0. Analog Devices Inc. (2014)
5. Durgun, A.C., Balanis, C.A., Birtcher, C.R., Allee, D.R.: Design, simulation, fabrication and testing of flexible bow-tie antennas. IEEE Trans. Antennas Propag. **59**(12), 4425–4435 (2011)
6. Communications Toolbox™ Support Package for Xilinx® Zynq®-Based Radio User's Guide, Version 20.2.0. The MathWorks Inc. (2020)
7. Communications Toolbox™ Support Package for Analog Devices® ADALM-Pluto Radio User's Guide, Version 20.2.0. The MathWorks Inc. (2020)
8. Yue, Z.Q., Liu, Z.Q.: Location determination of partial discharge sources in air-insulated substation (2021). in press

Signal Processing for Communications and Networking

Research on the Selection Method of Output Degree Distributions for LT Codes

Shuang Wu[1(✉)], Chen Cui[2], and Qingyang Guan[1]

[1] College of Engineering, Xi'an International University, Xi'an 710070, China
[2] School of Electronics and Information Engineering, Harbin Institute of Technology, Harbin 150001, China
cuichen@hit.edu.cn

Abstract. LT codes, the first class of rateless codes, are designed to overcome the packet lossy problems. LT codes with the dynamic overhead to measure the coding efficiency. It is necessary to note that the overhead can be considered as the reciprocal of code rate. Although the overhead is dynamic, but the different overhead corresponding different coding efficiency, which means that in the practical transmission scenarios, the overhead is the lower the better. For this reason, one need to find some other parameters to make the LT codes provide suitable error correction capabilities. As the error performances of LT codes also can be determined by output degree distributions, it is necessary to select the suitable distributions to provide the LT codes can provide enough erasure correcting capability on a lower overhead. For this reason, we derive the symbol error rates and the redundant probabilities of LT codes. Based on the derive results, two constraint conditions of selection of output degree distributions are proposed. And the simulation results shown that these two conditions are valid.

Keywords: LT codes · Rateless codes · Symbol error rate · Redundancy probability · Overhead

1 Introduction

LT codes, the first class of practical fountain codes, which are designed to overcome the packet lossy problems [1]. As LT codes can generate output symbols as much as needed, which code also been known as rateless codes, and have the potential to suit for various channel states [2]. As LT codes can overcome the packet lossy problems, which codes are originally to be used to take place of the transmission protocols based on Automatic Repeat-reQuest (ARQ) mechanisms. And LT codes is also can be used as the inner codes in the concatenate coding schemes [3].

Different with the traditional error correcting codes, LT codes cannot provide bit error correcting capability, but instead of erasure correcting property [4]. For

Q. Guo et al. (Eds.): WiSATS 2021, LNICST 410, pp. 325–334, 2022.
https://doi.org/10.1007/978-3-030-93398-2_32

this reason, there is no need to added too much redundancy data to provide error correcting capability, and the structure of LT codes can generated by using a random manner. For this reason, the LT codes can be flexibility to be the basis of many LT-based coding schemes. For example, as the inner codes of raptor codes [3], to be concatenated with other codes to provide rateless property [5], to provide unequal error protection property [6], and be re-designed to suit for the network communication systems [7,8], etc.

As the structures of LT and LT-based rateless codes are very flexible, which codes also be used in many research fields, for example the access technologies [9], network coding schemes [10] and to provide differentiates to recognize the user on Non-Orthogonal Multiple Access (NOMA) ensembles [11].

Because of the rateless property, the coding efficiency of LT codes are not measured by using the constant termed code rate. Actually, as the output symbols of LT codes can be generated as much as needed, the coding efficiency of LT codes are measured by using a dynamic ratio termed overhead, which is the reciprocal of code rate. As the transmission efficiency in one of the most important performances of the communication systems, which means that the overhead would as lower as better when the decoding processes are finished. Here is a paradox exists, on the one hand the overhead can be as large as needed, on the other hand the overhead would as lower as better. As the decoding processes are finished if the symbol error rate is lower enough to suit for the reliability requirements, and the error performances of LT codes are mainly determined by the output degree distributions, the output degree distributions need to be optimized to suit for the transmission conditions.

As the output degree distribution is the most important parameter in the construction of a LT codes, which have attracted much attentions. Luby have originally provide the idea and robust degree distributions [1], and some practical output degree distributions are given in [3]. Besides these classical studies, there are many other researcher studied on this issue [12,13]. Although there are so many studies focus on the output degree distributions, but in practice the encoders are mainly adopting the pre-designed output degree distributions from [3], because of which degree distributions are all can provide lower overhead performances. Unfortunately, the distributions proposed by [3] are designed on the code length 10^5, but in the practical scenarios, the code length are much lower, for this reason, it is necessary to select the suitable output degree distributions for such scenarios.

To resolve the above problems, by derive the relation between symbol error rate and average degrees, and quantized the redundancy probabilities of output symbols, we proposed two propositions to select output degree distributions, and the simulation results shown that these propositions are worked.

This paper is organized as following. In Sect. 2, we review the LT codes, some necessary definitions are also given. We derived the symbol error rate of LT codes by using a iterative expression, and quantized the corresponding overheads of given symbol error rates in Sect. 3. And the proposed propositions are simulated in Sect. 4, in which the results hold that the propositions are worked. At the lats, the conclusion of this paper is drawn in Sect. 5.

2 Preliminaries

In a LT code, the original data are framed into k input symbols. An output symbol with degree d is generated by XoR d randomly selected input symbols. The degree d is determined by given the output degree distribution $\Omega(x) = \sum_{d=1}^{D} \Omega_d x^d$. The input symbols which are selected in encoding process are termed as been *covered*, and each output symbol with degree 1 is called as *released* symbol. An output symbol and the selected input symbols in whose generating process are *neighbors* for each other. An input symbol can be *recovered* only if one of its neighbors is released. As there are n output symbols have been generated, the *overhead* is defined as $\gamma = \frac{n}{k}$. It can be found that the overhead is the reciprocal of the code rate.

The number of recovered input symbols are usually less than the number of collected output symbols, as an output symbol have the potential to recover at most one input symbol, there is an redundancy probability exists. Actually, an output symbol is redundant is caused of two reasons, the one is which symbol cannot to be processed into an released symbol, the other one is all the neighbors of an output symbols are recovered by their other output neighbors.Let$y$ and $P_{rdu,d}$ denote the symbol error requirement of communication systems and redundancy probability of an output symbol with degree d, then the redundancy probability can be given by following equations.

$$\begin{aligned} P_{rdu,d} &\triangleq P_{reason\ 1} + P_{reason\ 2} \\ &= 1 - d(1-y)^{d-1}y - (1-y)^d + (1-y)^d \\ &= 1 - d(1-y)^{d-1}y. \end{aligned} \tag{1}$$

The input symbols of LT codes are barely recovered if $\gamma < 1$. The *recovery rate* of rateless codes with $\gamma < 1$ is dubbed the *intermediate performance.* To improve the intermediate performance of rateless codes, Growth codes are proposed that have a dynamic output degree distribution $\Omega(x)$ [14]. The design of $\Omega(x)$ for Growth codes is intuitive. The partitions of output symbols with lower degrees would decrease with the increase in the number of output symbols.

3 The Differences Between the Degree of Output Symbols

It is well-known that the decoding performances of LT codes depending on their output degree distributions, but as the decoding performances of LT codes can be measured through various facets, the output degree distributions would affected on LT codes through different aspects.

3.1 On the Erasure Correcting Capability of LT Codes

In most of the scenarios, the data have to be transmitted based on a given reliability requirements. And the symbol error rate of LT codes is the performance which is used to measure the reliability of data, then the erasure correcting capability is the most important performance of a LT code.

By consider a given LT code, the symbol error rate can be measured by using the expected error rate of each input symbol. As the LT codes are processed over the symbol lossy channels, which means the bit error problems can be neglected, and an input symbol can be recovered only if one of its neighbor output symbols can be processed into with degree 1, and the only neighbor of this output symbol is the above input symbol.

As the input and output degree distributions $\Lambda(x) = \sum_{\hat{d}=1}^{D_{int}} \Lambda_d x^d$ and $\Omega(x) = \sum_{d=1}^{D_{out}} \Omega_d x^d$ are given, then for a randomly selected input symbol, whose degree is d', then the expected error rate of this symbol is

$$P_{error,\hat{d}} \triangleq \prod_{i=1}^{\hat{d}} (1 - P_{rel}^i), \tag{2}$$

where P_{error} is the expected error rate of the input symbol, and $P_{rel,i}$ is the probability that the ith output neighbor of the input symbol can be processed to released.

And for each output neighbor with degree d, the released probability can be computed by

$$P_{rel,d} = \binom{d}{d-1}(1 - P_{error})^{d-1} = d(1 - P_{error})^{d-1}, \tag{3}$$

where P_{error} is the average error rate of input symbols, and which can be calculated by

$$P_{error} = \sum_{\hat{d}}^{D_{int}} \Lambda_{\hat{d}} P_{error,\hat{d}}. \tag{4}$$

It can be found from Eq. (3), the released probability of an output symbol would decreases with its degree growth.

Assuming the encoding process is strictly random, then Eq. (2) can be rewritten as follows.

$$P_{error,\hat{d}} = \hat{d}(1 - P_{rel})^{\hat{d}-1}, \tag{5}$$

where P_{rel} denotes the expected released probability of output symbols, which can be expressed as

$$\begin{aligned} P_{rel} &= \sum_{d=1}^{D_{out}} \Omega_d P_{rel,d} \\ &= \sum_{d=1}^{D_{out}} \Omega_d d(1 - P_{error})^{d-1} \\ &= \Omega'(1 - P_{error}). \end{aligned} \tag{6}$$

The expected symbol error rate which shown in Eq. (4) can be rewritten by following.

$$\begin{aligned} P_{error} &= \sum_{\hat{d}}^{D_{int}} \Lambda_{\hat{d}} P_{error,\hat{d}} \\ &= \sum_{\hat{d}}^{D_{int}} \Lambda_{\hat{d}} \hat{d}(1 - P_{rel})^{\hat{d}-1} \\ &= \sum_{\hat{d}}^{D_{int}} \Lambda_{\hat{d}} \hat{d}\big(1 - \Omega'(1 - P_{error})\big)^{\hat{d}-1} \\ &= \Lambda'\big(1 - \Omega'(1 - P_{error})\big). \end{aligned} \tag{7}$$

It no hard to seen that for an input symbol, whose error rate would dramatically decreases with its degree growth. Assuming that the overhead in enough large such that $\Omega'(1 - P_{error})$ tends to its maximum value, then P_{error} would also tends to its minimum value.

As $\Lambda(x)$ is generating in the encoding process, and the average degrees of input symbol and output symbols $\bar{\hat{d}}$ and $\bar{d}$ satisfying $\bar{\hat{d}} = \gamma \bar{d}$, we can arbitrary given the following proposition.

Proposition 1. *For LT codes with the same average degree, if the overhead is enough large, which code would share the same symbol error rate.*

3.2 On the Coding Efficiency for LT Codes

As the reliability and efficiency are the two most important target of communication systems, the coding efficiency is also an important performance of LT codes besides the symbol error rate.

For the traditional error correcting codes, the coding efficiency are measured by using the code rate. But for LT code, as its rateless property, the code rates are not fixed, then the coding efficiency of LT codes (also for other rateless codes) are measured by overheads. The overhead is the ratio of the number of output symbols versus the number of input symbols, which is the reciprocal of the code rate. And it is worth to not that a code rate is a pre-designed constant, but the overhead is a dynamic ratio.

As LT codes are proposed to overcome the symbol lossy problems, which codes no need to correcting the bit errors, then the redundant informations are not necessary. For this reason, to obtain high coding efficiency performance, the target overheads of LT codes are as approach to 1.

Then consider a LT code with k input symbols, in which the symbol error rate is given by P_{error}. Then based on Eq. (1), the redundant probability of an output symbol with degree d can be rewritten as

$$P_{rdu,d} = 1 - d(1 - P_{error})^{d-1} P_{error}. \tag{8}$$

Furthermore, the expected redundant probability of this code is

$$P_{rdu} = \sum_{d=1}^{D_{out}} \Omega_d P_{rdu,d}. \tag{9}$$

By substitute Eq. (8) into Eq. (9), the redundant probability of the given LT code is

$$\begin{aligned} P_{rdu} &= \sum_{d=1}^{D_{out}} \Omega_d P_{rdu,d} \\ &= \sum_{d=1}^{D_{out}} \Omega_d - \sum_{d=1}^{D_{out}} \Omega_d d(1-P_{error})^{d-1} P_{error} \\ &= 1 - P_{error}\Omega'(1-P_{error}). \end{aligned} \tag{10}$$

As symbol error rate is a function of overhead γ, the redundancy probability is the function of functions of overhead. And the redundant problems would accumulated in the decoding process. Based on this reason, the following proposition can be given.

Proposition 2. *For a LT code, as the redundancy probability is a function of overhead* γ*, and the redundancy problem would accumulated in the decoding process, the output degree distributions should satisfying that in the region* $\gamma < 1$*, the* P_{rdu} *would as lower as possible.*

4 Simulation Results

In this section, the decoding performances of LT codes with various output degree distributions under asymptotic conditions are compared.

Firstly, we choose the output degree distributions proposed in [3], which are given in following

$$\begin{aligned} \Omega(x) = &\, 0.007969x^1 + 0.493570x^2 + 0.166220x^3 \\ &+ 0.072646x^4 + 0.082558x^5 + 0.056058x^8 \\ &+ 0.037229x^9 + 0.055590x^{19} + 0.025023x^{64} \\ &+ 0.003137x^{66}, \end{aligned} \tag{11}$$

$$\begin{aligned} \Omega(x) = &\, 0.007544x^1 + 0.493610x^2 + 0.166458x^3 \\ &+ 0.067900x^4 + 0.089209x^5 + 0.041731x^8 \\ &+ 0.050162x^9 + 0.038837x^{19} + 0.015537x^{20} \\ &+ 0.016298x^{66} + 0.012677^{67} \end{aligned} \tag{12}$$

and

$$\begin{aligned}\Omega(x) = 0.0.004807x^1 + 0.496472x^2 + 0.166912x^3 \\ + 0.073374x^4 + 0.082206x^5 + 0.057471x^8 \\ + 0.035951x^9 + 0.001167x^{18} + 0.054305x^{19} \\ + 0.018235x^{65} + 0.009100^{66}.\end{aligned} \tag{13}$$

As shown in Fig. 1, the above degree distributions are with average output degree values $\bar{d_1} = 5.8454$, $\bar{d_2} = 5.9708$, $\bar{d_3} = 5.8250$, the code 1 and code 3 with the nearly same symbol error rates in the region $\gamma > 1.1$, and as the code 2 with the average degree value little larger than the other two, which code with the better symbol error rate performance than the others. Which results proof that the Proposition 1 is valid.

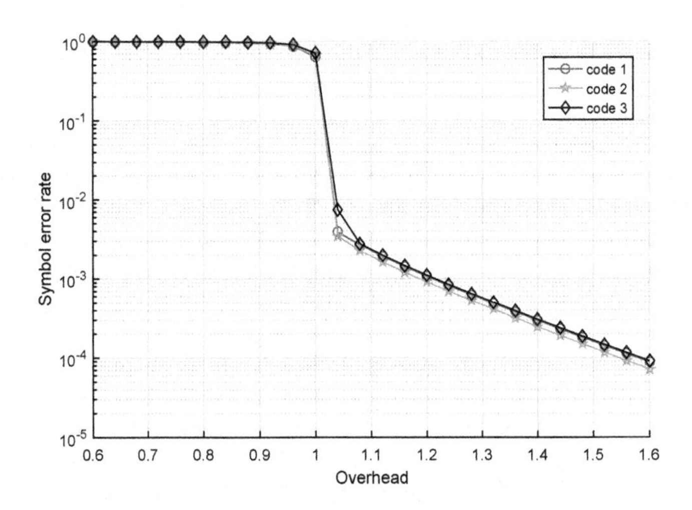

Fig. 1. The SER performances of LT codes with output degree distributions proposed in [3].

Then we consider the LT codes with robust degree distributions [1].As shown in Fig. 2, the codes are with the robust degree distributions with generating parameters ($k = 1000, c = 5, \delta = 0.5$), ($k = 1000, c = 2, \delta = 0.2$), ($k = 1000, c = 1, \delta = 0.1$) and ($k = 1000, c = 0.2, \delta = 0.1$). As the codes are with the average output degree values $\bar{d_1} = 5.9992$, $\bar{d_2} = 7.1500$, $\bar{d_3} = 7.9598$ $\bar{d_3} = 8.8662$, which values are increased with the sequence numbers growth, then symbol error rate performances of these codes are also as better as its sequence number goes. Which phenomenon also can be used to confirm the Proposition 1 in this paper.

By observing at Fig. 2, it can be found that although in the error flow region, the symbol error rate performances of the 4 codes are related to their average output degree values, but the error flow region are began from the different overhead which much larger than 1. By contrast, in the Fig. 1, the error flow

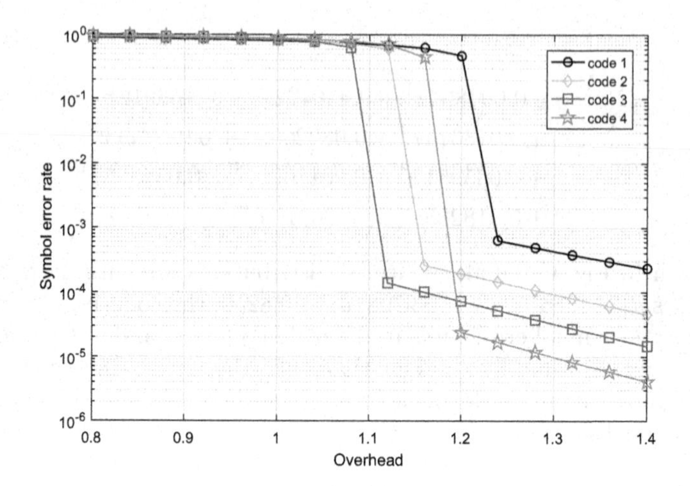

Fig. 2. The SER performances of LT codes with robust degree distributions.

region of the three given codes are began at the nearly same overhead which only a little larger than 1. Which because of the output degree distributions shown in Eqs. (11), (12) and (13) are carefully designed, but the above robust degree distributions are generated by random choosing the parameters. As the overheads are directly corresponding to the coding efficiency of LT codes, then we turned to focus on the redundancy probabilities of output symbols for LT codes.

Figure 3 illustrate the redundancy probabilities of output symbols for LT codes. In which the codes are with the output degree distributions shown in Eqs. (11) and (12), and the other codes are adopted the robust degree distributions with parameters $(k = 1000, c = 5, \delta = 0.5)$ and $(k = 1000, c = 1, \delta = 0.1)$. By observed at Fig. 1 and 2, it is can be found that the codes with output degree distributions proposed by [3] share the better overhead performances than which of the codes by using the robust degree distributions. And it can be found that for the codes share the same decoding overhead performances, whose redundancy probabilities are also the same, and it can be found that in the region $\gamma < 1$, the lower redundancy probabilities would leading to a lower decoding overhead, in other word leading to a higher coding efficiency. Which results illustrated that the redundant problems of output symbols would be accumulated in the decoding process, and the Proposition 2 can be holds.

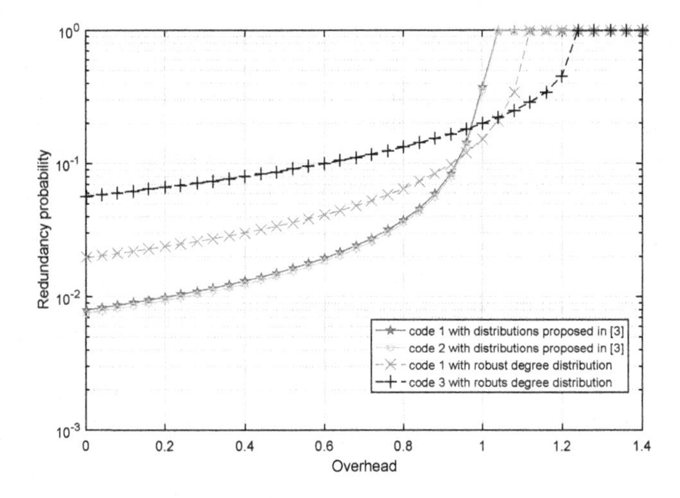

Fig. 3. The redundancy probabilities of output symbols for LT codes.

5 Conclusion

In this paper, we focus on the selection method of output degree distributions of LT codes. By derive the symbol error rate of input symbols and the redundancy of output symbols, two proposition are given. Firstly, the symbol error rate of input symbols would depending on the average degree values of output symbols. Secondly, the coding efficiency of LT codes would depending on the redundancy probabilities of output symbols in the decoding process, and the overall redundancy probability would be a accumulated value. The simulation results shown that the proposed proposition are valid, and which can be used to select an suitable output degree distributions before the encoding processes are beginning.

Acknowledgment. This work was supported by the Scientific Research Initiation Funds for the Doctoral Program of Xi'an International University (Grant No. XAIU2018070102 & XAIU2019002), Regional Innovation Capability Guidance Project (Grant No. 2021QFY01-08) and the General Project of science and Technology Department of Shaanxi Province (Grant No. 2020JM-638).

References

1. Luby, M.: LT codes. In: Proceedings of 43rd Annual IEEE Symposiumon Foundations of Computer Science, pp. 271–280 (2002)
2. Byers, J.W., Luby, M., Mitzenmacher, M.: A digital fountain approach to asynchronous reliable multicast. IEEE J. Sel. Areas Commun. **20**(8), 1528–1540 (2002)
3. Shokrollahi, A.: Raptor codes. IEEE Trans. Inf. Theory **52**, 2551–2567 (2006)

4. Mackay, D.J.C.: Fountain codes. IEE Proc. Commun. **152**(6), 1062–1068 (2005)
5. Li, B., Tse, D., Chen, K., Shen, H.: Capacity-achieving rateless polar codes. In: 2016 IEEE International Symposium on Information Theory (ISIT), pp. 46–50 (2016). https://doi.org/10.1109/ISIT.2016.7541258
6. Rahnavard, N., Vellambi, B., Fekri, F.: Rateless codes with unequal error protection property. IEEE Trans. Inf. Theory **53**(4), 1521–1532 (2007)
7. Talari, A., Rahnavard, N.: Distributed unequal error protection rateless codes over erasure channels: a two-source scenario. IEEE Trans. Commun. **60**(8), 2084–2090 (2012)
8. Gu, S., Yang, Y., Jian, J., Wei, X., Zhang, Q.: Distributed rateless coded collaboration for satellite relay networks. In: International Conference on Wireless Communications & Signal Processing (2015)
9. Shirvanimoghaddam, M., Li, Y., Dohler, M., Vucetic, B., Feng, S.: Probabilistic rateless multiple access for machine-to-machine communication. IEEE Trans. Wirel. Commun. **14**(12), 6815–6826 (2015). https://doi.org/10.1109/TWC.2015.2460254
10. Du, W., Li, Z., Liando, J.C., Li, M.: From rateless to distanceless: enabling sparse sensor network deployment in large areas. IEEE/ACM Trans. Netw. **24**(4), 2498–2511 (2016). https://doi.org/10.1109/TNET.2015.2476349
11. Chen, Y.M., Lai, W.M., Ueng, Y.L.: Rateless coded multiplexing for downlink transmission with two users: performance analysis and system design. IEEE Access **7**, 50440–50452 (2019). https://doi.org/10.1109/ACCESS.2019.2911214
12. Zhu, H., Zhang, G., Li, G.: A novel degree distribution algorithm of lt codes. In: 2008 11th IEEE International Conference on Communication Technology, pp. 221–224 (2008). https://doi.org/10.1109/ICCT.2008.4716207
13. Cheong, S.T., Fan, P.: Novel degree function over finite field for LT codes. In: 2016 International Conference on Electronics, Information, and Communications (ICEIC), pp. 1–5 (2016). https://doi.org/10.1109/ELINFOCOM.2016.7562925
14. Kamra, A., Misra, V.: Growth codes: maximizing sensor network data persistence. In: Proceedings of ACM Sigcomm, Pisa, Italy, pp. 255–266, September 2006

Novel Shaped 8/16PSK Modulation with Improved Spectral Efficiency

Xiao Chen[1], Yu Zhou[1(✉)], Fan Gao[1], Xuesong Shao[1], Yue Li[1], Zhuowen Mu[1], and Hao Lu[2]

[1] State Grid Jiangsu Electric Power Co., Ltd., Marketing Service Center, Nanjing, China
lygnr@aliyun.com
[2] Hohai University, Nanjing 210098, China
luhao@hhu.edu.cn

Abstract. In this work, a precoder based on 8/16 phase shift keying (PSK) is devised and used to generate a novel shaped 8/16 PSK modulation combining precoding and continuous phase modulation (CPM). Then, two receivers based on Bahl-Cocke-Jelinek-Raviv (BCJR) algorithm and I-Q coherent detection are derived. Theoretical analysis and simulation are provided to show that the proposed modulation achieves higher spectral efficiency than the conventional CPM.

Keywords: Continuous phase modulation · Phase shift keying · Spectrum efficiency

1 Introduction

Recently, as a spectrally efficient modulation scheme, continuous phase modulation (CPM) has attracted great attention. Constant envelop and continuous phase allow for robustness against nonlinear distortion and high spectral efficiency. Due to these properties, CPM has been widely used in a number of fields such as satellite communications, deep-space communications and telemetry [1–3]. As one of the components in CPM, the mapper converts original binary bits to the data symbol belonging to alphabet of $\{\pm 1, \pm 3, \cdots, \pm (M-1)\}$, where M denotes the modulation order.

The design of alphabet has great impact on the signal spectral property and bit error rate (BER). To the best of our knowledge, few contributions have been made on the improvement to the CPM mapper. One approach uses precoder to replace mapper. Using this approach, an improved multi-h CPM to increase the minimum Euclidean distance of binary CPMs was proposed in [4] without considering bandwidth. Reference [5] focused on constraining the signal phase evolution of low modulation order CPM via precoding. Reference [6] described a form of CPM known as shaped binary phase shift keying (SBPSK) which is a constant envelope waveform modulation with a controlled phase trajectory based

Q. Guo et al. (Eds.): WiSATS 2021, LNICST 410, pp. 335–344, 2022.
https://doi.org/10.1007/978-3-030-93398-2_33

on BPSK. Motivated by SBPSK, [7] designed a ternary precoder $\{-1, 0, 1\}$ based on the phase state of offset quadrature phase-shift keying (OQPSK). It combined the precoder with CPM modulator to form a special case of CPM, SOQPSK, with higher spectral efficiency. However, none of them has studied higher order PSK.

In this paper, we propose a novel precoder based on 8/16PSK. The precoder converts the original binary sequence to the data symbol drawn from $\{0, \pm1, \pm2, \cdots, \pm(M/2-1), M/2\}$. The data symbol sequence is then modulated by a CPM modulator to generate shaped 8PSK and shaped 16PSK. Theoretical analysis shows that the proposed precoder can obtain narrower bandwidth at the cost of decreasing Euclidean distance between symbols. Simulation shows that the shaped 8PSK and shaped 16PSK can achieve up to 29% and 39% spectral efficiency gains over the conventional CPM with the same modulation order and modulation index, respectively.

2 System Model

The complex envelope of a CPM signal can be written as

$$s(t) = \sqrt{\frac{E_s}{T_s}} e^{j\phi(t)}, \tag{1}$$

where the phase of the signal $\phi(t)$ is given by

$$\phi(t) = 2\pi h \sum_i b_i q(t - iT_s). \tag{2}$$

E_s is the average energy per information symbol, T_s is the symbol interval, $h = r/p$ is the modulation index (r and p are relatively prime integers), the symbols $\{b_i\}$ are drawn from the alphabet $\mathbb{F} \triangleq \{\pm1, \pm3, \cdots, \pm(M-1)\}$ with modulation order $M = 2^k$, $q(t)$ is the phase-smoothing response with duration LT_s and area 1/2 and its derivative is the frequency pulse $g(t)$

$$g(t) = \begin{cases} \frac{1}{2LT_s} & 0 \le t \le LT_s \\ 0 & \text{else}, \end{cases} \tag{3}$$

L is the memory length. The modulation scheme is shown in Fig. 1(a), In each symbol duration, the mapper maps k bits of original binary bits directly into $\mathbb{F}$ to generate data symbol sequence. Then, the CPM modulator modulates the data symbol sequence using (1) and (2). Noted that in this letter, we constrain our research to 1REC modulation, namely, $L = 1$. Given that the modulation index is rational, (2) is further expressed as

$$\phi(t) = \underbrace{2\pi h b_n q(t - nT_s)}_{\theta(t)} + \underbrace{\pi h \sum_{i=0}^{n-1} b_i}_{\theta_{n-1}}, \tag{4}$$

where $nT_s \leq t < (n+1)T_s, n \in \mathbb{N}, \theta_n$ is the phase state which takes on p values and can be recursively defined as

$$\theta_n = [\theta_{n-1} + \pi h b_n]_{2\pi}, \tag{5}$$

where $[\cdot]_{2\pi}$ denotes the "modulo 2π" operator.

3 Proposed Precoder

Our aim is to improve the spectral efficiency of the conventional CPM. Instead of using the mapper, we propose a precoder based on 8/16PSK to convert the original binary bits to data symbols drawn from a different alphabet. The role of the precoder is to orient the phase of the CPM signal to make it behave like the phase of a 8/16PSK signal. We use 8PSK as an example to illustrate the rules of the precoder. Assuming $t = nT_s$, firstly, we map $k = 3$ bits of the original binary bits $\{a_{3(n-1)+1}, a_{3(n-1)+2}, a_{3n}\}$ to one state θ_n drawn from the 8PSK phase constellation $\{0, \pi/4, 2\pi/4, 3\pi/4, 4\pi/4, 5\pi/4, 6\pi/4, 7\pi/4\}$ by Gray code. Then, as shown in (5), $\pi h b_n$ depends on the phase rotation between the adjacent time interval. To disperse the spectral energy in a more uniform manner, we constrain the phase rotation within π. In other words, in 8PSK constellation, when the phase rotation between adjacent symbols surpasses π in the clockwise, it toggles to rotate in the counter-clockwise direction and vice versa. Therefore, $\theta_n - \theta_{n-1}$ belongs to $\{0, \pm 1/4\pi, \pm 2\pi/4, \pm 3\pi/4, \pi\}$ and we correspond $\pi h b_n$ to this set. Intuitively, we can set $h = 1/4$ to give the final symbol alphabet $\{0, \pm 1, \pm 2, \pm 3, 4\}$. We use a similar method for 16PSK. The operation to convert binary sequence $\{a_n\}$ to data symbols $\{b_n\}$ is summarized in Algorithm 1. Due to the precoding operation, in any given symbol interval, $\{b_n\}$ is drawn from the alphabet $\mathbb{F}' \triangleq \{0, \pm 1, \pm 2, \cdots, \pm(M/2-1), M/2\}$. Compared to the conventional mapper in CPM, the spacing of data symbols in the proposed precoder is smaller. Besides, the precoder can constrain the phase change within π. Thus, we can expect a decreased bandwidth.

We show the generation step of shaped 8/16PSK in Fig. 1 (b). The precoder converts the original binary bits $\{a_n\}$ to final data symbol sequence $\{b_n\}$ according to Algorithm 1. After passing through the CPM modulator using (1) and (2), we can obtain the shaped 8/16PSK.

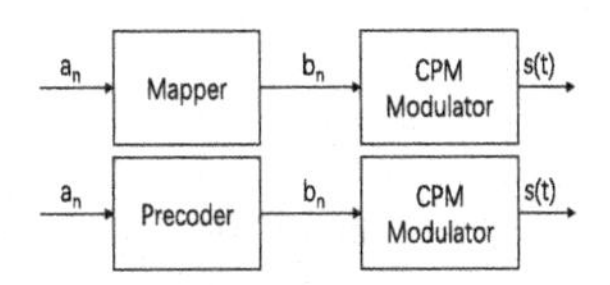

Fig. 1. Compared schemes. (a) Classical CPM, $b_n \in \{\pm 1, \pm 3, \cdots, \pm(M-1)\}$ and (b) Proposed Shaped 8/16PSK, $b_n \in \{0, \pm 1, \pm 2, \pm 3, 4\}$ for shaped 8PSK, $b_n \in \{0, \pm 1, \pm 2, \cdots, \pm 7, 8\}$ for shaped 16PSK.

Require: Original binary sequence $\{a_n\}$
Ensure: Final transmitted symbol sequence $\{b_n\}$
1: **for all** n **do**
2: Convert $\{a_{3(n-1)+1}, a_{3(n-1)+2}, a_{3n}\}$ to decimal form de_n;
3: $b_1 = de_1$;
4: **if** n>1 **then**
5: $b_n = de_n - de_{n-1}$;
6: **end if**
7: **if** $b_n \leq -M/2$ **then**
8: $b_n = b_n + M$;
9: **else if** $b_n \geq M/2$ **then**
10: $b_n = b_n - M$;
11: **end if**
12: **end for**

Algorithm 1: Shaped 8/16PSK Precoder

It is important to note that, in the following, we refer to the shaped 8/16PSK collectively as SMPSK due to the same precoding methods used. Where we need to distinguish, we will point out the different values of M. In addition, it is worth noting that to compare the difference between alphabets, the CPM in this letter is limited to 1REC CPM with the same modulation order M and modulation index h as SMPSK. In the following analysis, we denote it as MCPM.

In this letter, we use achievable spectral efficiency (ASE) to compare SMPSK with MCPM. ASE is one of the important metrics for communications. It represents the amount of information transmitted per unit of time and per unit of bandwidth. To compute ASE, we follow the approach in [8]

$$ASE = \frac{AIR}{BT_s} \tag{6}$$

where B is given by using the Carson's Rule measure, and achievable information rate (AIR) denotes the mutual information between channel input and channel output.

4 Receiver Design

Assuming that the channel is additive white Gaussian noise (AWGN) channel, the received signal is given by

$$y(t) = s(t) + w(t), \tag{7}$$

where $w(t)$ is a zero-mean white Gaussian process with two-sided power spectral density $N_0/2$. In this part, perfect frequency and time synchronization are assumed. Similar to SOQPSK, SMPSK can be constructed either by shaping the pulse of MPSK or connecting the proposed precoder with CPM modulator. Thus, we consider both the conventional MPSK I-Q coherent demodulator and optimal soft decision (i.e. BCJR) [9] could be deployed for reception.

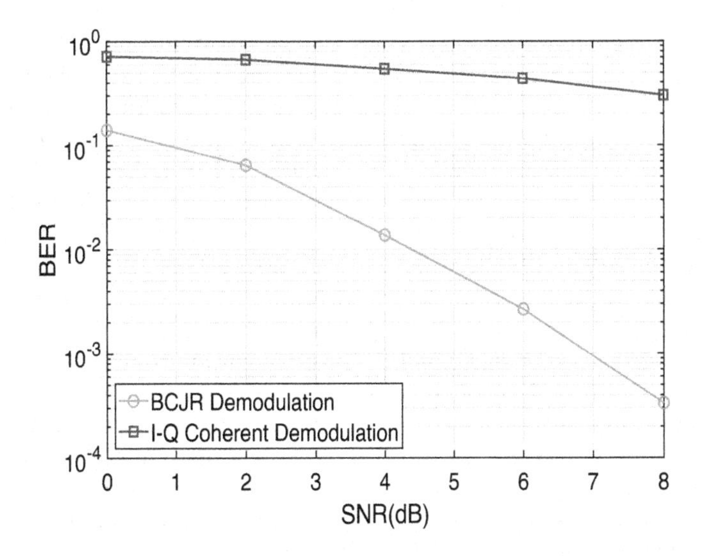

Fig. 2. BER comparison between two methods.

4.1 Receiver Based on BCJR

After the CPM modulation, the phase of SMPSK depends on the cumulative phase of previously transmitted symbols known as the phase memory. Therefore, we assume the SMPSK information sequence as finite and ergodic state Markov sequence, for which algorithms for maximum a posteriori symbol detection can be derived based on the BCJR algorithm. The transition between two states $c_{n-1} \rightarrow c_n$ corresponds to the transmitted symbol b_n.

Sampling the received signal $y(t)$ as $\boldsymbol{y} = (y_1, y_2, \cdots, y_N)$,

$$\hat{b}_n = \arg\max_{b_n \in \mathbb{F}'} p(b_n|\boldsymbol{y}), \tag{8}$$

where b_n denotes the symbol transmitted in the nth time index. $\hat{b}_n$ is the symbol estimate in the nth time index and $p(b_n|\boldsymbol{y})$ is the posteriori probability of the received signal. The transition between two states $c_{n-1} \rightarrow c_n$ corresponds to the transmitted symbol b_n. The conditional probability is given as,

$$p(b_n|\boldsymbol{y}) \propto \sum_{c_{n-1}} \alpha_{n-1}(c_{n-1})\beta(c_n)\gamma(c_n \rightarrow c_{n-1}, \boldsymbol{y})p(b_n). \tag{9}$$

The decision device chooses the symbol with the maximum a posteriori probability.

4.2 Receiver Based on I-Q Coherent Demodulation

As discussed in Sect. 3, each phase at symbol transition instants of SMPSK is drawn from the phase state in the MPSK constellation. Therefore, we adapt the detection techniques for MPSK systems to the detection of SMPSK. The received signal is first demodulated by multiplying it with two coherent carriers $cos(2\pi f_c t)$ and $sin(2\pi f_c t)$. The resultant inphase and quadrature waveforms are lowpass filtered and sampled synchronously at $t = nT_s$. The phase θ_{n-1} is computed from the inphase and quadrature samples, and then decoded as one of the phases of MPSK. The receiver stores the decoded phase and then subtracts it from θ_n. Finally, the transmitted symbol is estimated as one of the values in $\mathbb{F}'$. Figure 2 compares the two demodulation methods when $M = 8$ and $h = 1/4$. One sees that BCJR has better BER performance than I-Q coherent demodulation. This can be explained as follows. Firstly, BCJR pays attention to the distance between received signal and the possible transmitted signal in the whole time interval, while the I-Q coherent method only focuses on the distance in the symbol transition instants. Hence, the former can make full use of signal information. In addition, according to (5), if θ_n is incorrectly determined based on I-Q coherent demodulation, both b_n and b_{n+1} will be erroneously demodulated. In contrast, each symbol is independent in the BCJR demodulation. Hence, we use BCJR method to demodulate SMPSK in the following.

5 Performance Analysis

We derive the AIR of each modulation and then use it to compute the ASE. We consider AIR by modelling the SMPSK as finite and ergodic Markov chain. AIR denotes the mutual information between channel input and channel output over all possible input distributions $I(X, Y)$ as

$$\begin{aligned} I(X,Y) &= \lim_{N\to\infty} \frac{1}{N} I(X_1^N, Y_1^N) \\ &= \lim_{N\to\infty} [\frac{1}{N} H(X_1^N) - \frac{1}{N} H(X_1^N | Y_1^N)] \end{aligned} \tag{10}$$

where X_1^N is the transmitted data symbol sequence, and Y_1^N is the corresponding vector of the AWGN channel outputs. Then, we have

$$\frac{1}{N} H(X_1^N) = lbM, \tag{11}$$

and

$$\begin{aligned} \frac{1}{N} H(X_1^N | Y_1^N) &= \frac{1}{N} H(C_0^N | Y_1^N) \\ &= \frac{1}{N} \sum_{i=1}^{N} H(C_i | C_0^{i-1}, Y_1^N). \end{aligned} \tag{12}$$

Since C_0^{i-1} is already known and C_i is independent of C_0^{i-2},

$$\begin{aligned} & H(C_i | C_0^{i-1}, Y_1^N) \\ &= E_{p(C_{i-1}, Y_1^N)}(H(C_i | C_{i-1} = c_{i-1}, Y_1^N = y_1^N)) \\ &= E_{p(C_{i-1}, Y_1^N)} \\ &* \{E_{p(C_i | C_{i-1} = c_{i-1}, Y_1^N = y_1^N)}(lb(p(C_i | C_{i-1} = c_{i-1}, Y_1^N))\}, \end{aligned} \tag{13}$$

which can be computed from the BCJR algorithm.

We do not have a closed-form expression of AIR since BCJR algorithm is a simulation-based method. Nevertheless, state transition probability $p(C_i|C_{i-1} = c_{i-1}, Y_1^N)$ depends on the minimum squared Euclidean distance between different signals. Increasing the minimum Euclidean distance between different signals gives higher AIR. The minimum squared Euclidean distance d_{min}^2 of 1REC CPM is [10],

$$\begin{aligned} d_{min}^2 &= 2lbM[1 - sinc(\gamma h)] \\ &= 2lbM[1 - sinc(\gamma * (2/M))] \end{aligned} \tag{14}$$

where $\gamma = b_i - b_i^{'}$ denotes the error event between two symbols. Using the smallest value of γ, $\gamma = 2$ for MCPM and $\gamma = 1$ for SMPSK. Further, it can be concluded that the minimum squared Euclidean distance is $d_{min}^2 = 2lbM[1 - sinc(2/(M/2))]$ for MCPM and $d_{min}^{'2} = 2lbM[1 - sinc(1/(M/2))]$ for SMPSK.

We can see that the minimum squared Euclidean distance of proposed SMPSK is smaller and therefore it yields lower AIR than MCPM.

Secondly, for the denominator in (6), assuming that the same symbol interval T_s for both modulation schemes, we focus on the bandwidth B. Conventional research tends to deploy 99% power bandwidth. Unfortunately, this makes theoretical comparison of signals with different symbol alphabets difficult. The problem can be solved by using the Carson's Rule bandwidth measure, which decreases the complexity and still provides good bandwidth estimation for CPM. Denote the unmodulated signal as $m(t) \equiv \sum_i b_i g(t - iT_s)$. The definition of Carson's Rule bandwidth is [8]

$$B = 2h\sqrt{P_m} + 2f_m, \tag{15}$$

where P_m is the power of $m(t)$ and f_m is the one-sided effective bandwidth of $m(t)$. $f_m = P_m/(2S_m(0))$, where $S_m(f)$ is the power spectral density of $m(t)$. Since f_m can be derived from P_m, we only discuss P_m. According to [8], the power of the unmodulated MCPM signal is

$$P_m = \frac{M^2 - 1}{3T_s} \int_0^{T_s} g^2(t)dt. \tag{16}$$

The power of the unmodulated SMPSK signal is,

$$\begin{aligned} P_m^{'} &= (\frac{(\frac{M}{2} - 1)(M - 1)}{6T_s} + \frac{M}{4T_s}) \int_0^{T_s} g^2(t)dt \\ &= \frac{M^2 + 2}{12T_s} \int_0^{T_s} g^2(t)dt. \end{aligned} \tag{17}$$

Comparing (16) and (17), one sees that the final bandwidth occupied by SMPSK is narrower than that by MCPM as

$$\begin{aligned} \Delta B &= 2h\sqrt{P_m} + 2f_m - (2h\sqrt{P_m^{'}} + 2f_m^{'}) \\ &= 2h\sqrt{P_m} + \frac{6T_s P_m}{M^2-1} - (2h\sqrt{P_m^{'}} + \frac{24T_s P_m^{'}}{M^2+2}) \\ &= \frac{2}{MT_s} \cdot (\sqrt{\frac{M^2-1}{3}} - \sqrt{\frac{M^2+2}{12}}) \end{aligned} \tag{18}$$

where $f_m^{'}$ denotes the one-sided effective bandwidth of SMPSK. When $M > 1$, SMPSK requires less spectrum resources than MCPM.

These results show that the modulation providing higher information rate is not necessarily the one providing narrower bandwidth. Additionally, AIR depends on the simulation-based BCJR algorithm. Therefore, it is difficult to derive a closed-form result on which modulation method is more spectrally efficient.

6 Numerical Results and Discussion

In this section, we compare the two modulation schemes by simulation when AWGN channel is assumed. We set symbol number is 500 and $T_s = 40\,\mu\text{s}$. We use a Monte Carlo simulation to estimate the AIR and ASE of these modulations. Table 1 gives the occupied bandwidth of SMPSK and MCPM according to (15). Figure 3 shows how the AIR varies with SNR when different modulations are considered. First of all, it is natural that higher order modulation with higher transmission rate could achieve higher AIR for high SNRs. Second, for low SNRs, the larger minimum squared Euclidean distance is, the less sensitive it will be to noise, and therefore the higher the resulting AIR will be. It is clear that MCPM and low order modulation with larger minimum Euclidean distance can achieve higher AIR for low SNRs.

Table 1. Occupied bandwidth for different schemes.

Modulation scheme	Occupied bandwidth
S8PSK	$1.586/T_s$
8CPM 1REC, $h = 1/4$	$2.146/T_s$
S16PSK	$1.579/T_s$
16CPM 1REC, $h = 1/8$	$2.152/T_s$

On the other hand, Fig. 4 shows the ASE for the same modulations as in Fig. 3. These results show that the modulation providing the higher AIR is not

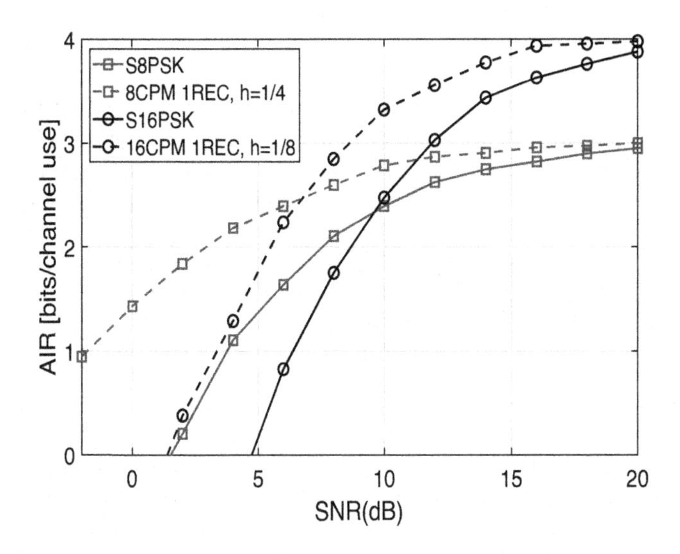

Fig. 3. Achievable information rate for different schemes.

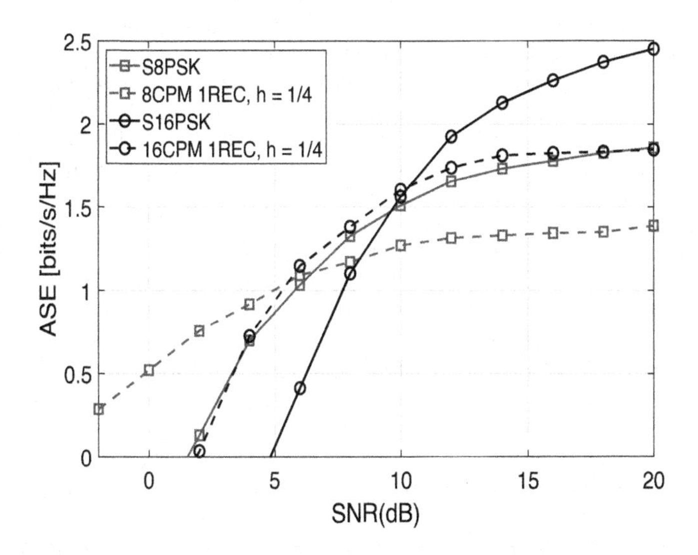

Fig. 4. Achievable spectal efficiency for different schemes.

that with higher ASE. Specifically, SMPSK outperforms MCPM by 29% to 39% in terms of spectral efficiency, when $M = 8$ and $M = 16$ at SNR is 20dB, respectively. This benefits from its precoder, which requires less bandwidth. Therefore, when the ASE is the key performance measure, we can trade degradation in AIR for improvement in ASE. We will consider SMPSK's unsatisfactory performance at low SNR in future work.

7 Conclusion

In this letter, we have proposed a novel precoder based on 8/16PSK. We connect the precoder with CPM modulator to generate a novel CPM modulation scheme shaped 8/16PSK. The analysis and simulation show that the proposed modulation scheme can obtain higher spectral efficiency than conventional CPM.

Acknowledgements. This work was supported by Research on Performance Evaluation and Optimization Technology of Local IOT for Client-side Metering Equipment under grant No. 5700-202118203A-0-0-00.

References

1. Bing, L., Aulin, T., Bai, B., Zhang, H.: Design and performance analysis of multiuser CPM with single user detection. IEEE Trans. Wirel. Commun. **15**(6), 4032–4044 (2016)
2. Wattamwar, R.R., Handore, P.S.: Comparison of bit error rate evaluation for SISO and MIMO system by CPM modulation technique using Matlab. In: 2018 International Conference On Advances in Communication and Computing Technology (ICACCT), pp. 269–272, February 2018
3. Xue, R., Yu, H., Cheng, Q.: Adaptive coded modulation based on continuous phase modulation for inter-satellite links of global navigation satellite systems. IEEE Access **6**, 20652–20662 (2018). https://doi.org/10.1109/ACCESS.2018.2825255
4. Fonseka, J.P.: Nonlinear continuous phase frequency shift keying. IEEE Trans. Commun. **39**(10), 1473–1481 (1991)
5. Messai, M., Piemontese, A., Colavolpe, G., Amis, K., Guilloud, F.: Binary CPMs with improved spectral efficiency. IEEE Commun. Lett. **20**(1), 85–88 (2016)
6. Dapper, M.J., Hill, T.J.: SBPSK: a robust bandwidth-efficient modulation for hard-limited channels. In: IEEE Military Communications Conference, MILCOM 1984, vol. 3, pp. 458–463, October 1984
7. DIS Agency: Department of Defense interface standard, interoperability standard for single-access 5-kHz and 25-kHz UHF satellite communications channels. Department of Defense, March 1999
8. Kuo, C.-H., Chugg, K.M.: On the bandwidth efficiency of CPM signals. In: IEEE MILCOM 2004, Military Communications Conference, vol. 1, pp. 218–224, October 2004
9. Bahl, L.R., Cocke, J., Jelinek, F., Raviv, J.: Optimal decoding of linear codes for minimizing symbol error rate (corresp.). IEEE Trans. Inf. Theory **20**(2), 284–287 (2003)
10. Ekanayake, N., Liyanapathirana, R.: On the exact formula for the minimum squared Euclidean distance of CPFSK. IEEE Trans. Commun. **42**(11), 2917–2918 (1994)

Location-Aided MmWave Hybrid Precoding Prediction for Air-to-Air Communications

Yipai Yan, Honglin Zhao(✉), Chengzhao Shan, Jiayan Zhang, and Yongkui Ma

School of Electronics and Information Engineering, Harbin Institute of Technology, Harbin 150001, China
hlzhao@hit.edu.cn

Abstract. Air-to-air (A2A) communications have been considered as one of the promising technologies to support various applications in future wireless communications, and high data rates are important for these applications such as aerial reconnaissance, remote sensing, and so on. However, it becomes unstable to provide high data rates services when suffering from the high mobility and huge path loss in A2A communications. Millimeter wave (mmWave) and massive MIMO precoding have been studied to meet the requirements of ever-increasing data rates. Unfortunately, MIMO full-digital precoding is considered by most of these methods, which leads to high complexity and power consumption. And the hybrid precoding is an efficient method to overcome this difficulty. Nevertheless, high relative velocity and long distance transmission are not taken into account in these researches. The objective of this paper is to develop an effective mmWave massive MIMO hybrid precoding strategy for high mobility and long distance communications. First, the A2A system is constructed and the evolution channel is derived. Second, a location-aided two-stage hybrid precoding prediction strategy is proposed. Moreover, the data transmission process of the system is discussed. Finally, numerical results manifest that the proposed strategy can effectively provide high data rates for A2A wireless communications.

Keywords: MmWave · Hybrid precoding · High mobility · A2A communications

1 Introduction

The dramatically growing demand for high data rates services, such as remoting sensing, disaster warnings, aerial photography and so on, has promoted the A2A communications attract much attention [1–3]. However, the A2A communications always need long range data transmission and are accompanied by the high relative velocity between the transceivers [4], which results in poor communication quality due to the significant path loss and high mobility. Millimeter wave (mmWave) and massive multiple-input-multiple-output (MIMO) are deemed as reliable techniques to meet the demand for high data rates [5]. MmWave offers

Q. Guo et al. (Eds.): WiSATS 2021, LNICST 410, pp. 345–356, 2022.
https://doi.org/10.1007/978-3-030-93398-2_34

huge available frequency spectrum ranging from 30 GHz to 300 GHz. And massive MIMO provides new spatial degrees of freedom communication capacity [6]. Thanks to the small wavelength of mmWave signals, the use of large-scale antenna arrays can be realized, and massive MIMO beamforming provides great array gain to combat the severe path loss at mmWave bands [7,8].

However, mmWave is rarely employed in long distance outdoor communications because of its severe free space attenuation and absorption loss [9]. Nevertheless, [10] explains the diminishing influence of the non-line-of-sight (NLOS) signal as altitude increases. Therefore, with the large-scale antenna arrays, highly directional beamforming can compensate huge path loss of LOS channel at mmWave. Meanwhile, for the high mobility systems, most of the researches [11–13] take MIMO full-digital architecture into consideration. On the one hand, to achieve the optimal spectral efficiency (SE), perfect channel state information (CSI) is needed [14], while it is quite challenging to be obtained in practice. On the other hand, the directional beamforming is rarely considered and resulting in short channel coherence time [15] under high mobility scenarios. In addition, due to the high complexity and power consumption of full-digital architecture [16], it is unaffordable for A2A system with limited energy supply. To overcome the above problems, the hybrid precoding architecture [16,17] has been proposed where the number of RF chains is far less than the number of antennas. It can perform both low-dimensional digital signal processing and directional beamforming. Unfortunately, most of the existing studies [18,19] assume the relative velocity between the transmitter and receiver is low, and the range of communication is short, which may not hold in A2A system.

In this paper, a novel hybrid precoding strategy is proposed to enhance SE in A2A wireless communications. First, the evolution channel for directional beamforming is derived. Then the location-aided analog precoder and combiner are obtained. Moreover, the Neumann series is utilized to predict digital precoder. Furthermore, hybrid precoding data transmission process is presented.

The remainder of this paper is organized as follows. In Sect. 2, an A2A system model and channel model for long distance mmWave communication are constructed, and some preliminaries of channel coherence time are revisited. In Sect. 3, a novel two-stage location-aided hybrid precoding prediction strategy is proposed, and data transmission process of the A2A system is also discussed. In Sect. 4, the performance of the proposed strategy is evaluated via simulations. Conclusions are drawn in Sect. 5.

We use the following notation throughout this paper: $\mathbf{A}^H$, $\mathbf{A}^{-1}$, $tr(\mathbf{A})$, $\|\mathbf{A}\|_F$ are the Hermitian, inverse, trace, Frobenius norm of the matrix $\mathbf{A}$; diag($\mathbf{a}$) is a diagonal matrix with the entries of $\mathbf{a}$ on its diagonal, blkdiag($\mathbf{a}_1, \ldots, \mathbf{a}_k$) is a block diagonal matrix with $\mathbf{a}_i$ is on its diagonal blocks.

2 System Model and Channel Model

2.1 System Model

We consider a mmWave massive MIMO A2A wireless system, as shown in Fig. 1, which consists of one control center and K mobile nodes. The altitude of control

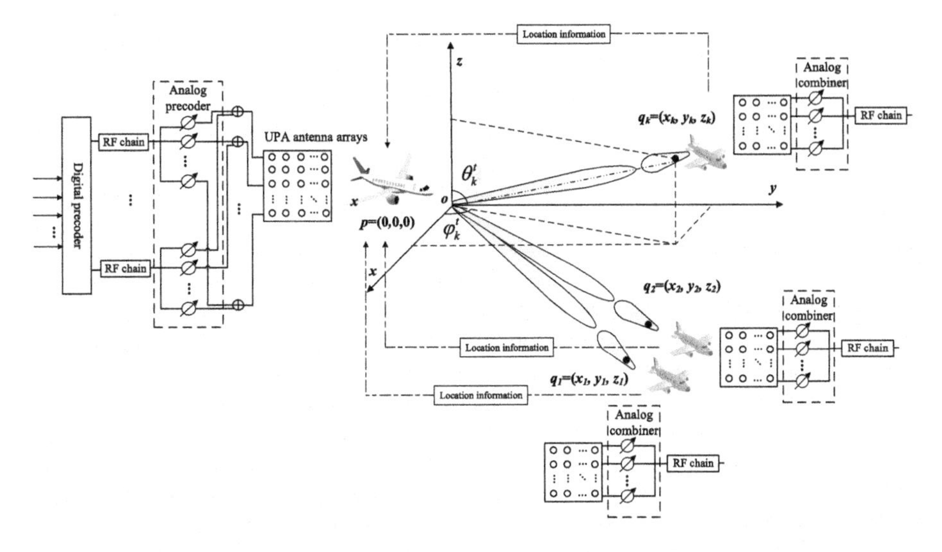

Fig. 1. A2A system model with hybrid precoding architecture.

is 3 km and mobile node is 8 km, and the relative velocity between control center and each mobile node is 300 m/s. It is assumed that the location information of the control center is $p = (0, 0, 0)$, and the location information of k-th mobile node is $q_k = (x_k, y_k, z_k)$. For control center, we assume a fully-connected architecture and it is equipped with K RF chains and $N_t = N_t^h \times N_t^v$ antennas, where $K \ll N_t$, while each mobile node is equipped with $N_r = N_r^h \times N_r^v$ antennas and has only one RF chain due to the limitation of energy supply. They both equipped with $\lambda/2$ spaced uniform planar array (UPA), N_t^h, N_t^v, N_r^h, and N_r^v are the number of antennas in horizontal and vertical directions for control center and each mobile node, respectively.

The transmitted symbols $\mathbf{s} \in \mathbb{C}^K$ with $\mathbb{E}\left[\mathbf{s}\mathbf{s}^H\right] = \mathbf{I}_K$ are first processed by a digital precoder $\mathbf{F}_{BB}$, then up-converted to RF domain and are processed with analog precoder $\mathbf{F}_{RF}$. With the k-th mobile node employs analog combiner $\mathbf{w}_{RF,k}$, the received signal at k-th mobile node is written as

$$y_k = \mathbf{w}_{RF,k}^H \mathbf{h}_k \mathbf{F}_{RF} \mathbf{F}_{BB} \mathbf{s} + \mathbf{w}_{RF,k}^H \mathbf{n}_k \tag{1}$$

where $\mathbf{h}_k$ is the channel matrix of k-th mobile node, $\mathbf{n}_k \in \mathbb{C}^{N_r}$ is the noise vector with $\mathcal{CN}\left(0, \sigma^2\right)$ entries.

The spectral efficiency at k-th mobile node is expressed as

$$R_k = \log_2(1 + \frac{\left|\mathbf{w}_{RF.k}^H \mathbf{h}_k \mathbf{F}_{RF} \mathbf{f}_{BB,k}\right|^2}{\sum_{i=1, i\neq k}^{K} \mathbf{w}_{RF.k}^H \mathbf{h}_k \mathbf{F}_{RF} \mathbf{f}_{BB,i} + \frac{\sigma^2 K}{P_t}}) \tag{2}$$

where P_t is the transmit power for per mobile node. To maximize the SE of the system, the objective function can be formulated as

$$
\begin{aligned}
(\mathbf{F}_{RF}, \mathbf{F}_{BB}) = \underset{\mathbf{F}_{RF}, \mathbf{F}_{BB}}{\arg\max} \sum_{k=1}^{K} R_k \\
s.t. \left|\mathbf{F}_{RF}(m,n)\right| = 1 \quad \forall m, n, \\
\left\|\mathbf{F}_{RF}\mathbf{F}_{BB}\right\|_F^2 = K.
\end{aligned} \tag{3}
$$

2.2 Channel Model

For high elevation and long distance communication, the mmWave A2A channel will consist of LOS propagation and few NLOS components [20]. The channel vector of the k-th mobile node at m-th time slot is modeled as

$$
\mathbf{h}_{m,k} = \sqrt{N_t N_r}\alpha_{m,k} e^{j2\pi f_d m T_c} \mathbf{a}_\mathbf{r}(\varphi_{m,k}^r, \theta_{m,k}^r)\mathbf{a}_t^H(\varphi_{m,k}^t, \theta_{m,k}^t) \tag{4}
$$

where $\alpha_{m,k}$ denotes the path gain, φ_k^t, θ_k^t, φ_k^r, θ_k^r, are azimuth angle of departure (AAoD), elevation angle of departure (EAoD), azimuth angle of arrival (AAoA), and elevation angle of arrival (EAoA), respectively. T_c is the symbol duration, f_d is the doppler shift, defined as

$$
f_d = \mathbf{\Psi}^T \mathbf{f}_{d\max} \tag{5}
$$

where $\mathbf{\Psi}$ is the spherical unit vector with φ_k^r and θ_k^r, defined as

$$
\mathbf{\Psi} = \left[\sin\theta_k^r \cos\phi_k^r, \sin\theta_k^r \cos\phi_k^r, \cos\theta_k^r\right]^T \tag{6}
$$

where $\mathbf{f}_{d\max}$ denotes the mobile node velocity vector and is given by

$$
\mathbf{f}_{d\max} = \frac{v_k}{\lambda}\left[\sin\theta_{k,v}\cos\phi_{k,v}, \sin\theta_{k,v}\sin\phi_{k,v}, \cos\theta_{k,v}\right]^T \tag{7}
$$

where v_k is the velocity of k-th mobile node, $\phi_{k,v}$ and $\theta_{k,v}$ denote the azimuth angle and elevation angle of travel direction, respectively. The array steering vector $\mathbf{a}(\varphi, \theta)$ is expressed as

$$
\begin{aligned}
\mathbf{a}(\varphi, \theta) &= \mathbf{a}_v(\varphi, \theta) \otimes \mathbf{a}_h(\varphi, \theta) \\
\mathbf{a}_v(\varphi, \theta) &= \frac{1}{\sqrt{N_v}}\left[1, \ldots, e^{j\pi(N_v-1)\cos\theta}\right]^T \\
\mathbf{a}_h(\varphi, \theta) &= \frac{1}{\sqrt{N_h}}\left[1, \ldots, e^{j\pi(N_h-1)\sin\theta\cos\varphi}\right]^T
\end{aligned} \tag{8}
$$

where N_v, N_h are the number of antennas in horizontal and vertical directions.

2.3 Temporal Correlations of the Channel

We assume perfect carrier and frame synchronization, and the effects of doppler have been mitigated by doppler pre-compensation process. The channel matrix at m-th time slot $\mathbf{H}_m$ is expressed as

$$\begin{aligned} \mathbf{H}_m &= \sqrt{N_t N_r}\mathbf{A}_r\left(\varphi_m^r, \theta_m^r\right)\boldsymbol{\alpha}_m \mathbf{A}_t^H\left(\varphi_m^t, \theta_m^t\right) \\ \boldsymbol{\alpha}_m &= \mathrm{diag}\left(\alpha_{m,1}, \ldots, \alpha_{m,K}\right) \\ \mathbf{A}_r\left(\varphi_m^r, \theta_m^r\right) &= \mathrm{blkdiag}\left[\mathbf{a}_r\left(\varphi_{m,1}^r, \theta_{m,1}^r\right), \ldots, \mathbf{a}_r\left(\varphi_{m,K}^r, \theta_{m,K}^r\right)\right] \\ \mathbf{A}_t\left(\varphi_m^t, \theta_m^t\right) &= \left[\mathbf{a}_t\left(\varphi_{m,1}^t, \theta_{m,1}^t\right), \ldots, \mathbf{a}_t\left(\varphi_{m,K}^t, \theta_{m,K}^t\right)\right] \end{aligned} \tag{9}$$

The channel matrix at $(m{+}1)$-th time slot $\mathbf{H}_{m+1}$ is expressed as

$$\mathbf{H}_{m+1} = \sqrt{N_t N_r}\mathbf{A}_r\left(\varphi_{m+1}^r, \theta_{m+1}^r\right)\boldsymbol{\alpha}_{m+1}\mathbf{A}_t^H\left(\varphi_{m+1}^t, \theta_{m+1}^t\right) \tag{10}$$

$$\begin{aligned} \boldsymbol{\alpha}_{m+1} &= \boldsymbol{\alpha}_m + \sqrt{1-\epsilon^2}\mathbf{W}_m \\ \mathbf{A}_r\left(\varphi_{m+1}^r, \theta_{m+1}^r\right) &= \mathbf{A}_r\left(\varphi_m^r + \Delta\varphi_m^r, \theta_m^r + \Delta\theta_m^r\right) \\ \mathbf{A}_t\left(\varphi_{m+1}^t, \theta_{m+1}^t\right) &= \mathbf{A}_t\left(\varphi_m^t + \Delta\varphi_m^t, \theta_m^t + \Delta\theta_m^t\right) \end{aligned} \tag{11}$$

where $\epsilon \in [0,1]$ is the channel correlation coefficient, and it is defined as [15]

$$\epsilon = \mathbb{E}\left[h_m h_{m+T_c}^*\right] = \int_{-\pi}^{\pi} e^{j2\pi f_d T_c} p\left(\alpha\right) \mathrm{d}\alpha \tag{12}$$

where the power angular spectrum (PAS) $p(\alpha)$ is normalized as $\int_{-\pi}^{\pi} p(\alpha)\,\mathrm{d}\alpha = 1$, α is the angle of incidence. For omnidirectional transmission, $p(\alpha) = (1/2\pi)$, while for directional beamforming, PAS can be modeled by a von Mises distribution function [15], the channel correlation coefficient becomes

$$\epsilon = \sqrt[4]{\frac{1}{\left(\frac{2\pi}{\lambda} v T_c \theta_b^2\right)^2 + 1}} \tag{13}$$

where θ_b is the beamwidth, λ is the signal wavelength.

$\mathbf{W}_m \in \mathbb{C}^{K\times K}$ denotes the additive white Gaussian noise with zero mean and unit variance, it is independent of $\boldsymbol{\alpha}_m$. $\Delta\varphi_m^t$, $\Delta\theta_m^t$, $\Delta\varphi_m^r$, $\Delta\theta_m^r$ are the angle variation of adjacent time slot in the direction of azimuth and elevation for AoD and AoA, respectively.

According to the Taylor expansion of the two variable functions, the matrix $\mathbf{A}_r\left(\varphi_m^r + \Delta\varphi_m^r, \theta_m^r + \Delta\theta_m^r\right)$ in (11) can be expressed as

$$\begin{aligned} &\mathbf{A}_r\left(\varphi_m^r + \Delta\varphi_m^r, \theta_m^r + \Delta\theta_m^r\right) \\ &= \mathbf{A}_r\left(\varphi_m^r, \theta_m^r\right) + \left(\Delta\varphi_m^r \frac{\partial}{\partial\varphi_m^r} + \Delta\theta_m^r \frac{\partial}{\partial\theta_m^r}\right)\mathbf{A}_r\left(\varphi_m^r, \theta_m^r\right) \\ &+ \frac{1}{2!}\left(\Delta\varphi_m^r \frac{\partial}{\partial\varphi_m^r} \Delta\theta_m^r \frac{\partial}{\partial\theta_m^r}\right)^2 \mathbf{A}_r\left(\varphi_m^r, \theta_m^r\right) + \mathbf{R}_n \end{aligned} \tag{14}$$

$\mathbf{A}_t\left(\varphi_m^t + \Delta\varphi_m^t, \theta_m^t + \Delta\theta_m^t\right)$ can be obtained by the similar way, which is omitted here.

For long distance A2A wireless communication, in a short time interval, $\Delta\varphi$ and $\Delta\theta$ are small [4], therefore, reserving the first term in (14), and other terms can be ignored. Then (14) can be rewritten as

$$\begin{aligned} \mathbf{A}_r\left(\varphi_m^r + \Delta\varphi_m^r, \theta_m^r + \Delta\theta_m^r\right) &\approx \mathbf{A}_r\left(\varphi_m^r, \theta_m^r\right) \\ \mathbf{A}_t\left(\varphi_m^t + \Delta\varphi_m^t, \theta_m^t + \Delta\theta_m^t\right) &\approx \mathbf{A}_t\left(\varphi_m^t, \theta_m^t\right) \end{aligned} \tag{15}$$

Combining (10), (11) and (15) yields

$$\begin{aligned} \mathbf{H}_{m+1} &= \epsilon\mathbf{H}_m + \sqrt{1-\epsilon^2}\mathbf{N}_m \\ \mathbf{N}_m &= \sqrt{N_t N_r}\mathbf{A}_r\left(\varphi_m^r, \theta_m^r\right)\mathbf{W}_m\mathbf{A}_t^H\left(\varphi_m^t, \theta_m^t\right) \end{aligned} \tag{16}$$

The following derivations are based on the evolution channel model in (16), and a more efficient hybrid precoding prediction strategy, which can exploit the temporal correlation of (16), is proposed for mmWave A2A system.

3 Two-Stage Hybrid Precoding Prediction Design for A2A Communications

In this section, a novel two-stage hybrid precoding strategy is presented for the system in Fig. 1. The proposed strategy leverages the channel correlation to maximize the spectral efficiency with reduced complexity.

3.1 Analog Precoder and Combiner Design

For long distance A2A communications, as discussed above, only LOS transmission is considered. Based on the geometrical relationship as illustrated in Fig. 1, AAoD and EAoD can be performed as

$$\begin{aligned} \varphi_k^t &= \arctan(\frac{y_k}{x_k}) \\ \theta_k^t &= \arctan(\frac{\sqrt{x_k^2+y_k^2}}{z_k}) \end{aligned} \tag{17}$$

φ_k^r and θ_k^r can be obtained by the similar way, which is omitted here.

It is assumed that the phase-only controlled phase shifter is employed, and we match the analog precoder and combiner with array steering vector and array response vector, respectively. For the given direction φ_k^t, θ_k^t, φ_k^r, θ_k^r, the ideal analog precoder and combiner of the k-th mobile node are formulated as

$$\begin{aligned} \mathbf{f}_{RF,k} &= \mathbf{a}_t(\varphi_k^t, \theta_k^t) \\ \mathbf{w}_{RF,k} &= \mathbf{a}_r(\varphi_k^r, \theta_k^r) \end{aligned} \tag{18}$$

We define the beam update period (BUP) T_b as the location information update time. At n-th BUP, location and relative velocity for the k-th user is

$$\begin{aligned} q_{n,k} &= (x_{n,k}, y_{n,k}, z_{n,k}) \\ v_{n,k} &= (v_{x_{n,k}}, v_{y_{n,k}}, v_{z_{n,k}}) \end{aligned} \tag{19}$$

It is assumed that the mobile node is moving uniformly and in a straight line during one BUP. The location information of $(n+1)$-th beam update period $q_{n+1,k} = (x_{n+1,k}, y_{n+1,k}, z_{n+1,k})$ can be written as

$$\begin{aligned} x_{n+1,k} &= x_{n,k} + T_b \cdot v_{x_{n,k}} \\ y_{n+1,k} &= y_{n,k} + T_b \cdot v_{y_{n,k}} \\ z_{n+1,k} &= z_{n,k} + T_b \cdot v_{z_{n,k}} \end{aligned} \tag{20}$$

Moreover, AoA and AoD of $(n+1)$-th BUP are obtained depending on (17).

3.2 Digital Precoder Design

Since the mobile nodes are not separated, there exists interference among data streams. Therefore, zero-forcing (ZF) approach to eliminate interference based on equivalent baseband channel is considered. We have the equivalent baseband channel matrix $\mathbf{H}_{eq}$ as follows

$$\mathbf{H}_{eq} = \mathbf{W}_{RF}^{H}\mathbf{H}\mathbf{F}_{RF} \tag{21}$$

where $\mathbf{F}_{RF} = [\mathbf{f}_{RF,1}, \mathbf{f}_{RF,2}, \dots, \mathbf{f}_{RF,K}]$, $\mathbf{W}_{RF} = \mathrm{blkdiag}[\mathbf{w}_{RF,1}, \mathbf{w}_{RF,2}, \dots, \mathbf{w}_{RF,K}]$.

As mentioned above, $\mathbf{H}_{eq}$ is a low dimensional square matrix. We define the channel update period (CUP, also called time slot) T_c as the equivalent channel update time. The ZF digital precoder of $(m+1)$-th CUP is derived as

$$\mathbf{F}_{BB,m+1} = \mathbf{H}_{eq,m+1}^{-1} \tag{22}$$

In general, $\mathbf{H}_{eq}$ should be updated at each time slot to maintain the spectral efficiency. However, when the relative velocity is high, there is insufficient time for channel estimation. Fortunately, in terms of (16), the channels of adjacent time slot are correlated, thus the outdated equivalent CSI can be utilized to predict the digital precoder.

Combining (16), (21) and (22) leads to

$$\mathbf{F}_{BB,m+1} = \left(\epsilon\mathbf{H}_{eq,m} + \sqrt{1-\epsilon^2}\mathbf{W}_m\right)^{-1} \tag{23}$$

The inverse of the sum of two arbitrary matrices is generally difficult to split. In particular, as ϵ is large, noise matrix can be considered as a disturbance term. Turn to Neumann series [21], it provides approximations of the inverse of the sum of two matrices when one of the matrix has entries of small magnitude. Then (23) can be rewritten as

$$\begin{aligned} \mathbf{F}_{BB,m+1} &= \sum_{k=0}^{\infty}(-1)^k\left((\epsilon\mathbf{H}_{eq,m})^{-1}\sqrt{1-\epsilon^2}\mathbf{W}_m\right)^k(\epsilon\mathbf{H}_{eq,m})^{-1} \\ &= (\epsilon\mathbf{H}_{eq,m})^{-1} - ((\epsilon\mathbf{H}_{eq,m})^{-1}\sqrt{1-\epsilon^2}\mathbf{W}_m)(\epsilon\mathbf{H}_{eq,m})^{-1} \\ &\quad + ((\epsilon\mathbf{H}_{eq,m})^{-1}\sqrt{1-\epsilon^2}\mathbf{W}_m)^2(\epsilon\mathbf{H}_{eq,m})^{-1} - \cdots \end{aligned} \tag{24}$$

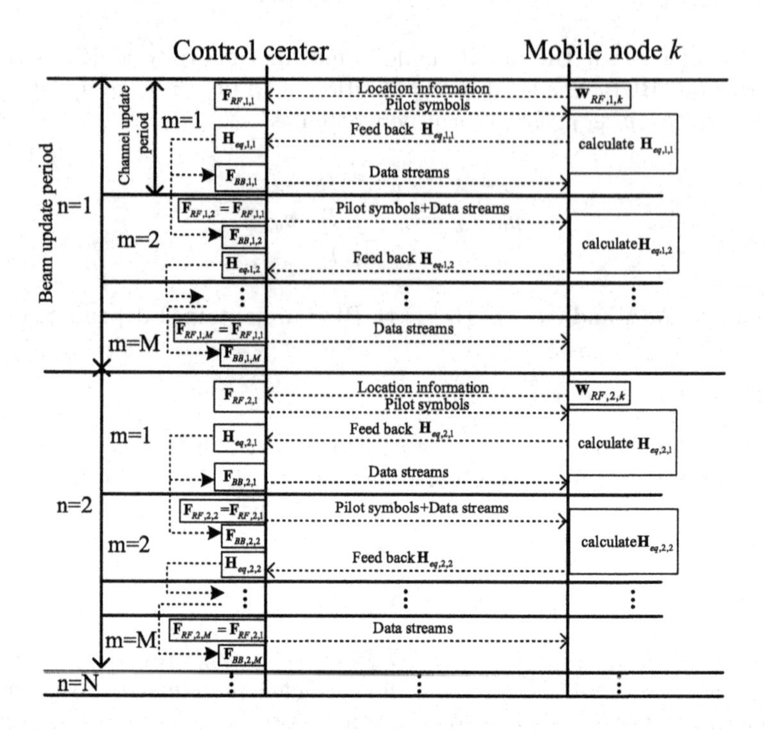

Fig. 2. Hybrid precoding data transmission process.

When ϵ is large enough, the first term in (24) is reserved, the following terms are ignored. The predictive digital precoder can be approximately written as

$$\widetilde{\mathbf{F}}_{BB,m+1} \approx (\epsilon \mathbf{H}_{eq,m})^{-1} \tag{25}$$

3.3 Hybrid Precoding Data Transmission Process

The definitions of BUP and CUP have been explained above, and it is assumed that one BUP contains several CUPs. The equivalent baseband channel almost unchanged within one CUP, but could vary for different CUP; AoA and AoD nearly do not vary in one BUP, they vary for different BUP. The steps involved in establishing a data transmission link during one BUP are summarized as follows:

- Step 1: At the initial CUP, each mobile node sends its location to the control center. Analog precoder and combiner are designed based on (17).
- Step 2: Control center transmits pilot symbols to calculate $\mathbf{H}_{eq}$.
- Step 3: $\mathbf{H}_{eq}$ is fed back to control center, then digital precoder is derived and data streams are transmitted to each mobile node.

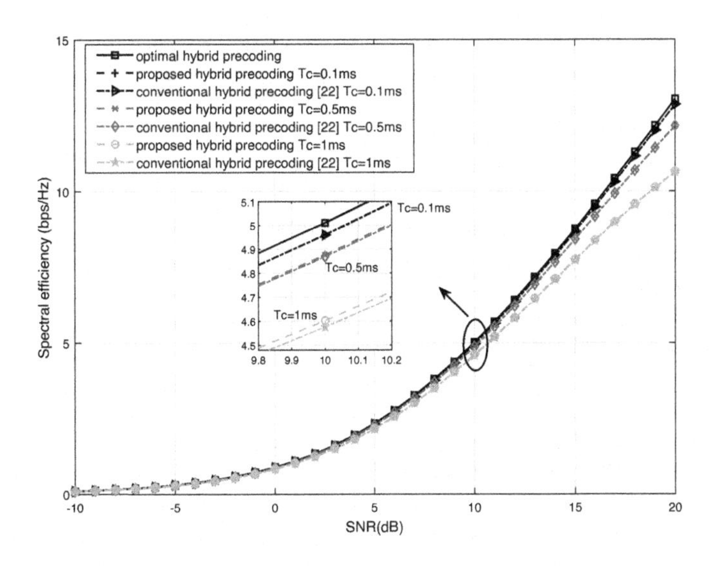

Fig. 3. Spectral efficiency versus SNR with different channel coherence time.

- Step 4: At the second CUP, analog precoder and combiner are unchanged, digital precoder is predicted according to (25). Then the pilot symbols and data streams are sent simultaneously by hybrid precoding. The pilot symbols are used to estimate the $\mathbf{H}_{eq}$ of current time slot.
- Step 5: $\mathbf{H}_{eq}$ of current time slot is fed back to control center to predict the digital precoder of next time slot.

Repeating step 4 to step 5 until the end of a BUP, and at the last CUP, only data streams need to be transfered. The details are shown in Fig. 2. For the second and subsequent BUP, if the mobile node moves uniformly in a straight line, location information does not need to be fed back, the analog precoder can be obtained according to (17) and (20). Otherwise, the location information should be fed back at the beginning of each BUP.

For conventional strategy [22], first the transmitter sends pilot symbols to estimate channel, then the receiver feeds back the estimated CSI to transmitter, next the precoder can be designed by transmitter to transmit data information. While for the proposed strategy, the channel estimation and data transmission are carried out together, thus the overhead of the system is considerably reduced.

4 Numerical Results

To confirm the validity of the derived digital precoder approximation in (25) and evaluate the performance of the proposed location-aided hybrid precoding strategy, we conduct simulations and discuss the numerical results in this section. We compare the proposed strategy to optimal strategy in (22) and the conventional strategy [22]. To provide a fair comparison, CSI is updated every BUP

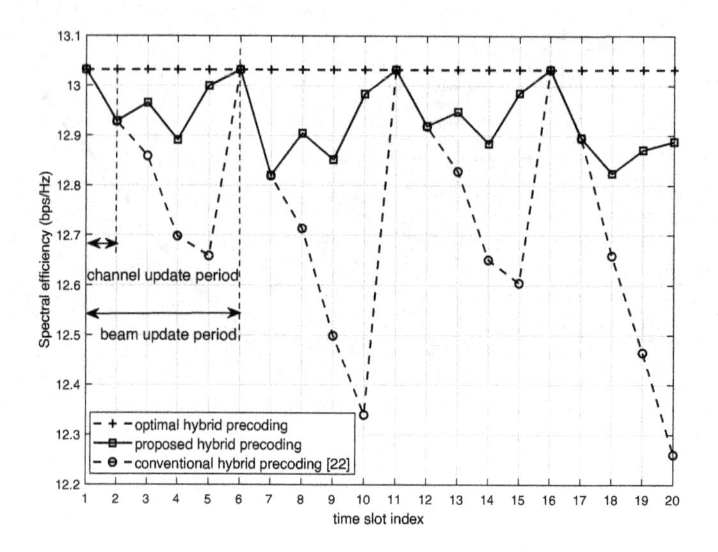

Fig. 4. Spectral efficiency versus time slot index with SNR = 20 dB and $T_c = 0.1$ ms.

for conventional strategy. We consider a mmWave massive MIMO system with carrier frequency $f_c = 30$ GHz where the control center is equipped with 64×64 UPA and each mobile node is equipped with 8×8 UPA. There are $K = 4$ mobile nodes. The initial distance between control center and mobile node is 44 km, 46 km, 47 km, 50 km, respectively, and the relative velocity is 300 m/s.

Figure 3 illustrates the spectral efficiency of the system with different channel coherence time versus SNR. As shown in Fig. 3, the SE of the proposed strategy is considerably close to the optimal hybrid precoding. As SNR goes large, a gap can be seen between the optimal strategy and the proposed strategy. In low SNR regime, the noise is the main influence for performance, while for high SNR regime, the interference between the data streams has a great impact on performance, residual interference due to approximation of Neumann series causes performance degradation. Figure 3 also shows that as channel coherence time becomes small, the proposed strategy approaches the optimal strategy closely. This is because the correlation between channel of adjacent time slot becomes stronger, thus the predicted precoder matrix has a smaller difference with the optimal precoder matrix. Therefore, we have to make a trade-off between SE and channel coherence time in practice. It can also be seen that the spectral efficiency of conventional strategy is close to the proposed strategy, the reason is that we only consider single time slot, without considering the influence of time on the performance.

Figure 4 depicts the spectral efficiency versus time slot index with different hybrid precoding strategy under the assumption that SNR = 20 dB, $T_c = 0.1$ ms. In simulation, 4 BUPs are considered here, and each BUP contains 5 CUPs. As shown in Fig. 4, the SE of the proposed strategy fluctuates under the optimal performance over time, in contrast, because of channel mismatch, the SE of conventional strategy continues to degrade as time goes on during one BUP, until it updates CSI. It is obvious that the proposed strategy can still maintain high spectral efficiency in high mobility scenarios compared with the conventional strategy.

5 Conclusions

In this paper, we have developed a novel hybrid precoding strategy for A2A system to overcome the problem of CSI outdated in high mobility scenarios and huge path loss for long distance transmission. We first derived the evolution channel for mmWave A2A communication. Then a two-stage hybrid precoding prediction strategy was proposed while the analog precoder and combiner were obtained based on location information, and the digital precoder was derived according to Neumann series approximation. Moreover, data transmission process of the proposed strategy was presented. Finally, numerical results verified the proposed strategy can achieve high spectral efficiency for A2A mmWave wireless communications.

References

1. Liu, W., Zang, X., Li, Y., Vucetic, B.: Over-the-air computation systems: optimization, analysis and scaling laws. IEEE Trans. Wirel. Commun. **19**(8), 5488–5502 (2020). https://doi.org/10.1109/TWC.2020.2993703
2. Ma, Z., Ai, B., He, R., Wang, G., Niu, Y., Zhong, Z.: A wideband non-stationary air-to-air channel model for UAV communications. IEEE Trans. Veh. Technol. **69**(2), 1214–1226 (2020). https://doi.org/10.1109/TVT.2019.2961178
3. Bai, L., Han, R., Liu, J., Yu, Q., Choi, J., Zhang, W.: Air-to-ground wireless links for high-speed UAVs. IEEE J. Sel. Areas Commun. **38**(12), 2918–2930 (2020). https://doi.org/10.1109/JSAC.2020.3005471
4. Li, Y., Miridakis, N.I., Tsiftsis, T.A., Yang, G., Xia, M.: Air-to-air communications beyond 5G: a novel 3D CoMP transmission scheme. IEEE Trans. Wirel. Commun. **19**(11), 7324–7338 (2020). https://doi.org/10.1109/TWC.2020.3010569
5. Bogale, T.E., Le, L.B.: Massive MIMO and mmWave for 5G wireless HetNet: potential benefits and challenges. IEEE Veh. Technol. Mag. **11**(1), 64–75 (2016). https://doi.org/10.1109/MVT.2015.2496240
6. Björnson, E., Larsson, E.G., Marzetta, T.L.: Massive MIMO: ten myths and one critical question. IEEE Commun. Mag. **54**(2), 114–123 (2016). https://doi.org/10.1109/MCOM.2016.7402270
7. Shan, C., Ma, Y., Zhao, H., Shi, J.: Joint radar-communications design based on time modulated array. Digit. Signal Process. **82**, 43–53 (2018). https://doi.org/10.1016/j.dsp.2018.07.013
8. Shan, C., Ma, Y., Zhao, H., Shi, J.: Time modulated array sideband suppression for joint radar-communications system based on the differential evolution algorithm. Digit. Signal Process. **97**, 102601 (2020). https://doi.org/10.1016/j.dsp.2019.102601
9. Xiao, M., et al.: Millimeter wave communications for future mobile networks. IEEE J. Sel. Areas Commun. **35**(9), 1909–1935 (2017). https://doi.org/10.1109/JSAC.2017.2719924

10. Goddemeier, N., Wietfeld, C.: Investigation of air-to-air channel characteristics and a UAV specific extension to the rice model. In: 2015 IEEE Globecom Workshops (GC Wkshps), pp. 1–5 (2015). https://doi.org/10.1109/GLOCOMW.2015.7414180
11. Kim, K., Kim, T., Love, D.J., Kim, I.H.: Differential feedback in codebook-based multiuser MIMO systems in slowly varying channels. IEEE Trans. Commun. **60**(2), 578–588 (2012). https://doi.org/10.1109/TCOMM.2011.012012.110051
12. Kim, T., Love, D.J., Clerckx, B.: MIMO systems with limited rate differential feedback in slowly varying channels. IEEE Trans. Commun. **59**(4), 1175–1189 (2011). https://doi.org/10.1109/TCOMM.2011.022811.090744
13. Chen, H., Lin, Y.: Differential feedback of geometrical mean decomposition precoder for time-correlated MIMO systems. IEEE Trans. Signal Process. **65**(14), 3833–3845 (2017). https://doi.org/10.1109/TSP.2017.2692741
14. Ayach, O.E., Rajagopal, S., Abu-Surra, S., Pi, Z., Heath, R.W.: Spatially sparse precoding in millimeter wave MIMO systems. IEEE Trans. Wirel. Commun. **13**(3), 1499–1513 (2014). https://doi.org/10.1109/TWC.2014.011714.130846
15. Va, V., Choi, J., Heath, R.W.: The impact of beamwidth on temporal channel variation in vehicular channels and its implications. IEEE Trans. Veh. Technol. **66**(6), 5014–5029 (2017). https://doi.org/10.1109/TVT.2016.2622164
16. Kutty, S., Sen, D.: Beamforming for millimeter wave communications: an inclusive survey. IEEE Commun. Surv. Tutor. **18**(2), 949–973 (2016). https://doi.org/10.1109/COMST.2015.2504600
17. Ahmed, I., et al.: A survey on hybrid beamforming techniques in 5G: architecture and system model perspectives. IEEE Commun. Surv. Tutor. **20**(4), 3060–3097 (2018). https://doi.org/10.1109/COMST.2018.2843719
18. Liu, J., Bentley, E.S.: Hybrid-beamforming-based millimeter-wave cellular network optimization. IEEE J. Sel. Areas Commun. **37**(12), 2799–2813 (2019). https://doi.org/10.1109/JSAC.2019.2947923
19. Sun, S., Rappaport, T.S., Shafi, M., Tataria, H.: Analytical framework of hybrid beamforming in multi-cell millimeter-wave systems. IEEE Trans. Wirel. Commun. **17**(11), 7528–7543 (2018). https://doi.org/10.1109/TWC.2018.2868096
20. Cuvelier, T., Heath, R.W.: MmWave MU-MIMO for aerial networks. In: 2018 15th International Symposium on Wireless Communication Systems (ISWCS), pp. 1–6 (2018). https://doi.org/10.1109/ISWCS.2018.8491045
21. Meyer, C.D.: Matrix Analysis and Applied Linear Algebra. Society for Industrial and Applied Mathematics (2000)
22. Zhou, S., Xu, W., Zhang, H., You, X.: Hybrid precoding for millimeter wave massive MIMO with analog combining. In: 2017 9th International Conference on Wireless Communications and Signal Processing (WCSP), pp. 1–5 (2017). https://doi.org/10.1109/WCSP.2017.8171060

Outage Probability for Device to Device and Cellular Heterogeneous Networks over Nakagami-m Channels

Yu Zhou[1(✉)], Xiao Chen[2], Xuesong Shao[1], Fan Gao[1], Qixin Cai[1], and Yue Li[1]

[1] Marketing Service Center, State Grid Jiangsu Electric Power Co. Ltd., Nanjing, China

[2] State Grid Jiangsu Electric Power Co. Ltd., Nanjing, China

Abstract. Device-to-Device (D2D) communication has become an important component in future communications, because it has the advantages of offloading traffic from the base stations and reducing the distance between the transmitter and receiver. In this paper, we derive a closed-form expression for the outage probability in uplink D2D communication and cellular heterogeneous networks over Nakagami-m fading, where the fading parameter m can vary among the devices with any positive value. The analytical results are presented and verified the simulation results.

Keywords: Device-to-Device communication · Nakagami-m fading · Outage probability

1 Introduction

Nowadays, the explosive growth of communication devices led to an increasing demand for higher capacity, data rates and radio spectrum resources. Device-to-Device (D2D) communication has been seen as a key component in the fifth generation (5G) wireless network, since it allows two adjacent D2D-enabled users, denoted as a D2D pair, to directly establish communication link without base stations assistance or control. Therefore, underlay D2D in conventional cellular networks has the potential to reduce the transmission distance, improve spectrum frequency and offload the traffic from base stations. It can be seen as a two tier heterogeneous networks (HetNets) where one tier is composed of the users that communicate over the D2D link and the other tier is the users transmitting signals through the base station. In D2D cellular heterogeneous networks, time/frequency resources can be portioned or shared by the D2D tier and cellular tier [1]. To avoid the inter-tier interference, the cellular and D2D users use the orthogonal resources by dividing the uplink frequency band into two non-overlapping portions [2,3]. However, sophisticated resource allocation schemes are required for system implement [4]. In [5] and [6], D2D pairs are allowed

Q. Guo et al. (Eds.): WiSATS 2021, LNICST 410, pp. 357–366, 2022.
https://doi.org/10.1007/978-3-030-93398-2_35

to reuse the uplink resource of the cellular users. Compared with the orthogonal spectrum scheme, the shared frequency network is efficient in the network resources and more flexible to perform.

However, the shared D2D-cellular HetNets improve the spectrum utilization with the cost of introducing the interference from both the D2D tier and cellular tier. In recent years, the performance analysis of the D2D communication in cellular networks has been extensively investigated. The outage probability is one of the key performance metrics to model the D2D communication underlaying the cellular network. The performance analysis of D2D-cellular HetNets under Rayleigh fading channel has been well studied in the literature. Reference [7] provided analytical coverage probability expression for such D2D and cellular users under Rayleigh fading channels. Outage probability for the uplink D2D cellular network with the multiple antennas has been analyzed by considering the Rayleigh faded channel [8]. It demonstrated that the performance was improved by equipping the D2D users with multiple antennas. Recently, Nakagami-m fading model has been proposed for modeling differing line-of-sight conditions and short-range communication i.e., D2D communication links. Nakagami-m channel is a generalization fading model which incorporates other fading scenarios by setting different fading parameters, i.e., Rayleigh for $m = 1$, the Rician shadow fading distribution for $m = (K+1)^2/(2K+1)$ and no fading for $m = \infty$ as special cases. Reference [9] derived the outage probability in an interference-limited system and assumed the Nakagami-m channels for desired signals and Rayleigh channels for interference signals. Reference [10] and [11] analyzed coverage performance of heterogeneous networks over the Nakagami-m channels by assuming the fading parameter m was a positive integer. However, to the best of the authors knowledge, none of these previous studies accommodate thermal noise and the outage performance of D2D-cellular HetNets over Nakagami-m fading channel with arbitrary fading parameters has not been reported in the literature.

In this paper, we derive an analytical expression for the outage probability of the D2D and cellular heterogeneous networks, where the uplink frequency resources are shared between the D2D users and cellular users. And we assume that the typical D2D transmitter is equipped with multiple antennas. The D2D transmitters and cellular users are equipped with single omnidirectional antennas. Both the desired signals and co-channel interference experience the Nakagami-m fading with different fading parameters. Then the closed-form expression for the outage probability is derived for D2D links over Nakagami-m channels, which is valid both for integer and non-integer values of m. By using the outage probability as the performance metric, we investigate impact of system parameters on the D2D communication.

The paper is organized as follows: the system model and Nakagami-m channel model are proposed in Sect. 2. The outage probability over D2D links has been derived in Sect. 3. The Monte-Carlo simulations and analytical results are compared in Sect. 4. Finally, some conclusions are given in Sect. 5.

2 System Model and Channel Model

2.1 System Model

In this section, we consider a two-tier uplink heterogeneous networks where the D2D users and cellular users are placed in the area as shown in Fig. 1. By reusing the uplink frequency band of cellular networks, D2D pair can directly communicate without the base stations assistance. The received signal at any receiving node is assumed to be corrupted by desired signal, additive white Gaussian noise (AWGN) and interference from both D2D transmitters and cellular users. Here, we consider a multi-antennas D2D communication system, where the typical D2D transmitter is equipped with N_R antennas. Other D2D devices and the cellular users are equipped with a single antenna. The channel fading of desired link and interference link are considered to follow the non-identically independent Nakagami-m fading distribution. Through this paper, we consider a typical

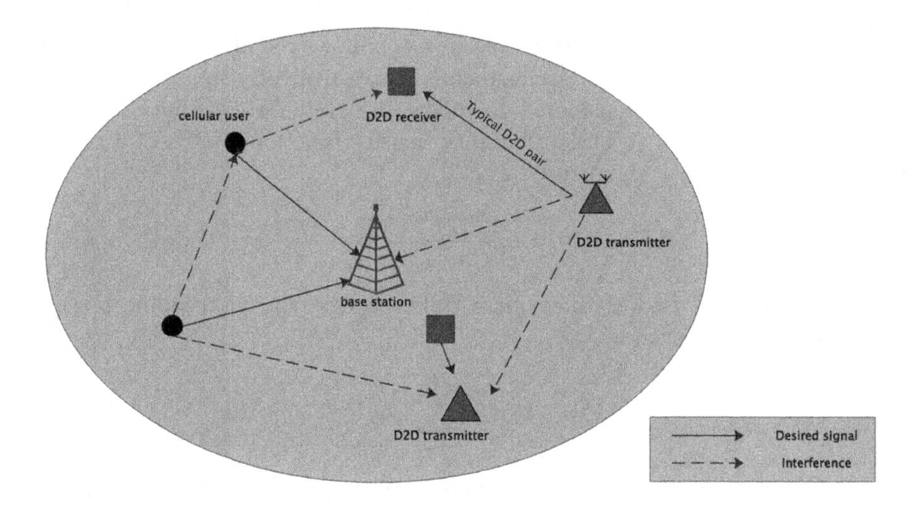

Fig. 1. System model of the D2D-cellular HetNets

D2D receiver at the center. Denoting d_0 and $\boldsymbol{h}_0$ are the distance and the channel fading between the D2D transmit and receive nodes in a typical D2D pair. The received signal at the D2D receive node is given as

$$y_0 = P_0 d_0^{-\alpha/2} \boldsymbol{h}_0^H \boldsymbol{x}_0 + \sum_{i=1}^{I} P_0 d_i^{-\alpha/2} \boldsymbol{h}_{D_i}^H \boldsymbol{x}_i + \sum_{j=1}^{J} P_1 d_j^{-\alpha/2} \boldsymbol{h}_{U_j}^H \boldsymbol{x}_j + \boldsymbol{n}_0 \tag{1}$$

where P_0 and I are the transmit power and number of the D2D transmitters, P_1 and J are the transmit power and number of cellular users, d_i and d_j are the distance from the D2D transmitters and cellular users to the typical D2D receiver, α is the path-loss parameter, $\boldsymbol{x}_i$ is the transmitted signal with

$\mathbb{E}[||x_i||^2] = 1$. $\boldsymbol{h_0} = \{h_{0,1}, h_{0,2}, \cdots, h_{0,N_R}\}$ is a $1 \times N_R$ vector and the envelope for each subchannel $||h_{0,n}||$ follows the i.i.d Nakagami-m distribution, denoted as $||h_{0,n}|| \sim Nakagami(m, \Omega)$. In addition, $\boldsymbol{h}_{D_i}$ and $\boldsymbol{h}_{U_j}$ are the $1 \times N_R$ channel vectors defining the channel fading between ith D2D transmitter and jth cellular users to the D2D receiver, i.e. $\boldsymbol{h}_{D_i} = \{h_{D_{i,1},D_{i,2},\cdots,D_{i,N_R}}\}$ and $\boldsymbol{h}_{U_i} = \{h_{U_{j,1},U_{j,2},\cdots,U_{j,N_R}}\}$. Similarly, we assume the envelope for each subchannel follows the Nakagami-m distribution, denoted as $||h_{D_{i,n}}|| \sim Nakagami(m_{D_i}, \Omega_{D_i})$ and $||h_{U_{j,n}}|| \sim Nakagami(m_{U_j}, \Omega_{U_j})$. n_0 is an $1 \times N_R$ vector of an i.i.d. zero mean complex AWGN with a variance σ_n^2.

From (1), the signal-to-interference-and-noise-ratio (SINR) at the typical D2D receiver is given as

$$\mathrm{SINR}_0 = \frac{P_0 d_0^{-\alpha} |\boldsymbol{h}_0|^2}{\sum_{i=1}^{I} P_0 d_i^{-\alpha} |\boldsymbol{h}_{D_i}|^2 + \sum_{j=1}^{J} P_1 d_j^{-\alpha} |\boldsymbol{h}_{U_j}|^2 + \sigma_n^2} \tag{2}$$

2.2 Channel Model

Since the element of fading channel vector $h_{0,n}$ follows the i.i.d Nakagami-m distribution, the fading power of each subchannel (equivalently squared Nakagami-m random variable) is considered to follow the Gamma distribution denoted as $||h_{0,n}||^2 \sim \mathcal{G}(m_0, \frac{\Omega_0}{m_0})$. The probability density function (PDF) of a Gamma random variable (RV) is expressed as [12]

$$f_{||h_{0,n}||^2}(x) = \frac{{m_0}^{m_0} x^{m_0-1}}{\Gamma(m_0)\Omega_0^m} \exp(-\frac{m_0}{\Omega_0}x) \tag{3}$$

Since the MRT used at the typical D2D receiver, the total fading power of N_R subchannels is given as [13]

$$\chi_0^2 = |\boldsymbol{h}_0|^2 = \sum_{n=1}^{N_R} ||h_{0,n}||^2, \tag{4}$$

where χ_0^2 is the summation of N_R subchannel gains. It is well known that the distribution of the sum of Gamma RVs can be approximated by single Gamma distribution [14]. By the moment matching method, the PDF of χ_0^2 is approximated by

$$f_{\chi_0^2}(x) \cong \frac{m_0^{m_0 N_R} x^{m_0 N_R - 1}}{\Gamma(m_0 N_R)\Omega_0^{m_0 N_R}} e^{-\frac{m_0}{\Omega_0}x}, \tag{5}$$

where $m_0 N_R$ is the shape parameter and Ω_0/m_0 is the scale parameter of the Gamma RV χ_0^2. And the complementary cumulative distribution function (CCDF) of χ_0^2 is given by [15]

$$\begin{aligned} F_{\chi_0^2}(x) &= 1 - \frac{\gamma(m_0 N_R, m_0 x/\Omega_0)}{\Gamma(m_0 N_R)} \\ &= \sum_{k=0}^{\lfloor t \rfloor} \frac{U_k (m_0 x/\Omega_0)^k}{k!} \exp(-m_0 x/\Omega_0) \end{aligned} \tag{6}$$

where $t = m_0 N_R$, $\lfloor\cdot\rfloor$ is the floor function, $U_k = 1$ when $k = 0, 1, \cdots, \lfloor t\rfloor - 1$ and $U_k = t - \lfloor t\rfloor$ when $k = \lfloor t\rfloor$. Note that, (6) is valid to characterize the CCDF of Gamma distribution for both integer and non-integer shape parameter. And it makes the performance analysis tractable in multi-antennas D2D communication over Nakagami-m channels. Similarly, the fading gains from the ith interfering D2D transmitter and jth cellular users follow the Gamma distribution, denoted as $\chi^2_{D_i} = ||h_{D_i}||^2 \sim \mathcal{G}(m_{D_i}, \frac{\Omega_{D_i}}{m_{D_i}})$ and $\chi^2_{U_j} = ||h_{U_j}||^2 \sim \mathcal{G}(m_{U_j}, \frac{\Omega_{U_i}}{m_{U_j}})$, respectively [11].

3 Outage Probability

Outage probability is defined to be the SINR falls than a predefined threshold, i.e. $P[\text{SINR} < \gamma]$. For a typical D2D receiver, the outage probability is given as

$$\begin{aligned}
P_0 &= P[\text{SINR}_0 < \gamma] \\
&= P[\frac{P_0 d_0^{-\alpha}\chi_0^2}{\sum_{i=1}^{I} P_0 d_i^{-\alpha}\chi^2_{D_i} + \sum_{j=1}^{J} P_1 d_j^{-\alpha}\chi^2_{U_j} + \sigma_n^2} < \gamma] \\
&= 1 - P[\chi_0^2 > \frac{\gamma d_0^{\alpha}}{P_0}(I + \sigma_n^2)] \\
&\overset{(a)}{=} 1 - \mathbb{E}_I[\sum_{k=0}^{\lfloor t\rfloor} \frac{U_k s^k (I+\sigma_n^2)^k}{k!} \exp(-s(I+\sigma_n^2))] \\
&\overset{(b)}{=} 1 - \sum_{k=0}^{\lfloor t\rfloor} \frac{U_k s^k \exp(-s\sigma_n^2)}{k!} \sum_{n=0}^{k} \binom{n}{k} (\sigma_n^2)^{n-k} \mathbb{E}_I[I^n e^{-sI}],
\end{aligned} \tag{7}$$

where $I = \sum_{i=1}^{I} P_0 d_i^{-\alpha}\chi^2_{D_i} + \sum_{j=1}^{J} P_1 d_j^{-\alpha}\chi^2_{U_j}$, $\binom{n}{k}$ is the binomial coefficient, $s = \gamma d_0^{\alpha} m_0 / P_0 \Omega_0$, (a) is derived from (6) and (b) is derived by the binomial expansion.

Now, we evaluate the PDF of the total interference I. Note that I is represented in terms of a series of weighted summations of Gamma RVs. As per the fact of weighted Gamma RV i.e. $X \sim \mathcal{G}(a, b)$ then $Y = cX \sim \mathcal{G}(a, cb)$. Since $\chi^2_{D_i}$ and $\chi^2_{U_j}$ are independent Gamma RVs, the distribution of the total interference received follows $I = \sum_{i=1}^{I} P_0 d_i^{-\alpha}\chi^2_{D_i} + \sum_{j=1}^{J} P_1 d_j^{-\alpha}\chi^2_{U_j} \sim \mathcal{G}(\beta, \lambda)$, where β and λ are the shape and scale parameters of the Gamma RV. Based on the second-order moment matching method, the parameters are given by [16]

$$\beta = \frac{(\sum_{i=1}^{I} P_0 m_{D_i}\Omega_{D_i} d_i^{-\alpha} + \sum_{j=1}^{J} P_1 m_{D_i}\Omega_{U_j} d_j^{-\alpha})^2}{\sum_{i=1}^{I} m_{D_i}(P_0\Omega_{D_i} d_i^{-\alpha})^2 + \sum_{i=1}^{I} m_{U_j}(P_1\Omega_{U_j} d_j^{-\alpha})^2} \tag{8}$$

and

$$\lambda = \frac{\sum_{i=1}^{I} m_{D_i}(P_0\Omega_{D_i} d_i^{-\alpha})^2 + \sum_{i=1}^{I} m_{U_j}(P_1\Omega_{U_j} d_j^{-\alpha})^2}{\sum_{i=1}^{I} P_0 m_{D_i}\Omega_{D_i} d_i^{-\alpha} + \sum_{j=1}^{J} P_1 m_{U_j}\Omega_{U_j} d_j^{-\alpha}} \tag{9}$$

With the shape and scale parameters in (8) and (9), we have

$$\begin{aligned}
&\mathbb{E}_I[I^n \exp(-sI)] \\
&= \int_0^{\infty} x^n \exp(-sx) \frac{x^{\beta-1}\exp(-\frac{x}{\lambda})}{\lambda^{\beta}\Gamma(\beta)} dx \\
&= \int_0^{\infty} \frac{x^{n+\beta-1}\exp(-(s+\frac{1}{\lambda})x)}{\lambda^{\beta}\Gamma(\beta)} dx \\
&= \frac{\lambda^n \Gamma(n+\beta)}{(1+s\lambda)^{n+\beta}\Gamma(\beta)},
\end{aligned} \tag{10}$$

Using the above expression, the outage probability defined in (7) is rewritten as

$$P_0 = 1 - \sum_{k=0}^{\lfloor t \rfloor} \sum_{n=0}^{k} \binom{n}{k} \frac{s^k \lambda^n U_k \exp(-s\sigma_n^2)\Gamma(n+\beta)}{k!\Gamma(\beta)(1+s\lambda)^{n+\beta}} \tag{11}$$

As in the case of integer parameter t, the outage probability is then simplified as

$$P_0 = 1 - \sum_{k=0}^{t-1} \frac{(s\lambda)^k \Gamma(k+\beta)}{k!\Gamma(\beta)(1+s\lambda)^{k+\beta}}. \tag{12}$$

4 Numerical Results and Analysis

In this section, we compares the analytical results of coverage probability for D2D link with the simulation results. The simulation results are derived by the Monte-Carlo simulation. Simulation parameter settings are given in Tabel 1.

Table 1. Simulation parameter settings

Parameter	Values
Cell radius R	500 m
Distance between the typical D2D pair d_0	30 m
Number of D2D transmitters I	4
Number of CUs J	4
Number of antennas at D2D transmitter N_R	3
Transmit power of cellular user P_1	28 dBm
Transmit power of D2D user P_0	25 dBm
Path loss exponent α	3.5
Thermal noise power σ_n^2	−174 dBm
Distance between interfering D2D users and typical D2D user	{40, 45, 50, 55} m
Distance between interfering cellular users and typical D2D user	{50, 80, 150, 200} m

Figure 2 shows the coverage probability of the D2D receiver with respect to SINR threshold in terms of different fading parameters. And we compare the analytical results with the Erlang approximation (denoted as EA) in [15]. From this illustration, we observe that the proposed expression in (11) closely match the corresponding simulation results compared to the Erlang approximation. Erlarng approximation is tight with the simulation results when the shaping parameter is an integer. We further observe that the outage probability of the D2D communication decreases as the fading parameter m_0 increases, due to the decrease in channel fading.

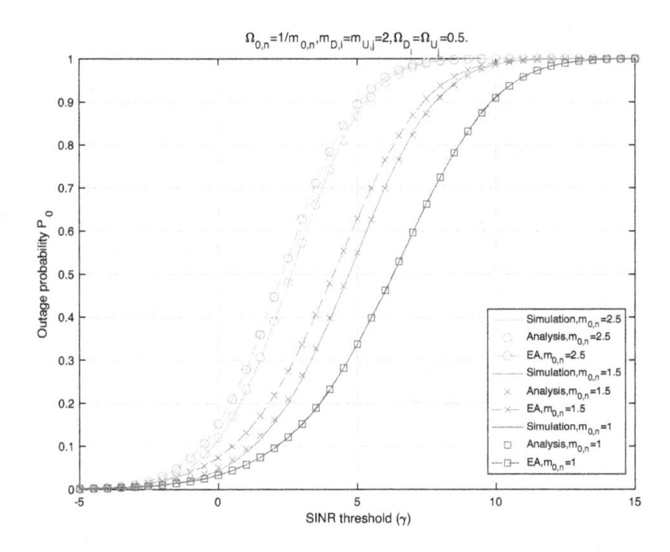

Fig. 2. Comparison the outage probability in terms of different fading parameters.

Figure 3 presents the outage probability with respect to the number of antennas at the receiver side N_R. Its clearly seen that the analytical results are very tight with the simulation results for different number of antennas, which implies that the approximation is reasonable. Furthermore, its clearly seen that, the outage probability decreases with the increasing of the number of antennas at the D2D receiver at low values of SINR. It demonstrates that the D2D with multiple antennas assure a great outage performance improvement.

Figure 4 presents the outage probability versus the average SINR for different threshold γ. One sees that average SINR in general degrades the outage performance due to increased desired signal. One also sees that the outage performance increase with the increase of SINR threshold.

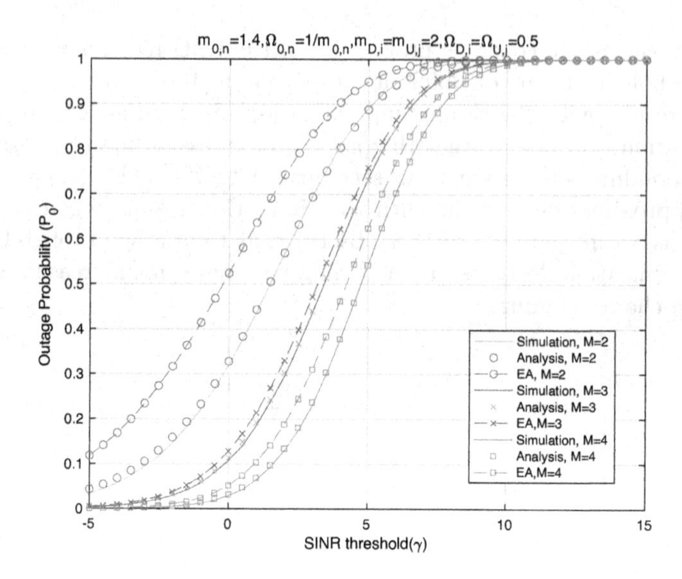

Fig. 3. Comparison the outage probability in terms of number of antennas at the D2D receiver.

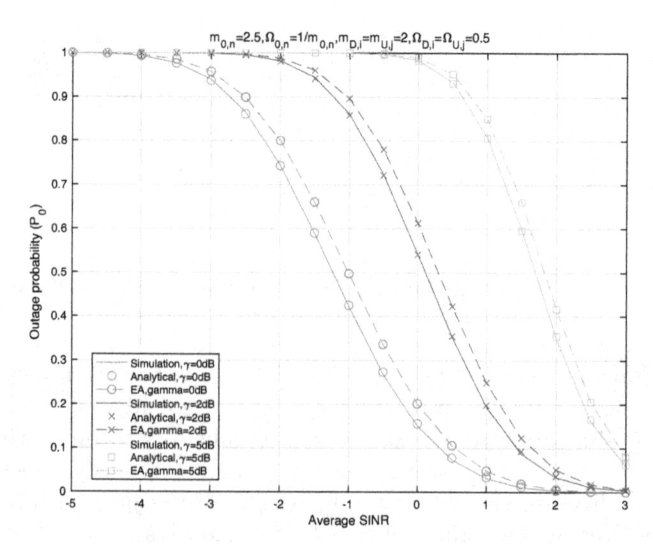

Fig. 4. Comparison the outage probability in terms of average SINR and threshold.

5 Conclusions

In this paper, outage probability for D2D links has been analyzed in uplink D2D communication and cellular HetNets. The closed-form expression for outage probability has been derived over the Nakagami-m fading. The expression is valid for both integer and non-integer values of fading parameters m and provides advantages in terms of less complexity in the numerical computations. Analytical results are very tight with the Monte-Carlo simulation. These results also show the impact of the number of D2D receiver antenna, Nakagami-m fading parameters and average SINR on outage performance.

Acknowledgments. This work was partially supported in part by the Study on Performance Evaluation and Optimization Technology of Local IoT of Customer Side Metering Equipment under grant 5700-202118203A-0-0-00.

References

1. Ye, Q., Al-Shalash, M., Caramanis, C., Andrews, J.G.: Resource optimization in device-to-device cellular systems using time-frequency hopping. IEEE Trans. Wirel. Commun. **13**(10), 5467–5480 (2014)
2. Min, H., Lee, J., Park, S., Hong, D.: Capacity enhancement using an interference limited area for device-to-device uplink underlaying cellular networks. IEEE Trans. Wirel. Commun. **10**(12), 3995–4000 (2011)
3. Chun, Y.J., Cotton, S.L., Dhillon, H.S., Ghrayeb, A., Hasna, M.O.: A stochastic geometric analysis of device-to-device communications operating over generalized fading channels. IEEE Trans. Wirel. Commun. **16**(7), 4151–4165 (2017)
4. Hoang, T.D., Le, L.B., Le-Ngoc, T.: Joint subchannel and power allocation for D2D communications in cellular networks. In: 2014 IEEE Wireless Communications and Networking Conference (WCNC), pp. 1338–1343 (2014)
5. Turgut, E., Gursoy, M.C.: Uplink Performance analysis in D2D-enabled mmWave cellular networks. In: 2017 IEEE 86th Vehicular Technology Conference (VTC-Fall), Toronto, ON, Canada, pp. 1–5 (2017)
6. Lu, B., Lin, S., Shi, J., Wang, Y.: Resource allocation for D2D communications underlaying cellular networks over Nakagami-m fading channel. IEEE Access **7**, 21816–21825 (2019)
7. Mustafa, H.A., Shakir, M.Z., Imran, M.A., Imran, A., Tafazolli, R.: Coverage gain and device-to-device user density: stochastic geometry modeling and analysis. IEEE Commun. Lett. **19**(10), 1742–1745 (2015)
8. Senadhira, N., Guo, J., Durrani, S.: Outage analysis of underlaid multi-antenna D2D communication in cellular networks. In: 2016 10th International Conference on Signal Processing and Communication Systems (ICSPCS), Surfers Paradise, QLD, Australia, pp. 1–7 (2016)
9. Singh, I., Jaiswal, R.K., Kumar, V., Verma, R., Singh, N.P., Singh, G.: Outage probability of device-to-device communication underlaying cellular network over Nakagami/Rayleigh fading channels. In: 2019 9th International Conference on Emerging Trends in Engineering and Technology - Signal and Information Processing (ICETET-SIP-19), Nagpur, India, pp. 1–5 (2019)
10. Joshi, S., Mallik, R.K.: Coverage probability analysis in a device-to-device network: interference functional and Laplace transform based approach. IEEE Commun. Lett. **23**(3), 466–469 (2019)
11. Atzeni, I., Arnau, J., Kountouris, M.: Downlink cellular network analysis with LOS/NLOS propagation and elevated base stations. IEEE Trans. Wirel. Commun. **17**(1), 142–156 (2018)
12. Salehi, M., Proakis, J.: Digital Communications. McGraw-Hill, New York (2007)

13. Magableh, A.M., Matalgah, M.M.: Capacity of SIMO systems over non-identically independent Nakagami-m channels. In: 2007 IEEE Sarnoff Symposium, Princeton, NJ, USA, pp. 1–5 (2007)
14. Nakagami, M.: The m-distribution-a general formula of intensity distribution of rapid fading. In: Hoffman, W.G. (ed.) Statistical Methods in Radio Wave Propagation. Pergamon, Oxford (1960)
15. Chen, J., Yuan, C.: Coverage and rate analysis in downlink L-tier HetNets with fluctuating Beckmann fading. IEEE Wirel. Commun. Lett. **8**(5), 1489–1492 (2019)
16. Heath, R., Kountouris, M., Bai, T.: Modeling heterogeneous network interference using Poisson point processes. IEEE Trans. Signal Process. **61**(16), 4114–4126 (2013)

On Codeword Bits Disparity in Polar Codes

Chen Cui[1], Wei Xiang[2], Zhenyong Wang[1(✉)], and Qing Guo[1]

[1] School of Electronics and Information Engineering, Harbin Institute of Technology, Harbin 150001, China
{cuichen,ZYWang,qguo}@hit.edu.cn

[2] School of Engineering and Mathematical Sciences, La Trobe University, Melbourne, VIC 3086, Australia
w.xiang@latrobe.edu.au

Abstract. Polar codes, the codes contributed based on channel polarization, are the first class of error-correction codes which can be proved to achieve the channel capacity of binary-input discrete memoryless channels. A set of bit-channels are generated after the channel combining and channel splitting operations. These bit-channels have different capacities. The idea of polar coding is to transmit information bits through bit-channels with higher capacities, while transmitting frozen bits through the ones with lower capacities. The generator matrix of polar codes is invertible. The input bits can also be represented as the codeword bits multiplied by the generator matrix. Based on these observations, we make an assumption that there are significance disparities existing in the codeword bits of polar codes and the significance level of codeword bits are relevant to their bit indices. To analyze the significance level of codeword bits, the codeword bits are transmitted over non-uniform channels. Some of the codeword bits are transmitted through pure noisy channels, which can be equalized as punctured bits. Simulation results show that there exist significance disparities in the codeword bits.

Keywords: Polar codes · Significance disparity · Codeword bits · SC decoding

1 Introduction

Polar codes proposed by Arikan are the first class of error-correction codes which can be proved to achieve the capacity of binary-input discrete memoryless channels (B-DMC) [1]. They have recursive structures and simple encoding and decoding complexity . They can achieve good error-correction performance even with short code lengths. Due to these characteristics, polar codes have been adopted as coding schemes for the 5G communication systems [2].

The main idea of polar codes is to generate a set of coordinate channels called bit-channels and transmit information through the bit-channels with

Q. Guo et al. (Eds.): WiSATS 2021, LNICST 410, pp. 367–377, 2022.
https://doi.org/10.1007/978-3-030-93398-2_36

higher capacity. For a polar code with code length N and code dimension k, the construction of polar codes is to select k good bit-channels out of all the N bit-channels. In [3,4], density evolution (DE) is employed to construct polar codes. In [5], the authors use upgraded and degraded channels to approximate the bit-channels. In [6], Gaussian approximation (GA) is employed to decrease the computation complexity of DE. In [7], fast construction of polar codes based on polarization weight (PW) is proposed.

Many efficient decoding algorithms of polar codes have been studied. In [8], successive cancellation list (SCL) decoding algorithm is proposed to improve the error-correction performance of polar codes. The cyclic redundancy check (CRC) aided SCL decoding can substantially improve the performances of polar codes [8,9]. An adaptive CRC aided SCL decoder which enlarges the list size progressively is proposed in [10]. Simplified-SC (SSC) decoder [11] and fast-simplified-SC (FSSC) [12] decoder are proposed to reduce decoding latency. Belief propagation (BP) decoding is proposed in [13] for pipelined implementations.

The code lengths of polar codes are the power of 2 with a kernel of $F = \begin{bmatrix} 1 & 0 \\ 1 & 1 \end{bmatrix}$. To achieve arbitrary code lengths, many puncturing and shortening methods have been proposed. Some of the coded bits are not transmitted. For punctured polar codes, these bits are unknown bits. In the receiving end, the log-likelihood ratios (LLRs) of the punctured bits are set to zeros. For shortened polar codes, these bits are known bits and the LLRs of the shortened bits are set to infinity. In [14], quasi-uniform puncturing (QUP) algorithm is proposed. The first N_p coded bits are punctured. In [15], the last N_p coded bits are shortened. The information sets are reselected after puncturing and shortening in [14,15]. In [16], punctured bits are selected according to the information and frozen sets.

The generator matrix of polar codes is invertible. There may be some similarity between the codeword bits and the input bits. In [17], the authors make a conjecture that the codeword bits can be sorted according to the reliability of information bits. The various performances of these puncturing and shortening methods also indicate that codeword bits of different positions have different effects on the performance of polar codes. In [18], the importance differences of input bits have been studied. This paper will study the importance differences of codeword bits.

The remainder of the paper is organized as follows. Section 2 presents a brief review of polar codes. In Sect. 3, the various importance levels of codeword bits of polar codes are studied. Simulation results are presented in Sect. 4. Finally, concluding remarks are drawn in Sect. 5.

2 Preliminaries

2.1 Polar Codes

Given a B-DMC channel $(\mathcal{X}, \mathcal{Y}, W(y \mid x))$, where $\mathcal{X} \in \{0, 1\}$ is the input alphabet, $\mathcal{Y}$ is the output alphabet, and $W(y \mid x)$ is the transition probabilities of channel W. Use a_i^j to denote $(a_i, a_{i+1}, \cdots, a_j)$. When $j < i$, $a_i^j = \varnothing$. Let W^N denote N independent uses of W, we have

$$W^N\left(y_1^N \mid x_1^N\right) = \prod_{i=1}^{N} W\left(y_i \mid x_i\right), \tag{1}$$

These N independent channels can be combined into a single synthesized channel W_N. The relationship between W_N and W^N can be represented as

$$W_N\left(y_1^N \mid u_1^N\right) = W^N\left(y_1^N \mid u_1^N G_N\right), \tag{2}$$

where u_1^N are the input bits of W_N.

The channel W_N can be split into a series of bit-channels $W_N^{(i)} : \mathcal{X} \rightarrow \mathcal{Y}^N \times \mathcal{X}^{i-1}, 1 \leq i \leq N$, each channel $W_N^{(i)}$ with single input bits u_i and its channel transmission probabilities are given by

$$W_N^{(i)}\left(y_1^N, u_1^{i-1} \mid u_i\right) \triangleq \sum_{u_{i+1}^N \in \mathcal{X}^{N-i}} \frac{1}{2^{N-1}} W_N\left(y_1^N \mid u_1^N\right). \tag{3}$$

As N tends to infinity, the capacity of the bit-channels $W_N^{(i)}$ will tend to 0 or 1. In other words, this set of bit-channels will be polarized into noiseless and noisy channels. This phenomenon is called *channel polarization.* The idea of polar coding is to transmit the information bits through the noiseless bit-channels and assign fixed bits over the noisy bit-channels.

Let u_1^N and x_1^N denote the input bits and codeword bits of polar codes. The encoding process of polar codes can be written as

$$x_1^N = u_1^N G_N, \tag{4}$$

where G_N is the generator matrix of order N, and

$$G_N = B_N F^{\otimes n}, \tag{5}$$

where B_N is the bit-reversal matrix, $\otimes n$ is the Kronecker power and $F \triangleq \begin{bmatrix} 1 & 0 \\ 1 & 1 \end{bmatrix}$.

Let $\mathcal{A}$ denote the information set, and $\mathcal{A}^c$ be its complementary set. The codeword x_1^N can be rewritten as

$$x_1^N = u_{\mathcal{A}} G_{\mathcal{A}} + u_{\mathcal{A}^c} G_{\mathcal{A}^c}, \tag{6}$$

where $u_{\mathcal{A}}$ denote the input bits with indices corresponding to the information set, $G_{\mathcal{A}}$ denote the submatrix of G_N formed by the rows with indices in $\mathcal{A}$ and $u_{\mathcal{A}^c}$ are referred as frozen bits.

2.2 Successive Cancellation Decoding Algorithm

The successive cancellation (SC) decoding algorithm of polar codes decodes input bits sequentially from u_1 to u_N. Define the log-likelihood ratio (LLR) as

$$L_N^{(i)}\left(y_1^N, \hat{u}_1^{i-1}\right) \triangleq \log \frac{W_N^{(i)}\left(y_1^N, \hat{u}_1^{i-1} \mid u_i = 0\right)}{W_N^{(i)}\left(y_1^N, \hat{u}_1^{i-1} \mid u_i = 1\right)}, \tag{7}$$

The LLRs can be calculated recursively as follows

$$\begin{aligned} &L_N^{(2i-1)}\left(y_1^N, \hat{u}_1^{2i-2}\right) \\ &= 2\tanh^{-1}\left(\tanh\left(L_{N/2}^{(i)}\left(y_1^{N/2}, \hat{u}_{1,e}^{2i-2} \oplus \hat{u}_{1,o}^{2i-2}\right)/2\right)\right. \\ &\left.\times \tanh\left(L_{N/2}^{(i)}\left(y_{N/2+1}^N, \hat{u}_{1,e}^{2i-2}\right)/2\right)\right), \end{aligned} \tag{8}$$

$$\begin{aligned} L_N^{(2i)}\left(y_1^N, \hat{u}_1^{2i-1}\right) &= L_{N/2}^{(i)}\left(y_{N/2+1}^N, \hat{u}_{1,e}^{2i-2}\right) \\ &+ (-1)^{\hat{u}_{2i-1}} L_{N/2}^{(i)}\left(y_1^{N/2}, \hat{u}_{1,e}^{2i-2} \oplus \hat{u}_{1,o}^{2i-2}\right). \end{aligned} \tag{9}$$

The ith decision element $\hat{u}_i$ can be represented as

$$\hat{u}_i \triangleq \begin{cases} u_i, & \text{if } i \in \mathcal{A}^c \\ h_i\left(y_1^N, \hat{u}_1^{i-1}\right), & \text{if } i \in \mathcal{A} \end{cases} \tag{10}$$

where $h_i : \mathcal{Y}^N \times \mathcal{X}^{i-1} \rightarrow \mathcal{X}, i \in \mathcal{A}$ is the decision function, and

$$h_i\left(y_1^N, \hat{u}_1^{i-1}\right) \triangleq \begin{cases} 0, & \text{if } L_N^{(i)}\left(y_1^N, \hat{u}_1^{i-1}\right) \geq 0 \\ 1, & \text{otherwise} \end{cases} \tag{11}$$

2.3 Construction of Polar Codes

The construction of polar codes focuses on choosing the bit-channels with low error probabilities to transmit the information bits. However, the exact value of error probabilities of bit-channels $P_e(W_N^{(i)})$ are not easy to calculate. Many approximation methods have been proposed to calculate these error probabilities, including density evolution [3,4], Tal and Vardy method [5], Gaussian approximation (GA) method [6] and the construction method based on polarization weight [7]. In this subsection, the GA method of polar codes is introduced.

When the codeword bits of polar codes are transmitted over additive white Gaussian noise (AWGN) channels, the intermediate values of SC decoding can be seen as Gaussian values [6]. Assume that all zero bits are transmitted. The signals are BPSK modulated with $s_i = 1 - 2x_i, i = 1, \cdots, N$, where s_i is the signal after modulation, and x_i is the codeword bit. Let y_i denote the received signal. Then $y_i = s_i + n_i, i = 1, \cdots, N$, where $(n_1, n_2, \cdots, n_N)$ is an i.i.d. set of Gaussian random variables with mean zero and variance σ^2. Let $L_1^{(i)}(y_i)$ denote the LLRs of y_i. Then $L_1^{(i)}(y_i)$ is a Gaussian value with the distribution function of $\mathcal{N}\left(\frac{2}{\sigma^2}, \frac{4}{\sigma^2}\right)$.

The LLR values given by (8)–(9) can also be considered as Gaussian random variables with $\mathbf{D}\left[L_N^{(i)}\right] = 2\mathbf{E}\left[L_N^{(i)}\right]$, where $\mathbf{E}$ and $\mathbf{D}$ are the mean and variance, respectively. Calculate the expected values of both sides of (8)–(9), the following expressions can be derived

$$\mathbf{E}\left[L_N^{(2i-1)}\right] = \phi^{-1}\left(1 - \left(1 - \phi\left(\mathbf{E}\left[L_{N/2}^{(i)}\right]\right)\right)^2\right), \tag{12}$$

$$\mathbf{E}\left[L_N^{(2i)}\right] = 2\mathbf{E}\left[L_{N/2}^{(i)}\right], \tag{13}$$

where

$$\phi(x) = \begin{cases} 1 - \frac{1}{\sqrt{4\pi x}} \int_{-\infty}^{\infty} \tanh\left(\frac{u}{2}\right) \mathrm{e}^{-\frac{(u-x)^2}{4x}}\, du, & x > 0 \\ 1, & x = 0. \end{cases} \tag{14}$$

The error probabilities of bit-channels [19] are given by

$$P_e(W_N^{(i)}) = Q\left(\sqrt{\mathbf{E}\left[L_N^{(i)}\right]/2}\right), \quad 1 < i < N. \tag{15}$$

3 The Various Important Levels of Different Codeword Bits in Polar Codes

3.1 BLER Performance of Polar Codes over Non-uniform Channels

When the coreword bits are transmitted over non-uniform channels, the Gaussian approximation method can be generalized. For the basic unit of encoding and decoding in [1] with input bits (u_1, u_2) and received signals (y_1, y_2), we have

$$\mathbf{E}\left[L\left(u_1\right)\right] = \phi^{-1}\left(1 - \left(1 - \phi\left(\mathbf{E}\left[L\left(y_1\right)\right]\right)\right)\left(1 - \phi\left(\mathbf{E}\left[L\left(y_2\right)\right]\right)\right)\right), \tag{16}$$

and

$$\mathbf{E}\left[L\left(u_2\right)\right] = \mathbf{E}\left[L\left(y_1\right)\right] + \mathbf{E}\left[L\left(y_2\right)\right]. \tag{17}$$

$\mathbf{E}\left[L_N^{(i)}\right]$ can be calculated recursively using (16) and (17), and $P_e(W_N^{(i)})$ can be calculated accordingly.

The block error rate of polar codes is calculated as

$$P(\varepsilon) = 1 - \prod_{i \in \mathcal{A}} \left(1 - P_e(W_N^{(i)})\right). \tag{18}$$

3.2 The Proposed Scheme About Significance Disparity of Codeword Bits in Polar Codes.

The generator matrix G_N of polar codes is invertible, and $G_N^{-1} = G_N$, so that (4) can be rewritten as $u_1^N = x_1^N G_N$. This equality shows that there may be some similarity between input bit u_i and codeword bit x_i. As is known to us, u_i has different error probabilities for different bit indices. To research the significance disparity of the codeword bits with different positions, we transmit the codeword bits through non-uniform channels. Some of the codeword bits are transmitted through the pure noisy channels, which can be seen as punctured bits, and the other codeword bits will be transmitted over AWGN channels.

Consider a codeword bit x_i which is over the pure noisy channel. The LLR of x_i can be calculated as

$$L(y_i) = \log \frac{\Pr(y_i \mid x_i = 0)}{\Pr(y_i \mid x_i = 1)} = 0. \tag{19}$$

It is the same as when x_i is punctured. Note that for puncturing schemes of polar codes, the information sets are often redesigned after a puncturing pattern is set. In this paper, the information sets are fixed, and we study the effects of different codeword bits on the BLER performances.

4 Simulation Results

To illustrate the various significance disparity of codeword bits in polar codes, we use the punctured bits to equalize the codeword bits transmitted over the pure noisy channels.

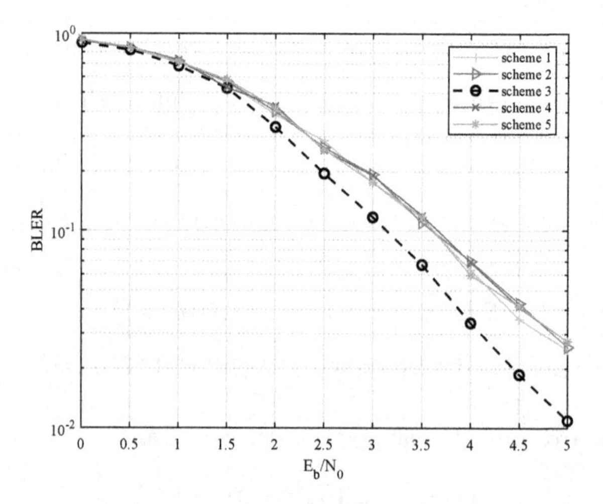

Fig. 1. The BLER performances of polar codes with 30 punctured codeword bits.

Firstly, we consider the polar codes with code length 256, code dimension 93. Figure 1 illustrates the BLER performances of polar codes by randomly puncturing 30 codeword bits in different sets. For puncturing schemes 1 to 5, the punctured codeword bits are randomly selected from the sets $[1, 64]$, $[65, 128]$, $[97, 160]$, $[129, 192]$ and $[193, 256]$, respectively. It can be found that the BLER

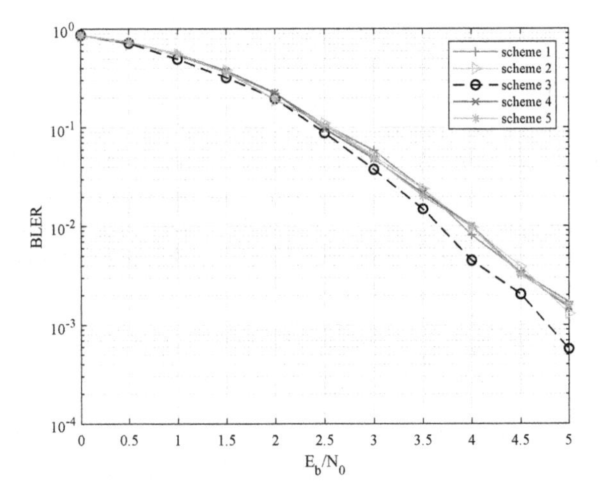

Fig. 2. The BLER performances of polar codes with 20 punctured codeword bits.

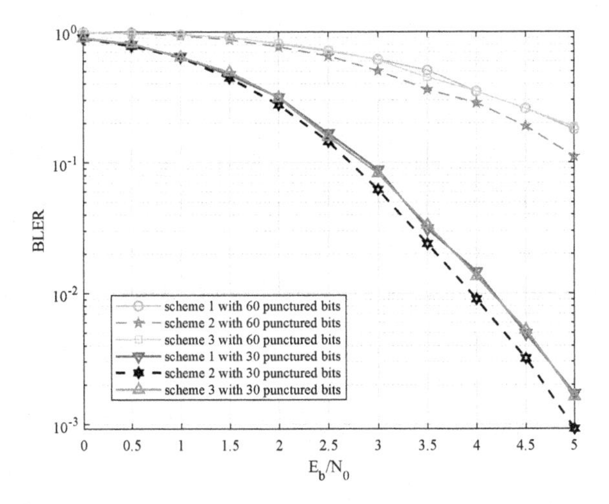

Fig. 3. The BLER performances of polar codes with randomly punctured codeword bits in different sets.

performances of scheme 3 are better than the other schemes, which implies that the average significance levels of codeword bits in the 3rd set are lower than the codeword bits in other sets. We also provide the BLER performances of polar codes by randomly puncturing 20 codeword bits in the aforementioned sets, as shown in Fig. 2. It can be seen in Fig. 2 that scheme 3 also has better BLER performances than the other schemes when there are 20 punctured codeword bits. It can be shown that there are various significance levels between the codeword bits, and the significance levels of the codeword bits will be affected by their positions.

Following the phenomena shown in Fig. 1 and Fig. 2, we make an assumption that the codeword bits in the middle set are with the lower average significance levels than the others. We define three puncturing schemes, in which the codeword bits are punctured from the sets [1, 128], [65, 192] and [129, 256], respectively. The BLER performances of the three schemes are given in Fig. 3 with 60 and 30 punctured codeword bits in each set. It can be seen that scheme 2 has better performances than the other schemes both with 60 and 30 punctured codeword bits. The puncture set of scheme 2 [65, 192] is still in the middle sets.

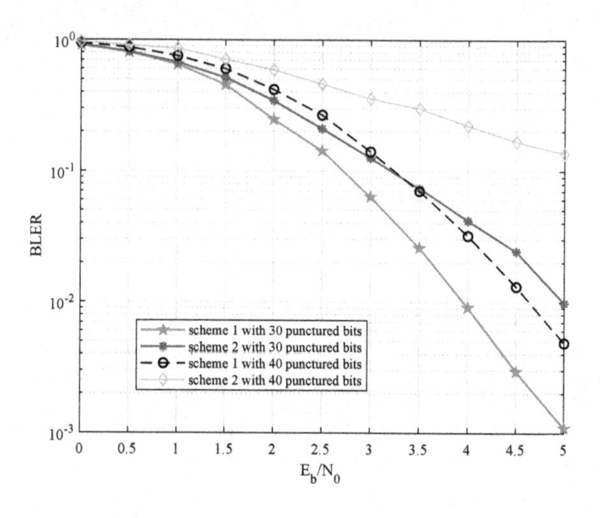

Fig. 4. The BLER performances of polar codes with randomly punctured bits in the middle part of bit indices.

Observing the BLER curves in the previous figures, all these results imply that the codeword bits in the middle of the index are with lower average significance levels than the others. Then we focus on the schemes with better performances than the others in the previous results, which means we focus on the puncturing range [65, 192] and [97, 160]. For simplicity, the scheme with puncture sets [65, 192] and [97, 160] are termed as scheme 1 and 2, respectively. From Fig. 4, it can be seen that although there are the same number of codeword bits being punctured, scheme 1 has better BLER performances than that of scheme 2. As both the schemes are puncturing codeword bits from the middle set, and the puncturing set of scheme 1 is larger than that of scheme 2, we conclude that the continuous impulse noise would lead to much more negative affections on BLER performances of polar codes than that of the discrete impulse noise.

Figure 5 illustrates the BLER performances of puncturing codeword bits from different sets with different probabilities. The code length of polar codes is 512 and code rate is 0.375. The codeword bits are divided into 2 sets, the first set is $[1, 128] \cup [385, 512]$, and the second set is [129, 384]. For all the schemes 1 to 6, there are 128 codeword bits being punctured. For scheme 1 to 6, there are 16, 32,

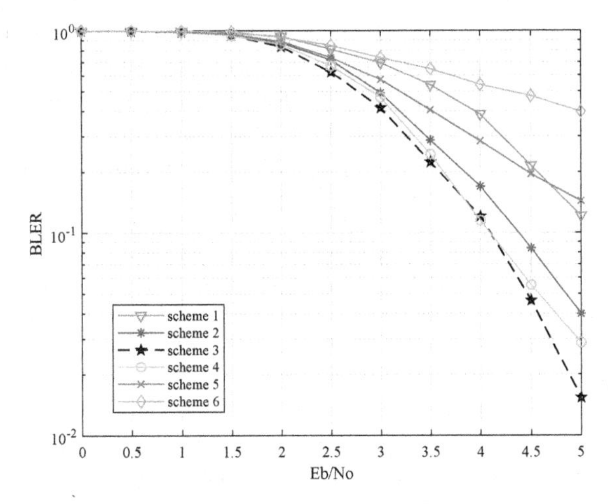

Fig. 5. The BLER performances of polar codes with punctured bits from different sets.

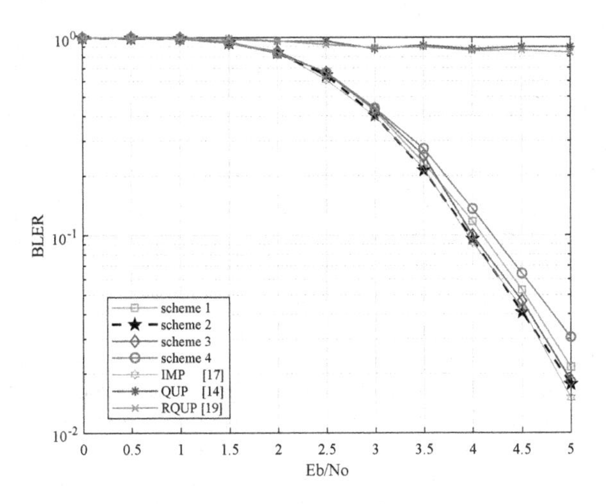

Fig. 6. The BLER performance comparison between the proposed schemes and other existing schemes.

48, 64, 80 and 96 codeword bits randomly punctured from the first set, and the remaining codeword bits are randomly punctured from the second set. It can be found that scheme 4 can be equalized as randomly puncturing 128 bits without dividing sets, and scheme 3 has better BLER performance than scheme 4, which means the randomly puncturing cannot be optimal. Figure 5 also illustrates that the codeword bits in the middle positions are with the lower average significance levels than the codeword bits in other positions.

Finally, we compare the proposed schemes with other existing puncturing schemes, which are shown in Fig. 6. The proposed schemes are with the same divided sets as shown in Fig. 5. For scheme 1 to 4, there are 40, 48 and 56 and 64 bits being punctured from the first set, respectively. It can be found that the proposed scheme 1 to 3 have better BLER performances than scheme 4, which is equalized to randomly puncturing. The performances of the proposed schemes are similar to that of the method in [17], where the importance levels are sorted according to the error probabilities of input bits with the same indices. The QUP [14] and RQUP [20] schemes are also compared, and it can be found that the QUP and RQUP schemes cannot provide satisfactory performance with fixed information sets.

5 Conclusion

In this paper, to research the significance disparity between the codeword bits of polar codes, we adopted non-uniform channels to transmit the codeword bits. By analyzing the log-likelihood ratio of a codeword bit affected by pure noisy channel, we equalized this bit as a punctured bit. By studying the previous research, we made an assumption that the significance disparity of codeword bits would follow a symmetrically distributed law based on their positions. The simulation results proved that the significance disparities of codeword bits exist.

References

1. Arikan, E.: Channel polarization: a method for constructing capacity-achieving codes for symmetric binary-input memoryless channels. IEEE Trans. Inf. Theory **55**(7), 3051–3073 (2009)
2. Evaluation on channel coding candidates for eMBB control channel. 3GPP TSG RAN WG1, R1-1611109 (2017)
3. Mori, R., Tanaka, T.: Performance of polar codes with the construction using density evolution. IEEE Commun. Lett. **13**(7), 519–521 (2009)
4. Mori, R., Tanaka, T.: Performance and construction of polar codes on symmetric binary-input memoryless channels. In: Proceedings of IEEE International Symposium on Information Theory, Seoul, Korea, pp. 1496–1500 (2009)
5. Tal, I., Vardy, A.: How to construct polar codes. IEEE Trans. Inf. Theory **59**(10), 6562–6582 (2013)
6. Trifonov, P.: Efficient design and decoding of polar codes. IEEE Trans. Commun. **60**(11), 3221–3227 (2012)
7. He, G., et al.: Beta-expansion: a theoretical framework for fast and recursive construction of polar codes. In: GLOBECOM 2017–2017 IEEE Global Communications Conference, Singapore, pp. 1–6 (2017)
8. Tal, I., Vardy, A.: List decoding of polar codes. IEEE Trans. Inf. Theory **61**(5), 2213–2226 (2015)
9. Niu, K., Chen, K.: CRC-aided decoding of polar codes. IEEE Commun. Lett. **16**(10), 1668–1671 (2012)
10. Li, B., Shen, H., Tse, D.: An adaptive successive cancellation list decoder for polar codes with cyclic redundancy check. IEEE Commun. Lett. **16**(12), 2044–2047 (2012)

11. Alamdar-Yazdi, A., Kschischang, F.R.: A simplified successive-cancellation decoder for polar codes. IEEE Commun. Lett. **15**(12), 1378–1380 (2011)
12. Sarkis, G., Giard, P., Vardy, A., Thibeault, C., Gross, W.J.: Fast polar decoders: algorithm and implementation. IEEE J. Sel. Areas Commun. **32**(5), 946–957 (2014)
13. Arikan, E.: Polar codes: a pipelined implementation. In: Proceedings of 4th International Symposium on Broadband Communications (ISBC2010), Melaka, Malaysia, pp. 11–14 (2010)
14. Niu, K., Chen, K., Lin, J.: Beyond turbo codes: rate-compatible punctured polar codes. In: 2013 IEEE International Conference on Communications (ICC), pp. 3423–3427 (2013)
15. Wang, R., Liu, R.: A novel puncturing scheme for polar codes. IEEE Commun. Lett. **18**(12), 2081–2084 (2014)
16. Zhang, L., Zhang, Z., Wang, X., Yu, Q., Chen, Y.: On the puncturing patterns for punctured polar codes. In: 2014 IEEE International Symposium on Information Theory, Honolulu, HI, USA, pp. 121–125 (2014)
17. Saber, H., Marsland, I.: An incremental redundancy hybrid ARQ scheme via puncturing and extending of polar codes. IEEE Trans. Commun. **63**(11), 3964–3973 (2015)
18. Cui, C., Xiang, W., Wang, Z., Guo, Q.: Polar codes with the unequal error protection property. Comput. Commun. **123**, 116–125 (2018)
19. Proakis, J.G.: Digital Communications. McGraw Hill, London (1995)
20. Niu, K., Dai, J., Chen, K., Lin, J., Zhang, Q.T., Vasilakos, A.V.: Rate-compatible punctured polar codes: Optimal construction based on polar spectra. arXiv:1612.01352 (2016)

Generalized Simplified Successive-Cancellation Decoding of Multi-kernel Polar Codes

Yanlong Zhao, Zhilu Wu, Zhendong Yin(✉), and Qingzhi Liu

School of Electronics and Information Engineering, Harbin Institute of Technology, Harbin, China
{16B905027,17B905015}@stu.hit.edu.cn, {wuzhilu,yinzhendong}@hit.edu.cn

Abstract. Multi-Kernel (MK) offers more flexibility code length selections for polar codes compared to size-2 kernel proposed by Arikan. In this paper, a generalized Simplified Successive-Cancellation List (SSCL) decoding algorithm is introduced. We first provide sufficient conditions and corresponding proofs to perform SSC and SSCL decoding on MK polar codes. These simplifications are proven valid for MK polar codes whose transform matrix are constructed by kernels who satisfy certain conditions. Time-complexity reduction of introducing generalized simplification decoding is discussed. Numerical results shows that our proposed methods can reduce time-step while preserving the error-correction performance.

Keywords: Polar codes · Successive-cancellation · List decoding · Multi-kernel

1 Introduction

Polar codes are the first error-correcting codes introduced by Arikan in [1] who are proven to be able to reach channel capacity for a large number of channels. In recent years, polar codes were adopted as channel coding for uplink and downlink control information for the enhanced mobile broadband communication service of 5G. Arikan proposes a 2×2 kernel $\boldsymbol{T}_2 = \left(\begin{smallmatrix} 1 & 0 \\ 1 & 1 \end{smallmatrix}\right)$ as transform matrix to construct polar codes, which only allows for code lengths power of 2. This constraint limits polar codes for typical 5G applications. Therefore, puncturing, shortening [2] and extending [3] are studied for polar codes 5G typical scenarios.

Authors in [4] exhibit that other kernels except $\boldsymbol{T}_2$ could also achieve polarization phenomenon, which means code length of polar codes is no longer limit to power of 2. This coding method allows to conjunct different size of polarizing matrices to build a polar code with more flexibility in code length. MK polar codes are introduced in [5] which provide a practical approach for constructing more flexible polar codes. To improve the error-correction performance,

Q. Guo et al. (Eds.): WiSATS 2021, LNICST 410, pp. 378–391, 2022.
https://doi.org/10.1007/978-3-030-93398-2_37

minimum-distance profile is proposed in [6]. Then, a hybrid design combining reliability and distance properties is further explored in [7]. [8] also designs a kernel substitution coding method which further improves the performance of MK polar codes.

Successive-Cancellation (SC) decoding is first used to decode polar codes when they were proposed. However, affecting by the serial nature, SC decoding suffers great latency and undesirable throughput. To overcome this drawback, Simplified Successive-Cancellation (SSC) has been proposed in [9] who introduced parallel decoder to SC decoding. And then, more special nodes have been raised in [10] and [11] etc. Recently, [12] provides a generalized approach to further reduce SC-based decoding latency. In order to fill the gap between the performance of SC and maximum-likelihood decoding, Successive-Cancellation List (SCL) proposed in [13] offers performance competitive with many other channel codes. The simplification algorithms of SCL are consecutively studied in [14] and [15]. Due to the coding structure, SC decoding is also performed for multi-kernel polar codes when they were proposed in [4]. Then, [16] explores general procedure to marginalize kernels of any size. Further, fast simplified SC decoding was also be presented in [17] to decode ternary kernel based polar codes.

This paper explores a generalized approach to SSC and SSCL decoding of MK polar codes to further reduce computation complexity. We provide sufficient conditions for kernels who are suitable for simplification calculations of SC and SCL decoding of multi-kernel polar codes. Then, the relative conditions are verified by deriving $\boldsymbol{T}_3$ kernel who was proposed in [16]. At next, we offer proofs for generalized simplified decoding methods of MK polar codes. We focus on Rate-1 and Rate-0 nodes which have much more simpler code structures. And the exact and approximation formulas of Path Metric (PM) are both considered and proven. The proofs and following numerical results indicate that these simplifications are valid for any kernels who are generalized for MK polar codes without incurring any performance degradation.

The rest of this letter is organized as follows. Section 2 reviews MK polar codes and the conventional decoding methods. Section 3 describes the generalize simplified decoding of MK polar codes. The sufficient conditions and corresponding proofs are provided, and time-step requirements are discussed separately. Section 4 presents numerical results which indicate our proposed simplification do not damage error-correction performance. Section 5 draws the conclusions.

2 Preliminaries

In this section, we briefly introduce the basic knowledge of multi-kernel polar codes and their corresponding decoding method. A N length polar code is presented by $\mathcal{P}(N, K)$, which carries K bits information. The process of polar coding can be denoted as a matrix multiplication as $\boldsymbol{x} = \boldsymbol{u}\boldsymbol{G}$. $\boldsymbol{u} = \{u_0, u_1, ..., u_{N-1}\}$ is the input sequence. $\boldsymbol{x} = \{x_0, x_1, ..., x_{N-1}\}$ is the polar coded sequence. As Arikan first introducing polar codes in [1], $\boldsymbol{G}$ is the n-th Kronecker product of the polarizing matrix $\boldsymbol{G} = \boldsymbol{T}_2^{\otimes n}$.

2.1 Multi-kernel Polar Codes

Multi-kernel polar codes are a generalization of Arikan's polar codes. $\boldsymbol{T}_3 = \left(\begin{smallmatrix}1&0&0\\1&1&0\\1&0&1\end{smallmatrix}\right)$ was introduced in [16] which is invertible and can be used for systematic encoding and decoding. MK polar codes provide more flexible length in conventional polar coding whose transform matrix defines $\boldsymbol{G} = \boldsymbol{T}_{k_0} \otimes ...\boldsymbol{T}_{k_i} ... \otimes \boldsymbol{T}_{k_{S-1}}$, where $\boldsymbol{T}_{k_i}$ is a $N_{k_i} \times N_{k_i}$ kernel matrix. The size of decoder v's output is $N_v = \prod_{i=0}^{S-1} N_{k_i}$, where S is the number of kernels. The design and arrangement of $\boldsymbol{T}_{k_i}$ influence MK polar codes performance. We collect set $\mathcal{T}_v = \{\boldsymbol{T}_{k_i} | 0 \leq i < S\}$ to represent the corresponding kernel matrices of decoder v.

2.2 SC and SSC Decoding

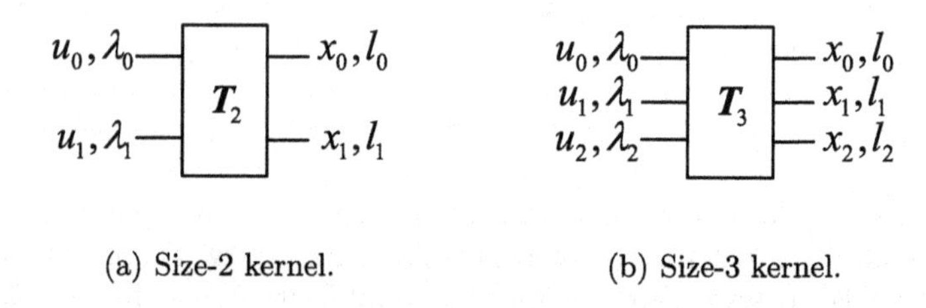

(a) Size-2 kernel. (b) Size-3 kernel.

Fig. 1. Block in Tanner graph, corresponding to $N_k \times N_k$ kernel $\boldsymbol{T}_k$.

SC decoding could also be used in MK polar codes on the Tanner graph of the code. Figure 1 describes the Tanner graph of corresponding MK polar codes. Denoting l_i the input Log-Likelihood Ratios (LLRs) of decoder and λ_i the output LLRs.

$$\begin{aligned} x_0 &= u_0 \oplus u_1, \\ x_1 &= u_1, \end{aligned} \tag{1}$$

describes coding relation defined in $\boldsymbol{T}_2$ corresponding Tanner graph. Based on the inverse of the update rule, the estimation of received code $\hat{u}_i$ is calculated by

$$\begin{aligned} \hat{u}_0 &= \mathrm{h}(\lambda_0) = \mathrm{h}(l_0 \boxplus l_1), \\ \hat{u}_1 &= \mathrm{h}(\lambda_1) = \mathrm{h}((-1)^{\hat{u}_0} l_0 + l_1). \end{aligned} \tag{2}$$

In Eq. (2), $\mathrm{h}(x)$ is hard decision function, which equals 0 when $x > 0$, and otherwise, 1. $\boxplus$ defines as

$$a \boxplus b = \ln(\frac{1 + e^{a+b}}{e^a + e^b}) \approx \min(a, b)(\mathrm{sgn}(a)\mathrm{sgn}(b)). \tag{3}$$

$\mathrm{sgn}(x)$ is sign function, which equals 1 when $x > 0$, and otherwise, -1. $\boldsymbol{T}_3$'s coding schemes are defined as Eq. (4),

$$\begin{aligned} x_0 &= u_0 \oplus u_1 \oplus u_2, \\ x_1 &= u_1, \\ x_2 &= u_2. \end{aligned} \tag{4}$$

SSC decoding was proposed to exclude redundant calculation of conventional SC decoding. The basic idea is to perform parallel decoder on SC decoding tree. Identifying special nodes from binary tree of polar codes, and deducing corresponding equivalent formulas, which saves the time-step of traversing from the special nodes to the leaf nodes. In general, Rate-1 and Rate-0 nodes are widely used in SSC decoding. Rate-1 node represent special node whose corresponding leaf nodes all carry information bits. On the contrary, Rate-0's leaf nodes are all frozen bits. According to SSC proposed in [9], a Rate-1 node v of binary tree decoder's candidate $\boldsymbol{\beta}_v$ can be calculated via

$$\boldsymbol{\beta}_v = \mathrm{h}(\boldsymbol{\alpha}_v), \tag{5}$$

$\boldsymbol{\alpha}_v$ is the input LLRs of local decoder v. As for Rate-0 node, its candidate code $\boldsymbol{\beta}_v = 0$. At the top of decoding tree, the corresponding codeword $\boldsymbol{x}$ is calculated via $\boldsymbol{\beta}_v \boldsymbol{G}$.

2.3 SCL and SSCL Decoding

SCL decoding significantly improves the error-correction performance of polar codes at short and medium code length. It considers and stores both possible values 0 and 1, at each step a information bit is estimated. By introducing path metric, it maintains L paths of possible candidates with the most possibility. $\mathrm{PM}_{i,l}$ is the i-th step of l-th path's path metric value, defined as

$$\mathrm{PM}_{i,l} = \mathrm{PM}_{i-1,l} + \ln(1 + e^{-(1-2\hat{u}_{i,l})\alpha_{i,l}}). \tag{6}$$

Every i-th step estimating bit produces $2L$ new candidates, half of which with larger PM values will be discarded. Equation (7) is a Hardware Friendly (HWF) version calculation of Eq. (6), reducing calculation complexity at the expense of decoding performance.

$$\mathrm{PM}_{i,l} = \begin{cases} \mathrm{PM}_{i-1,l}, & \text{if } \hat{u}_{i,l} = \frac{1}{2}(1 - \mathrm{sng}(\alpha_{i,l})), \\ \mathrm{PM}_{i-1,l} + |\alpha_{i,l}|, & \text{otherwise.} \end{cases} \tag{7}$$

SSCL performs simplified method on list decoding of polar codes. The key is to prove and induct PM calculation can still be deduced and performed on special nodes proposed in SSC decoding. [14] offered relevant proofs and defined Rate-1 nodes' $\mathrm{PM}_{v,l}$ is calculated via

$$\mathrm{PM}_{v,l} = \begin{cases} \sum_{i=0}^{N_v-1} \ln(1 + e^{-(1-2\beta_{v(i),l})\alpha_{v(i),l}}), & \text{Exact,} \\ \frac{1}{2}\sum_{i=0}^{N_v-1} \mathrm{sgn}(\alpha_{v(i),l})\alpha_{v(i),l} & \text{HWF.} \\ -(1-2\beta_{v(i),l})\alpha_{v(i),l}, & \end{cases} \tag{8}$$

And the PM value of Rate-0 node is calculated through

$$\mathrm{PM}_{v,l} = \sum_{i=0}^{N_v-1} \ln(1 + e^{-\alpha_{v(i),l}}). \tag{9}$$

3 Generalized Simplified Decoding Algorithms

In this paper, we only focus on Rate-1 and Rate-0 which have much simpler representations under MK polar coding scheme. We extend the simplification conclusion of size-2 based polar codes to MK. In order to reduce the calculation complexity of MK polar codes decoding, the generalized conditions for kernels who could be adopted to perform simplified SC and SCL are explored in our proposed algorithms. At the end of each algorithm, the time-step reduction effects are discussed separately.

3.1 Generalized SSC Decoding for MK Polar Codes

In this section, we first derive the equivalent expression of $\boldsymbol{T}_3$ kernel's Rate-1 decoder. Then we propose the sufficient conditions and proof for kernels of MK polar codes who are suitable for generalized SSC decoding. The corresponding time-step analysis is mentioned at the end.

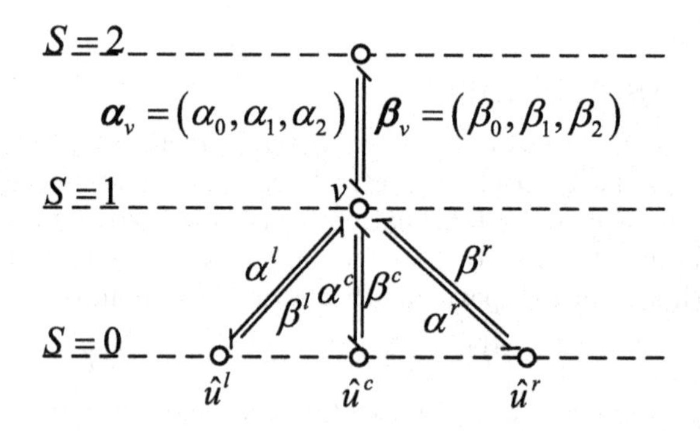

Fig. 2. SC decoding tree which $\boldsymbol{G} = \boldsymbol{T}_3$.

Rate-1 Nodes. Figure 2 depicts the decoding tree defined by $\boldsymbol{T}_3$ kernel. Stage $S = 0$ represents leaf nodes of decoding trees. Their corresponding input messages are represented as α^l, α^c and α^r, which are relative to their farther node's input message $\boldsymbol{\alpha}_v$. For $\boldsymbol{T}_3$ defined Tanner graph, according to Eq. (4), we have

$$\begin{aligned} \beta^l &= \beta_0 \oplus \beta_1 \oplus \beta_2, \\ \beta^c &= \beta_1 = \hat{u}^l \oplus \beta_0 \oplus \beta_2, \\ \beta^r &= \beta_2 = \hat{u}^l \oplus \hat{u}^c \oplus \beta_0. \end{aligned} \tag{10}$$

$\hat{u}^l$, $\hat{u}^c$ and $\hat{u}^r$ are the estimation bits. In Fig. 2, β^l, β^c and β^r represent the output candidate code of their corresponding nodes. For leaf nodes, $\beta^l = \hat{u}^l$ and etc. Therefore, the corresponding soft messages are calculated as

$$\begin{aligned} \alpha^l &= \alpha_0 \boxplus \alpha_1 \boxplus \alpha_2, \\ \alpha^c &= \alpha_1 + (-1)^{\hat{u}^l} \alpha_0 \boxplus \alpha_2, \\ \alpha^r &= \alpha_2 + (-1)^{\hat{u}^l \oplus \hat{u}^c} \alpha_0. \end{aligned} \tag{11}$$

Now, assuming that $\alpha_i \neq 0$ and performing hard decision function on both sides of the equal sign of Eq. (11), we have

$$\mathrm{h}(\alpha^l) = \mathrm{h}(\alpha_0) \oplus \mathrm{h}(\alpha_1) \oplus \mathrm{h}(\alpha_2),$$

$$\begin{aligned}\mathrm{h}(\alpha^c) &= \mathrm{h}\left(\alpha_1 + (1 - 2\mathrm{h}(\alpha^l))(\alpha_0 \boxplus \alpha_2)\right)\\ &= \mathrm{h}(\alpha_1 + (1 - 2(\mathrm{h}(\alpha_0) \oplus \mathrm{h}(\alpha_1) \oplus \mathrm{h}(\alpha_2)))(\alpha_0 \boxplus \alpha_2))\\ &= \mathrm{h}(\alpha_1 + (1 - 2(\mathrm{h}(\alpha_0 \boxplus \alpha_2) \oplus \mathrm{h}(\alpha_1)))(\alpha_0 \boxplus \alpha_2))\\ &= \mathrm{h}(\alpha_1),\end{aligned}$$

$$\begin{aligned}\mathrm{h}(\alpha^r) &= \mathrm{h}(\alpha_2 + (1 - 2\mathrm{h}(\alpha^l \boxplus \alpha^c))\alpha_0)\\ &= \mathrm{h}(\alpha_2 + (1 - 2(\mathrm{h}(\alpha^l) \oplus \mathrm{h}(\alpha^c)))\alpha_0)\\ &= \mathrm{h}(\alpha_2 + (1 - 2(\mathrm{h}(\alpha_0) \oplus \mathrm{h}(\alpha_2)))\alpha_0)\\ &= \mathrm{h}(\alpha_2).\end{aligned}$$

According to that and Eq. (4), we have

$$\begin{aligned}\beta_0 &= \beta^l \oplus \beta^c \oplus \beta^r = \mathrm{h}(\alpha^l) \oplus \mathrm{h}(\alpha^c) \oplus \mathrm{h}(\alpha^r)\\ &= \mathrm{h}(\alpha_0) \oplus \mathrm{h}(\alpha_1) \oplus \mathrm{h}(\alpha_2) \oplus \mathrm{h}(\alpha_1) \oplus \mathrm{h}(\alpha_2)\\ &= \mathrm{h}(\alpha_0),\\ \beta_1 &= \beta^c = \mathrm{h}(\alpha^c) = \mathrm{h}(\alpha_1),\\ \beta_2 &= \beta^r = \mathrm{h}(\alpha^r) = \mathrm{h}(\alpha_2).\end{aligned} \tag{12}$$

Equation (12) indicates $\boldsymbol{\beta}_{\boldsymbol{T}_3} = \mathrm{h}(\boldsymbol{\alpha}_{\boldsymbol{T}_3})$, where $\boldsymbol{\beta}_{\boldsymbol{T}_3}$ is $\boldsymbol{T}_3$ defined $N_v = 3$ length decoder's output candidate and $\boldsymbol{\alpha}_{\boldsymbol{T}_3}$ is the relative input message vector.

Here, we propose Theorem 1.

Theorem 1. *If $\boldsymbol{T}_k \in \mathcal{T}_v$ and $\boldsymbol{\beta}_{\boldsymbol{T}_k} = h(\boldsymbol{\alpha}_{\boldsymbol{T}_k})$, decoder v is a Rate-1 node. Then $\boldsymbol{\beta}_v = h(\boldsymbol{\alpha}_v)$.*

Proof (Proof of Theorem 1).

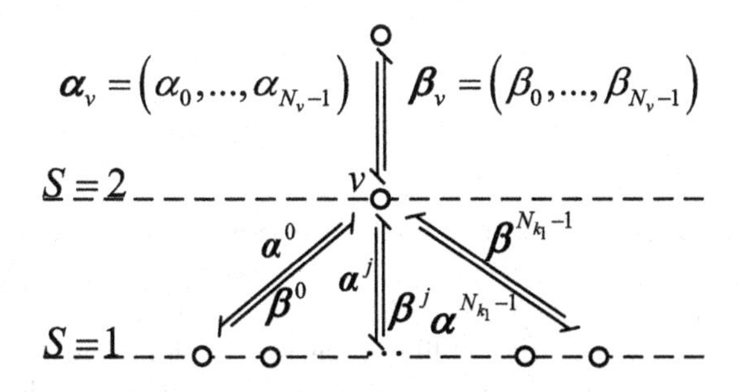

Fig. 3. SC decoding tree which $\boldsymbol{G} = \boldsymbol{T}_{k_0} \otimes \boldsymbol{T}_{k_1}$.

Figure 3 shows stage 2's corresponding soft massage and candidate code updates. $\boldsymbol{\alpha}^i$ and $\boldsymbol{\beta}^i$ are node v's i-th son's corresponding input and output vectors.

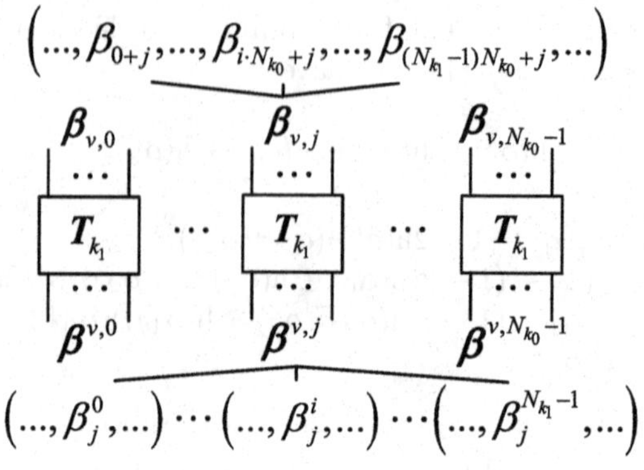

Fig. 4. SC decoding tree's stage 1 equivalent decoders.

By rearranging stage 2's input and output candidates, we divide node v into N_{k_0} encoders. And the j-th encoder's input code is denoted as $\boldsymbol{\beta}^{v,j}$, where $\boldsymbol{\beta}^{v,j} = \{\beta_j^0, ..., \beta_j^i, ...\beta_j^{N_{k_1}-1}\}$ and $0 \leq j < N_{k_0}$. And the corresponding output code is $\boldsymbol{\beta}_{v,j}$, where $\boldsymbol{\beta}_{v,j} = \{\beta_{0+j}, ..., \beta_{i \cdot N_{k_0}+j}, ..., \beta_{(N_{k_1}-1)N_{k_0}+j}\}$, $0 \leq j < N_{k_0}$, which is rearranged by $\boldsymbol{\beta}_v$.

According to polar coding scheme, we have $\boldsymbol{\beta}_{v,j} = \boldsymbol{\beta}^{v,j}\boldsymbol{T}_{k_1}$. And we suppose $\boldsymbol{T}_{k_0}$ satisfies $\boldsymbol{\beta}_{\boldsymbol{T}_{k_0}} = \mathrm{h}(\boldsymbol{\alpha}_{\boldsymbol{T}_{k_0}})$. That indicates that $\boldsymbol{\beta}^{v,j} = \mathrm{h}(\boldsymbol{\alpha}^{v,j})$, which is the same with $\hat{u} = \mathrm{h}(\alpha)$. Hence, we have $\boldsymbol{\beta}_{v,j} = \mathrm{h}(\boldsymbol{\alpha}_{v,j})$. It should be noted that the rearrangement of v's input and output does not change the relative order of corresponding vectors. Then, we have $\boldsymbol{\beta}_v = \mathrm{h}(\boldsymbol{\alpha}_v)$. For $S = 2$, the theorem is proven. From Fig. 3, it's easy to see that the proof is still hold when $S > 2$. Therefore, Theorem 1 is proven for MK polar codes' Rate-1 node with any stage.

Rate-0 Nodes. A Rate-0 node is decoder whose corresponding leaf nodes are frozen bits. Take $\boldsymbol{u}_v$ as the corresponding information vector of node v, where $\boldsymbol{u}_v = 0$. For local decoder v, we know that $\boldsymbol{\beta}_v = \boldsymbol{u}_v\boldsymbol{G}$. Hence, $\boldsymbol{\beta}_v = 0$, which is irrelevant to $\boldsymbol{G}$. In a word, the output candidate for a Rate-0 node of MK polar codes is always 0, no matter what stage and transition matrix it is.

Time-Step Analysis of Generalized SSC. Here, we analyze time-step spending of MK polar codes SSC decoding. The original SC decoding algorithm [1] takes $2N_v - 1$ time steps to decode N_v length code, which contains one step to calculate the corresponding likelihood ratios. As for local decoder v who is not root node, it costs $2N_v - 2$ time steps for $\boldsymbol{T}_2$ kernel SC decoding. Follow the same rule, MK polar code SC decoding takes $\sum_{j=0}^{S-1}\prod_{i=0}^{j} N_{\boldsymbol{T}_{k_i}}$ time steps. Due to the deletion of the complete tree traversal, SSC decoding for Rate-1 and Rate-0 nodes only need 1 time step.

3.2 Generalized SSCL for MK Polar Codes

To collect the conditions for kernels of MK polar codes which could perform generalized SSCL decoding, we study $\boldsymbol{T}_3$ as example. The exact and hardware friendly calculation formulas of PMs are derived. Then we provide the sufficient conditions and proofs for generalized simplified list decoding of MK polar codes. Time-step requirements are also discussed.

Rate-1 Nodes. The $\boldsymbol{T}_2$ defined polar codes' list decoding simplification has been proved in [14]. In this paper, we attempt to generalize this simplification to $\boldsymbol{T}_3$ and other kernels who satisfied certain conditions. For simplicity consideration, we use η to represent $1-2\beta$.

For Rate-1 polar code of length $N_v=3$ and $\boldsymbol{G}=\boldsymbol{T}_3$, based on Eq. (11), we have

$$\begin{aligned} \alpha^l &= \alpha_0 \boxplus \alpha_1 \boxplus \alpha_2, \\ \alpha^c &= \alpha_1 + \eta^l(\alpha_0 \boxplus \alpha_2), \\ \alpha^r &= \alpha_2 + \eta^l\eta^c\alpha_0. \end{aligned} \tag{13}$$

And analyze the relationship defined by Eq. (4), we have

$$\begin{aligned} \eta^l &= \eta_0\eta_1\eta_2, \\ \eta^c &= \eta_1, \\ \eta^r &= \eta_2. \end{aligned} \tag{14}$$

Take it to Eq. (13), we have

$$\begin{aligned} \alpha^l &= \alpha_0 \boxplus \alpha_1 \boxplus \alpha_2, \\ \alpha^c &= \alpha_1 + \eta_0\eta_1\eta_2(\alpha_0 \boxplus \alpha_2), \\ \alpha^r &= \alpha_2 + \eta_0\eta_2\alpha_0. \end{aligned} \tag{15}$$

Now, with Eq. (14) and (15), we build relationship of input soft messages and output candidate codes between a Rate-1 node and its descendants nodes. Now, we take Eq. (14) and (15) into (8), the path metric associated with $\boldsymbol{T}_3$ is

$$\begin{aligned} \mathrm{PM}_{\boldsymbol{T}_3} &= \ln\left(e^{-\eta_0\eta_1\eta_2(\alpha_0\boxplus\alpha_1\boxplus\alpha_2)}+1\right) \\ &\quad+\ln\left(e^{-\eta_1(\alpha_1+\eta_0\eta_1\eta_2(\alpha_0\boxplus\alpha_2))}+1\right) \\ &\quad+\ln\left(e^{-\eta_2(\alpha_2+\eta_0\eta_2\alpha_0)}+1\right) \\ &= \ln\left(\left(\frac{e^{\alpha_0}+e^{\alpha_1}+e^{\alpha_2}+e^{\alpha_0+\alpha_1+\alpha_2}}{1+e^{\alpha_0+\alpha_1}+e^{\alpha_0+\alpha_2}+e^{\alpha_1+\alpha_2}}\right)^{-\eta_0\eta_1\eta_2}+1\right) \\ &\quad+\ln\left(e^{-\eta_1\alpha_1}\left(\frac{1+e^{\alpha_0+\alpha_2}}{e^{\alpha_0}+e^{\alpha_2}}\right)^{-\eta_0\eta_2}+1\right) \\ &\quad+\ln\left(e^{-\eta_0\alpha_0-\eta_2\alpha_2}+1\right). \end{aligned} \tag{16}$$

For the sake of brevity, we introduce Δ to replace $\ln((e^{\alpha_0}+1)(e^{\alpha_1}+1)(e^{\alpha_2}+1))$, after expand and simplification Eq. (16) we have

$$\mathrm{PM}_{\boldsymbol{T}_3} = \begin{cases} \Delta - \alpha_0 - \alpha_1 - \alpha_2, & \text{when } \eta_0, \eta_1, \eta_2 = 1, 1, 1, \\ \Delta - \alpha_0 - \alpha_1, & \text{when } \eta_0, \eta_1, \eta_2 = 1, 1, -1, \\ \Delta - \alpha_0 - \alpha_2, & \text{when } \eta_0, \eta_1, \eta_2 = 1, -1, 1, \\ \Delta - \alpha_0, & \text{when } \eta_0, \eta_1, \eta_2 = 1, -1, -1, \\ \Delta - \alpha_1 - \alpha_2, & \text{when } \eta_0, \eta_1, \eta_2 = -1, 1, 1, \\ \Delta - \alpha_1, & \text{when } \eta_0, \eta_1, \eta_2 = -1, 1, -1, \\ \Delta - \alpha_2, & \text{when } \eta_0, \eta_1, \eta_2 = -1, -1, 1, \\ \Delta, & \text{when } \eta_0, \eta_1, \eta_2 = -1, -1, -1, \end{cases}$$
$$= \ln(1+e^{-\eta_0\alpha_0}) + \ln(1+e^{-\eta_1\alpha_1}) + \ln(1+e^{-\eta_2\alpha_2})$$
$$= \sum_{i=0}^{N_{T_3}-1} \ln(1+e^{-\eta_i\alpha_i}). \tag{17}$$

Furthermore, we will introduce the HWF version path metric calculation's simplification of MK polar codes list decoding. For simplicity, we introduce $s_i = \mathrm{sgn}(\alpha_i)$. And based on the character of sign function, we have $\mathrm{sgn}(x)x = |x|$. Take the approximation of Eq. (13) and (14), we have

$$\begin{aligned} \alpha^l &= s_0 s_1 s_2 \min(|\alpha_0|, |\alpha_1|, |\alpha_2|), \\ \alpha^c &= \alpha_1 + \eta^l s_0 s_1 \min(|\alpha_0|, |\alpha_2|) \\ &= \alpha_1 + \eta_0\eta_2 s_0 s_2 \min(|\alpha_0|, |\alpha_2|), \\ \alpha^r &= \alpha_2 + \eta^l \eta^c \alpha_0 = \alpha_2 + \eta_0\eta_2\alpha_0. \end{aligned} \tag{18}$$

According to Eq. (8)'s HWF version, we have $\boldsymbol{T}_3$ kernel's corresponding path metric formulation Eq. (19).

$$\begin{aligned} \mathrm{PM}_{\boldsymbol{T}_3} = \tfrac{1}{2} (& \mathrm{sgn}(\alpha^l)\alpha^l - \eta^l\alpha^l + \mathrm{sgn}(\alpha^c)\alpha^c - \eta^c\alpha^c \\ & + \mathrm{sgn}(\alpha^r)\alpha^r - \eta^r\alpha^r). \end{aligned} \tag{19}$$

From Eq. (19), substituting Eq. (18), the path metric associated with $\boldsymbol{T}_3$ would be computed as

$$\begin{aligned} 2\mathrm{PM}_{\boldsymbol{T}_3} = & \underbrace{(1 - \eta_0\eta_1\eta_2 s_0 s_1 s_2)\min(|\alpha_0|, |\alpha_1|, |\alpha_2|)}_{p_0} \\ & + \underbrace{|\alpha_1 + \eta_0\eta_1\eta_2 s_0 s_2 \min(|\alpha_0|, |\alpha_2|)|}_{p_1} - \eta_1\alpha_1 \\ & \underbrace{-\eta_0\eta_2 s_0 s_2 \min(|\alpha_0|, |\alpha_2|) + |\alpha_2 + \eta_0\eta_2\alpha_0|}_{p_2} \\ & - \eta_2\alpha_2 - \eta_0\alpha_0. \end{aligned}$$

Then, we analyze this equation with several situations:

First, we calculate $p_0 + p_1$. When $\eta_0\eta_1\eta_2 s_0 s_1 s_2 = 1$, we have

$$p_0 + p_1 = |\alpha_1 + s_1 \min(|\alpha_0|, |\alpha_2|)| = |\alpha_1| + \min(|\alpha_0|, |\alpha_2|).$$

When $\eta_0\eta_1\eta_2 s_0 s_1 s_2 = -1$, we have

$$p_0 + p_1 = 2\min(|\alpha_0|, |\alpha_1|, |\alpha_2|) + |\alpha_1 - s_1 \min(|\alpha_0|, |\alpha_2|)|\,.$$

In this situation, when $|\alpha_1| \le \min(|\alpha_0|, |\alpha_2|)$,

$$\begin{aligned} p_0 + p_1 &= 2|\alpha_1| + |\min(|\alpha_0|, |\alpha_2|) - |\alpha_1|| \\ &= |\alpha_1| + \min(|\alpha_0|, |\alpha_2|). \end{aligned}$$

Otherwise, we have

$$\begin{aligned} p_0 + p_1 &= 2\min(|\alpha_0|, |\alpha_2|) + ||\alpha_1| - \min(|\alpha_0|, |\alpha_2|)| \\ &= |\alpha_1| + \min(|\alpha_0|, |\alpha_2|). \end{aligned}$$

Therefore, we have

$$p_0 + p_1 = |\alpha_1| + \min(|\alpha_0|, |\alpha_2|).$$

Then we focus on the calculation of $p_0 + p_1 + p_2$. When $\eta_0\eta_2 s_0 s_2 = 1$,

$$\begin{aligned} p_2 &= -\min(|\alpha_0|, |\alpha_2|) + |s_2|\alpha_2| + \eta_0\eta_2 s_0|\alpha_0|| \\ &= -\min(|\alpha_0|, |\alpha_2|) + |s_2|\alpha_2| + s_2|\alpha_0|| \\ &= -\min(|\alpha_0|, |\alpha_2|) + |\alpha_2| + |\alpha_0|. \end{aligned}$$

Hence, we have

$$p_0 + p_1 + p_2 = |\alpha_0| + |\alpha_1| + |\alpha_2|.$$

When $\eta_0\eta_2 s_0 s_2 = -1$,

$$\begin{aligned} p_2 &= \min(|\alpha_0|, |\alpha_2|) + |s_2|\alpha_2| - s_2|\alpha_0|| \\ &= \min(|\alpha_0|, |\alpha_2|) + ||\alpha_2| - |\alpha_0||. \end{aligned}$$

If $|\alpha_2| \ge |\alpha_0|$,

$$p_0 + p_1 + p_2 = 2|\alpha_0| + |\alpha_2| - |\alpha_0| + |\alpha_1|.$$

Else, we have

$$p_0 + p_1 + p_2 = 2|\alpha_2| - |\alpha_2| + |\alpha_0| + |\alpha_1|.$$

Finally, we have

$$\begin{aligned} \mathrm{PM}_{T_3} &= \tfrac{1}{2}(|\alpha_0| + |\alpha_1| + |\alpha_2| - \eta_0\alpha_0 - \eta_1\alpha_1 - \eta_2\alpha_2) \\ &= \textstyle\sum_{i=0}^{N_{T_3}-1}(\mathrm{sgn}(\alpha_i)\alpha_i - \eta_i\alpha_i). \end{aligned} \tag{20}$$

Before we introduce Theorem 2, we define function $pm(\eta, \alpha)$, which is

$$\mathrm{pm}(\eta, \alpha) = \begin{cases} \ln(1 + e^{-\eta\alpha}), & \text{Exact}, \\ \frac{1}{2}\left(\mathrm{sgn}(\alpha)\alpha - \eta\alpha\right), & \text{HWF}. \end{cases} \tag{21}$$

Theorem 2. *If* $\forall \boldsymbol{T}_k \in \mathcal{T}_v$ *all satisfy* $PM_{\boldsymbol{T}_k} = \sum_{i=0}^{N_k-1} pm(\eta_i, \alpha_i)$, v *is a Rate-1 node. Then* $PM_v = \sum_{i=0}^{N_v-1} pm(\eta_i, \alpha_i)$.

Proof (Proof of Theorem 2). Figure 5 shows stage 2's path metric updating procedure.

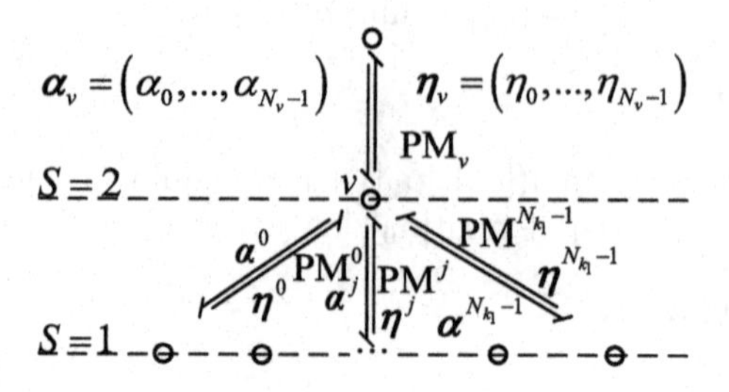

Fig. 5. Path metrics of SC decoding tree

Node v's path metric is represented as PM_v. The corresponding candidate is replaced by $\boldsymbol{\eta}_v$ who satisfies $\boldsymbol{\eta}_v = 1 - 2\boldsymbol{\beta}_v$. PM^i is node v's i-th son's corresponding path metric and vector $\boldsymbol{\eta}^i$ is the corresponding output candidate. The path metric at a node v is the summation of the path metrics calculated at its sons. Hence, we have

$$\begin{aligned}\mathrm{PM}_v &= \sum_{i=0}^{N_{k_1}-1} \mathrm{PM}^i \\ &= \sum_{i=0}^{N_{k_1}-1} \sum_{j=0}^{N_{k_0}-1} \mathrm{pm}(\eta_j^i, \alpha_j^i) \\ &= \sum_{j=0}^{N_{k_0}-1} \sum_{i=0}^{N_{k_1}-1} \mathrm{pm}(\eta_j^i, \alpha_j^i).\end{aligned} \tag{22}$$

According to Fig. 4, we know that $\boldsymbol{\beta}_{v,j} = \boldsymbol{\beta}^{v,j}\boldsymbol{T}_{k_1}$. Hence, we have $\mathrm{PM}_{v,j} = \sum_{i=0}^{N_{k_1}-1} \mathrm{pm}(\eta_j^i, \alpha_j^i)$, where $\mathrm{PM}_{v,j}$ is the j-th PM value of the new rearranged decoder. $\boldsymbol{T}_{k_1}$ defines the new decoder and satisfies above conditions which means $\mathrm{PM}_{v,j} = \sum_{i=0}^{N_{k_1}-1} \mathrm{pm}(\eta_{i\cdot N_{k_0}+j}, \alpha_{i\cdot N_{k_0}+j})$. Then we have

$$\begin{aligned}\mathrm{PM}_v &= \sum_{j=0}^{N_{k_0}-1} \sum_{i=0}^{N_{k_1}-1} \mathrm{pm}(\eta_{i\cdot N_{k_0}+j}, \alpha_{i\cdot N_{k_0}+j}) \\ &= \sum_{i=0}^{N_{k_0}N_{k_1}-1} \mathrm{pm}(\eta_i, \alpha_i).\end{aligned} \tag{23}$$

For $S = 2$, the theorem is proven. From Fig. 5, it's easy to see that the proof is still hold when $S > 2$. Hence, Theorem 2 is valid for any stage of tree decoder.

Rate-0 Nodes

Theorem 3. *The PM value of a Rate-0 node can be calculated as:*

$$PM_v = \begin{cases} \sum_{i=0}^{N_v-1} \ln(1+e^{-\alpha_i}), \; \textit{Exact}, \\ \sum_{i=0}^{N_v-1} \frac{1}{2}\left(|\alpha_i| - \alpha_i\right), \; \textit{HWF}, \end{cases} \tag{24}$$

where α_i is the input soft message of the Rate-0 node tree.

Proof (Proof of Theorem 3). To proof this theorem, it worth to note that Rate-0 node is only a special case of Rate-1 node. That means decoder already know the output candidate of Rate-0 node tree without traversal the whole tree from top to the leaves. Then we revisit the calculation of Rate-1 decoder fast list decoding path metric Eq. (21). If we take Rate-0 as a normal Rate-1 node. Due to the prior knowledge, η_v equals to 1 at this situation. Hence, we have $\text{PM}_v = \sum_{i=0}^{N_v-1} \ln(1+e^{-\alpha_i})$ in the exact calculation of path metric. And for the approximation version, $\text{PM}_v = \sum_{i=0}^{N_v-1} \frac{1}{2}\left(|\alpha_i| - \alpha_i\right)$.

Time-Step Analysis of Generalized SSCL. The conventional SCL needs $3N_v - 2$ to decode a N_v length code. According to [14] and [15], a Rate-1 node needs $\min(L-1, N_v)$ time steps and a Rate-0 node takes one step for $\boldsymbol{T}_2$ kernel based polar codes. When performing conventional SCL decoding algorithm on MK polar codes, a N_v length decoder will cost $\sum_{j=0}^{S-1}\prod_{i=0}^{j} N_{\boldsymbol{T}_{k_i}} + N_v$ time steps. In this paper, the generalize SSCL-MK reduces the necessary time steps to $\min(L-1, N_v)$ and 1.

4 Numerical Results

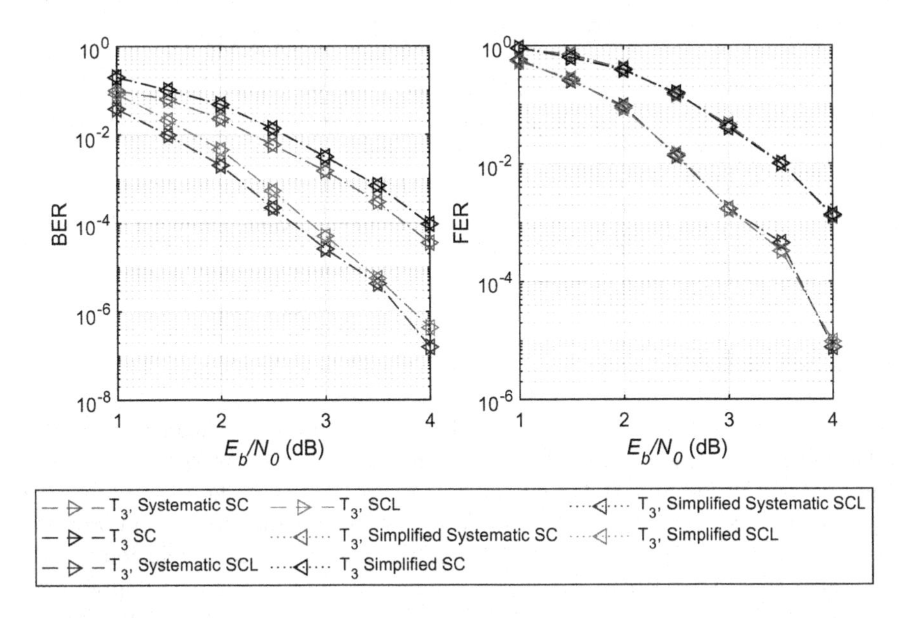

Fig. 6. Bit Error Rate (BER) and Frame Error Rate (FER) curves for $\boldsymbol{T}_3$ kernel MK polar codes with coding rate 0.5.

In this section, we demonstrate our proposed generalized simplified SC and SCL MK decoding algorithms with the conventional decoding algorithms. We assume binary phase shift key transmission over the additive white Gaussian noise channel. MK polar codes in Fig. 6 and 7 are build by the method proposed in [18]. $\boldsymbol{T}_3$ kernel MK polar codes in Fig. 6 are 729 length. $\boldsymbol{T}_2$ and $\boldsymbol{T}_3$ kernels MK polar codes are 648 length. Systematic and unsystematic coding methods are used in $\boldsymbol{T}_3$ kernel MK polar codes. SCL method is performed with list size 8. The results in Fig. 6 and 7 show our proposed generalized simplified decoding holds the error-correction performance of conventional SC and SCL methods.

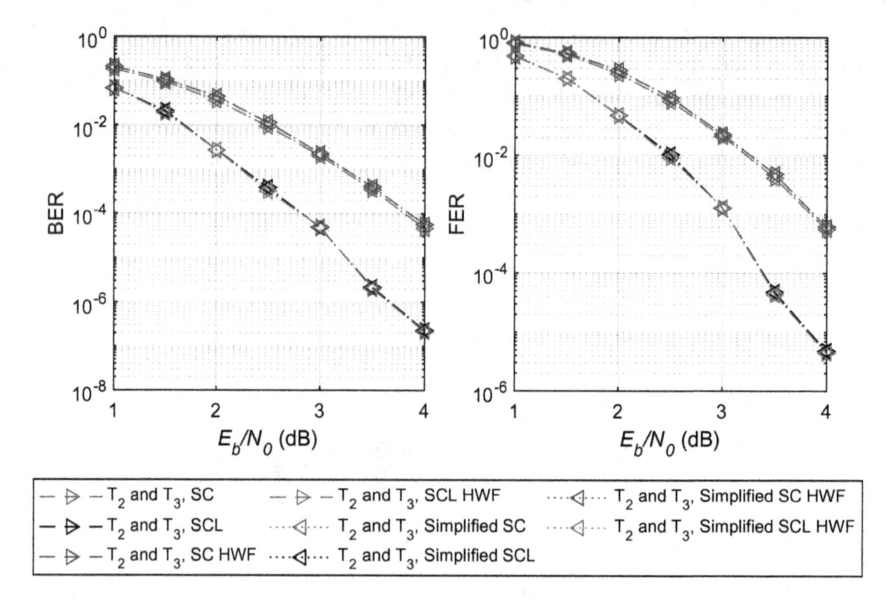

Fig. 7. Bit Error Rate (BER) and Frame Error Rate (FER) curves for $\boldsymbol{T}_2$ and $\boldsymbol{T}_3$ kernels MK polar codes with coding rate 0.5.

5 Conclusion

In this paper, we have explored generalized SSC and SSCL algorithm for MK polar codes. We provide the sufficient conditions for kernels who are suitable for simplification of SC and SCL calculations. We proved that these simplifications do not incur error-correction performance loss which means that Arikan proposed kernel $\boldsymbol{T}_2$ is a special case of multi-kernel polar codes. This conclusion is valid for any kernel-based polar codes decoding with both exact and approximate calculations. The proposed method saves time step requirements of Rate-1 and Rate-0 nodes from $\sum_{j=0}^{S-1}\prod_{i=0}^{j}N_{\boldsymbol{T}_{k_i}}+N_v$ to $\min(L-1,N_v)$ and 1, which would remarkably reduce the time complexity of conventional SCL.

References

1. Arikan, E.: Channel polarization: a method for constructing capacity-achieving codes for symmetric binary-input memoryless channels. IEEE Trans. Inf. Theory **55**(7), 3051–3073 (2009)
2. Bioglio, V., Gabry, F., Land, I.: Low-complexity puncturing and shortening of polar codes. In: IEEE Wireless Communications and Networking Conference Workshops (WCNCW). IEEE 2017, pp. 1–6 (2017)
3. El-Khamy, M., Lin, H.-P., Lee, J., Kang, I.: Circular buffer rate-matched polar codes. IEEE Trans. Commun. **66**(2), 493–506 (2017)
4. Korada, S.B., Şaşoğlu, E., Urbanke, R.: Polar codes: characterization of exponent, bounds, and constructions. IEEE Trans. Inf. Theory **56**(12), 6253–6264 (2010)
5. Gabry, F., Bioglio, V., Land, I., Belfiore, J.-C.: Multi-kernel construction of polar codes. In: 2017 IEEE International Conference on Communications Workshops (ICC Workshops), pp. 761–765. IEEE (2017)
6. Bioglio, V., Gabry, F., Land, I., Belfiore, J.-C.: Minimum-distance based construction of multi-kernel polar codes. In: GLOBECOM 2017–2017 IEEE Global Communications Conference, pp. 1–6. IEEE (2017)
7. Bioglio, V., Gabry, F., Land, I., Belfiore, J.-C.: Multi-kernel polar codes: concept and design principles. IEEE Trans. Commun. **68**(9), 5350–5362 (2020)
8. Xia, C., Tsui, C.-Y., Fan, Y.: Construction of multi-kernel polar codes with kernel substitution. IEEE Wirel. Commun. Lett. **9**(11), 1879–1883 (2020)
9. Alamdar-Yazdi, A., Kschischang, F.R.: A simplified successive-cancellation decoder for polar codes. IEEE Commun. Lett. **15**(12), 1378–1380 (2011)
10. Sarkis, G., Giard, P., Vardy, A., Thibeault, C., Gross, W.J.: Fast polar decoders: algorithm and implementation. IEEE J. Sel. Areas Commun. **32**(5), 946–957 (2014)
11. Hanif, M., Ardakani, M.: Fast successive-cancellation decoding of polar codes: identification and decoding of new nodes. IEEE Commun. Lett. **21**(11), 2360–2363 (2017)
12. Condo, C., Bioglio, V., Land, I.: Generalized fast decoding of polar codes. In: IEEE Global Communications Conference (GLOBECOM) 2018, pp. 1–6 (2018)
13. Tal, I., Vardy, A.: List decoding of polar codes. IEEE Trans. Inf. Theory **61**(5), 2213–2226 (2015)
14. Hashemi, S.A., Condo, C., Gross, W.J.: Simplified successive-cancellation list decoding of polar codes. In: IEEE International Symposium on Information Theory (ISIT), pp. 815–819. IEEE (2016)
15. Hashemi, S.A., Condo, C., Gross, W.J.: Fast simplified successive-cancellation list decoding of polar codes. In: IEEE Wireless Communications and Networking Conference Workshops (WCNCW), IEEE, pp. 1–6 (2017)
16. Bioglio, V., Land, I.: On the marginalization of polarizing kernels. In: IEEE 10th International Symposium on Turbo Codes & Iterative Information Processing (ISTC), pp. 1–5 (2018)
17. Cavatassi, A., Tonnellier, T., Gross, W.J.: Fast decoding of multi-kernel polar codes. In: IEEE Wireless Communications and Networking Conference (WCNC), pp. 1–6. IEEE (2019)
18. Trifonov, P.: Efficient design and decoding of polar codes. IEEE Trans. Commun. **60**(11), 3221–3227 (2012)

Multiple Dimensional Scaling Hybrid Precoding in Millimeter Wave MIMO System

Jiaqi Su[1], Wenbin Zhang[1,2](✉), Liyan Zhang[1], Jinwei Huang[1], and Shaochuan Wu[1]

[1] Harbin Institute of Technology, Harbin 150001, People's Republic of China
zwbgxy1973@hit.edu.cn

[2] Science and Technology on Communication Networks Laboratory, Shijiazhuang 050050, Hebei, People's Republic of China

Abstract. In most wireless systems, the channel transmission loss of millimeter wave signals is often greater than that of traditional signals. Hybrid precoding can leverage large antenna arrays to compensate this loss and further improve the spectral efficiency of millimeter wave. This paper explores the sparsity of mmWave channel—only a few main paths are useful for the precoding procedure. This sparsity makes it possible to use a limited feedback in channel estimation. We propose a novel hybrid precoding scheme in this paper, which utilizes the main vector of the channel. We call this scheme Multiple Dimensional Scaling (MDS) hybrid precoding. On the one hand, comparing with many traditional hybrid precoding schemes with full channel information feedback, our scheme can decrease feedback overheads significantly. On the other hand, comparing with some hybrid precoding with limited channel feedback information, our scheme can improve spectral efficiency. Moreover, the simulation results show that the system spectral efficiency increases with the number of data streams significantly.

Keywords: Millimeter wave communications · Hybrid precoding · Limited feedback · Multiple Dimensional Scaling

1 Introduction

The principle of millimeter wave (mmWave) communication to achieve high data transmission rate is to utilize the potential available large bandwidth in the high frequency band [1–4]. However, because of the high frequency of mmWave and the small wavelength, the path loss is much higher than that of microwave band. The solution to compensate for the loss caused by the increase of frequency, is to

Supported by National Nature Science Foundation of China (NSFC) under Grant 62071148.

Q. Guo et al. (Eds.): WiSATS 2021, LNICST 410, pp. 392–404, 2022.
https://doi.org/10.1007/978-3-030-93398-2_38

use precoding technology and multi-antenna technology. Future cellular network base stations will deploy hundreds of antennas to form mmWave massive MIMO. However, the design of precoding matrix and combining matrix in mmWave MIMO system is not as simple as that in the low frequency scheme. It is mainly due to different hardware constraints, higher cost and higher power consumption. Therefore, adjusting the number of antennas and radio frequency (RF) chains, reducing hardware complexity and improving energy efficiency become the main problems in Massive MIMO technology [5–7]. Thus, research on the precoding scheme in mmWave MIMO systems has become a very important field.

In this context, analog-digital hybrid precoding is proposed, which achieves the purpose of reducing the number of RF chains by dividing the precoding process into analog domain and digital domain [2,3]. Digital precoding can achieve higher accuracy and analog precoding can reduce the originally high power consumption, which makes hybrid precoding have better performance. It can achieve a compromise and balance between hardware complexity and system performance. In analog-digital hybrid precoding, the transmitter needs to know the channel information. In FDD (frequency division duplex) mode, the mobile station estimates the downlink channel by receiving the pilot transmitted by the base station, and then transmits the channel information to the base station. In [2,3], the hybrid architecture under lower frequency is studied. In mmWave systems, the concept of hybrid precoding is similar to that of the lower frequency. In [8], considering the sparsity of mmWave channel, the precoder and combiner of mmWave system with large antenna array are combined to design the system. This precoding problem is transformed into a sparse reconstruction problem by utilizing the spatial sparsity of mmWave channels. In this paper, based on the principle of base tracking, a system is designed that can approximately achieve the optimal result under all-digital precoding. [9] proposes a low complexity reconstruction algorithm based on [8], which avoids the process of matrix inversion by querying the orthogonal codebook, thereby reducing the computational complexity. Similar to [8], in [10], the author proposes a greedy algorithm to design hybrid analog/digital precoders, which has lower complexity and is approximately orthogonal. The principle of the algorithm is to greedily select the RF beamforming vectors through the Gram-Schmidt orthogonalization process. Besides, [14] proposes a situation where only partial channel information is known. Some other heuristic algorithms that do not include the orthogonal matching pursuit process can also be seen in [12,13] to design hybrid precoding, which need to meet the condition of having perfect channel information at the transmitter. In [14] the combined channel matrix is fed back to the base station. The base station uses a simplified effective channel matrix with little training and feedback overhead to design the hybrid precoding. The performance analysis of the above mentioned algorithms is based on large dimensional regime and single-path channels.

This paper proposes an algorithm that can use limited channel feedback to achieve hybrid precoding, which compresses the channel information at the receiver and then feedback the compressed information to reduce the feedback overhead and power consumption. Then, the base station uses the compressed feedback information to design precoding. The simulation results show that although we have reduced the computational complexity, the performance of the algorithm in some cases has not been affected.

The following text structure of the paper is as follows. Section 2 briefly introduces the system model of mmWave MIMO hybrid precoding. Section 3 analyzes the hybrid precoding problem. Section 4 explains the proposed Multiple Dimensional Scaling hybrid precoding. Section 5 shows the results of simulation at the achievable rate. At last, the conclusion of the article is described in Sect. 6.

Notation: boldface uppercase, boldface lowercase, and lowercase letter $\mathbf{A}$, $\mathbf{a}$, a denote a matrix, vector, and scalar variable respectively. $\mathbf{A}^T$, $\mathbf{A}^*$ and $\mathbf{A}^{-1}$ represent the transpose, the conjugate transpose and the inverse of a matrix, respectively. $|\ |$ denotes absolute value, *tr()* denotes trace of a matrix, *diag* denotes a diagonal matrix.

2 System Model

2.1 System Model

First, Fig. 1 shows the mm Wave system model, where a base station (BS) with N_t antennas and N_{tRF} RF chains will establish communication relationship with a single mobile station (MS) with N_r antennas and N_{rRF} RF chains. The communication process between BS and MS utilizes N_s data streams, and the number N_s meets the condition that $N_s \leq N_{tRF} \leq N_t$ and $N_s \leq N_{rRF} \leq N_r$.

At the transmitter, we first use a $N_{tRF} \times N_s$ digital precoding matrix $\mathbf{F}_{BB}$ to precode the $N_s \times 1$ data symbols $\mathbf{s}$, then the symbols will be applied to $N_t \times N_{tRF}$ RF precoding $\mathbf{F}_{RF}$, such that $E[\mathbf{s}\mathbf{s}^*] = \frac{1}{N_s}\mathbf{I}_{N_s}$. Therefore, the transmitted discrete-time signal can be expressed by $\mathbf{x} = \mathbf{F}_{RF}\mathbf{F}_{BB}\mathbf{s}$. Since we use analog phase shifters to realize $\mathbf{F}_{RF}$, it is required to meet the condition that $(\mathbf{F}_{RF}{}^{(i)}\mathbf{F}_{RF}{}^{(i)*})_{l,l} = \frac{1}{N_t}$, where $()_{l,l}$ represents the l-th diagonal element of a matrix. Due to the total power limitations of the transmitters, we must normalize $\mathbf{F}_{BB}$ to satisfy $\| \mathbf{F}_{BB}\mathbf{F}_{RF} \|_F^2 = N_s$.

We discuss a narrowband block-fading propagation channel as in [4–7] for simplicity, which generates a received signal

$$\mathbf{y} = \sqrt{\rho}\mathbf{H}\mathbf{F}_{RF}\mathbf{F}_{BB}\mathbf{s} + \mathbf{n} \tag{1}$$

where $\mathbf{y}$ denotes the $N_r \times 1$ received vector, $\mathbf{H}$ represents the $N_r \times N_t$ channel matrix which meets the condition that $E[\| \mathbf{H} \|_F^2] = N_t N_r$, ρ is the average received power, and $\mathbf{n}$ indicates the noise vector of i.i.d $\mathbb{CN}(0, \sigma_n^2)$. The received signal at the MS after combination can be represented as the following formula

$$\tilde{\mathbf{y}} = \sqrt{\rho}\mathbf{W}_{BB}^*\mathbf{W}_{RF}^*\mathbf{H}\mathbf{F}_{RF}\mathbf{F}_{BB}\mathbf{s} + \mathbf{W}_{BB}^*\mathbf{W}_{RF}^*\mathbf{n} \tag{2}$$

where $\mathbf{W}_{RF}$ denotes the $N_r \times N_{rRF}$ RF combining matrix, and $\mathbf{W}_{BB}$ represents the $N_{rRF} \times N_s$ baseband combining matrix. Like the transmitter, the receiver need to satisfy the power constrain such that $(\mathbf{W}_{RF}{}^{(i)}\mathbf{W}_{RF}{}^{(i)*})_{l,l} = \frac{1}{N_r}$. When the signal transmitted on the mmWave channel satisfies the Gaussian distribution, the system spectral efficiency can be expressed by

$$R = log_2(|\mathbf{I}_{N_s} + \frac{\rho}{N_s} \frac{\mathbf{R}_s}{\sigma_n^2 \mathbf{W}_{BB}^* \mathbf{W}_{RF}^* \mathbf{W}_{RF} \mathbf{W}_{BB}}|) \tag{3}$$

where $\mathbf{R}_s = \mathbf{W}_{BB}^* \mathbf{W}_{RF}^* \mathbf{H} \mathbf{F}_{RF} \mathbf{F}_{BB} \times \mathbf{F}_{BB}^* \mathbf{F}_{RF}^* \mathbf{H}^* \mathbf{W}_{RF} \mathbf{W}_{BB}$.

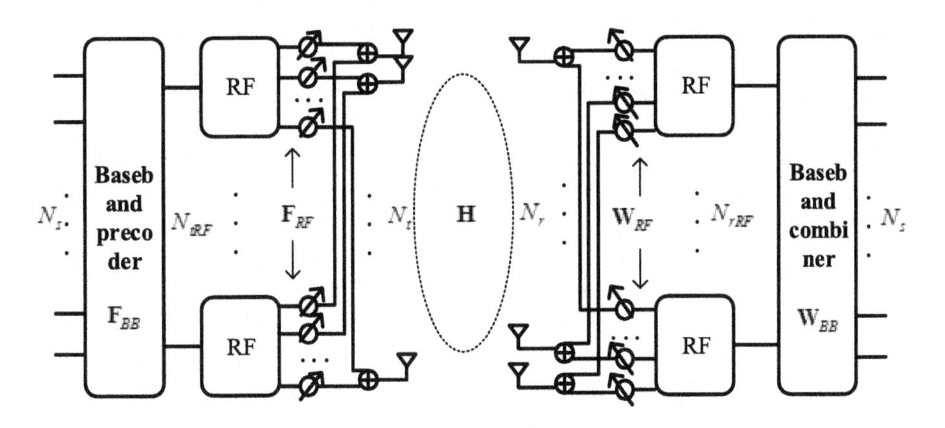

Fig. 1. A mmWave downlink system model with digital baseband precoding followed by a module using RF phase shifters to decrease radio frequency chains.

2.2 Channel Model

Millimeter wave transmission loses some degrees of freedom, which will cause limited spatial selectivity and limited scattering. We model a 2D channel through a widely used geometric mmWave channel model, which is a narrowband clustered channel expression in the light of the extended Saleh-Valenzuela model. The advantage of this model is that it enables us to accurately know the mathematical structure of the mmWave channel [8].

According to the narrowband clustered channel model, the channel matrix $\mathbf{H}$ is the set of all scattered transmission paths of N_{cl} scattering clusters, and each N_{ray} propagation paths of each scattering cluster is the composition of the channel matrix. Thus, the narrowband discrete-time channel matrix $\mathbf{H}$ can be expressed as the following formula

$$\mathbf{H} = \gamma \sum_{i,l} \alpha_{il} \mathbf{a}_r(\phi_{il}^r, \theta_{il}^r) \mathbf{a}_t^*(\phi_{il}^t, \theta_{il}^t) \tag{4}$$

where γ represents a normalization factor and equals to $\sqrt{\frac{N_r N_t}{N_{cl} N_{ray}}}$, α_{il} is the complex gain of i.i.d $\mathbb{CN}(0, \sigma_{\alpha,i}^2)$, which represents the gain of the lth ray in the ith scattering cluster, $\sigma_{\alpha,i}^2$ is the average power of the ith cluster. The average cluster powers satisfy the equation that $\sum_{i=1}^{N_{cl}} \sigma_{\alpha,i}^2 = \gamma$. For the lth ray in the ith scattering cluster, $\phi_{il}^r, \theta_{il}^r, \phi_{il}^t, \theta_{il}^t$ stand for its azimuth/elevation angles of arrival and departure respectively. The vector $\mathbf{a}_r(\phi_{il}^r, \theta_{il}^r)$ and $\mathbf{a}_t(\phi_{il}^t, \theta_{il}^t)$ denote the normalized receive and transmit array response vectors at an azimuth/elevation angle of $\phi_{il}^r, \theta_{il}^r, \phi_{il}^t, \theta_{il}^t$ respectively. We make an assumption that the azimuth and elevation angles of departure $\phi_{il}^t, \theta_{il}^t$, are distributed with a uniformly-random mean cluster angle of ϕ_i^t, θ_i^t and angular spread of σ_{ϕ^t}, σ_{θ^t} in the cluster i respectively. The azimuth and elevation angles of arrival $\phi_{il}^r, \theta_{il}^r$ are again randomly distributed with mean cluster angles of ϕ_i^r, θ_i^r and angular spreads σ_{ϕ^r}, σ_{θ^r}.

$\mathbf{a}_r(\phi_{il}^r, \theta_{il}^r)$ and $\mathbf{a}_t(\phi_{il}^t, \theta_{il}^t)$ are the receiving and transmitting antenna array response vectors of MS and BS respectively, and have nothing to do with the antenna element properties. The following two illustrative examples of generally-accepted antenna arrays can both be applied to the algorithms and simulation results in this paper. We give the conclusion that the array response vector for an N-element uniform linear array (ULA) on the y-axis can be expressed as the following formula

$$\mathbf{a}_{ULA}(\phi) = \frac{1}{N}[1, e^{jkdsin(\phi)}, ..., e^{j(N-1)kdsin(\phi)}]^T \tag{5}$$

where $k = \frac{2\pi}{\lambda}$, λ denotes the signal wavelength, and d represents the inter-element spacing. It should be pointed out that we do not consider θ in the discuss of a ULA, because the array's response remains constant in the elevation domain. Then, consider another antenna array distribution situation, which is a uniform planar array (UPA) that has W and H elements on the y and z axes respectively in the yz-plane, the array response vector is expressed by

$$\begin{aligned}\mathbf{a}_{UPA}(\phi, \theta) = \frac{1}{N}[1, ..., e^{jkd(msin(\phi)sin(\theta)+ncos(\theta))} \\ ..., e^{jkd((W-1)sin(\phi)sin(\theta)+(H-1)cos(\theta))}]^T\end{aligned} \tag{6}$$

where the parameters meet the condition that $0 \le m \le W$, $0 \le n \le H$, and $N = WH$. Through the above analysis, we can know that uniform planar arrays (UPA) have the following advantages, the UPA has smaller antenna array dimensions, has the ability to perform the beamforming process in the elevation domain, and packs more components in a reasonably sized antenna array. Therefore, it is of great importance to study uniform planar arrays in mmWave beamforming.

3 Problem Formulation

In order to improve the system spectral efficiency and reduce computational complexity, the goal of this paper is to propose a feasible hybrid precoding algorithm. In Sect. 2, we have introduced the system model in details. In order to make the subsequent research more simple, we assume that the optimal nearest neighbor decoding algorithm adopted by the receiver satisfies the following conditions. First, the received signal is N_s-dimensional, and secondly, it is transmitted by all-digital hardware. The above characteristics allow us to decouple the design of the transceiver and focus on the design of hybrid precoding. The amount of mutual information of the system after the hybrid precoding process can be maximized [10], which we define as

$$I(\mathbf{F}_{RF}, \mathbf{F}_{BB}) = log_2(|\mathbf{I}_{N_s} + \frac{\rho}{N_s\sigma_n^2}\mathbf{H}\mathbf{F}_{RF}\mathbf{F}_{BB} \times \mathbf{F}_{BB}^*\mathbf{F}_{RF}^*\mathbf{H}^*|) \tag{7}$$

where $SNR = \frac{\rho}{\sigma_n^2}$. Because the all-digital hardware design scheme is not feasible in practice, the design problem of hybrid precoding will become complicated and need to be reconsidered. However, the hybrid precoding design ideas proposed in this paper can directly construct a hybrid combining matrix. The design process of $\mathbf{W}_{RF}$ and $\mathbf{W}_{BB}$ is almost similar to the process of designing hybrid precoders. Due to the limitation of the length of the article, we will not repeat them here. We assume that the RF beamforming vector is directly obtained through the F_{RF} codebook, which meets the constraints of RF hardware design, then the maximum mutual information under the hybrid precoding model proposed in the article is given by the following formula

$$\begin{aligned} (\mathbf{F}_{RF}^*, \mathbf{F}_{BB}^*) &= \underset{\mathbf{F}_{RF}, \mathbf{F}_{BB}}{argmax}\, I(\mathbf{F}_{RF}, \mathbf{F}_{BB}) \\ s.t. \quad & \mathbf{F}_{RF} \in \mathscr{F}_{RF} \\ & \| \mathbf{F}_{BB}\mathbf{F}_{RF} \|_F^2 = N_s \end{aligned} \tag{8}$$

where F_{RF} represents the RF precoding domain, which is, the set of $N_t \times N_{tRF}$ matrices which have constant-magnitude elements.

4 Multiple Dimensional Scaling Hybrid Precoding

Generally, hybrid precoding systems need to obtain channel information to realize precoding design at the transmitting end. In large-scale antenna settings, channel information fed back from the receiver is huge, which will bring high feedback overhead and power consumption. Therefore, we propose a multi-dimensional scaling hybrid precoding scheme, which reduces the channel feedback by reducing the information dimension. The limited feedback scheme is shown in Fig. 2.

The commonly used dimension of information is tens of thousands. In addition, many computing methods involve distance calculation, and high-dimensional space will bring great trouble to distance calculation. When the dimension of information is high, it is no longer easy to calculate the inner product. As the number of antennas continues to increase, the computational complexity of the algorithm increases too. These computational obstacles in high dimensions are called "dimension disaster".

An important way to alleviate the "dimension disaster" is to dimension-reduction. The specific method is to transform the original high-dimensional

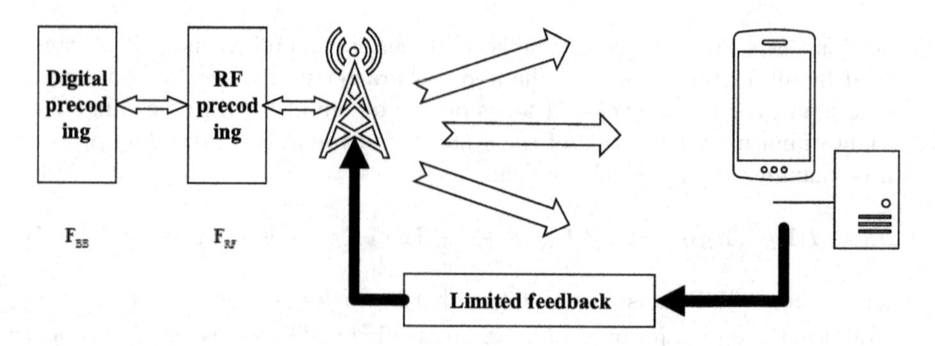

Fig. 2. A mmWave downlink system model with mobile station has a limited feedback channel to the base station.

attribute space into a low-dimensional "subspace" through some mathematical transformation. In this space, the sample density is greatly increased, and the distance calculation becomes easier. The principle of dimension-reduction is that: in many cases, the observed data samples are high-dimensional, however, only a low-dimensional distribution are closely related to task, that is, a low-dimensional embedding in high-dimensional space. As shown in Fig. 3, a three-dimensional graphic is simplified to a two-dimensional graphic. Sample points are easier to calculate in this low-dimensional subspace. But dimension-reduction also brings a defect, which means it loses some useful information of system and leads to the degradation of system performance.

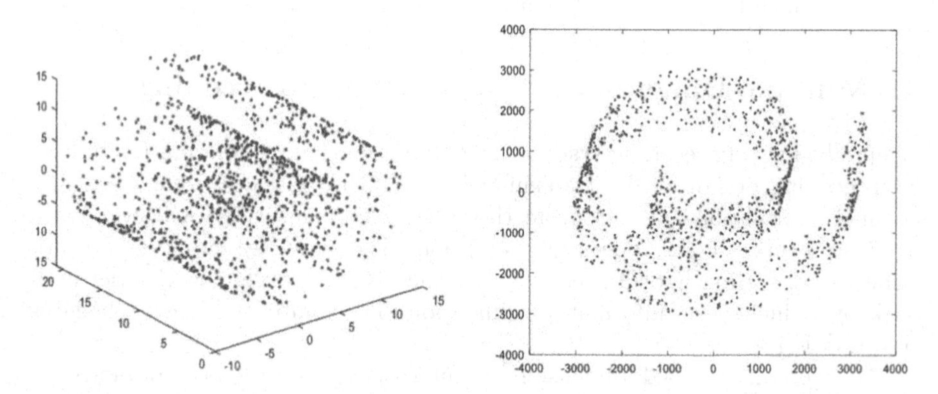

Fig. 3. A brief introduction to Multiple Dimensional Scaling (MDS), using MDS to compress data from 3D to 2D.

The starting point of this method—Multiple Dimensional Scaling (MDS) is to keep the distance between different points in the original data space unchanged as far as possible in the transformed space. Assuming m samples, it can be seen here as a channel vector whose distance matrix in the original channel space is

$\mathbf{D} \in \mathbb{R}^{m \times m}$, and it's ith row, jth column element $dist_{ij}$ is between the channel vector $\mathbf{h}_i$ and $\mathbf{h}_j$, that is

$$\| \mathbf{h}_i - \mathbf{h}_j \| = dist_{ij} \tag{9}$$

Our goal is to obtain the representation of samples in the d' dimension space $\mathbf{Z} \in \mathbb{R}^{d' \times m}$, $d' \leq d$, and the Euclidean distance of any two samples in the low-dimensional space equals the distance in the original space, that is, $\| \mathbf{z}_i - \mathbf{z}_j \| = dist_{ij}$. Let $\mathbf{B} = \mathbf{Z}^T\mathbf{Z} \in \mathbb{R}^{m \times m}$, which is the inner product matrix $\mathbf{B}$ of the reduced dimension sample $b_{ij} = \mathbf{z}_i^T \mathbf{z}_j$, we have

$$dist_{ij}^2 = \| \mathbf{z}_i \|^2 + \| \mathbf{z}_j \|^2 - 2\mathbf{z}_i^T \mathbf{z}_j \tag{10}$$

easily, we can obtain:

$$\sum_{i=1}^{m} dist_{ij}^2 = tr(\mathbf{B}) + mb_{jj} \tag{11}$$

$$\sum_{j=1}^{m} dist_{ij}^2 = tr(\mathbf{B}) + mb_{ii} \tag{12}$$

$$\sum_{i=1}^{m}\sum_{j=1}^{m} dist_{ij}^2 = 2mtr(\mathbf{B}) \tag{13}$$

where $tr(\mathbf{B}) = \sum_{i=1}^{m} \| \mathbf{z}_i \|^2$, let

$$dist_{i.}^2 = \frac{1}{m}\sum_{j=1}^{m} dist_{ij}^2 \tag{14}$$

$$dist_{.j}^2 = \frac{1}{m}\sum_{i=1}^{m} dist_{ij}^2 \tag{15}$$

$$dist_{..}^2 = \frac{1}{m^2}\sum_{i=1}^{m}\sum_{j=1}^{m} dist_{ij}^2 \tag{16}$$

From the above formula, we can get that:

$$b_{ij} = \frac{1}{2}(dist_{ij}^2 - dist_{i.}^2 - dist_{.j}^2 + dist_{..}^2) \tag{17}$$

Thus, through keeping the distance matrix $\mathbf{D}$ unchanged before and after dimension reduction, we can obtain the inner product matrix $\mathbf{B}$.

The matrix $\mathbf{B}$ is decomposed into $\mathbf{B} = \mathbf{V}\Lambda\mathbf{V}^T$, in which the diagonal matrix $\Lambda = diag(\lambda_1, ..., \lambda_d)$, $\lambda_1 \geq \lambda_1 \geq, ..., \geq \lambda_d)$ is composed of eigenvalues, and V is the eigenvector matrix. Supposing that there are d^* non-zero eigenvalue which form a diagonal matrix $\Lambda_* = diag(\lambda_1, ..., \lambda_d^*)$, so that the corresponding eigenvector matrix can be represented as $\mathbf{V}_*$, then $\mathbf{Z}$ can be expressed as

$$\mathbf{Z} = \Lambda_*^{1/2}\mathbf{V}_*^T \in \mathbf{R}^{d^* \times m} \tag{18}$$

In order to reduce dimension effectively in practical applications, the distance after dimension reduction is often as close as possible to the distance in the original space, rather than strictly equal. In this case, a diagonal matrix consisting of the largest eigenvalues $\tilde{\Lambda} = diag(\lambda_1, ..., \lambda_{d'})$, $d' \leq d$ can be taken to represent the corresponding eigenvector matrix $\tilde{\mathbf{V}}$, which can be expressed as:

$$\mathbf{Z} = \tilde{\Lambda}^{1/2}\tilde{\mathbf{V}}^T \in \mathbb{R}^{d' \times m} \tag{19}$$

Now, we have the reduced channel vector $\mathbf{Z}$. We suppose that the receiver sends the reduced matrix $\mathbf{Z}$ as feedback to the transmitter. The transmitter can use the reduced information to design the precoding matrix of the transmitter. We can use the orthogonal matching pursuit to design the corresponding hybrid precoding. The algorithm is summarized in Algorithm 1. The following is procedure of this algorithm. $\mathbf{V}_*$ is first considered as the optimal precoding matrix. Then all the analog beam vectors $\mathbf{A}_t$ are projected to $\mathbf{V}_*$ in turn, where $\mathbf{A}_t = [\mathbf{a}_t(\phi^t_{i,l}, \theta^t_{i,l}), ..., \mathbf{a}_t(\phi^t_{N_{cl},N_{ray}}, \theta_{N_{cl},N^t_{ray})}]$. According to the maximum projection, the optimal analog precoding vector is found, and then the vector is set as the analog vector $\mathbf{F}_{RF(:,i)}$. After finding the principal eigenvector, the digital precoding matrix can be calculated by the least square method. Then the optimal analog precoding vectors are found in a loop, and the process will not stop until all the N_{tRF} analog vectors are found. After the N_{tRF} times iteration, the RF precoding matrix $\mathbf{F}_{RF}$ and the digital precoding matrix $\mathbf{F}_{BB}$ can be found. Finally, we need to normalize the digital precoding vector so that the normalized vector meets the power constraint

$$\mathbf{F}_{BB} = \sqrt{N_s}\frac{\mathbf{F}_{BB}}{\| \mathbf{F}_{RF}\mathbf{F}_{BB} \|_F} \tag{20}$$

5 Simulation Results

The parameter settings in the simulation process are as follows: we use the multipath extended Saleh-Valenzuela channel given in Sect. 2-B. Each channel has $N_{cl} = 8$ cluster and each cluster has $N_{ray} = 8$ scattering path. For simplicity, it is assumed that each cluster has equal power allocation. In the simulation, uniform linear array (ULA) is adopted, so only plane angle ϕ needs to be considered. Receiving angle and transmitting angle ϕ_r, ϕ_t are randomly uniformly distributed in $[0, \pi]$. All precoding schemes have the same SNR, which is defined as $SNR = \frac{\rho}{\sigma_n^2}$. Each simulation implements through 50 channel realizations. Receiver and transmitter arrays have $N_r = N_t = 32$ antennas and $N_{rRF} = N_{tRF} = 8$ RF chains.

Algorithm 1. Multiple dimensional scaling hybrid precoding scheme

Input: $d^* = N_s$, $\mathbf{H}$, $\mathbf{A}_t$
Output: $\mathbf{F}_{RF}$, $\mathbf{F}_{BB}$

1: compute matrix $\mathbf{D}$ based on (9)
2: compute matrix $\mathbf{B}$ based on (17)
3: decompose $\mathbf{B} = \mathbf{V}\Lambda\mathbf{V}^T$
4: we have $\mathbf{Z} = \Lambda_*^{1/2}\mathbf{V}_*^T \in \mathbb{R}^{d^* \times m}$
5: feed back $\mathbf{V}$ to BS for precoding
6: $\mathbf{F}_{RF} = []$
7: $\mathbf{F}_{res} = \mathbf{V}_*^T$
8: **for** $i \leq N_{tRF}$ **do**
9: $\quad \Psi = \mathbf{A}_t^*\mathbf{F}_{res}$
10: $\quad k = argmax_{l=1,\ldots,N_{cl}N_{ray}}(\Psi\Psi^*)_{(l,l)}$
11: $\quad \mathbf{F}_{RF} = [\mathbf{F}_{RF}|\mathbf{A}_t^{(k)}]$
12: $\quad \mathbf{F}_{BB} = (\mathbf{F}_{RF}^*\mathbf{F}_{RF})^{-1}\mathbf{F}_{RF}^*\mathbf{V}_*^T$
13: $\quad \mathbf{F}_{res} = \frac{\mathbf{V}_*^T - \mathbf{F}_{RF}\mathbf{F}_{BB}}{\|\mathbf{V}_*^T - \mathbf{F}_{RF}\mathbf{F}_{BB}\|_F}$
14: $\mathbf{F}_{BB} = \sqrt{N_s}\frac{\mathbf{F}_{BB}}{\|\mathbf{F}_{RF}\mathbf{F}_{BB}\|_F}$
15: **return** $\mathbf{F}_{RF}$, $\mathbf{F}_{BB}$

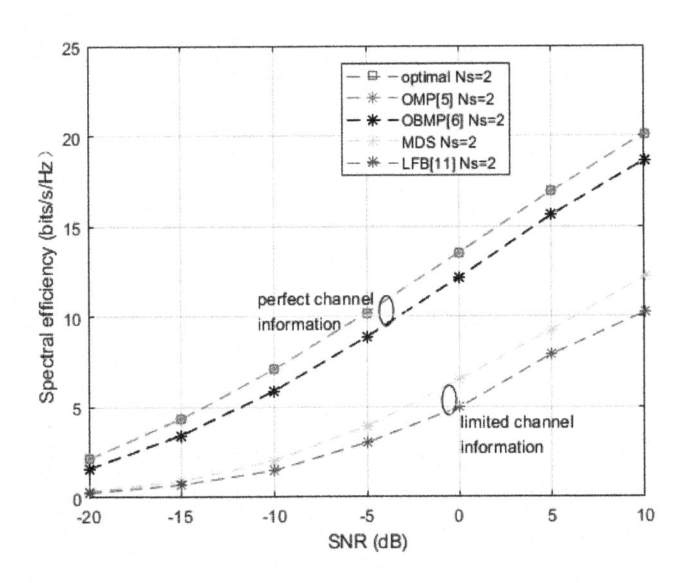

Fig. 4. When $N_s = 2$, performance comparison between Multiple Dimensional Scaling (MDS) hybrid precoding scheme and other schemes.

Experimental summary: from Fig. 4, it can be seen that under the same conditions and the number of data streams equals to two, the proposed Multiple Dimensional Scaling (MDS) hybrid precoding scheme outperforms the finite feedback scheme [13], and our scheme performs better than the finite feedback scheme [14] in the case of varying SNR.

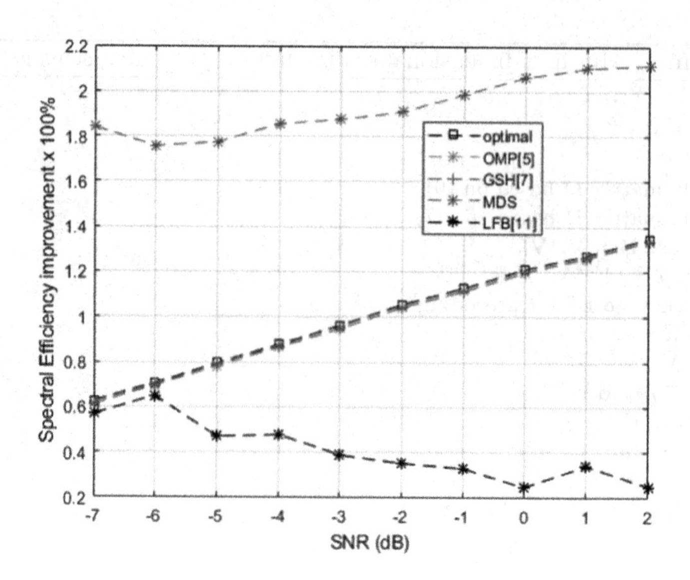

Fig. 5. When $N_s = 2$ and $N_s = 8$, performance improvement between Multiple Dimensional Scaling (MDS) hybrid precoding scheme and other schemes.

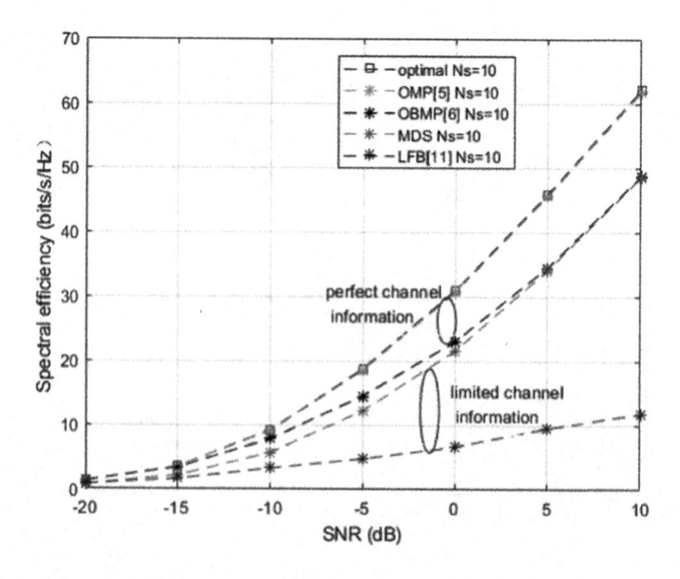

Fig. 6. When $N_s = 10$, $N_{tRF} = 16$, performance comparison between Multiple Dimensional Scaling (MDS) hybrid precoding scheme and other schemes.

In Fig. 5, when we increase the number of data streams from $N_s = 2$ to $N_s = 8$, we compare the performance and capability of various schemes. The formula in Fig. 5 is as follows: $imp = (SE(N_s = 8) - SE(N_s = 2))/SE(N_s = 2) \times 100\%$. From Fig. 5, we can see that when the number of data streams increases, the performance of the MDS precoding scheme increases the most, nearly 200%, while the performance of other schemes increases slightly more than 100%, and the performance of the limited feedback scheme [11] even decreases slightly.

In Fig. 6, when we continue to increase the number of data streams to $N_s = 10$, we get the simulation results of Fig. 6. We can see that when the number of data streams is large, our scheme always performs better than the limited feedback proposal [11], and is close to some full feedback schemes, so our scheme is more suitable for systems with large number of data streams.

6 Conclusion

This paper explores the sparsity of mmWave channel—only a few main paths are useful to help the precoding procedure. We study a limited feedback scenario for mmWave massive MIMO system, we propose a hybrid precoding scheme called Multiple Dimensional Scaling (MDS) hybrid precoding. Instead of the most paper which uses perfect channel information, our scheme only feeds back main vector of the channel which have a significant decrease of feedback overhead. Compared with other limited feedback scheme, our plan has a good performance. When the data streams improve, our scheme improves most of them. Our scheme always performs better than the limited feedback scheme, and is close to some full feedback schemes, and when the system increases the number of data streams, the system performance improves significantly. Therefore, in conclusion, the proposed scheme in this paper can be applied to the limited feedback scheme, and is more suitable for systems with large number of data streams. However, the future communication technology is mainly affected by two factors, namely spectrum efficiency and energy efficiency. This article mainly studies and simulates the spectrum efficiency of our proposed scheme. Therefore, research on energy efficiency of our scheme will be carried out in the future.

References

1. Andrews, J.G., et al.: What will 5G be? IEEE J. Sel. Areas Commun. **32**(6), 1065–1082 (2014)
2. Heath, R.W., González-Prelcic, N., Rangan, S., Roh, W., Sayeed, A.M.: An overview of signal processing techniques for millimeter wave MIMO systems. IEEE J. Sel. Topics Signal Process. **10**(3), 436–453 (2016)
3. Rusek, F., et al.: Scaling up MIMO: opportunities and challenges with very large arrays. IEEE Signal Process. Mag. **30**(1), 40–60 (2013)
4. Chen, C., Tsai, C., Liu, Y., Hung, W., Wu, A.: Compressive sensing (CS) assisted low-complexity beamspace hybrid precoding for millimeter-wave MIMO systems. IEEE Trans. Signal Process. **65**(6), 1412–1424 (2017)
5. Zhang, W., Xia, X., Fu, Y., Bao, X.: Hybrid and full-digital beamforming in mmWave massive MIMO systems: a comparison considering low-resolution ADCs. China Commun. **16**(6), 91–102 (2019). https://doi.org/10.23919/JCC.2019.06.008
6. Zhang, Y., Du, J., Chen, Y., Li, X., Rabie, K.M., Kharel, R.: Near-optimal design for hybrid beamforming in mmWave massive multi-user MIMO systems. IEEE Access **8**, 129153–129168 (2020). https://doi.org/10.1109/ACCESS.2020.3009238

7. Xue, X., Wang, Y., Yang, L., Shi, J., Li, Z.: Energy-efficient hybrid precoding for massive MIMO mmWave systems with a fully-adaptive-connected structure. IEEE Trans. Commun. **68**(6), 3521–3535 (2020). https://doi.org/10.1109/TCOMM.2020.2979139
8. Ayach, O.E., Rajagopal, S., Abu-Surra, S., Pi, Z., Heath, R.W.: Spatially sparse precoding in millimeter wave MIMO systems. IEEE Trans. Wirel. Commun. **13**(3), 1499–1513 (2014)
9. Hung, W.L., Chen, C.H., Liao, C.C., et al.: Low-complexity hybrid precoding algorithm based on orthogonal beamforming codebook. In: IEEE Workshop on Signal Processing Systems. IEEE (2015)
10. Alkhateeb, A., Heath, R.W.: Gram schmidt based greedy hybrid precoding for frequency selective millimeter wave MIMO systems. In: IEEE International Conference on Acoustics. IEEE (2016)
11. Alkhateeb, A., El Ayach, O., Leus, G., et al.: Channel estimation and hybrid precoding for millimeter wave cellular systems. IEEE J. Sel. Topics Signal Process. **8**(5), 831–846 (2014)
12. Sohrabi, F., Yu, W.: Hybrid digital and analog beamforming design for large-scale MIMO systems. In: Proceedings of the IEEE International Conference on Acoustics, Speech and Signal Processing (ICASSP), Brisbane, Australia, April 2015
13. Chen, C.-E.: An iterative hybrid transceiver design algorithm for millimeter wave MIMO systems. IEEE Wirel. Commun. Lett. **4**(3), 285–288 (2015)
14. Alkhateeb, A., Leus, G., Heath, R.W., Jr.: Limited feedback hybrid precoding for multi-user millimeter wave systems. IEEE Trans. Wirel. Commun. **14**(11), 6481–6494 (2014)

A Novel Spatial Modulation Based on Cosine Function Pattern

Liyan Zhang[1], Wenbin Zhang[1,2](✉), Jiaqi Su[1], and Shaochuan Wu[1]

[1] Harbin Institute of Technology, Harbin 150001, People's Republic of China
zwbgxy1973@hit.edu.cn
[2] Science and Technology on Communication Networks Laboratory, Shijiazhuang 050050, Hebei, People's Republic of China

Abstract. Spatial modulation (SM) is a special type of Massive MIMO technology, which utilizes the index of transmit antenna to carry some information bits. Recently, a novel spatial modulation called ABPM appears, which uses the index of antenna beam pattern to carry some information bits. When the receivers are located at the cross points of all beam patterns, ABPM is better than traditional spatial modulation in term of bit error rate (BER). However, the probability that receivers located at the position above is low. Thus ABPM can not be widely used in application. In order to solve this problem, this paper proposes a new method of beam pattern modulation based on cosine function pattern (M-CBPM), the scheme uses the Woodword-Lawson algorithm to calculate the weighted vectors and is more suitable for mobile communication scenarios. The results of theory analysis and simulation show that M-CBPM is better than traditional spatial modulation in term of BER and transmission rate.

Keywords: MIMO · Spatial modulation · Antenna beam pattern modulation · Massive MIMO · M-CBPM

1 Introduction

Massive MIMO is a key technology of 5G, which covers many techniques in application [1]. One of these techniques is spatial modulation (SM), which has many advantages such as simple transmitter structure, low power consumption with few activated RF chains [2,3]. Thus, spatial modulation gradually becomes one of the technologies that attracts much attention in Massive MIMO area. According to the number of activated antennas, spatial modulation is classified into traditional SM and general SM, which activates one antenna and many antennas respectively [4–6].

Recently, a novel SM called Antenna Beam Pattern Modulation (ABPM) is proposed [7], which transmits a part of information bits by generating the pattern

Supported by National Nature Science Foundation of China (NSFC) under Grant 62071148.

Q. Guo et al. (Eds.): WiSATS 2021, LNICST 410, pp. 405–417, 2022.
https://doi.org/10.1007/978-3-030-93398-2_39

of antenna array. In the receiving part, the maximum likelihood(ML) algorithm is used to recover the transmitted bits. The ML algorithm brings the best BER performance to the receiver, but results in the highest computational complexity. To tackle with this problem, [8] used lattice reduction and linear detection jointly. This receiver structure reduces the computational complexity efficiently, at the same time, achieving the same BER performance as ML algorithm.

In ABPM, there are a lot of available patterns of antenna array. If one receiver lies in the non-intersection area among different patterns, it can not receive the electromagnetic wave from certain patterns, which results in some not recognized patterns in this receiver. To solve this intrinsic problem of ABPM, this paper proposes a scheme based on cosine pattern of antenna array.

The contents of this paper are organized as follows. First, we introduce the model of ABPM including transmitter, wireless channel and receiver. Second, we propose modified cosine beam pattern modulation, which makes receivers can identify each beam pattern anywhere. Third, we give numerical results of our scheme to show each factor's effect on BER and spectral efficiency. At last, the conclusion is drawn.

Notation: $\mathbf{A}^{\mathrm{T}}$, $\mathbf{A}^{\mathrm{H}}$ and a^* represent the transpose, the conjugate transpose and the conjugate respectively.

2 The Model of ABPM

The model of ABPM is shown in Fig. 1, which includes transmitter, wireless channel and receiver. At the transmitter, the input bit stream is divided into many groups, each group includes k bits which can be expressed as $\mathbf{b} = [b_1\, b_2 \cdots b_k]$. Next, we convert each bit-group into two parts by serial-to-parallel conversion. The first m bits are divided into the first part, which are used to select a weighted vector of transmit antenna by some mapping relationship. The weighted vector can be expressed as $\mathbf{w} = \{w_1, w_2, \cdots w_{N_t}\}$, where N_t is the number of transmit antennas. The next $k - m$ bits are divided into the second part, these bits are mapped to many points of constellation graph according to the modulation type. The constellation point collection can be expressed as s. Then, multiplying the selected weighted vector w by the selected constellation point collection s, we can obtain the transmission vector $\mathbf{x} = \mathbf{w} \cdot \mathbf{s} = \left[x_1\, x_2 \cdots x_{N_t}\right]^{\mathrm{T}}$ This vector will be transmitted to wireless channel described by a $N_t \times N_r$ matrix $\mathbf{H}$, whose arbitrary element $H_{i,j}, i \in \{1, 2, \cdots N_t\}, j \in \{1, 2, \cdots N_r\}$ follows the i.i.d circular symmetry complex Gaussian distribution with $\mathcal{CN}(0, 1)$. After transmission, the signal vector arriving at receiving antenna array can be expressed as $\mathbf{y} = \mathbf{Hx} + \mathbf{v}$, where $\mathbf{v} = \left[v_1\, v_2 \cdots v_{N_r}\right]^{\mathrm{T}}$ is the additive white Gaussian noise (AWGN) vector introduced by receiving antennas. The elements of $\mathbf{v}$ follow the i.i.d complex Gaussian distribution with $\mathcal{CN}\left(0, \sigma_v^2\right)$. Assuming that the channel is quasi-static, that is, the elements of $\mathbf{H}$ do not change during the transmission of $\mathbf{b}$, and $\mathbf{H}$ in different transmission during are independent. At the receiver, we use maximum likelihood (ML) algorithm to detect the transmitted k bits from $\mathbf{y}$.

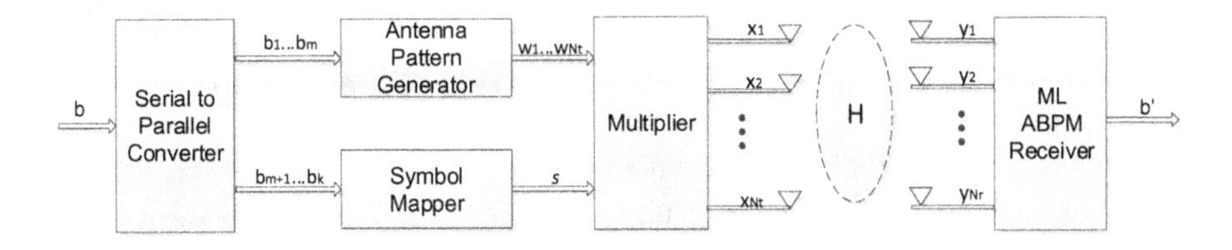

Fig. 1. The model of ABPM system.

In order to understand the principle of ABPM, we take a 2×2 MIMO system as an example. Each group of bits $\mathbf{b} = [b_1\ b_2\ b_3]$ includes three information bits, the first bit b_1 is used to select a weighted vector w , the next two bits b_2, b_3 are mapped to a constellation point of QPSK. The weighted vector w is determined by

$$w_1^{\mathrm{H}} [\mathbf{a}(120°)\ \mathbf{a}(60°)]^{\mathrm{H}} = [1\ 0]^{\mathrm{T}} \tag{1}$$

$$w_2^{\mathrm{H}} [\mathbf{a}(60°)\ \mathbf{a}(120°)]^{\mathrm{H}} = [1\ 0]^{\mathrm{T}} \tag{2}$$

Where, $\mathbf{a}(120°) = [1\ e^{-j2\pi d\cos(120°)/\lambda}]$ and $\mathbf{a}(60°) = [1\ e^{-j2\pi d\cos(60°)/\lambda}]$ are array response vectors at 120° and 60° respectively. When $b_1 = 0$, the beam pattern 1 is selected, the gain of this pattern at 120° is 1, and the gain at 60° is 0. When $b_1 = 1$, the beam pattern 2 is selected, the gain of this pattern at 60° is 1, and the gain at 120° is 0. Generally, d is the distance between adjacent antenna elements of array, which is equal to $\lambda/2$. λ is the wavelength. After solving the w_1 and w_2, two beam patterns corresponding to them can be drawn in Fig. 2. Obviously, there is large difference between two beam patterns. If the receiver can discriminate them, it will recovery one bit information correctly. However, if the receiver locates at the red point in Fig. 2, it can not receive the signal carried by beam pattern 2. This is an intrinsic weak point of ABPM. In this paper, our proposed scheme can solve this problem efficiently.

3 Modified Cosine Beam Pattern Method (M-CBPM)

Considering one cell in cellular network, where a base station is located in the center of circular area and some user equipment are uniformly located near the base station. First, in Fig. 3, we divide the circle into n sectors equally and allocate different frequencies to different sectors. Thus, there are no interference among different sectors. The base station has many antenna arrays, each of them is allocated to one sector. The desired beam pattern is described by the following cosine function.

$$\begin{gathered} p\{\xi\} : F(\theta, \xi) = \left|\cos\left[(\Omega_0 + \xi \cdot \Delta\omega)(\theta - \alpha)\right]\right|, \xi = 1, 2, \cdots 2^m, \theta \in [0, \pi] \\ \Omega_0 + 2^m \cdot \Delta\omega = \Omega_m \end{gathered} \tag{3}$$

Where, α stands for the central angle of a sector, m represents the number of bits used to select weighted vector, in other words, there are 2^m weighted

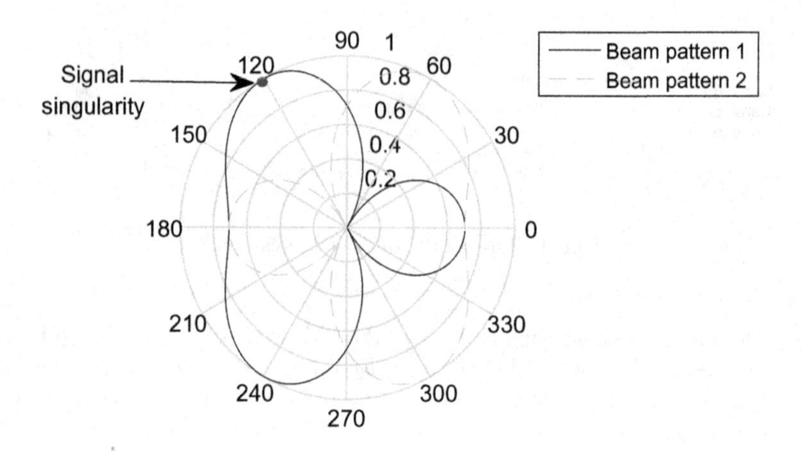

Fig. 2. Two beam patterns corresponding to w_1 and w_2. (Color figure online)

vectors available. Ω_m is the maximum cosine angular frequency that can be synthesized by N_t transmit antennas, it is equal to $N_t/2\pi$ by spatial sampling theory and determines the narrowest width of main lobe $\theta_{\min} = \left(2\arccos\left(\frac{\sqrt{2}}{2}\right)\right)/\Omega_m$. Thus, the number of sectors n is equal to $n = \pi/\theta_{\min}$. When m, $\Delta\omega$ and Ω_m are fixed, as an additional angle frequency, Ω_0 ensures $\Omega_0 + 2^m\Delta\omega = \Omega_m$. Figure 4 is an example of cosine pattern with $m = 3$and $\alpha = \pi/2$. By observing (3), we can find that when Ω_m is fixed, $\Delta\omega$ determines the similar degree among beam patterns, which reduces with the increase of $\Delta\omega$. So the receiver can identify different patterns more easily and the system has a better BER performance. It is worth noting that when Ω_m is fixed, the increase of $\Delta\omega$ also results in the reduction of m, which means the reduction of number of information bits carried by beam pattern.

Next, we will design the weighted vectors corresponding to the cosine beam patterns. Obviously, we can not utilize method introduced in (1) and (2) to calculate the weighted vectors. Here, we introduce Woodword-Lawson algorithm [9] to calculate the weighted vectors from a desired cosine beam pattern. First, we supply basic principle of this algorithm. Taking the central point of axis of uniform linear array as a reference point, the response vector of a uniform linear array can be written as

$$\begin{aligned}\mathbf{v}(\theta) = \Big[& e^{j\left(0-\frac{N_t-1}{2}\right)\frac{2\pi d\cos\theta}{\lambda}} \; e^{j\left(1-\frac{N_t-1}{2}\right)\frac{2\pi d\cos\theta}{\lambda}} \; \cdots \\ & e^{j\left(N_t-2-\frac{N_t-1}{2}\right)\frac{2\pi d\cos\theta}{\lambda}} \; e^{j\left(N_t-1-\frac{N_t-1}{2}\right)\frac{2\pi d\cos\theta}{\lambda}} \Big]^{\mathrm{T}}\end{aligned} \tag{4}$$

Where, λ is the wavelength of electric-magnetic wave. N_t is the number of transmit antennas, without loss of generality, we assume it is an odd. d is the distance between adjacent antennas, generally speaking, $d = \lambda/2$. θ is the angle between

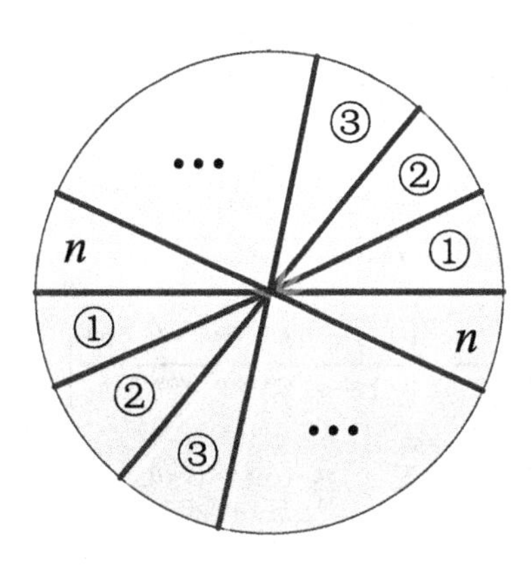

Fig. 3. The cell is divided into n sectors equally.

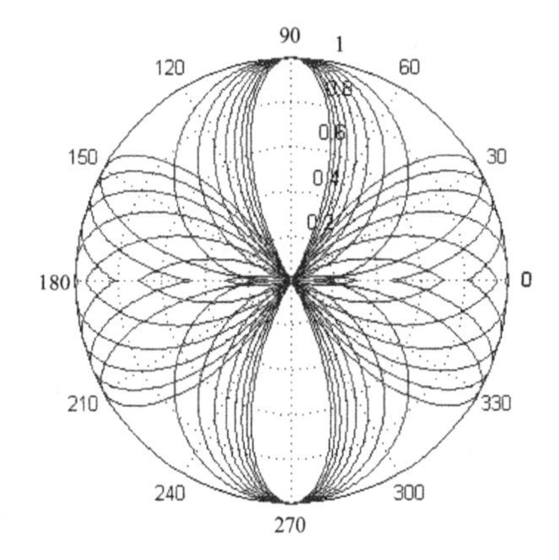

Fig. 4. Beam pattern set with $m = 3$ and $\alpha = \pi/2$.

direction of electric-magnetic wave and axis of array. Thus, the pattern of uniform linear array with N antenna elements can be written as

$$F(\theta) = \mathbf{w}^H \mathbf{v}(\theta) = \sum_{n=0}^{N_t-1} w_n^* e^{j\left(n-\frac{N_t-1}{2}\right)\pi\cos\theta} \tag{5}$$

We use the following sampling function to fit the pattern of (5).

$$\frac{\sin\left[N_t\pi(\cos\theta - \cos\theta_m)/2\right]}{N\sin\left[\pi(\cos\theta - \cos\theta_m)/2\right]}, 0 \le \theta \le \pi \tag{6}$$

Where, θ_m denotes the m-th sampling angle, the sampling is uniformly, which means $\cos\theta_{m+1} - \cos\theta_m = 2/N_t$, $\forall m \in \{0, 1, 2, \cdots N_t - 1\}$. Therefore, the pattern in (5) can be rewritten as

$$F(\theta) = \sum_{m=0}^{N_t-1} F_D(\theta_m) \frac{\sin\left[N_t\pi(\cos\theta - \cos\theta_m)/2\right]}{N_t\sin\left[\pi(\cos\theta - \cos\theta_m)/2\right]} \tag{7}$$

Where, $F_D(\theta_m)$ stands for the discrete pattern value on sampling point θ_m. Assuming that sampling points are symmetry about $\cos\theta = 0$. According to the sampling criterion of space, which is similar to time-frequency one, sampling interval equals $2/N_t$ in term of $\cos\theta$. So, the sampling angle must meet the following condition.

$$\cos\theta_m = \frac{2}{N_t}\left(m - \frac{N_t - 1}{2}\right), m = 0, 1, \cdots, N_t - 1 \tag{8}$$

Next, we derive the expression of weighted vector from (7) utilizing $\sin x = (e^{jx} - e^{-jx})/2j$ and $\sum_{k=0}^{N_t-1} x^k = \frac{1-x^N}{1-x}$, we can rewrite (7) as

$$\begin{aligned} F(\theta) &= \sum_{m=0}^{N_t-1} F_D(\theta_m)\frac{1}{N_t} \cdot \frac{e^{j\frac{N_t}{2}\pi(\cos\theta-\cos\theta_m)} - e^{-j\frac{N_t}{2}\pi(\cos\theta-\cos\theta_m)}}{e^{-j\frac{\pi}{2}(\cos\theta-\cos\theta_m)} - e^{-j\frac{\pi}{2}(\cos\theta-\cos\theta_m)}} \\ &= \sum_{m=0}^{N_t-1} F_D(\theta_m)\frac{1}{N_t} \cdot \frac{-e^{-j\frac{N_t}{2}\pi(\cos\theta-\cos\theta_m)}}{-e^{-j\frac{\pi}{2}(\cos\theta-\cos\theta_m)}} \cdot \left[\frac{1 - e^{jN_t\pi(\cos\theta-\cos\theta_m)}}{1 - e^{j\pi(\cos\theta-\cos\theta_m)}}\right] \\ &= \sum_{m=0}^{N_t-1} F_D(\theta_m)\frac{1}{N_t} \cdot e^{-j\frac{N_t-1}{2}\pi(\cos\theta-\cos\theta_m)} \cdot \sum_{k=0}^{N_t-1} e^{jk\pi(\cos\theta-\cos\theta_m)} \\ &= \sum_{m=0}^{N_t-1} F_D(\theta_m)\frac{1}{N_t} \cdot \sum_{k=0}^{N_t-1} e^{j(k-\frac{N_t-1}{2})\pi(\cos\theta-\cos\theta_m)} \\ &= \sum_{k=0}^{N_t-1} e^{j(k-\frac{N_t-1}{2})\pi\cos\theta} \left(\sum_{m=0}^{N_t-1} F_D(\theta_m)\frac{1}{N_t} \cdot e^{-j(k-\frac{N_t-1}{2})\pi\cos\theta_m}\right) \end{aligned} \tag{9}$$

Comparing (5) with (9), we can obtain the k-th element of weighted vector.

$$w_k^* = \frac{1}{N_t}\sum_{m=0}^{N_t-1} F_D(\theta_m) \cdot e^{-j(k-\frac{N_t-1}{2})\pi\cos\theta_m}, k = 0, 1, \cdots, N_t - 1 \tag{10}$$

Now, we have derived the weighted vectors of general array pattern. Subsequently, we propose an algorithm to obtain the weighted vector based on cosine beam pattern.

Algorithm 1. Weighted vector calculation Algorithm for M-CBPM

Input: The number of transmit antennas N_t.
Output: The weighted vector $\mathbf{w}^H$.

1: Substituting N_t into (8), and calculating the sampling point θ_m;
2: Calculating $F_D(\theta_m)$ according to the desired pattern (3) and sampling point θ_m;
3: Substituting θ_m, N_t and $F_D(\theta_m)$ into (7), and synthesizing actual pattern $F(\theta)$;
4: Substituting θ_m, N_t and $F_D(\theta_m)$ into (10), and calculating the weighted vector $\mathbf{w}^H$;
5: **return** The weighted vector $\mathbf{w}^H$.

---

4 Performance Analysis of M-CBPM

First, we analyze the number of bits carried by weighted vectors in M-CBPM. We have known $\Omega_0 + 2^m\Delta\omega = \Omega_m$ and $\Omega_m = N_t/2\pi$. Once Ω_m and $\Delta\omega$ are determined, m can be maximized when $\Omega_0 = 0$. Thus, the weighted vectors can carry $\log_2(\Omega_m/\Delta\omega)$ bits. Second, we analyze the BER performance of M-CBPM in theory. Without loss of generality, we select maximize likelihood(ML) to detect the received signal. Considering there are innumerable pattern schemes, we only analyze the up-bound of BER. Assuming that transmitter sends $\mathbf{x}_m$, and receiver judge $\mathbf{x}_{\hat{m}}$ by ML. The pairwise error probability (PEP) between $\mathbf{x}_m$ and $\mathbf{x}_{\hat{m}}$ can be written as

$$\begin{aligned} P\left(\mathbf{x}_m \rightarrow \mathbf{x}_{\hat{m}}\right) &= p\left(\left\|\mathbf{y}-\mathbf{H}\mathbf{x}_{\hat{m}}\right\|^2 - \left\|\mathbf{y}-\mathbf{H}\mathbf{x}_m\right\|^2 \leq 0\right) \\ &= \mathbb{E}\left[Q\left(\sqrt{\frac{1}{2\sigma_v^2}\sum_m\sum_{\hat{m}}\left\|\mathbf{H}\left(\mathbf{x}_m-\mathbf{x}_{\hat{m}}\right)\right\|^2}\right)\right] \end{aligned} \tag{11}$$

Utilizing the union-bounding technique given by [10], the union bound on BER of M-CBPM can be written as

$$\begin{aligned} P_{e,bit} &\leq \mathbb{E}_{\mathbf{x}_m}\left[\sum_{\hat{m}} N(m,\hat{m}) P\left(\mathbf{x}_m \rightarrow \mathbf{x}_{\hat{m}}\right)\right] \\ &\leq \sum_{m}^{L}\sum_{\hat{m},\hat{m}\neq m}^{L} \frac{N(m,\hat{m})}{kL} P(\mathbf{x}_m \rightarrow \mathbf{x}_{\hat{m}}) \end{aligned} \tag{12}$$

Where, k denotes the number of information bits carried by each transmitted symbol of M-CBPM. L denotes the number of transmitted symbols of M-CBPM. $N(m,\hat{m})$ represents the number of different bits between $\mathbf{x}_m$ and $\mathbf{x}_{\hat{m}}$. $\mathbb{E}_{\mathbf{x}_m}$ stands for the expectation on $\mathbf{x}_m$. According to [10], $P\left(\mathbf{x}_m \rightarrow \mathbf{x}_{\hat{m}}\right)$ is given by

$$P\left(\mathbf{x}_m \rightarrow \mathbf{x}_{\hat{m}}\right) = \left(\frac{1-\frac{1}{p}}{2}\right)^{\Lambda}\sum_{q=0}^{\Lambda-1} 2^{-q}\binom{\Lambda-1+q}{q}\left(1+\frac{1}{p}\right)^q \tag{13}$$

Where, $p=\sqrt{1+1/E_s d^2(m,\hat{m})/4N_t\sigma_v^2}$, E_s is the average energy of symbols of M-CBPM, $\Lambda = N_t N_r$, σ_v^2 is the variance of AWGN, $d(m,\hat{m})$ is the Euclidean distance between $\mathbf{x}_m$ and $\mathbf{x}_{\hat{m}}$. Substituting (13) into (12) and deriving the up-bound of BER of M-CBPM.

$$P_{e,bit} \leq \sum_{m}^{L}\sum_{\hat{m},\hat{m}\neq m}^{L} \frac{N(m,\hat{m})}{kL}\left(\frac{1-\frac{1}{p}}{2}\right)^{\Lambda}\sum_{q=0}^{\Lambda-1} 2^{-q}\binom{\Lambda-1+q}{q}\left(1+\frac{1}{p}\right)^q \tag{14}$$

Observing the above expression of p, we find that $p \geq 1$. If $p=1$, the result of (14) is zero. If $p \neq 1$, the smaller p is benefit to the BER of M-CBPM. So it is expected that the Euclidean distance between $\mathbf{x}_m$ and $\mathbf{x}_{\hat{m}}$ is as possible as large. Next, we analyze the effect of $\Delta\omega$ on the Euclidean distance between $\mathbf{x}_m$ and $\mathbf{x}_{\hat{m}}$. Assuming $\mathbf{x}_m = \mathbf{w}_m s_m$ and $\mathbf{x}_{\hat{m}} = \mathbf{w}_{\hat{m}} s_{\hat{m}}$. The event $\mathbf{x}_m \neq \mathbf{x}_{\hat{m}}$ happens when any one of the following three conditions holds, which are $\mathbf{w}_m = \mathbf{w}_{\hat{m}}, s_m \neq s_{\hat{m}}$, $\mathbf{w}_m \neq \mathbf{w}_{\hat{m}}, s_m = s_{\hat{m}}$ and $\mathbf{w}_m \neq \mathbf{w}_{\hat{m}}, s_m \neq s_{\hat{m}}$. Our analyses do not involve in any condition. For simplicity, assuming $|s_m|^2 = |s_{\hat{m}}|^2 = c$. The squared Euclidean distance between $\mathbf{x}_m$ and $\mathbf{x}_{\hat{m}}$ is written as

$$\begin{aligned}
d^2(m,\hat{m}) &= \|\mathbf{w}_m s_m - \mathbf{w}_{\hat{m}} s_{\hat{m}}\|^2 \\
&= \sum_{n=1}^{N_t} \left[(w_{m,n}s_m - w_{\hat{m},n}s_{\hat{m}}) \cdot (w_{m,n}s_m - w_{\hat{m},n}s_{\hat{m}})^* \right] \\
&= \sum_{n=1}^{N_t} \left[w_{m,n}s_m w_{m,n}^* s_m^* + w_{\hat{m},n}s_{\hat{m}} w_{\hat{m},n}^* s_{\hat{m}}^* - (w_{\hat{m},n}s_{\hat{m}} w_{m,n}^* s_m^* + w_{m,n}s_m w_{\hat{m},n}^* s_{\hat{m}}^*) \right] \\
&= \sum_{n=1}^{N_t} \left[|w_{m,n}|^2 c + |w_{\hat{m},n}|^2 c - (w_{\hat{m},n}w_{m,n}^* s_{\hat{m}} s_m^* + w_{m,n}w_{\hat{m},n}^* s_m s_{\hat{m}}^*) \right] \\
&\geq \sum_{n=1}^{N_t} \left[c|w_{m,n}|^2 + c|w_{\hat{m},n}|^2 - c(w_{\hat{m},n}w_{m,n}^* + w_{m,n}w_{\hat{m},n}^*) \right] \\
&\geq c \sum_{n=1}^{N_t} \|w_{m,n} - w_{\hat{m},n}\|^2 \\
&\geq c\|\mathbf{w}_m - \mathbf{w}_{\hat{m}}\|^2 = c d^2(\mathbf{w}_m, \mathbf{w}_{\hat{m}})
\end{aligned} \tag{15}$$

From (15), we can find that the Euclidean distance between $\mathbf{x}_m$ and $\mathbf{x}_{\hat{m}}$ depends on the Euclidean distance between $\mathbf{w}_m$ and $\mathbf{w}_{\hat{m}}$. According to the expression of the weighted vectors in (10), we can rewrite the $d(\mathbf{w}_m, \mathbf{w}_{\hat{m}})$ as

$$\begin{aligned}
d(\mathbf{w}_m, \mathbf{w}_{\hat{m}}) &= \left(\sum_{n=1}^{N_t} |w_{m,n} - w_{\hat{m},n}|^2 \right)^{\frac{1}{2}} \\
&= \left(\sum_{n=1}^{N_t} \left(\begin{array}{l} \sum\limits_{h=0}^{N_t-1} \left((F_m(\theta_h) - F_{\hat{m}}(\theta_h)) e^{-j\left(n - \frac{N_t-1}{2}\right)\pi\cos\theta_h} \right) \cdot \\ \left((F_m(\theta_h) - F_{\hat{m}}(\theta_h)) e^{-j\left(n - \frac{N_t-1}{2}\right)\pi\cos\theta_h} \right)^* \end{array} \right) \right)^{\frac{1}{2}}
\end{aligned} \tag{16}$$

Substitute (3) into $F_m(\theta_h) - F_{\hat{m}}(\theta_h)$ and obtain

$$F_m(\theta_h) - F_{\hat{m}}(\theta_h) = \cos\left[m \cdot \Delta\omega(\theta_h - \alpha)\right] - \cos\left[\hat{m} \cdot \Delta\omega(\theta_h - \alpha)\right], \theta_h \in [0, \pi] \tag{17}$$

From (3), we have known that the larger $\Delta\omega$ is, the less the number of weighted vectors available is. It means that the average distance between $m \cdot \Delta\omega$ and $\hat{m} \cdot \Delta\omega$ becomes larger with the increase of $\Delta\omega$. Consequently, $F_m(\theta_h) - F_{\hat{m}}(\theta_h)$ in (17) increases in average meaning, and $d(\mathbf{w}_m, \mathbf{w}_{\hat{m}})$ in (16) also increases.

5 Numerical Results

Due to $\Omega_m = N_t/2\pi$, the number of information bits carried by each M-CBPM symbol is $\log_2\left[N_t/\left(2\pi\Delta\omega\right)\right]$. However, the number of information bits carried by each ABPM symbol is $\log_2 N_t$. So the actual number of transmitted information bits carried by M-CBPM symbol depends on $\Delta\omega$. In order to show the effect of $\Delta\omega$ clearly, the Fig. 5 is drawn, which shows the comparison of the results of SM and M-CBPM with different $\Delta\omega$. By observing the fig, we can see that once $\Delta\omega \geq 0.001$, M-CBPM excels SM in term of the number of information bits carried by each transmitted symbol. Whether it is SM or M-CBPM, the number of information bits carried by each symbol increases with the number of transmit antennas. Next, we show the effect of $\Delta\omega$ on BER of M-CBPM.

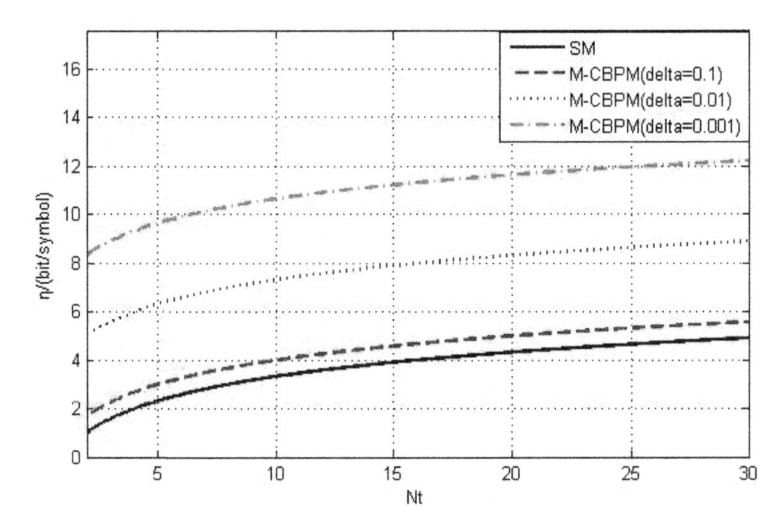

Fig. 5. Under the condition of different $\Delta\omega$, the number of information bits carried by each symbol varies with the number of transmit antennas.

First, we set the system parameters in the simulation and list them in Table 1. The receiver uses maximize likelihood algorithm to detect the received signal, and Fig. 6 shows the curves of BER with different m. By observing this fig, we can find that when the SNR is low, m has little effect on BER. However, with the increase of m, the number of information bits carried by each transmitted symbol increases. Subsequently, the number of transmitted pattern increases, which leads to the increase of transmit antennas according to the space sampling theory. Therefore, there is a trade-off between the number of transmit antennas and transmission rate. It is lucky that there are a very large number of transmit antennas in massive MIMO system, which ensures both reliability and validity of transmission.

Table 1. System parameters in simulation.

Parameter of system	Value
The number of transmit antennas N_t	25
The number of receive antennas N_r	5
Wavelength of the carrier	0.116 m
The distance between adjacent antennas in ULA	0.058 m
Ω	3
$\Delta\omega$	0.3

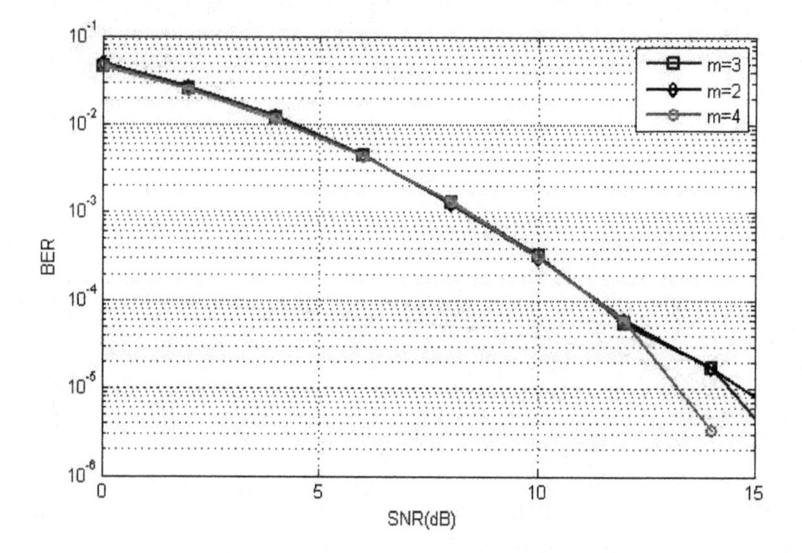

Fig. 6. The BER performance of M-CBPM with different m.

We have known that $\Delta\omega$ has an effect on the Euclidean distance between weighted vectors. The larger the $\Delta\omega$, the smaller the correlation between beam patterns. Therefore, the receiver can distinguish different patterns more easily, and the BER performance of the receiver can be improved. To validate the conclusion above, we implement the simulation of BER on different $\Delta\omega$, the simulation results are shown in Fig. 7. From this fig, we can see that the BER performance of M-CBPM improves efficiently with the increase of $\Delta\omega$.

In order to show the effect of Woodword-Lawson algorithm, we give the beam patterns with different $\Delta\omega$ in Fig. 8. These patterns are synthesized by this algorithm. Comparing this fig with Fig. 4, we find that the pattern synthesized by Woodword-Lawson algorithm is close to the ideal pattern.

The simulations above are based on maximize likelihood algorithm, which is known to have the best BER performance. However, the computational complexity is the highest. In order to obtain the trade-off between the BER and complexity, we implement the simulation about MMSE algorithm, and the results of simulation are shown in Fig. 9.

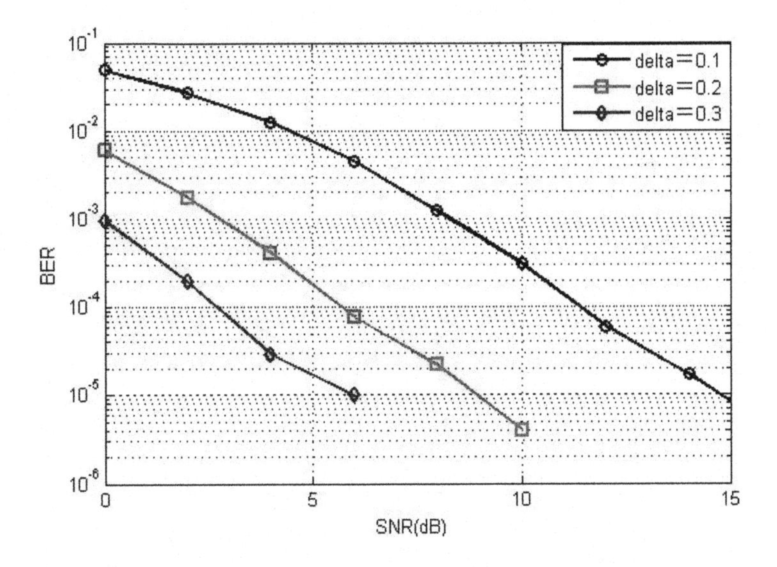

Fig. 7. The BER performance of M-CBPM with different $\Delta\omega$.

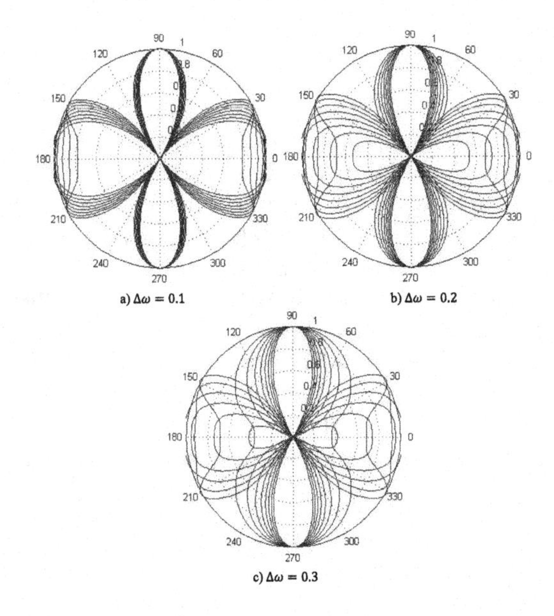

Fig. 8. The pattern synthesized by Woodword-Lawson algorithm with different $\Delta\omega$.

By observing this fig, we can find that although ML is better than MMSE in term of BER, as SNR increase, the BER performance of MMSE with different N_r will gradually improve, especially for the larger N_r. If the requirement for the BER of the receiver is not high, the MMSE algorithm with a large number of receive antennas is a suitable choice.

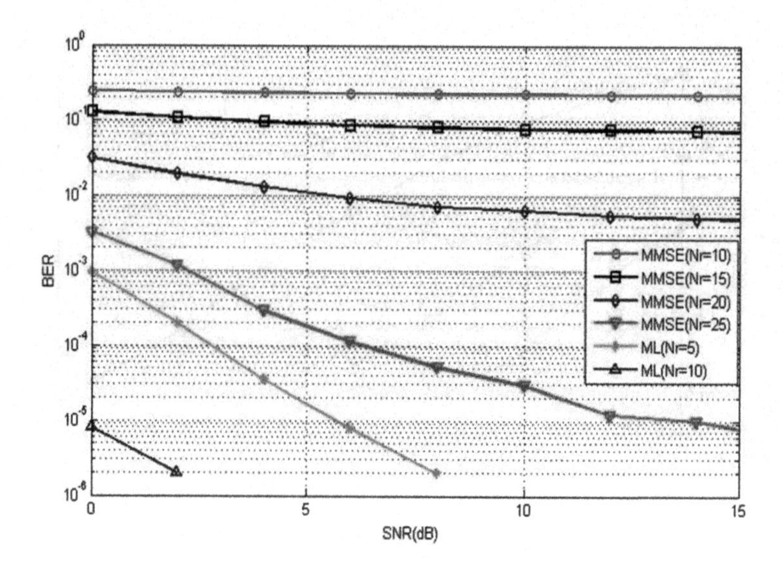

Fig. 9. The BER performance of M-CBPM with MMSE algorithm.

6 Conclusion

Theoretical analysis and simulation results show that our proposed M-CBPM is superior to traditional spatial modulation in term of BER and transmission rate. Moreover, compared with ABPM, this scheme overcomes the defect that the signal can not be received in some zero lobe directions, so it is more suitable for cellular system. It is worth noting that due to the limitation of the number of antennas in current base station, M-CBPM cannot work efficiently. However, with the prevalence of 5G network, more and more base stations will deploy massive MIMO. Then, by virtual of a large number of antennas available, we can choose a smaller $\Delta\omega$ to improve the BER and transmission rate of M-CBPM, and choose a low-complexity algorithm to detect the received signal. Therefore, M-CBPM must have some potential applications in base stations equipped with massive MIMO.

References

1. Lu, L., Li, G.Y., Swindlehurst, A.L., Ashikhmin, A., Zhang, R.: An overview of massive MIMO: benefits and challenges. IEEE J. Sel. Top. Signal Process. **8**(5), 742–758 (2014). https://doi.org/10.1109/JSTSP.2014.2317671
2. Huang, F., Zhan, Y.: Design of spatial constellation for spatial modulation. IEEE Wirel. Commun. Lett. **9**(7), 1097–1100 (2020). https://doi.org/10.1109/LWC.2020.2981612
3. Di Renzo, M., Haas, H., Ghrayeb, A., Sugiura, S., Hanzo, L.: Spatial modulation for generalized MIMO: challenges, opportunities, and implementation. Proc. IEEE **102**(1), 56–103 (2014). https://doi.org/10.1109/JPROC.2013.2287851

4. Wang, J., Jia, S., Song, J.: Generalised spatial modulation system with multiple active transmit antennas and low complexity detection scheme. IEEE Trans. Wirel. Commun. **11**(4), 1605–1615 (2012). https://doi.org/10.1109/TWC.2012.030512.111635
5. Yang, P., Di Renzo, M., Xiao, Y., Li, S., Hanzo, L.: Design guidelines for spatial modulation. IEEE Commun. Surv. Tutor. **17**(1), 6–26 (2015). https://doi.org/10.1109/COMST.2014.2327066
6. Zhao, Y., Hu, J., Xie, A., Yang, K., Wong, K.-K.: Receive spatial modulation aided simultaneous wireless information and power transfer with finite alphabet. IEEE Trans. Wirel. Commun. **19**(12), 8039–8053 (2020). https://doi.org/10.1109/TWC.2020.3019011
7. Ramirez-Gutierrez, R., Zhang, L., Elmirghani, J., Almutairi, A.: Antenna beam pattern modulation for MIMO channels. In: Wireless Communications and Mobile Computing Conference (IWCMC), pp. 591–595 (2012). https://doi.org/10.1109/TWC.2020.3019011
8. Ramirez-Gutierrez, R., Zhang, L., Elmirghani, J.: Antenna beam pattern modulation with lattice reduction aided detection. IEEE Trans. Veh. Technol. **65**, 2007–2015 (2015). https://doi.org/10.1109/TVT.2015.2422299
9. Ghayoula, E., Bouallegue, A., Ghayoula, R., Fattahi, J., Pricop, E., Chouinard, J.-Y.: Radiation pattern synthesis using hybrid Fourier-Woodward-Lawson-neural networks for reliable MIMO antenna systems. In: IEEE International Conference on Systems, Man, and Cybernetics (SMC), Banff, AB, Canada, pp. 3290–3295 (2017). https://doi.org/10.1109/SMC.2017.8123136
10. Lin, Z., Erkip, E., Stefanov, A.: Exact pairwise error probability for the MIMO block fading channel. In: Proceedings of International Symposium on Information Theory and Its Applications (2004)

Low Complexity Hybrid Precoding for Millimeter Wave MIMO Systems

Mengying Jiang, Jiayan Zhang, and Honglin Zhao(✉)

Harbin Institute of Technology, Harbin, China
hlzhao@hit.edu.cn

Abstract. Massive multiple input multiple output (MIMO) technology and millimeter wave (mmWave) system, as the key technologies of the new generation of mobile communications, can effectively increase channel capacity and relieve spectrum resources. Because the mmWave has a short wavelength, the transceiver can be composed of a large antenna array to reduce severe signals attenuation. Furthermore, the use of hybrid precoding technology can improve system performance and reduce system hardware complexity. The classic hybrid precoding algorithm that based on simultaneous orthogonal matching pursuit (SOMP) requires matrix inversion, which leads to high complexity, and its performance depends on the accuracy of channel estimation. In this paper, by modeling the mmWave MIMO system, we compare three improved algorithms, which are orthogonality based matching pursuit algorithm (OBMP), matrix-inversion-bypass simultaneous orthogonal matching pursuit algorithm (MIB-SOMP) and residual matrix-singular value decomposition algorithm (RM-SVD). We analyze the performance of the algorithms, such as complexity, spectrum efficiency, bit error rate, as well as the advantages and disadvantages of the algorithms.

Keywords: Massive MIMO · Millimeter wave · Hybrid precoding · Low complexity

1 Introduction

The millimeter wave (mmWave) multiple input multiple output (MIMO) system is a promising technology to achieve data rates above 10Gbit/s [1]. The most prominent feature of mmWave system is their huge bandwidth, which can support high data rates communications. However, compared with the frequency bands of existing cellular communications, mmWave signals have severe path loss. But because of the shorter wavelength of mmWave, more antennas can be integrated on the same physical size to improve antenna gain [2].

An effective solution to achieve high link quality and support high data rates is to transmit multiple data streams through precoding technology [3]. In 4G mobile communications, traditional digital precoding technology has been widely used [4]. However, it is not practical to use traditional digital precoding on a large-scale antenna array, because digital precoding requires a separate radio frequency (RF) chain for each antenna, which will greatly increase the cost and power consumption of the system.

Q. Guo et al. (Eds.): WiSATS 2021, LNICST 410, pp. 418–431, 2022.
https://doi.org/10.1007/978-3-030-93398-2_40

Generally, hybrid precoding technology is used in mmWave MIMO system, which combines baseband precoding with RF precoding to solve the problems of high hardware cost and power consumption.

The hybrid precoding design can be solved by simultaneous orthogonal matching pursuit (SOMP) [5], an algorithm requiring matrix inversion and assuming that the channel state information (CSI) is perfectly known. Hence, several approaches are proposed to simplify the matrix inversion computation and avoid the dependence on channel estimation. Based on the singular value decomposition (SVD), a hybrid precoding algorithm based on the SVD of the residual matrix is proposed [6], which reduces the dependence on channel estimation. And in [7], the DFT codebook is used as a candidate matrix for its orthogonality, which can avoid the iterations in matching pursuit. In addition, an approach is proposed to simplify the matrix inversion computation by applying Schur-Banachiewicz blockwise inversion [8].

On this basis, we compare the three improved algorithms, quantitatively analyze the complexity and the performance of these algorithms, as well as the existing problems.

This paper is organized as follows. The system model, channel model and the classic hybrid precoding algorithm are introduced in Sect. 2. The RM-SVD, OBMP and MIB-SOMP algorithms are presented in Sect. 3. The simulation analysis and performance comparison of the three improved algorithms are presented in Sect. 4. Finally, the conclusions about the performance analysis are presented in Sect. 5.

2 System Model

2.1 System Model

Consider the single-user mmWave MIMO system shown in Fig. 1 [9].

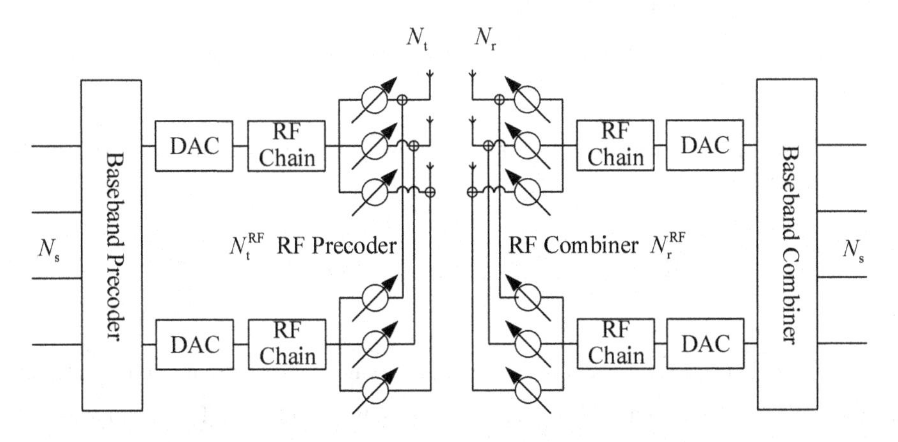

Fig. 1. MmWave MIMO hybrid precoding system block diagram

The base station equipped with N_t antennas transmit N_s data streams to the user equipment equipped with N_r antennas. There are N_t^{RF} and N_r^{RF} RF chains at the transmitter and the receiver to enable multi-stream transmission such that $N_s \leq N_t^{RF} \leq N_t$, $N_s \leq N_r^{RF} \leq N_r$. In the system, the signal processed by the baseband precoder $\mathbf{F}_{BB} \in \mathbb{C}^{N_t^{RF} \times N_s}$ and RF precoder $\mathbf{F}_{RF} \in \mathbb{C}^{N_t \times N_t^{RF}}$ is transmitted to the channel $\mathbf{H} \in \mathbb{C}^{N_r \times N_t}$. The transmitted signal at the base station is:

$$\mathbf{x} = \mathbf{F}_{RF}\mathbf{F}_{BB}\mathbf{s} \tag{1}$$

where $\mathbf{s} = [s_1, s_2, \ldots, s_{N_s}]^T$ denotes the data stream such that $\mathbb{E}[\mathbf{s}\mathbf{s}^H] = \frac{1}{N_s}\mathbf{I}_{N_s}$. The total transmit power constraint satisfies $\|\mathbf{F}_{RF}\mathbf{F}_{BB}\|_F^2 = N_s$. The received signal at the user equipment is:

$$\mathbf{y} = \sqrt{\rho}\mathbf{H}\mathbf{F}_{RF}\mathbf{F}_{BB}\mathbf{s} + \mathbf{n} \tag{2}$$

where ρ represents the average received power, $\mathbf{n} \sim \mathcal{CN}\left(0, \sigma_n^2\mathbf{I}_{N_r}\right)$ is the noise, ρ/σ_n^2 is the received SNR, and $\mathbf{H}$ is the channel matrix such that $\mathbb{E}\left[\|\mathbf{H}\|_F^2\right] = N_t N_r$. After being processed by the RF combiner $\mathbf{W}_{RF} \in \mathbb{C}^{N_r \times N_r^{RF}}$ and baseband combiner $\mathbf{W}_{BB} \in \mathbb{C}^{N_r^{RF} \times N_s}$, the received signal is:

$$\hat{\mathbf{y}} = \sqrt{\rho}\mathbf{W}_{BB}^H\mathbf{W}_{RF}^H\mathbf{H}\mathbf{F}_{RF}\mathbf{F}_{BB}\mathbf{s} + \mathbf{W}_{BB}^H\mathbf{W}_{RF}^H\mathbf{n} \tag{3}$$

Since $\mathbf{F}_{RF}$ and $\mathbf{W}_{RF}$ is implemented using analog phase shifters, its elements are constrained to satisfy $\left|\mathbf{F}_{RF}^{(i,j)}\right| = 1/\sqrt{N_t}$ and $\left|\mathbf{W}_{RF}^{(i,j)}\right| = 1/\sqrt{N_r}$.

2.2 Millimeter Wave Channel Model

Considering that the mmWave has the characteristics of high path loss and line-of-sight transmission, the narrowband scattering cluster channel model is used to model the smmWave channel [10]. Assuming that the mmWave channel contains N_{cl} scattering cluster, each cluster contains N_{ray} propagation path, the system channel can be described as:

$$\mathbf{H} = \sqrt{\frac{N_t N_r}{N_{cl} N_{ray}}} \sum_{i,l} \alpha_{i,l}\mathbf{a}_r\left(\phi_{il}^r, \theta_{il}^r\right)\mathbf{a}_t\left(\phi_{il}^t, \theta_{il}^t\right)^H \tag{4}$$

where $\alpha_{i,l} \sim \mathcal{CN}\left(0, \sigma_{\alpha,i}^2\right)$ is the complex path gain of the lth ray in the ith scattering cluster, and $\sigma_{\alpha,i}^2$ represents the average power of the ith cluster. The average cluster powers are such that $\sum_{i=1}^{N_{cl}} \sigma_{\alpha,i}^2 = \gamma$ where γ satisfies $\mathbb{E}\left[\|\mathbf{H}\|_F^2\right] = N_t N_r$. ϕ_{il}^r and θ_{il}^r represent the azimuth and elevation angles of the arrival angle. In the same way, ϕ_{il}^t and θ_{il}^t

represent the azimuth and elevation angles of the departure angle. $\mathbf{a}_{\mathrm{r}}(\phi_{il}^{\mathrm{r}}, \theta_{il}^{\mathrm{r}})$ and $\mathbf{a}_{\mathrm{t}}(\phi_{il}^{\mathrm{t}}, \theta_{il}^{\mathrm{t}})$ are the antenna array response vector at the receiver and the transmitter.

For the convenience of description, formula (4) can be written as:

$$\mathbf{H} = \mathbf{A}_{\mathrm{r}}\mathbf{H}_{\mathrm{a}}\mathbf{A}_{\mathrm{t}}^{H} \tag{5}$$

where $\mathbf{H}_{\mathrm{a}}=\sqrt{N_{\mathrm{t}}N_{\mathrm{r}}/N_{\mathrm{cl}}N_{\mathrm{ray}}}\,\mathrm{diag}\left(\alpha_{1,1}, \alpha_{1,2}, \cdots, \alpha_{cl,ray}\right)$ represents the path gain of all propagation paths in the channel. $\mathbf{A}_{\mathrm{t}}$ and $\mathbf{A}_{\mathrm{r}}$ represent the transmit candidate matrix and receive candidate matrix.

In this paper, we adopt uniform linear arrays and then the array response vector is:

$$\mathbf{a}(\phi) = \sqrt{\frac{1}{N}}\left[1, e^{-jkd\sin\phi}, e^{-j2kd\sin\phi}, \ldots, e^{-j(N-1)kd\sin\phi}\right]^{T} \tag{6}$$

where N is the number of antennas, d is the antenna spacing, and $k = 2\pi/\lambda$.

2.3 Hybrid Precoder Designs

The target of hybrid precoding is to maximize the spectral efficiency over all possible solutions of $(\mathbf{F}_{\mathrm{RF}}, \mathbf{F}_{\mathrm{BB}}, \mathbf{W}_{\mathrm{RF}}, \mathbf{W}_{\mathrm{BB}})$, which is written as:

$$R = \log_2\left|\mathbf{I}_{N_s} + \frac{\rho}{N_s}\mathbf{R}_{\mathrm{n}}^{-1}\mathbf{W}_{\mathrm{BB}}^{H}\mathbf{W}_{\mathrm{RF}}^{H}\mathbf{H}\mathbf{F}_{\mathrm{RF}}\mathbf{F}_{\mathrm{BB}}\mathbf{F}_{\mathrm{BB}}^{H}\mathbf{F}_{\mathrm{RF}}^{H}\mathbf{H}^{H}\mathbf{W}_{\mathrm{RF}}\mathbf{W}_{\mathrm{BB}}\right| \tag{7}$$

where $\mathbf{R}_{\mathrm{n}} = \sigma_{\mathrm{n}}^{2}\mathbf{W}_{\mathrm{BB}}^{H}\mathbf{W}_{\mathrm{RF}}^{H}\mathbf{W}_{\mathrm{RF}}\mathbf{W}_{\mathrm{BB}}$ represents the noise covariance matrix processed by the user equipment. According to mathematical derivation, the joint optimization problem can be decoupled. Then we only consider the design of the hybrid precoder $\mathbf{F}_{\mathrm{RF}}\mathbf{F}_{\mathrm{BB}}$ and assume the receiver can decode perfectly.

In general, the precoder design problem can be rewritten as:

$$\begin{aligned}
\left(\mathbf{F}_{\mathrm{RF}}^{\mathrm{opt}}, \mathbf{F}_{\mathrm{BB}}^{\mathrm{opt}}\right) &= \underset{\mathbf{F}_{\mathrm{RF}},\mathbf{F}_{\mathrm{BB}}}{\arg\min}\left\|\mathbf{F}_{\mathrm{opt}} - \mathbf{F}_{\mathrm{RF}}\mathbf{F}_{\mathrm{BB}}\right\|_{F} \\
\text{s.t.} \quad & \mathbf{F}_{\mathrm{RF}} \in \varphi, \quad \left\|\mathbf{F}_{\mathrm{RF}}\mathbf{F}_{\mathrm{BB}}\right\|_{F}^{2} = N_{\mathrm{s}}
\end{aligned} \tag{8}$$

where $\mathbf{F}_{\mathrm{opt}}$ represents the optimal full-digital precoder, which consists of right singular vectors associated with the largest N_{s} eigenvalues of $\mathbf{H}$.

Similarly, considering the fixed hybrid precoder $\mathbf{F}_{\mathrm{RF}}$ and $\mathbf{F}_{\mathrm{BB}}$, we adopt the minimum mean square error (MMSE) criterion to design the combiner $\mathbf{W}_{\mathrm{RF}}$ and $\mathbf{W}_{\mathrm{BB}}$. According to derivation, the design problem of the combiner can be equivalent to [5]:

$$\begin{aligned}
\left(\mathbf{W}_{\mathrm{RF}}^{\mathrm{opt}}, \mathbf{W}_{\mathrm{BB}}^{\mathrm{opt}}\right) &= \underset{\mathbf{W}_{\mathrm{RF}},\mathbf{W}_{\mathrm{BB}}}{\operatorname{argmin}}\left\|\mathbb{E}\left[yy^{H}\right]^{1/2}\left(\mathbf{W}_{\mathrm{MMSE}} - \mathbf{W}_{\mathrm{RF}}\mathbf{W}_{\mathrm{BB}}\right)\right\|_{F} \\
\text{s.t.} \quad & \mathbf{W}_{\mathrm{RF}} \in \varphi
\end{aligned} \tag{9}$$

Therefore, the design of the MMSE combiner is similar to the design of the hybrid precoder. The differences are that there is a weighting of the received signal power $\mathbb{E}[yy^H]$ in the optimization objective function and there is no power limitation for the combiner. In this paper, we adopt the MMSE criterion to convert the precoding algorithm to the receiver to solve the design problem of the combiner.

The hybrid precoding algorithm based on SOMP is shown in Table 1.

Table 1. The Hybrid precoding algorithm based on SOMP

Algorithm 1: SOMP
Input $\mathbf{F}_{\text{opt}}$ and $\mathbf{A}_{\text{t}}$
Output $\mathbf{F}_{\text{RF}}$ and $\mathbf{F}_{\text{BB}}$
1: $\mathbf{F}_{\text{RF}} = [\ ]$
2: $\mathbf{F}_{\text{res}} = \mathbf{F}_{\text{opt}}$
3: for $i = 1 : N_{\text{t}}^{\text{RF}}$ do
4: $\boldsymbol{\Psi}_i = \mathbf{A}_{\text{t}}^H \mathbf{F}_{\text{res}}$
5: $k = \arg\max_{l=1,\ldots,N_{\text{cl}}N_{\text{ray}}} \left(\boldsymbol{\Psi}_i \boldsymbol{\Psi}_i^H\right)_{l,l}$
6: $\mathbf{F}_{\text{RF}} = \left[\mathbf{F}_{\text{RF}} \middle\| \mathbf{A}_{\text{t}}^{(k)}\right]$
7: $\mathbf{F}_{\text{BB}} = \left(\mathbf{F}_{\text{RF}}^H \mathbf{F}_{\text{RF}}\right)^{-1} \mathbf{F}_{\text{RF}}^H \mathbf{F}_{\text{opt}}$
8: $\mathbf{F}_{\text{res}} = \dfrac{\mathbf{F}_{\text{opt}} - \mathbf{F}_{\text{RF}}\mathbf{F}_{\text{BB}}}{\left\Vert \mathbf{F}_{\text{opt}} - \mathbf{F}_{\text{RF}}\mathbf{F}_{\text{BB}} \right\Vert_F}$
9: end for
10: $\mathbf{F}_{\text{BB}} = \sqrt{N_{\text{s}}} \dfrac{\mathbf{F}_{\text{BB}}}{\left\Vert \mathbf{F}_{\text{RF}}\mathbf{F}_{\text{BB}} \right\Vert_F}$

The algorithm mainly has two issues: 1) High complexity of matrix inversion. SOMP algorithm requires matrix inversion for updating the baseband precoder with least squares method. When the number of selected base vectors increases, the dimension of matrix inversion will also increase, and matrix inversion is complicated for hardware implementation, which may cause more long calculation delay and higher power consumption. 2) High dependence on the channel estimation. In this algorithm, the RF precoder is the premise of the baseband precoder. And because of the constant modulus constraints and power control, the design of RF precoder is the main factor affecting system performance. This algorithm uses the antenna array response vector as a candidate matrix for RF precoder, and the system performance will be affected by the accuracy of the channel estimation.

3 Hybrid Precoding Algorithm

3.1 RM-SVD

The RM-SVD algorithm is mainly composed of two parts: the design of the initial RF precoding matrix and the update of the initial precoding matrix. The initial RF precoding matrix is constructed by performing SVD on the optimal full-digital precoding matrix $\mathbf{F}_{\mathrm{opt}}$, as shown in Eq. (10):

$$\mathbf{F}_{\mathrm{opt}} = \mathbf{SVD}^{H} \tag{10}$$

where $\mathbf{SV} \in \mathbb{C}^{N_{\mathrm{t}} \times N_{\mathrm{s}}}$. In order to determine the initial RF precoding matrix, a $N_{\mathrm{t}} \times \left(N_{\mathrm{t}}^{\mathrm{RF}} - N_{\mathrm{s}}\right)$ dimensional matrix is constructed so that the phase of its elements obeys a uniform random distribution on $[0, 2\pi)$. At the same time, it is forced to limit its amplitude to $1/\sqrt{N_{\mathrm{t}}}$ for meeting the constant modulus constraints of $\mathbf{F}_{\mathrm{RF}}$. Therefore, the Eq. (10) has the following equivalent form:

$$\mathbf{F}_{\mathrm{opt}} = [\mathbf{SVF}_{\mathrm{R}}]\begin{bmatrix} \mathbf{D}^{H} \\ \mathbf{0} \end{bmatrix} \tag{11}$$

where $\mathbf{0}$ represents an all-zero matrix with $\left(N_{\mathrm{t}}^{\mathrm{RF}} - N_{\mathrm{s}}\right) \times N_{\mathrm{s}}$ dimensions. Under unrestricted conditions, according to the SVD of $\mathbf{F}_{\mathrm{opt}}$, a global optimal solution can be obtained:

$$\mathbf{F}_{\mathrm{RF}}^{*} = [\mathbf{SVF}_{\mathrm{R}}],\ \mathbf{F}_{\mathrm{BB}}^{*} = \begin{bmatrix} \mathbf{D}^{H} \\ \mathbf{0} \end{bmatrix} \tag{12}$$

But the $\mathbf{SV}$ does not meet the constant modulus constraints. Therefore, the phases of all elements of the $\mathbf{SV}$ are preserved, and the amplitude of all elements is forced to be $1/\sqrt{N_{\mathrm{t}}}$. The processed $\mathbf{F}_{\mathrm{RF}}^{*}$ is used as the initial RF precoding matrix.

The hybrid precoding algorithm based on RM-SVD is shown in Table 2.

Table 2. The Hybrid precoding algorithm based on RM-SVD

Algorithm 2: RM-SVD
Input $\mathbf{F}_{\text{opt}}$ and initial $\mathbf{F}_{\text{RF}}$
Output $\mathbf{F}_{\text{RF}}$ and $\mathbf{F}_{\text{BB}}$
1: for $i = 1 : N_{\text{t}}^{\text{RF}}$ do
2: $\mathbf{F}_{\text{RF}}(:,1) = [\ ]$
3: $\mathbf{F}_{\text{BB}} = \left(\mathbf{F}_{\text{RF}}^{H}\mathbf{F}_{\text{RF}}\right)^{-1}\mathbf{F}_{\text{RF}}^{H}\mathbf{F}_{\text{opt}}$
4: $\mathbf{F}_{\text{res}} = \mathbf{F}_{\text{opt}} - \mathbf{F}_{\text{RF}}\mathbf{F}_{\text{BB}}$
5: $\mathbf{F}_{\text{res}} = \mathbf{USV}^{H}$
6: $\mathbf{n} = 1/\sqrt{N_{\text{t}}} \cdot \mathbf{e}^{j\arg(\mathbf{U}(:,1))}$
7: $\mathbf{F}_{\text{RF}} = [\mathbf{F}_{\text{RF}}\ \mathbf{n}]$
8: end for
9: $\mathbf{F}_{\text{BB}} = \left(\mathbf{F}_{\text{RF}}^{H}\mathbf{F}_{\text{RF}}\right)^{-1}\mathbf{F}_{\text{RF}}^{H}\mathbf{F}_{\text{opt}}$
10: $\mathbf{F}_{\text{BB}} = \sqrt{N_{\text{s}}}\dfrac{\mathbf{F}_{\text{BB}}}{\left\|\mathbf{F}_{\text{RF}}\mathbf{F}_{\text{BB}}\right\|_F}$

Compared with the SOMP algorithm, this algorithm does not require a candidate matrix, which reduces the dependence on channel estimation. Furthermore, through the SVD of the residual matrix, the information of the residual matrix can be better used, so that the hybrid precoder $\mathbf{F}_{\text{RF}}\mathbf{F}_{\text{BB}}$ is closer to $\mathbf{F}_{\text{opt}}$. However, due to matrix inversion and SVD, the algorithm complexity is relatively high.

3.2 OBMP

To avoid the issues of the SOMP algorithm, the candidate matrix must satisfy the orthogonality of the column vectors. And considering the sparse characteristics of the mmWave channel, the DFT codebook is used as the candidate matrix, denoted as $\mathbf{A}_{\text{t,DFT}}$

$$\mathbf{A}_{\text{t,DFT}}(:,k) = \frac{1}{\sqrt{N_{\text{t}}}}\left[1, e^{\frac{-j2\pi(k-1)}{N_{\text{t}}}}, \ldots, e^{\frac{-j2\pi(k-1)(N_{\text{t}}-1)}{N_{\text{t}}}}\right]^{T} \tag{13}$$

where $k = 1, \ldots, N_{\text{t}}$. It is well known that the DFT codebook is the basis for the space that is spanned by array response vectors.

Based on the orthogonality of the DFT codebook, as shown in Fig. 2, by calculating the correlation matrix of $\mathbf{F}_{\text{opt}}$ and $\mathbf{A}_{\text{t,DFT}}$, the power contribution distributed in N_{t}

beamforming directions can be calculated in parallel. The correlation matrix can be expressed as:

$$\mathbf{\Psi}_0 = \mathbf{A}_{\mathrm{t,DFT}}^H \mathbf{F}_{\mathrm{opt}} \tag{14}$$

The power contribution of each beamforming direction can be obtained according to formula (18):

$$\boldsymbol{\beta} = \mathrm{diag}\left(\mathbf{\Psi}_0 \mathbf{\Psi}_0^H\right) \tag{15}$$

where the ith element of $\boldsymbol{\beta} \in \mathbb{C}^{N_t \times 1}$ represents the contribution of the ith beamforming direction in the DFT codebook. Because of the orthogonality of the DFT beamforming vectors, the N_t^{RF} column vectors can be selected in parallel in the DFT codebook without iteration to form $\mathbf{F}_{\mathrm{RF}}$:

$$\mathbf{F}_{\mathrm{RF}} = \mathbf{A}_{\mathrm{t,DFT}}(:, \mathcal{V}) \tag{16}$$

where $\mathcal{V}$ is the index set of the largest N_t^{RF} elements selected in $\boldsymbol{\beta}$. Since $\mathbf{F}_{\mathrm{RF}} = \mathbf{A}_{\mathrm{t,DFT}}(:, \mathcal{V})$ is composed of orthogonal column vectors in the DFT codebook, the inverse matrix of $\mathbf{F}_{\mathrm{RF}}^H \mathbf{F}_{\mathrm{RF}}$ is the identity matrix. Then $\mathbf{F}_{\mathrm{BB}}$ can be expressed as:

$$\begin{aligned} \mathbf{F}_{\mathrm{BB}} &= \left(\mathbf{A}_{\mathrm{t,DFT}}(:, \mathcal{V})^H \mathbf{A}_{\mathrm{t,DFT}}(:, \mathcal{V})\right)^{-1} \mathbf{A}_{\mathrm{t,DFT}}(:, \mathcal{V})^H \mathbf{F}_{\mathrm{opt}} \\ &= \mathbf{A}_{\mathrm{t,DFT}}(:, \mathcal{V})^H \mathbf{F}_{\mathrm{opt}} = \mathbf{\Psi}_0(\mathcal{V}, :) \end{aligned} \tag{17}$$

As shown in Fig. 2, the least squares method can be simplified as selecting rows in $\mathbf{\Psi}_0$ with $\mathcal{V}$ as the index set.

The hybrid precoding algorithm based on OBMP is shown in Table 3.

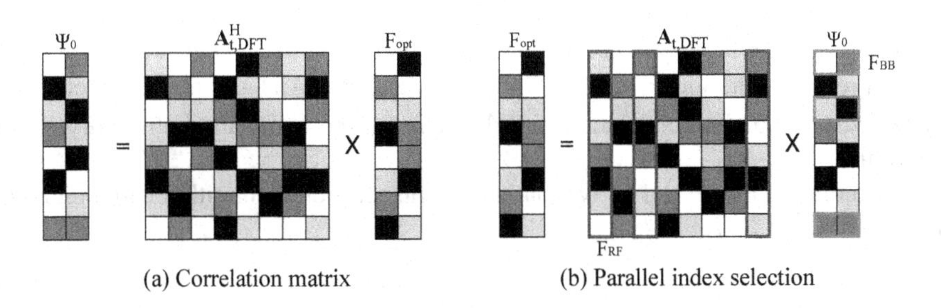

(a) Correlation matrix (b) Parallel index selection

Fig. 2. The explanation of OBMP algorithm

Table 3. The Hybrid precoding algorithm based on OBMP

Algorithm 3: OBMP
Input $\mathbf{F}_{\text{opt}}$ and $\mathbf{A}_{\text{t,DFT}}$
Output $\mathbf{F}_{\text{RF}}$ and $\mathbf{F}_{\text{BB}}$
1: $\boldsymbol{\Psi}_0 = \mathbf{A}_{\text{t,DFT}}^H \mathbf{F}_{\text{opt}}$
2: $\boldsymbol{\beta} = \text{diag}\left(\boldsymbol{\Psi}_0 \boldsymbol{\Psi}_0^H\right)$
3: $\mathcal{V}_0 = [\ ]$
4: for $i = 1 : N_{\text{t}}^{\text{RF}}$ do
5: $k = \arg\max_{l=1,\ldots,N_{\text{t}}} \boldsymbol{\beta}(l)$
6: $\mathcal{V}_i = [\mathcal{V}_{i-1}\ k]$
7: $\boldsymbol{\beta}(k) = 0$
8: end for
9: $\mathbf{F}_{\text{RF}} = \mathbf{A}_{\text{t,DFT}}(:, \mathcal{V})$
10: $\mathbf{F}_{\text{BB}} = \boldsymbol{\Psi}_0(\mathcal{V}, :)$
11: $\mathbf{F}_{\text{BB}} = \sqrt{N_{\text{s}}} \dfrac{\mathbf{F}_{\text{BB}}}{\left\Vert \mathbf{F}_{\text{RF}} \mathbf{F}_{\text{BB}} \right\Vert_F}$

3.3 MIB-SOMP

In algorithm MIB-SOMP, Schur-Banachiewicz blockwise inversion is used to avoid complex matrix inversion. The algorithm divides a high-dimensional matrix into a low-dimensional matrix, and uses the result of the previous iteration to perform the inversion operation of the block matrix in the current iteration, which reduce the complexity of the SOMP algorithm.

In this algorithm, we can use auxiliary variables $\mathbf{A}, V, \mathbf{M}$ and the calculation result of the previous iteration to update $\mathbf{G}_{\mathcal{I}_i,\mathcal{I}_i}^{-1}, \mathbf{F}_{\text{BB}_i}$ and $\boldsymbol{\Psi}_i$. And the matrix inversion is converted to matrix multiplication, which improves the hardware realizability of the algorithm.

According to theoretical derivation, the hybrid precoding algorithm based on MIB-SOMP is shown in Table 4.

Table 4. The Hybrid precoding algorithm based on MIB-SOMP

Algorithm 4: MIB-SOMP
Input $\mathbf{F}_{\mathrm{opt}}$ and $\mathbf{A}_{\mathrm{t}}$
Output $\mathbf{F}_{\mathrm{RF}}$ and $\mathbf{F}_{\mathrm{BB}}$
1: $\mathbf{G} = \mathbf{A}_{\mathrm{t}}^{H}\mathbf{A}_{\mathrm{t}}\left(\mathbf{G}_{\mathcal{I},\mathcal{J}} = \mathbf{A}_{\mathrm{t}_{\mathcal{I}}}^{H}\mathbf{A}_{\mathrm{t}_{\mathcal{J}}}\right)$
2: $\mathbf{G}_{\mathcal{I}_0,\mathcal{I}_0}^{-1} = \mathbf{F}_{\mathrm{BB}_0} = [\ ]$, $\mathbf{\Psi}_0 = \mathbf{A}_{\mathrm{t}}^{H}\mathbf{F}_{\mathrm{opt}}$
3: $\mathcal{I}_0 = [\ ]$, $\overline{\mathcal{I}}_0 = \{i,\ 1 \le i \le N_{\mathrm{cl}}N_{\mathrm{ray}}\}$
4: for $i = 1 : N_{\mathrm{t}}^{\mathrm{RF}}$ do
5: $k = \arg\max_l \left(\mathbf{\Psi}_{i-1}\mathbf{\Psi}_{i-1}^{H}\right)_{l,l}$
6: $\mathbf{A} = \mathbf{G}_{k,\mathcal{I}_{i-1}}\mathbf{G}_{\mathcal{I}_{i-1},\mathcal{I}_{i-1}}^{-1}$
7: $V = 1/\left(\mathbf{G}_{k,k} - \mathbf{A}\mathbf{G}_{\mathcal{I}_{i-1},k}\right)$
8: $\mathbf{M} = \mathbf{A}\mathbf{\Psi}_0(\mathcal{I}_{i-1},:) - \mathbf{\Psi}_0(k,:)$
9: $\mathcal{I}_i = [\mathcal{I}_{i-1}\ k]$, $\overline{\mathcal{I}}_i = \overline{\mathcal{I}}_{i-1} - \{k\}$
10: $\mathbf{G}_{\mathcal{I}_i,\mathcal{I}_i}^{-1} = \begin{bmatrix} \mathbf{G}_{\mathcal{I}_{i-1},\mathcal{I}_{i-1}}^{-1} + V\mathbf{A}^{H}\mathbf{A} & -V\mathbf{A}^{H} \\ -V\mathbf{A} & V \end{bmatrix}$
11: $\mathbf{F}_{\mathrm{BB}_i} = \begin{bmatrix} \mathbf{F}_{\mathrm{BB}_{i-1}} + V\mathbf{A}^{H}\mathbf{M} \\ -V\mathbf{M} \end{bmatrix}$
12: $\mathbf{\Psi}_i = \mathbf{\Psi}_{i-1}(\overline{\mathcal{I}}_i,:) - \mathbf{G}_{\overline{\mathcal{I}}_i,\mathcal{I}_i}\begin{bmatrix} V\mathbf{A}^{H}\mathbf{M} \\ -V\mathbf{M} \end{bmatrix}$
13: end for
14: $\mathbf{F}_{\mathrm{RF}} = \mathbf{A}_{\mathrm{t}_{\mathcal{I}_{N_{\mathrm{t}}^{\mathrm{RF}}}}}$
15: $\mathbf{F}_{\mathrm{BB}} = \sqrt{N_{\mathrm{s}}}\dfrac{\mathbf{F}_{\mathrm{BB}}}{\lVert \mathbf{F}_{\mathrm{RF}}\mathbf{F}_{\mathrm{BB}} \rVert_F}$

4 Performance Analysis

4.1 Complexity Analysis

Compared with addition, multiplication calculation is the main factor of the complexity of the algorithm. Therefore, in this paper we only consider the number of complex multiplications in each iteration.

The number of complex multiplications during the ith iteration of the hybrid precoding algorithm of SOMP and RM-SVD is shown in Table 5 and Table 6.

Table 5. Number of complex multiplications in i^{th} the iteration of SOMP

Computation	SOMP
$\boldsymbol{\Psi}_i$	$L \times N_t \times N_s$
$\boldsymbol{\Psi}_i \boldsymbol{\Psi}_i^H$	$L \times N_s \times L$
$\mathbf{F}_{RF}\mathbf{F}_{BB}$	$N_t \times N_t^{RF} \times N_s$
$\Vert \mathbf{F}_{opt} - \mathbf{F}_{RF}\mathbf{F}_{BB} \Vert_F$	$N_t \times N_s$

Table 6. Number of complex multiplications in i^{th} the iteration of RM-SVD

Computation	RM-SVD
$\mathbf{F}_{RF}\mathbf{F}_{BB}$	$N_t \times N_t^{RF} \times N_s$
$\mathbf{USV}^H$	$N_t \times N_s \times N_t$
$\mathbf{n} = 1/\sqrt{N_t} \cdot \mathbf{e}^{j\arg(\mathbf{U}(:,1))}$	N_t

where L is the number of basis vectors in the candidate matrix, N_t is the number of antennas at the transmitter, N_s is the number of data streams, and N_t^{RF} is the number of RF chains at the transmitter. The complexity of the SOMP algorithm is mainly concentrated in the calculation of the correlation matrix and the residual matrix. Furthermore, the complexity of the RM-SVD algorithm is mainly concentrated in the SVD of the residual matrix. Due to the sparsity of mmWave channels, N_t is much larger than L in mmWave MIMO systems, so the RM-SVD algorithm requires more complex multiplications.

The number of complex multiplications during the ith iteration of the hybrid precoding algorithm of OBMP and MIB-SOMP is shown in Table 7 and Table 8.

Table 7. Number of complex multiplications in i^{th} the iteration of OBMP

Computation	OBMP
$\boldsymbol{\Psi}_0$	$N \times N \times N_s$
$\boldsymbol{\beta}$	$N \times N_s/2$

Table 8. Number of complex multiplications in i^{th} the iteration of MIB-SOMP

Computation	MIB-SOMP	Computation	MIB-SOMP
$\boldsymbol{\Psi}_0$	$L \times N \times N_s$	$\mathbf{M}$	$(i-1) \times N_s$
$\mathbf{G}$	$L \times N \times L$	$\mathbf{A}^H\mathbf{V}$	$i-1$
$(\boldsymbol{\Psi}_{i-1}\boldsymbol{\Psi}_{i-1}^H)_{l,l}$	$[L-(i-1)] \times N_s/2$	$(\mathbf{A}^H\mathbf{V})\mathbf{A}$	$(i-1) \times (i-1)$
$\boldsymbol{\Psi}_i$	$(L-i) \times i \times N_s$	$(\mathbf{A}^H\mathbf{V})\mathbf{M}$	$(i-1) \times N_s$
$\mathbf{A}$	$(i-1) \times (i-1)$	$\mathbf{VM}$	N_s
$\mathbf{V}$	$i-1$	$\mathbf{G}_i^{-1}$	0

The OBMP algorithm uses the DFT codebook with orthogonal column vectors as the candidate matrix and selects N_t^{RF} column vectors with the largest power contribution in parallel. In addition, the algorithm converts the least square method into a simple indexing process so that matrix inversion is not required, which greatly reduces the complexity of the algorithm.

The MIB-SOMP algorithm applies Schur-Banachiewicz blockwise inversion and achieves hybrid precoding through iterative inversion. This procedure greatly reduces the computation complexity and enable efficient hardware implementation.

4.2 Spectral Efficiency

The simulation parameters in the performance analysis are shown in Table 9.

Table 9. The simulation parameters

Parameters	Values
$N_t \times N_r$	128×32
d	0.5λ
Modulation scheme	QPSK
N_{cl}	5
N_{ray}	10

In Fig. 3(a), we compare the spectral efficiency of all hybrid precoding algorithms. This simulation sets $N_s = 2$ and $N_t^{RF} = N_r^{RF} = 4$. From this figure we observe that MIB-SOMP algorithm exhibits same spectral efficiency as the SOMP algorithm, this is because the MIB-SOMP algorithm only changes the method of matrix inversion, but the hybrid precoding matrix is the same as that of the SOMP algorithm. The spectral efficiency of the RM-SVD algorithm is higher than that of the SOMP algorithm, which is closer to the optimal full-digital precoding algorithm. This algorithm better extracts the information in the optimization target matrix by performing singular value decomposition on the residual matrix. The OBMP algorithm has the lowest spectral efficiency, because the candidate matrix is defined in advance without CSI.

In Fig. 3(b), we compare the spectral efficiency of all hybrid precoding algorithms when the number of RF chains varies from 2 to 6. This simulation sets $N_s = 2$ and $\text{SNR} = -5\text{dB}$. As the number of RF chains increases, the spectral efficiency of the SOMP algorithm and improved hybrid precoding algorithm is closer to the optimal full-digital precoding algorithm. With the same SNR, the spectral efficiency of the RM-SVD algorithm is about 5%–10% higher than that of SOMP algorithm. And the spectral efficiency of the MIB-SOMP algorithm is the same as the spectrum efficiency of the SOMP algorithm. The OBMP algorithm has the lowest spectrum efficiency, which is about 80% of that of the SOMP algorithm.

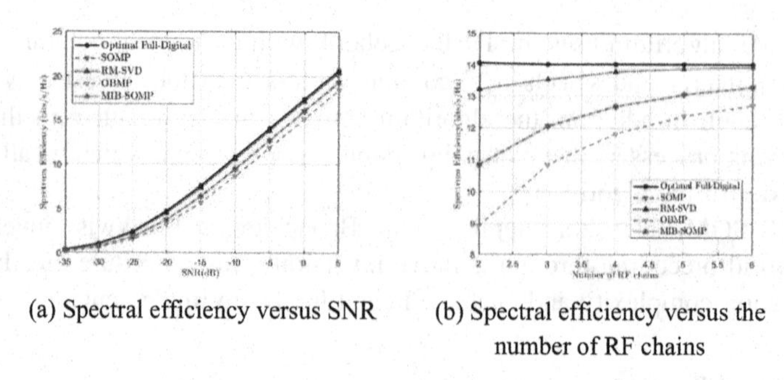

(a) Spectral efficiency versus SNR (b) Spectral efficiency versus the number of RF chains

Fig. 3. Spectral efficiency of hybrid precoding algorithms

4.3 Bit Error Rate

In Fig. 4(a), we compare the bit error rate (BER) of the algorithms. This simulation sets $N_s = 2$ and $N_t^{RF} = N_r^{RF} = 4$. Consistent with the spectral efficiency of algorithms, MIB-SOMP and SOMP have the same BER. The RM-SVD has a lower BER than the SOMP algorithm. The OBMP algorithm has the highest BER.

In Fig. 4(b), we compare the BER of the algorithms when the number of RF chains varies from 4 to 10. This simulation sets $N_s = 4$ and SNR $= -15$ dB. With the same SNR, we observe that the BER of the OBMP algorithm is increased by about 35%–40%, the BER of the RM-SVD algorithm is reduced by about 30%–40%. And the BER of the MIB-SOMP algorithm is the same as that of the SOMP algorithm.

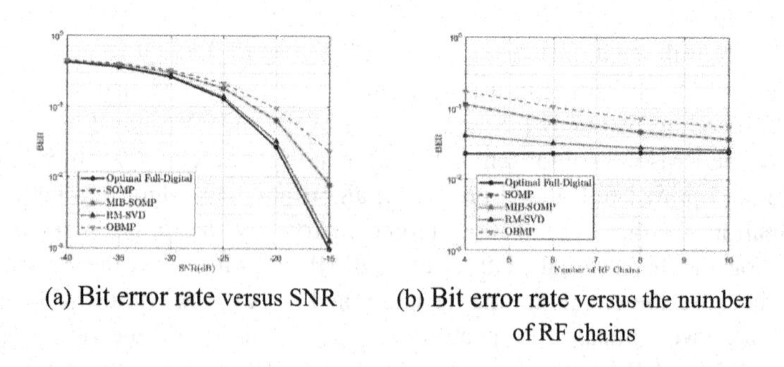

(a) Bit error rate versus SNR (b) Bit error rate versus the number of RF chains

Fig. 4. Bit error rate of hybrid precoding algorithms

5 Conclusions

In this paper, we analyze the main issues of the SOMP algorithm, which are high complexity of matrix inversion and high dependence on the channel estimation. On this basis, we compare and analyze the performance and complexity of the three improved hybrid precoding algorithms. The RM-SVD algorithm is more complicated than the SOMP algorithm due to the SVD of the residual matrix, but the spectrum efficiency is approximately improved 5%–10%, and the bit error rate is approximately reduced

30%–40%. The OBMP algorithm uses the DFT codebook as a candidate matrix to completely avoid matrix inversion. The complexity is about 20% of the MIB-SOMP algorithm, the spectral efficiency is reduced by about 20%, and the bit error rate is increased by about 35%–40%. The MIB-SOMP algorithm uses the Schur-Banachiewicz blockwise inversion and replaces the matrix inversion with iterative inversion. The algorithm exhibits same performance as the SOMP algorithm via simulation results with reduced computational complexity.

References

1. Huang, H., Liu, K.P., Wen, R.: Joint channel estimation and beamforming for millimeter wave cellular system. In: IEEE Global Communications Conference, San Diego (2015)
2. Doan, C.H., Emami, S., Sobel, D.A., Niknejad, A.M.: Design considerations for 60 GHz CMOS radios. IEEE Commun. Mag. **42**, 132–140 (2004)
3. Vu, M., Paulraj, A.: MIMO wireless linear precoding. IEEE Signal Process. Mag. **24**, 86–105 (2007)
4. Marzetta, T.L.: Noncooperative cellular wireless with unlimited numbers of base station antennas. IEEE Trans. Wirel. Commun. **9**(11), 3590–3600 (2010)
5. Ayach, O.E., Rajagopal, S., Abu-Surra, S.: Spatially sparse precoding in millimeter wave MIMO systems. IEEE Trans. Wirel. Commun. **13**(3), 1499–1513 (2014)
6. Xiang, J.W., Yu, X.L., Jing, X.R.: Low complexity hybrid precoding method in mmWave massive MIMO system. Telecommun. Sci. **32**(09), 10–15 (2016)
7. Chen, C.H., Tsai, C.R., Liu, Y.H.: Compressive sensing assisted low-complexity beamspace hybrid precoding for millimeter-wave MIMO systems. IEEE Trans. Signal Process. **65**(6), 1412–1424 (2017)
8. Lee, Y.Y., Wang, C.H.: A hybrid RF/baseband precoding processor based on parallel-index-selection matrix-inversion-bypass simultaneous orthogonal matching pursuit for millimeter wave MIMO systems. IEEE Trans. Signal Process. **63**(2), 305–317 (2015)
9. Ayach, O.E., Heath, R.W., Abu-Surra, S.: Low complexity precoding for large millimeter wave MIMO systems. In: IEEE International Conference Communications (ICC), pp. 3724–3729 (2012)
10. Alkhateeb, A., El Ayach, O., Leus, G., Heath, R.W.: Heath: Hybrid precoding for millimeter wave cellular systems with partial channel knowledge. In: Information Theory and Applications Workshop (ITA), San Diego, CA, USA (2013)

Cloud Prediction Based on the Combination of Optical Flow and Deep Learning

Peng Muzi(✉), Zhao Kanglian, Dai Zheng, and Li Wenfeng

NanJing University, 163 Xianlin Street, Qixia Distirct,
Nanjing 210023, Jiangsu, China

Abstract. Satellite-to-ground laser communication has a problem of being susceptible to specific atmospheric environments, which will attenuate the laser transmission signal severely. To solve this problem, we have to know important prior information about whether the construction of a specific laser communication link is suitable. In this paper, in order to predict future images of cloud clusters around the laser links in advance, we propose a cloud prediction model based on the combination of optical flow and deep learning. Our model is based on Deep Voxel Flow (DVF), an end-to-end CNN designed for video frame synthesis. The 3D optical flow vector across space and time in the input cloud images is used to form an intermediate layer in DVF. By using DVF for multiple times to iterate the input cloud images at t second and t + 25 s, we can get the predicted cloud images during the next 100 s. Our experimental results show that, compared to the optical flow extrapolation method which is a typical method used for nowcast, our cloud prediction model can predict future cloud images with higher quality and accuracy.

Keywords: Laser communication · Cloud prediction model · Deep learning

1 Introduction

Laser communication is a kind of wireless communication which uses optical signal as the carrier to transmit information directly in the atmosphere. With the advantages of large communication capacity and strong confidentiality, laser communication has a wide application potential in satellite-to-ground communication [1].

However, satellite-to-ground laser communication has a problem of being susceptible to specific atmospheric environments [2]. Compared to satellite-to-satellite laser links, satellite-to-ground laser links have to pass through the atmosphere, during which the laser transmission signal would be easily affected by atmospheric environments such as clouds, fog, haze and so on. These complex atmospheric environments will attenuate the laser transmission signal severely, and even cause communication interruption. As a result, in satellite-to-ground laser communication, we must take the influence of different atmospheric environments into consideration.

In this case, if the condition of cloud clusters around the laser links can be obtained in advance, we will get to know important prior information about whether the

Q. Guo et al. (Eds.): WiSATS 2021, LNICST 410, pp. 432–444, 2022.
https://doi.org/10.1007/978-3-030-93398-2_41

construction of a specific laser communication link is suitable and we will be able to predict the quality of the link, thus guaranteeing the stability of the uninterrupted satellite-to-ground laser communication. Such a task of extrapolating future cloud condition from the past trends of cloud change can be termed as cloud prediction.

2 Related Work

In weather forecast domain, radar echo image extrapolation [3] is a main approach of nowcast. Many typical methods of radar echo extrapolation have been proposed, including the centroid tracking method, the cross-correlation extrapolation method and the optical flow extrapolation method. The core idea of all these three typical methods is to find the corresponding relationship between the frames at the adjacent time.

The centroid tracking method tries to identify the monomers, and then scans the images at the adjacent time to match and track the target monomers [4]. However, this algorithm is only suitable for images with easily identified target monomers. As the target monomers on most of our cloud images are complex and hard to be identified, the centroid tracking method has limitations on our cloud prediction.

The cross-correlation extrapolation method [5] first divides the image region into several small regions, and then calculates the correlation coefficient between the small regions at the adjacent time. The corresponding relationship of the regions at the adjacent time is determined by the maximum correlation coefficient. However, the cross-correlation extrapolation method often fails when the image motion is fast. As the motion of cloud tends to be fast, this method also has limitations on our cloud prediction.

The optical flow extrapolation method makes use of the optical flow to track the image motion. Optical flow is the instantaneous velocity field of a moving object. It can be used to calculate the next position of the target point. It has been proved that the optical flow method has advantages over the other two methods above. Gunnar Farneback's algorithm [6] is used to calculate dense optical flow—the optical flow of all pixels in the image is calculated. Farneback optical flow method is a gradient based method. In this method, the image gradient is assumed to be constant and the local optical flow is assumed to be constant. Farneback optical flow method is a very suitable optical flow extrapolation method for weather radar echo extrapolation at present. However, the optical flow extrapolation method also has limitations on our cloud prediction. The optical flow method requires the image to follow the assumption of gray invariance, while the brightness of the target point is constantly changing because the actual cloud change with time is often accompanied by generation, development, weakening and dissipation. As a result, when the moving speed of clouds is fast and the time interval is long, the prediction error rate can still be large.

3 Our Cloud Prediction Model

In this paper, we decide to improve the optical flow extrapolation method by combining it with convolutional neural networks (CNN), which is one of the most representative networks in deep learning used for tasks of prediction. Our cloud prediction model is based on Deep Voxel Flow (DVF) [7]. DVF is an end-to-end CNN designed for video frame synthesis. Compared to optical flow extrapolation method and simple CNN-based model without using the optical flow [8], DVF can synthesize the next video frame with higher quality and accuracy with the input of former two consecutive frames. The structure of DVF is composed of two parts. The first part is the convolutional encoder-decoder used to predict voxel flow, as is show in Fig. 1. And the second part of DVF is the volume sampling layer used to synthesize the predicted frame by bilinear interpolation, with the predicted voxel flow and the previous two frames, as is shown in Fig. 2.

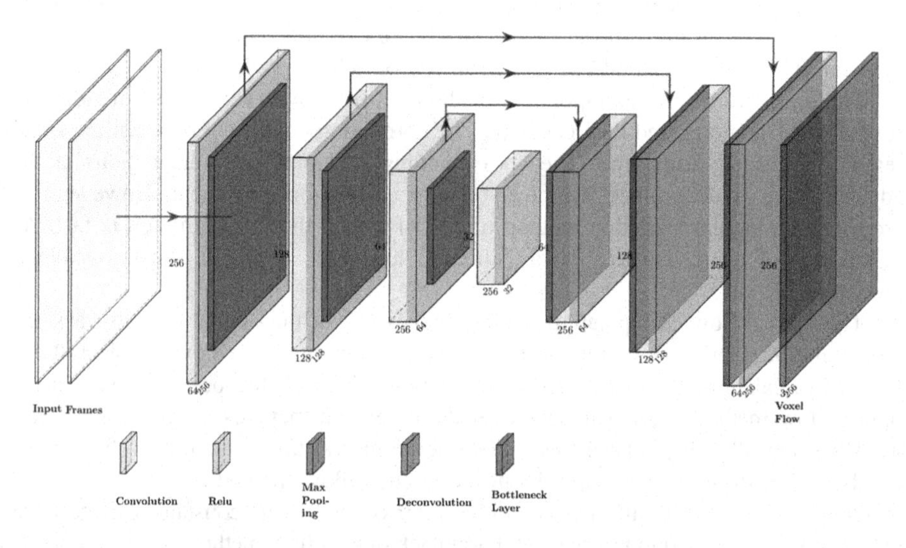

Fig. 1. The convolutional encoder-decoder in DVF

Instead of using CNN to predict and output the optical flow itself as FlowNet [9]do, the voxel flow vector across space and time in the input cloud images is used to form an intermediate layer in DVF, which means the correctness of the optical prediction will never be directly tested. The output of DVF is the predicted frame and we only have to directly consider the correctness of the predicted frame. Because of the superiority of DVF in the video prediction task, we adopt DVF to our prediction model. The structure of our cloud prediction model is shown in Fig. 3. By using DVF for multiple times to iterate the input cloud images at t second and t + 25 s, we can get the predicted cloud images during the next 100 s.

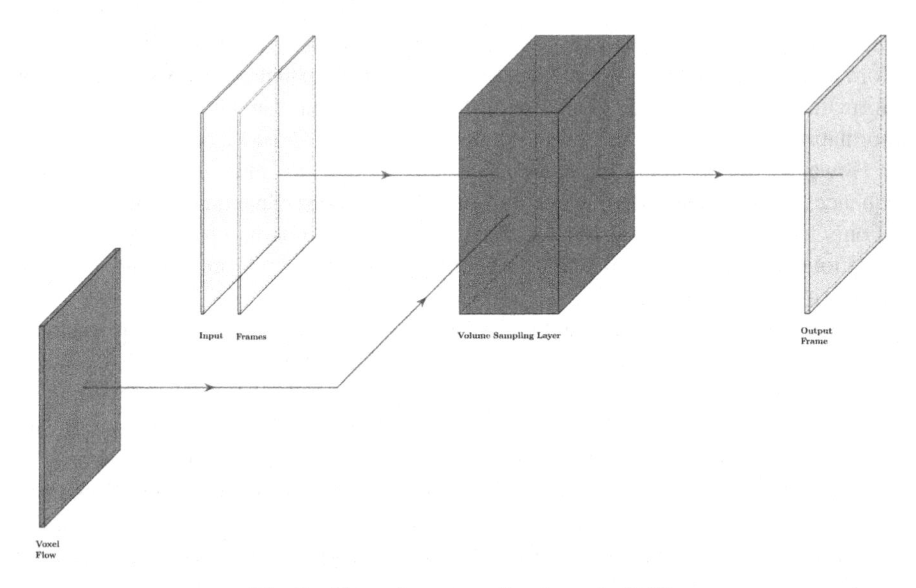

Fig. 2. The volume sampling layer in DVF

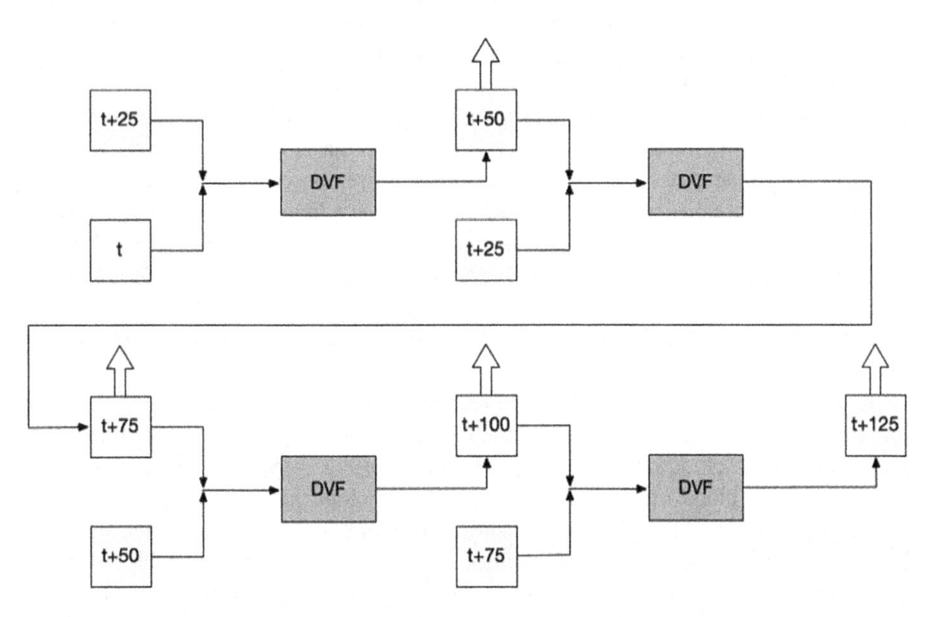

Fig. 3. Our cloud prediction model

3.1 Voxel Flow

The optical flow is a 2D vector describing instantaneous velocity of a point. Accordingly, the optical flow field is a 2D vector field, which reflects the gray change trend of each point in the image, and can be regarded as the instantaneous velocity field generated by the pixel points with gray level moving on the image plane. In a word, the information contained in the 2D vector field is the instantaneous velocity vector information of each image point, and such information can be termed as spatial information.

However, the spatial information is not enough to measure spatiotemporal sequences, such as the video frames and our cloud images. Spatiotemporal sequences not only have the spatial information, but also have the temporal information. The task of spatiotemporal sequence prediction has to take both spatial information and temporal information into consideration.

The voxel flow adds the temporal dimension to the original two-dimensional optical flow. It is a per-pixel 3D vector across space and time. The voxel flow field is on a 2D grid of integer target pixel location. Let the two video frames as the input of DVF be $\mathbf{X} \in \mathbb{R}^{H \times W \times 2}$, and the predicted video frame be $\mathbf{Y} \in \mathbb{R}^{H \times W}$, where H and W are the height and width of the frame, then the voxel flow field $\mathbf{F}$ can be expressed as $\mathbf{F} = (\Delta x, \Delta y, \Delta t)$. The first and second dimension of $\mathbf{F}$ represents the optical flow from the target frame to the next frame. It can be understood as the spatial component of the voxel flow. Especially, the optical flow is assumed to be locally linear and temporally symmetric between two consecutive frames. In this case, we will be able to find the location of a target pixel point in previous two frames. Let the coordinates of the target pixel point be (x, y), then we can get its coordinates in previous two frames as $(x - 2\Delta x, y - 2\Delta y)$ and $(x - \Delta x, y - \Delta y)$. Moreover, based on the assumption of local linearity and temporal symmetry between the two consecutive frames, the temporal component of voxel flow F is a linear blend weight between the previous two frames. In computer vision, this temporal component is called mask. When we use the selected image to cover the processed image to control the region of image processing. The selected image is called a mask. The mask is a binary image composed of 0 and 1. When a mask is applied to a processed image, the 1-value region is processed, while the 0-value region which is covered will not be processed. The volume sampling layer which uses the voxel flow to synthesize the predicted frame by trilinear interpolation will help us to better understand the function of the mask. This will be discussed in Sect. 3.3.

In conclusion, the voxel flow is composed of the spatial information F_{motion} and the temporal information F_{mask}. To reflect the concept of the voxel flow more vividly, we visualize it on our cloud image dataset, as is shown in Fig. 4. Specifically, we visualize the ground truth cloud images which have different cloud forms and their F_{motion} and F_{mask} which predicted by DVF. Visualization of the optical flow is achieved by using flow field color coding [10]. In this method, flow direction is encoded with color and magnitude is encoded with color intensity, as is shown in Fig. 5. And we use heatmap to visualize F_{mask}, so the connection between F_{mask} and previous two frames can be vividly showed, as is shown in Fig. 6.

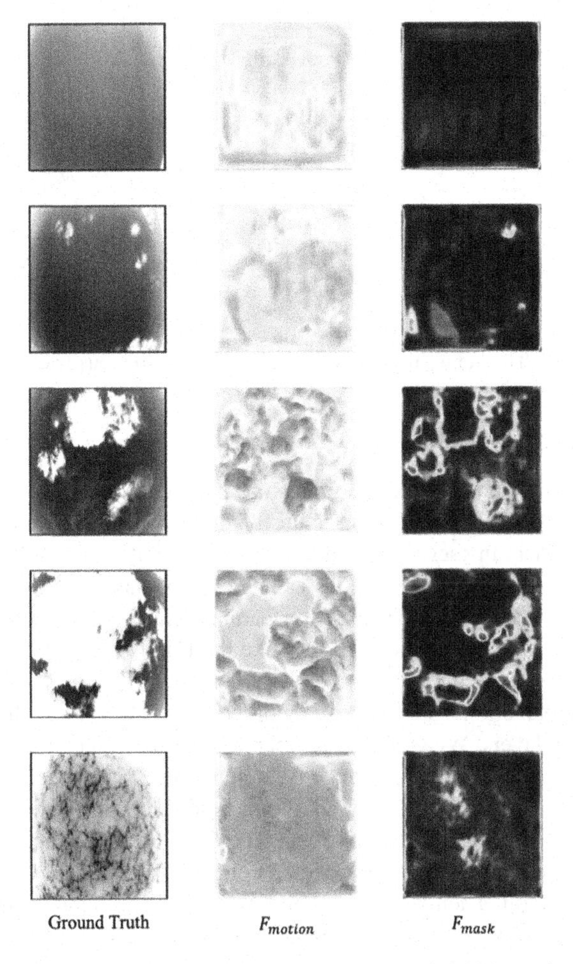

Fig. 4. Visualization of the voxel flow on our cloud image dataset

Fig. 5. The flow field color coding method

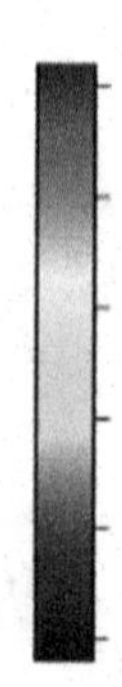

Fig. 6. The connection between F_{mask} and the previous two frames

3.2 Convolutional Encoder-Decoder

The convolutional encoder-decoder architecture is used to predict voxel flow with the input of two previous frames. The structure of this convolutional encoder-decoder is shown in Fig. 1. It is in fact a kind of U-Net [11] architecture. The U-Net was first proposed to solve the problem of medical image segmentation. U-Net is composed of two parts. The first part of U-Net is used for feature extraction and the second part is for up-sampling. As the whole structure of it is like the English letter U, it is termed as U-Net.

The feature extraction part of the network contains four convolution layers and three max pooling layers. We will get four feature maps of different sizes as 256×256, 128×128, 64×64 and 32×32. The convolution kernel sizes are 5×5, 5×5, 3×3 respectively. Then, in the up-sampling part, we first do deconvolution on the 32×32 feature map to get the 64×64 feature map. The 64×64 feature map is concatenated with the previous 64×64 feature map. Such concatenating action will better maintain the spatial information. Then we do convolution and up-sampling on the concatenated feature map to get the 128×128 feature map. And then we do the same thing to get the 256×256 feature map. Finally, through a bottleneck layer, we get the predicted voxel flow, which has a size of $3 \times 256 \times 256$.

3.3 Volume Sampling Layer

The volume sampling layer in DVF is used to synthesize the predicted frame by trilinear interpolation. The structure of it is shown in Fig. 2. The inputs of the volume sampling layer are the voxel flow generated by the convolutional encoder-decoder and the two previous frames, and the output is the predicted frame. The volume sampling function samples colors by interpolating within an optical-flow-aligned video volume computed from input $\mathbf{X}$.

In the paper Video Frame Synthesis using Deep Voxel Flow [7], the author has used mathematical expressions to explain the volume sampling function in detail. Here we use Python code to interpret the function of this volume sampling layer in a brief way, as is shown below. Firstly, we have to use F_{motion} to find the relationship between the output grids and their corresponding locations in the inputs. Then we fill the pixel value of the corresponding position in the inputs into the output grid. As there are two input frames, so we will get two interpolation results. Finally, we can use the F_{mask} to combine two interpolation results. Specifically, we can multiply the interpolation result of the first frame with *mask* and multiply the interpolation result of the second frame with $(1 - mask)$.

```
coor_x_1 = grid_x - flow[0, :, :] * 2
coor_y_1 = grid_y - flow[1, :, :] * 2
coor_x_2 = grid_x - flow[0, :, :]
coor_y_2 = grid_y - flow[1, :, :]


output_1 = torch.nn.functional.grid_sample(
    input[0:3, :, :],
    torch.stack([coor_x_1, coor_y_1], dim=2),
    padding_mode='border')

output_2 = torch.nn.functional.grid_sample(
    input[3:6, :, :],
    torch.stack([coor_x_2, coor_y_2], dim=2),
    padding_mode='border')


prediction = mask * output_1 + (1.0 - mask) * output_2
```

4 Experiments

4.1 Our Cloud Image Dataset

We use the cloud image dataset collected by our laboratory. Using the infrared imager independently developed, our laboratory collected cloud image sequences over a specific area of Nanjing in March 2019. Cloud images were taken every 5 s and cloud images collected are all gray images with a resolution of 720×480. In the data preprocessing part, we resize them to 256×256.

In this experiment of cloud prediction, we sample the cloud sequences every 5 images, which means the time interval between two cloud images in our sampled image sequences is 25 s. We do this because the cloud motion change is relatively slow when it is compared to the image motion in many life scenes. By lengthening the time interval between sampled cloud images, we can get more obvious cloud motion changes for model training, although this will make it more challenging to train the DVF.

For sampled image sequences in the training set, every 3 consecutive images make up a training data. The first and second image serve as the input of DVF and the third one serves as the label. For sampled image sequences in the testing set, every 6 consecutive images make up a testing data. The first and the second cloud images serve as the input of our cloud prediction model and the following 4 cloud images serve as the ground truth. Finally, we get 42240 data for training and 7848 data for testing.

4.2 Model Training

We use MSE loss as our loss function. And we use Adam as the optimizer. The original learning rate is 0.0001, and we adjust it dynamically in the training process. The batch size is set to be 32.

4.3 Model Evaluation

We use MSE, PSNR, SSIM to evaluate the testing result of our cloud prediction model. These three indexes are the main indexes for image quality evaluation. We compare the results of our cloud prediction model with the optical extrapolating method using Gunnar Farneback's algorithm. Specifically, we make use of the calcOpticalFlowFarneback function in OpenCV library to implement the optical flow extrapolating method.

Average MSE, PSNR and SSIM on the testing set are shown in Table 1. The change trend of these three indices as the prediction time increases is shown in Fig. 7. From the results we can see that our cloud prediction model is better than the optical extrapolating method in all these three indices. As the prediction time step increases, our model keeps a higher prediction accuracy and quality than the optical flow extrapolation method. Moreover, we can see from the line charts that with the increase of the prediction time step, the advantage of our model becomes bigger and bigger compared to the optical flow extrapolation method.

Table 1. Average MSE, PSNR and SSIM on the testing set

Time step	Our cloud prediction model	Optical flow extrapolation method
Average MSE		
t + 50	0.010	0.022
t + 75	0.017	0.041
t + 100	0.024	0.057
t + 125	0.030	0.072
Average PSNR		
t + 50	23.673	20.830
t + 75	20.601	17.097
t + 100	18.801	15.086
t + 125	17.604	13.790
Average SSIM		
t + 50	0.874	0.844
t + 75	0.836	0.776
t + 100	0.814	0.728
t + 125	0.799	0.692

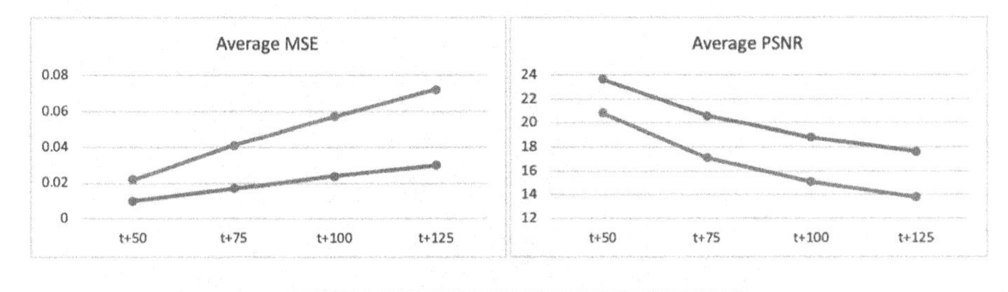

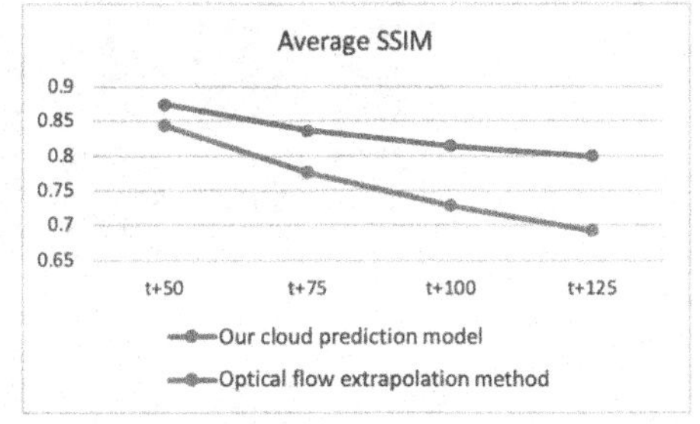

Fig. 7. The change trend of the three indices as the prediction time increases

Then we select prediction results of different cloud condition in the testing set and visualize them, as is shown in Fig. 8. In fact, the classification of cloud is a complex task. When a cloud is classified, cloud height, external characteristics, formation process and other principles are all needed to be considered. Here we mainly list several typical cloud conditions according to the amount and density of cloud.

As is shown in Fig. 8, prediction by our cloud prediction model is in the line of Prediction1, prediction by the optical flow extrapolation is in the line of Prediction2. Ground truth is in the third line. Also, to better show the difference between the prediction and ground truth, we visualize the difference images between the prediction result and ground truth in the fourth and fifth line. Difference1 is the difference between Prediction1 and the ground truth, and Difference2 is the difference between Prediction2 and the ground truth.

According to the visualized testing results, we can intuitively observe that our cloud prediction model can predict the location of cloud clusters with higher quality and accuracy than the optical flow extrapolation method. In particular, when the distribution of clouds is dense and messy, our cloud prediction model is obviously better than the optical flow extrapolation method. Our cloud prediction model is more accurate in the estimation of cloud location. Also, our model does better in the prediction of the edge information of future cloud images. In contrast, the optical flow extrapolation method brings much distortion. The position of the cloud cluster sometimes deviated greatly. Moreover, the optical flow tracking sometimes fails especially when the cloud condition changes rapidly, so we can see that there are obvious black spots on some cloud images predicted by the optical flow extrapolation method.

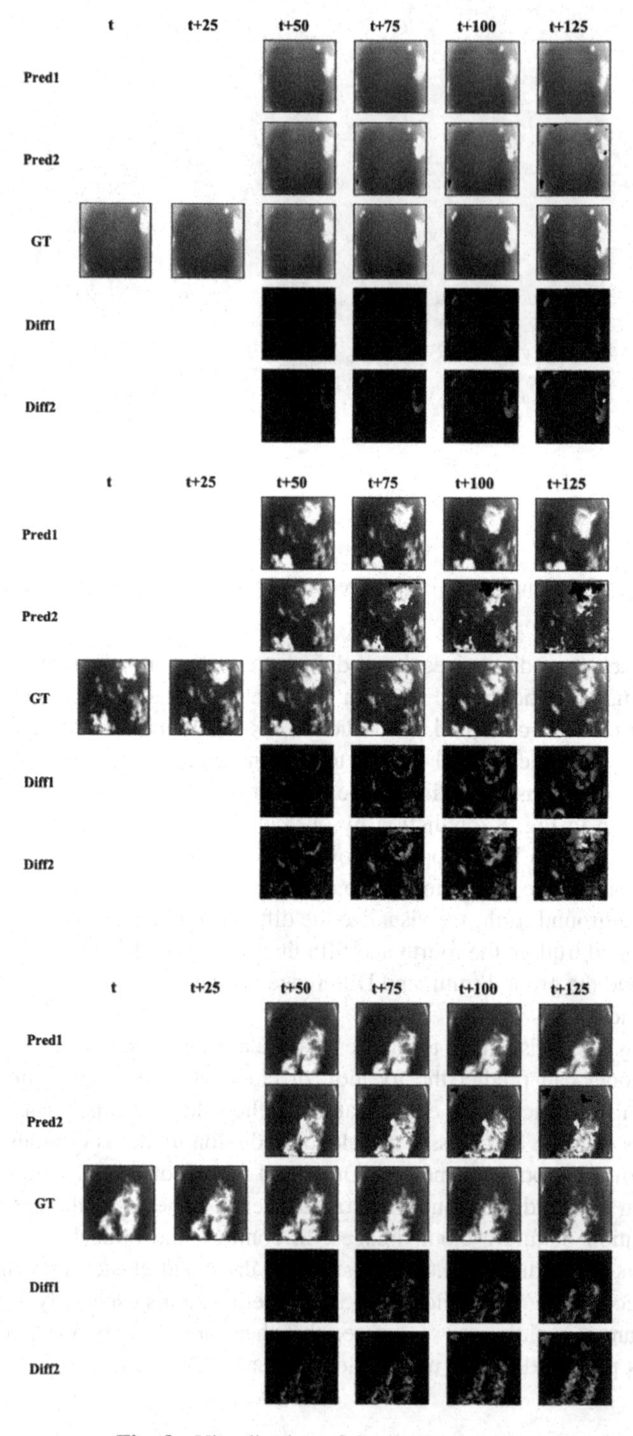

Fig. 8. Visualization of the testing results

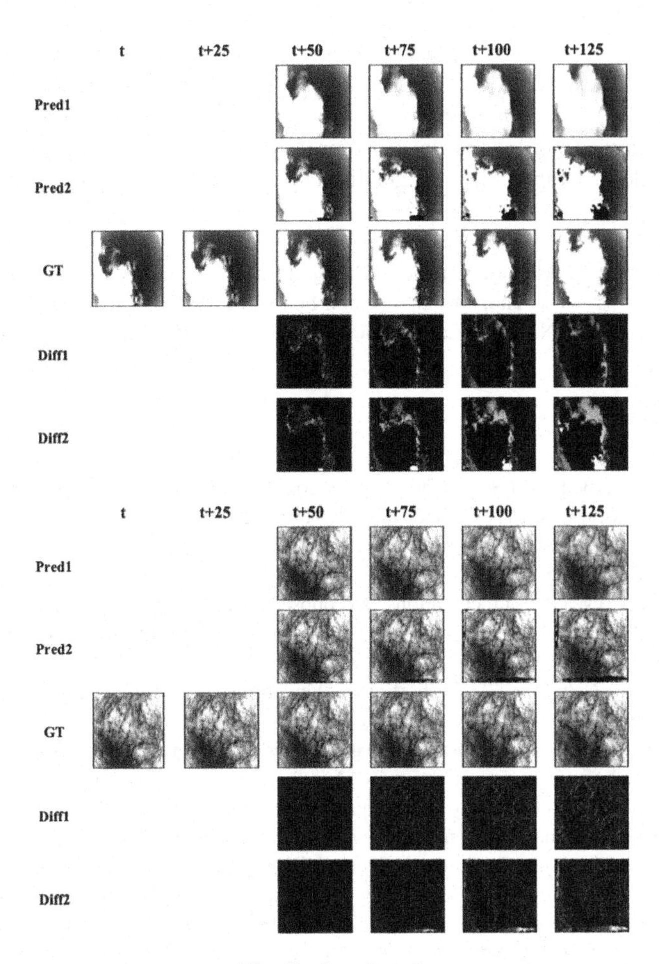

Fig. 8. (*continued*)

5 Conclusion and Future Work

In this paper, in order to deal with the problem of satellite-to-ground laser communication being susceptible to specific atmospheric environments, we propose a cloud prediction model based on the combination of optical flow and deep learning to predict future cloud images. Our model is based on Deep Voxel Flow (DVF), a CNN structure designed for video synthesis. By using DVF for multiple times to iterate the input cloud images at t second and t + 25 s, we can predict cloud images during the next 100 s. We train and test our cloud prediction model on the cloud image dataset collected by our laboratory. Our experimental results show that, compared to the optical flow extrapolation method which is a typical method used for nowcast, our cloud prediction model can predict future cloud images with higher quality and accuracy.

For future work, firstly we consider adding a cloud image pre-classification model. As the cloud conditions are complex, actually it will be challenging to train a single model to apply to all types of cloud condition. As a result, it will be sensible to add a cloud image pre-classification model and then train the cloud prediction model for different types of cloud conditions respectively. Moreover, we will do more research to improve the structure of the present cloud prediction model. We expect that our model can predict cloud images for a longer time span and with a higher accuracy in the future.

Acknowledgement. This work is supported by the 13th Five-Year Civil Aerospace Technology Pre Research Project, the Fundamental Research Funds for the Central Univesities under Grant 021014380187 and the National Natural Sciences Foundation of China under Grant 62131012.

References

1. Arnon, S., Kopeika, N.S.: Laser satellite communication network-vibration effect and possible solutions. Proc. IEEE **85**(10), 1646–1661 (1997)
2. Ricklin, J.C., et al.: Atmospheric channel effects on free-space laser communication. J. Opt. Fiber Commun. Rep. (2006)
3. Austin, K.: Nowcasting precipitation—a proposal for a way forward. J. Hydrol. (2000)
4. Lei, H.: Review on Development of Radar-based Storm Identification, Tracking and Forecasting. Meteorological Monthly (2007)
5. Hamill, T.M., Nehrkorn, T.: A short-term cloud forecast scheme using cross correlations. Weather Forecast. **8**(4), 401–411 (1993)
6. Farnebäck, G.: Two-frame motion estimation based on polynomial expansion. In: Bigun, J., Gustavsson, T. (eds.) Image Analysis. SCIA 2003. LNCS, vol. 2749, pp. 363–370. Springer, Berlin, Heidelberg (2003). https://doi.org/10.1007/3-540-45103-X_50
7. Liu, Z., et al.: Video frame synthesis using deep voxel flow. IEEE (2017)
8. Goodfellow, I.J., et al.: Generative Adversarial Nets. MIT Press, Cambridge (2014)
9. Fischer, P., et al.: FlowNet: learning optical flow with convolutional networks. In: 2015 IEEE International Conference on Computer Vision (ICCV). IEEE (2016)
10. Butler, D.J., et al.: A naturalistic open source movie for optical flow evaluation. In: Fitzgibbon, A., Lazebnik, S., Perona, P., Sato, Y., Schmid, C. (eds.) Computer Vision – ECCV 2012. ECCV 2012. LNCS, vol. 7577, pp. 611–625. Springer, Berlin, Heidelberg (2012). https://doi.org/10.1007/978-3-642-33783-3_44
11. Ronneberger, O., Fischer, P., Brox, T.: U-Net: convolutional networks for biomedical image segmentation. In: Navab, N., Hornegger, J., Wells, W., Frangi, A. (eds.) Medical Image Computing and Computer-Assisted Intervention – MICCAI 2015. MICCAI 2015. LNCS, vol. 9351, pp. 234–241. Springer, Cham (2015). https://doi.org/10.1007/978-3-319-24574-4_28

Chaotic Sea Clutter Modeling Based on Gray Wolf Algorithm

Jurong Hu(✉), Bao Zeng, and Xujie Li

Hohai University, Nanjing 210098, China

Abstract. This paper mainly verifies the chaotic characteristics of sea clutter and studies the prediction of chaotic sea clutter. The chaotic phase space was reconstructed by calculating the delay time and embedding dimension pairs. This paper provide a sea clutter prediction method based on the Grey Wolf Optimizer (GWO). The iterative results of the Grey Wolf algorithm were used as the weights of the RBF neural network. The simulation shows the predicted value of GWO-RBF network is close to the real value, and the prediction error is small, so as to verify the effectiveness of GWO-RBF network for chaotic sea clutter prediction.

Keywords: Chaos · Sea clutter · Phase space reconstruction · Grey Wolf algorithm · RBF neural network

1 Introduction

Sea clutter is the backscattered echo of a small area of ocean covered by the transmitted radar signal, and this signal is a time series with chaotic properties. Chaotic time series are a part of chaos theory and a representation of the time change of some states of a chaotic system. Thus the properties of chaotic time series are also properties of chaos theory. Using chaotic time series to reconstruct chaotic system is the first step to study chaotic system. Phase space is reconstructed according to Takens theorem [1], and chaotic system is reconstructed by combining neural network, deep learning and other methods, so as to realize the analysis of chaotic system. Therefore, the study of chaotic time series becomes one of the important research directions in the field of radar signal processing [2].

Leng et al. [3], found for the first time that the sea clutter in the X-band had chaotic characteristics by calculating the correlation dimension of sea clutter in 1990. Later, a large number of literatures showed that sea clutter had chaotic characteristics. In 2013, R. A. Simon et al. [4] proved that the randomness of sea clutter data was generated by deterministic chaos rather than random process through a large number of experiments. Meanwhile, Lyapunov exponent is calculated, and the result is greater than 0, which further verifies the chaotic characteristics of sea clutter.

This work was supported in part by the Provincial Key Research and Development Program of Jiangsu under Grant BE2019017, in part by the Provincial Water Science and Technology Program of Jiangsu under Grant 2020028.

Q. Guo et al. (Eds.): WiSATS 2021, LNICST 410, pp. 445–455, 2022.
https://doi.org/10.1007/978-3-030-93398-2_42

In recent years, through in-depth study of sea clutter, it is found that sea clutter is not a complete random time series. For the most part, it contains some features that are evident in chaotic systems. Because chaotic systems belong to deterministic systems, the trajectories of chaotic attractors can be predicted in a very short time [5, 6]. Early researchers considered sea clutter as a random signal and proposed a variety of modeling methods. The common problems of these modeling methods are complex establishment process, poor universality and low accuracy. These traditional sea clutter prediction methods adopt statistical process, which essentially cannot make accurate prediction. The new method combines neural network, machine learning and deep learning methods to model sea clutter using chaotic characteristics. Therefore, through in-depth research on the chaotic sea clutter can provide a new development direction for sea clutter prediction and sea target detection. Neural network and some other nonlinear algorithms have also been applied to target detection in the marine environment [7, 8].

At present, most of the research on sea clutter mainly focuses on the construction of the probability model of sea clutter [9], or the classification of clutter and target by using neural network classifier [10]. The application of these new technologies and algorithms makes the target detection technology under the background of sea clutter develop continuously and innovate, and makes the detection level reach a new height.

In 2013, Mirjalili et al. [11], put forward the Gray Wolf algorithm based on the cooperative hunting mechanism of simulated wolves in accordance with the inspiration of group hunting of gray wolves. Grey Wolf algorithm has a rigorous hierarchical system, adaptive convergence factor and information feedback system, and can effectively improve global search and local optimization. Therefore, this algorithm has a good solution effect in terms of accuracy and convergence speed.

2 Chaotic Characteristics of Sea Clutter

2.1 Phase Space Reconstruction

The first step in studying chaos theory is to reconstruct phase space. In the basic theory of phase space reconstruction, the change of any variable in space is associated with other variables, and they interact with each other. According to the theory proposed by Packard et al. [12], the phase space can be reconstructed by using the delay coordinate of a variable in the dynamical system. Takens proved that under appropriate embedding dimensions [1], the dimensions of the prime dynamic system can be restored.

The first step in reconstructing phase space is to calculate two parameters, One is the embedding dimension and the other is the delay time. The mutual information method is used to calculate the delay time and the False Nearest Neighbors method is used to calculate the embedding dimension [13]. Set time series $x(i) i = 1, 2, 3, 4, ..., N$, the delay time is τ, and the embedding dimension is m, then the phase space reconstruction vector is:

$$Y(i) = \{x(i), x(i+\tau), \cdots, x(i+(m-1)\tau)\} \tag{1}$$

2.2 Delay Time

Let the sampling data set $X(j) = x(t_0 + jT)$, where t_0 is the sampling time, T is the sample period, $j = 1, 2, \cdots, N$. Then its correlation function is:

$$C(\tau) = \frac{\sum_{j=1}^{N} [x(t_0 + jT + \tau) - \langle x \rangle][x(t_0 + jT) - \langle x \rangle]}{\sum_{j=1}^{N} [x(t_0 + jT) - \langle x \rangle]^2} \tag{2}$$

Where,

$$\langle x \rangle = \frac{1}{N} \sum_{j=1}^{N} x(t_0 + jT) \tag{3}$$

If the autocorrelation function has a zero crossing at time τ, then τ can be used as the optimal solution. Otherwise, the first local minimum value is used as the best delay time.

2.3 Embedding Dimension

In an m-dimensional phase space, the vector of the phase space attractor is $X(j) = \{x(j), x(j+\tau), \ldots, x(j+(m-1)\tau)\}, j = 1, 2, \cdots, N$. Chaotic time series can be regarded as the projection of the trajectory of the attractor in the high-dimensional space into the low-dimensional space, and the two adjacent points projected are false adjacent points. Therefore, when the dimension m changes to $m+1$, it is necessary to investigate which adjacent points of the trajectory line $x(j)$ are real adjacent points and which are false adjacent points. Set $X^{NN}(j)$ is the neighbor of $X(j)$, and the distance between them be:

$$R_m(j) = \left\| X(j) - X^{NN}(j) \right\| \tag{4}$$

When the dimension is incremented from m to $m+1$, the $R_{m+1}(j)$ is:

$$R_{m+1}(j) = \sqrt{R_l^2(j) + \| X(j+\tau m) - X^{NN}(j+\tau m) \|} \tag{5}$$

$$a_1(j, m) = \frac{|X(j+\tau m) - X^{NN}(j+\tau m)|}{R_m(j)} > R_r \tag{6}$$

$$a_2(j, m) = \frac{\left\| X_{m+1}(j) - X_{m+1}^{NN}(j) \right\|}{\left\| X_m(j) - X_m^{NN}(j) \right\|} \tag{7}$$

If $R_{m+1}(j)$ is much larger than $R_m(j)$, it can be considered that these two points are not the nearest neighbors, but the higher-dimensional attractors are projected onto the lower-dimensional trajectories. So they're not really the nearest neighbor.

Among Eq. (6) (7), $X_m(j)$ and $X_m^{NN}(j)$ of m dimensional space of J th vector and its adjacent points, $X_{m+1}(j)$ and $X_{m+1}^{NN}(j)$ of m + 1 dimensional space of the J th vector and its adjacent points, $R_r \in [10, 50]$. Define:

$$\begin{cases} E(m) = \frac{1}{N-m\tau} \sum_{i=1}^{N-m\tau} a_2(i, m) \\ E_2(m) = E(m+1)/E(m) \end{cases} \quad (8)$$

If the time series is deterministic, then $E(m)$ will unchanged after m is greater than a value. If the time series is a random, then $E(m)$ should increase gradually. However, in the application, it is difficult to judge whether E(m) of finite time series changes slowly or has been stable, so supplement to a condition:

$$\begin{cases} E(m) = \frac{1}{N-m\tau} \sum_{i=1}^{N-m\tau} X(j+\tau m) - X^{NN}(j+\tau m) \\ E_2(m) = E(m+1)/E(m) \end{cases} \quad (9)$$

For a deterministic time series, the correlation between data points is dependent on the change of embedding dimension m values, and there are always some m values where $E_2(m)$ does not equal 1.

2.4 Correlation Dimension

The correlation dimension describes the distribution of points in dynamic space. The correlation dimension is usually calculated by the G-P (Grassber-Procaccia) method. $X_1, X_2, \ldots, X_N$ is the point in the phase space, and given a distance r, the proportion of the phase points whose two points are more than r is:

$$C(r) = \frac{1}{N(N-1)} \sum_{i=1, i \neq j}^{N} \sum_{j=1}^{N} H(r - \|X_i - X_j\|) \quad (10)$$

Among them, $\|\cdot\|$ is Euclidean norm, H is Heaviside step function:

$$H(x) = \begin{cases} 1 & x \geq 0 \\ 0 & x < 0 \end{cases} \quad (11)$$

So, D_2 is:

$$D_2 = \lim_{r \to 0} \frac{\ln(C(r))}{\ln(r)} \quad (12)$$

When, $r \to 0$ $C(r) \approx r^{D_2}$:

$$\lim_{r\to 0} C(r) = \lim_{r\to\infty} \frac{1}{N(N-1)} \sum_{i=1,i\neq j}^{N} \sum_{j=1}^{N} H(r - \|X_i - X_j\|) \tag{13}$$

2.5 Lyapunov Exponents

Lyapunov exponent is a quantitative index to describe the state evolution of a dynamical system. It measures the average divergence or convergence rate of adjacent orbits in the phase space of the system. Small data volume method is a method to calculate Lyapunov exponent with less data and low complexity. Its basic principle is as follows: after the reconstruction of time series, look for the nearest neighbor $X_{\hat{j}}$ of the JTH point X_j on the orbit, that is:

$$d_j(0) = \min\left\|X_j - X_{\hat{j}}\right\|,\ \left|j - \hat{j}\right| > p \tag{14}$$

Among them, p is the period of X_j. And calculate the average dispersion of each point to get largest Lyapunov exponent. So largest Lyapunov exponent is:

$$\lambda_1(i) = \frac{1}{i\Delta t} \frac{1}{(N-i)} \sum_{j=1}^{N-i} \ln \frac{d_j(i)}{d_j(0)} \tag{15}$$

Where, Δt is the sample period, $d_j(i)$ is the distance of the JTH pair of nearest neighbors on the orbit after i discrete time steps. The above equation is rewritten as follows:

$$\lambda_1(i,k) = \frac{1}{k\Delta t} \frac{1}{(N-k)} \sum_{j=1}^{N-k} \ln \frac{d_j(i+k)}{d_j(i)} \tag{16}$$

Where, K is constant,

$$d_j(i) \approx C_j e^{\lambda_1(i\cdot\Delta t)},\ C_j = d_j(0) \tag{17}$$

$$\ln(d_j(i)) = \ln(C_j) + \lambda_1(i\cdot\Delta t), j = 1, 2, \cdots, N-(m-1)\tau \tag{18}$$

The value of the largest Lyapunov exponent can be obtained by approximating the slope of Eq. (17) by the least square method:

$$x(j) = \frac{1}{\Delta t}\langle \ln(d_j(t)) \rangle \tag{19}$$

Where, $\langle\cdot\rangle$ is taking the average of all about j points.

3 Grey Wolf Algorithm

3.1 Grey Wolf Algorithm

In 2013, Mirjalili proposed a new group intelligence algorithm based on the hunting and inter-population relations of gray wolves. The Gray Wolf optimization algorithm mainly imitates the hunting behavior of natural Gray wolves to solve complex optimization problems [11, 14].

In the Gray Wolf algorithm, there are four different levels of Wolf α, β, δ, ω, and the different levels of Wolf represent different solutions. α represents the optimal solution, β represents second optimal solution, δ represents the third optimal solution, ω Wolf is the solution that updates α, β, δ.

During each iteration, the position of ω is actively updated, replacing the corresponding α, β, δ with the position of ω only when the solution of ω is better than that of α, β, δ. The position update formula of the three wolves is as follows:

$$d_\alpha = |Y_\alpha \cdot C_1 - Y|, d_\beta = |Y_\beta \cdot C_2 - Y|, d_\delta = |Y_\delta \cdot C_3 - Y| \tag{20}$$

$$Y_1 = Y_\alpha - d_\alpha \cdot A_1, Y_2 = Y_\beta - d_\beta \cdot A_2, Y_3 = Y_\delta - d_\delta \cdot A_3 \tag{21}$$

$$Y(t+1) = (Y_1 + Y_2 + Y_3)/3 \tag{22}$$

Where, Y_α, Y_β, Y_δ are the position vectors of α, β and δ, and Y is the position vector of ω in iteration t, $Y(t+1)$ is the position vector of ω in iteration $t+1$, The formula for A and C is:

$$A = \mathbf{r_1} \cdot 2\alpha - \alpha, \quad C = 2 \cdot \mathbf{r_2} \tag{23}$$

$\boldsymbol{r}_1$ and $\boldsymbol{r}_2$ are random numbers in [0, 1], as the number of iterations of the algorithm increases, α gradually decreases from 2 to 0.

$x_i^j(t)$ is jth dimensional position value of the ith wolf in the current iteration, x_{max} is maximum position, x_{min} is minimum position.

3.2 RBF Neural Network Based on Gray Wolf Optimization Algorithm

RBF neural network has excellent performance characteristics such as simple structure, self-learning, self-organization, self-adaptation, easy to overcome local minimum, etc. [8], and has been widely used in online identification, model approximation and parameter learning of nonlinear systems. Because of this, this paper will use RBF neural network as the basis to achieve the prediction of chaotic sea clutter. Meanwhile, Gray Wolf algorithm was used to train the parameters of RBF neural network to improve the prediction accuracy.

In Fig. 1, the input vector of the RBF network is $\boldsymbol{x} = [x_1, x_2, \cdots, x_i]$ input of the network. Let RBF neural network have N hidden layers, the output vector of the hidden layer is h, $h = [h_1, h_2, \cdots, h_N]$, the calculation formula of h_i is:

$$h_j = \exp(\frac{\|\mathbf{x} - \varphi_j\|^2}{2\theta_j^2}), j = 1, \cdots, N \tag{24}$$

$$y_i = \sum_{j=1}^{N} p_{ij} h_j, j = 1, \cdots, N \tag{25}$$

Where, $\boldsymbol{x}$ is input value, $\boldsymbol{\varphi}_j$ is central vector of the JTH hidden layer neuron, $\boldsymbol{\varphi}_j = [\varphi_{1,j}, \varphi_{2,j}, \cdots, \varphi_{r,j}]$. N central vectors constitute a central vector matrix $\boldsymbol{\Phi} = [\varphi_1, \varphi_2, \cdots, \varphi_N]$, θ_j is the Gaussian RMS width value of the jth hidden layer neuron $\boldsymbol{\theta} = [\theta_1, \theta_2, \cdots, \theta_N]^{\mathrm{T}} (\theta_j > 0)$, Y is final output of RBF network $Y = [y_1, y_2, \cdots, y_n]$, p_{ij} is transmission weights between neurons in the JTH hiding layer and the ITH output neuron and weight matrix is $\boldsymbol{p} = [p_1, p_2, \cdots, p_N]^T$.

In the GWO-RBF network, $\boldsymbol{\theta}$, $\boldsymbol{\Phi}$ and $\boldsymbol{p}$ are the wolf position vectors, and the optimal objective function is selected as follows:

$$E = \frac{1}{2n} \sum_{i=1}^{n} (y_i - y_{mi})^2 \tag{26}$$

Where, the number of training samples is N, y_{mi} is the expected output value of the ith $\boldsymbol{Y_m} = [y_{m1}, y_{m2}, \cdots, y_{mn}]$, y_i is the actual value of the ITH $\boldsymbol{Y} = [y_1, y_2, \cdots, y_n]$. The steps of chaotic sea clutter prediction by GWO-RBF network are as follows:

Step1: The network input data is reconstructed sea clutter data.
Step2: Compared to the expected value and output data.
Step3: Output data put in GWO.
Step4: Repeat Step1 ~ Step3.until the end of the iteration.

4 Simulation and Modeling

The validity of the proposed GWO-RBF network for sea clutter prediction is verified through a group of comparative experiments. The first group of experiments used RBF neural network to predict the sequence, and the second group of experiments used GWO-RBF network to predict the sequence, and the data were all sea clutter data.

The sea clutter data came from IPIX radar data collected by McMaster University in Canada in 1993, which spanned 14 range cells. We select the data in one of the unit doors as the test data.

The experimental data is affected by a variety of errors, and the data should be preprocessed before analysis to reduce these errors and eliminate the influence of different attributes of data with different orders of magnitude.

Figure 2 shows the embedding dimension. The intersecting point of $E_1(d)$ and $E_2(d)$ is the embedding dimension calculated by CAO method, and $E_1(d)$ gradually

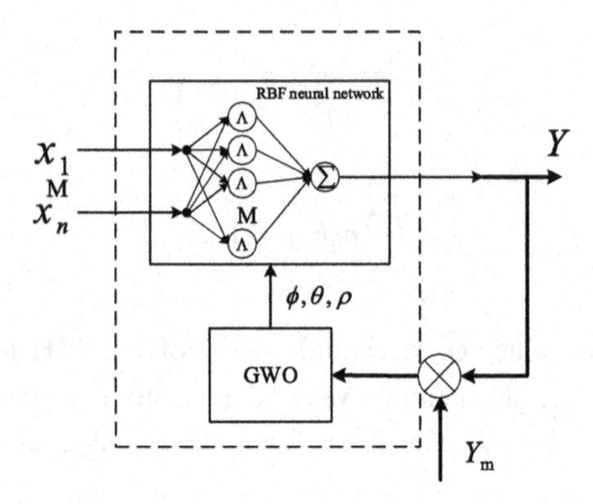

Fig. 1. RBF-GWO network structure

approaches to 1 with the increase of dimension, while $E_2(d)$ slowly changes around 1 with the increase of dimension. According to the figure, the embedding dimension is 8.

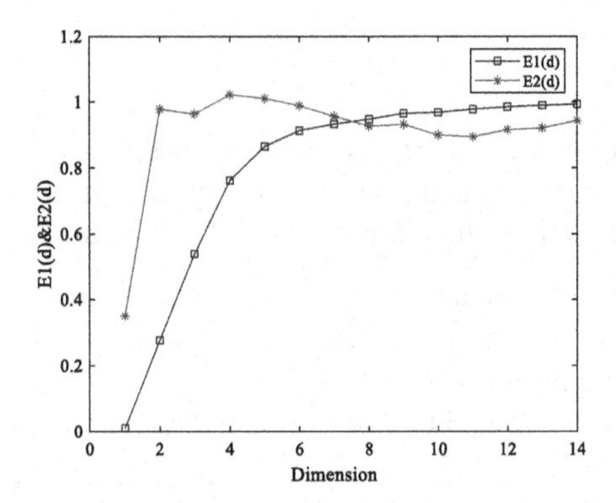

Fig. 2. Embedding dimension

Figure 3 shows the correlation dimensions under different embedded dimension change curve of the figure shows, with the increase of critical distance $log(r)$ correlation integral $log\,(C(r))$ increase gradually and tends to a stable value, the correlation integral every curve fitting, the slope is the correlation dimension of different number of embedded, correlation dimension drawing curve as shown in Fig. 3(a). As the embedding dimension increases, the correlation dimension is gradually becoming larger and tends to be stable. This stable value is the correlation dimension requested. In Fig. 3(b), the correlation dimension is approximately 4. When the image slope is a positive value, the correlation dimension is a chaotic invariant, which proves that one characteristic of sea clutter sequence is chaos.

In this paper, we choose small data sets method as the method to calculate largest Lyapunov exponents, using formula (13)–(16) to calculate relationship between $\mathrm{ln}d_j(i)$ and $\lambda_1(i)$. The slope is calculated by the least square method. And then we get largest

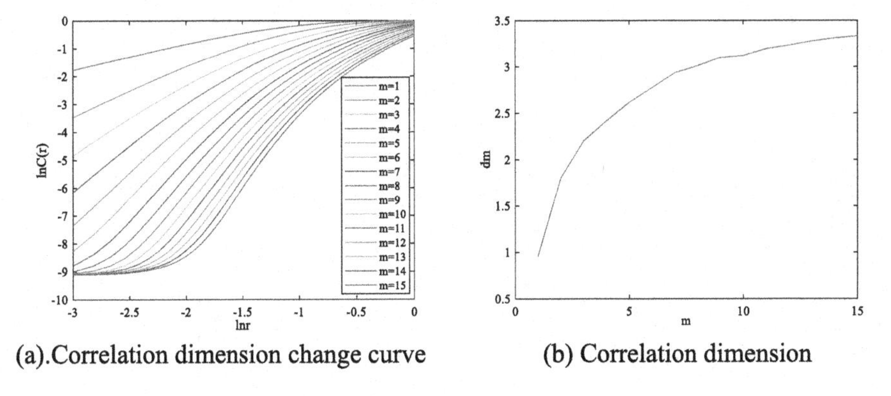

(a).Correlation dimension change curve (b) Correlation dimension

Fig. 3. Correlation dimension

Lyapunov exponent is equal to 0.0181. In Fig. 4 (a), shown the largest Lyapunov exponent is 0.0181, and Fig. 4(b) is the Lyapunov exponent curve drawn by the sea clutter data of four different distances. As can be seen from the figure, the two graphs have certain upward fluctuations in the initial stage, which is caused by the chaotic superposition of the actual sea clutter books. But the overall trend is upward.

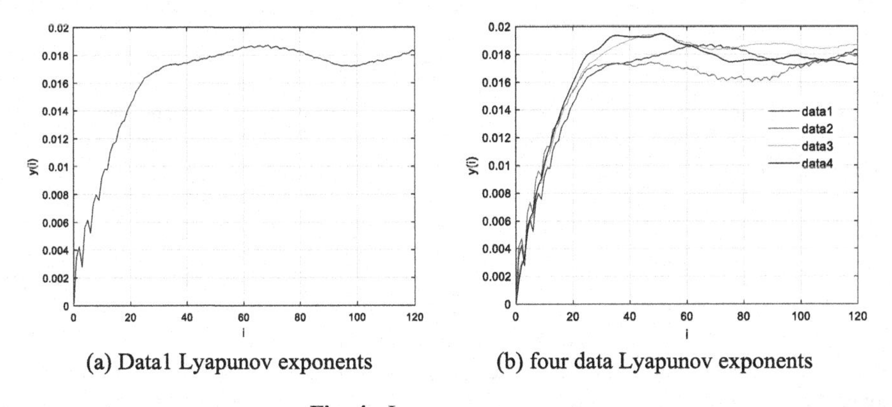

(a) Data1 Lyapunov exponents (b) four data Lyapunov exponents

Fig. 4. Lyapunov exponents

By combining the correlation dimension and the calculation results of the maximum A, it can be concluded that the clutter data has chaotic characteristics.

The reconstructed data were input to the input end of the neural network. The first 1800 data were selected as the test data, and the last 200 data were selected as the detection data.

Therefore, two sets of different results were obtained. Figure 5(a) is the comparison between the predicted and true sea clutter values of GWO-RBF network. The figure shows, Only using RBF neural network to predict sea clutter is not effective, but using GWO-RBF network predict sea clutter series, the predicted results are very close to the actual values.

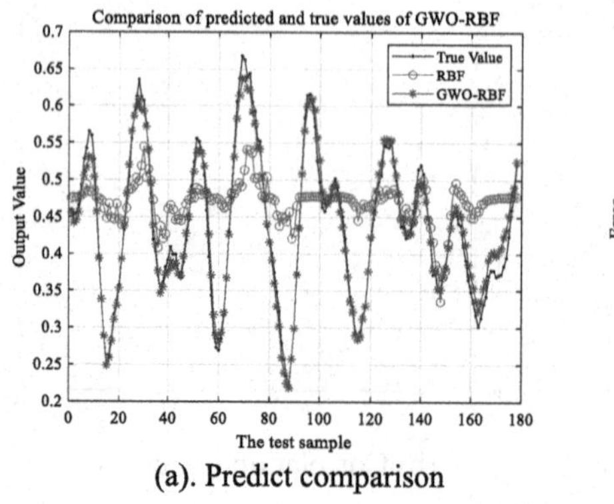

(a). Predict comparison

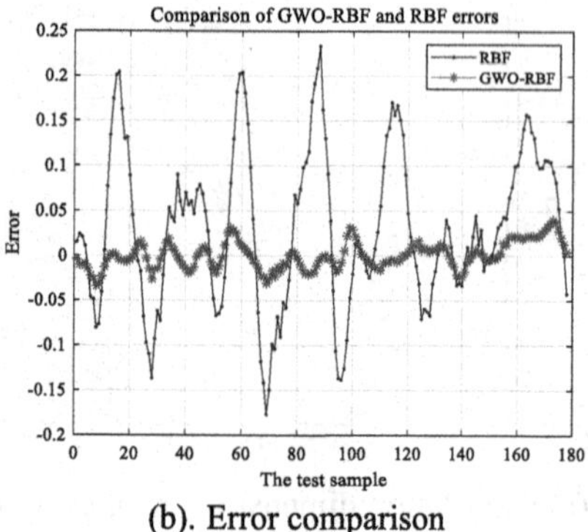

(b). Error comparison

Fig. 5. Prediction model and error

Figure 5(b) is the error comparison diagram of RBF and GWO-RBF network. According to the figure, the error of GWO-RBF network hovers around 0, while the error of RBF is a large fluctuation of neural network. In terms of error, the error of GWO-RBF network is smaller than that of RBF network. It is proved that the prediction result of GWO-RBF network is more accurate and the error is less when it is used in sea clutter prediction.

Table 1. Error analysis

	Mean square error	Root of mean square error	Mean absolute error	Mean absolute percentage error
Before	0.0083811	0.091548	0.072779	19.56%
After	4.7054e-05	0.0068596	0.005162	1.16%

Table 1 is the error recording table of before and after optimization. It can be seen from the table that the mean square error before optimization is much larger than after optimization, and the mean absolute error is also greater than the result after optimization.

5 Conclusion

This paper mainly verifies whether sea clutter has chaotic characteristics. First of all, by calculating delay time (τ) and embedding dimension (m), reconstructing phase space, The second, calculating correlation dimension of sea clutter by G-P algorithm. On the basis of the slope of image correlation dimension is greater than 0. The last calculating the largest Lyapunov exponent by small data sets method, and largest Lyapunov exponent is 0.0181, According to the calculated correlation dimension and the maximum Lyapunov exponent, it can be concluded that the clutter sequence has chaotic characteristics.

The gray wolf algorithm is combined with RBF neural network to predict the sea clutter sequence after phase space reconstruction. The simulation results show that the GWO-RBF network can effectively reduce the prediction error, the prediction result of sea clutter sequence is very close to the real value, and the prediction effect is obviously better than that of the RBF neural network.

References

1. Takens, F.: Detecting strange attractors in turbulence. In: Rand, D., Young, L.S. (eds.) Dynamical Systems and Turbulence, Warwick 1980. Lecture Notes in Mathematics, vol. 898, pp. 366–381. Springer, Berlin, Heidelberg (1981). https://doi.org/10.1007/BFb0091924
2. Callaghan, D., Burger, J., Mishra, A.K.: A machine learning approach to radar sea clutter suppression. In: IEEE Radar Conference, pp. 1222–1227 (2017)
3. Leung, H., Haykin, S.: Is there a radar clutter attractor? Appl. Phys. Lett. **56**(6), 593–595 (1990)
4. Simon, R.A., Kumar, P.V.: A nonlinear sea clutter analysis using chaotic system. In: 2013 Fourth International Conference on Computing, Communications and Networking Technologies (ICCCNT) (2013)
5. Haykin, S., Puthusserypady, S.: Chaotic dynamics of sea clutter. Chaos **7**(4), 777–802 (1997)
6. Leung, H.: Applying chaos to radar detection in an ocean environment: an experimental study. IEEE J. Oceanic Eng. **20**(1), 56–64 (1995)
7. Shen, Y., Li, G.: The chaotic neural network is used to predict the sea clutter signal. In: 2009 International Conference on Artificial Intelligence and Computational Intelligence, pp. 25–30 (2009)
8. Haykin, S., Puthusserypady, S.: Chaos, sea clutter, and neural networks. In: Conference Record of the Thirty-First Asilomar Conference on Signals, Systems and Computers, pp. 1224–1227 (1998)
9. Jayaprakash, A., Reddy, G.R., Prasad, N.S.S.R.K.: Small target detection within sea clutter based on fractal analysis. Procedia Technol. **24**, 988–995 (2016)
10. Pathak, J., et al.: Using machine learning to replicate chaotic attractors and calculate Lyapunov exponents from data. Chaos **27**(12) (2017)
11. Mirjalili, S., Mirjalili, S.M., Lewis, A.: Grey Wolf optimizer. Adv. Eng. Softw. **69**, 46–61 (2014)
12. Packard, N.H., Crutchfield, J.P.: Geometry from a time series. Phys. Rev. Lett. 712–716 (1980)
13. Cao, L.: Practical method for determining the minimum embedding dimension of a scalar time series. Phys. D Nonlinear Phenom. **110**(1) (1997)
14. Mirjalili, S.: How effective is the Grey Wolf optimizer in training multi-layer perceptrons. Appl. Intell. **43**(1), 150–161 (2015)

Low Overhead Growth Degree Coding Scheme for Online Fountain Codes with Limited Feedback

Pengcheng Shi[1], Zhenyong Wang[1,2(✉)], and Dezhi Li[1]

[1] School of Electronics and Information Engineering, Harbin Institute of Technology, Harbin 150001, China
ZYWang@hit.edu.cn
[2] Shenzhen Academy of Aerospace Technology, Shenzhen 518057, China

Abstract. A new growth degree encoding scheme (GDS) for online fountain codes is proposed to achieve a low overhead when the feedback is limited. When the feedback points are determined at the completion phase, the encoder sends coded symbols with growth degrees between the two feedback points, rather than symbols with fixed degrees. This increases the effective probability of the coded symbols, thereby reducing the overall overhead. We analyze the overhead of the proposed scheme to demonstrate the performance. Simulation results show that our proposed scheme has better overhead performance compared to the conventional online fountain codes with limited feedback.

Keywords: Online fountain codes · Feedback · Overhead analysis · Rateless codes

1 Introduction

Fountain codes, also called rateless codes, are a class of erasure correction codes that can generate an infinite number of coded symbols from a limited number of original symbols through the eXclusive-OR (XOR) operation. In 2002, Luby proposed the first practical realization of digital fountain, named LT codes [1]. In 2006, Raptor codes are proposed by Shokrollahi through cascading fixed-rate codes and LT codes [2]. Later, spinal codes [3] and online fountain codes (OFC) [4] were proposed as new classes of fountain codes to transmit data efficiently and reliably.

Fountain codes were initially designed to transmit information without feedback. However, as it evolved, it made sense to use a small amount of feedback to enhance its performance. In 2015, online fountain codes (OFC) [4] were first proposed by Cassuto and Shokrollahi. The online property means that once given an instantaneous decoding state, the encoder can find the optimal coding strategy efficiently. Online fountain codes as a class of fountain codes, which feedback the current decoding state of the receiver to the transmitter over the feedback

Q. Guo et al. (Eds.): WiSATS 2021, LNICST 410, pp. 456–467, 2022.
https://doi.org/10.1007/978-3-030-93398-2_43

channel, so that the transmitter can adjust the degree value of coding symbols according to the feedback information, thereby efficiently transmitting encoding symbols and reducing decoding overhead. Later, in [5] and [6], the full recovery performance of online fountain codes is improved by non-random selection of original symbols. The intermediate performance is important in many applications such as audio and video streaming. Therefore, the intermediate performance of online fountain codes were studied in recent years [7–9]. In [10] and [11], the unequal error protection (UEP) property and unequal recovery time (URT) property are studied in online fountain codes.

With the research on online fountain codes, they have been applied to many applications such as wireless sensor networks [12] and satellite broadcast system [13]. However, the existing methods are designed for unlimited feedback. In practical applications, due to the shortage of feedback resources, the feedback times are usually limited. Therefore, it is a meaningful research direction to study how to efficiently transmit data with online fountain codes under limited feedback scenario. In [14], the authors first studied the feedback strategies for online fountain codes with limited feedback and proposed two schemes to select the optimal feedback points. However, since the degree value of the symbols sent between the two feedback points is fixed, when there are too many optimal degree values are skipped by the adjacent feedback points, the overhead will be high. This explains why the feedback strategy has a large overhead when the number of feedbacks is small. To address this problem, we propose the growth degree encoding scheme (GDS), which sends coded symbols with growth degrees between adjacent feedback points. This increases the effective probability of coded symbols, thus reducing the overall overhead.

The rest of this paper is organized as follows. Section 2 introduces the online fountain codes with limited feedback. The proposed growth degree encoding scheme are provided in Sect. 3. Section 4 provides the overhead analysis for the proposed scheme. The simulation results are given in Sect. 5. Section 6 concludes this paper.

2 Online Fountain Codes with Limited Feedback

In this section, we briefly review the online fountain codes (OFC) and the online fountain scheme with limited feedback. For the online fountain codes, the uni-partite graph are introduced to represent the decoding state. As shown in Fig. 1, an example of the bi-partite and the corresponding uni-partite graph is presented, the original symbols, or called source symbols, are represented by circle nodes, while the output symbols, or called encoded symbols, are represented by square nodes. In the bi-partite graph, if a square node are neighbored with two circle nodes by edges, it means the corresponding encoded symbol are generated by the XOR of these two source symbols. While in the uni-partite graph, the nodes are only used to represent the source symbols. The edge between two nodes indicates that an encoded symbol generated by the XOR of the two source symbols is received by the decoder. Blacked node represents the source symbol that have been recovered. A component means a set that any two circle nodes in this set are connected by an edge. Obviously, a component is decoded when one input symbol in this set is colored black. Note that the size of a component is the number of source symbols in this component.

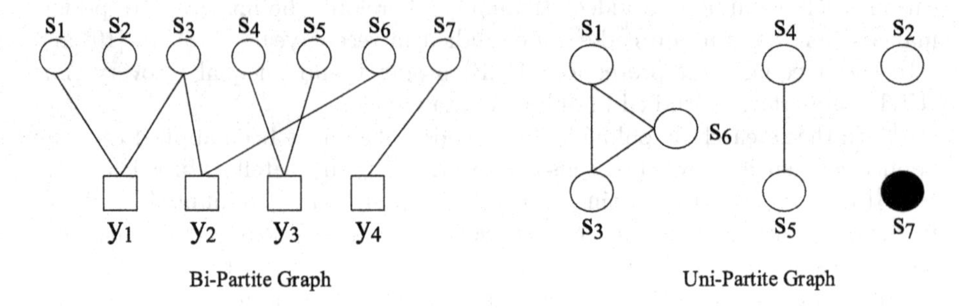

Fig. 1. A bi-partite graph for online fountain codes and the corresponding uni-partite graph.

The decoder receives coded symbols and updates the uni-partite graph. According to the uni-partite graph, the decoder can clear the current decoding state and get the optimal degree value at this time. When the optimal degree value changes, the decoder feeds back the decoding state to the transmitter. Then the transmitter modifies the degree value to transmit encoded symbols. At the decoder, an coded symbol is useful when it belongs to the following two cases.

- **Case 1:** A received symbol that degree m is the XOR of a single white symbol and $m-1$ black symbols.
- **Case 2:** A received symbol that degree m is the XOR of two white symbols and $m-2$ black symbols.

We assume there are k source symbols. The encoding process of online fountain codes is divided into two phases: build-up phase and completion phase.

Build-up Phase: The phase is consist of 2 steps. In the first step, the transmitter generate coded symbols with degree 2 continuously. The decoder receives the coded symbols and updates the uni-partite graph. Until the size of the largest component is $k\beta_0$, where β_0 is a predefined value that satisfies $0 < \beta_0 < 1$, the decoder sends feedback to inform the transmitter, then the first step ends at this time. In the second step, the transmitter sends the coded symbols with degree 1 until the largest component is colored black, which means all the source symbols in the largest component is recovered. Then the build-up phase ends.

Completion Phase: The transmitter generates degree-$\hat{m}$ coded symbols based on β until the decoding process is successful. β represents the recovery ratio of source symbols. The encoder selects an optimal value of $\hat{m}$ to maximize the probability of the coded symbol becoming a useful symbol. In other words, to maximize the sum of probabilities of Cases 1 and 2. Then the value of $\hat{m}$ satisfies:

$$\hat{m} = \arg\max_{m}[P_1(m,\beta) + P_2(m,\beta)] \tag{1}$$

The $P_1(m,\beta)$ and $P_2(m,\beta)$ can be calculated as follows:

$$P_1(m,\beta) = \binom{m}{1}\beta^{m-1}(1-\beta) \tag{2}$$

$$P_2(m,\beta) = \binom{m}{2}\beta^{m-2}(1-\beta)^2 \tag{3}$$

For the online fountain codes with limited feedback, the authors in [14] proposed two strategies to select optimal feedback points in all the feedback points. When the number of feedback times is determined, the heuristic table-lookup algorithm based on effective probability (HTLEP) and the heuristic table-lookup algorithm based on overhead (HLTO) are applied to select appropriate feedback points to reduce the overhead. However, since a fixed value of degree is selected between two feedback points, when the number of feedback times is small, the optimal degree value skipped between the two feedback points is too much. So the effective probability of the fixed value of degree will gradually decrease. This will lead to a high overhead.

3 Growth Degree Encoding Scheme for Online Fountain Codes

In this section, we will introduce the proposed growth degree encoding scheme for online fountain codes with limited feedback. For convenience, we refer to the limited feedback online fountain codes with fixed degree value between two feedback points as conventional scheme. At the completion phase, the degree value of the coded symbols sent by the conventional scheme is from the feedback, which is fixed until the next feedback. However, in our proposed scheme, the degree value between two feedback points is gradually growing. We use a set V to store the number of coded symbols that need to be sent for each skipped optimal degree, then the transmitter sends a fixed number of coded symbols according to the set V until the next feedback point is reached. If all the symbols in the set V are sent before the next feedback point is reached, the degree value is randomly selected according to the degree distribution for transmission until the next feedback is received. The specific design method is as follows.

We assume the ratio of decoded symbols to all source symbols as β. With the increase of β, from Eq. (1), we can calculate the optimal degree $\hat{m}$ corresponding to each β that maximizes the useful probability. We call the β that changes the value of $\hat{m}$ each time as the degree transition points. Note that in the conventional scheme, since there is no limit to the number of feedback times, the set of degree transition points are the set of feedback points.

In the proposed GDS scheme, the transmitter get the selected feedback points from the set of degree transition points according to the HTLEP or HTLO algorithm. We denote by B and M the set of selected feedback points and the corresponding feedback degrees, respectively. When the number of feedback times is determined to be N, we can get:

$$B = \{\beta_0, \beta_1, \ldots, \beta_N, 1\} \tag{4}$$

$$M = \{m_0, m_1, \ldots, m_N, m_{max}\} \tag{5}$$

where β_i and m_i represents the ith feedback point and feedback degree, $1 \leq i \leq N$, m_0 represents the corresponding optimal degree when β is β_0. m_{max} represents the maximum degree in the scheme.

It can be seen from the above description that the transmitter sends coded symbols between every two adjacent feedback points. Some transition points will be skipped between any two adjacent feedback points, and the optimal degree value will change at these transition points. Without loss of generality, we assume the transmitter receives the feedback and knows the decoded symbols ratio β is β_i. Between the feedback points (β_i, β_{i+1}), the skipped transition points are denoted by $\{\beta_{i,1}, \beta_{i,2}, \ldots, \beta_{i,j}\}$ and the corresponding transition degree are denoted by $\{m_{i,1}, m_{i,2}, \ldots, m_{i,j}\}$. By combining the set of feedback points and transition points, we can get the following two sets:

$$B^i = \{\beta_{i,0}, \beta_{i,1}, \ldots, \beta_{i,j}, \beta_{i+1,0}\} \tag{6}$$

$$M^i = \{m_{i,0}, m_{i,1}, \ldots, m_{i,j}, m_{i+1,0}\} \tag{7}$$

where $\beta_{i,0}$ and $m_{i,0}$ are the β_i and m_i in the set of B and M, respectively. With these two sets, we can derive the set N^i. It is the set of coded symbols that needs to be sent between the two feedback points. The dimension of N^i is $|N^i| = |B^i| - 1 = j + 1$. We denote by $N^i(x)$ the xth element in this set. Then $N^i(x)$ can be get by the following equation.

$$N^i(x) = \frac{k[B^i(x+1) - B^i(x)]}{P(M^i(x), \frac{B^i(x+1)}{k})} \tag{8}$$

where $P(m, \beta)$ represents the sum of probabilities of the received coded symbol with degree m belongs to Cases 1 and 2. So it can be calculated as follows.

$$\begin{aligned} P(m, \beta) &= P_1(m, \beta) + P_2(m, \beta) \\ &= \binom{m}{1} \beta^{m-1}(1-\beta) + \binom{m}{2} \beta^{m-2}(1-\beta)^2 \end{aligned} \tag{9}$$

The value of $N^i(x)$ represents the theoretical upper bound of the number of coded symbols with degree $M^i(x)$ when the value of β is between the transition points $B^i(x)$ and $B^i(x+1)$.

After obtaining N^i and M^i, the transmitter first sends $N^i(1)$ coded symbols of degree $M^i(1)$. Then it increases the value of degrees and sends $N^i(2)$ coded symbol of degree $M^i(2)$. Continue in this way until $N^i(j+1)$ coded symbols of degree $M^i(j+1)$ have been sent or the feedback β_{i+1} is received.

If all symbols in the set N^i are sent and still no feedback is received, the calculated degree distribution is used to select the degree values to generate a new coded symbol until the next feedback is reached. We denote by $\Omega^i(x)$ the degree distribution between the feedback points β_i and β_{i+1}. The probability of being selected for a degree value $M^i(t)$ is the ratio of the number of symbols of degree $M^i(t)$ in N^i to the number of all symbols in N^i, i.e., $\frac{N^i(t)}{sum(N^i)}$, where $sum(N^i)$ represents the sum of the numbers in the set N^i. So the degree distribution can be calculated as follows.

$$\Omega^i(x) = \sum_{t=1}^{j+1} \frac{N^i(t)}{sum(N^i)} x^{M^i(t)} \tag{10}$$

4 Overhead Analysis

In this section, we provide analysis for the performance of conventional online fountain codes with limited feedback and the proposed GDS scheme based on the theoretical analysis framework in [5].

4.1 Performance Analysis of Online Fountain Codes with Limited Feedback

In this subsection, we analyze the performance of online fountain codes with limited feedback. First we introduce a lemma as follows.

Lemma 1 [5]. *Denote by $P(n)$ the probability that a received coded symbol belongs to Case 1 or Case 2 when the degree of the symbol is optimal, where n represents the number of recovered symbols. Then we can calculate the value of $P(n)$ as below.*

$$P(n) = P_1(\hat{m}, \beta_0 + \frac{n}{k}) + P_2(\hat{m}, \beta_0 + \frac{n}{k}) \tag{11}$$

where $\hat{m}$, P_1 and P_2 satisfy the Eqs. (1), (2) and (3), respectively.

To give the relationship between the number of received coded symbols and the number of recovered symbols, we introduce a new lemma.

Lemma 2 [5]. *Denote $T(s)$ as the number of coded symbols to transmit with unlimited feedback when s source symbols have been recovered. Then the value of $T(s)$ can be evaluated as follows.*

$$T(s) = \frac{1}{2}kc + \frac{1}{\beta_0} + (1 - \frac{1}{2}(1-\beta_0)c) \sum_{i=1}^{s-k\beta_0} \frac{1}{P(i)} \tag{12}$$

where $k\beta_0 < s \leq k$. c is the average degree of the source symbols when the build-up phase is over, which satisfies the following:

$$c = \frac{\ln(1-\beta_0)}{-\beta_0} \tag{13}$$

From Lemma 2, we can get the overhead performance analysis of the online fountain codes with limited feedback as follows.

Corollary 1. *Denote $T^N(s)$ as the number of coded symbols to transmit with N times of feedback when s source symbols have been recovered. The set of feedback points and the corresponding feedback degrees are B and M. Then the value of $T^N(s)$ can be evaluated as follows.*

$$T^N(s) = \frac{1}{2}kc + \frac{1}{\beta_0} + (1 - \frac{1}{2}(1 - \beta_0)c) \sum_{i=1}^{s-k\beta_0} \sum_{m \in M} \frac{f(i,m)}{P(m, \beta_0 + \frac{i}{k})} \tag{14}$$

where $P(m, \beta_0 + \frac{i}{k})$ satisfies Eq. (9) and $f(i,m)$ is a function that takes a value of 1 when the value of β is between the feedback points and the degree value m is the corresponding degree point at the same time. For convenience, we assume the value of i and m that meet the conditions as the event E. Then we can get the value of $f(i,m)$ as follows.

$$f(i,m) = \begin{cases} 1 & (i,m) \in E \\ 0 & else \end{cases} \tag{15}$$

Proof. Since online fountain codes with limited feedback is designed for the completion phase, the relationship between the number of coded symbols sent by the transmitter at the build-up phase and the recovery rate is the same as the conventional online fountain codes. From Lemma 2, we know at the end of build-up phase, the transmitter needs to send $\frac{1}{2}kc + \frac{1}{\beta_0}$ coded symbols. And it can be seen from [5] that at the completion phase, n useful symbols, including the build-up edges, Case 1 symbols and Case 2 symbols, can recover n source symbols. Therefore, at the completion phase, we denote by $T_{ca1,2}(n)$ the number of Case 1 and Case 2 symbols required to recover n source symbols. Then it can be calculated as follows:

$$T_{ca1,2}(n) = (1 - \frac{1}{2}(1 - \beta_0)c)n \tag{16}$$

In the online fountain codes with limited feedback, different from the conventional online fountain codes, the decoder only feeds back the corresponding degree value at a few fixed feedback points. The degree value between the two feedback points is constant. Therefore, the number of Cases 1 and 2 symbols n needs to be calculated separately between every two adjacent feedback points. Without loss of generality, we assume that the value of s/k is between the feedback points β_i and β_{i+1}. Then we first calculate the number of Cases 1 and 2 symbols required between the feedback points β_0 and β_1. We represent the value as n_0. Same as the analysis in [5], we can get:

$$n_0 = \sum_{t=1}^{k(\beta_1 - \beta_0)-1} \frac{1}{P(m_0, \beta_0 + \frac{t}{k})} \tag{17}$$

We denote by n_i the number of Cases 1 and 2 symbols required between the feedback points β_i and β_{i+1}. Based on the Eq. (17), we can get $n_1, n_2, \ldots, n_{i-1}$ according to the above analysis. In addition, the value of n_i can be calculated by

$$n_i = \sum_{t=k(\beta_i-\beta_0)}^{s-k\beta_0} \frac{1}{P(m_i, \beta_0 + \frac{t}{k})} \tag{18}$$

Therefore, the number of Cases 1 and 2 symbols needed to recover s symbols is the sum of $n_0, n_1, \ldots, n_i$. From Eq. (16), at the completion phase, the number of coded symbols required to recover s symbols is

$$T^N(s) = \frac{1}{2}kc + \frac{1}{\beta_0} + (1 - \frac{1}{2}(1-\beta_0)c)\sum_{t=0}^{i} n_t \tag{19}$$

Through derivation, it is obvious that Eq. (14) and Eq. (19) are equivalent. □

4.2 Performance Analysis of the Proposed GDS Scheme

In this subsection, we analyze the theoretical performance of the proposed GDS scheme over lossy channels. We denote by ϵ the channel erasure probability. Note that when the channel is lossless channels, the value of ϵ is 0.

We present the performance analysis of our proposed GDS scheme through the following corollary.

Corollary 2. *Denote $T_p^N(s)$ as the number of coded symbols to transmit in the GDS scheme with N times of feedback when s source symbols have been recovered. The set of feedback points and the corresponding feedback degrees are B and M. The set B^i, M^i and N^i, Ω^i have been calculated when $0 \le i \le N$. Then the value of $T_p^N(s)$ can be evaluated as follows.*

$$\begin{aligned} T_p^N(s) =& \frac{kc}{2(1-\epsilon)} + \frac{1}{\beta_0(1-\epsilon)} + (1 - \frac{1}{2}(1-\beta_0)c) \\ &\cdot [\sum_{\beta_i=B(1)}^{B(b-1)} G_{\beta_i}(k\beta_{i+1} - k\beta_i) + G_{B(b)}(s - k\beta_{B(b)})] \end{aligned} \tag{20}$$

where $B(b)$ represents that the value of s/k is between the value of $B(b)$ and $B(b+1)$ in B, i.e., $s/k \in [B(b), B(b+1)]$. We assume $G_{\beta_i}(x)$ is a function that satisfies the equation: $G_{\beta_i}(x) = \sum_{t=1}^{x} g(\beta_i, t)$. And $g(\beta_i, t)$ is a function that satisfies the following cases:

- *The value is $\frac{1}{P(M^i(1), \beta_i + \frac{t}{k})}$, when $0 \le G_{\beta_i}(t-1) \le N^i(1)(1-\epsilon)$.*
- *The value is $\frac{1}{P(M^i(r), \beta_i + \frac{t}{k})}$, when $\sum_{h=1}^{r} N^i(h)(1-\epsilon) < G_{\beta_i}(t-1) \le \sum_{h=1}^{r+1} N^i(h)(1-\epsilon)$ $\forall r > 1$.*

– *The value is* $\sum_{h=1}^{j+1} \frac{\Omega^i(h)}{P(M^i(h),\beta_i+\frac{t}{k})}$, *when* $G_{\beta_i}(t-1) > sum(N^i)(1-\epsilon)$.

Proof. Because the proposed GDS scheme is designed for the completion phase, the performance analysis of the build-up phase is similar to the conventional online fountain codes. And based on the Eq. (16) in the proof of Corollary 1, we know n represents the number of Cases 1 and 2 symbols, which is the focus of the analysis.

In the proposed GDS scheme, we need to analyze the number of Cases 1 and 2 symbols for every two adjacent feedback points. Without loss of generality, we analyze the number for β_i and β_{i+1}. The value of N^i can be calculated by (8). The transmitter sends $N^i(1)(1-\epsilon)$ coded symbols with degree $M^i(1)$, the probability of being Cases 1 and 2 symbols is $\frac{1}{P(M^i(1),\beta_i+\frac{t}{k})}$. So we can prove the first condition of $g(\beta_i, t)$. In the same way, we can prove the second condition.

For the third condition, when the symbols in set N^i have been sent, the transmitter sends symbols according to the degree distribution Ω^i. For every degree value, we get the product of the number of symbols with the degree value becomes Cases 1 and 2 symbols and the probability that the degree value is selected. Then we sum all the products to get the expected value. So we prove the third condition. □

5 Simulation Results

In this section, we first verify the proposed analysis by comparing with the simulation results. We assume the $k = 1000$ and $\beta_0 = 0.5$. We compare the performance for the three feedback points selection strategies, HTLEP, HTLO and EVEN. The EVEN strategy is a simple feedback point selection strategy that equals the number of degrees skipped between every two feedback points.

Figure 2 shows the overhead performance of the OFC with limited feedback and the proposed GDS scheme. For the feedback points selection strategies, HTLEP, HTLO and EVEN, we perform analysis and simulations over the lossless channels when the number of feedback times is 1 to 5. As shown in Fig. 2, the theoretical analysis matches with the simulation results, which demonstrate the accuracy of our proposed analysis. When the channel is lossless channels, we also observe that compared with the conventional encoding scheme, the GDS scheme can effectively reduce the overhead, especially when the number of feedback times is extremely small.

In Tables 1 and 2, we compare the overhead performance of conventional scheme and GDS scheme over the lossy channel. We set the channel erasure $\epsilon = 0.1$ and $\epsilon = 0.2$. As can been seen from the tables, the proposed GDS scheme still has good overhead performance over lossy channel. Even with the simple EVEN feedback selection strategy, online fountain codes with limited feedback can achieve very low overhead by using the GDS scheme.

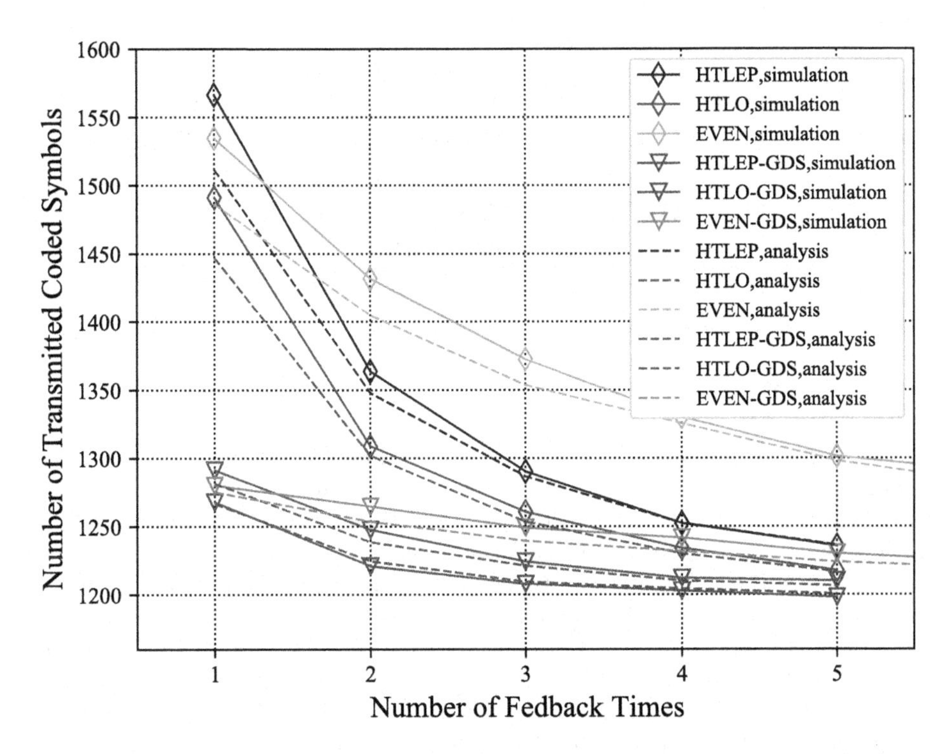

Fig. 2. Analysis and simulation results of the conventional OFC with limited feedback and the proposed GDS scheme. We set $k = 1000$, $\beta_0 = 0.5$ and $\epsilon = 0$. We simulate the three feedback strategies HTLEP, HTLO and EVEN when the number of feedback times is 1 to 5.

Table 1. Overhead performance of the conventional OFC with limited feedback and the GDS scheme when $\epsilon = 0.1$.

	$N = 1$	$N = 2$	$N = 3$	$N = 4$	$N = 5$
HTLEP	1754.34	1507.09	1433.26	1391.31	1379.14
HTLO	1651.6	1458.16	1398.66	1370.12	1353.07
EVEN	1720.37	1587.89	1516.88	1475.83	1451.15
HTLEP-GDS	1409.42	1369	1349.03	1343.61	1338.21
HTLO-GDS	1391.22	1352.81	1337.48	1336.77	1334.86
EVEN-GDS	1400.67	1384.88	1376.45	1365.33	1355.98

Table 2. Overhead performance of the conventional OFC with limited feedback and the GDS scheme when $\epsilon = 0.2$.

	$N = 1$	$N = 2$	$N = 3$	$N = 4$	$N = 5$
HTLEP	1955.83	1701.6	1613.87	1570.31	1551.93
HTLO	1859.36	1634.92	1576.42	1548.86	1524.73
EVEN	1930.94	1793.89	1722.36	1664.51	1635.85
HTLEP-GDS	1555.27	1525.12	1517.77	1506.44	1502.45
HTLO-GDS	1545.87	1516.67	1503.21	1500.56	1499.83
EVEN-GDS	1557.6	1529.9	1527.61	1521.84	1517.53

6 Conclusions

In this paper, we proposed growth degree encoding scheme for online fountain codes with limited feedback to achieve a low overhead. Between the two feedback points, the transmitter initially sends a fixed number of coded symbols with growth degree values. When all the coded symbols are transmitted, if no feedback is received at this time, the coded symbols are generated according to the calculated degree distribution until the feedback information is received. We also analyzed the overhead performance of the conventional OFC with limited feedback and the proposed scheme. Both the theoretical analysis and the simulation results showed that the proposed scheme can achieve a significant overhead reduction when the number of feedback times is small.

Acknowledgements. This work was supported in part by Beijing Natural Science Foundation under Grant L182032, by the National Natural Science Foundation of China (No. 61601147), and by the Development Foundation of CECT54 (Grant No. XX17641C008).

References

1. Luby, M.: LT codes. In: the 43rd Annual IEEE Symposium on Foundations of Computer Science, pp. 271–280 (2002)
2. Shokrollahi, A.: Raptor codes. IEEE Trans. Inf. Theory **52**(6), 2551–2567 (2006)
3. Perry, J., Iannucci, P.A., Fleming, K.E., Balakrishnan, H., Shah, D.: Spinal codes. SIGCOMM Comput. Commun. Rev. **42**(4), 49–60 (2012)
4. Cassuto, Y., Shokrollahi, A.: Online fountain codes with low overhead. IEEE Trans. Inf. Theory **61**(6), 3137–3149 (2015)
5. Huang, J., Fei, Z., Cao, C., Xiao, M., Jia, D.: Performance analysis and improvement of online fountain codes. IEEE Trans. Commun. **66**(12), 5916–5926 (2018)
6. Huang, T., Yi, B.: Improved online fountain codes based on shaping for left degree distribution. AEU-Int. J. Electron. C. **79**, 9–15 (2017)
7. Zhao, Y., Zhang, Y., Lau, F.C.M., Yu, H., Zhu, Z.: Improved online fountain codes. IET Commun. **12**(18), 2297–2304 (2018)

8. Shi, P., Wang, Z., Li, D., Xiang, W.: Zigzag decodable online fountain codes with high intermediate symbol recovery rates. IEEE Trans. Commun. **68**(11), 6629–6641 (2020)
9. Huang, J., Fei, Z., Cao, C., Xiao, M.: Design and analysis of online fountain codes for intermediate performance. IEEE Trans. Commun. **68**(9), 5313–5325 (2020)
10. Huang, J., Fei, Z., Cao, C., Xiao, M., Jia, D.: On-line fountain codes with unequal error protection. IEEE Commun. Lett. **21**(6), 1225–1228 (2017)
11. Cai, P., Zhang, Y., Pan, C., Song, J.: Online fountain codes with unequal recovery time. IEEE Commun. Lett. **23**(7), 1136–1140 (2019)
12. Yi, B., Xiang, M., Huang, T., Huang, H., Qiu, K., Li, W.: Data gathering with distributed rateless coding based on enhanced online fountain codes over wireless sensor networks. AEU-Int. J. Electron. Commun. **92**, 86–92 (2018)
13. Huang, J., Fei, Z., Cao, C., Xiao, M., Xie, X.: Reliable broadcast based on online fountain codes. IEEE Commun. Lett. **25**(2), 369–373 (2021)
14. Cai, P., Zhang, Y., Wu, Y., Chang, X., Pan, C.: Feedback strategies for online fountain codes with limited feedback. IEEE Commun. Lett. **24**(9), 1870–1874 (2020)

Ad Hoc Networks; Optical Communications and Networks

HFSWR Polarization DOA Estimation Based on Quaternion

Shao Shuai[1], Liu Aijun[1](✉), Yu Changjun[1], Lv Zhe[1], and Zhao Quanrui[2]

[1] Harbin Institute of Technology, Harbin, Shandong Province, China
liuaijun@hit.edu.cn
[2] Shanghai Radio Equipment Research Institute, Shanghai, China

Abstract. This paper examines the direction of arrival (DOA) estimation for polarized signals impinging on a vector sensor array for high frequency surface wave radar (HFSWR). The vector array effectively utilizes the polarization domain information of incident signals, and the quaternion model is adopted for signals polarization characteristic maintenance and computational burden reduction. The quaternion data model based on vector uniform linear array (ULA) is established. Based on the model, a quaternion MUSIC algorithm based on HFSWR signal processing is proposed for HFSWR DOA estimation. The algorithm combines the HFSWR signal processing and vector ULA to enhance the DOA estimation performance. Analytical simulations are operated to certify the capability of the algorithm.

Keywords: Direction of arrival · Vector uniform linear array · Polarization · Quaternion model

1 Introduction

Array signal processing is a basic theory in the fields of radar, sonar, navigation, etc. [1, 2]. With the increasing reliability of vector sensors, polarization is added to the DOA estimation as a basic information attribute. Therefore, many researchers have proposed multi-component data processing algorithms. For polarization vector sensor arrays, MUSIC-like algorithms are introduced in [3] and [4] and ESPRIT techniques in [5–7]. [8] and [9] have studied the Cramer–Rao bound for the vector-sensor arrays. However, these methods presume that the complex-valued data model represents incident signals frequency domain samples. A data covariance matrix is next described as second-order statistical magnitude between all sensor components. In the recent few decades, algorithms based on quaternion were introduced [10] for array signal processing. A multidimensional complex signals hypercomplex version [11] was also introduced by Bülow and Sommer [12]. In [13], Sebastian Miron proposes a quaternion data model algorithm based on vector-sensor arrays providing an estimation of DOA and polarization parameters, which decreases the data covariance model representation memory size leading to an efficient algorithm.

In HFSWR signal processing, the polarization domain information of the incident signal is effectively utilized by vector array. Quaternion data model is adopted to increase

Q. Guo et al. (Eds.): WiSATS 2021, LNICST 410, pp. 471–481, 2022.
https://doi.org/10.1007/978-3-030-93398-2_44

the orthogonality constraint of received data and maintain the polarization characteristics of the signal. Since the shift from the scalar sensor to the polarization vector sensor increases the amount of data to be processed, the quaternion model can reduce the amount of computation. For HFSWR DOA estimation, a quaternion data model based on vector array is firstly established, and then combined with HFSWR signal processing, a quaternion DOA estimation algorithm suitable for HFSW is proposed. The purpose is to improve the DOA estimation performance by using the polarization information and to improve the operation rate and convergence rate by combining the quaternion model. Finally, the effectiveness of the algorithm is verified by computer simulation.

This paper consists of the following sections. In Sect. 2, a short quaternions description is introduced and a polarized signal model is given. In Sect. 3, a description of the quaternion DOA estimation based on vector array is given. In Sect. 4, the algorithm performances are evaluated by computer simulations. Finally, the conclusions of this paper are presented.

2 Polarized Quaternion Signal Model

2.1 Quaternion

Quaternions discovered by Hamilton in 1843 are four-dimensional (4-D) hypercomplex numbers, which are an extension of complex numbers into 4-D space. A quaternion is defined via one real and three imaginaries, whose Cartesian form can be described as

$$q = w + x\mathrm{i} + y\mathrm{j} + z\mathrm{k}, \tag{1}$$

where $\mathrm{i}^2 = \mathrm{j}^2 = \mathrm{k}^2 = \mathrm{ijk} = -1$, $\mathrm{ij} = \mathrm{k}, \mathrm{ji} = -\mathrm{k}, \mathrm{ki} = \mathrm{j}, \mathrm{ik} = -\mathrm{j}, \mathrm{jk} = \mathrm{i}, \mathrm{kj} = -\mathrm{i}$.

Several complex numbers properties may be extended into quaternions.

(1) the conjugate of q, q*, is expressed as $q^* = w - x\mathrm{i} - y\mathrm{j} - z\mathrm{k}$;.
(2) a pure quaternion is a null real part quaternion: $q = x\mathrm{i} + y\mathrm{j} + z\mathrm{k}$;.
(3) the quaternion modulus q is $\|q\| = \sqrt{qq^*} = \sqrt{q^*q} = \sqrt{w^2 + x^2 + y^2 + z^2}$;
(4) its inverse is expressed as

$$q^{-1} = q^*/\|q\|; \tag{2}$$

(5) a null quaternion is given by $w = x = y = z = 0$;
(6) the quaternions set, noted $\mathbb{H}$, forms a noncommutative normed algebra, meaning

$$q_1 q_2 \neq q_2 q_1; \tag{3}$$

(7) conjugation over $\mathbb{H}$ is an anti-involution

$$(q_1 q_2)^* = q_2^* \cdot q_1^*. \tag{4}$$

In [16], a Cayley–Dickson form quaternion is expressed as: $q = q_1 + q_2\text{j}$, where $q_1 = w + x\text{i}$ and $q_2 = y + z\text{i}$. That is

$$q = \begin{bmatrix} 1 & \text{j} \end{bmatrix}\begin{bmatrix} q_1 & q_2 \end{bmatrix}^{\text{T}}. \tag{5}$$

2.2 Polarization Model

In this section, the concept of electromagnetic wave polarization is firstly expounded, then the joint expression of spatial domain and polarization domain of signal is modeled and the geometric structure of electromagnetic vector sensor array is modeled, and finally, the signal receiving a model of the electromagnetic vector sensor array is established. This article mainly studies fully polarized electromagnetic waves.

Assume that there is a point signal source with elevation angle θ and azimuth angle ϕ at a certain point in space, which is mutually excited by the magnetic field and the electric field. The ideal transmission medium is transmitted to the array along the direction of the Poynting vector, as shown in Fig. 1.

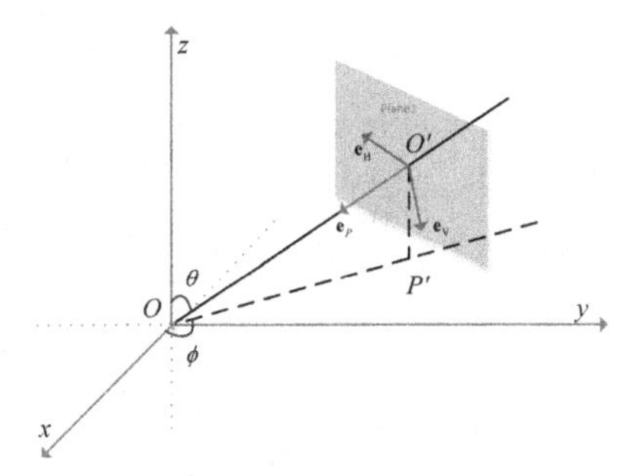

Fig. 1. Characterization of fully polarized electromagnetic waves

The unit vectors pointing in the positive direction of x, y, and z axis are denoted as $\mathbf{e}_x$, $\mathbf{e}_y$, $\mathbf{e}_z$ respectively. The coordinate origin serves as the reference phase point of the electromagnetic vector sensor array. As can be seen from Fig. 1, the unit vector of the incident direction of the signal is the propagation vector of the signal:

$$\mathbf{e}_P = \begin{bmatrix} -\sin\theta\cos\phi & -\sin\theta\sin\phi & -\cos\theta \end{bmatrix}^{\text{T}}. \tag{6}$$

The horizontal vector $\mathbf{e}_{\text{H}}$ and the vertical vector $\mathbf{e}_{\text{V}}$ constitute a set of standard orthogonal bases perpendicular to the plane of the propagation direction. The three form a right-handed coordinate system, which is related to the angle of arrival of the signal. According to Fig. 1:

$$\mathbf{e}_{\text{H}} = -\mathbf{e}_P\left(\phi + \frac{\pi}{2}, \frac{\pi}{2}\right) = \begin{bmatrix} -\sin\phi & \cos\phi & 0 \end{bmatrix}^{\text{T}}, \tag{7}$$

$$\mathbf{e}_V = -\mathbf{e}_P\left(\phi, \theta + \frac{\pi}{2}\right) = [\cos\theta\cos\phi \quad \cos\theta\sin\phi \quad -\sin\theta]^{\mathrm{T}}. \tag{8}$$

The signal polarization information may be described via the instantaneous ratio of the electric field amplitude and phase in the H and V directions, and recorded $\tan\gamma = E_{\mathrm{Vm}}/E_{\mathrm{Hm}}, \gamma \in [0, \pi/2]$ is polarization assist angle, $\eta = \varphi_{\mathrm{V}} - \varphi_{\mathrm{H}}, \eta \in [0, 2\pi]$ is polarization phase difference. For a fully polarized wave, its endpoint polarization trajectory is an ellipse with a fixed long and short axial ratio and inclination. The coordinates of a complete six-dimensional electric and magnetic field in a plane rectangular coordinate system are:

$$\begin{bmatrix} \mathbf{E}_x(t) \\ \mathbf{E}_y(t) \\ \mathbf{E}_z(t) \\ \mathbf{H}_x(t) \\ \mathbf{H}_y(t) \\ \mathbf{H}_z(t) \end{bmatrix} = \begin{bmatrix} -\sin\phi & \cos\theta\cos\phi \\ \cos\phi & \cos\theta\sin\phi \\ 0 & -\sin\theta \\ \cos\theta\cos\phi & \sin\phi \\ \cos\theta\sin\phi & -\cos\phi \\ -\sin\theta & 0 \end{bmatrix} \begin{bmatrix} \cos\gamma \\ \sin\gamma e^{i\eta} \end{bmatrix} E_c(t) \tag{9}$$

where

$$\mathbf{A}_P(\phi, \theta, \gamma, \eta) = \begin{bmatrix} -\sin\phi & \cos\theta\cos\phi \\ \cos\phi & \cos\theta\sin\phi \\ 0 & -\sin\theta \\ \cos\theta\cos\phi & \sin\phi \\ \cos\theta\sin\phi & -\cos\phi \\ -\sin\theta & 0 \end{bmatrix} \begin{bmatrix} \cos\gamma \\ \sin\gamma e^{i\eta} \end{bmatrix} \tag{10}$$

$\mathbf{A}_P(\phi, \theta, \gamma, \eta)$ is polarization steering vector, and **E** is the electric field component and **H** is the magnetic field component.

There is a uniform linear electromagnetic vector sensor array composed of N electromagnetic vector sensors arranged in sequence along the positive direction of the y axis. The electric dipoles of the electric field are orthogonal to each other, and the distance between the electromagnetic vector sensor array elements is d. The y-axis coordinate y_n of the nth electromagnetic vector sensor is $(n-1)d$; $n = 1, 2, \ldots, N$ in Fig. 2. And make $M (M < N)$ far-field expected signals (fully polarized waves) incident on the array and elevation angle $\theta = 90°$. Let the azimuth of the desired signal, the spatial phase delay between the nth element and the reference element ($d = \lambda/2$) is

$$\alpha_n = -\pi(n-1)\sin\phi. \tag{11}$$

The steering vector

$$\mathbf{A}_s = \begin{bmatrix} e^{i\alpha_1} & e^{i\alpha_2} & \cdots & e^{i\alpha_N} \end{bmatrix}^{\mathrm{T}}. \tag{12}$$

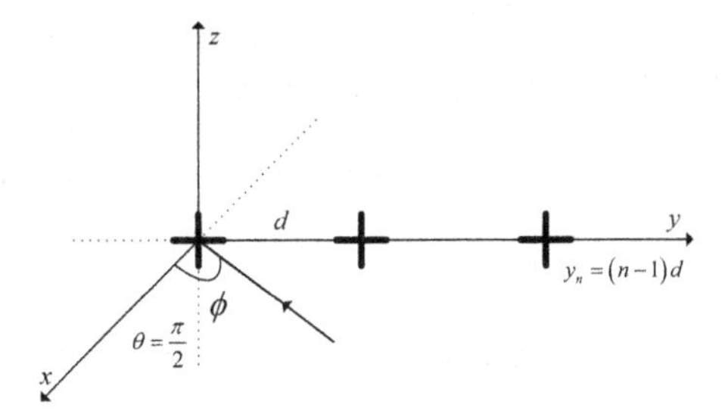

Fig. 2. Uniform linear array

The data received by the entire array is

$$\mathbf{X}(t) = \begin{bmatrix} e^{i\alpha_1}\mathbf{A}_P s(t) \\ e^{i\alpha_2}\mathbf{A}_P s(t) \\ \vdots \\ e^{i\alpha_N}\mathbf{A}_P s(t) \end{bmatrix} + \begin{bmatrix} \mathbf{n}_1(t) \\ \mathbf{n}_2(t) \\ \vdots \\ \mathbf{n}_N(t) \end{bmatrix} = \mathbf{A}_s(\phi,\theta) \otimes \mathbf{A}_P(\phi,\theta,\gamma,\eta)s(t) + \mathbf{N}(t) \tag{13}$$

where $\otimes$ represents tensor product and $\mathbf{n}_n(t) = \begin{bmatrix} n_{ny}(t) & n_{nz}(t) \end{bmatrix}^{\mathrm{T}}, n = 1, 2, \ldots, N$ is noise vector in directions of yz. It is assumed that the noise is independent between each array element, and the noise is independent between the vector components within each array element, and the signal and noise are relatively independent. Noise is an independent Gaussian white noise with zero mean and σ^2 variance.

3 HFSWR Polarized Quaternions DOA Estimation Algorithm

3.1 HFSWR Signal Processing

This section describes target azimuth acquisition based on the Range Doppler spectrum. The LFM continuous-wave transmitting signal with an FM period is set as:

$$S_{\mathrm{T}}(t) = \cos\left[2\pi\left(f_c t + 0.5kt^2\right)\right] = \cos[\phi_T(t)], \tag{14}$$

where f_c is center frequency and k is frequency modulation slope. After passing through the radio frequency amplifier at the receiver, the received echo signal is mixed with the transmitted signal, that is:

$$\begin{aligned} S_{\mathrm{d}}(t) &= A\cos\left(2\pi\left(f_c t + 0.5kt^2 - f_c(t - t_{\mathrm{d}}) - 0.5k(t - t_{\mathrm{d}})^2\right)\right) \\ &= A\cos\left(2\pi\left(f_c t_{\mathrm{d}} + kt_{\mathrm{d}}t - 0.5kt_{\mathrm{d}}^2\right)\right) \end{aligned} \tag{15}$$

where t_d is time delay. Based on the phase of the difference frequency signal of the nth frequency modulation period, the Fourier transform result is:

$$V_n(f) = \frac{AT}{2}\left\{\begin{array}{l} \frac{\sin[2\pi(f-f_{dn})T/2]}{2\pi(f-f_{dn})T/2} e^{-j\phi_0 + j2\pi f_c \frac{2v}{c} nT_r} \\ + \frac{\sin[2\pi(f+f_{dn})T/2]}{2\pi(f+f_{dn})T/2} e^{+j\phi_0 - j2\pi f_c \frac{2v}{c} nT_r} \end{array}\right\} \tag{16}$$

where v is speed and T_r is Frequency sweep cycle. Take out the positive frequency part of V, a total of M sampling points. The process of obtaining the range spectrum from the differential frequency signal from Eq. (16) is called the solution range transformation in the HFSWR signal processing.

Continuous transmitting N FM signals, and the difference frequency signal of each frequency modulation period do the distance transformation, can get $N \times M$ samples

$$\mathbf{V} = \begin{bmatrix} V_{1,1} & V_{1,2} & \cdots & V_{1,M} \\ V_{2,1} & V_{2,2} & \cdots & V_{2,M} \\ \vdots & \vdots & \ddots & \vdots \\ V_{N,1} & V_{N,2} & \cdots & V_{N,M} \end{bmatrix} \tag{17}$$

The mth (m = 1, 2, …, M) column is N samples taken from the mth distance gate, namely the time series on the distance gate. The Fourier transform of each column is the solution velocity transform. The Result of the Fourier transform is:

$$Q_{nm} = KNT_r \frac{\sin[2\pi(f - 2f_c v/c)NT_r/2]}{2\pi(f - 2f_c v/c)NT_r/2} \tag{18}$$

After the signal processing solution velocity transformation, the Range-Doppler matrix can be obtained:

$$\boldsymbol{Q} = \begin{bmatrix} Q_{1,1} & Q_{1,2} & \cdots & Q_{1,M} \\ Q_{2,1} & Q_{2,2} & \cdots & Q_{2,M} \\ \vdots & \vdots & \ddots & \vdots \\ Q_{N,1} & Q_{N,2} & \cdots & Q_{N,M} \end{bmatrix} \tag{19}$$

The Range-Doppler (RD) spectrum of the target can be drawn from the matrix. It can be seen from the theoretical derivation that the Range-Doppler spectrum is the 2-dimensional envelope form of the sinc function.

After Range processing and Doppler processing, the echo data can be represented as a 3-dimensional data block containing Range-Doppler-Array information. Based on the RD spectrum, the points with the same range and Doppler are extracted to form time series. Replace the previous polarization signal with this time series:

$$\mathbf{X}(t) = \mathbf{A}_s(\phi, \theta) \otimes \mathbf{A}_P(\phi, \theta, \gamma, \eta) s_{\mathrm{RD}}(t) + \mathbf{N}(t) \tag{20}$$

3.2 DOA Estimation Algorithm

Take the $\mathbf{e}_y$ and $\mathbf{e}_z$ in formula (16) to obtain the polarization domain steering vector of the mth signal received by this electromagnetic vector sensor array:

$$\mathbf{A}_P(\phi_m, \theta_m, \gamma_m, \eta_m) = \begin{bmatrix} \cos\phi_m & \cos\theta_m \sin\phi_m \\ 0 & -\sin\theta_m \end{bmatrix} \begin{bmatrix} \cos\gamma_m \\ \sin\gamma_m e^{\mathrm{i}\eta_m} \end{bmatrix} \tag{21}$$

The elevation $\theta = 90°$, after simplification:

$$\mathbf{A}_P(\phi_m, \gamma_m, \eta_m) = \begin{bmatrix} \cos\phi_m \cos\gamma_m \\ -\sin\gamma_m e^{\mathrm{i}\eta_m} \end{bmatrix} \tag{22}$$

Applying formula (5) here, the quaternion Cayley-Dickson representation synthesizes the polarization domain steering vector in the complex number domain into a polarization domain steering vector in the quaternion domain, namely:

$$p(\phi_m, \gamma_m, \eta_m) = \begin{bmatrix} 1 & \mathrm{j} \end{bmatrix} \begin{bmatrix} \cos\phi_m \cos\gamma_m \\ -\sin\gamma_m e^{\mathrm{i}\eta_m} \end{bmatrix} \in \mathbb{H} \tag{23}$$

after simplification:

$$p = \cos\phi_m \cos\gamma_m - \sin\gamma_m \cos\eta_m \mathrm{j} - \sin\gamma_m \sin\eta_m \mathrm{k} \tag{24}$$

According to formula (20), it can be known that the mth signal received by the entire electromagnetic vector sensor array is:

$$\mathbf{X}_m(t) = \mathbf{A}_s(\phi_m) p(\phi_m, \gamma_m, \eta_m) s_{\mathrm{RD}m}(t) + \mathbf{N}(t) \tag{25}$$

where

$$\mathbf{A}_s(\phi_m) = \begin{bmatrix} e^{\mathrm{i}\alpha_{m,1}} & e^{\mathrm{i}\alpha_{m,2}} & \cdots & e^{\mathrm{i}\alpha_{m,N}} \end{bmatrix}^{\mathrm{T}} \tag{26}$$

The quaternion noise is synthesized from the complex noise vector by Cayley-Dickson representation as:

$$n = \begin{bmatrix} 1 & \mathrm{j} \end{bmatrix} \begin{bmatrix} n_{ny}(t) \\ n_{nz}(t) \end{bmatrix} = n_{ny}(t) + n_{nz}(t)\mathrm{j} \tag{27}$$

All signals received by the entire array can be expressed as:

$$\mathbf{X}(t) = \sum_{m=1}^{M} \mathbf{A}_s(\phi_m) p(\phi_m, \gamma_m, \eta_m) s_{\mathrm{RD}m}(t) + \mathbf{N}(t) \tag{28}$$

where

$$\mathbf{N}(t) = \begin{bmatrix} n_1 & n_2 & \cdots & n_N \end{bmatrix}^{\mathrm{T}} \tag{29}$$

Make $\mathbf{a}_m = \mathbf{A}_s(\phi_m)p(\phi_m, \gamma_m, \eta_m)$ write formula (26) as a matrix:

$$\mathbf{X}(t) = \begin{bmatrix} \mathbf{A}_{s1} & \mathbf{A}_{s2} & \cdots & \mathbf{A}_{sM} \end{bmatrix} \cdot \begin{bmatrix} p_1 & & & \\ & p_2 & & \\ & & \ddots & \\ & & & p_M \end{bmatrix} \begin{bmatrix} s_1(t) \\ s_2(t) \\ \vdots \\ s_M(t) \end{bmatrix} + \mathbf{N}(t) \tag{30}$$

A is called the spatial-polarization joint steering vector matrix. In this way, a signal receiving data model of the electromagnetic vector sensor array in the quaternion domain is established.

The equivalent second order representation for a vector-sensor array using Quaternion spectral matrix (QSM)

$$\mathbf{\Omega} = \mathrm{E}\{\mathbf{X}\mathbf{X}^{\triangleleft}\} \tag{31}$$

where $\mathrm{E}\{\bullet\}$ is the mathematical expectation operator and $^{\triangleleft}$ represents quaternion conjugate transpose.

Assuming the decorrelation between sources themselves and between the noise and the sources, the quaternion spectral matrix becomes

$$\mathbf{\Omega} = \mathrm{E}\{\mathbf{A}\mathbf{S}\mathbf{S}^{\triangleleft}\mathbf{A}^{\triangleleft}\} + \mathrm{E}\{\mathbf{N}\mathbf{N}^{\triangleleft}\} = \mathbf{\Omega}_S + \mathbf{\Omega}_N \tag{32}$$

where

$$\mathbf{\Omega}_S = \mathbf{A}\mathrm{E}\{\mathbf{S}\mathbf{S}^{\triangleleft}\}\mathbf{A}^{\triangleleft} = \sum_{m=1}^{M} \sigma_m^2 \mathbf{A}\mathbf{A}^{\triangleleft} = \sum_{m=1}^{M} \sigma_m^2 \|p_m\|^2 \mathbf{A}_{sm}\mathbf{A}_{sm}^{\triangleleft} \tag{33}$$

and $\mathbf{\Omega}_N = \mathrm{E}\{\mathbf{N}\mathbf{N}^{\triangleleft}\}$ is a matrix containing noise second order statistics. In (33), $\mathbf{\Omega}_S$ is the signal part and $\sigma_m^2\|p_m\|^2$ is mth source antenna power. Let the eigenvalue decomposition (EVD) be given by

$$\mathbf{\Omega} = \mathbf{U}\mathbf{D}\mathbf{U}^{\triangleleft} \tag{34}$$

with $\mathbf{U} = [\mathbf{u}_1, \cdots, \mathbf{u}_N] \in \mathbb{H}^{N \times N}$ containing the N eigenvectors and $\mathbf{D}$ is the diagonal matrix of its eigenvalues.

Define two matrices $\mathbf{U}_S \in \mathbb{H}^{N \times M}$ and $\mathbf{U}_G \in \mathbb{H}^{N \times (N-M)}$, such as

$$\mathbf{U}_S = [\mathbf{u}_1, \cdots, \mathbf{u}_M] \tag{35}$$

$$\mathbf{U}_G = [\mathbf{u}_{M+1}, \cdots, \mathbf{u}_N] \tag{36}$$

$\mathbf{U}_{\mathrm{S}}$ contains the signal subspace eigenvectors and $\mathbf{U}_{\mathrm{G}}$ the noise subspace eigenvectors. Because

$$\mathbf{A}^{\triangleleft}\mathbf{U}_{G} = 0 \tag{37}$$

If (37) is multiplied on the right by $(\mathbf{A}^{\triangleleft}\mathbf{U}_{G})^{\triangleleft}$, (37) can be expressed using columns of $\mathbf{A}$ as

$$\mathbf{a}_{m}^{\triangleleft}(\phi_{m}, \theta_{m}, \gamma_{m}, \eta_{m})\mathbf{U}_{G}\mathbf{U}_{G}^{\triangleleft}\mathbf{a}_{m}(\phi_{m}, \theta_{m}, \gamma_{m}, \eta_{m}) = 0 \tag{38}$$

for all sets of $\{\phi_{m}, \theta_{m}, \gamma_{m}, \eta_{m}\}$ corresponding to parameters of M signal sources. $\mathbf{\Pi}_{N} = \mathbf{U}_{G}\mathbf{U}_{G}^{\triangleleft} \in \mathbb{H}^{N \times N}$ represents the noise subspace projector.

The DOA estimator based on quaternion is then achieved by projecting the quaternion steering vector $\mathbf{a}_{m}(\phi_{m}, \theta_{m}, \gamma_{m}, \eta_{m}) \in \mathbb{H}^{N}$

$$\mathbf{a}_{m} = \begin{bmatrix} e^{\mathrm{i}\alpha_{m,1}} & e^{\mathrm{i}\alpha_{m,2}} & \cdots & e^{\mathrm{i}\alpha_{m,N}} \end{bmatrix}^{\mathrm{T}} p(\phi_{m}, \theta_{m}, \gamma_{m}, \eta_{m}) \tag{39}$$

on the noise subspace as:

$$SP_{\mathrm{Q}}(\phi_{m}, \theta_{m}, \gamma_{m}, \eta_{m}) = \frac{1}{\mathbf{a}_{m}^{\triangleleft}\hat{\mathbf{\Pi}}_{N}\mathbf{a}_{m}} \tag{40}$$

The functional in (40) has maxima for sets of $\{\phi_{m}, \theta_{m}, \gamma_{m}, \eta_{m}\}$ corresponding to sources present in the signal. In this article, the DOA estimation of the signal is mainly concerned and it is assumed that the polarization of the signal is known.

4 Simulation Analysis

In this section, numerical examples illustrate the polarization DOA estimation performance superiority based on quaternion in HFSWR. Assume that the number of the sources is known and all sources are equal power. RD spectrum is obtained after range and velocity processing in HFSWR, as shown in Fig. 3. Figure 3 shows the single-channel result, and the RD spectrum of other channels is similar. As the RD spectrum acquisition period is relatively long, the number of snapshots should not be too much. To combine the HFSWR signal processing, the snapshot number is about ten.

The next simulations consider the RMSE performance versus the input SNR, the number of snapshots. The fixed parameter setting is following: SNR is from –10 dB to 10 dB because SNR varies under different conditions, T are 10 snapshots. Figure 4 shows the RMSE of the DOA estimates versus the SNR. As the SNR increases, all the RMSEs decrease. Moreover, the Quaternion MUSIC achieves smaller RMSE than Long vector MUSIC and MUSIC across a wide range of the SNR. Figure 5 illustrates the RMSE of the DOA estimates versus the number of the snapshot, all DOA estimates become more accurate and stabilized as snapshots increase, and Quaternion MUSIC behaves better than Long vector MUSIC and MUSIC.

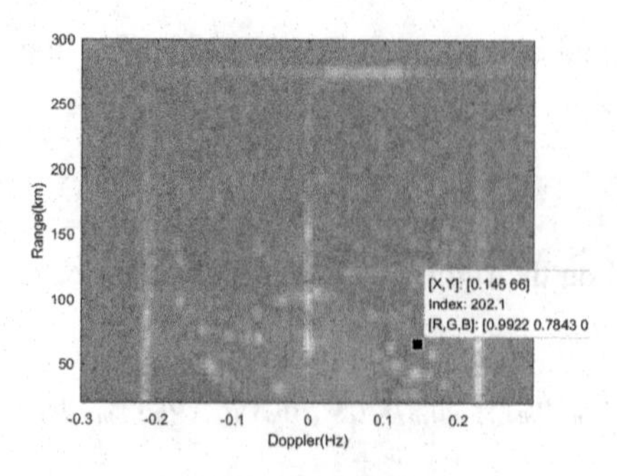

Fig. 3. RD spectrum

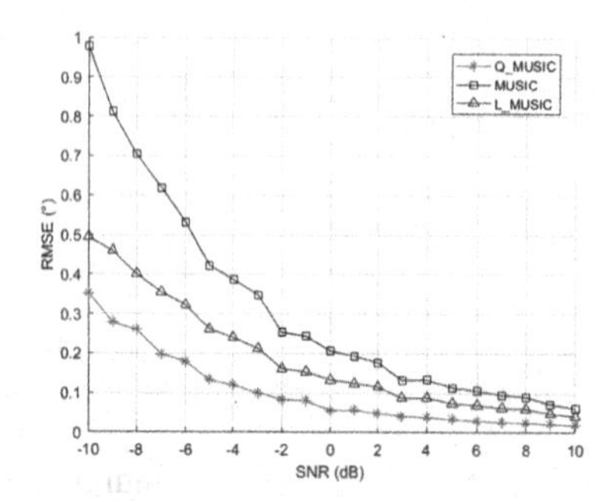

Fig. 4. RMSE versus the SNR

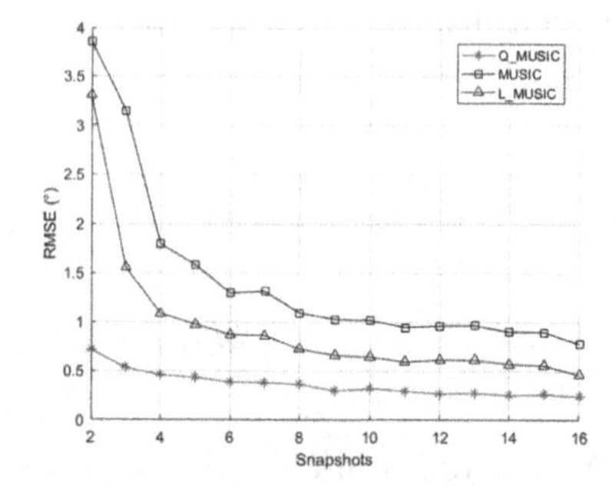

Fig. 5. RMSE versus the number of snapshots

5 Conclusion

In this paper, combining HFSWR signal processing with a quaternion data model, we propose a quaternion-based polarization DOA estimation algorithm for HFSWR. The proposed method has been compared to the classical MUSIC algorithm and the long vector MUSIC algorithm. Q-MUSIC is more accurate than them, reducing the computational burden the memory required. The results indicate the quaternion potential to model polarized signals and the possibility of describing more than two complex-valued components signals using higher-dimensional hypercomplex algebras in HFSWR signal processing.

Acknowledgement. Thanks to National Natural Science Foundation of China 62031015 and Harbin Institute of Technology at Weihai Scientific Research Innovation Fund 2019KYCXJ-JYB07 for funding.

References

1. Krim, B., Viberg, M.: Two decades of array signal processing research. IEEE Sig Proc Mag. **13**(4), 67–94 (1996)
2. Trees, H.: Detection, Estimation, and Modulation Theory, Optimum Array Processing. Publishing House of Electronics Industry, Beijing (2013)
3. Wong, K., Zoltowski, M.: Self-initiating MUSIC-based direction finding and polarization estimation in spatio-polarizational beamspace. IEEE Trans. Antennas Propag. **48**(8), 1235–1245 (2000)
4. Wong, K., Zoltowski, M.: Diversely polarized Root-MUSIC for azimuth-elevation angle-of-arrival estimation. In: Antennas and Propagation Society International Symposium (1996)
5. Zoltowski, M., Wong, K.: ESPRIT-based 2-D direction finding with a sparse uniform array of electromagnetic vector sensors. IEEE Trans. Signal Process. **48**(8), 2195–2204 (2000)
6. Wong, K., Zoltowski, M.: Uni-vector-sensor ESPRIT for multisource azimuth, elevation, and polarization estimation. Antennas Propag. IEEE Trans. **45**(10), 1467–1474 (1997)
7. Li, J., Compton, R.: Angle and polarization estimation using ESPRIT with a polarization sensitive array. IEEE Trans. Antennas Propag. **39**(9), 1376–1383 (1991)
8. Weiss, A., Friedlander, B.: Performance analysis of diversely polarized antenna arrays. IEEE Trans. Signal Process. **39**(7), 1589–1603 (1991)
9. Nehorai, A., Paldi, E.: Vector-sensor array processing for electromagnetic source localization. Signal Process. IEEE Trans. **42**(2), 376–398 (1994)
10. Schutte, H., Wenzel, J.: Hypercomplex numbers in digital signal processing. In: IEEE International Symposium on Circuits & Systems (1990)
11. Hahn, S.: Multidimensional complex signals with single-orthant spectra. Proc. IEEE **80**(8), 1287–1300 (1992)
12. Bulow, T., Sommer, G.: Hypercomplex signals-a novel extension of the analytic signal to the multidimensional case. IEEE Trans. Signal Process. (2001)
13. Miron, S., Bihan, N., Mars, J.: Quaternion-MUSIC for vector-sensor array processing. IEEE Trans. Signal Process. **54**(4), 1218–1229 (2006)
14. Kantor, I., Solodovnikov, A.: Hypercomplex Numbers: An Elementary Introduction to Algebras. Springer, Heidelberg (1989)
15. Ward, J.: Quaternions and cayley numbers (1997)
16. Lee, H.: Eigenvalues and canonical forms of matrices with quaternion coefficients. Proc. R. Irish Acad. **52**, 253–260 (1948)

Trajectory Optimization in UAV Communication System Based on User-QoS

Yikun Zou[1], Gang Wang[1(✉)], Jinlong Wang[1], Haoyang Liu[1], and Yinghua Wang[2]

[1] Communication Research Center, Harbin Institute of Technology, Harbin, China
gwang51@hit.edu.cn

[2] Nanjing Research Institute of ZTE Corporation, Nanjing, China

Abstract. Due to the mobility and flexibility of unmanned aerial vehicles (UAVs), the UAV communication networks can effectively support information transmission in the cellular networks. The purpose of this paper is to propose a scheme to ensure the fairness of information transmission for the whole users on the premise of taking into account the quality of service (QoS) of some users in the cellular networks. While considering the QoS of the users, we ensure that information transmission is not monopolized by some users by maximizing the minimum effective capacity of the whole users in the networks. By jointly optimizing user scheduling and UAV's trajectory, an iterative algorithm based on the block coordinate descent method and successive convex optimization technique is proposed. Numerical analysis results show that our proposed fairness scheme based on the QoS can improve the minimum effective capacity of the users compared with the existing schemes.

Keywords: UAV communication · User quality of service · Maximum and minimum fairness · Capacity optimization

1 Introduction

With the popularity of 5G and the grand concept of 6G, the integrated air-space-earth-sea networks have played an important role in it [1]. In the integrated air-space-earth-sea networks, unmanned aerial vehicles(UAVs), as an important part of airborne mobile wireless networks, have been widely used in a variety of military and civilian scenes [2]. Among them, the emergency communication of post-disaster rescue based on UAVs and the UAV-assisted cellular networks have been studied in depth [3]. Due to the high mobility and high flexibility, UAVs can provide high-quality line-of-sight communication links when the ground environment is not complex [4,5]. At the same time, the fixed deployment infrastructure

This work is supported in part by National Natural Science Foundation of China (No. 62071146 and No. 62071147).

Q. Guo et al. (Eds.): WiSATS 2021, LNICST 410, pp. 482–495, 2022.
https://doi.org/10.1007/978-3-030-93398-2_45

cannot meet the requirements of resource overload and high dynamic, which provides a great advantage for UAV-assisted cellular networks [6].

At present, the researches of UAV-assisted communication mainly focus on several aspects: such as UAV-assisted coverage, UAV-assisted communication, UAV-assisted Internet of Things, etc. In terms of UAV-assisted coverage, UAVs served as the aerial base stations to provide ground users with the uplink/downlink information transmission [7,8]. Lyu *et al.* focused on the efficient deployment algorithm of multi-UAV, which minimized the number of UAVs while completely covering a group of ground users in the system [7]. A cyclical time-division multiple access scheme based on the UAV communication system was proposed, in which the relationship between throughput and access delay was discussed [9]. The UAV trajectory design of a multi-UAV communication system was studied. The minimum throughput of all ground users in the downlink was maximized by optimizing multi-user communication scheduling and association as well as UAV trajectory and power control [10]. In addition, some researchers proposed various methods to improve the quality of service (QoS) of the networks in the research of the UAV-assisted cellular networks [11–13].

In this paper, we propose a UAV-assisted communication scheme based on the QoS, in which a single UAV serves each user on the ground through a cyclical time-division multiple access scheme. In this scheme, we propose several conditions to guarantee the QoS. In particular, we introduce the delay-constrained QoS index to ensure the delay requirements of the users. For the downlinks between the single-antenna UAV and the single-antenna users, we propose the optimization problem of maximizing the minimum effective capacity of each user under the delay-constrained condition to increase capacity to ensure all users' delay requirements and fairness. We consider an iterative algorithm in which UAV trajectory and user scheduling are jointly optimized and the block coordinate descent method and successive convex optimization technique are used to solve the non-convex problem. Simulation results will verify the effectiveness and convergence of the proposed algorithm.

2 System Model

As shown in Fig. 1, we consider the process of information transmission between a single UAV with a single antenna and multi-users with a single antenna on the ground. We assume that the maximum flight speed of the UAV is V_{max}, and the period T is divided into N time slots. In this paper, the three-dimensional Cartesian coordinate system is used, in which the position of the kth user and the discretized position of the UAV are expressed as $\mathbf{u}_k = [x_k, y_k, z_k]^T \in \mathbb{R}^{3\times1}$ and $\mathbf{q}(n) = [x(n), y(n), z(n)]^T \in \mathbb{R}^{3\times1}$ at the time $0 < t < T$. During the period T, the position of the UAV shall meet the following conditions

$$\mathbf{q}(1) = \mathbf{q}(N), \tag{1}$$

$$||\mathbf{q}[n+1] - \mathbf{q}[n]||^2 \leqslant \left(\frac{V_{max}T}{N}\right)^2, \forall n. \tag{2}$$

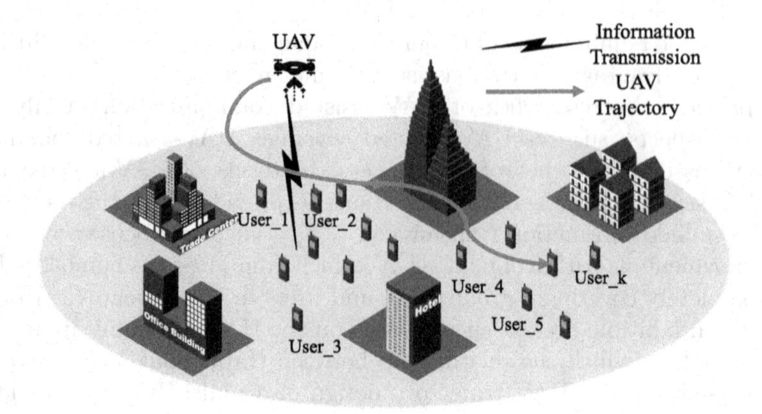

Fig. 1. UAV-enabled communication system.

Equation (1) indicates that the UAV returns to its starting point during the period T. Equation (2) states that the speed of the UAV cannot exceed its maximum flight speed V_{max} in each time slot.

In this paper, the variable $\mathbf{X} = \{x_k[n], \forall k, n\}$ is introduced when considering the connectivity between the UAV and the users. $x_k[n] = 0$ indicates that user k is not connected with the UAV in the nth time slot. $x_k[n] = 1$ indicates that user k is connected with the UAV in the nth time slot. Since the UAV can only connect to one user in each time slot, there are the following constraints

$$\sum_{k=1}^{K} x_k[n] \leqslant 1, \tag{3}$$

$$x_k[n] = \{0, 1\}. \tag{4}$$

In this paper, the communication links dominated by LoS links are adopted. In the nth time slot, the channel gain between the UAV and the kth user is expressed as

$$h_k[n] = \frac{\rho_0}{H^2 + ||\mathbf{q}[n] - u_k||^2}, \tag{5}$$

where ρ_0 represents the channel power at the reference distance $d_0 = 1$ m. The maximum transmission power of the UAV is defined as $P_{\max}$. To maximize the transmission capacity of the system of a single UAV, we assume that the UAV always transmits data at the maximum transmission power $P_{\max}$. Therefore, in the nth time slot, the received signal-to-noise ratio and the achievable rate of the kth user are

$$\gamma_k[n] = \frac{P_{\max} h_k[n]}{\sigma^2}, \tag{6}$$

$$R_k[n] = x_k[n]\log_2(1 + \gamma_k[n]), \tag{7}$$

where σ^2 is the power of the additive white Gaussian noise (AWGN) at the receiver. To ensure the delay requirements in information transmission, the effective capacity based on QoS index is introduced in this paper. When the service rate sequence $R[k]$ is stationary and time-independent, the effective capacity is written as

$$C(\theta) = -\frac{1}{\theta}\log\left(\mathbb{E}\left\{e^{-\theta R[k]}\right\}\right), \tag{8}$$

where $\mathbb{E}\{\cdot\}$ represents expectation. And θ represents the QoS index. Therefore, the effective capacity of the kth user in the period T is

$$C_k = -\frac{1}{\theta_k}\log\left(\frac{1}{N}\sum_{n=1}^{N} e^{-\theta_k R_k[n]}\right). \tag{9}$$

UAV trajectory and user scheduling are $\mathbf{Q} = \{\mathbf{q}[n], \forall n\}$ and $\mathbf{X} = \{x_k[n], \forall k, n\}$, respectively. $C_{\min} = \min(C_k)$ is defined as the minimum effective capacity among the whole users, then the objective function is to maximize the minimum effective capacity among the whole users. Assuming that the location of the ground user is known, the optimization problem is expressed as

$$\max_{\mathbf{X},\mathbf{Q},C_{\min}} \; C_{\min}, \tag{10a}$$

$$\text{subject to} \quad -\frac{1}{\theta_k}\log\left(\frac{1}{N}\sum_{n=1}^{N} e^{-\theta_k R_k[n]}\right) \geqslant C_{\min}, \tag{10b}$$

$$||\mathbf{q}[n+1] - \mathbf{q}[n]||^2 \leqslant \left(\frac{V_{max}T}{N}\right)^2, \forall n, \tag{10c}$$

$$\mathbf{q}(1) = \mathbf{q}(N), \tag{10d}$$

$$\sum_{k=1}^{K} x_k[n] \leqslant 1, \tag{10e}$$

$$x_k[n] = \{0, 1\}, \forall k, n. \tag{10f}$$

Based on (10b) and (10f), problem (10) is non-convex. The left side of (10b) is neither convex nor concave for variable $\mathbf{Q}$ or variable $\mathbf{X}$. Meanwhile, the variable $\mathbf{X}$ in (10f) is a binary discrete variable, thus (10f) has the same phenomenon with (10b). Based on this, we try to adopt the successive convex optimization technique and block coordinate descent method to solve these problems in the next section.

3 The Proposed Algorithm

To solve the problem that the variable $\mathbf{X}$ is a binary discrete variable, we can relax it into $\mathbf{X} = \{x_k[n] \in [0, 1], \forall k, n\}$. Therefore, (11f) is transformed into

$0 \leqslant x_k[n] \leqslant 1, \forall k, n$. We propose an effective iterative algorithm using the block coordinate descent method to solve problem (11).

3.1 User Scheduling Optimization

For any given UAV trajectory $\mathbf{Q}$, we can simplify it into an optimization problem (11) as follows

$$\max_{\mathbf{X},\mathbf{Q},C_{\min}} C_{\min}, \tag{11a}$$

$$\text{subject to} \quad -\frac{1}{\theta_k}\log\left(\frac{1}{N}\sum_{n=1}^{N} e^{-\theta_k x_k[n]\log_2(1+\gamma_k[n])}\right) \geqslant C_{\min}, \tag{11b}$$

$$\sum_{k=1}^{K} x_k[n] \leqslant 1, \tag{11c}$$

$$0 \leqslant x_k[n] \leqslant 1, \forall k, n. \tag{11d}$$

In the problem (11), (11a), (11c) and (11d) are convex for the variable $\mathbf{X}$, so we need to explore whether (11b) is convex for the variable $\mathbf{X}$. The first-order derivation and the second-order derivation of the effective capacity C_k in the period T with respect to any variable $x_k[n]$ are

$$C_k{}'(x_k[n]) = \frac{\beta_k[n]}{(\ln 2)^2}\frac{(1+\gamma_k[n])^{-\beta_k[n]-1}}{\sum\limits_{n=1}^{N}(1+\gamma_k[n])^{-\beta_k[n]}}, \tag{12}$$

$$C_k{}''(x_k[n]) = \frac{1}{(\ln 2)^2 (F(x_k[n]))^2}\left[\frac{\theta_k}{\ln 2}\beta_k^2[n](1+\gamma_k[n])^{-2\beta_k[n]-2}\right.$$
$$\left. + \frac{\theta_k}{\ln 2}(1+\gamma_k[n])^{-\beta_k[n]-2}\left[(1+\gamma_k[n]) - \beta_k[n]\left(\beta_k[n]+1\right)\right]F(x_k[n])\right], \tag{13}$$

where $F(x_k[n]) = \sum\limits_{n=1}^{N}(1+\gamma_k[n])^{-\beta_k[n]}$ and $\beta_k = \frac{\theta_k x_k[n]}{\ln 2}$. In Eq. (13), we find that the concavity and convexity of C_k for any variable $x_k[n]$ depends on the term $(1+\gamma_k[n]) - \beta_k[n]\left(\beta_k[n]+1\right)$. Since this article aims at the general users, the QoS index θ_k is usually in the range of $(0, 0.1]$. Therefore, the numerical value of $(1+\gamma_k[n]) - \beta_k[n]\left(\beta_k[n]+1\right)$ must be larger than zero, which means that C_k is convex for any variable $x_k[n]$. Finally, problem (13) is a standard convex optimization problem for the variable $\mathbf{X}$, which can be effectively solved by existing optimization tools such as CVX.

3.2 UAV Trajectory Optimization

For any given user scheduling $\mathbf{X}$, we can simplify it into an optimization problem (14) as follows

$$\max_{\mathbf{Q},C_{\min}} C_{\min} \tag{14a}$$

$$\text{subject to } -\frac{1}{\theta_k}\log_2\left(\frac{1}{N}\sum_{n=1}^{N}\left(1+\frac{P_{\max}\rho_0}{\sigma^2\left(H^2+||\mathbf{q}[n]-u_k||^2\right)}\right)^{-\beta_k}\right) \geqslant C_{\min}, \tag{14b}$$

$$||\mathbf{q}[n+1]-\mathbf{q}[n]||^2 \leqslant \left(\frac{V_{max}T}{N}\right)^2, \forall n, \tag{14c}$$

$$\mathbf{q}(1) = \mathbf{q}(N). \tag{14d}$$

In problem (14), Eqs. (14a), (14c) and (14d) are all convex for the variable $\mathbf{Q}$, but the left side of (14b) is neither convex nor concave for any variable $\mathbf{q}[n]$. Further, assuming that $z_k[n] = H^2 + ||\mathbf{q}[n] - u_k||^2$, the effective capacity C_k in the period T for any variable $z_k[n]$ has the following first-order derivation and second-order derivation

$$C_k{}'(z_k[n]) = -\frac{\beta_k[n]\alpha}{\theta_k \ln 2}\frac{(1+\frac{\alpha}{z_k[n]})^{-\beta_k[n]-1}}{(z_k[n])^2\sum\limits_{n=1}^{N}(1+\frac{\alpha}{z_k[n]})^{-\beta_k[n]}}, \tag{15}$$

$$\begin{aligned} C_k{}''(z_k[n]) = &-\frac{\alpha\beta_k[n](1+\frac{\alpha}{z_k[n]})^{-\beta_k[n]-2}}{\theta_k \ln 2(z_k[n])^4}\left[(1+\frac{\alpha}{z_k[n]})^{-\beta_k[n]}(-2z_k[n]-\alpha)\right. \\ &\left.+\left((\beta_k[n]-1)\alpha-2z_k[n]\right)\sum_{j=1,j\neq n}^{N}(1+\frac{\alpha}{z_k[j]})^{-\beta_k[n]}\right], \end{aligned} \tag{16}$$

where $\alpha = \frac{P_{\max}\rho_0}{\sigma^2}$ and $\beta_k[n] = \frac{\theta_k x_k[n]}{\ln 2}$. According to Eq. (16), since the positive and negative of other terms are known, the positive and negative of the second-order derivation of C_k directly depends on $((\beta_k[n]-1)\alpha - 2z_k[n])$. In the previous section, we mentioned that the targeted users are general users, so the value of second-order derivation of C_k must be larger than zero in the range of $(0, 0.1]$. Furthermore, C_k is convex for any variable $z_k[n]$. At the same time, we can see that $z_k[n] = H^2 + ||\mathbf{q}[n] - u_k||^2$ is also convex for any variable $\mathbf{q}[n]$. In this way, we can adopt successive convex optimization technique. Since the first-order Taylor expansion of an arbitrary point of an arbitrary convex function is its global lower bound, we will use the first-order Taylor expansion to approximate the original function C_k at a point $\mathbf{Z}^l$. (where $\mathbf{Q} = \mathbf{q}[\text{n}], \forall\text{n}\}$ is given). Assuming that the UAV trajectory $\mathbf{Q}^l = \mathbf{q}^l[\text{n}], \forall\text{n}\}$ and $\mathbf{Z}^l = \{H^2 + ||\mathbf{q}^l[n] - u_k||^2, \forall k, n\}$

are given in the lth round of the iteration, the objective function C_k can be approximated to the following form

$$\begin{aligned} C_k &= -\frac{1}{\theta_k}\log_2\left(\frac{1}{N}\sum_{n=1}^{N}\left(1+\frac{P_{\max}\rho_0}{\sigma^2 z_k[n]}\right)^{-\beta_k[n]}\right) \\ &\geqslant C_k(\mathbf{Z}^l) + C_k'(\mathbf{Z}^l)(\mathbf{Z}-\mathbf{Z}^l) \\ &= A - \sum_{n=1}^{N} B_k[n]\left(||\mathbf{q}[n]-u_k||^2 - ||\mathbf{q}^l[n]-u_k||^2\right) \\ &\triangleq C_k^{lb}, \end{aligned} \tag{17}$$

where

$$A = -\frac{1}{\theta_k}\log_2\left(\frac{1}{N}\sum_{n=1}^{N}\left(1+\frac{P_{\max}\rho_0}{\sigma^2(H^2+||\mathbf{q}^l[n]-u_k||^2)}\right)^{-\beta_k[n]}\right), \tag{18}$$

$$B_k[n] = \frac{\alpha\beta_k[n]}{\theta_k \ln 2(H^2+||\mathbf{q}^l[n]-u_k||^2)^2}\frac{(1+\frac{\alpha}{H^2+||\mathbf{q}^l[n]-u_k||^2})^{-\beta_k[n]-1}}{\sum_{n=1}^{N}(1+\frac{\alpha}{H^2+||\mathbf{q}^l[n]-u_k||^2})^{-\beta_k[n]}}. \tag{19}$$

Problem (14) is transformed into an optimization problem (20) in the following form

$$\max_{\mathbf{Q},C^r_{\min\_\mathbf{Q}}} \; C^r_{\min\_\mathbf{Q}} \tag{20a}$$

$$\text{subject to } C_k^{lb} \geqslant C^r_{\min\_\mathbf{Q}}, \tag{20b}$$

$$||\mathbf{q}[n+1]-\mathbf{q}[n]||^2 \leqslant \left(\frac{V_{max}T}{N}\right)^2, \forall n, \tag{20c}$$

$$\mathbf{q}(1) = \mathbf{q}(N). \tag{20d}$$

Since (20c) and (20d) are convex for any variable $\mathbf{q}[n]$, so problem (20) is a convex optimization problem, which can be effectively solved by standard convex optimization solvers (such as CVX). It is worth noting that the target value obtained by the approximation problem (20) is the lower bound of the optimal target value of the optimization problem (14).

3.3 Global Algorithm and Convergence

According to the contents of the previous two sections, we adopt the block coordinate descent method to optimize this problem, and the proposed algorithm is as follows.

$C_{\min}\left(\mathbf{X}^l,\mathbf{Q}^l\right)$ is defined as the minimum effective capacity among all users in the lth iteration. $C^{lb}_{\min}\left(\mathbf{X}^l,\mathbf{Q}^l\right)$ is defined as the lower bound of the minimum

Algorithm 1. Block coordinate descent algorithm for problem (10).

1: **Initialize** $\mathbf{Q}^0$ and $\mathbf{X}^0$. Let $l = 0$.
2: **Repeat**
3: Given $\left\{\mathbf{Q}^l, \mathbf{X}^l\right\}$ to solve the problem (11), and define the optimal case as $\left\{\mathbf{X}^{l+1}\right\}$.
4: Given $\left\{\mathbf{Q}^l, \mathbf{X}^{l+1}\right\}$ to solve the problem (20), and define the optimal case as $\left\{\mathbf{Q}^{l+1}\right\}$.
5: Update $l = l + 1$.
6: **Until** The gain of the objective function is less than a threshold $\varepsilon > 0$.

effective capacity among all users obtained in the lth iteration. For step 3 of Algorithm 1, we can get

$$C_{\min}\left(\mathbf{X}^l, \mathbf{Q}^l\right) \leqslant C_{\min}\left(\mathbf{X}^{l+1}, \mathbf{Q}^l\right). \tag{21}$$

Since $C_{\min}\left(\mathbf{X}^{l+1}, \mathbf{Q}^l\right)$ is the optimal value in the l+1 iteration, and the optimal value after optimization must be larger than or equal to the value before optimization.

For step 4 of Algorithm 1, we can get

$$\begin{aligned} C_{\min}\left(\mathbf{X}^{l+1}, \mathbf{Q}^l\right) &\overset{(a)}{=} C_{\min}^{lb}\left(\mathbf{X}^{l+1}, \mathbf{Q}^l\right) \\ &\overset{(b)}{\leqslant} C_{\min}^{lb}\left(\mathbf{X}^{l+1}, \mathbf{Q}^{l+1}\right) \\ &\overset{(c)}{\leqslant} C_{\min}\left(\mathbf{X}^{l+1}, \mathbf{Q}^{l+1}\right), \end{aligned} \tag{22}$$

where (a) indicates that the first-order Taylor expansion is tightened at a given local point, and both the left term and the right term have the same target value. Maximizing the objective function in (b) results in that the target value after optimization must be larger than or equal to that before optimization. In (c), $C_{\min}^{lb}\left(\mathbf{X}^{l+1}, \mathbf{Q}^{l+1}\right)$ is the lower bound of $C_{\min}\left(\mathbf{X}^{l+1}, \mathbf{Q}^{l+1}\right)$. Thus, we can obtain

$$C_{\min}\left(\mathbf{X}^l, \mathbf{Q}^l\right) \leqslant C_{\min}\left(\mathbf{X}^{l+1}, \mathbf{Q}^{l+1}\right). \tag{23}$$

In the above convergence analysis, we find that although the approximate function takes the place of the objective function in the process of trajectory optimization, the overall target value is not reduced after each iteration. Since the target value of the optimization problem (10) is the upper bound of the finite value, Algorithm 1 ensures convergence. Meanwhile, the initialization scheme usually affects the convergence speed for iterative algorithms. In this section, we simplify the existing initialization scheme in [10] to make it more suitable for a single UAV. The detailed scheme is shown in Algorithm 2.

Algorithm 2. Trajectory initialization scheme for a single UAV.

1: Determine the geometric centers of multiple ground users $c_u = \frac{1}{K}\sum_{k=1}^{K} U_k$.
2: Determine the minimum radius of the UAV trajectory that can cover all users $r_u = \max_{k\in\mathcal{K}} \|u_k - c_u\|$.
3: The maximum flight radius that ensures that the UAV can return to the starting point within the cycle T $r_{\max} = \frac{V_{\max}T}{2\pi}$.
4: Determine the final flight radius of the UAV $r = \min\left(r_{\max}, \frac{r_u}{2}\right)$.
5: Determine the initial trajectory of the UAV $q^0[n] = [x + rcos\left(2\pi\frac{(n-1)}{(N-1)}\right), y + rsin\left(2\pi\frac{(n-1)}{(N-1)}\right)], \forall n$.

4 Numerical Results

In this section, we provide simulation to illustrate the effectiveness and convergence of Algorithm 1. We consider a UAV communication system with $K = 4$ ground users, where the positions of all ground users are randomly generated within an area of $1 \times 1\,\text{km}^2$. The randomly generated location of the ground users is shown in Fig. 2. The UAV is assumed to fly at a fixed altitude $H = 100\,\text{m}$. The channel power gain at the reference distance $d_0 = 1\,\text{m}$ is set to $\rho_0 = -60\,\text{dB}$. The noise power at the receiver is assumed to be $\sigma^2 = -110\,\text{dBm}$. The maximum transmission power and maximum flight speed of the UAV are $P_{\max} = 0.1\,\text{W}$ and $V_{\max} = 50\,\text{m/s}$, respectively. The convergence threshold of Algorithm 1 is $\varepsilon = 10^{-5}$.

Before analyzing the simulation, we first explain the initialization scheme of user scheduling. In this scheme, the UAV position of each time slot is connected with the nearest user, and it is guaranteed that only one user is connected to one UAV in each time slot. Figure 2 shows the flight trajectory of the UAV generated by the proposed Algorithm 1 with different periods, where the user QoS is set to $\theta = [0.1, 0.05, 0.01, 0.05]$. We find that with the increase of the flight period of the UAV, the trajectory of the UAV is also gradually increasing. This shows that the trajectory of the UAV is subject to the flight period T. Comparing the flight trajectory with a period of T=5 s and the flight trajectory with T=15 s, we can find that the longer the period of the trajectory is, the bigger the gap between the initial trajectory and the trajectory. Further, the trajectory with a longer period can arrange the flight trajectory more adaptively to maximize the minimum effective capacity in the network. From this point, we can infer that when the period is large enough, the UAV will stay directly above some users for a while to increase the minimum effective capacity, thereby increasing the overall system capacity.

Figure 3 shows the UAV flight trajectories of different user QoS, of which the two QoS schemes are $\theta_1 = [0.1, 0.05, 0.01, 0.05]$ and $\theta_2 = [0.001, 0.005, 0.08, 0.01]$. Table 1 shows the max-min effective capacity comparison of different user QoS. The higher the QoS index, the more stringent the delay requirements of the

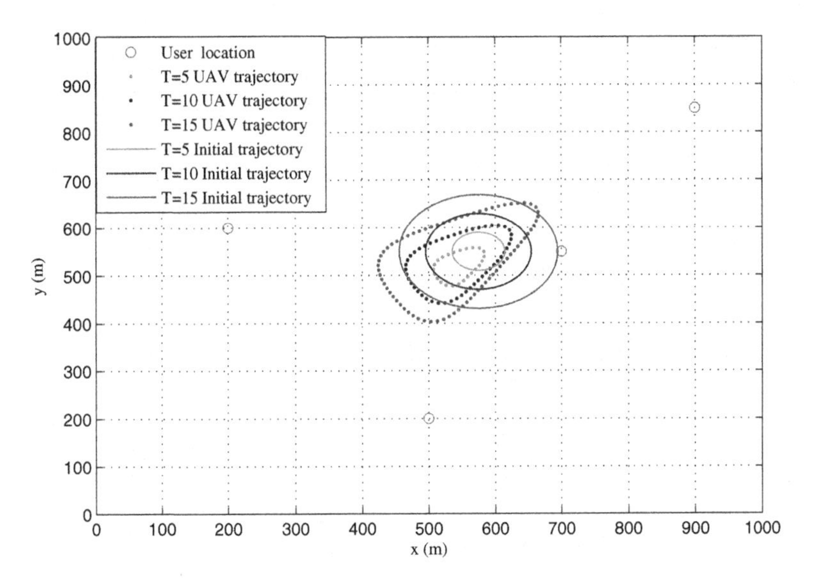

Fig. 2. The optimized trajectory of UAVs with the periods of $T = 5\,\mathrm{s}$, $T = 10\,\mathrm{s}$ and $T = 15\,\mathrm{s}$.

Table 1. Comparison of the max-min effective capacity with different user QoS

User	QoS scheme 1	QoS scheme 2
User 1	0.036618947541554	0.037037338823121
User 2	0.036618947541777	0.037037338823389
User 3	0.036618947548263	0.037037338828709
User 4	0.036618947541154	0.037037338822966

corresponding users. On the contrary, the smaller the QoS, the looser the delay requirements of users. The overall QoS of scheme 1 is larger than that of scheme 2, which shows that the users of scheme 1 have relatively strict requirements for delay. The target value of scheme 2 in Table 1 shows an increasing trend relative to scheme 1. This is because the system needs to allocate more bandwidth for users with stricter delay requirements to adapt to the effective capacity of downlink transmission. However, we strictly regulate the allocation of bandwidth, which leads to the reduction of the effective capacity of the downlink. The increase of the effective capacity of scheme 2 User 3 means that it occupies part of the effective capacity increase of other users. Comparing the trajectories corresponding to the two QoS schemes, we can see that the UAV trajectory of scheme 2 is closer to user 3 and user 4 compared to scheme 1. Since the delay requirements of user 3 and user 4 of solution 2 have become more stringent in comparison with solution 1, the UAV will approach them to meet the delay requirements.

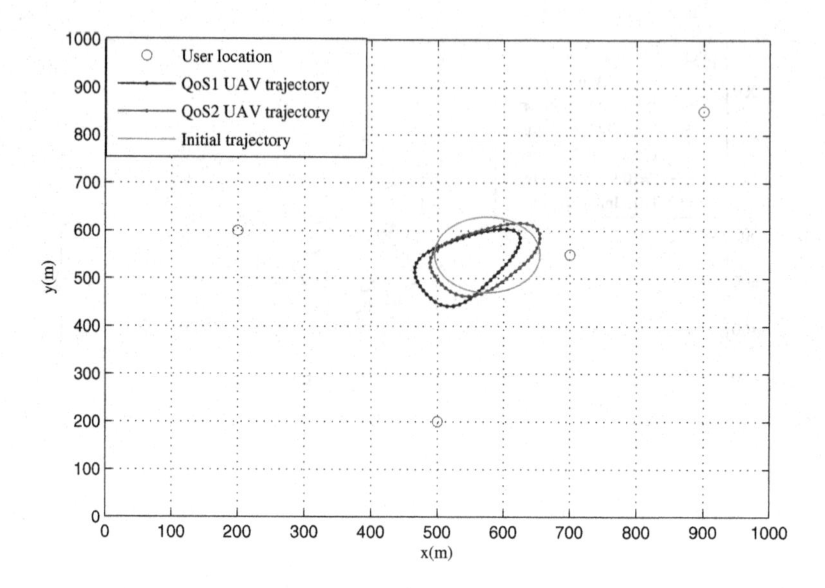

Fig. 3. The UAV optimized trajectory corresponding to different QoS when the period T=10 s.

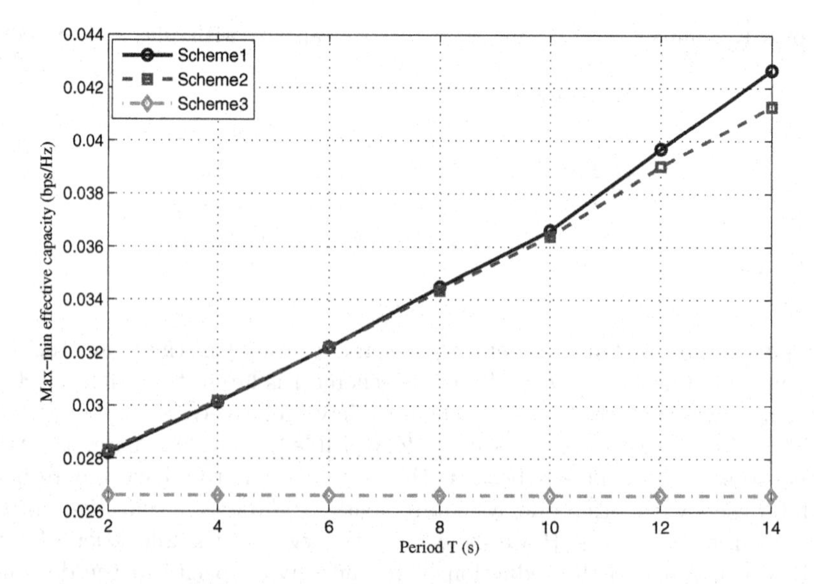

Fig. 4. The max-min effective capacity of different trajectory designs in period T.

Figure 4 shows the comparison trend of the optimal value of the proposed algorithm and the comparison algorithm in different periods of T. 1) Scheme 1: Algorithm 1 proposed in this paper, namely, joint optimization of trajectory and user scheduling; 2) Scheme 2: Optimizing the initial trajectory of user scheduling;

3) Scheme 3: UAV hovering at the geometric center. It can be seen from Fig. 4 that the target value of Scheme 3 remains unchanged and stays at a very low level. The curve of scheme 1 always stays above scheme 2. As the period increases, the gap between the two schemes gradually increases. From this, we can infer that the gap between scheme 1 and scheme 2 will increase significantly with the increase of the period. Since the flight radius of scheme 2 will not increase when the period T increases to a specific value. Even if the user scheduling is optimized, the fixed trajectory of scheme 2 will limit the communication distance between the UAV and the user to restrict the increase of the effective capacity of each user. In contrast, the UAV ensures the hovering time above each user and increases the effective capacity of each user by adaptively planning the trajectory in scheme 1 when the period T is large enough. Finally, we find that the target value in Fig. 4 is significantly smaller for two reasons: 1) the period T selected in the simulation is small, and the target value of the scheme will increase obviously with the increase of period T. 2) It is impossible for each user to connect N time slots with the UAV, and the joint optimization of user scheduling and the number of connection time slots will lead to the inability to solve the problem. Therefore, although the optimal solution obtained by scheme 1 is small, the effectiveness and convergence of Algorithm 1 can still be verified by comparing the effective capacity of all users after the compromise.

Table 2. Comparison of the maximum total capacity scheme and the max-min effective capacity scheme

User	Maximum total capacity scheme	Maxi-min effective capacity scheme
User 1	0	0.0344727255844699
User 2	0	0.0344727669486103
User 3	1	0.0344727310850233
User 4	0	0.0344728034346694

Figure 5 shows the convergence behavior of the proposed algorithm with a period of $T = 12$ s. We find that the target value will increase after each iteration and convergence will be reached quickly within about 10 iterations. Table 2 shows the user situation of maximizing the minimum effective capacity and maximizing the total capacity when the period $T = 8$ s. In the maximum total capacity scheme, the objective function is

$$C = \sum_{k=1}^{K} -\frac{1}{\theta_k} \log\left(\frac{1}{N}\sum_{n=1}^{N} e^{-\theta_k R_k[n]}\right). \tag{24}$$

It can be seen from Table 2 that the effective capacity is concentrated on one user in the maximum total capacity scheme. Since the goal is to maximize the total capacity, the UAV will find the user with the largest effective capacity in the iterative process, and then hover over the user to maximize the total system capacity. This behavior is fatal to other users' information transmission. The effective

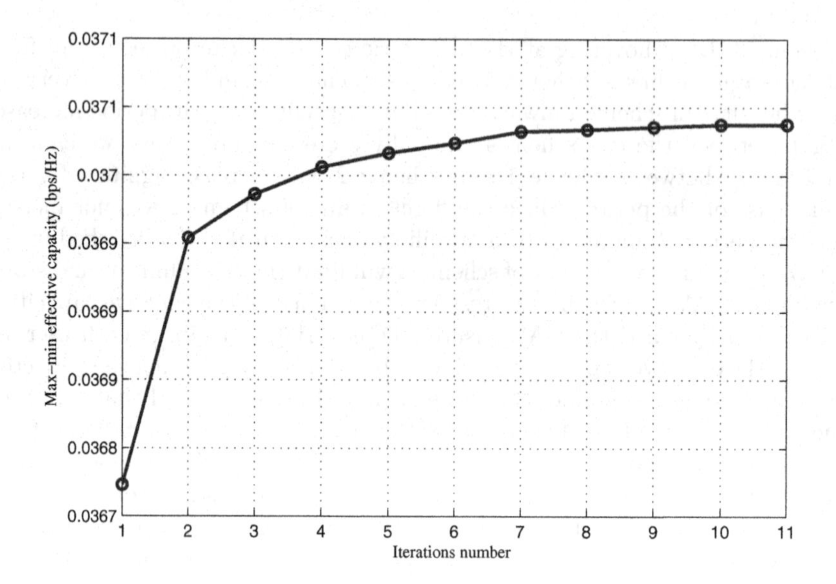

Fig. 5. Convergence behavior of proposed Algorithm 1.

capacity of each user corresponding to the objective function mentioned in this paper is very balanced, with a maximum difference of $7.78501995 \times 10^{-8}$ bps/Hz. It is verified that this paper guarantees the fairness of users. It is worth noting that the fairness of users comes at the expense of reducing the total effective capacity.

5 Conclusion

This paper studied the fairness of information transmission among users in UAV communication systems based on QoS. The main goal of this paper was to design an effective scheme for jointly optimizing user scheduling and UAV trajectory. The proposed scheme could satisfy different users' QoS requirements while ensuring that a single user could not monopolize information transmission to indirectly promote the fairness of information transmission. To this end, we proposed an iterative algorithm that adopted the block coordinate descent method and successive convex optimization technology. Compared with the circular trajectory scheme and fixed-point trajectory scheme, the convergence and effectiveness of the proposed algorithm were verified.

References

1. Cao, X., Wu, C., Lan, J., Yan, P., Li, X.: Vehicle detection and motion analysis in low-altitude airborne video under urban environment. IEEE Trans. Circuits Syst. Video Technol. **21**(10), 1522–1533 (2011)

2. Zhang, S., Liu, J., Sun, W.: Stochastic geometric analysis of multiple unmanned aerial vehicle-assisted communications over Internet of Things. IEEE Internet Things J. **6**(3), 5446–5460 (2019)
3. Liu, M., Yang, J., Gui, G.: DSF-NOMA: UAV-assisted emergency communication technology in a heterogeneous Internet of Things. IEEE Internet Things J. **6**(3), 5508–5519 (2019)
4. Jiang, B., Yang, J., Xu, H., Song, H., Zheng, G.: Multimedia data throughput maximization in Internet-of-Things system based on optimization of cache-enabled UAV. IEEE Internet Things J. **6**(2), 3525–3532 (2019)
5. Yao, H., Wang, L., Wang, X., Lu, Z., Liu, Y.: The space-terrestrial integrated network: an overview. IEEE Commun. Mag. **56**(9), 178–185 (2018)
6. Jeong, S., Simeone, O., Kang, J.: Mobile edge computing via a UAV-mounted cloudlet: optimization of bit allocation and path planning. IEEE Trans. Veh. Technol. **67**(3), 2049–2063 (2018)
7. Lyu, J., Zeng, Y., Zhang, R., Lim, T.J.: Placement optimization of UAV-mounted mobile base stations. IEEE Commun. Lett. **21**(3), 604–607 (2017)
8. Zeng, Y., Zhang, R., Lim, T.J.: Throughput maximization for UAV-enabled mobile relaying systems. IEEE Trans. Commun. **64**(12), 4983–4996 (2016)
9. Lyu, J., Zeng, Y., Zhang, R.: Cyclical multiple access in UAV-aided communications: a throughput-delay tradeoff. IEEE Wirel. Commun. Lett. **5**(6), 600–603 (2016)
10. Wu, Q., Zeng, Y., Zhang, R.: Joint trajectory and communication design for multi-UAV enabled wireless networks. IEEE Trans. Wirel. Commun. **17**(3), 2109–2121 (2018)
11. Zhang, X., Cheng, W., Zhang, H.: Heterogeneous statistical QoS provisioning over airborne mobile wireless networks. IEEE J. Sel. Areas Commun. **36**(9), 2139–2152 (2018)
12. Zhu, S., Gui, L., Cheng, N., Sun, F., Zhang, Q.: Joint design of access point selection and path planning for UAV-assisted cellular networks. IEEE Internet Things J. **7**(1), 220–233 (2020)
13. Bejaoui, A., Park, K., Alouini, M.: A QoS-oriented trajectory optimization in swarming unmanned-aerial-vehicles communications. IEEE Wirel. Commun. Lett. **9**(6), 791–794 (2020)

Research on Dynamic Timeslot Reservation Handover Algorithm Based on Remaining Time in Beam Hopping System

Hongyu Pan[1,2](✉), Chuang Wang[3], and Lifeng Jiang[4]

[1] College of Telecommunications and Information Engineering, Nanjing University of Posts and Telecommunications, Nanjing 210003, China
[2] National Engineering Research Center for Communication and Network Technology, Nanjing University of Posts and Telecommunications, Nanjing 210003, China
[3] Military Representative Office in Zhenjiang District of Some Department, Zhenjiang 212000, China
[4] Xi'an Institute of Space Radio Techonology, Xi'an 710000, China

Abstract. Recently, beam hopping (BH) technology has been considered as the key technology of the next generation High Throughput Satellite (HTS) system for its flexibility. In order to ensure the communication continuity of the terminal in the beam hopping satellite system, it is particularly important to study efficient beam handover technology. This paper proposes a dynamic timeslot reservation handover algorithm based on remaining time. Firstly, the terminal obtains remaining time of terminal in the current beam position. Then this paper simulates load imbalance through Poisson random distribution, and then clusters the beam positions to calculate the probability density function of the timeslot occupied in the cluster. Finally the algorithm dynamically reserves timeslot resources for the terminal. The simulation results show that, compared with the traditional handover algorithm, when the timeslot resources are insufficient, the algorithm can dynamically reserve timeslot resources to reduce the handover failure probability, and improve QoS.

Keywords: High Throughput Satellite · Beam Hopping (BH) · Beam handover · Timeslot reservation

1 Introduction

The core concept of beam hopping [1] is to employ time-slicing method: not all beams are illuminated at the same time, only part of them are activated on demand. Compared with the traditional multi-beam satellite system, the beam hopping technology is more able to meet the scenarios with unbalanced traffic requirements [2–5]. On the basis of this technology, in order to ensure the continuity of the communication of high-speed mobile terminals between beam positions, it is particularly important to study beam handover technology.

Many scholars have done a lot of research on beam handover algorithm. Literature [6] proposed a hard handover algorithm, which is based on the received signal strength

Q. Guo et al. (Eds.): WiSATS 2021, LNICST 410, pp. 496–507, 2022.
https://doi.org/10.1007/978-3-030-93398-2_46

and the speed of the mobile terminal. The algorithm does not take into account the current geographical location the terminal, and hard handover is more likely to cause communication interruption. A conventional handover algorithm was proposed in literature [7], which is based on the distance between the terminal and the center of each beam position. This algorithm does not consider the speed of the mobile terminal and follows the first-come first-served principle, which has great limitations. A handover algorithm was proposed based on remaining time [8, 9]. This algorithm takes into account information about geographic location and terminal speed to obtain remaining time at the current beam position, which solves the urgency problem. Mul-priority channel reservation allocation strategy was proposed in literature [10]. This algorithm classifies call types and reserves channels for handover calls to reduce handover failure probability. The beam handover strategy should be a multi-factor integration problem, so literature [11] proposes a handover algorithm based on fuzzy logic, combining multiple variables into a cost function for judgment. All the above algorithms are based in multi-beam satellite system. In beam hopping satellite system, the success of handover mainly depends on the occupancy of timeslot resources.

In view of the lack of research on beam hopping handover algorithm, with the goal of optimizing handover failure probability and improve the quality of service, this paper focuses on minimizing the failure of handover calls by reserving timeslot resources in the case of random distribution. Firstly, the terminal obtains the remaining time of terminals at the current beam position. Secondly, the beam positions are clustered. This paper simulates load imbalance through random distribution and the probability density function of the occupied timeslots is obtained. Finally, according to the handover failure probability, the timeslot resources are dynamically reserved for handover call terminals: when the handover failure probability is greater than the threshold, increase the number of timeslot reservations, otherwise the number of timeslot reservations remains unchanged. The simulation results show that under high arrival rate, the dynamic timeslot reservation handover algorithm based on remaining time has a lower handover failure probability, and improves QoS.

The rest of this paper is arranged as follows: Sect. 2 gives the beam hopping system model; Sect. 3 mainly explains the handover algorithm and handover process; Sect. 4 analyzes the simulation results; This paper is summarized in Sect. 5.

2 Beam Hopping System Model

The beam hopping model is shown in Fig. 1. The core concept of beam hopping is to employ time-slicing method: not all beams are illuminated at the same time, only part of them are activated on demand. Each cluster has timeslot period tables, and the beam positions are illuminated according to the timeslot period table. When the gateway receives the handover request, according to the usage of timeslot resources, beam is switched by the beam hopping controller on the satellite to ensure the continuity of mobile terminal communication.

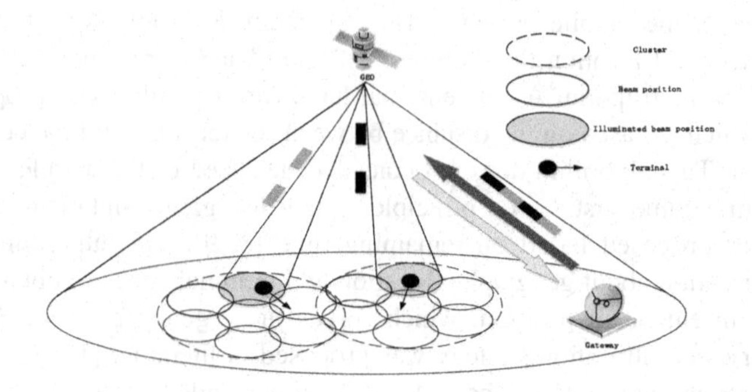

Fig. 1. Beam hopping system model

For ease of analysis, we assume that the system is a single beam hopping satellite system. Firstly, the beam positions are clustered. N_b beam positions form a cluster. Mobile terminals are randomly distributed in the cluster. At the same time, at most one beam position in each cluster is illuminated, and the number of timeslots in each cycle is the same, and the window length is W. Assuming that each terminal needs at least one timeslot for access, at most W mobile terminals are allowed to access at the same time in a cluster. In practice, the experience of handover failure is worse than waiting for new connection request access, so this paper sets the priority of handover request higher than new connection request. K timeslots are reserved for initialization, and the reserved timeslots are only allocated to the terminal that handover request. The remaining (W-K) timeslots are called by the handover and the new connection calls compete together.

3 Handover Algorithm and Analysis Model

3.1 Handover Mechanism

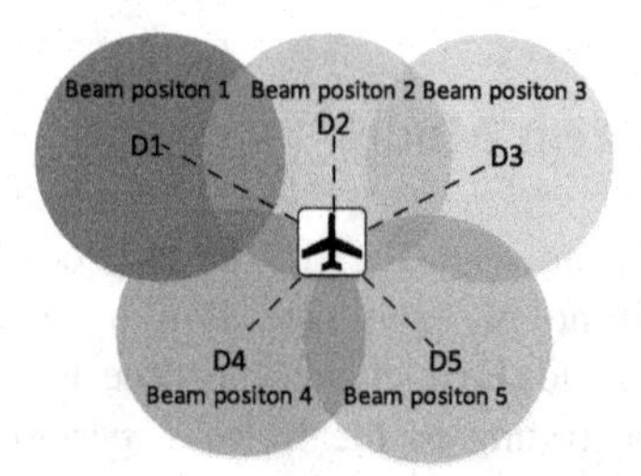

Fig. 2. Schematic diagram of the movement of the terminal in the multi-beam

Handover Mechanism Based on Location. In the handover mechanism [7, 8] based on location, the location of the terminal is used as a condition for triggering handover. As shown in Fig. 2, **D1–D5** are the distance between the mobile terminal and the beam position center respectively. Assuming that the current beam position No. 2 is serving the mobile terminal, the terminal periodically calculates the distance D1–D5 from the center of each beam position. Then D1–D5 compare with D2. The distance difference **para** $= D_i - D_2$. We set the terminal's mobility management requirement to 0, and the terminal sends the position information to the gateway via satellite. When para < 0, the handover request is met, the gateway prepares timeslot resources for the terminal on the target beam position. If the occupied time window is smaller than the timeslot window W, then the gateway execute handover. If the timeslot window is occupied, the terminal enters the waiting timeslot resource state, and continuously send the current location information within the overlap range until it leaves the overlap area.

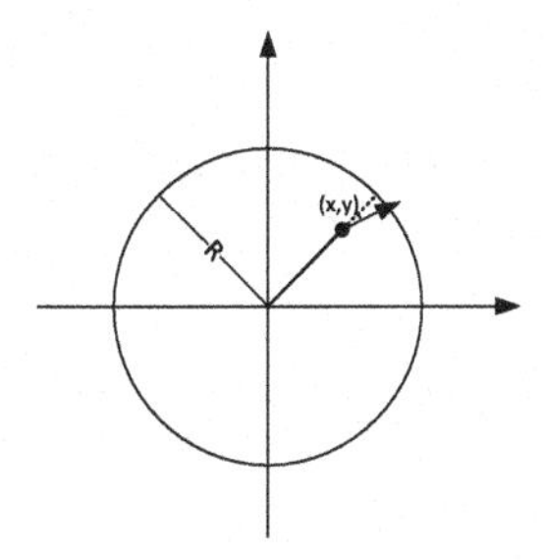

Fig. 3. The mobile station in the beam residence time calculation chart

Handover Mechanism Based on Remaining Time. Under the same satellite system, high-speed terminals such as airplanes and trains compete for the same channel or time slot resources. Assuming that in a overlap area, terminals with different speeds and different trajectories make handover requests to the gateway. According to the Mechanism in 3.1.1, the gateway will handle the handover requests according to the first-come first-served principle. For systems that include terminals of different speeds, this Mechanism has great limitations. This section proposes the handover mechanism [4, 7] based on remaining time that takes the remaining time of the terminal in the current beam position as the decision factor for handover. As shown in Fig. 3, the current position of the terminal is (x, y), the distance from the center of the beam position is $\mathbf{d} = \sqrt{x^2 + y^2}$. Assuming that the angle is θ, the remaining distance in the current beam position is $D = \sqrt{R^2 - d^2 \sin^2\theta} - d\cos\theta$. The remaining time in the current beam position is $\mathbf{T} = \mathbf{D}/\mathbf{V}$. When it is less than the remaining time threshold, all mobile terminal handover requests will be queued according to the remaining time, and the gateway will give priority to the terminal with the least remaining time to allocate timeslot resources. It can be seen that under this handover mechanism, some urgent terminals can be processed preferentially, which reduces the handover failure probability.

3.2 Analysis Model

For a general connection, the channel holding time of the spot beam can be derived from [12]:

$$T_{hold} = \{T_{res}, T_{un}\} \tag{1}$$

Where T_{res} is the time interval between the moment of reaching the beam position and the moment of reaching the boundary of the adjacent beam position. T_{un} is the unencumbered connection duration. For these two types of terminals, airplanes(i = a) and trains(i = t), we assume that T_{un} is an exponentially distributed random variable with parameter τ. When deriving the T_{hold} [13], the position where the terminal enters the beam position cannot be ignored, that is, T_{hold} is related to the original position of the terminal [7, 14].

The terminals are always randomly distributed within the beam position, and then average channel holding times are derived from [9]:

$$E\left[T_{hold,i}^{new}\right] = \frac{1}{\tau}\left(1 - \frac{1-e^{-\tau R_{max,i}}}{\tau R_{max,i}}\right) \tag{2}$$

$$E\left[T_{hold,i}^{handover}\right] = \frac{1}{\tau}\left(1 - e^{-\tau R_{max,i}}\right) \tag{3}$$

Where $R_{max,i}$ is the time required for the terminal to cross the beam position. $E\left[T_{hold,i}^{new}\right]$ is average channel holding time of the terminal in original beam position. $E\left[T_{hold,i}^{handover}\right]$ is average channel holding time of the terminal in handover beam position.

The new connection arrival rate and handover request arrival rate of different terminals are independent Poisson processes, and arrival rates are $\lambda_{new,i}$ and $\lambda_{handover,i}$ respectively. Their relationship is [9]:

$$\lambda_{handover,i} = \frac{\lambda_{new,i}(1-P_{nb})P_{new,i}}{1-\left(1-P_{hf}\right)P_{handover,i}} \tag{4}$$

where P_{nb} is new connections blocking probability. P_{hf} is handover failure probability. $P_{new,i}$ is the probability that a certain terminal needs to switch in the original beam position. $P_{handover,i}$ is the probability that a certain terminal needs to switch in the handover beam position. In order to simplify our analysis, a traditional approximation is made to the above formulas.

The average connection time of the terminal $\frac{1}{\tau_i}$ is derived from the above formula [9]:

$$\begin{aligned}\frac{1}{\tau_i} &= \frac{\lambda_{new,i}(1-P_{nb})}{\lambda_{n,i}(1-P_{nb}) + \lambda_{h,i}\left(1-P_{hf}\right)} E\left[t_{hold,i}^{new}\right] \\ &+ \frac{\lambda_{handover,i}\left(1-P_{hf}\right)}{\lambda_{new,i}(1-P_{nb}) + \lambda_{handover,i}\left(1-P_{hf}\right)} E\left[t_{hold,i}^{handover}\right]\end{aligned} \tag{5}$$

The number of occupied timeslots in the cluster is C, and the probability density function $f_C(z)$ of variable C is derived through the algorithm proposed in [15], and $z = 1, 2, 3, 4 \ldots W$:

$$f_C(z) = \frac{1}{z}\sum_{j=1}^{4}\frac{\lambda_j}{\tau_j} f_C(z-1)\Delta_j(z-1) \tag{6}$$

$$\Delta_j(z) = \begin{cases} 1 & z \le threshold_j - 1 \\ 0 & otherwise \end{cases} \tag{7}$$

Where $threshold_j$ is the threshold of each connection type. $j = 1, 2, 3, 4$ are the new connection request of train, the new connection request of airplane, the handover request of the train and the handover request of airplane respectively. Since trains and airplanes share the same timeslot resources, only the type of connection is distinguished, not the type of terminal. Then $\Delta_{new}(j = 1, 2)$, $\Delta_{handover}(j = 3, 4)$, $threshold_{\text{new}}(j = 1, 2)$, $threshold_{\text{handover}}(j = 3, 4)$ will be considered below.

After normalizing $f_C(z)$, the new connection blocking probability is derived from [9]:

$$P_{nb} = 1 - \sum_{z=0}^{W} f_C(z-1)\Delta_{new}(z-1) \tag{8}$$

According to $f_C(z)$, the probability P_{all} that all timeslots are occupied is derived:

$$P_{all} = 1 - \sum_{z=0}^{W} f_C(z-1)\Delta_{handover}(z-1) \tag{9}$$

Equation (9) does not consider the queuing strategy. For this reason, we assume that the maximum queuing time of the *i-th* terminal is an exponentially distributed random variable with parameter τ_i^P. At the same time, the results given in (9) are extended to a single type of terminal:

$$P_{hf} = P_{all}\left(\sum_{i=a,t} P_i P_i^{active} P_i^{occupied}\right) \tag{10}$$

Where P_i is the probability that the terminal requesting handover is the *i-th* type of terminal, P_i^{active} is the probability that the terminal remains connected in the overlapping area and $P_i^{active} = \frac{\tau_i^P}{\tau_i + \tau_i^P}.P_i^{occupied}$ is the probability that the timeslot is not released within the maximum queuing time, and $P_i^{occupied} = \frac{\tau_i^P}{\tau_a Com_{\frac{a}{tot}} + \tau_t Com_{\frac{t}{tot}} + \tau_i^P}$, $Com_{\frac{a}{tot}}$ and $Com_{\frac{t}{tot}}$ respectively are the ratio of the number of timeslots occupied by planes and trains, and $Com_{\frac{a}{tot}} + Com_{\frac{t}{tot}} \le 1$. This method allows us to implement the handover mechanism by changing the τ_i^P value: Under the mechanism based location, $\frac{1}{\tau_t^P} = \frac{1}{\tau_a^P} = 550$ s, and under the mechanism based remaining time, $\frac{1}{\tau_t^P} = 275$ s, $\frac{1}{\tau_a^P} = 80$ s.

For a certain beam position, the interval between two adjacent illuminations cannot be too long [16]. Otherwise, the transmission delay will be too long and the communication will be interrupted. A condition is needed here:

$$mint_{slot} \geq 10\,ms$$

Where $t_{slot} = T_{slot} \times N.t_{slot}$ is the total timeslot duration of a certain beam position, T_{slot} is duration of one timeslot. N is the number of timeslots allocated to a certain beam position.

3.3 Dynamic Timeslot Reservation Handover

For beam hopping system, users in the same cluster are served by the same beam, and do not need to switch at network layer. At the same time, the handover signaling overhead is low, but it puts forward high requirements on resource allocation: the corresponding timeslots need to be allocated when the terminals arrive and are released when the terminals leave. For multi-beam system, each beam resource is relatively fixed, there are many handover signaling interactions, and beam handover often involves network layer protocols.

The handover failure probability P_{hf} derived in the previous section is a handover algorithm based on static timeslot reservation, and does not reflect the superiority of the handover algorithm based on dynamic timeslot reservation. The handover algorithm based on dynamic timeslot reservation proposed in this paper, which based on static timeslot reservation, the number of timeslot reservations is dynamically adjusted according to the handover failure probability: when the handover failure probability is greater than the threshold, increase the number of timeslot reservations, otherwise the number of timeslot reservations remains unchanged. Dynamic timeslot reservation is shown in Fig. 4:

1. Initialize the terminal arrival rate λ_1, λ_2.
2. Initialization. For different types of terminals, $P_{hf} = P_{nb} = P_{all} = 0$, and the number of remaining timeslots in the cluster is W, then $f_C(0) = 1$.
3. According to the arrival rates of airplanes and trains λ_1, λ_2, the handover arrival rates of airplanes and trains λ_3, λ_4 are derived.
4. According to (5), the average connection time $\frac{1}{\tau_i}$ of a certain terminal is derived.
5. The probability density function $f_C(z)$ of the number of occupied timeslots in the cluster is derived through (6), and then normalized.
6. Select the handover mechanism, get different τ_i^P, and derive $NewP_{hf}$, $NewP_{nb}$, $NewP_{all}$.
7. If $NewP_{hf}$ is greater than the $threshold_{Phf}$, the number of timeslot reservations is increased by one, and repeat 5.
8. If $mint_{slot}$ is greater than 10 ms and $\sum_k |NewP - P| < \alpha$, output handover failure probability P_{hf} and new connection blocking probability P_{nb}, otherwise, repeat 3.

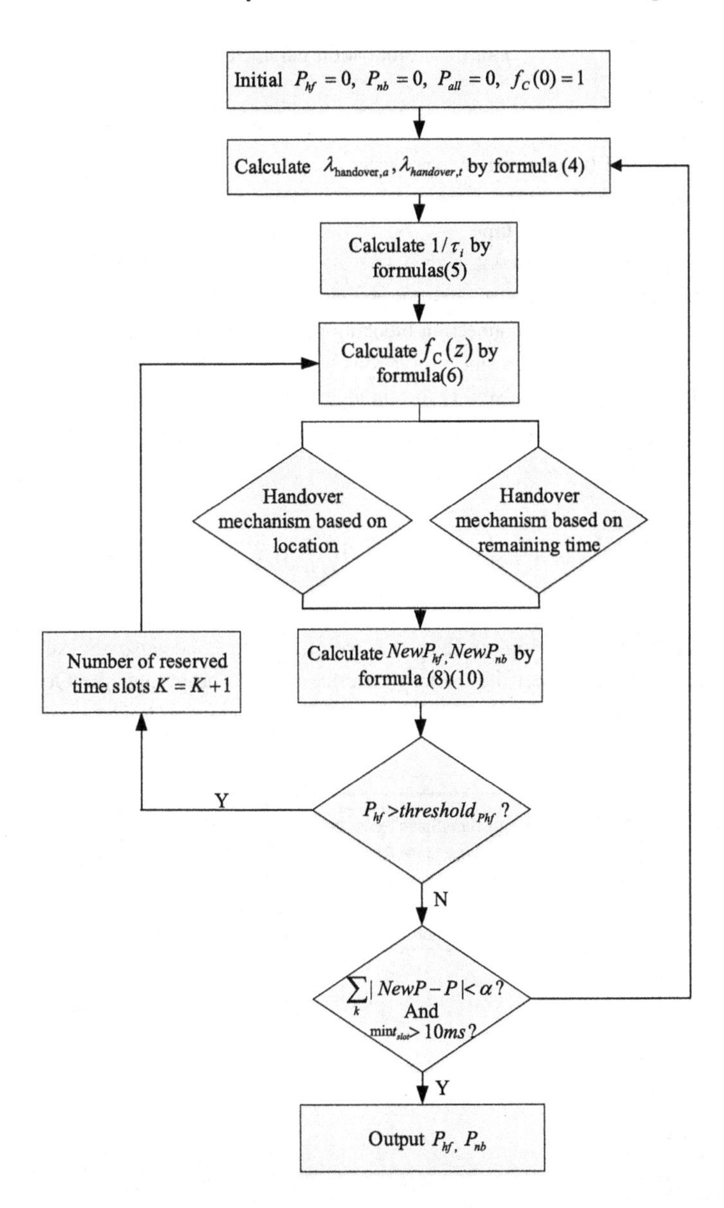

Fig. 4. Dynamic reservation handover flow chart

4 Simulation and Analysis

The simulation parameters are shown in Table 1. In addition, in order to better compare the simulation results, P_{hf}, P_{nb}, and QoS are used as indicators, and QoS is defined as:

Table 1. Simulation parameters

Parameters	Value
Beam hopping period	W = 140 T_{slot}
Number of beam positions per cluster	$N_b = 7$
Arrival rate of terminal	λ from 1 to 20
Average connection time	$\tau = 1$ h
Duration of one timeslot	$T_{slot} = 2$ ms
Importance of handover failure probability	$\alpha_{hf} = 0.8$
Importance of new connection blocking probability	$\alpha_{nb} = 0.2$
Number of initial timeslot reservations	K = 7
Handover failure probability threshold	$threshold_{Phf} = 0.01$
Convergence threshold	$\alpha = 0.01$

$$QoS = 1 - \left(0.7 \sum_{i=a,i=t} \alpha_{hf} P_{hf} + 0.3 \sum_{i=a,i=t} \alpha_{nb} P_{nb}\right) \tag{11}$$

The weight of P_{hf} is higher than the weight of P_{nb}, because the success of the terminal handover can affect the user experience more. The greater the QoS, the better the experience in this paper.

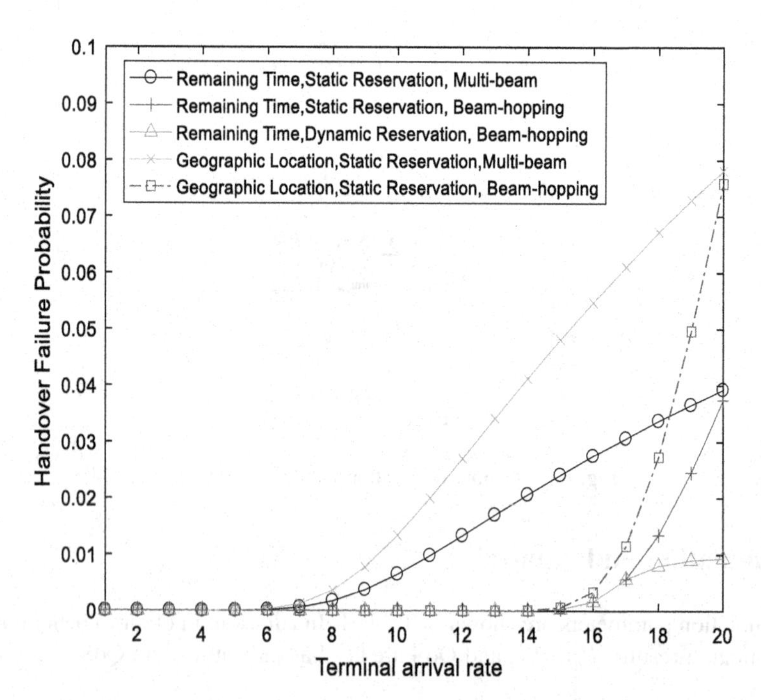

Fig. 5. The comparison of handover failure probability

As shown in Fig. 5, in the same scenario, it can be seen that the higher the arrival rate, the timeslot reservation handover algorithm based on remaining time has a significant improvement in handover failure probability. Under the same handover algorithm, there is a lower handover failure probability in beam hopping. Because beam hopping has more flexibility in timeslot allocation. This paper proposes a dynamic timeslot reservation handover algorithm based on remaining time. When the arrival rate is large enough, more timeslot resources need to be reserved to ensure that handover failure probability is within $threshold_{Phf}$.

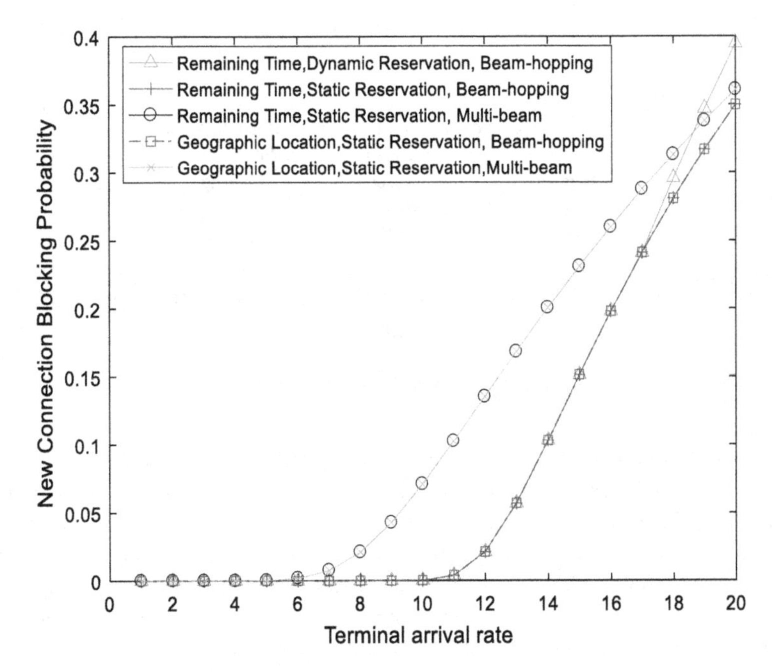

Fig. 6. The comparison of new connection blocking probability

As shown in Fig. 6, in the same scenario, the static timeslot reservation handover algorithm based on location and the static timeslot reservation handover algorithm based on remaining time have the same new connection blocking probability, indicating that the difference between the two handover mechanisms is not the new connection access, so different handover mechanisms have little effect on the new connection blocking probability. Under the same handover mechanism, there is a lower new connection blocking probability in the beam hopping, because beam hopping has more flexibility in timeslot allocation. This paper proposes a dynamic timeslot reservation handover algorithm based on remaining time. As handover failure probability approaches $threshold_{Phf}$, more timeslots are reserved for handover calls, which increases the new connection blocking probability.

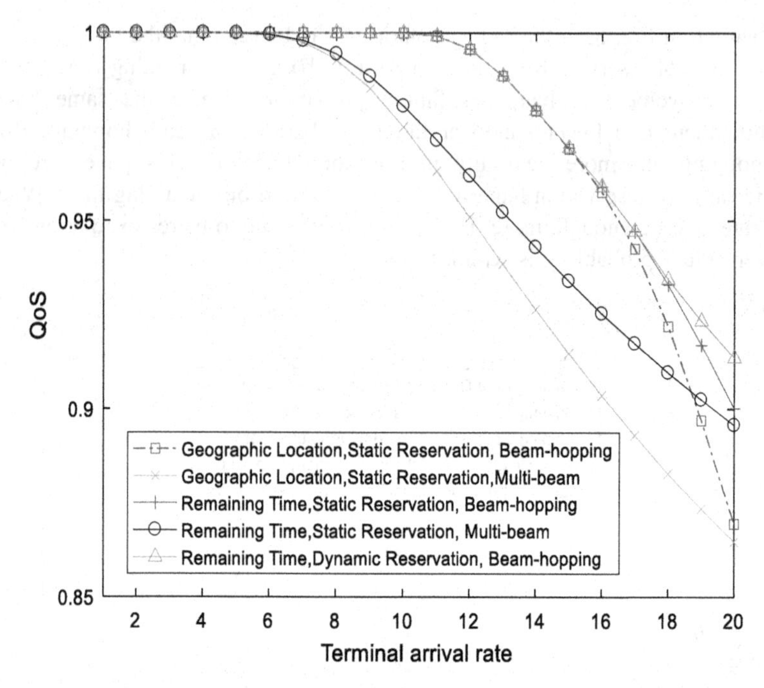

Fig. 7. The comparison of QoS

Successful handover can improve the user experience more, so the handover success probability is far more important than the new connection success probability. As shown in Fig. 7, in the same scenario, the QoS of the timeslot reservation handover algorithm based on remaining time goes down more slowly. For the same handover algorithm, the QoS goes down more slowly in beam hopping. In the case of high arrival rate or insufficient resources, the beam hopping technology uses can effectively reduce the handover failure probability for its flexibility. The dynamic timeslot reservation handover algorithm based on remaining time proposed in this paper reserves more timeslots to ensure the handover success probability. So that the QoS goes down the slowest.

5 Conclusion

In this paper, random distribution is used to simulate load imbalance. In order to reduce the handover failure probability of the system, a dynamic timeslot reservation handover algorithm based on remaining time is proposed: when the handover failure probability is greater than the threshold, increase the number of timeslot reservations, otherwise the number of timeslot reservations remains unchanged. The simulation results show that the algorithm can update the number of timeslot reservations according to the handover failure probability, which effectively guarantees the high-priority call service in the case of insufficient timeslot resources, and improves the user's quality of service.

References

1. Tang, J., Li, G., Bian, D., Hu, J.: Review on resource allocation for beam-hopping satellite. Mob. Commun. **43**(05), 26–31 (2019)
2. Yi, K., Li, Y., Sun, C., Nan, C.: Recent development and its prospect of satellite communications. J. Commun. **36**(06), 161–176 (2015)
3. Zhao, X., Zhang, C., Zhang, G.: Summary of beam-hopping technology of broadband satellite communication system. In: The 15th Annual Satellite Communication Conference, p. 7 (2019)
4. Feng, S., Tang, H., Xu, Z., Li, H.: The development and standardization process of broadband multimedia satellite system. Space Int. **12**, 20–25 (2012)
5. Chen, T., Liu, H., Wang, D., Hao, X.: Research on airborne satellite communication technology based on Ka high-throughput satellite communication system. Sci. Technol. Vis. (21), 219–221+143 (2019)
6. Song, L., Liu, A., Tian, X., Xi, G.: Dynamic handoff algorithm for GEO mobile satellite system. J. Syst. Simul. **21**(11), 3411–3415 (2009)
7. Liu, M., Shi, Y., Cheng, Z.: Performance analysis of beam handover mechanism in high throughput satellite communication system. Electron. Des. Eng. **27**(02), 154–159+165 (2019)
8. Deng, Z., Long, B., Lin, W., Wang, J.: GEO satellite communications system soft handover algorithm based on residence time. In: 3rd International Conference on Computer Science and Network Technology, Dalian, pp. 12–13, October (2013)
9. Lattanzi, F.,Acar, G., Evans, B.: Performance study of a lightweight DVB- RCS handover scheme for vehicular GEO networks. In: IEEE International Workshop on Satellite and Space Communications, Toulouse (2008)
10. Gui, Y., Zhu, L.: Multi-priority channel reservation allocation strategy for satellite airborne terminals. Radio Commun. Technol. **45**(01), 62–67 (2019)
11. Wang, J., Yu, H.: Research of Spotbeam handoff strategy based on fuzzy logic for GEO Multibeam satellite mobile communication system. Commun. Technol. **50**(10), 2266–2273 (2017)
12. Lin, Y., Chang, L., Noerpel, A.: Modeling hierarchical microcell/Macrocell PCS architecture. In: IEEE International Conference, vol. 1, pp. 405–409 (1995)
13. Del Re, E., Fantacci, R.: Different queuing policies for handover requests in low earth orbit mobile satellite systems. IEEE Trans. Veh. Technol. **48**(2), 448–458 (1999)
14. Tani, S., Uchida, S., Okamura, A.: Overlapping clustering for beam-hopping systems. In: 36th International Communications Satellite Systems Conference. Niagara Falls, pp. 14–18, October (2018)
15. Kyrgiazos, A., Evans, B., Thompson, P.: Smart gateways designs with time switched feeders and beam hopping user links. In: Advanced Satellite Multimedia Systems Conference and the Signal Processing for Space Communications Workshop, pp. 1–6 (2016)
16. Zhang, C., Zhang, G., Wang, X.: Design of next generation high throughtput satellite communication system based on beam-hopping. J. Commun. **41**(07), 59–72 (2020)

Equal Proportional Probability Caching Policy for Heterogeneous Network Based on Stochastic Geometry Theory

Haoyang Liu(✉), Gang Wang, Wenchao Yang, Jinlong Wang, Yao Xu, and Donglai Zhao

Communication Research Center, Harbin Institute of Technology, Harbin, China
{gwang51,wenchaoy}@hit.edu.cn, xuyao_hit@sina.com

Abstract. In this paper, we compare different caching policies in a cache-enabled heterogeneous network (HetNet). The network is equipped with a tier of macro cell base stations (MBSs) with backhaul link and a tier of small cell base stations (SBSs) with cache memories. The two-dimensional spatial distribution of these two-tier network is modeled by stochastic geometry theory, the probability of the files being requested is modeled by Zipf distribution and the probability of the files being cached on the SBS are modeled by different particular distributions corresponding to different caching policies. We use average outage probability (AOP) to measure the performance of the system. In order to decilne the AOP of a user request, we propose a new caching policy with equal proportional probability. The performance of the new caching policy, the affect caused by SBS density changing and the affect caused by cache skewness changing are verified by Monte-Carlo simulation.

Keywords: Caching policy · Heterogeneous network · Stochastic geometry · Average outage probability · Equal proportional probability

1 Introduction

With the explosive growth of mobile data services, the user's demand for high-speed data streams is rapidly growing. Since requesting contents from the centralized cloud usually generates significant backhaul link pressure, the recent research tends to cache the contents on the edge storage deployed on the SBSs of HetNet [1].

Focusing on the minimization of resource consumption, by combining the caching memory and backhaul link as a unified resource with balance coefficient, the optimal memory size has been obtained in closed-form, and the optimal density of SBSs has been found numerically, which can reduce the required capacity of backhaul but keep the wireless QoS effectively [2].

This work is supported in part by National Natural Science Foundation of China (No. 62071146, 62071147).

Q. Guo et al. (Eds.): WiSATS 2021, LNICST 410, pp. 508–519, 2022.
https://doi.org/10.1007/978-3-030-93398-2_47

Focusing on different performance metrics, to reduce time delay, by jointly modeling the backhaul and wireless transmission phases, a theoretical framework for partially connected edge cache network was derived to characterize the trade-off relations between time delay and deployment parameters, and the optimal deployment strategy is obtained [3]. To maximize cache hit probability, by modeling the network with limited number of BSs, cached files and users, [4] relaxed the problem to a convex optimization problem and obtained the optimal caching policy with randomized rounding algorithm. When considering outage probability of the service, [5] compared the performance of different classical caching policies and analytically derived the outage probability of serving the requested content by jointly considering spectrum allocation and storage constraints. To maximize the success probability and area spectral efficiency (ASE), [6] takes SBS idling into consideration and obtained the optimal caching probability respectively maximizing the success probability and ASE, which concluded that each SBS tends to cache the most popular files to maximize ASE.

Focusing on special network structure, with extra D2D communication inserted, [7] proposed a near optimal MOAC algorithm and a low-complexity MRAC algorithm, which demonstrated that the proposed algorithms solved the problem of user selfishness through the natural social efficiency and individual rationality of the proposed auction model.

This paper characterizes the AOPs of user request corresponding to different caching policies by modeling the spatial distribution of the BSs as several independent Poisson Point Processes (PPPs). Unlike [5], we propose a new caching policy which could avoid the shortcomings of the classical policies and provide lower outage probability than the classical policies. Furthermore, we investigate the relationship between the density of SBS and AOP and the relationship between the sliding parameter and AOP.

2 System Model

Consider the downlink transmission scenario of HetNet with cache-enabled SBSs and backhaul link-deployed MBSs as shown in Fig. 1. The two dimensional spatial distribution of the MBSs and SBSs follows two independent PPPs denoted by Φ_{M} and Φ_{S}, respectively, which represent the spatial location sets of MBSs and SBSs. The density values of MBS, SBS are λ_{M} (nodes per square meter) and λ_{S}, respectively. Since we only consider the performance from the user side, we assume the spectrum allocation of both SBSs and MBSs and the same. The transmit power of MBS, SBS is denoted by P_{M} and P_{S}, respectively. The radium of service area for each SBS is denoted by $R_{\max}$.

We assume that each user requests the content from a library which contains N files, and the cache space of each SBS is $M\,(M \leq N)$, where each file is assumed to follow unit size. The files are indexed according to their popularity, ranking from the most popular (the 1st file) to the least popular (the Nth file). The probability of each file being requested follows the Zipf distribution,

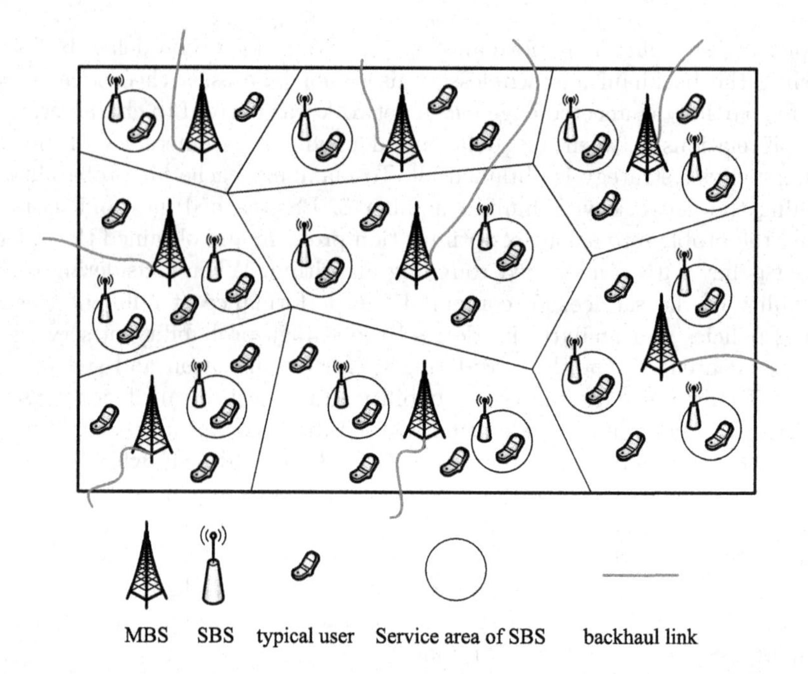

Fig. 1. HetNet with cache-enabled SBSs and backhaul-equipped MBSs.

according to which the requesting probability of the ith popular file is

$$p_i = \frac{i^{-\delta}}{\sum\limits_{j=1}^{N} j^{-\delta}}, \tag{1}$$

where δ ($\delta > 0$) is the skew parameter of the popularity of different files, as δ grows larger, the gap between the high-popularity files and the low-popularity files becomes larger.

We assume each SBS selects the files to be cached independently. The probability of the ith popular file to be cached on SBS is denoted as q_i [8], since each file follows unit size, the average space occupied by the ith popular file is q_i, thus, the summation of q_i is

$$\sum_{i=1}^{N} q_i = M \tag{2}$$

The most commonly used caching policies are uniform caching policy (UCP) and popularity-based caching policy (PCP), where UCP makes SBSs cache each file with equal probability and PCP only cache the most M popular files, denoted as

$$q_{i,\mathrm{UCP}} = \frac{M}{N}(i = 1, 2, ..., N) \tag{3}$$

$$q_{i,\mathrm{PCP}}=\begin{cases}1 & i=1,2,...,M\\ 0 & i=M+1,M+2,...,N\end{cases} \tag{4}$$

Since each SBS caches the files independently, the spatial location set of the SBS which caches the ith file is denoted as $\Phi_{\mathrm{S}i}$, with the density value thinned by value q_i, i.e., $\lambda_{\mathrm{S}i}=\lambda_{\mathrm{S}}q_i$. On the contrary, $\Phi_{\mathrm{S}i'}$ and $\lambda_{\mathrm{S}i'}=\lambda_{\mathrm{S}}(1-q_i)$ denote the spatial location set of SBSs which do not cache the ith file and their density value, respectively.

The downlink signal-to-interference-plus-noise ratio (SINR) for a typical user requesting the ith file when associating with SBS (MBS) is

$$\mathrm{SINR}_{\mathrm{S}i}=\frac{P_{\mathrm{S}}h_{\mathrm{S}i}R_{\mathrm{S}i}^{-\alpha}}{I_{\mathrm{S}i}+\sigma^2}, \tag{5}$$

$$\mathrm{SINR}_{\mathrm{M}}=\frac{P_{\mathrm{M}}h_{\mathrm{M}}R_{\mathrm{M}}^{-\alpha}}{I_{\mathrm{M}i}+\sigma^2}, \tag{6}$$

where $h_{\mathrm{S}i}$ and h_{M} are the equivalent channel power from the associated SBS (MBS) i.e., Si0 (M0) to the typical user, we assume both SBS-to-user and MBS-to-user channels are modeled as Rayleigh fading channel. Thus, both h_{Si} and h_{M} follow the exponential distribution with unit mean, i.e., $h_{\mathrm{Si}}\sim exp(1)\,(i=1,2,...,N)$, $h_{\mathrm{M}}\sim exp(1)$ [9], $R_{\mathrm{S}i}$ is the distance between the typical user and the nearest SBS which caches the ith file, R_{M} is the distance between the typical user and the nearest MBS, α is the path loss exponent, for the ease of the calculation below, the value of α is set to 4. The interference generating to the typical user which is associating with SBS is $I_{\mathrm{S}i}=I_{\mathrm{S}i\mathrm{M}}+I_{\mathrm{S}i\mathrm{S}i}+I_{\mathrm{S}i\mathrm{S}i'}$, where $I_{\mathrm{S}i\mathrm{M}}=\sum_{j\in\Phi_{\mathrm{M}}}P_{\mathrm{M}}h_{\mathrm{M}j}R_{\mathrm{M}j}^{-\alpha}$, $I_{\mathrm{S}i\mathrm{S}i}=\sum_{j\in\Phi_{\mathrm{S}i}\setminus\mathrm{S}i0}P_{\mathrm{S}}h_{\mathrm{S}ij}R_{\mathrm{S}ij}^{-\alpha}$ and $I_{\mathrm{S}i\mathrm{S}i'}=\sum_{j\in\Phi_{\mathrm{S}i'}}P_{\mathrm{S}}h_{\mathrm{S}ij}R_{\mathrm{S}ij}^{-\alpha}$ respectively represent the interference produced by the MBSs, the SBSs which cache the ith file, the SBSs which do not cache the ith file. The interference to the typical user associating with MBS is $I_{\mathrm{M}i}=I_{\mathrm{M}i\mathrm{M}}+I_{\mathrm{M}i\mathrm{S}i}+I_{\mathrm{M}i\mathrm{S}i'}$, where $I_{\mathrm{M}i\mathrm{M}}=\sum_{j\in\Phi_{\mathrm{M}}\setminus\mathrm{M}0}P_{\mathrm{M}}h_{\mathrm{M}j}R_{\mathrm{M}j}^{-\alpha}$, $I_{\mathrm{M}i\mathrm{S}i}=\sum_{j\in\Phi_{\mathrm{S}i}}P_{\mathrm{S}}h_{\mathrm{S}ij}R_{\mathrm{S}ij}^{-\alpha}$ and $I_{\mathrm{M}i\mathrm{S}i'}=\sum_{j\in\Phi_{\mathrm{S}i'}}P_{\mathrm{S}}h_{\mathrm{S}ij}R_{\mathrm{S}ij}^{-\alpha}$, which represent the same as above. Since the HetNets are usually interference-limited [10], we neglect the effect caused by the additive noise, i.e., $\sigma^2=0$.

3 Caching Policies and Performance Analysis

In this section, we propose a equal proportional probability caching policy and a connection strategy to increase the usage ratio of SBSs, then we use the outage probability as the performance metric to analyze the validity of the system.

3.1 Caching Policies and Connecting Strategy

When using UCP, if the request probability gap between each file is large (when δ is large), the high-ranking files possess the same cache probability as others. On the contrary, when using PCP, if the request probability gap between each file is little (when δ is little), they low-ranking files still won't be cached although the gap of the requesting probability between the low-ranking files and the high-ranking files is not large, resulting in the low cache hit probability.

The primary reason of this problem is the cache probability of a typical file cannot match its corresponding request probability, to solve these problem, the equal proportional probability caching policy (EPPCP) is proposed. In this policy, the 1st popular file is most likely to be cached, i.e., $\arg\max_i q_i = 1$. Then q_i is equally proportional to q_{i-1} with proportionality factor d $(0 < d < 1)$, i.e., $q_i = q_{i-1}d$. So if we sets q_1 $((M/N) < q_1 < 1)$ as a function of sliding parameter β $(0 < \beta < 1)$ which makes q_1 slide from M/N to 1, i.e., $q_1 = \beta + (1-\beta)(M/N)$, according to (2), the summation of their average space is

$$\begin{aligned} \sum_{i=1}^{N} q_i &= \sum_{i=1}^{N} q_1 d^{i-1} \\ &\overset{(a)}{=} \frac{q_1(1-d^N)}{1-d} \\ &= \frac{[\beta + (1-\beta)(M/N)](1-d^N)}{1-d} \\ &= M, \end{aligned} \tag{7}$$

where step (a) follows the summation formula of geometric sequence, by solving this Nth degree polynomial equation with dichotomy, we can obtain a unique solution of d within (0,1), then we can obtain the caching probability of each file q_1, $q_2 = q_1 d$, $q_3 = q_1 d^2$, ..., $q_N = q_1 d^{N-1}$, the above is the achievement method of EPPCP.

To release the load of the backhaul link, we use the SBS-priority connecting strategy (SPCS). When a typical user requests the ith file, it firstly searches and associates with the nearest SBS which caches the ith file within $R_{\max}$, if there is no SBS which satisfies the requirement, the nearest MBS would fulfill the request and obtain the requested file from the core network through backhaul link. By this strategy, the backhaul link would be occupied only when the SBSs cannot provide service, resulting in link pressure to be released.

The probability of the user associating with SBS tier when requesting the ith file is denoted as $p_{\mathrm{S}i}$, which is equivalent with the probability that there's at least one SBS within the circle area centered on the typical user of radius $R_{\max}$.

$$p_{\mathrm{S}i} = 1 - e^{-q_i \lambda_{\mathrm{S}} \pi R_{\max}^2}, \tag{8}$$

When a typical user connects with SBS tier, the probability density function of user-SBS distance is

$$\begin{aligned} f_{R_{\mathrm{S}i}|R_{\mathrm{S}i}<R_{\max}}(r) &= \frac{d(F_{R_{\mathrm{S}i}|R_{\mathrm{S}i}<R_{\max}}(r))}{dr} \\ &= \frac{2\pi q_i \lambda_{\mathrm{S}} r e^{-q_i \lambda_{\mathrm{S}} \pi r^2}}{1 - e^{-q_i \lambda_{\mathrm{S}} \pi R_{\max}^2}}, \end{aligned} \tag{9}$$

similar to (9), when a typical user connects with MBS tier, the probability density function of user-MBS distance is

$$f_{R_{\mathrm{M}}} = 2\pi \lambda_{\mathrm{M}} r e^{-\lambda_{\mathrm{M}} \pi r^2} \tag{10}$$

3.2 Outage Probability

We use AOP, i.e. $\mathbb{P}_{\mathrm{out}}$ as the performance metric to reflect the quality of service, which is the average probability that the downlink channel SINR be inferior to the given threshold γ, which is

$$\mathbb{P}_{\mathrm{out}} = \sum_{i=1}^{N} p_i [p_{\mathrm{S}i} \mathbb{P}_{out,\mathrm{S}i} + (1 - p_{\mathrm{S}i}) \mathbb{P}_{out,\mathrm{M}i}] \tag{11}$$

where $\mathbb{P}_{out,\mathrm{S}i}$ represents the outage probability for a SBS-connected user when requesting the ith file, which is

$$\begin{aligned} \mathbb{P}_{out,\mathrm{S}i} &= \mathbb{E}_{R_{\mathrm{S}i}}[p(\mathrm{SINR}_{\mathrm{S}i} < \gamma | R_{\mathrm{S}i} = r)] \\ &= 1 - \int_0^{R_{\max}} p(\mathrm{SINR}_{\mathrm{S}i} > \gamma | R_{\mathrm{S}i} = r) f_{R_{\mathrm{S}i}|R_{\mathrm{S}i}<R_{\max}}(r) dr \\ &= 1 - \int_0^{R_{\max}} \mathcal{L}_{I_{\mathrm{SiM}}}(\frac{\gamma}{P_{\mathrm{S}} r^{-\alpha}}) \mathcal{L}_{I_{\mathrm{SiSi}}}(\frac{\gamma}{P_{\mathrm{S}} r^{-\alpha}}) \mathcal{L}_{I_{\mathrm{SiSi'}}}(\frac{\gamma}{P_{\mathrm{S}} r^{-\alpha}}) \frac{2\pi q_i \lambda_{\mathrm{S}} r e^{-q_i \lambda_{\mathrm{S}} \pi r^2}}{1 - e^{-q_i \lambda_{\mathrm{S}} \pi R_{\max}^2}} dr, \end{aligned} \tag{12}$$

where $\mathcal{L}_{I_{\mathrm{SiM}}}(\frac{\gamma}{P_{\mathrm{S}} r^{-\alpha}})$, $\mathcal{L}_{I_{\mathrm{SiSi}}}(\frac{\gamma}{P_{\mathrm{S}} r^{-\alpha}})$ and $\mathcal{L}_{I_{\mathrm{SiSi'}}}(\frac{\gamma}{P_{\mathrm{S}} r^{-\alpha}})$ respectively represent the Laplace transform of the probability density function of I_{SiM}, I_{SiSi} and $I_{\mathrm{SiSi'}}$ with parameter $\frac{\gamma}{P_{\mathrm{S}} r^{-\alpha}}$.

Proof: Since $I_{\mathrm{S}i\mathrm{M}}$, $I_{\mathrm{S}i\mathrm{S}i}$ and $I_{\mathrm{S}i\mathrm{S}i'}$ are independent of each other, so the probability density function of their summation is their convolution, i.e., $f_{I_{\mathrm{S}i}}(u) = f_{I_{\mathrm{SiM}}}(u) * f_{I_{\mathrm{S}i\mathrm{S}i}}(u) * f_{I_{\mathrm{S}i\mathrm{S}i'}}(u)$, then we get

$$
\begin{aligned}
p(\mathrm{SINR}_{\mathrm{S}i} > \gamma | R_{\mathrm{S}i} = r) &= p(\frac{P_{\mathrm{S}} h_{\mathrm{S}i} r^{-\alpha}}{I_{\mathrm{S}i}} > \gamma) \\
&= \mathbb{E}_{I_{\mathrm{S}i}}[p(h_{\mathrm{S}i} > \frac{\gamma I_{\mathrm{S}i}}{P_{\mathrm{S}} r^{-\alpha}})|I_{\mathrm{S}i}] \\
&\overset{(a)}{=} \mathbb{E}_{I_{\mathrm{S}i}}[e^{-\frac{\gamma I_{\mathrm{S}i}}{P_{\mathrm{S}} r^{-\alpha}}}|I_{\mathrm{S}i}] \\
&= \int_0^\infty e^{-\frac{\gamma u}{P_{\mathrm{S}} r^{-\alpha}}} f_{I_{\mathrm{S}i}}(u) du \\
&= \mathcal{L}_{I_{\mathrm{S}i}}(\frac{\gamma}{P_{\mathrm{S}} r^{-\alpha}}) \\
&\overset{(b)}{=} \mathcal{L}_{I_{\mathrm{S}i\mathrm{M}}}(\frac{\gamma}{P_{\mathrm{S}} r^{-\alpha}}) \mathcal{L}_{I_{\mathrm{S}i\mathrm{S}i}}(\frac{\gamma}{P_{\mathrm{S}} r^{-\alpha}}) \mathcal{L}_{I_{\mathrm{S}i\mathrm{S}i'}}(\frac{\gamma}{P_{\mathrm{S}} r^{-\alpha}}),
\end{aligned}
\tag{13}
$$

step (a) is from $h_{\mathrm{M}} \sim exp(1)$, step (b) follows the theorem of convolution, i.e., $\int_0^\infty e^{-su}(f_a(u) * f_b(u))du = \int_0^\infty e^{-su} f_a(u)du \int_0^\infty e^{-su} f_b(u)du$.

The Laplace transform of $\mathcal{L}_{I_{\mathrm{S}i\mathrm{M}}}(\frac{\gamma}{P_{\mathrm{S}} r^{-\alpha}})$ is

$$
\begin{aligned}
\mathcal{L}_{I_{\mathrm{S}i\mathrm{M}}}(\frac{\gamma}{P_{\mathrm{S}} r^{-\alpha}}) &= \mathbb{E}_{I_{\mathrm{S}i\mathrm{M}}}[e^{-\frac{\gamma I_{\mathrm{S}i\mathrm{M}}}{P_{\mathrm{S}} r^{-\alpha}}}|I_{\mathrm{S}i\mathrm{M}}] \\
&= \mathbb{E}_{\Phi_{\mathrm{M}}, \mathrm{h}_{\mathrm{M}}}[\exp(-\frac{\gamma}{P_{\mathrm{S}} r^{-\alpha}}(\sum_{j \in \Phi_{\mathrm{M}}} P_{\mathrm{M}} h_{\mathrm{M}j} R_{\mathrm{M}j}^{-\alpha}))] \\
&\overset{(a)}{=} \mathbb{E}_{\Phi_{\mathrm{M}}}[\prod_{j \in \Phi_{\mathrm{M}}} \frac{P_{\mathrm{S}} r^{-\alpha}}{P_{\mathrm{S}} r^{-\alpha} + \gamma P_{\mathrm{M}} R_{\mathrm{M}j}^{-\alpha}}] \\
&\overset{(b)}{=} \exp(2\pi\lambda_{\mathrm{M}} \int_0^\infty (\frac{P_{\mathrm{S}} r^{-\alpha}}{P_{\mathrm{S}} r^{-\alpha} + \gamma P_{\mathrm{M}} x^{-\alpha}} - 1) x dx) \\
&= \exp(\pi\lambda_{\mathrm{M}} \int_0^\infty (\frac{P_{\mathrm{S}} r^{-\alpha}}{P_{\mathrm{S}} r^{-\alpha} + \gamma P_{\mathrm{M}} u^{-\frac{\alpha}{2}}} - 1) du) \\
&\overset{(c)}{=} \exp(-\frac{\pi^2}{2} \lambda_{\mathrm{M}} \sqrt{\frac{\gamma P_{\mathrm{M}}}{P_{\mathrm{S}}}} r^2),
\end{aligned}
\tag{14}
$$

step (a) is from $h_{\mathrm{M}} \sim exp(1)$, step (b) is from the probability generating function of PPP [11] and step (c) is from the assumption that the path-loss parameter $\alpha = 4$.

Similar to (14), $\mathcal{L}_{I_{\text{S}i\text{S}i}}(\frac{\gamma}{P_{\text{S}}r^{-\alpha}})$ and $\mathcal{L}_{I_{\text{S}i\text{S}i'}}(\frac{\gamma}{P_{\text{S}}r^{-\alpha}})$ are

$$\begin{aligned}\mathcal{L}_{I_{\text{S}i\text{S}i}}(\frac{\gamma}{P_{\text{S}}r^{-\alpha}}) &= \mathbb{E}_{\Phi_{\text{S}i}}[\prod_{j\in\Phi_{\text{S}i}}\frac{r^{-\alpha}}{r^{-\alpha}+\gamma R_{\text{S}ij}^{-\alpha}}]\\ &= \exp(2\pi q_i\lambda_{\text{S}}\int_r^{\infty}(\frac{r^{-\alpha}}{r^{-\alpha}+\gamma x^{-\alpha}}-1)xdx)\\ &= \exp(-\pi q_i\lambda_{\text{S}}\sqrt{\gamma}r^2(\frac{\pi}{2}-\arctan(\frac{1}{\sqrt{\gamma r^2}})))\end{aligned} \tag{15}$$

$$\begin{aligned}\mathcal{L}_{I_{\text{S}i\text{S}i'}}(\frac{\gamma}{P_{\text{S}}r^{-\alpha}}) &= \mathbb{E}_{\Phi_{\text{S}i'}}[\prod_{j\in\Phi_{\text{S}i'}}\frac{P_{\text{S}}r^{-\alpha}}{P_{\text{S}}r^{-\alpha}+\gamma P_{\text{S}}R_{\text{S}ij}^{-\alpha}}]\\ &= \exp(2\pi(1-q_i)\lambda_{\text{S}}\int_0^{\infty}(\frac{r^{-\alpha}}{r^{-\alpha}+\gamma x^{-\alpha}}-1)xdx)\\ &= \exp(-\frac{\pi^2}{2}(1-q_i)\lambda_{\text{S}}\sqrt{\gamma}r^2)\end{aligned} \tag{16}$$

Similar to (12), the outage probability of the typical user request when associating with MBS (i.e., $\mathbb{P}_{out,\text{M}i}$) is

$$\begin{aligned}\mathbb{P}_{out,\text{M}i} &= \mathbb{E}_{R_{\text{M}i}}[p(\text{SINR}_{\text{M}i}<\gamma|R_{\text{M}i}=r)]\\ &= \int_0^{\infty}(1-\mathcal{L}_{I_{\text{M}i\text{M}}}(\frac{\gamma}{P_{\text{M}}r^{-\alpha}})\mathcal{L}_{I_{\text{M}i\text{S}i}}(\frac{\gamma}{P_{\text{M}}r^{-\alpha}})\mathcal{L}_{I_{\text{M}i\text{S}i'}}(\frac{\gamma}{P_{\text{M}}r^{-\alpha}}))2\pi\lambda_{\text{M}}re^{-\lambda_{\text{M}}\pi r^2}dr,\end{aligned} \tag{17}$$

where $\mathcal{L}_{I_{\text{M}i\text{S}i}}(\frac{\gamma}{P_{\text{M}}r^{-\alpha}})$, $\mathcal{L}_{I_{\text{M}i\text{S}i}}(\frac{\gamma}{P_{\text{M}}r^{-\alpha}})$, $\mathcal{L}_{I_{\text{M}i\text{S}i'}}(\frac{\gamma}{P_{\text{M}}r^{-\alpha}}))$ are

$$\begin{aligned}\mathcal{L}_{I_{\text{M}i\text{M}}}(\frac{\gamma}{P_{\text{M}}r^{-\alpha}}) &= \mathbb{E}_{\Phi_{\text{M}}\backslash\text{M}0}[\prod_{j\in\Phi_{\text{M}}\backslash\text{M}0}\frac{r^{-\alpha}}{r^{-\alpha}+\gamma R_{\text{M}j}^{-\alpha}}]\\ &= \exp(\frac{1}{\sqrt{\gamma r^2}}(\frac{\pi}{2}-\arctan(\frac{1}{\sqrt{\gamma r^2}})))\end{aligned} \tag{18}$$

$$\begin{aligned}\mathcal{L}_{I_{\text{M}i\text{S}i}}(\frac{\gamma}{P_{\text{M}}r^{-\alpha}}) &= \mathbb{E}_{\Phi_{\text{S}i}}[\prod_{j\in\Phi_{\text{S}i}}\frac{P_{\text{M}}r^{-\alpha}}{P_{\text{M}}r^{-\alpha}+\gamma P_{\text{S}}R_{\text{S}ij}^{-\alpha}}]\\ &= \exp(-\pi q_i\lambda_{\text{S}}\sqrt{\frac{\gamma P_{\text{S}}}{P_{\text{M}}}}r^2(\frac{\pi}{2}-\arctan(\frac{R_{\max}}{\sqrt{\frac{\gamma P_{\text{S}}r^4}{P_{\text{M}}}}})))\end{aligned} \tag{19}$$

$$\begin{aligned}\mathcal{L}_{I_{\text{M}i\text{S}i'}}(\frac{\gamma}{P_{\text{M}}r^{-\alpha}}) &= \mathbb{E}_{\Phi_{\text{S}i'}}[\prod_{j\in\Phi_{\text{S}i'}}\frac{P_{\text{M}}r^{-\alpha}}{P_{\text{M}}r^{-\alpha}+\gamma P_{\text{S}}R_{\text{S}ij}^{-\alpha}}]\\ &= \exp(-\pi\lambda_{\text{M}}\sqrt{\gamma}r^2(\frac{\pi}{2}-\arctan(\frac{1}{\sqrt{\gamma}r})))\end{aligned} \tag{20}$$

Thus, we can obtain the AOP of different caching policies by (11).

4 Numerical Results

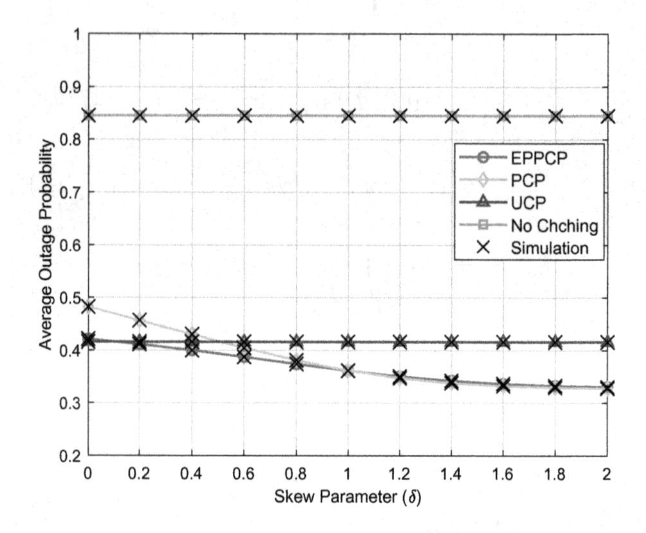

Fig. 2. Outage probability v.s. skew parameter. $\lambda_S = 1000\,\text{km}^{-1}$, $\beta = 1$.

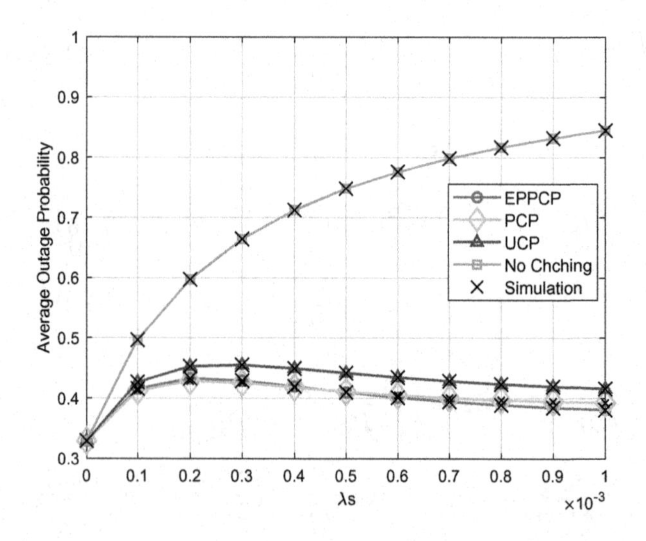

Fig. 3. Outage probability v.s. λ_S. $\delta = 1$, $\beta = 1$.

In this section, we present the numerical analysis of the outage probability, and verify the analysis results with 10,000-time Monte-Carlo simulations of a $5\,\text{km} \times 5\,\text{km}$ size network. γ is 20 dBm, P_M is 50 dBm, P_S is 30 dBm, $M = 70$, $N = 100$, $R_{\max} = 30\,\text{m}$, $\lambda_M = 10\,\text{km}^{-1}$.

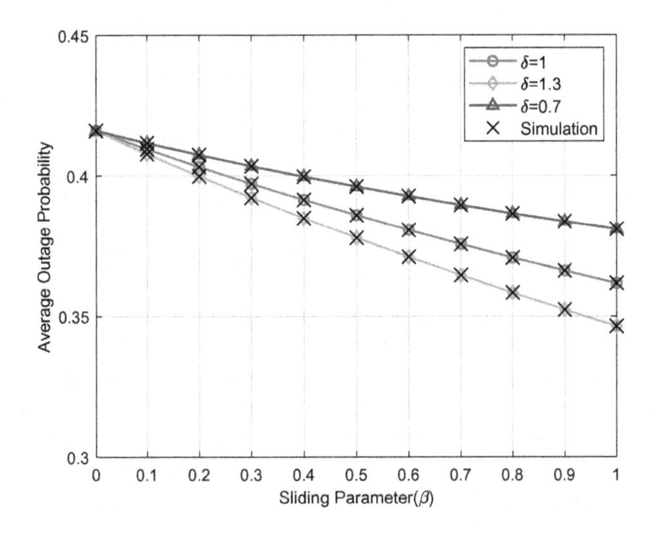

Fig. 4. Outage probability v.s. sliding parameter. $\lambda_S = 1000\,\text{km}^{-1}$.

Figure 2 shows the AOP for different caching policies as the function of skew parameter δ. We can see the results of the performance analysis match the simulation results, the red line, i.e., the AOP of EPPCP is lower than others at the range of (0.2, 1.0). When there's no cache deployed on SBSs (i.e. $q_i = 0$, $i = 1, 2, ..., N$), from (12) we have $\mathbb{P}_{out,Si} = 1$, $i = 1, 2, ..., N$, since the typical user associates with the nearest MBS regardless of the file it requests and the interference produced by SBSs keeps unchanged with i, the SINR does not change with δ, resulting in the AOP keeps stable. When using UCP, the outage probability when requesting the ith file keeps unchanged, resulting in the AOP stabilizes with δ. Since the requesting probability gap between the high-ranking files and the low-ranking files is not large between (0,1), the EPPCP has lower $\mathbb{P}_{out,Si}$ when requesting low-ranking files than PCP, resulting in lower AOP. On the contrary, the AOP of EPPCP is higher than UCP between (0,0.2) for the same reason. Since δ has the typical value of 0.5$\sim$1.0 [12], EPPCP would be a better option than other caching policies.

Figure 3 shows the AOP as the function of λ_S. We find that for different caching policies, the AOP firstly grows larger, that's because when $\lambda_S = 0$ (i.e. there is no SBS), there's no interference produced by SBS, which is the main part of the total interference, resulting in large SINR. As λ_S grows larger, the AOP turns to fall since the average distance to the associating SBS gets shorter, resulting in larger signal power received and larger SINR. Thus, when we build ultra-dense HetNet with cache-enabled SBS, we would sacrifice a little bit of AOP in exchange for the release of backhaul link pressure and lower time delay. Unlike the above, when there's no cache deployed, the AOP could only rise with λ_S since there's only interference increasing caused by higher densed SBSs.

Figure 4 shows the AOP as the function of β when using EPPCP. We find that the AOP becomes smaller with β, which means the performance becomes better with a more skewed caching policy.

5 Conclusion

In this paper, we applied an equal proportional probability caching policy with SBS-priority connecting strategy on cache-enabled heterogeneous celluar network. By modeling the spatial distribution of the BSs as two independent Poisson Point Processes, we obtained the average outage probability for a typical user's request. The simulation result showed that the proposed policy resulted in lower AOP than the classical policies when δ is in the interval (0.2, 1.0). In addition, we investigated the relationship between the density of SBS and AOP, which concluded the high-densed SBSs would sacrifice a little bit of AOP in exchange for the release of backhaul link and lower time delay. Lastly, we investigated the relationship between the sliding parameter and AOP, which concluded the performance becomes better with a more skewed caching policy. This paper highlights that the new caching policy performs better than the conventional policies under cache-enabled HetNets, underscoring the fact that using high-densed network in conjunction with a more skewed caching policy would obtain a better system performance.

References

1. Mao, Y., You, C., Zhang, J., Huang, K., Letaief, K.B.: A survey on mobile edge computing: the communication perspective. IEEE Commun. Surv. Tutor. **19**(4), 2322–2358 (2017)
2. Song, J., Choi, W.: Minimum cache size and backhaul capacity for cache-enabled small cell networks. IEEE Wirel. Commun. Lett. **7**(4), 490–493 (2018)
3. Yu, T., Zhang, S., Chen, X., Xu, S.: An analytical framework for delay optimal mobile edge deployment in wireless networks. IEEE Wirel. Commun. Lett. **9**(12), 2149–2153 (2020)
4. Krishnendu, S., Bharath, B.N., Bhatia, V.: Cache enabled cellular network: algorithm for cache placement and guarantees. IEEE Wirel. Commun. Lett. **8**(6), 1550–1554 (2019)
5. Tamoor-ul Hassan, S., Bennis, M., Nardelli, P.H.J., Latva-aho, M.: Caching in wireless small cell networks: a storage-bandwidth tradeoff. IEEE Commun. Lett. **20**(6), 1175–1178 (2016)
6. Liu, D., Yang, C.: Caching policy toward maximal success probability and area spectral efficiency of cache-enabled hetnets. IEEE Trans. Commun. **65**(6), 2699–2714 (2017)
7. Zhang, T., Fang, X., Liu, Y., Li, G.Y., Xu, W.: D2d-enabled mobile user edge caching: a multi-winner auction approach. IEEE Trans. Veh. Technol. **68**(12), 12314–12328 (2019)
8. Blaszczyszyn, B., Giovanidis, A.: Optimal geographic caching in cellular networks. In: 2015 IEEE International Conference on Communications (ICC), pp. 3358–3363 (2015)

9. Zhang, J., Kountouris, M., Andrews, J.G., Heath, R.W.: Multi-mode transmission for the mimo broadcast channel with imperfect channel state information. IEEE Trans. Commun. **59**(3), 803–814 (2011)
10. Jo, H.S., Sang, Y.J., Xia, P., Andrews, J.G.: Heterogeneous cellular networks with flexible cell association: a comprehensive downlink sinr analysis. IEEE Trans. Wirel. Commun. **11**(10), 3484–3495 (2012)
11. Andrews, J.G., Baccelli, F., Ganti, R.K.: A tractable approach to coverage and rate in cellular networks. IEEE Trans. Commun. **59**(11), 3122–3134 (2011)
12. Breslau, L., Cao, P., Fan, L., Phillips, G., Shenker, S.: Web caching and zipf-like distributions: evidence and implications. In: IEEE INFOCOM, vol. 1, pp. 126–134 (1999)

Momentum-Based Adversarial Attacks Against End-to-End Communication Systems

Qiuna Zhang, Yongkui Ma, Honglin Zhao(✉), Chengzhao Shan, and Jiayan Zhang

Communication Research Center, Harbin Institute of Technology, Harbin, China
{yk_ma,hlzhao,czshan,jyzhang}@hit.edu.cn

Abstract. Deep learning (DL) based communication system is a promising novel architecture to implement end-to-end optimization compared with conventional block-separated optimization schemes. However, the vulnerability to adversarial examples of deep neural networks poses significant security concern on the end-to-end communication systems. Adversarial attacks serve as a fundamental surrogate to evaluate the robustness of the DL-based communication systems before they are deployed. Specifically, we propose a new adversarial attack method with momentum iterative gradient against the end-to-end communication systems. For targeted attacks, embedding the momentum term in the iterative process can help loss function stabilize the update direction and avoid getting stuck in saddle points and poor local minima. Therefore, the momentum-based method can enhance the effectiveness without losing the transferability of adversarial attacks. Numerous simulation results illustrate that the proposed method can achieve superior block error rate compared with traditional jamming attacks and no momentum accumulated adversarial attacks.

Keywords: Momentum · Adversarial attacks · End-to-end communication systems · Deep learning · Wireless security · Model robustness

1 Introduction

Due to powerful nonlinear approximation and optimization capabilities, deep learning (DL) has become a promising technique to satisfy the growing demands of the fifth-generation (5G) wireless communications and beyond, such as high reliability, ultra-high capacity and low-latency. Recently, abundant concepts and applications of deep learning have been deployed in the field of communications. One of the most novel concepts is an end-to-end communication system based on an autoencoder framework [1], and a series of studies have been conducted around it [2–5].

Q. Guo et al. (Eds.): WiSATS 2021, LNICST 410, pp. 520–532, 2022.
https://doi.org/10.1007/978-3-030-93398-2_48

Whereas, deep neural networks (DNNs) are exceedingly susceptible to adversarial examples [6,7], these examples can fool classifiers by adding small and human-imperceptible perturbations to legitimate examples. This unique characteristic of DNN raises significant security and robustness concern about implying DL method in the field of communications, especially for the end-to-end communication system based on the autoencoder architecture, whose transmitter, channel and receiver are composed of DNNs. Compared with conventional jamming attacks, adversarial attacks are more destructive and unnoticeable, because adversarial perturbations are essentially optimization vectors deliberately crafted in feature space, which can change the optimization of loss gradients in wrong directions. For deep learning based communication systems, owing to the openness of wireless communication channel, an illegitimate attacker can add a small perturbation to the transmitted signal when it passes through the channel, thus confuses the receiver [8]. As a consequence, how to improve the robustness of the DL-based communication systems is an urgent challenge. Therefore, for the DL-based communication systems, it is necessary to conduct further research from the perspective of adversarial attacks, since it inspires the study of defense methods, as well as facilitates the robustness assessment.

For the DL-based communication systems, despite broad studies on various aspects, poor discussions of adversarial attacks are conducted around them. In [8] and [9], one-step gradient based perturbation generating algorithm is adopted to craft white-box and black-box attacks. More transferable adversarial examples can be generated by one-step gradient based attack methods, however, they usually have a low success rate for attacking white-box models [10]. Therefore, we consider applying momentum into adversarial attacks against DL-based communication systems, aiming to improve the effectiveness of crafting white-box attacks without reducing their transferability.

In this paper, we investigate momentum-based adversarial attacks against end-to-end communication systems. The main contributions of our work are: (i) We propose a momentum-based adversarial perturbation generation algorithm to generate adversarial perturbations for end-to-end communication systems, in which we utilize velocity vector to stabilize the optimization direction of loss gradient, so as to escape from local minima and saddle points. (ii) Through crafting targeted attacks against autoencoders with different structures, we verify that momentum-based adversarial attacks can achieve better block error rate (BLER) than traditional jamming attacks, which demonstrates their powerful destructiveness. (iii) By crafting white-box and black-box attacks, we show that adversarial attacks with momentum are more destructive than attacks without momentum accumulation in [8], which reveals that momentum-based attacks can increase the attack effectiveness of white-box models without losing the transferability of attacking black-box models.

2 System Model

We consider a DL-based communication system consisting of transmitter, channel and receiver, which can be represented as an autoencoder since its optimization object is to reconstruct its input at the output [1]. The characteristic of the

DL-based communication system is that all its transmitter, channel and receiver are composed of deep neural networks, in which both transmitter and receiver are trainable. Nevertheless, in most structures, the channel layer is untrainable, unless the channel state information (CSI) is unknown, in this setting, a channel modeling method based on generative adversarial network (GAN) can be used to train the channel layer [3].

The DL-based communication system tries to learn the channel characteristic (noise jamming, fading and distortion) mapped from the transmitter to the receiver, so that the massage sent by the transmitter can be recovered at the receiver side with minimum error rate. Corresponding to the components of the autoencoder, the transmitter and receiver are represented by the encoder and decoder, respectively.

In the end-to-end communication system, k bit message $s \in \{1, 2, \ldots, M\}$ first passes through the transmitter (encoder) to the channel, where it may be attacked by illegitimate attackers using adversarial attack methods, and then the receiver (decoder) tries to recover it. Before input to the encoder, s needs to be represented as a one-hot vector of dimension $M = 2^k$, and then the encoder transforms it by applying $f : \mathbb{R}^M \to \mathbb{R}^{2n}$ to generate the transmitted signal $\mathbf{x} = f(s) \in \mathbb{R}^{2n}$ for n complex channel uses [2]. However, in practice, $2n$ real channel uses are usually used to replace the n complex channel uses, since there is no complex operation in the actual operation of DNN [1]. The last layer of the encoder is a normalization layer constrained by $\mathbb{E}\left[x_i^2\right] \leq 0.5,\ \forall i.$, where $\mathbb{E}\left[\cdot\right]$ is the expectation of the elementwise square of $\mathbf{x}$, which ensures the power constraint before $\mathbf{x}$ is transmitted to the channel [1].

When $\mathbf{x}$ is transmitted to the wireless channel, it may face various channel environments. In this work, as set in [8], we consider the additive white Gaussian noise (AWGN) channel realized by a noise layer, in which illegitimate attackers could craft adversarial perturbations to attack the transmitted signal $\mathbf{x}$. Therefore, the output of the channel, namely, the received signal of the receiver $\mathbf{y}$ is given by

$$\mathbf{y} = \mathbf{x} + \mathbf{n} + \mathbf{p} \tag{1}$$

where $\mathbf{n}$ is a Gaussian distribution vector and $\mathbf{p}$ is an adversarial perturbation. The noise vector $\mathbf{n} \sim \mathcal{N}(0, \sigma^2 \mathbf{I}_n)$, whose variance $\sigma^2 = (2RE_b/N_0)^{-1}$. For σ^2, R represents the data rate, E_b denotes the energy per bit, and N_0 is the noise power spectral density. As for the adversarial perturbation $\mathbf{p}$, it is delicately crafted for the characteristic of DNN by illegitimate attackers.

After the channel, the receiver (decoder) tries to achieve the transformation $g : \mathbb{R}^{2n} \to \mathbb{R}^M$ for the received signal $\mathbf{y}$, to generate a reconstructed message $\hat{s}$. The last layer of the decoder realizes softmax activation operation [11], and the estimated message $\hat{s}$ is the index of the highest probability element in the M dimensional probability vector b, which is the output of the softmax activation function. The composition of the end-to-end communication system under adversarial attacks is shown in Fig. 1.

To express the reconstruction problem of communication message as a classification mission, the cross-entropy loss function is adopted to optimize the distance between s and b. Thus, the cross-entropy loss function L is written as

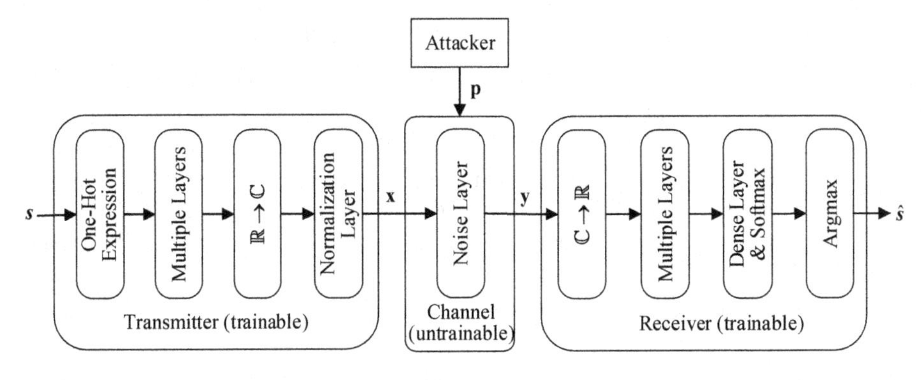

Fig. 1. An end-to-end communication system represented as an autoencoder under an adversarial attack.

$$L = -\sum_{i} s_i \log(b_i) \tag{2}$$

where s_i and b_i are corresponding ith element of s and b, respectively.

Generally, block error rate (BLER) P_e is used to measure the performance of the DL-based communication systems, defined as

$$P_e = \frac{1}{M} \sum_{s} \Pr(\hat{s} \neq s | s) \tag{3}$$

It is worth noting that the BLER in the end-to-end communication systems is equivalent to the symbol error rate (SER) in conventional communication systems.

3 A Momentum-Based Adversarial Perturbation Generation Algorithm

The concept of momentum was first proposed in [12], as a tool to help improve gradient descent algorithms by accumulating the gradients of loss function in previous iterations as a velocity vector. [13] applies the concept of momentum to generate adversarial examples for image adversarial attacks, inspired by this, we incorporate momentum into adversarial perturbation generation of the end-to-end communication systems. [14] shows that the accumulation of previous gradients could help DNN to avoid drawbacks of local minimization, so in targeted attacks, we can leverage momentum to help the current gradient to barrel through the critical points of loss surface, including saddle points and poor local minima or maxima. Moreover, from [15] we konw that the momentum-based gradient iteration method could also achieve better stability for the update process of stochastic gradient descent, which plays an important role in the effectiveness of adversarial attacks. Therefore, based on the analysis mentioned above, we propose a momentum-based adversarial perturbation generation method, which is formulated in this section.

3.1 Momentum-Based Iterative Gradient Method

Compared with untargeted adversarial attacks, targeted adversarial attacks are easier to implement and more efficient, so we focus on targeted attacks. In order to generate a targeted adversarial example $\mathbf{x}^{adv}$ based on a real example $\mathbf{x}$, it should be constrained by L_p norm within accuracy constraint ε, to make sure the adversarial attack is unnoticeable. Therefore, the gradient-based methods solving the constrained optimization problem as below to seek the optimal adversarial example

$$\begin{aligned} &\arg\min_{\mathbf{x}^{adv}} J(\mathbf{x}^{adv}, y^{target}) \\ &\text{s.t.} \left\| \mathbf{x}^{adv} - \mathbf{x} \right\|_p \leq \varepsilon \end{aligned} \tag{4}$$

The gradient iteration methods can be easily generalized to the attack setting of explicit norm bound constrains, such as L_1, L_2, L_∞ norm bounds. Using the cumulative gradient of all previous steps to substitute current gradient, we can extend any gradient iteration approach to its momentum gradient variant. However, in the field of wireless communications, compared to other norm bounds, L_2 norm seems to be a more appropriate choice since it usually represents the power of the adversarial perturbation [9]. Therefore, we only introduce targeted attack methods for generating adversarial perturbations in terms of L_2 norm bound.

For targeted attacks, to seek an adversarial perturbation within the vicinity of a real example subject to L_2 distance, $\left\| \mathbf{x}^{adv} - \mathbf{x} \right\|_2 \leq \varepsilon$, the update process of the momentum-based gradient method incorporates accumulated gradient of all previous steps into current gradient, which can be written as

$$\mathbf{g}_t = \mu \cdot \mathbf{g}_{t-1} + \alpha \cdot \frac{\nabla J(\mathbf{x}_t^{adv}, y^{target})}{\left\| \nabla J(\mathbf{x}_t^{adv}, y^{target}) \right\|_2} \tag{5}$$

where μ is the decay factor, $\mathbf{g}_{t-1}$ denotes the accumulated gradient up to the current iteration, y^{target} is defined as the targeted misclassification label, α is the stepsize.

In consequence, the targeted momentum-based iterative gradient within a L_2 norm constraint is

$$\mathbf{x}_t^{adv} = \mu \cdot \mathbf{x}_{t-1}^{adv} - \alpha \cdot \frac{\mathbf{g}_t}{\|\mathbf{g}_t\|_2} \tag{6}$$

3.2 Generative Algorithm of Momentum-Based Adversarial Perturbation

In order to evaluate the robustness of the wireless communication systems based on DL, we propose a momentum-based adversarial attack method against the end-to-end communication systems. Adversarial attack method with momentum generates adversarial perturbations, which can fool the receiver (decoder), resulting in superior increasing of BLER. Algorithm 1 presents how to generate an adversarial perturbation based on the momentum iterative gradient. In Algorithm 1, we use bisection search to find an appropriate value of the stepsize α

among all possible targeted categories as in [8]. Furthermore, for a specific target class, the momentum-based algorithm tries to minimize $J(\mathbf{x}_t^*, y^{target})$, and then use the bisection search to find appropriate gradient after T iterations with momentum.

Algorithm 1. Generating a Momentum-based Adversarial Perturbation

Input: A real input $\mathbf{x}$ and its ground-truth label y_{true}, the pretrained model f with loss function J, desired perturbation accuracy ε_{acc}, the number of iterations T, maximum allowed perturbation norm p_{max} and decay factor μ.

Output: An adversarial perturbation $\mathbf{p}_{\mathbf{x}}^{adv}$ of input $\mathbf{x}$, an adversarial example $\mathbf{x}^{adv}$ with constraint $\left\|\mathbf{x}^{adv} - \mathbf{x}\right\|_2 \leq \varepsilon$.

1: **Initialization:** Set the initial value $\mathbf{g}_0 = 0$ and $\mathbf{x}_0^* = \mathbf{x}$.
2: **for** *class-index* in *range*(C) **do**
3: $\quad\varepsilon_{\max} \leftarrow p_{\max}$, $\varepsilon_{\min} \leftarrow 0$, $\alpha \leftarrow \varepsilon_{\max} + \varepsilon_{\min}/2$
4: $\quad$Input $\mathbf{x}_t^*$ to pretrained model f for obtaining the gradient $\nabla_{\mathbf{x}} J(\mathbf{x}_t^*, y^{target})$
5: $\quad$**for** $t = 1$ to T **do**
6: $\quad\quad$Update $\mathbf{g}_t$ by accumulating the velocity vector in the gradient direction as
7: $\quad\quad\mathbf{g}_t = \mu \cdot \mathbf{g}_{t-1} + \alpha \cdot \nabla_{\mathbf{x}} J(\mathbf{x}_t^*, y^{target})(\left\|\nabla_x J(\mathbf{x}_t^*, y^{target})\right\|_2)^{-1}$
8: $\quad\quad$**while** $\varepsilon_{\max} - \varepsilon_{\min} > \varepsilon_{acc}$ **do**
9: $\quad\quad\quad\varepsilon_{ave} = \varepsilon_{\max} + \varepsilon_{\min}/2$ and $\alpha = \varepsilon_{ave}$
10: $\quad\quad\quad$Update $\mathbf{x}_t^*$ by applying the momentum gradient as
11: $\quad\quad\quad\mathbf{x}_t^* = \mathbf{x}_{t-1}^* - \alpha \cdot \mathbf{g}_t(\|\mathbf{g}_t\|_2)^{-1}$
12: $\quad\quad\quad$**if** $f(\mathbf{x}_t^*) = y_{true}$ **then**
13: $\quad\quad\quad\quad\varepsilon_{\min} \leftarrow \varepsilon_{ave}$
14: $\quad\quad\quad$**else**
15: $\quad\quad\quad\quad\varepsilon_{\max} \leftarrow \varepsilon_{ave}$
16: $\quad\quad\quad$**end if**
17: $\quad\quad$**end while**
18: $\quad\quad\varepsilon_{class-index} = \varepsilon_{\max}$
19: $\quad$**end for**
20: $\quad\mathbf{x}_{class-index}^* = \mathbf{x}_T^*$
21: **end for**
22: $target - class_{\varepsilon} = \operatorname{argmin} \varepsilon_{class-index}$ and $\alpha^* = \min \varepsilon_{class-index}$
23: $target - class_{\mathbf{g}} = \operatorname{argmin} \mathbf{g}_{class-index}$ and $\mathbf{g}_T^* = \min \mathbf{g}_{class-index}$
24: **return** $\mathbf{x}^{adv} = \mathbf{x}_{target-class_{\mathbf{g}}}^*$, $\mathbf{p}_{\mathbf{x}}^{adv} = \alpha^* \cdot \mathbf{g}_T^*(\|\mathbf{g}_T^*\|_2)^{-1}$

In our experiments, we only report the results of targeted attacks using momentum-based perturbation generation algorithm. We set the decay factor μ to 1.0 and the number of iterations T to 20. The decay factor μ plays an instrumental role in increasing the BLER of momentum-based algorithms. In order to find the appropriate value of the decay factor, we attack the end-to-end communication systems based on CNN and MLP respectively with PSR $= -6$ dB, $E_b/N_0 = 8$ dB, while the decay factor ranging from 0.0 to 2.0 with a granularity 0.1. In Fig. 2, the curve of BLER for end-to-end communications with different values of μ does not change much, but it can achieve stable performance when $\mu = 1.0$ for different network structures, so we choose $\mu = 1.0$ as one of our hyper-parameters, just like in [13], which means we simply add up all gradients of previous iterations to update the current gradient.

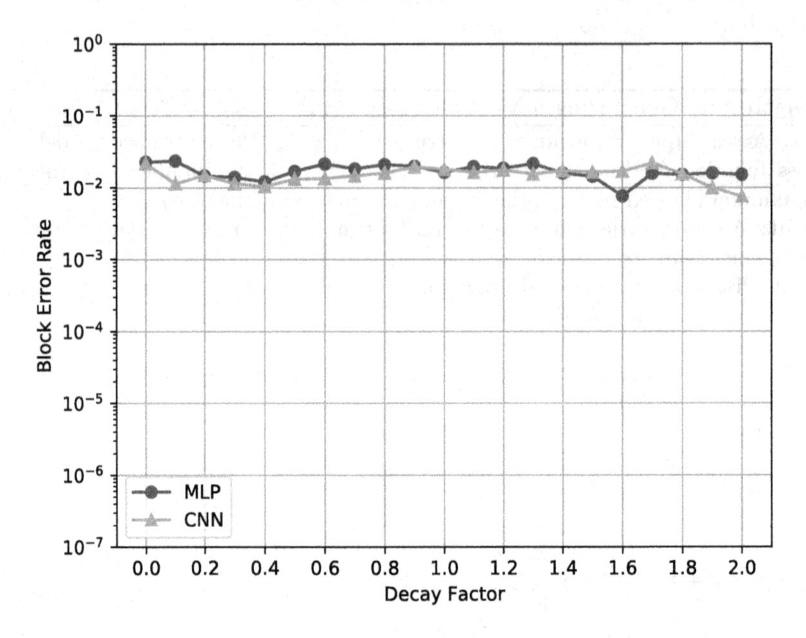

Fig. 2. BLER versus the decay factor μ under the momentum-based attacks against the MLP autoencoder and CNN autoencoder.

We also investigate the effect of the number of iterations T on BLER when using the momentum-based adversarial attacks. We adopt the hyper-parameter $\mu = 1$ to attack end-to-end communication systems with T ranging from 1 to 20. We evaluate the BLER of adversarial perturbations against autoencoders in Fig. 3. It can be observed that the BLER of momentum-based algorithm against autoencoders with different structures maintains a stable value. It proves that the adversarial perturbations generated by momentum iterative methods are difficult to overfit a white-box model and maintain stable effectiveness for different models. However, from the result of simulation, it is obvious that when $T = 20$, the BLER of different network structures are better than others, so we choose $T = 20$ as the value of another hyper-parameter.

4 Crafting Momentum-Based Adversarial Attacks Against End-to-End Communication Systems

Equation (1) shows how an adversarial perturbation is added to the transmitted signal in the wireless channel, therefore, the optimization objective of

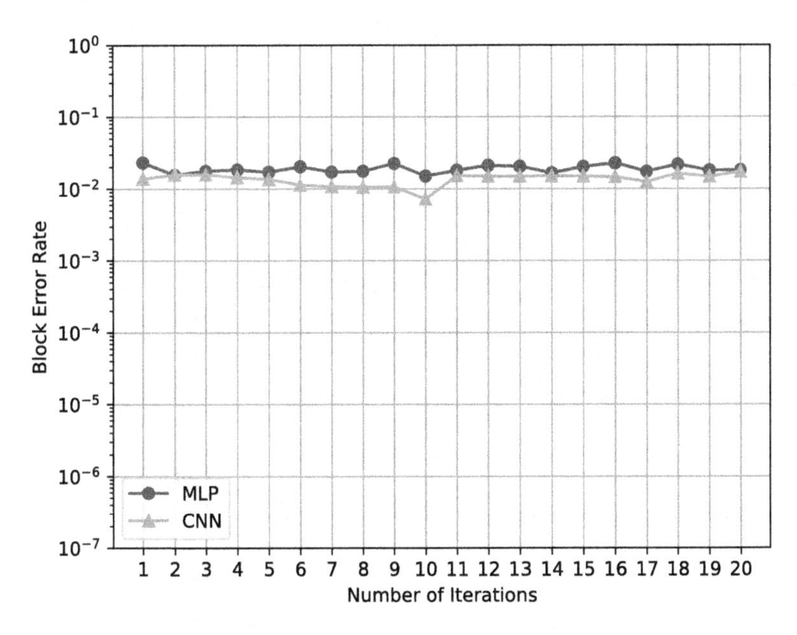

Fig. 3. BLER versus the number of iterations T under the momentum-based attacks against the MLP autoencoder and CNN autoencoder.

the attacker is to craft an adversarial perturbation to confuse the considered receiver(decoder) as in [8]

$$\begin{aligned} &\min_{\mathbf{p}} \|\mathbf{p}\|_2 \\ &\text{s.t. } g(\mathbf{x}+\mathbf{n}+\mathbf{p}) \neq g(\mathbf{x}+\mathbf{n}) \end{aligned} \tag{7}$$

From [6], we know that the Eq. (7) does not belong to convex function because the receiver mapping g does not have a convex structure. [8] uses fast gradient method without momentum [6,7] to approximate the optimum solution of Eq. (7), which would face the dilemma of being trapped by poor local minima or saddle points. By using the momentum-based gradient method, we expect to provide a velocity vector in the gradient direction of the loss surface across iterations, which can feed the demand of stabilizing the update direction as well as escaping from poor local minima and saddle points.

The structures of autoencoder generally used as the benchmarks in end-to-end communication systems are multi-layer perceptron (MLP) and convolutional neural network (CNN). To enable a performance comparison for the adversarial attacks against the end-to-end communication systems without momentum, we adopt the same autoencoder structures and dimensions as in [8], namely a MLP-based autoencoder and a CNN-based autoencoder which are shown in Table 1 and Table 2 respectively. The results of BLER are obtained by Monte Carlo simulation, and the number of simulation is 1,000,000. All the simulations are performed using TensorFlow on an NVIDIA GeForce GTX 1080 Ti graphic processing unit.

Table 1. The structure and dimension of the considered autoencoder based on MLP.

Block structure	Layer	Output dimension
Encoder	One-Hot Input	M
	Dense + ReLU	M
	Dense + Linear	2n
	Power Normalization	2n
Channel	AWGN Layer (+Adversarial Attack)	2n
Decoder	Dense + ReLU	M
	Dense + Softmax	M

Table 2. The structure and dimension of the considered autoencoder based on CNN.

Block structure	Layer	Output dimension
Encoder	One-Hot Input	M
	Dense + eLU	M
	Conv1D + eLU	16 × M
	Conv1D + eLU	2n
	Power Normalization	2n
Channel	AWGN Layer (+Adversarial Attack)	2n
Decoder	Conv2D	$16 \times 2n$
	Conv2D+Flattening	$8 \times 2n$
	Dense + ReLU	2M
	Dense + Softmax	2M

4.1 Crafting a White-Box Adversarial Attack with Momentum

Under the framework of the end-to-end communication system, we adopt the white-box and black-box attack crafting method in [8] to craft momentum-based white-box and black-box attacks. Then we compare the performance of the proposed momentum-based attacks with pure jamming attacks and adversarial attacks proposed in [8]. In a white-box attack, we assume the attacker has full knowledge of the autoencoder, including the network structure and parameters; while in a black-box attack, the attacker has no knowledge or limited knowledge of the autoencoder. First of all, Fig. 4 and Fig. 5 show the BLER performance of the MLP autoencoder and CNN autoencoder (described in Table 1 and Table 2) under the proposed momentum-based attacks respectively. We adopt perturbation-to-signal ratio (PSR) as metric, which is the power ratio of the received perturbation to the received signal [9]. To evaluate the effectiveness of the momentum-based adversarial attacks, we also use jamming attacks (Gaussian noise) as counterparts with the same PSR. The values of PSR we selected are –2 dB, –6 dB and –10 dB. Figure 4 and Fig. 5 illustrate that the proposed momentum-based attacks are more destructive than conventional jamming

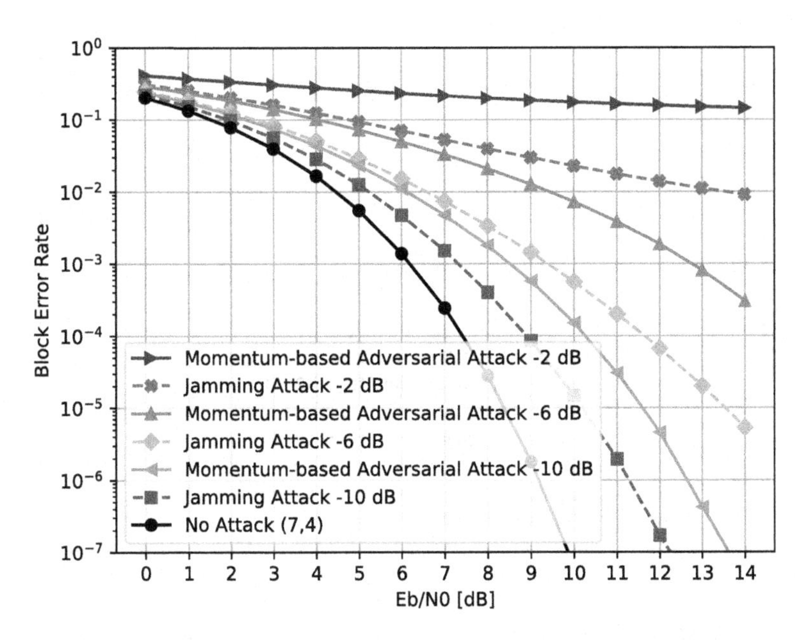

Fig. 4. BLER versus E_b/N_0 under the momentum-based adversarial and jamming attacks against the MLP autoencoder.

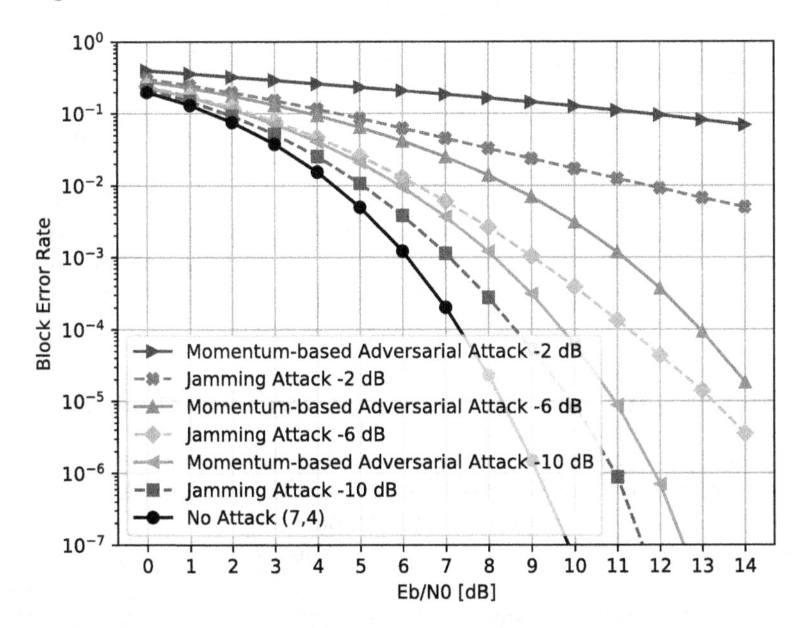

Fig. 5. BLER versus E_b/N_0 under the momentum-based adversarial and jamming attacks against the CNN autoencoder.

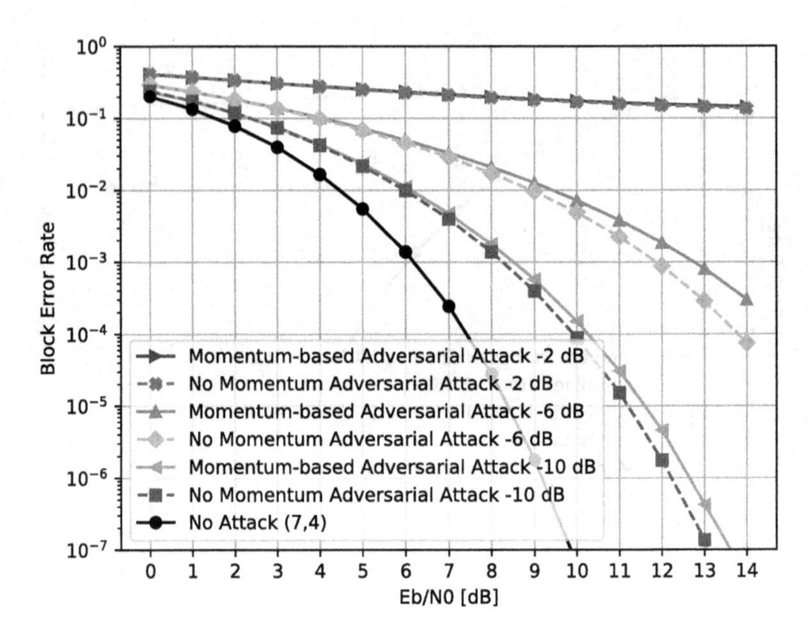

Fig. 6. BLER versus E_b/N_0 under the momentum-based and no momentum white-box attacks.

attacks even for small PSR values, and the performance advantages are become more obvious as the signal-to-ratio (SNR) increases.

Secondly, we compare the performance of the proposed momentum-based method with the adversarial attack method without momentum in Fig. 6. It can be seen that the performance of the proposed momentum-based algorithm is significantly better than the benchmark algorithm in [8] when crafting white-box adversarial attacks. This is due to the momentum-based algorithm accumulates the gradient of the loss function at each iteration to achieve stabilized optimization and get rid of poor local minima and saddle points.

4.2 Crafting a Black-Box Adversarial Attack with Momentum

For verifying the transferability, we also craft black-box attacks for the proposed momentum-based adversarial attack method and the benchmark in [8]. In the black-box attacks, we first use the MLP autoencoder (Table 1) as the substitute model to obtain shift-invariant perturbations as in [8], whereas, based on momentum. Due to the transferability of adversarial attacks, we then attack the CNN autoencoder (Table 2), whose structure and parameters are unknown for the attackers. We use the black-box attack crafting method (without momentum) in [8] as a comparison, and the results of BLER performance are presented in Fig. 7. Comprehensively considering the effect of the white-box and black-box attacks, even the performance improvement of the momentum-based attacks in the black-box attacks are not obvious enough, it can be concluded that the pro-

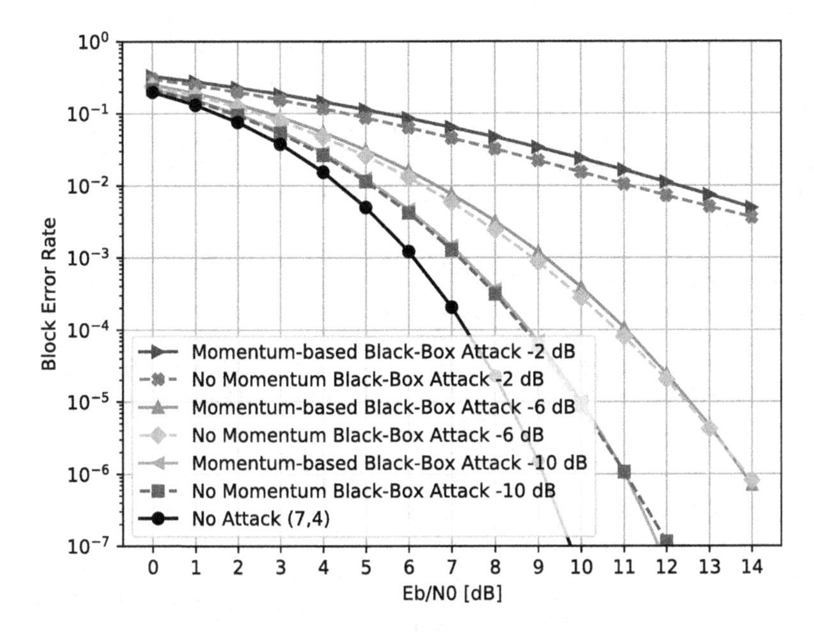

Fig. 7. BLER versus E_b/N_0 under the momentum-based and no momentum black-box attacks.

posed momentum-based method could increase the effectiveness of white-box attacks and alleviate the cost of transferability at the same time.

5 Conclusion

In this paper, we proposed a momentum-based adversarial attack method against the end-to-end communication systems, which incorporates momentum into the adversarial perturbation generative algorithm for better performance. Numerical results showed that, compared with conventional jamming attacks, the adversarial attacks based on momentum can effectively increase the BLER for the DL-based communication models with various network structures. Meanwhile, we also illustrated that the proposed method is more destructive than adversarial attacks without momentum accumulation. Therefore, we demonstrated that the adversarial attack incorporated with momentum can alleviate the trade-off between effectiveness and transferability of the adversarial perturbation, which raises new security issues and robustness evaluation methods for building more reliable end-to-end communication systems.

References

1. O'Shea, T., Hoydis, J.: An introduction to deep learning for the physical layer. IEEE Trans. Cognit. Commun. Netw. **3**(4), 563–575 (2017). https://doi.org/10.1109/TCCN.2017.2758370

2. Dörner, S., Cammerer, S., Hoydis, J., Brink, S.t.: Deep learning based communication over the air. IEEE J. Sel. Top. Signal Process. **12**(1), 132–143 (2018). https://doi.org/10.1109/JSTSP.2017.2784180
3. Ye, H., Liang, L., Li, G.Y., Juang, B.H.: Deep learning-based end-to-end wireless communication systems with conditional gans as unknown channels. IEEE Trans. Wirel. Commun. **19**(5), 3133–3143 (2020). https://doi.org/10.1109/TWC.2020.2970707
4. Chen, X., Cheng, J., Zhang, Z., Wu, L., Dang, J., Wang, J.: Data-rate driven transmission strategies for deep learning-based communication systems. IEEE Trans. Commun. **68**(4), 2129–2142 (2020). https://doi.org/10.1109/TCOMM.2020.2968314
5. Ye, H., Li, G.Y., Juang, B.H.F.: Deep learning based end-to-end wireless communication systems without pilots. IEEE Trans. Cognit. Commun. Netw. 1 (2021). https://doi.org/10.1109/TCCN.2021.3061464
6. Goodfellow, I.J., Shlens, J., Szegedy, C.: Explaining and harnessing adversarial examples. arXiv preprint arXiv:1412.6572 (2014)
7. Szegedy, C., et al.: Intriguing properties of neural networks. arXiv preprint arXiv:1312.6199 (2013)
8. Sadeghi, M., Larsson, E.G.: Physical adversarial attacks against end-to-end autoencoder communication systems. IEEE Commun. Lett. **23**(5), 847–850 (2019). https://doi.org/10.1109/LCOMM.2019.2901469
9. Sadeghi, M., Larsson, E.G.: Adversarial attacks on deep-learning based radio signal classification. IEEE Wirel. Commun. Lett. **8**(1), 213–216 (2019). https://doi.org/10.1109/LWC.2018.2867459
10. Kurakin, A., Goodfellow, I., Bengio, S.: Adversarial machine learning at scale. arXiv preprint arXiv:1611.01236 (2016)
11. Goodfellow, I., Bengio, Y., Courville, A., Bengio, Y.: Deep Learning, vol. 1. MIT press, Cambridge (2016)
12. Polyak, B.T.: Some methods of speeding up the convergence of iteration methods. USSR Comput. Math. Math. Phys. **4**(5), 1–17 (1964)
13. Dong, Y., et al.: Boosting adversarial attacks with momentum. In: Proceedings of the IEEE Conference on Computer Vision and Pattern Recognition, pp. 9185–9193 (2018)
14. Duch, W., Korczak, J.: Optimization and global minimization methods suitable for neural networks. Neural Comput. Surv. **2**, 163–212 (1998)
15. Sutskever, I., Martens, J., Dahl, G., Hinton, G.: On the importance of initialization and momentum in deep learning. In: International Conference on Machine Learning, pp. 1139–1147. PMLR (2013)

A Routing Algorithm Based on Node Utility and Energy in Opportunistic Networks

Peiyan Yuan[1,2(✉)] and Xiaoyan Huang[1]

[1] College of Computer and Information Engineering, Henan Normal University, Xinxiang 453007, China
peiyan@htu.cn
[2] Engineering Lab of Intelligence Business and Internet of Things, Xinxiang, Henan, China

Abstract. There does not exit a complete transmission path in the opportunistic network. In order to further improve the delivery rate and transmission delay, hybrid routing algorithms with node utility and redundancy was proposed, but they face the problem of higher network overhead. In addition, data transmission consumes energy while the energy of node is limited. Therefore, efficient nodes may lead to energy depletion due to excessive data transmission, aggravating the network disconnection. Considering this fact, a routing algorithm based on node utility and energy is proposed, which takes into account the influence of self-difference and dynamic variation of node relationship on routing packets, and makes full use of social relations to calculate the social utility of nodes, and synthesizes the node's residual energy to evaluate the node's forwarding capability, so as to make balance between communication overhead and energy consumption. Finally, compared with other algorithms, the proposed routing scheme can achieve better packet delivery rate and transmission delay, while network overhead and energy balance are greatly improved.

Keywords: Opportunistic network · Data forwarding · Social utility · Energy balance · Routing algorithm

1 Introduction

Nodes in traditional networks can find an end-to-end communication link before routing, and then complete the packet forwarding task according to the determined path. This means that the traditional network topology is connected in most of the time and there is at least one connected path. However, in some practical application scenarios, there is usually no end-to-end multi-hop wireless link between the source and destination node pairs, resulting in the traditional network routing strategy ineffectively. In order to solve the communication problem in disconnected environment, the opportunistic network has been widely concerned by researchers. The opportunistic network takes advantage of the encounter opportunities formed in the process of node movement and adopts the way of "store-carry-forward" for data transmission [1], which has a strong adaptability to the characteristics of node random movement, uncertain distribution density and limited communication range. Therefore, the

Q. Guo et al. (Eds.): WiSATS 2021, LNICST 410, pp. 533–553, 2022.
https://doi.org/10.1007/978-3-030-93398-2_49

opportunistic network has a wide application prospect in the fields of vehicle network, wildlife information collection, large gatherings (such as large sports events, concerts, etc.), remote areas and deep space communications.

The network topology of the opportunistic network changes frequently and the duration of links is uncertain, so the network is divided into several disconnected sub-regions, which can also be called "intermittently connected networks" [2]. The forwarding path cannot be determined in advance for data transmission in such a network environment. It is necessary to select node dynamically. Usually, the data packet will be sent to the destination after forwarding with multi-hop nodes. Obviously, routing decision-making problem has always been the focus of research in opportunistic networks, and its essence is how to select appropriate relay nodes to complete the data transmission task. So far, the routing protocols in opportunistic networks can be divided into two categories according to the way of selecting relays. One is zero-information routing, which does not need complex network information, and only uses the encounter opportunities generated by node movement to complete data transmission. The other is information-assisted routing, which employs additional information to make a forwarding decision, that is, calculate the utility value of the node according to the node information, and further select high-utility nodes to route packets, which is also called utility routing.

Zero-information routing does not consider the heterogeneity of nodes in the network, and uses a relatively single routing method to complete the exchange and delivery of data packets. Information-assisted routing is the focus of the current work, and the evaluation function judges the ability of nodes to route packets according to different types of parameter information. Literature [3–5] use the historical contact information of nodes to evaluate node's utility. The Prophet algorithm [3] is a classical probability algorithm based on the contact frequency of nodes. The IPRA algorithm [4] uses the historical contact information (contact times) of the direct encounter node and the two-hop neighbor node to calculate the contact probability, and then makes the forwarding decision. The HPR algorithm [5] calculates the probability that the intermediate node can successfully deliver the data packet to the destination node based on the encounter frequency and contact duration. Literature [6–11] uses the social information of nodes to evaluate node's utility. The PageRank algorithm further makes the forwarding decision by calculating the centrality of neighbors [6]. The PeopleRank algorithm [7] sorts the centrality of the nodes on the basis of the PageRank algorithm, and selects the node with high global centrality as the next-hop relay. The Bubblerap algorithm [8] uses the number of times that act as a relay to calculate the centrality of the node, and selects the appropriate relay node based on its centrality ranking in the local community and global environment. The CMTR algorithm [9] is a social awareness protocol based on throw-boxes, which uses the number of encounters of nodes to calculate the degree of centrality, and deploys static and dynamic throw-boxes to forward data between remote communities. The SAPC algorithm [10] calculates the social utility of nodes based on the degree of social activity and physical contact factors of nodes. The HiBOp algorithm [11] mainly uses the current and historical information of the node (including the node's own information and neighbor information) to calculate the similarity probability between the node and the destination node as the basis for forwarding. Literature [12–15] uses the location information of nodes to evaluate node's utility. The LOOP algorithm [12] learns the historical movement trajectory by building a Bayesian model and predicts the future movement of the node, and finally forwards the packet to the destination area rather than to a specific node. The EDR algorithm [13] dynamically determines the next-hop forwarding node by using the ratio of encounter parameters to distance parameters, and selects a better relay node by maximizing the number of encounters with the destination node and minimizing the distance between the packet and the destination node. The MLProph algorithm [14] uses decision tree and neural network to calculate the probability of successful delivery of nodes by training multiple state information such as position, energy, speed and so on. The Geo-social algorithm [15] uses the geographical location history of users to mine the similarity between users, and makes forwarding decisions according to the similarity between users. Literature [16–18] uses node energy information to evaluate node's utility. The routing algorithm proposed in reference [16] relies on the energy of each node to calculate the forwarding probability of the node in order to maximize the network lifetime under the energy consumption constraint of each node. The ESW algorithm [17] establishes an evaluation function based on the speed and residual energy of the node, and calculates the forwarding utility of the node. The ProphetEA algorithm [18] forwards according to the residual energy of the node and the delivery rate defined by the Prophet algorithm. Many routing protocols in opportunistic networks are intersected and related, such as the above literature [8, 9, 14]. The ultimate goal is to integrate the various information of the node and measure the forwarding ability of the node in order to improve the network performance according to the different network environment.

The nodes in the opportunistic network are in a state of frequent movement, and each node divides the network into several sub-networks with independent communication opportunities. When the node cannot transmit the data packet to the destination node in time, it can only forward the data between the intermediate nodes, which may cause a large packet transmission delay. In order to further improve the delivery rate and transmission speed of the network, on the basis of information-assisted routing, researchers propose a hybrid routing scheme of utility and redundancy, but this scheme may have high network overhead. In addition, the packet is always transmitted in the direction of the high-utility node, which will cause the energy of the high-utility node to be consumed too much. Especially in the case of limited energy, high-utility nodes often participate in data reception and forwarding, the greater the energy consumption. The energy of high-utility will even be exhausted, which will aggravate the network intermittence and lead to a sharp decline in network performance. It has been proved that better routing performance can be achieved by selecting relays based on the social attribute information of nodes, so the focus of this paper is to make full use of the social utility of nodes to guide routing and further reduce network overhead. Secondly, balance the energy consumption of nodes in the data transmission process, so as to avoid the rapid exhaustion of energy of high-utility nodes.

Based on the above analysis, this paper proposes a multi-copy routing algorithm based on node utility and energy (ProEnergy algorithm), which makes full use of node utility to reduce network overhead and balance node energy consumption under the premise of guaranteeing better network performance. In Sect. 2, the design idea based on utility and energy algorithm is given and a network model based on social attributes is constructed. In Sect. 3, the calculation process of social utility of nodes is introduced. In Sect. 4, the main steps of the algorithm are given. In Sect. 5, the network performance of the four routing algorithms is compared by simulation. Finally, we summarize the full study in Sect. 6.

2 Design Idea of ProEnergy Algorithm

2.1 Motivation

The opportunistic network relies on the communication opportunities brought by the node movement to complete the data transmission task. In general, most of the nodes in the network are composed of intelligent terminal devices carried by people, and their social activities may follow some social characteristics [19]. Different routing protocols use different social metrics such as centrality, similarity, and community attributes to select appropriate relay nodes [20–22]. The relationship between people in the network is complex, the equipment is in a state of continuous movement with people's social activities, and the relationship between nodes has human characteristics. In the previous routing work, the influence of self-difference and dynamic variability of node relationship on routing packets is ignored, so a new utility metric is proposed to measure the social relations of nodes. In order to avoid the problem of excessive network overhead in multi-backup routing, the focus of the routing algorithm is to study how to make full use of node relationships to further reduce network overhead. In addition, in practical application scenarios, considering the fact that node communication needs energy support but node energy is limited, in the process of data forwarding, the social utility and residual energy of the node are jointly used to judge the forwarding capability of the node.

The design idea of the routing algorithm proposed in this paper is shown in Fig. 1. Firstly, the network model is constructed to maintain the contact information and information update mechanism of nodes, so as to provide sufficient preparation for data forwarding. Secondly, a routing scheme based on node utility and energy is designed, which mainly includes three steps: identifying intimate nodes, building relationship model and routing decision.

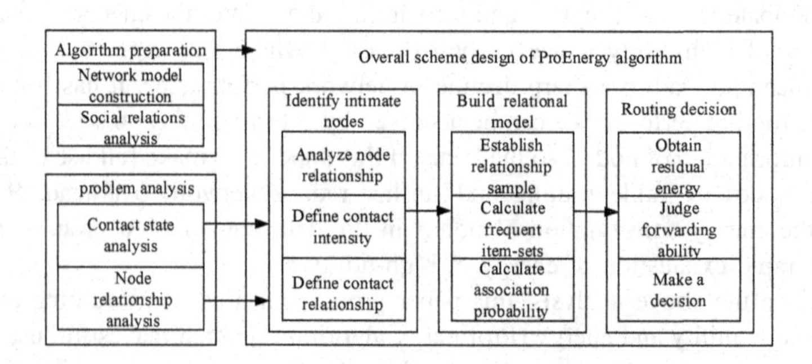

Fig. 1. Overall scheme design of ProEnergy algorithm

Each link of ProEnergy algorithm is described in detail in the following section. In Sect. 2.2, the node relationship is firstly analyzed, and then the contact relationship between nodes is defined to build a network model.

2.2 The Opportunistic Network Model Based on Social Attributes

In the opportunistic network based on social characteristics, the area of node activity is affected by human social activities, and human social activities are determined by social relations or the same social attributes. For example, nodes with certain social relationships (such as family, friends, colleagues) often appear in an environment with similar mobility and more contact time than unfamiliar relationships. Secondly, the node relies on social environmental life that there are many social relationships, and the relationship between nodes is a relative state. However, considering the influence of personality differences on the state of nodes, the boundaries of the relationship between nodes are not the same. In addition, with the increase of network time, the node relationship may change due to the influence of some force majeure factors (living environment and social pressure). Based on the above analysis of node relationship, the contact diagram with time attributes $G_t(V,E)$ is used to model the opportunistic network. Assuming that there are n nodes in the network, $V = \{v_1, v_2, ..., v_n\}$ represents the node set, and t_x represents a current moment when the network is running. The relevant definitions are as follows:

Definition 1 Neighbor Node Set: When the distance between nodes v_i and v_j is less than a certain threshold d, then the two nodes are neighbors of each other. The neighbor set of node v_i is represented as:

$$Nb(v_i) = \{v_j | D(v_i, v_j) < d\} \tag{1}$$

Where $D(v_i, v_j)$ is the communication distance between v_i and v_j, and d is the maximum communication distance between two nodes.

Definition 2 Intra and Inter Contact Time: The cumulative intra-contact time of nodes v_i and v_j is defined as $CT_{v_i}^{t_x}(v_j)$, and the cumulative inter-contact time of nodes v_i and v_j is defined as $OT_{v_i}^{t_x}(v_j)$.

The total duration of the network is represented by $T = \{t_0, t_1, t_2, ..., t_n\}$, where T is evenly divided into m different moments, the time gap between every two adjacent moments is $\Delta t. \delta_{\Delta t}^x(v_i, v_j)$ represents the contact situation between v_i and v_j in the x-th time gap.

$$\delta_{\Delta t}^x(v_i, v_j) = \begin{cases} 1 & v_j \in Nb(v_i) \\ 0 & v_j \notin Nb(v_i) \end{cases} \tag{2}$$

When $\delta_{\Delta t}^x(v_i, v_j) = 1$, it means that v_i and v_j communicate with each other and can send data packets to each other. On the contrary, the communication link between them has been disconnected. $CT_{v_i}^{t_x}(v_j)$ and $OT_{v_i}^{t_x}(v_j)$ of nodes v_i and v_j are automatically updated according to the contact status of the node, that is:

$$\begin{cases} CT_{v_i}^{t_x}(v_j) = CT_{v_i}^{t_{x-1}}(v_j) + \Delta t & \delta_{\Delta t}^{x}(v_i, v_j) = 1 \\ OT_{v_i}^{t_x}(v_j) = OT_{v_i}^{t_{x-1}}(v_j) + \Delta t & \delta_{\Delta t}^{x}(v_i, v_j) = 0 \end{cases} \tag{3}$$

Definition 3 Contact Intensity: The contact intensity of nodes v_i and v_j is defined as the ratio of the intra-contact time to the inter-contact time.

$$CS_{v_i}^{t_x}(v_j) = \frac{CT_{v_i}^{t_x}(v_j)}{OT_{v_i}^{t_x}(v_j)} \tag{4}$$

The greater the contact intensity of the two nodes, the greater the chance for the two nodes to establish contact. It should be noted that because the contact between the two nodes is mutual, that is, $CS_{v_i}^{t_x}(v_j) = CS_{v_j}^{t_x}(v_i)$.

Definition 4 Average Contact Intensity: The average contact intensity of the node v_i at t_x is expressed as the average of the sum of the current contact intensity with the historical communication node, which represents the average level of the contact condition of the node at t_x.

$$ACS_{v_i}^{t_x} = \frac{1}{|N(v_i)|} \sum_{v_j \in N(v_i)} CS_{v_i}^{t_x}(v_j) \tag{5}$$

where $N(v_i)$ and $|N(v_i)|$ respectively represent the historical contact neighbor node set of v_i and the number of neighbor nodes in the set. $N(v_i)$ updates according to the contact condition of v_i at t_x.

$$N(v_i) = N(v_i)_{old} \cup \{v_j | D(v_i, v_j) < d \cap v_j \notin N(v_i)_{old}\} \tag{6}$$

$N(v_i)_{old}$ represents the neighbor node set of v_i at t_{x-1}. Therefore, the historical neighbor node set of the node is composed of the historical neighbor node set of the previous moment and the newly joined neighbor nodes at the current time.

Definition 5 Contact Relationship: when $CS_{v_i}^{t_x}(v_j) > ACS_{v_i}^{t_x}$, it indicates that there is a strong contact relationship between v_i and v_j. On the contrary, when $CS_{v_i}^{t_x}(v_j) < ACS_{v_i}^{t_x}$, it indicates that there is a weak contact relationship between v_i and v_j.

3 Relation Model

3.1 Contact Intensity Prediction

The contact intensity of nodes is constantly updated with the increase of network time, and this value is a statistical value under the current time of the network. According to the average contact level of nodes in the network, the nodes with more contact opportunities with themselves are further identified. Based on the global time, the contact intensity between different nodes can be compared better, but it cannot better reflect the instantaneous change of the contact intensity between a single node pair. Therefore, we try to predict the change trend of contact intensity between single node pairs, excavate the change rule of contact intensity in a short time, and predict the future contact intensity of the two nodes, so as to guide the routing work more pertinently.

GM (1,1) prediction model is a small sample prediction model, which can use a small amount of historical data to predict the data of the next moment, and it has a better prediction effect on uncertainty [23–25]. Due to the mobility of nodes, the contact between nodes is constantly changing, and the contact or disconnection state of nodes is random. Figure 2 shows the contact situation of a node pair in the network. It is assumed that the node pair is in the contact state at t_x, while it may be in the contact state or disconnected state at t_{x-1}, and has no relation with the contact or disconnection state of the node at the past t_{x-1}, t_{x-2}, t_{x-3},..., t_{x-n}. Due to the sudden change and irregularity of node movement over a long period of time, with the increase of network time, the contact intensity of the two nodes at time t_{x-n} and t_{x+n} may change greatly. However, the contact intensity of the node in a short period of time fluctuates slightly, which may show a certain regularity. Therefore, using the changing law of contact intensity in a short period of time, predict the contact intensity in a short period of time in the future.

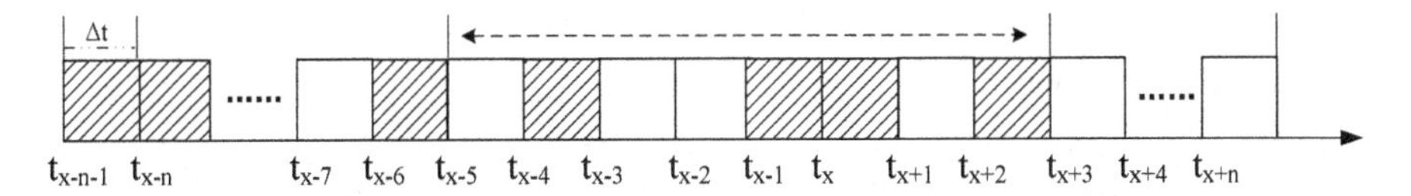

Fig. 2. The change diagram of contact and disconnection between nodes over time

The prediction process of contact intensity based on GM (1,1) is as follows:

(a) The contact intensity of the current time t_x and its previous history $w - 1$ time is obtained, and the total contact intensity of w time is obtained, from which the original contact intensity sequence is given.

$$CS^{(0)}_{v_i,v_j} = \left\{CS^{(0)}_{v_i,v_j}(1), CS^{(0)}_{v_i,v_j}(2), \cdots, CS^{(0)}_{v_i,v_j}(w)\right\} \tag{7}$$

(b) Accumulate the original contact intensity sequence to generate $CS^{(1)}_{v_i,v_j}$ sequence, and call $CS^{(1)}_{v_i,v_j}$ the 1-AGO (Accumulated Generating Operation) sequence of $CS^{(0)}_{v_i,v_j}$.

$$CS^{(1)}_{v_i,v_j} = \left\{CS^{(1)}_{v_i,v_j}(1), CS^{(1)}_{v_i,v_j}(2), \cdots, CS^{(1)}_{v_i,v_j}(w)\right\} \tag{8}$$

Where $CS^{(1)}_{v_i,v_j}(k) = CS^{(1)}_{v_i,v_j}(1) + CS^{(1)}_{v_i,v_j}(2) + \cdots + CS^{(1)}_{v_i,v_j}(k)$. A new sequence $Z^{(1)}_{v_i,v_j}$ is generated by processing the nearest neighbor mean of the cumulative sequence $CS^{(1)}_{v_i,v_j}$.

$$Z^{(1)}_{v_i,v_j} = \left\{ Z^{(1)}_{v_i,v_j}(2), Z^{(1)}_{v_i,v_j}(3), \cdots, Z^{(1)}_{v_i,v_j}(w) \right\} \tag{9}$$

$Z^{(1)}_{v_i,v_j}$ is called background value, $z^{(1)}_{v_i,v_j}(k) = \frac{z^{(1)}_{v_i,v_j}(k) + z^{(1)}_{v_i,v_j}(k-1)}{2}$.

(c) The first-order differential equation of the GM(1,1) model is established for the accumulated contact intensity sequence, and the corresponding whitening differential equation is as follows:

$$\frac{dCS^{(1)}_{v_i,v_j}}{dt} + aCS^{(1)}_{v_i,v_j}(t) = b \tag{10}$$

Where a is called the development coefficient, and b is called the gray effect, both of which are constant. The least square method is used to obtain the values of a and b, which satisfy the following equation:

$$[a, b]^T = \left(B^T B\right)^{-1} B^T Y \tag{11}$$

The value of Y and B is:

$$Y = \begin{bmatrix} CS^{(0)}_{v_i,v_j}(2) \\ CS^{(0)}_{v_i,v_j}(3) \\ \vdots \\ CS^{(0)}_{v_i,v_j}(w) \end{bmatrix}, B = \begin{bmatrix} -Z^{(1)}_{v_i,v_j}(2) & 1 \\ -Z^{(1)}_{v_i,v_j}(3) & 1 \\ \vdots & \vdots \\ -Z^{(1)}_{v_i,v_j}(w) & 1 \end{bmatrix} \tag{12}$$

(d) Expand the matrix to get the expression of a and b.

$$a = \frac{\sum_{k=2}^{n} CS^{(0)}_{v_i,v_j}(k) \sum_{k=2}^{n} Z^{(1)}_{v_i,v_j}(k) - (w-1) \sum_{k=2}^{n} CS^{(0)}_{v_i,v_j}(k) Z^{(1)}_{v_i,v_j}(k)}{(w-1) \sum_{k=2}^{n} (Z^{(1)}_{v_i,v_j}(k))^2 - (\sum_{k=2}^{n} Z^{(1)}_{v_i,v_j}(k))^2} \tag{13}$$

$$b = \frac{\sum_{k=2}^{n} CS^{(0)}_{v_i,v_j}(k) \sum_{k=2}^{n} (Z^{(1)}_{v_i,v_j}(k))^2 - \sum_{k=2}^{n} Z^{(1)}_{v_i,v_j}(k) \sum_{k=2}^{n} Z^{(1)}_{v_i,v_j}(k) CS^{(0)}_{v_i,v_j}(k)}{(w-1) \sum_{k=2}^{n} (Z^{(1)}_{v_i,v_j}(k))^2 - (\sum_{k=2}^{n} Z^{(1)}_{v_i,v_j}(k))^2} \tag{14}$$

(e) Solve the differential equation and get the discrete solution of the GM(1,1) model:

$$CS^{(1)}_{v_i,v_j}(k+1) = (CS^{(0)}_{v_i,v_j}(1) - \frac{b}{a})e^{-ak} + \frac{b}{a} \tag{15}$$

(f) Reverted to the original sequence, the prediction model is:

$$CS^{(0)}_{v_i,v_j}(k+1) = CS^{(1)}_{v_i,v_j}(k+1) - CS^{(1)}_{v_i,v_j}(k) \tag{16}$$

Bring Eq. 15 into Eq. 16 to get:

$$CS^{(0)}_{v_i,v_j}(k+1) = (1-e^{a})(CS^{(0)}_{v_i,v_j}(1) - \frac{b}{a})e^{-ak}, k = 1, 2, \cdots, w \tag{17}$$

When $k = w$, $CS^{(0)}_{v_i,v_j}(k+1)$ is the predicted value of the contact intensity at the time t_{x+1} in the future.

The prediction method of contact intensity takes full account of the motion characteristics of nodes and is based on the motion rule of nodes in a short time. Mining the dynamic characteristics of contact intensity between nodes with the passage of time is more suitable for predicting the contact intensity between nodes in a short time in the future. GM (1,1) prediction model has high prediction accuracy for data samples with small scale, small fluctuation and persistence in the short term.

3.2 Build Relationship Samples

Nodes rely on the social environment to survive rather than independent individuals, and each node will have a relatively close social relationship. There may be more opportunities for contact between nodes and nodes with close relationships. Therefore, we believe that nodes and nodes with strong contact relationships have relatively close social relationships with a higher probability. In order to further analyze the node relationship to help reduce network overhead, first use the node contact relationship to establish the node relationship matrix.

Definition 6 Relational Model: Maintain a relationship matrix R based on the average contact intensity of the nodes, which is as follows:

$$\begin{pmatrix} r_1^{t_x} & r_{2\to1} & \cdots & r_{n\to1} \\ r_{1\to2} & r_2^{t_x} & \cdots & r_{n\to2} \\ \vdots & \vdots & r_j^{t_x} & \vdots \\ r_{n\to1} & r_{n\to2} & \cdots & r_n^{t_x} \end{pmatrix}$$

The diagonal position of R is $r_i^{t_x}$, where $r_i^{t_x}$ represents the average contact intensity $\widehat{ACS}_{v_i}^{t_x}$ of node v_i calculated according to the predicted contact intensity at t_x. The intersection of the i-th row and the j-th column represents the contact relationship $r_{j \to i}$ between node v_j and node v_i, and its value is as follows:

$$r_{j \to i} = \begin{cases} 1 & CS_{v_i}^{t_x}(v_j) > ACS_{v_i}^{t_x} \\ 0 & CS_{v_i}^{t_x}(v_j) < ACS_{v_i}^{t_x} \end{cases} \tag{18}$$

Where, $CS_{v_i}^{t_x}(v_j)$ is the predicted contact intensity between v_i and v_j at t_x. When $r_{j \to i} = 1$, it means that v_i and v_j have some kind of intimate relationship. When $r_{j \to i} = 0$, it means that there is no intimate relationship between v_i and v_j. It should be noted that the relational matrix R is an asymmetric matrix, $r_{i \to j} \neq r_{j \to i}$, the value of $r_{i \to j}$ is based on $ACS_{v_j}^{t_x}$, i.e.,

$$r_{i \to j} = \begin{cases} 1 & CS_{v_j}^{t_x}(v_i) > ACS_{v_j}^{t_x} \\ 0 & CS_{v_j}^{t_x}(v_i) < ACS_{v_j}^{t_x} \end{cases} \tag{19}$$

3.3 Relational Model

The Apriori algorithm was proposed by Agrawal in 1994. It is a classic algorithm for mining data association rules [26]. The most famous case is the ‘supermarket shopping basket’ case, which optimizes the placement of supermarket items by analyzing and mining supermarket shopping data to find out the set of items that people frequently buy [27]. Figure 3 shows the workflow of the model. It is mainly divided into two major steps. The first step is to find frequent item-sets. The purpose is to find frequent item-sets from the dataset. The second step is to discover association rules, that is, use conditional probabilities to find association rules in frequent item sets, and use potential rules to make predictions. Scan is called dataset selection process, and its main function is to filter the support, delete item-sets that do not meet the minimum support, and retain item-sets that meet the minimum support.

Aiming at the network environment proposed in this paper, a relationship model is established based on the idea of Apriori algorithm. The relationship matrix R of the node is used as the input of the model, and the social forwarding utility of the node is finally obtained to help routing to make forwarding decisions. The following describes the process of mining potential relationships between nodes based on the relationship model shown in Fig. 3. According to the input relationship matrix R, the model saves the close relationship of each node in the corresponding record of the relationship transaction sample Ω, that is, the name of the node with the close relationship is saved in the record. Each element in the candidate item-sets (C_1 and C_2) and frequent item-sets (L_1 and L_2) is a set, and the elements in the set are nodes in the network. Each element in C_1 and L_1 is called a 1-item-set. In the same way, each element in C_2 and L_2 is called a 2-item-set. 1-item-sets and 2-item-sets are also collectively called item-sets.

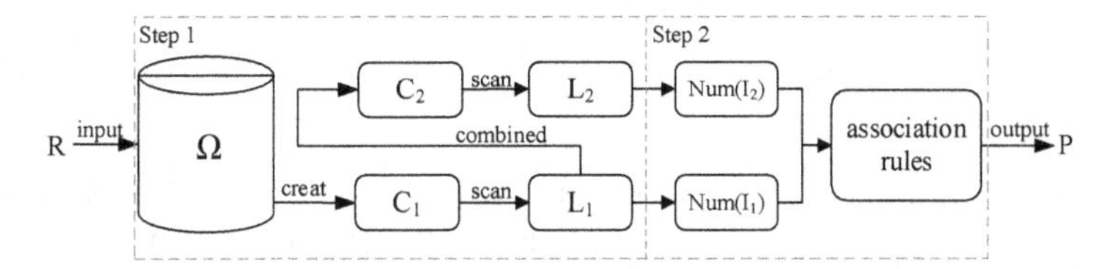

Fig. 3. Workflow chart of relationship mining model based on Apriori algorithm

It can be seen from Fig. 3 that the candidate 1-item-set is generated according to Ω, then $C_1=\{\{v_1\},\{v_2\},\cdots,\{v_n\}\}$, which contains all the node items in the transaction sample, that is, contains a total of *n* 1-item-sets. The Scan process is divided into two stages. The first stage is to calculate the item-set frequency, which is based on candidate item-set *C*. Assuming that the item-set is represented by *I*, and the frequency of item-set *I* at t_x is defined as:

$$Fre_I^{t_x} = \partial^{t_x}(I) \tag{20}$$

$$\partial^{t_x}(I) = \sum_{k=1}^{n} F(I,k) \tag{21}$$

Where $F(I,k)$ is expressed as the relation function between the k-th node in the network and the nodes contained in the item set *I*. *I* is divided into 1-item-set I_1 and 2-item-set I_2. When $I = I_1$, the frequency of 1-item-set is calculated to obtain the number of intimate relationships of nodes at t_x. Assume that $I_1=\{v_i\}$,where v_i is the *i*-th node in the network, then $F(I,k)=F(I_1,k)$ and $k \neq i$, its value is as follows:

$$F(I_1,k) = \begin{cases} 1, & R[k][i] = 1 \\ 0, & R[k][i] = 0 \end{cases} \tag{22}$$

When $I = I_2$, the frequency of 2-item-set is calculated to obtain the number of common close nodes of the two nodes in the item-sets at t_x. Assume that $I_2 = \{v_i, v_j\}$, where v_i and v_j are respectively the *i*-th and *j*-th nodes in the network, then $F(I,k) = F(I_2,k)$ and $k \neq i$ or $k \neq j$, its value is as follows:

$$F(I_2,k)=\begin{cases} 1, & R[k][i] = 1, R[k][j] = 1 \\ 0, & otherwise \end{cases} \tag{23}$$

In the first stage of the Scan process, the frequency of all candidate item-sets is finally obtained. In the second stage, the frequent item-sets *L* is obtained according to the minimum support T_M. At the current network time t_x, according to the relationship between the frequency of the item-set $Fre_I^{t_x}$ in the candidate item-sets *C* and the minimum support T_M, the frequent set *L* is obtained, and the basis is as follows:

$$\begin{cases} I \in L^{t_x} & Fre_I^{t_x} > T_M \\ I \notin L^{t_x} & Fre_I^{t_x} < T_M \end{cases} \tag{24}$$

L^{t_x} includes $L_1^{t_x}$ and $L_2^{t_x}$, which represent frequent 1-item-sets and frequent 2-item-sets at the current time t_x. The nodes included in the frequent 1-item-sets are called active nodes in the network. If a node has a close relationship with many nodes, then the node is more likely to contact other nodes, so the node can transmit the message to the destination node with a higher probability. In this process, T_M helps to reduce unnecessary data transmission and increase the number of valid data packets received by neighbor nodes. The corresponding two nodes included in the frequent 2-item-sets are called frequent contact node pairs in the network. If two nodes have more intimacy in common, the more likely it is to establish a connection between the two nodes. Therefore, the two nodes can establish a connection through a common intimacy with a higher probability. In this process, T_M helps to eliminate the node pairs with a lower degree of association in the network, and finally make the data packets flow to the node that has a higher probability of establishing contact with the destination node, achieving the purpose of further reducing network overhead and improving network performance.

The first step of the Apriori model is to finally obtain a pair of nodes in the network that can establish potential connections with a higher probability. Therefore, in the second step, it is necessary to further calculate the probability that a node can establish contact through a common close node. Assuming that at t_x, 2-item-set $\{v_i, v_j\} \in L_2^{t_x}$, it is necessary to calculate the probability that nodes v_i and v_j can establish a connection through the potential relationship. Let $p_{(v_i \Rightarrow v_j)}^{t_x}$ denote the probability that the node v_i establishes a connection with the node v_j through the latent relationship at t_x. The formula is defined as follows:

$$p_{(v_i \Rightarrow v_j)}^{t_x} = \frac{Num^{t_x}(v_i \cup v_j)}{Num^{t_x}(v_i)} = \frac{Fre_{\{v_i, v_j\}}^{t_x}}{Fre_{\{v_i\}}^{t_x}} \tag{25}$$

where $Fre_{\{v_i, v_j\}}^{t_x}$ represents the frequency of 2-item-set $\{v_i, v_j\}$ in $L_2^{t_x}$, and $Fre_{\{v_i\}}^{t_x}$ represents the frequency of 1-item-set $\{v_i\}$ in $L_1^{t_x}$.$Num^{t_x}(v_i \cup v_j)$ is the number of transactions that include v_i and v_j in the transaction sample, and $Num^{t_x}(v_i)$ is the number of transactions that include node v_i in the transaction sample. $p_{(v_i \Rightarrow v_j)}^{t_x}$ represents the conditional probability of node v_j in the transaction containing node v_i at t_x. This probability represents the possibility that the node v_i and the node v_j will establish a connection through a potential relationship, which can also be called the social forwarding utility of v_i being able to forward the data packet to v_j. It is worth noting that, according to the relational model, at t_x, the social forwarding utility $p_{(v_j \Rightarrow v_i)}^{t_x}$ of node v_j that can forward the data packet to node v_i and the social forwarding utility $p_{(v_i \Rightarrow v_j)}^{t_x}$ of node v_i that can forward the data packet to node v_j are not equal. The formula is as follows:

$$p^{t_x}_{(v_j \Rightarrow v_i)} = \frac{Num^{t_x}(v_i \cup v_j)}{Num^{t_x}(v_j)} = \frac{Fre^{t_x}_{\{v_i, v_j\}}}{Fre^{t_x}_{\{v_j\}}} \tag{26}$$

That is, the conditional probability of the occurrence of node v_i in the transaction containing node v_j is used as the forwarding utility of node v_j being able to forward the data packet to node v_i. When the 2-item-set $\{v_i, v_j\} \notin L^{t_x}_2$, then $p^{t_x}_{(v_i \Rightarrow v_j)} = 0$, it is considered that the probability of establishing potential association between node v_i and v_j is small and can be ignored.

4 ProEnergy Routing

4.1 Energy Weight Factor

Considering that the relay node has an association relationship with the destination node of multiple data packets at the same time, it can be called the key node in the network. The node needs to continuously receive and forward data, which will inevitably consume a lot of energy. In practical application scenarios, the communication between nodes requires energy support and the energy of the nodes is limited. If the energy factor of the node is not considered, it will stop working when the energy of the node is exhausted, resulting in the all paths through the node to be interrupted, which will greatly increase the transmission delay of the message. Therefore, in the ProEnergy routing scheme, the initial value of the node energy is set to 500. It is assumed that the node will consume one unit of energy when receiving and forwarding data packets, and the energy consumption required by the node when it is moving is ignored. Determine whether to participate in data forwarding according to the remaining energy of the node. When the energy of the node is less than a certain threshold, it only receives data packets that need to be transmitted to itself. Only when the energy of the node is greater than the threshold, the social forwarding utility of the node and the remaining energy need to be integrated to calculate the forwarding capacity of the node, as shown in the following formula:

$$P(v_i, v_j) = \lambda \times p^{t_x}_{(v_i \Rightarrow v_j)} + (1 - \lambda) \times C_i \tag{27}$$

Where $\lambda(\lambda \in [0, 1])$ is expressed as a weighting factor, $p^{t_x}_{(v_i \Rightarrow v_j)}$ is the social forwarding utility from v_i to v_j and $p^{t_x}_{(v_i \Rightarrow v_j)} < 1$. C_i is the remaining energy of node v_i. In addition, $P(v_i, v_j) \in (0, 1)$ and $p^{t_x}_{(v_i \Rightarrow v_j)} \in (0, 1)$, in order to eliminate the dimensional influence between the indicators, the dispersion standardization is used to standardize the remaining energy C_i of the node v_i, as shown in the following formula:

$$C_i = \frac{C_i - C_{min}}{C_{max} - C_{min}} \tag{28}$$

Where C_{min} represents the minimum remaining energy of the node in the network, and C_{max} represents the maximum remaining energy of the node in the network. This method maps the remaining energy of the node to [0,1] by linearly transforming the remaining energy of the node. After the energy is standardized, the forwarding capacity of the intermediate node is jointly calculated based on the remaining energy and social utility to guide the completion of the routing work. In this way, the energy consumption of nodes can be balanced to a certain extent, and nodes with too low energy can be prevented from participating in data packet forwarding. In order to better balance the problems of network performance and resource consumption, the value of the weighting factor λ will be explained in detail in the experimental part.

4.2 ProEnergy Routing Algorithm

The main steps of the ProEnergy routing algorithm proposed in this paper are shown in Table 1.

Table 1. Main steps of ProEnergy routing algorithm

ProEnergy Algorithm
Input: The packet *M*, the destination node d_M, the contact intensity of node pair GM(1,1)
Output: Whether to forward *M*
1. **FOR** each node $a, b, d_M \in V$ **DO**
2. **IF** *a* and *b* communicate with each other and *a* is the sender of *M* **THEN**
3. calculate $CS_a^{t_x}(b)$ based on $CT_a^{t_x}(b)$ and $OT_a^{t_x}(b)$,update GM(1,1)
4. **IF** *b* does not carry the copy of *M* **THEN**
5. **IF** the energy of *a* and *b* are greater than 50 **THEN**
6. **IF** *b* is the destination node of *M* **THEN**
7. update packet information, *a* forwards a copy of *M* to *b*
8. **ELSE**
9. calculate $CS_a^{t_x}(d_M)$ and $CS_b^{t_x}(d_M)$, update GM(1,1), calculate $CS_a^{t_x}(d_M)$, $CS_b^{t_x}(d_M)$, $CS_a^{t_x}(b)$, ACS_a, ACS_b, ACS_{d_M}, update *R*
10. calculate L_1 and L_2 based on *R*, calculate $p_{(a \Rightarrow d_M)}^{t_x}$ and $p_{(b \Rightarrow d_M)}^{t_x}$, normalize C_a and C_b, calculate $P(a,d_M)$ and $P(b,d_M)$
11. **IF** $P(a,d_M) < P(b,d_M)$ **THEN**
12. update packet information, *a* forwards a copy of *M* to *b*
13. **END IF**
14. **END IF**
15. **END IF**
16. **END IF**
17. **END IF**
18. **END FOR**

5 Simulation Experiment

5.1 Simulation Environment and Parameter Setting

For performance evaluation, the experiment in this paper is based on the mobile opportunistic network simulator platform, which is described in detail in [28, 29]. The movement trajectory of the node uses the KAIST data set, which is provided by the Korea Institute of Science and Technology. The detailed introduction of the data set can be found in [30]. Other simulation parameters are shown in Table 2. This paper is based on the same parameters for simulation experiments. Where the Apriori algorithm is used to calculate the associated probability, the T_M value is set to 45. We use packet delivery rate, transmission delay, network overhead, energy consumption, and energy consumption to evaluate the performance of the routing algorithm, and compares it with the Prophet algorithm, the PageRank algorithm, and the GeoSocial algorithm.

Table 2. Experimental parameter setting

Parameter	Value
Simulation field size (m^2)	600 × 600
Simulation time (s)	15000
Number of packets	200
Number of nodes	90
Maximum communication distance (m)	250
Send packets rate (ms)	100
Neighborhood search cycle (ms)	100

The influence of λ on network performance is analyzed in first. Figure 4(a) shows the delivery rate of λ with different values from 0.5 to 0.8. Observation shows that the delivery rate trends in the four cases are similar, and the delivery rate from $\lambda = 0.8$ to $\lambda = 0.5$ decreases sequentially, because the social utility influence of nodes gradually decreases. ProEnergy uses a multi-copy transmission method, so it has a higher transmission probability under different λ values. Figure 4(b) shows the transmission delay of λ with different values from 0.5 to 0.8. Observation shows that the transmission delay is the lowest when $\lambda = 0.7$. When $\lambda = 0.5$, the higher transmission delay is because when the value is small, more data packets are transmitted to nodes with relatively high energy but low correlation probability, which causes the delay to increase. Figure 4(c) shows the network overhead of λ with different values from 0.5 to 0.8. Observation shows that the network overhead of nodes from $\lambda = 0.5$ to $\lambda = 0.8$ decreases sequentially. When λ increases, the chances of those nodes with lower correlation probability but higher energy carrying data packets will be smaller, so the number of network copies will decrease. Through comprehensive comparative analysis, we believe that based on the above network parameters, when $\lambda = 0.7$, routing can obtain the best network performance.

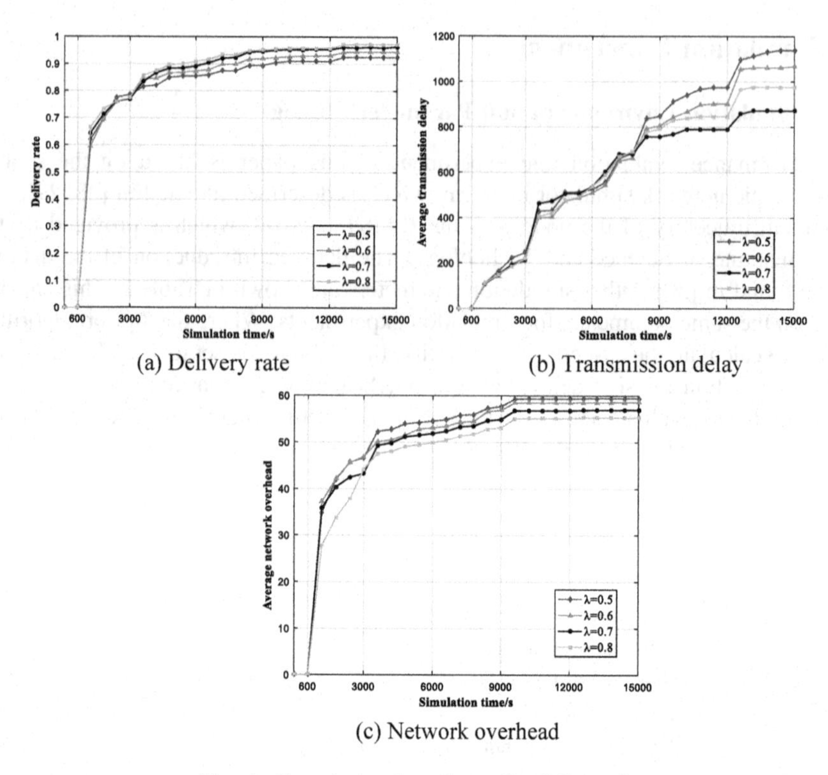

(a) Delivery rate

(b) Transmission delay

(c) Network overhead

Fig. 4. Experimental results under different λ

5.2 Experimental Results Analysis

Figure 5(a) shows the total energy consumption of 90 nodes in the network. It can be seen from the figure that the energy consumption of the four algorithms increases with the increase of simulation time. Before the simulation time is 3200 s, the total energy consumption of ProEnergy algorithm is greater than Prophet algorithm. Because the number of contacts of the nodes is limited at the beginning of the simulation, the Prophet algorithm only calculates the utility of the nodes within the communication range for forwarding. The ProEnergy algorithm calculates the social utility of nodes based on historical communication records. Due to the unstable contact between nodes in the initial period of the network, and the sufficient energy of the nodes, more nodes participate in data forwarding, resulting in higher energy consumption. However, as the simulation time increases, the total energy consumed by the ProEnergy algorithm increases slowly and tends to stabilize, while the Prophet algorithm continues to increase. At the end of the simulation, the energy consumption of the PageRank algorithm reached about 35,000, which was the highest consumption among the routing algorithms. The energy consumption of the GeoSocial algorithm is about 34,000, that of the Prophet algorithm is about 29,000, and the ProEnergy algorithm is about 21,000.

Figure 5(b) shows the variation of the standard deviation of node energy consumption as the simulation time increases. The remaining energy of the 90 nodes is output every 600 s, and then the standard deviation of the energy consumption of the 90 nodes in each 600 s is finally calculated. It can be seen from the figure that all algorithms have large fluctuations in the standard deviation at the beginning of the simulation ($0 \sim 3000$ s), which is caused by the uneven distribution of messages in the network. However, after 3000 s, it can be clearly seen that the standard deviation of ProEnergy algorithm is smaller than the other three routing algorithms, and the fluctuations are small and relatively stable, which indicates that ProEnergy algorithm can balance the energy consumption of the nodes, and does not overuse a certain node for data forwarding during data transmission, thus avoiding premature death of some efficient nodes in the network due to energy exhaustion.

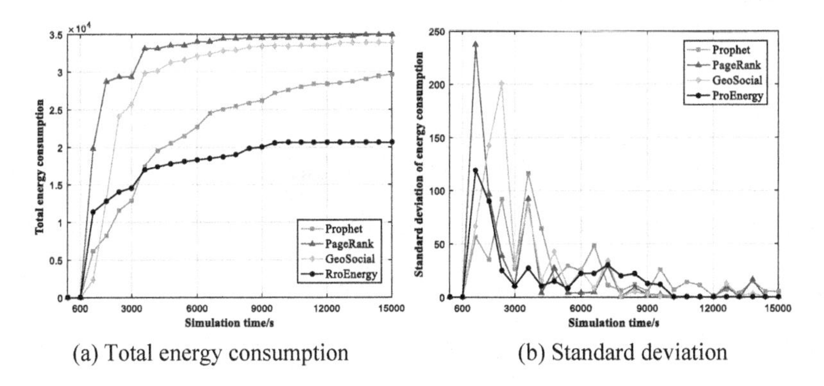

(a) Total energy consumption (b) Standard deviation

Fig. 5. Energy experiment results

The delivery rate of the four algorithms is shown in Fig. 6(a). All of the four algorithms complete routing work in the way of multiple copies, so they all get higher delivery rates. It can be seen that PageRank algorithm is obviously higher than the other three algorithms before 7000 s. At the end of the simulation experiment, the delivery rate of PageRank is 0.98, GeoSocial is 0.97, and the ProEnergy algorithm is 0.96. The delivery rates of the three are similar. The delivery rate of Prophet algorithm is 0.89.

The average transmission delay of the four algorithms is shown in Fig. 6(b). It can be seen that the transmission delay increases with the increasing simulation time. In the initial simulation stage ($0 \sim 3200$ s), the transmission delay of Geosocial algorithm is the largest, but then the transmission delay of the Prophet algorithm increases with the increasing of the simulation time, because the Prophet algorithm updates the node utility when the node is in contact, and the high-utility node carries the data packet, which does not consider the energy problem of the node. After transferring the node to the high-utility node, the high-utility node may consume too much energy and aggravate the network intermittence, resulting in high network overhead. At the end of the simulation, the delay of Prophet is the largest, followed by Geosocial. ProEnergy is about 870 s. Compared with Prophet and Geosocial, the ProEnergy algorithm can effectively reduce the transmission delay.

The average network overhead is shown in Fig. 6 (c). The average network overhead represents the ratio of the number of copies of all data packets generated in the network to the total number of packets. In the simulation experiment before 3200 s, the network overhead of ProEnergy algorithm is higher than Prophet algorithm, but after that, the ProEnergy algorithm is gradually stable, and the reason is similar to energy. At the end of the experiment, the number of copies of PageRank, GeoSocial and Prophet algorithms is higher than ProEnergy. Because PageRank, GeoSocial, and Prophet algorithms are based on different utility indicators, they forward when they encounter nodes with higher utility, and they do not consider the energy factor of the node, so it brings a higher number of data packet copies. In the ProEnergy algorithm, the relational utility of the node is used to mine the association probability of the node, and the energy of the node is also considered as the forwarding factor to effectively reduce the number of copies of data packets.

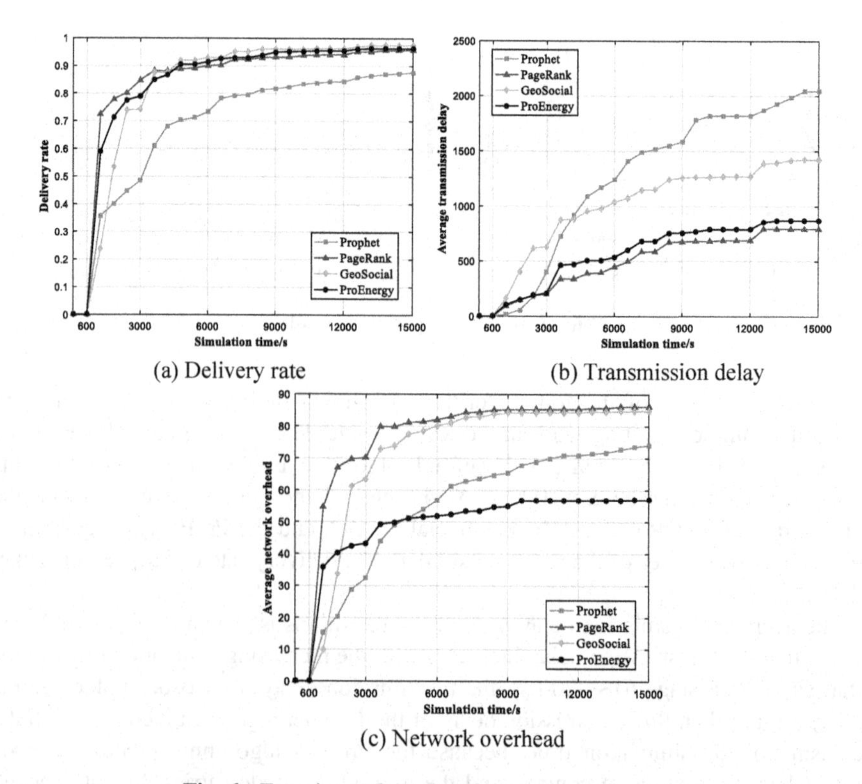

(a) Delivery rate (b) Transmission delay

(c) Network overhead

Fig. 6. Experimental results of network performance

6 Conclusion

The nodes in the opportunistic network are in a state of continuous movement. When they cannot be delivered to the destination node, the packet will be transmitted between the intermediate nodes, which may bring high transmission delay. In order to further improve the network delivery rate and reduce the transmission delay, researchers proposed a hybrid routing mechanism of utility and redundancy, but this mechanism has the problem of high network overhead. In addition, the routing mechanism is based on the multi-copy transmission mode. In the actual environment, the energy of the node is limited, and the high-utility node will experience premature energy exhaustion due to excessive data transmission. In order to solve the above problems, the goal of the routing algorithm based on utility and energy proposed in this paper is to reduce the network overhead and balance the energy consumption of nodes while ensuring better network performance. First of all, the relationship between nodes is analyzed, the network model is constructed, and the Apriori algorithm is introduced to establish the relationship model to calculate the social forwarding utility of nodes. Secondly, add energy factors to balance the energy consumption of nodes and optimize routing performance. Finally, the proposed algorithm is simulated. The experimental results show that while ensuring a higher delivery rate and a lower average delay, it effectively reduces network overhead and balances the energy consumption of nodes, which provides a reference for future research on multiple backup routing algorithms.

Acknowledgements. This work was supported in part by the National Natural Science Foundation of China under Grants U1804164, 62072159 and U1404602.

References

1. Wu, D., Zhang, F., Wang, H., et al.: Security-oriented opportunistic data forwarding in mobile social networks. Future Gener. Comput. Syst. **87**(10), 803–815 (2017)
2. Yuan, P., Fan, L., Liu, P., et al.: Recent progress in routing protocols of mobile opportunistic networks: a clear taxonomy, analysis and evaluation. J. Network Comput. Appl. 62(C), 163–170 (2016)
3. Patel, D., Shah, R.: Improved PROPHET routing protocol in DTN. Int. Res. J. Eng. Technol. 503–509 (2016)
4. Pan, H., Chaintreau, A., Scott, J., et al.: Pocket switched networks and human mobility in conference environments. In: Proceedings of the 2005 ACM SIGCOMM Workshop on Delay-Tolerant Networking, pp. 244–251. ACM, Philadelphia (2005)
5. Wang, X., Lin, Y., Zhang, S., et al.: A social activity and physical contact-based routing algorithm in mobile opportunistic networks for emergency response to sudden disasters. Enterp. Inf. Syst. 1–30 (2015)
6. Yuan, P., Ma, H., Fu, H.: Hotspot-entropy based data forwarding in opportunistic social networks. Pervasive Mob. Comput. **16**, 136–154 (2015)
7. Ayyat, S.A., Harras, K.A., Aly, S.G.: Interest aware PeopleRank: towards effective social-based opportunistic advertising. In: IEEE Wireless Communications and Networking Conference. IEEE (2013)

8. Hui, P., Crowcroft, J., Yoneki, E.: BUBBLE Rap: social-based forwarding in delay-tolerant networks. IEEE Trans. Mob. Comput. **10**(11), 1576–1589 (2011)
9. Qirtas, M.M., Faheem, Y., Rehmani, M.H.: A cooperative mobile Throwbox-based routing protocol for social-aware delay tolerant networks. Wirel. Netw. **2**, 1–13 (2020)
10. Wang, X., Lin, Y., Zhang, S., et al.: A social activity and physical contact-based routing algorithm in mobile opportunistic networks for emergency response to sudden disasters. Enterp. Inf. Syst. **11**(1–5), 597–626 (2015)
11. Boldrini, C., Conti, M., Jacopini, J., et al.: HiBOp: a history based routing protocol for opportunistic networks. In: IEEE International Symposium on World of Wireless, Mobile and Multimedia Networks (2007)
12. Lindgren, A., Doria, A.: Probabilistic routing in intermittently connected networks. ACM SIGMOBILE Mobile Comput. Commun. Rev. **7**(3), 19 (2003)
13. Dhurandher, S.K., et al.: EDR: an encounter and distance based routing protocol for opportunistic networks. In: IEEE International Conference on Advanced Information Networking and Applications IEEE (2016)
14. Sharma, D.K., Dhurandher, S.K., Woungang, I., et al.: A machine learning-based protocol for efficient routing in opportunistic networks. IEEE Syst. J. **12**, 2207–2213 (2016)
15. Ying, Z., Zhang, C., Li, F., et al.: Geo-social: routing with location and social metrics in mobile opportunistic networks. In: IEEE International Conference on Communications, pp. 3405–3410 (2015)
16. Jang, K., Lee, J., Kim, S.K., et al.: An adaptive routing algorithm considering position and social similarities in an opportunistic network. Wirel. Netw. **22**(5), 1537–1551 (2016)
17. Jia, W.U., Chen, Z.: Reducing energy consumption priority selection of node transmission routing algorithm in opportunistic network. Adv. Inf. Sci. Serv. Sci. (2014)
18. Sobin, C.C., Raychoudhury, V., Saha, S.: An energy-efficient and buffer-aware routing protocol for opportunistic smart traffic management. In: Proceedings of the 18th International Conference on Distributed Computing and Networking, pp. 1–8 (2017)
19. Basaras, P., Iosifidis, G., Katsaros, D., et al.: Identifying influential spreaders in complex multilayer networks: a centrality perspective. IEEE Trans. Netw. Sci. Eng. **6**(1), 31–45 (2017)
20. Freeman, L.C.: Centrality in social networks conceptual clarification. Soc. Netw. **1**(3), 215–239 (1978)
21. Wei, K., Xiao, L., Ke, X.: A survey of social-aware routing protocols in delay tolerant networks: applications, taxonomy and design-related issues. IEEE Commun. Surv. Tutor. **16** (1), 556–578 (2014)
22. Ahmad, T., Li, X.J., Seet, B.C., et al.: Social network analysis based localization technique with clustered closeness centrality for 3D wireless sensor networks. Electronics **9**(5), 738 (2020)
23. Ding, S., Hipel, K.W., Dang, Y.G.: Forecasting China's electricity consumption using a new grey prediction model. Energy **149**(4), 314–328 (2018)
24. Yu, Q., Lyu, J., Jiang, L., et al.: Traffic anomaly detection algorithm for wireless sensor networks based on improved exploitation of the GM (1,1) Model. Int. J. Distrib. Sens. Netw. **12**(7), 2181256 (2016)
25. Kun, G., Qishan, Z.: Privacy preserving method based on GM (1,1) and its application to clustering. Grey Syst. Theor. Appl. **2**(2), 157–165 (2012)
26. Pratima, G., Pardasani, K.R.: A fast algorithm for mining multilevel association rule based on Boolean matrix. Int. J. Comput. Sci. Eng. **2**(3), 746–752 (2010)
27. Soni, A., Saxena, A., Bajaj, P.: A methodological approach for mining the user requirements using Apriori algorithm. J. Cases Inform. Technol. **22**(4), 1–30 (2020)

28. Yuan, P., Song, M.: MONICA one simulator for mobile opportunistic networks. In: 11th EAI International Conference on Mobile Multimedia Communications, pp. 21–32 (2018)
29. Ma, H., Zhao, D., Yuan, P.: Opportunities in mobile crowd sensing. Infocommunications J. **7** (2), 32–38 (2015)
30. Rhee, I., Shin, M., Hong, S., et al.: On the levy-walk nature of human mobility. IEEE/ACM Trans. Netw. (TON) **19**(3), 630–643 (2011)

A Design Scheme and Security Analysis of Unmanned Aerial Vehicle

Dongyu Yang[1,2,3](✉), Yue Zhao[1,2,3], Kaijun Wu[1,2,3], Zhongqiang Yi[1,2,3], and Haiyang Peng[1,2,3]

[1] Science and Technology on Communication Security Laboratory, Chengdu 610041, China

[2] No. 30 Research Institute of China Electronics Technology Group Corporation, Chengdu 610041, China

[3] China Electronics Technology Cyber Security Co., Ltd., Chengdu 610041, China

Abstract. Since the 21st century, informatization, modernization and intellectualization have become an important direction of national science and technology development. Especially in recent years, with the continuous improvement and perfection of artificial intelligence, 5G, edge computing and autonomous unmanned technology, the UAV industry has made unprecedented development and has been applied to many fields of military and civil, such as: information collection, investigation and combat integration, geological mapping, border defense inspection, emergency rescue, power inspection, traffic supervision, intelligent logistics, pesticide spraying, meteorological monitoring, etc. However, with the continuous development of UAV business, more and more researchers realize that UAV depends on security and effective information system and network connection, and the cyber security problem of UAV is becoming increasingly prominent. The cyber security research of UAV has become a new subject. This paper first elaborates the development of UAV through a large number of researches. Secondly, according to the current development status of UAV, the future development trend of UAV is analyzed. Then, focusing on the communication, network and system of UAV, the security threats faced by UAV are elaborated and analyzed in detail. After that, the new cyber security problems caused by UAV are analyzed from three aspects of important infrastructure security, public security and privacy security. Finally, the paper gives suggestions on the future development of UAV from the perspective of cyber security, which provides a strong support for solving the cyber security problems of UAV.

Keywords: UAV · Cyber security · Communication security · Privacy security

1 Introduction

UAV is the abbreviation of unmanned aerial vehicle, which refers to the unmanned aircraft controlled by radio remote control or self-contained program. It can fly autonomously or remotely, carry a variety of mission equipment, perform a variety of tasks, and can be used once or repeatedly [1].

Q. Guo et al. (Eds.): WiSATS 2021, LNICST 410, pp. 554–567, 2022.
https://doi.org/10.1007/978-3-030-93398-2_50

With the development of unmanned technology, the concept of UAV is constantly enriched and evolved. In the 1920s, the newly emerging UAV was known as pilotless aircraft. In the 1930s, UAV was more used as target aircraft, known as drone. In the early 1950s, UAV controlled by radio signal, known as radio controlled aerial target (RCAT). In the mid-1950s, unmanned aerial vehicle (UAV) was equipped with reconnaissance function, known as surveillance drone. In the 1960s, UAV was equipped with remote control function, known as remote piloted vehicle (RPV). In the 1980s, unmanned aerial vehicle (UAV) can perform autonomous or preset tasks, which is called unmanned aircraft (UMA),unmanned aerial vehicle (UAV) has been widely used since 1990s [2].

With the rapid development of information technology, UAV related technology is becoming more and more mature. According to report [3], in recent years, the scale of UAV market has maintained a rapid growth trend of 50% every year, and the annual sales volume of UAV is expected to reach 4.33 million by 2020. It is estimated that sales of UAVs will exceed $12 billion by 2021. With the continuous development of UAV business, UAV is more and more advanced, and the cyber security problem of UAV is very prominent. Due to the characteristics of UAV and its communication structure, when the attacker intercepts the communication information or hijacks the UAV itself, it will bring serious consequences to the user and the surrounding environment. Furthermore, in addition to being vulnerable to attack, UAV itself can also be used as a new means of cyber attack.

This paper is divided into six sections, the second section introduces the development of UAV in detail. The third section analyzes the development trend of UAV. In the fourth section, the cyber security threats of UAV are described in detail from three aspects of communication, network and system. In the fifth section, the cyber security problems caused by UAV as a new means of cyber attack are analyzed in detail from three aspects of important infrastructure, public security and privacy security. The sixth section summarizes the full text and some suggestions are put forward for the development of UAV from the perspective of cyber security.

2 Overview of UAV Development

The global UAV industry is in the stage of vigorous development, and all countries in the world are aware of the huge application potential and broad application prospects of UAV in military and civil fields.

The development of UAV has gone through the following stages: 1920–1960 is the initial stage, and UAV was often used as target aircraft. 1960–1980 is the practical stage, and UAV was used in the battlefield. Since 1990 is the high-speed development stage, and a large number of modern and intelligent UAVs have emerged.

2.1 Initial Stage

The development of UAV can be traced back to 1917, when the Royal Aircraft Establishment (RAE) developed the world's first unmanned aircraft. In 1918, France's first radio controlled aircraft was successfully tested. In 1921, Britain developed the world's first practical unmanned target aircraft, which can fly at a speed of 160km/h at an altitude of nearly 2 km. In September 1931, Fairey Company refitted the "Queen" manned biplane into a "Fairey Queen" target aircraft, as shown in Fig. 1. In 1933, Britain developed the famous "Queen Bee" drone, which produced 420 drones from 1934 to 1943. In 1948, the American Ryan Aeronautical Company began to develop a high subsonic, jet propelled target aircraft, which was later known as the "Fire Bee" target aircraft. Due to the successful design, mass production began in 1953, and soon 1280 early "Fire Bee" Q-2A and KDA were in service in the US armed forces and the Royal Canadian air force.

At this stage, in addition to the United Kingdom and the United States, France, Italy, Australia, Canada, Israel, Japan and Germany have also developed many target aircraft. The development of unmanned target aircraft drives the development of key technologies of UAV, such as remote control and telemetry technology, flight control and guidance technology, small engine technology, launch and recovery technology and special equipment for UAV. In the development process of UAV, UAV technology has broken through the speed limits of low speed, high subsonic speed and supersonic speed, as well as the airspace flight limits of ultra-low altitude, low altitude, medium altitude and high altitude, laying a foundation for the comprehensive development of UAV in the future [4].

Fig. 1. Fire bee

2.2 Practical Stage

In 1960, the American Ryan Aeronautical Company began to try to transform "Fire Bee" into a kind of unmanned reconnaissance aircraft 147A with low radar detectability, longer voyage and better maneuverability. Later, it quickly improved and completed the 147B with longer voyage, and then developed the famous 147D "Firefly" unmanned reconnaissance aircraft.

Military UAV is the first large-scale application in Vietnam battlefield. During the Vietnam War, the "Fire Bee" series of unmanned high altitude reconnaissance aircraft were used for as many as 3435, and had carried out such tasks as high altitude and ultra-low altitude photo reconnaissance, electronic eavesdropping, jamming the communication of Vietnam radio station, scattering metal chaff in the air corridor to escort the bombers, of which 2873 sorties returned safely, with a battle damage rate of only 16%.

The outstanding performance of "Fire Bee" reconnaissance UAV in Vietnam battlefield makes people realize the new value of UAV, and also makes UAV used in actual combat for the first time as combat equipment, opening up a new stage of UAV application and development. During the Middle East War, Israel successively developed two kinds of unmanned reconnaissance aircraft: "Scout" and "Mastiffs", which are used to collect radar signals and conduct photoelectric composite reconnaissance. These two kinds of unmanned reconnaissance aircraft can be deployed flexibly and have all-weather working ability. Since then, Pakistan, India, Singapore, Iraq, Iran and other countries have carried out the development of unmanned reconnaissance aircraft, and made great progress.

2.3 High-Speed Development Stage

Since the 1980s, the military value of UAV has been gradually valued by the military of all countries. After the 1990s, several high-tech local wars have provided a broader stage for UAV to show its combat capability. Guided by the needs of war, UAV has entered a stage of rapid rise and rapid development.

Since the 1990s, many countries have placed UAV development in an important strategic position, and the investment has increased year by year. At present, there are nearly 1000 kinds of UAV systems developed by 57 countries in the world, among which nearly 400 have become UAV products. The United States has occupied the technological commanding height of UAV development. Israel started early and has characteristics and advantages in Tactical UAV and long endurance UAV. Russia has never relaxed the development and application of advanced technology. European countries and Asian countries have also accelerated the pace of UAV development and set off a climax of UAV development in the world.

In the civil field, UAV is more and more used in all walks of life. The agricultural UAV uses the high-precision camera to realize the real-time monitoring of crop growth and the surrounding soil moisture, and accordingly sowing, watering, fertilizing and spraying pesticides. Through the aerial survey and aerial exploration of UAV, mineral deposits and other resources can be found, and the local geological conditions can be monitored at any time to guide the development of resources. In daily use, UAVs can patrol important public facilities such as roads, railways, high-voltage wires and oil and gas pipelines to reduce accidents. The application of UAV in many industries in the civil field has promoted social progress and gradually become an important growth point to promote social and economic development.

3 UAV Development Trend

Looking forward to the future, UAV technology will continue to improve, mission will continue to expand, the number of equipment will continue to grow. At the same time, UAVs will continue to develop in the direction of diversification, intelligence and swarming, so as to better adapt to various complex environments.

3.1 System Performance Level

With the development of new power and energy, diversified detection, identification technology, advanced communication and control technology, the capability of UAV system such as mission duration, situation awareness, information transmission and autonomous control will be greatly increased in the future, and higher, farther, larger, smaller and faster UAV Systems will continue to be applied.

3.2 Application Fields

In the military aspect, it will continue to improve the application scope, flexibility, efficiency and adaptability of the unmanned system, and ultimately cover all mission fields such as ground, sea, air, missile defense and network power attack and defense. In the civil aspect, it will have broad market development prospects in the fields of Agriculture, forestry, animal husbandry and sideline fishing, entertainment, logistics, emergency rescue, public services and other industries.

3.3 The Level of Intelligent Autonomy

Mastering autonomous ability is the ultimate goal of unmanned system development. With the gradual improvement of the autonomy of the unmanned system, the human intervention required by the unmanned system in the process of completing the task will be greatly reduced in the future. Finally, the unmanned system will have the ability of autonomous learning and adapting to the environment, and be able to make decisions independently and provide suggestions to human beings, so as to achieve a higher level of more autonomous "man-in-the-loop" or even "man out of the loop". In addition, the intelligent technology of unmanned system will develop rapidly in the direction of aggregating many single agents to realize swarm intelligence, robust architecture and more efficient cost ratio.

3.4 Human Machine Cooperation

It is an important development direction for all kinds of unmanned systems to realize the cooperative combat capability between manned and unmanned systems. The U.S. military has further emphasized the coordinated development and joint application of unmanned equipment in its latest unmanned system roadmap. The U.S. Army plans to build a modern force composed of manned unmanned system teams. The U.S. Air Force has verified the ability of manned and UAV formation to independently attack targets. The U.S. Navy is vigorously promoting the coordinated development of air, surface and underwater unmanned systems. With the development of human-computer force, the cross domain cooperative combat capability of underwater, surface and air man-machine formation has been verified, trying to build a new maritime combat system with high efficiency and cooperation.

3.5 Swarm

Unmanned system swarm is a kind of low-cost unmanned system with an index of 10 or 100. It is like a bee colony performing tasks in groups and rapidly assembling in local areas to form large-scale equipment advantages. It has the characteristics of swarm substitution mobility, quantity improvement ability and cost creation advantages. It is an important development direction of unmanned system. The United States has carried out swarm research of unmanned systems, conducted dozens of swarm tests of UAVs and unmanned boats, and conducted grouping and maneuvering flight tests. In the past two years, China and the United States have refreshed the scale record of UAV swarm flight for four times. In 2017, China completed 119 UAV swarm flight tests, which broke the world record of UAV swarm test again. The competition between China and the United States in this field is becoming increasingly fierce.

4 UAV: The Target of Cyber Attacks

UAV is mainly includes the following parts: UAV system, ground control station and communication link to transmit information, as shown in Fig. 2. The UAV system includes power system, main controller, communication link module, sensor and task execution unit. The ground control station includes remote controller, intelligent terminal and communication link module [5]. The control command is transmitted to the UAV platform through the ground control station, and the data collected by the UAV platform and its operation data are also transmitted to the ground control station [6]. UAV relies on security and effective information system and network connection, which makes UAV become a new network attack target. These network attacks may destroy the control system of UAV, hijack UAV or sneak into the data collected by UAV and take it away. With the continuous development of UAV business, the cyber security problems of UAV become increasingly prominent [7, 8]. At present, the cyber security problems of UAV mainly focus on communication security, cyber security and system security.

4.1 Security Threats of UAV Communication

The UAV completes the flight mission under the guidance and control of the ground control station. The communication link between the UAV and the ground control station is used to exchange control commands and data. Generally, Ku band tactical communication data link (TCDL) is used in satellite communication, and C-band radio signal, 2.4 GHz wireless signal or wireless data transmission are used in communication between UAV and ground control station [9]. Because wireless signal is the main communication mode between UAV and controller, UAV receives command or

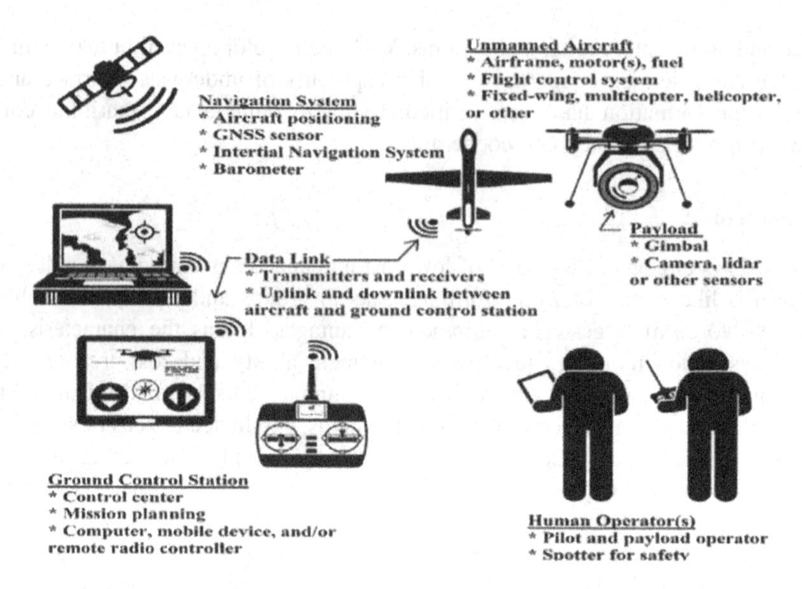

Fig. 2. UAV components

transmits information through communication link. Once the communication link is attacked, it is like closing the "ears" of UAV.

The attack mode of UAV communication link can be divided into three layers: signal layer, protocol layer and system layer. Signal layer attack is mainly against the tracking of transmission channel and the cracking of signal receiving and sending. Protocol layer attack is against the communication protocol by identifying and analyzing the protocol itself and the loopholes existing in the implementation. System layer attack is mainly about the attack methods against the operating system and various application software. The countermeasure of signal layer needs accurate communication protocol, while the countermeasure of protocol layer needs the modulation of signal layer and the control of system layer. The system layer needs the other two layers to provide physical access. At present, the main attacks on UAV communication link are as follows:

1. Eavesdropping attack
   Due to the lack of encryption and other protection mechanisms, UAV information interacting in an open environment can be directly accessed by the enemy.
2. Information injection attack
   If there is no proper authentication scheme, the opponent can disguise as a legal entity to inject false information or commands. One of them is man in the middle attack. As the intermediary between UAV and remote control, attacker can steal and tamper with communication information.
3. Denial of service attack (DOS)/Distributed denial of service attack (DDoS)
   If there is no proper DOS/DDoS Defense mechanism, single or many hijacked systems attack the target UAV, resulting in the UAV refusing to provide services to its legitimate users.

4. Flooding attack
   Usually send a large number of SYN, UDP, ICMP and Ping packets, causing network congestion. Buffer overflow attack forces the buffer overflow of network devices, making the network card unable to accept new requests.
5. Interference attack
   Transmit high-power multi band radio to interfere with the communication of UAV, or use the vulnerability of WIFI communication to remove the communication connection between UAV and ground control station.

The attack on UAV communication link can directly affect the normal operation of UAV, and even gain control of the UAV. Successfully invade the UAV and obtain full authority, not only can control the flight function of the UAV, but also browse, copy, and tamper with the data stored on the UAV at will.

4.2 Security Threats of UAV Network

In some application scenarios, multiple UAVs or cooperation between UAVs and other facilities may be required to complete a task. UAV network can be seen as a flying wireless network [10]. Each UAV can be a network data transceiver node or a network relay The UAV network can be self-organized or based on ground or satellite infrastructure support [11]. The air communication network composed of multiple UAVs is called Mobile ad hoc Network (MANET), which mostly adopts ad hoc mode architecture [12]. According to the topology type, Manet can be divided into planar topology and hierarchical topology. In the planar topology, each node in the network plays an equal role in routing calculation, message sending and receiving. Planar Topology Routing Protocols generally include AODV, DSR and OLSR [13]. In the hierarchical topology, the UAV nodes in the network are divided into multiple groups, each group has a group head, which is responsible for the topology management and communication between groups. The group heads of multiple groups can form a higher level network, and the higher level group heads are selected to form a multi-level network. The group head in each group is selected by the group members, which is dynamic and self-organized. Hierarchical Topology Routing Protocols generally include ZRP, LANMAR and CGSR.

As the UAV network is a subclass of mobile ad hoc network, the existing attacks of traditional mobile ad hoc network will also threaten the UAV ad hoc network, such as denial of service attack, black hole attack and so on. However, there are some special cyber security requirements in UAV ad hoc network.

1. Denial of service attack
   The traditional denial of service attack is that the attacker makes the network service abnormal by occupying a large number of system resources, so as to make the legitimate users unable to use the service normally. Similarly, the same idea can be adopted for UAV, but it has a greater impact on UAV ad hoc network, and even leads to the fall and destruction of UAV, threatening ground facilities and the masses. For example, Vasconcelos [14] and others used tools such as hping3 to implement a denial of service attack on drone 2.0, which made the legitimate operators unable to control the UAV and eventually led to the UAV crash.

2. Black hole attack
   The attack mode of black hole attack is that the malicious node disguises itself as the best path node by using the defects of routing protocol, so when forwarding packets, it will give priority to the malicious node as a relay node, that is, change the routing table. However, when the normal node in the UAV network forwards the packet to the malicious node, the malicious node will not transmit the packet to the next node in the routing table, but directly discard it, which leads to the failure of control command transmission. When the UAV at the edge of UAV group wants to receive the command sent by the controller, it needs the relay forwarding of the middle UAV. When the malicious UAV enters the UAV network, the malicious UAV will not continue to forward the data, which makes the edge UAV unable to receive the control command.

4.3 Security Threats of UAV System

UAV system refers to the sum of the various parts that make up the UAV system, which is usually composed of aircraft subsystem, remote control and navigation subsystem, mission equipment subsystem, data communication subsystem and support equipment. The security threats of UAV system mainly focus on sensors, positioning system and system software.

1. Sensor security threats
   The sensors of UAV collect environment data from physical domain and feed back to UAV control system to assist UAV to issue control instructions. When the sensor is attacked, the wrong data will be collected, making the control system unable to give correct instructions, resulting in the UAV flying out of control and even crashing.
2. Positioning system security threat
   GPS is an important sensor of UAV, which is responsible for providing accurate position information for UAV. When the attacker collects some public information of GPS system, such as signal definition, communication link, communication protocol. They can calculate the location, time and other information of each GPS transmission according to the attack target, and through high-power transmission of forged GPS signal with specific direction, they can cheat the UAV to choose the false signal with higher receiving strength, so as to achieve the purpose of deceptive attack. For example, the attacker can send the jamming signal with the same frequency as GPS, so that the GPS receiver cannot receive the normal signal, and can also send the high-power forged GPS signal, so that the UAV GPS can receive the forged GPS signal and get the wrong location information.
3. Software security threats
   Both the underlying software and the flight control software of the UAV have certain security vulnerabilities, and hackers can use these security vulnerabilities to launch attacks on the UAV. For example, the UAV flight control software Maldrone has a security loophole. Attackers can enter the UAV system through this loophole, or install a backdoor program on the control end to steal UAV data or perform

remote control. In addition, ZigBee chip threat and keyboard Trojan horse threat are also the means of attack against UAV software.

5 UAV: A New Means of Cyber Attack

From the perspective of cyber security, UAV has two sides. On the one hand, UAV becomes a new target of network attack, which causes data leakage and hijacking by means of UAV equipment vulnerability or network attack. On the other hand, UAV has become a new way of network attack. UAV may provide a new way for network attack. By injecting worms into the target's data and network, UAV can carry out data penetration attack or other attacks, thus destroying its key infrastructure. In addition, as shown in Table 1, the widespread use of UAV also brings new problems to public security and data privacy.

Table 1. Part of the UAV security events

NO.	Event description
1	On February 24, 2015, five UAVs of unknown origin hovered over the US consulate in Paris
2	On September 15, 2013, German President Angela Merkel was at a campaign rally when a drone crashed in front of a number of senior executives and Merkel herself
3	In April 2015, a drone carrying radioactive material fell on the Japanese Prime Minister's residence
4	On June 30, 2013, a man used a remote-controlled UAV to take aerial photos of letoria hospital
5	From 1995 to 2014, there were hundreds of terrorist attacks against politicians' UAVs carrying explosives, chemical or biological weapons, of which only 16 were cracked
6	In November 2014, 20 UAVs flew over many nuclear facilities in France
7	At the end of 2014, after a damaged nuclear reactor in Belgium was reopened, it was visited by a micro UAV the next day
8	On January 28, 2015, some light UAVs flew to the anchorage of Brest military port, which is a French nuclear military base with four nuclear submarines capable of submarine launched nuclear missiles
9	On January 26, 2015, a UAV broke into the no fly zone of the White House and fell in the zone. In this process, the air defense radar system did not find the whereabouts of the aircraft
10	On August 14, 2015, a drone crashed while flying near HMP Pentonville prison in the United States. Police intercepted a large number of drugs and mobile phones it was carrying
11	On January 22, 2015, an unmanned aerial vehicle (UAV) on the U.S. - Mexico border crashed due to overloading and tried to transport drugs and weapons
12	Mike Tassey and rich Perkins, former members of the U.S. air force, have successfully built an unmanned reconnaissance plane, which can perform Wi-Fi password cracking, phone tapping, SMS interception, etc

5.1 Cyber Attack Performance of UAV

In addition to being vulnerable to attack, UAV itself can also be used as a weapon to launch network attacks. According to its network attack performance, UAV can be divided into jamming UAV, eavesdropping UAV and defensive UAV.

1. Jamming UAV
   Selex Glileo, a US Italian technology military contractor, has designed a small UAV for electronic warfare and cyber attacks. The main purpose of its design is to interfere with the surface to air missile system. UAV can interfere with the communication system of the target, such as Bluetooth or Wi-Fi signal. It can also control suicide missiles, electronic warfare equipment will automatically delete its stored data before it is destroyed.
2. Eavesdropping UAV
   Septier Communications, an Israeli company, launched its first drone in 2017 to eavesdrop on phone calls and transmit data from smart phones. The UAV is equipped with a network listener, which can monitor the data of 2G, 3G and 4G networks. The maximum monitoring range of the UAV is one kilometer, which means that it cannot be physically detected by its monitoring target. This UAV is very likely to use proximity degradation attack to force devices in high security network to reduce the security level, such as from 4G to 2G. At this time, the UAV is not an unmanned vehicle, but a high-end camera for physical monitoring or attack.
3. Defensive UAV
   With the rise of the military UAV market, the anti UAV market, as a balancing force, also rises. The Cyber-Box UAV sold by Israel company in 2015 can detect and control the UAVs around the equipment boundary. By maintaining a kind of radio frequency jamming weapon, DoS attack or zero day vulnerability attack can be carried out on the UAV until the enemy UAV is completely controlled.

5.2 The Threat of UAV to Important Infrastructure

The vulnerability of UAV can make physical access to the network and equipment of important infrastructure. Extract information from systems that are not otherwise accessible due to scope constraints. UAVs can also cover up the identity of intruders to some extent. In addition, in their recent report, the researchers highlighted the risk of UAVs penetrating critical infrastructure with high security. Researchers have shown that UAVs can be used for wireless intrusion access points, insecure networks and devices. In 2016, for example, Israeli researchers manipulated a drone near an office building and used a flaw in a radio protocol called Zigbee to invade a smart light bulb inside the building.

5.3 The Threat of UAV to Public Security

For different application scenarios, different types of UAVs play their own strengths. Military UAVs are used in battlefield and enemy intelligence collection, while civilian UAVs are used in some illegal espionage and criminal activities. For example, in 2015, drones broke into the White House and made headlines in major media. The literature describes that UAVs may be used by criminals in terrorism, drug smuggling, unconventional weapons (such as biological and chemical weapons) attacks, etc. All of these pose a great threat to public security and cause irreparable losses to people's property security and national interests.

5.4 The Threat of UAV to Privacy

Flexible mobility makes the UAV like a "thief" can intrude into the forbidden or private airspace, equipped with high-definition cameras may pry into other people's privacy [15, 16]. Unmanned aerial vehicles (UAVs) with cameras fly to the house above the courtyard or near the window, and become "voyeurism" in the high-tech field. Even if UAVs are used for monitoring on specific occasions, there are privacy violations, such as video monitoring of parking lots, roads, parks, and so on. The privacy issues are also mentioned in [17–20] of the literature. Another example: hackers use dedicated Wi-Fi similar to fake base stations to attack passersby's mobile phones and steal personal privacy. Even the U.S. government uses spy planes with fake base stations to monitor millions of U.S. smartphones.

6 Summary

UAV brings convenience to people's life, but also faces increasingly serious cyber security problems. As a complete physical information system, UAV is used in the uncontrolled environment. UAV itself is constantly changing, and its security threats are also constantly changing. Therefore, the security protection measures of UAV must be updated. It is necessary to continuously evaluate the cyber security of UAV and find new countermeasures. Aiming at the cyber security problem of UAV, this paper gives the following suggestions:

1. From the top-level design of UAV, a complete UAV cyber security protection system should be established.
2. Researchers should pay more attention to the cyber security related topics of large UAV, and constantly pay attention to the latest technological achievements in related fields.
3. UAV, like unmanned vehicle, unmanned ship and other unmanned systems, need security systems such as firewall, and become a part of product safety standards.

Acknowledgements. This work was supported by Sichuan Science and Technology Program 2021JDRC0072.

References

1. Department of Defense (DoD). U.S. Army "Unmanned Aircraft Systems Roadmap 2010–2035". Office of the Secretary of Defense. US Fort Rucker, Alabama (2010)
2. Suraj, G., Mangesh, M., Jawandhiya, M.: Review of unmanned aircraft system (UAS). Int. J. Adv. Res. Comput. Eng. Technol. (IJARCET) **4**(2) (2013). ISSN 2278–1323
3. Schmidt, M., Shear, M.: A drone too small for radar to detect rattles the White House. http://www.suasnews.com/2015/01/a-drone-too-small-for-radar-to-detect-rattles-the-white-house/
4. Villasenor, J.: Drones and the future of domestic aviation. Proc. IEEE **102**(3), 235–238 (2014)
5. Abhishek, S., Pankhuri, V., Nikhil, P., et al.: Communication and networking technologies for UAVs: a survey. J. Netw. Comput. Appl. **168** (2020)
6. Imad, J., Nader, M., Jameela, A., et al.: Communication and networking of UAV-based systems: classification and associated architectures. J. Netw. Comput. Appl. **84**, 93–108 (2017)
7. Rodday, N., Schmidt, R., Pras, A.: Exploring security vulnerabilities of unmanned aerial vehicles. In: Network Operations & Management Symposium, pp. 993–994 (2016)
8. Javaid, Y., Sun, W., Devabhaktuni, K., et al.: Cyber security threat analysis and modeling of an unmanned aerial vehicle system. In: 2012 IEEE Conference on Technologies for Homeland Security (HST), Waltham, MA, pp. 585–590 (2012)
9. Hartmann, K., Steup, C.: The vulnerability of UAVs to cyber attacks: an approach to the risk assessment. In: 2013 5th International Conference on the Cyber Conflict (CyCon), Tallinn, Estonia, pp. 1–23 (2013)
10. Mansfield, K., Eveleigh, T., Holzer, T., et al.: Unmanned aerial vehicle smart device ground control station cyber security threat model. In: The IEEE International Conference on the Technologies for Homeland Security (HST), Waltham, USA, pp. 722–728 (2013)
11. He, D., Chan, S., Guizani, M.: Communication security of unmanned aerial vehicles. IEEE Wirel. Commun. 2–7 (2017)
12. Muhammad, A., Alamgir, S., Ijaz, M., et al.: Flying ad-hoc networks (FANETs): a review of communication architectures, and routing protocols. In: 2017 First International Conference on Latest trends in Electrical Engineering and Computing Technologies (INTELLECT), Karachi, Pakistan (2017)
13. Deng, H., Li, W., Agrawal, D.: Routing security in wireless ad hoc networks. IEEE Commun. Mag. **40**(10), 70–75 (2002)
14. Vasconcelos, G., Carrijo, G., Miani, R., et al.: The impact of DoS attacks on the AR. Drone 2.0. In: 2016 XIII Latin American Robotics Symposium and IV Brazilian Robotics Symposium (LARS/SBR), Recife, pp. 127–132 (2016)
15. Korshunov, P., Ebrahimi, T.: Using warping for privacy protection in video surveillance. In: The 18th IEEE International Conference on Digital Signal Processing (DSP), Fira, pp. 1–6 (2013)
16. Korshunov, P., Ebrahimi, T.: Using face morphing to protect privacy. In: The 10th IEEE International Conference on Advanced Video and Signal Based Surveillance (AVSS), Krakow, pp. 208–213 (2013)
17. Pitt, J., Perailslis, C., Michael, H.: Drones humans introduction to the special issue. Technol. Soc. Mag. **33**(2), 38–39 (2014)

18. Wilson, R.L.: Ethical issues with use of drone aircraft. In: The 2014 IEEE International Symposium on Ethics in Science, Technology and Engineering, Chicago, pp. 1–4 (2014)
19. Finn, R.L., Wright, D.: Unmanned aircraft systems: surveillance, ethics and privacy in civil applications. Comput. Law Secur. Rev. **28**(2), 184–194 (2012)
20. Villasenor, J.: Observations from above: unmanned aircraft systems and privacy. Harv. J. Law Public Policy **36**, 457 (2013)

On the Performance of Diversified Hybrid Carrier System Based on the Extended WFRFT

Ge Song, Xiaojie Fang, and Xuejun Sha(✉)

School of Electronics and Information Engineering,
Harbin Institute of Technology, Harbin 150000, China
shaxuejun@hit.edu.cn

Abstract. In this paper, an extended hybrid carrier system based on extended weighted fractional Fourier transform is proposed to improve the confidentiality of wireless communication. We illustrate the principles and put forward a fast algorithm to ensure the feasibility of the scheme. By extending the existing forms of WFRFT, the proposed scheme achieves a great improvement of the parameter dimensions and constructs diversified anti-interception signals, which significantly reduces the interception probability of eavesdroppers without consuming more physical layer resources or power. The validity of the EWFRFT scheme is verified by theoretical analysis and numerical simulations, and the performance of the system is evaluated in terms of the secrecy capacity, bit error rate, and computational complexity. Simulation results and analysis show that the proposed scheme remedies the defects of the existing hybrid carrier system and safeguards the physical layer security of the system effectively at the cost of a slight increase in computational complexity.

Keywords: WFRFT · Physical layer security · Computational complexity · Hybrid carrier

1 Introduction

With the rapid development of information technology, the demand for security of communication is increasing, and related issues have been widely concerned and studied [1]. The current research is dominated by the information layer encryption system with cryptography as the core, which is mainly based on the encryption and decryption algorithms deployed in the upper layer of the

Supported by the Natural Science Foundation of China under Grant 61901140, supported by China Postdoctoral Science Foundation Funded Project under Grant 2019M650067, supported by Science and Technology on Communication Networks Laboratory under Grant 6142104190203.

Q. Guo et al. (Eds.): WiSATS 2021, LNICST 410, pp. 568–582, 2022.
https://doi.org/10.1007/978-3-030-93398-2_51

protocol stack to meet the requirements of communication confidentiality and authentication. However, due to the insufficient consideration on the physical layer of wireless communication, the existing security schemes still have room for improvement [2,3]. The physical layer security technologies protect the communication signal and the device itself based on some essential characteristics of the physical layer in the wireless transmission of the signal, rather than pursuing the computational security that cryptography relies on [4,5]. Researchers have proposed artificial noise, transmit precoding, cooperative jamming, and the combination of multiple physical layer technologies to enhance the security capacity of the system and ensure the confidentiality of communication, which have achieved rich research result [6–8]. The development of physical layer security systems has made up for the lack of the underlying secrecy performance of traditional cryptography, which has the advantages of good compatibility and saving resources as well as challenges such as power consumption, hardware conditions, and engineering implementation.

In recent years, it is considered a worthy research direction that protecting physical layer waveforms through transform domain signal processing. In particular, the weighted fractional Fourier transform (WFRFT) has been paid more and more attention to the anti-detection and anti-interception of the signal due to its unique four-component superposition structure and beneficial changes to the signal characteristics. [9] introduces WFRFT into the communication system and reveals its physical significance. It points out that the hybrid carrier system based on WFRFT realizes the fusion of single carrier system and multi-carrier system. On this basis, a large number of studies demonstrate that because of the time-frequency backup structure of WFRFT, the hybrid carrier system shows robust anti-channel interference capability in different scenarios [10,11]. On the other hand, the feasibility and advantages of WFRFT in the field of communication security are also valued. By changing the constellation characteristics of the signal through the hybrid carrier modulation, the signal with Gaussian-like distribution can be constructed to realize the concealment of communication signals. Meanwhile, transform parameters can be designed to make the eavesdroppers misjudge the form of the transformation. In addition, the sensitivity of transform parameters can also be utilized to deteriorate the equivalent signal-to-noise ratio of the eavesdropper to reduce the probability of interception. Based on the four-component structure and good security features of WFRFT, a series of secure transmission schemes are proposed and analyzed to cope with the rapid development of targeted interception means by eavesdroppers [12,13]. However, some potential safety hazards have gradually emerged. As the security capability of WFRFT is constrained by the transform parameters and the transformation degree of freedom is limited, the signal is at risk of being cracked. With the extensive research of WFRFT theory, the eavesdroppers gradually master the same prior knowledge of the fraction domain as the partner, which makes the transformation in the form of a single parameter unable to effectively guarantee the security of the system, especially facing eavesdroppers with strong computing capacity.

To cope with the defects of the existing hybrid carrier system, on the one hand, researchers try to combine WFRFT with spread spectrum, antenna array, chaos interference, and other physical layer security technologies to improve the overall secrecy performance of the system [14–16]. On the other hand, there are also some studies on the extension of the transformation form. MP-WFRFT shows the possibility of parameter extension through the construction of coefficient parameter vectors [17]. Furthermore, GWFRFT is proposed in [18], which has more abundant multi-component energy distribution schemes. Compared with WFRFT, GWFRFT can improve the dimension partly due to the relaxation of the boundary conditions, but its research mainly focuses on combating channel fading through parameter design. The potential safety hazards of the existing HC system have not been effectively remedied. Therefore, it is of great research value that extending the WFRFT form to construct diversified anti-intercepted signal forms.

In this paper, we propose a new form of extended weighted fractional Fourier transform and illustrate its advantages in the anti-interception of the signal. The proposed transformation greatly improves the diversity and confidentiality of the signal through extension of the basic operator and the construction of the weighting coefficient vector without consuming more physical resources or power. This paper presents the mathematical principles of the EWFRFT scheme, and a fast implementation process is put forward to simplify the computational complexity of partner communication in ordet to ensure the feasibility of the scheme. Theoretical analysis and simulation verification demonstrates that the proposed scheme can improve the security of the system effectively.

The remainder of this paper is organized as follows: The basic theory of the WFRFT is presented in Sect. 2. The mechanism and implementation process of the proposed EWFRFT scheme are described in detail in Sects. 3, and the improvement of system performance is analyzed. In Sect. 4, the numerical simulations of the proposed scheme are provided. Finally, Sect. 5 concludes this paper.

2 Preliminary

According to [19], the classical weighted fractional Fourier transform can be defined as

$$F^{a}[x] = \sum_{l=0}^{3} \omega_l\left(\alpha\right) F^{l} x \tag{1}$$

where F is the normalized DFT matrix, whose elements in row m-th and column n-th satisfying as $[F]_{m,n} = \frac{1}{\sqrt{N}} \exp\left[\frac{-2\pi mn}{N} i\right]$. $\omega_l\left(\alpha\right), l = 0, 1, 2, 3$ is the weighted coefficient generated by the transform order α, and its calculation method is shown as follows

$$\omega_l\left(\alpha\right) = \frac{1}{4} \sum_{k=0}^{3} \exp\left[\frac{2\pi(\alpha - l)k}{4} i\right] \tag{2}$$

The WFRFT operator satisfies periodicity, unitarity, and additivity, as shown in (3)–(5).

$$F^{a+4}[x] = F^{a}[x] \tag{3}$$

$$F^{\alpha}\left[F^{-\alpha}[x]\right] = x \tag{4}$$

$$F^{a+\beta}[x] = F^{a}\left[F^{\beta}[x]\right] \tag{5}$$

MP-WFRFT is an extended form of WFRFT, whose weighting coefficients can be expressed as follows

$$\omega_p^{\alpha,m_k,n_k} = \frac{1}{4}\sum_{k=0}^{3} \exp\left\{\frac{2\pi j}{4}\left[(4m_k+1)\,\alpha\,(k+4n_k) - pk\right]\right\} \tag{6}$$

where $\{m_k, n_k\} \in Z^{+}, k = 0,1,2,3$. Further, [18] proposes the GWFRFT with relaxed boundary conditions, and its weighting coefficients can be expressed as

$$\begin{cases} \omega_0^{\theta} = \frac{1}{4}\left(e^{\theta_0 i} + e^{\theta_1 i} + e^{\theta_2 i} + e^{\theta_3 i}\right) \\ \omega_1^{\theta} = \frac{1}{4}\left(e^{\theta_0 i} - e^{\theta_1 i} - e^{\theta_2 i} + e^{\theta_3 i}\right) \\ \omega_2^{\theta} = \frac{1}{4}\left(e^{\theta_0 i} - e^{\theta_1 i} + e^{\theta_2 i} - e^{\theta_3 i}\right) \\ \omega_3^{\theta} = \frac{1}{4}\left(e^{\theta_0 i} + e^{\theta_1 i} - e^{\theta_2 i} - e^{\theta_3 i}\right) \end{cases} \tag{7}$$

where $\theta_k, k = 0,1,2,3$ are the transform parameters of period 2π. Due to $F^0 = \mathrm{I}, F^2 = \Pi$, where I is the unit matrix identity and Π is the shift matrix satisfying $[\Pi]_{m,n} = \delta\left(\langle m+n\rangle_N\right)$ where $\langle\cdot\rangle_N$ denotes modulo-N calculation, these WFRFT forms can be implemented by the FFT module and inversion module, and their computational complexity is slightly increased compared with FFT.

3 System Description

3.1 Extended Weighted Fractional Fourier Transform

Weighted fractional Fourier transform has the feature of dispersing the signal energy into four components in proportion. For eavesdroppers, since the specific dispersing method is unknown, the original signal can be recovered only by parameter scanning. On this basis, an extended weighted fractional Fourier transform (EWFRFT) is proposed. By extending the basic operator and constructing the weighting coefficient vector, the signal energy can be further distributed to more components, which greatly increases the diversity of the transformation and the difficulty of intercepting. The proposed transformation can make up for the defects of the existing hybrid carrier system and effectively improve the physical layer security of the communication system. The EWFRFT scheme can be expressed as follows

$$Y = \sum_{l=0}^{N-1} H_l T^l x \tag{8}$$

x is the original modulation signal, and T is the basic operator matrix with period N. To better realize the purpose of energy dispersion, elements in N are taken to satisfy $[T]_{m,n} = \delta(\langle n-m\rangle_N - 1)$, where $[\cdot]_{m,n}$ represents the m-th row and n-th column elements in the matrix. The diagonal matrix H_l is the weighting coefficient matrix of the transformation. The inverse transformation of with coefficient matrix ${H_l}^{-1}$ can be expressed as $\tilde{x} = \sum_{l=0}^{N-1} {H_l}^{-1} T^l Y = \sum_{\substack{l=0 \\ (l+p) \bmod N=0}}^{N-1} {H_l}^{-1} T^l H_p T^p x + \sum_{(l+p) \bmod N=1}^{N-1} \sum_{l=0}^{N-1} {H_l}^{-1} T^l H_p T^p x$. In order to satisfy the reliability of the partner communication, $\tilde{x} = x$ should be required. Since the basic operator matrix T satisfies $\left[T^l H_p T^p\right]_{m,n} = 0, m \neq n, (l+p) \bmod N = 0$, $\left[T^l H_p T^p\right]_{m,n} = 0, m = n, (l+p) \bmod N = q$, the transformation should satisfy the following relation.

$$\begin{cases} \sum\limits_{\substack{l=0 \\ (l+p) \bmod N=0}}^{N-1} {H_l}^{-1} T^l H_p T^p = I \\ \sum\limits_{\substack{l=0 \\ (l+p) \bmod N=q}}^{N-1} {H_l}^{-1} T^l H_p T^p = 0_{N\times N}, q = 1, \ldots, N-1 \end{cases} \tag{9}$$

A set of weighting coefficient is presented in (10), which can be obtained through the iterative method as shown in (11).

$$h_n^m = \prod_{k=0}^{\log_2 N - 1} h_{k,n}^m \tag{10}$$

where k is the iteration series, h_n^m is the element in H_l, which can be expressed as $[H_l]_{m,m} = h_n^m, (n-m) \bmod N = l$ or $[H_l]_{n,n} = h_n^m, (m-n) \bmod N = l$, and $h_{k,n}^m$ is the iteration coefficient of order k, which can be obtained as follows

$$h_{k,n}^m = \begin{cases} \frac{1}{2}\left(e^{\theta_{k,0}^{\left[\frac{z_k^m}{2^{k+1}}\right]} i} + e^{\theta_{k,1}^{\left[\frac{z_k^m}{2^{k+1}}\right]} i}\right), \left[\frac{z_k^m}{2^k}\right] = \left[\frac{n}{2^k}\right] \\ \frac{1}{2}\left(e^{\theta_{k,0}^{\left[\frac{z_k^m}{2^{k+1}}\right]} i} - e^{\theta_{k,1}^{\left[\frac{z_k^m}{2^{k+1}}\right]} i}\right), others \end{cases} \tag{11}$$

where $\theta_{k,0}^t, \theta_{k,1}^t, t = 0, 1, \ldots, \frac{N}{2^{k+1}} - 1$ are the transform parameters and $[\cdot]$ represents rounding down. When $k = \log_2 N - 1$, $z_k^m = m$, and when $k = 0, 1, \ldots, \log_2 N - 2$, z_k^m can be obtained by iteration of the following method.

$$z_k^m = \begin{cases} z_{k+1}^m, h_{k+1,n}^m = \frac{1}{2}\left(e^{\theta_{k+1,0}^t i} + e^{\theta_{k+1,1}^t i}\right), t = 0, 1, \ldots, \frac{N}{2^{k+1}} - 1 \\ 2^{k+3}\left[\frac{z_{k+1}^m}{2^{k+2}}\right] + 2^{k+2} - z_{k+1}^m - 1, others \end{cases} \tag{12}$$

In this way, we get the expression of EWFRFT, which is an extension of the existing weighted transformation. Through the design of the weighting coefficients, the proposed scheme greatly increases the parameter dimension of WFRFT without losing the excellent characteristics of the existing HC signals, which is suitable for the communication system and has the advantage of security. As for eavesdroppers, the probability of accurately judging the form of transformation will be greatly reduced due to the diversity of transformation. To achieve a high probability of signal interception by carrying out the correct inverse transformation, it is necessary to periodically scan all transform parameters on the basis of an accurate judgment of the transformation form, which has to paid unacceptable cost of computational complexity. Therefore, the security of the physical layer waveform is strongly guaranteed.

3.2 System Model

In this section, we propose the EWFRFT-based secure transmission scheme to enhance the confidentiality of communication. The system model is shown in Fig. 1. At the transmitter, EWFRFT is carried out for each code block of the signal. At the receiver, due to the shared transform parameters, the partners can perform a corresponding inverse transformation to recover the signal accurately. While for eavesdroppers, without loss of generality, it is assumed that the mechanism of weighted-type transformation is public, but the specific scheme is unknown. Besides, since we mainly concern the security of the system, the anti-fading process such as equalization is omitted in the block diagram.

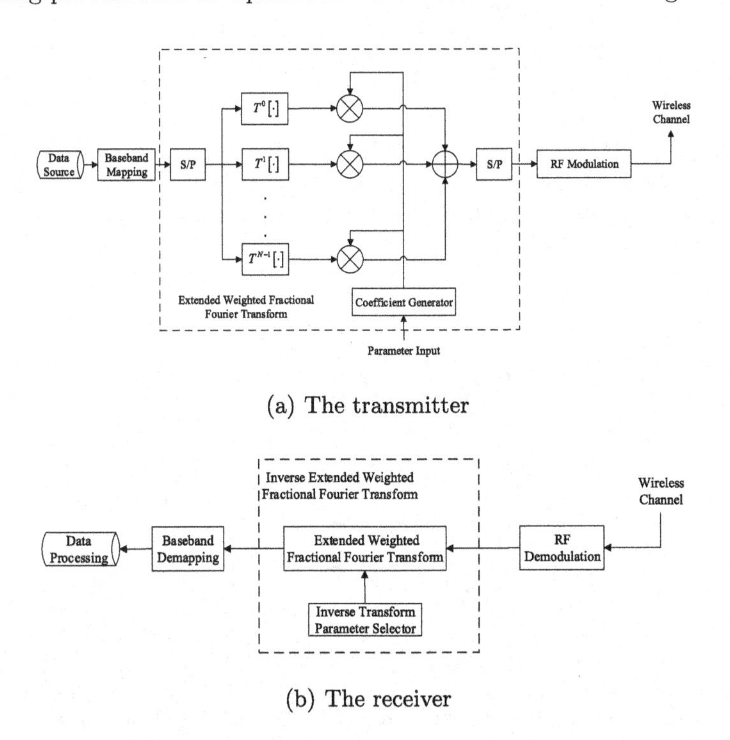

(a) The transmitter

(b) The receiver

Fig. 1. The framework of the proposed scheme.

As shown in Fig. 1 - (a), the proposed scheme generates the weighting coefficients according to the transform parameters and utilizes the parallel structure to carry out the weighted summation of the components processed by the periodic operator to obtain the EWFRFT signal. In this case, the computational complexity of EWFRFT can be expressed as $o\,(N \times N)$, which obviously has room for optimization. To decrease the cost of cooperative communication, we give a fast implementation scheme just like simplifying the calculation of DFT by FFT, which can ensure the feasibility of the proposed scheme. For briefness, it can be expressed in matrix form as follows

$$Y = \mathscr{F}^{\theta}\,[x] = T\,[F_k]\,x, k = 0, 1, \ldots, \log_2 N - 1 \tag{13}$$

where $\mathscr{F}^{\theta}$ represents the EWFRFT operator with parameter θ, $F_E = T\,[F_k]$ is the EWFRFT matrix, which carries out the multiplicative operation on F_k, that is $F_E = F_0 F_1 \ldots F_{N-1}$ or $F_E = F_{N-1} F_{N-2} \ldots F_0$. F_k is the block diagonal matrix of size $N \times N$, where the j-th subblock can be expressed as

$$[F_k]_j = \omega_0^{k,j} I_{2^{k+1}} + \omega_1^{k,j} \Pi_{2^{k+1}}, j = 0, 1, \ldots, \frac{N}{2^{k+1}} - 1 \tag{14}$$

I is the unit matrix identity and Π is the shift matrix, whose elements satisfying $[\Pi]_{m,n} = \delta\left(\langle n + m + 1 \rangle_N\right)$, where 2^{k+1} is the size of subblocks. $\omega_0^{k,j}, \omega_1^{k,j}$ is the weighting coefficients expressed as

$$\begin{cases} \omega_0^{k,j} = \frac{1}{2}\left(e^{\theta_{k,0}^{j} i} + e^{\theta_{k,1}^{j} i}\right) \\ \omega_1^{k,j} = \frac{1}{2}\left(e^{\theta_{k,0}^{j} i} - e^{\theta_{k,1}^{j} i}\right) \end{cases} \tag{15}$$

where $\theta_{k,0}^{j}, \theta_{k,1}^{j}$ is the $2N - 2$ transform parameters. At this point, the computational complexity of the system can be expressed as $o\,(2N\log_2 N)$. By contrast, the fast algorithm greatly reduces the cost of EWFRFT, which makes the proposed scheme feasible.

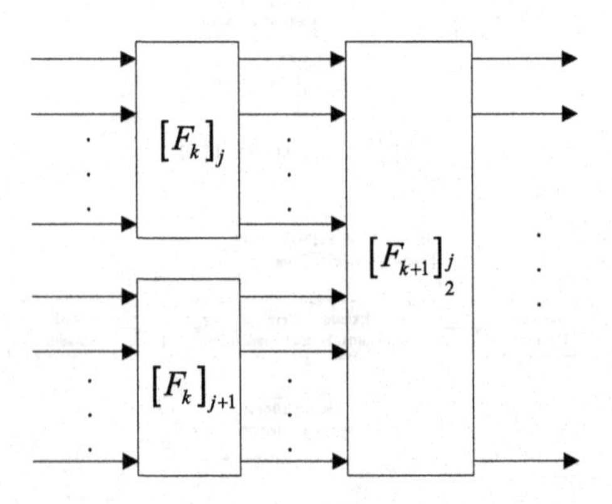

Fig. 2. Basic iterative structure of the EWFRFT fast algorithm.

Figure 2 shows a basic iterative structure schematic diagram of the simplified implementation process, which can be combined step by step to realize the EWFRFT. It is worth noting that due to the unitarity of the iteration function, the EWFRFT simplified algorithm can realize the undistorted recovery of the signal as long as the reverse transformation is carried out step by step in the opposite order to the forward one, thus there are diversified iteration methods. In addition, the iterative function is not unique, but can also be extended to the generalized weighted fractional Fourier transform to further increase the parameter dimension, while its computational complexity will also increase to a certain extent. It can be found that due to the diversity of basic operators, the proposed scheme has strong design flexibility, which is beneficial to resist the eavesdropping of unauthorized receivers. In practical application, it should be reasonably selected according to the system complexity limit and the security requirements to ensure the overall performance of the system.

Next, we describe the structure of the receiver, as shown in Fig. 1 - (b). The inverse transform module can also be implemented in a simplified multi-level iterative process. For the partners, since all parameters are shared, the original signal can be recovered by performing the corresponding inverse transformation shown as follows

$$Y = \mathscr{F}_{-1}^{\theta}[x] = T^{-1}\left[F_k^{-1}\right]x \tag{16}$$

where $T^{-1}[\cdot] = \left(T\left[(\cdot)^T\right]\right)^T$ is the multiplication in the order opposite to the operator $T[\cdot]$, and the k-th inverse transformation matrix F_k^{-1} is the block diagonal matrix, where the j-th subblock can be expressed as

$$\left[F_k^{-1}\right]_j = \frac{1}{2}\left[\left(e^{\left[\theta_{k,0}^j\right]^r i} + e^{\left[\theta_{k,1}^j\right]^r i}\right)I + \left(e^{\left[\theta_{k,0}^j\right]^r i} - e^{\left[\theta_{k,1}^j\right]^r i}\right)\Pi\right] \tag{17}$$

where the inverse transform parameters satisfying $\left[\theta_{k,0}^j\right]^r = -\theta_{k,0}^j$, $\left[\theta_{k,1}^j\right]^r = -\theta_{k,1}^j$. Through the inverse transformation, the original signal can be restored without reduction of the equivalent signal-to-noise ratio, which ensures the reliability of the communication of the partner. Besides, EWFRFT is compatible with the existing communication system and various physical layer security schemes. The proposed scheme has good universality in the field of anti-interception.

3.3 Analysis of Anti-interception Performance

In this section, we will analyze the anti-interception performance of the system in detail. For eavesdroppers, the existence and carrier scheme of the signal need to be accurately detected first. Similar to the HC signal, the EWFRFT has the advantage of changing the signal characteristics. From the point of signal detection, the signal with Gaussian-like distribution can effectively resist the detection. Due to the diversity of EWFRFT, rich transformation can be obtained to make the signal show Gaussian-like statistical property through parameter designs to realize the hiding or camouflage of the signal. Thus, the signal can not be accurately detected rely on prior knowledge by eavesdroppers, which can effectively make up for the defect caused by the simple change of signal characteristics with the parameter in the existing HC system.

Next, we focus on the receiving performance of eavesdroppers. To reassemble the original signal from the energy dispersed signal, it should accurately judge the signal processing means adopted by the transmitter and select the correct inverse transformation operator to realize the signal reconstruction. When the operator can not be exactly chosen, the eavesdropper will hardly obtain any useful information in high probability. In fact, due to the extension of the transformation form, EWFRFT has the extensibility and design flexibility of the basic operator, which will create obstacles for detecting the signal and ensure the physical layer security of the system. On this basis, the eavesdroppers not only need to master the concrete implementation process of the EWFRFT fast algorithm to realize the inverse transformation of the appropriate complexity but also need to determine all transform parameters. Otherwise, it will lead to a decline in receiving performance. We assume that the eavesdroppers adopt the same fast algorithm as the partners to process the signal, which can be expressed as

$$y = \mathscr{F}_{-1}^{\theta_e}\left[\mathscr{F}^{\theta}\left[x\right]\right] = T^{-1}\left[F_k^e\right]T\left[F_k\right]x \tag{18}$$

The inverse transform parameters of the eavesdroppers can be expressed as $\left[\theta_{k,0}^j\right]_e^r = -\theta_{k,0}^j + \sigma_k$, $\left[\theta_{k,1}^j\right]_e^r = -\theta_{k,1}^j$, since the weighting coefficient of transformation satisfies $\sum\limits_{\substack{0 \le u,v \le 1 \\ u+v=l}} \omega_u^{k,j}(\theta_p)\omega_v^{k,j}(\tau_p) = \omega_l^{k,j}(\theta_p + \tau_p)$, we can obtain $F_k^e F_k = diag\left(F_{k,0}^e F_{k,0}, F_{k,1}^e F_{k,1}, \ldots, F_{k,\frac{N}{2^{k+1}}-1}^e F_{k,\frac{N}{2^{k+1}}-1}\right) = F_k^{\sigma_k}$. In this case, the received signal can be expressed as

$$Y = T\left[F_k^{\sigma_k}\right]x = F^{\Delta}x \tag{19}$$

where $\Delta = \left[\sigma_0, \sigma_1, \ldots, \sigma_{\log_2 N-1}\right]$ is the error parameter vector, and $F_k^{\sigma_k}$ is the block diagonal matrix satisfying $\left[F_k^{\sigma_k}\right]_q = \frac{1}{2}\left[\left(e^{\sigma_k i} + 1\right)I + \left(e^{\sigma_k i} - 1\right)\Pi\right]$, $q = 0, 1, \ldots, \frac{N}{2^{k+1}} - 1$. It can be seen that due to the existence of parameter errors, the unitarity of EWFRFT is destroyed, which introduces additional signal distortion and decline the receiving performance. At this point, the received signal under the fading channel can be expressed as

$$Y_e = \left|g_e\right|^2 F^{\Delta}X + T^{-1}\left[F_k^e\right]g_e^* V^e \tag{20}$$

where g_e is the random channel gains and V^e is the AWGN noise vector with variance σ_e^2. It can be seen that the distortion of the received signal is jointly determined by channel parameters and the error of inverse transform parameters. With the increase of the error, the equivalent signal-to-noise ratio of the eavesdropper decreases rapidly, which harms the receiving reliability. Then, expected secrecy capacity of the proposed system can be expressed as

$$
\begin{aligned}
C_s &= \underset{g_p, g_e}{E}\left[\log_2(1+\eta) - \log_2(1+\eta_e)\right]^+ \\
&= \underset{g_p, g_e}{E}\left[\log_2\left(\frac{\sigma_e^2 |g_p|^2 P_s + \sigma_s^2 \sigma_e^2}{\sigma_s^2 |g_e|^2 P_s + \sigma_e^2 \sigma_s^2}\right)\right. \\
&\left. + \log_2\left(1 + \frac{P_s}{\sigma_e^2}|g_e|^2\left(1 - \frac{1}{N^3}\sum_{c=0}^{N-1}\left|\prod_{k=0}^{\log_2 N-1}\left(\exp\left(\theta_{k,0}^{\left[\frac{c}{2^{k+1}}\right]} i\right) + 1\right)\right|^2\right)\right)\right]^+
\end{aligned} \tag{21}
$$

where $E[\cdot]$ represents the expectation. $\eta = \frac{|g_p|^2 P_s}{\sigma_s^2}$ is the signal-to-noise ratio of the legitimate receiver. While the equivalent SNR of eavesdroppers can be expressed as $\eta_e = \frac{|H_0|^2 P_s}{\sum_{e=1}^{N-1} |H_e|^2 P_s + \sigma_e^2}$. Since $\frac{1}{4}\left|\sum_{s=0}^{1} \exp\left[\left(\theta_{k,s}^{\left[\frac{c}{2^{k+1}}\right]}\right) i\right]\right|^2 \leq 1$, there is $C_s \geq C_0 = \underset{g_p, g_e}{E}\left[\log_2\left(1+\frac{|g_p|^2 P_s}{\sigma_s^2}\right) - \log_2\left(1 + \frac{|g_e|^2 P_s}{\sigma_e^2}\right)\right]$. It can be seen that the introduction of EWFRFT provides an additional gain aside from the channel for the enhancement of the secrecy performance, and improve the expected secrecy capacity. On the other hand, the transform parameters with higher dimensions greatly improve the diversity of signals and guarantee the flexibility of the weighted transformation design, which is beneficial to the physical layer waveform anti-interception. The system is easy to obtain better security. For the eavesdroppers, to achieve a better reception effect, the secrecy capacity caused by EWFRFT should be reduced as far as possible to make it meet $C_s \approx C_0$. In this case, the error parameter vector should satisfy

$$
\left[\theta_{k,q}^{j}\right]_e - \left[\theta_{k,q}^{j}\right]_r = \sigma_k \leq \psi_k, j = 0, 1, \ldots, \tfrac{N}{2^{k+1}} - 1, q = 0, 1 \tag{22}
$$

where $\Psi = \left[\psi_0 \; \psi_1 \; \ldots \; \psi_{\log_2 N-1}\right]$ is the tolerance of parameter sensitivity, which should be reasonably selected according to communication requirements. To obtain the inverse transformation satisfying the sensitivity tolerance, the eavesdropper can only scan each transform parameter within the period 2π by traversal. In this way, the computational complexity of the eavesdropper can be expressed as

$$
o\left(\prod_{k=0}^{\log_2 N-1}\left(\frac{2\pi}{\psi_k}\right)^{\frac{N}{2^k}} 2N\log_2 N\right) \tag{23}
$$

Compared with the computational complexity of the eavesdropper in the WFRFT scheme expressed as $o\left[\frac{4}{\Delta_\alpha}(N\log_2 N + 4N)\right]$ and the GWFRFT scheme expressed as $o\left[\left(\frac{2\pi}{\Delta_G}\right)^4 (N\log_2 N + 4N)\right]$, it can be seen that the EWFRFT scheme greatly increases the cost of eavesdropping due to the extension of the transformation form, which reduces the probability of interception signals through parameter cracking, thus effectively makes up the drawbacks of the existing WFRFT schemes and ensures the anti-interception performance of the system. In this case, the partner computational complexity of the proposed scheme $o\,(2N\log_2 N)$ is slightly higher than that of the existing WFRFT scheme $o\,(N\log_2 N + 4N)$ and GWFRFT scheme $o\,(N\log_2 N + 4N)$, but still in the same magnitude, which is acceptable without adversely affecting the communication of partners. The EWFRFT scheme gains a great improvement of secrecy performance with a small increase in computational complexity.

4 Simulation Results and Discussion

In this section, the performance of the proposed EWFRFT-based extended hybrid carrier security system will be illustrated through numerical simulations. First, we verify the feasibility of the scheme, as shown in Fig. 3.

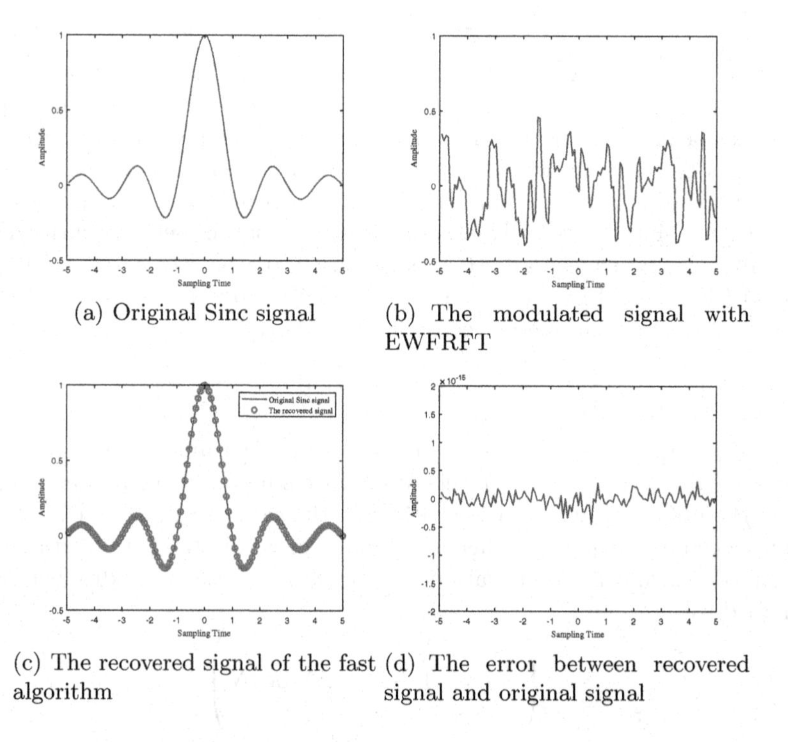

(a) Original Sinc signal (b) The modulated signal with EWFRFT

(c) The recovered signal of the fast algorithm (d) The error between recovered signal and original signal

Fig. 3. The feasibility of the EWFRFT scheme.

Sinc function is used as the original signal in the simulation as shown in Fig. 3 - (a). First, we perform EWFRFT modulation on the original Sinc signal as (8) with weighting coefficients generated according to (10), and the result is shown in Fig. 3 - (b). At this point, if the transformation can guarantee the unitarity, the original signal can be recovered by the corresponding inverse transformation. The fast algorithm as shown in (13) is adopted to demodulate the modulated signal, and the result is shown in Fig. 3 - (c). The error between the demodulation result and the original signal is given in Fig. 3 - (d). As can be seen from the simulation results, the undistorted recovery of the EWFRFT modulated signal is realized, which proves the unitarity of EWFRFT and the equivalence of the fast algorithm with the original transform, thus guaranteeing the feasibility of the proposed scheme in anti-interception.

On the premise of verifying its mathematical mechanism, we concern about the anti-detection feature of EWFRFT signals. The degree of Gaussian-like distribution of the signal can be measured by kurtosis $K_X = \frac{E[X-E(X)]^4}{E^2[X-E(X)]^2} - 3$. The simulation results show that there are rich parameter combinations of EWFRFT that can achieve $K_X \approx 0$ regardless of the modulation mode. That is, different from the WFRFT signal with the transform parameter α, due to the extension of parameter dimensions, the proposed scheme has design flexibility and can achieve diversified Gaussian-like distribution, which has advantages in signal hiding and anti-recognition as well as good universality.

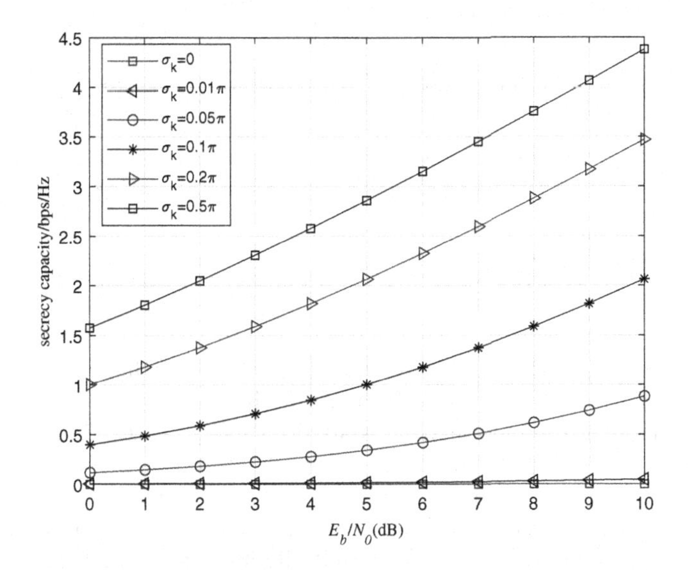

Fig. 4. Secrecy capacity of the proposed scheme.

Then, we analyze the expected secrecy capacity of the system, which is intuitively presented in Fig. 4. It can be seen that, when parameters cannot be correctly selected, the expected secrecy capacity shows a steep growth trend with the increase of SNR as well as that of parameters error, indicating that the system has achieved great secrecy performance. It should be noted that to clearly describe the influence of EWFRFT on the eavesdropper, we assume that the channel conditions of the cooperative receiver and the eavesdropper are the same in the simulation. That is, the expected secrecy capacity of the proposed scheme as shown in the Fig. 4 depends on the energy dispersion and aggregation process provided by EWFRFT, rather than the superiority or confidentiality of channel gain. On this basis, because of its good compatibility, it is no obstacle to combine the EWFRFT scheme with various traditional physical layer security technologies that can selectively degrade the eavesdropper's channel, which will obtain higher secrecy capacity obviously.

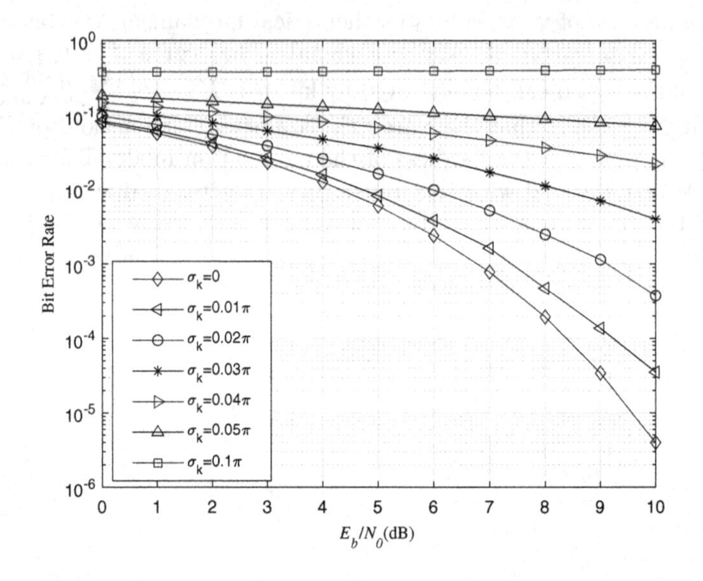

Fig. 5. BER performance of the proposed scheme.

Next, we illustrate the effectiveness of the proposed scheme through the bit error rate. Figure 5 shows the BER curve of the eavesdropper under different parameter errors. QPSK modulation is adopted. We pay attention to the receiving performance in the AWGN channel to show that the signal distortion caused by parameter mismatch reduces the equivalent SNR of the eavesdropper and worsens BER performance, which is the main reason for the improvement of the security. The simulation results show that the increase of σ_k will decrease the BER performance of the eavesdropper. More specifically, under the condition of $\mathrm{Eb/N0} = 10\,\mathrm{dB}$, the increase of bit error rate can be clearly observed when $\sigma_k = 0.01$, while BER increases by more than 4 magnitudes when $\sigma_k = 0.05$, which means that the eavesdropper can hardly obtain any useful information compared with the partner. The core advantage of the EWFRFT scheme is that the extension of the transformation greatly increases the dimension of the parameters as well as the cost of interception. According to the tolerance determined by the analysis results of sensitivity, the eavesdroppers have to scan all the transform parameters to obtain the signal. In the communication scenarios with real-time requirements, especially considering more uncertainty brought about by dynamic changes of parameters jointly, it will inevitably lead to the decline of receiving performance when the eavesdropper cannot pay such a high computational complexity, which effectively guarantees the confidentiality of communication.

In addition, it should be pointed out that in the previous analysis, we assumed that the eavesdroppers master the same knowledge as the partner, as well as accurately select the basic operator of transformation and the iterative scheme of the fast algorithm, but there might be errors in the judgment of parameters. In fact, as an extension of WFRFT, basic operators and parameter generation schemes of EWFRFT have great diversity. The eavesdropper cannot get correct inverse transformation results with the wrong operator even if all parameters are scanned in full cycle, which is the additional security brought by the diversity of EWFRFT. The secrecy performance is effectively improved compared to existing HC systems.

5 Conclusion

This paper focuses on the anti-interception of physical layer waveform in the wireless communication system. To remedy the potential safety hazards of the existing hybrid carrier system, we propose a new EWFRFT scheme and illustrate its superiority. In the proposed scheme, the anti-interception advantages brought by the diversity of EWFRFT signal are fully utilized, and greatly increase the difficulty and cost of intercepting signals for eavesdroppers without affecting the communication of the partner, which effectively guarantees the robust non-zero secrecy capacity. Theoretical analysis and numerical simulations show that, compared with the traditional hybrid carrier system, the proposed scheme can significantly improve the confidentiality of communication with a slight computational complexity increase. Future work will focus on combination with traditional physical layer security schemes to deal with the increasingly sophisticated interception methods of eavesdroppers.

References

1. Cao, J., et al.: A survey on security aspects for 3GPP 5G networks. IEEE Commun. Surv. Tutor. **22**(1), 170–195 (2020)
2. Liang, X., Zhang, K., Shen, X., Lin, X.: Security and privacy in mobile social networks: challenges and solutions. IEEE Wirel. Commun. **21**(1), 33–41 (2014)
3. Choo, K.R., Gritzalis, S., Park, J.H.: Cryptographic solutions for industrial internet-of-things: research challenges and opportunities. IEEE Trans. Ind. Inf. **14**(8), 3567–3569 (2018)
4. Zou, Y., Zhu, J., Wang, X., Hanzo, L.: A survey on wireless security: technical challenges, recent advances, and future trends. Proc. IEEE **104**(9), 1727–1765 (2016)
5. Wang, D., Bai, B., Zhao, W., Han, Z.: A survey of optimization approaches for wireless physical layer security. IEEE Commun. Surv. Tutor. **21**(2), 1878–1911 (2019). Second quarter
6. Jameel, F., Wyne, S., Kaddoum, G., Duong, T.Q.: A comprehensive survey on cooperative relaying and jamming strategies for physical layer security. IEEE Commun. Surv. Tutor. **21**(3), 2734–2771 (2019). thirdquarter

7. Li, L., Hu, Y., Zhang, H., Liang, W., Gao, A.: Deep learning based physical layer security of D2D underlay cellular network. China Commun. **17**(2), 93–106 (2020)
8. Zhang, Y., Shen, Y., Jiang, X., Kasahara, S.: Secure millimeter-wave ad hoc communications using physical layer security. IEEE Trans. Inf. Forensics Secur. 1 (2021)
9. Mei, L., Sha, X., Zhang, N.: The approach to carrier scheme convergence based on 4-weighted fractional Fourier transform. IEEE Commun. Lett. **14**(6), 503–505 (2010)
10. Hui, Y., Li, B., Tong, Z.: 4-weighted fractional Fourier transform over doubly selective channels and optimal order selecting algorithm. Electron. Lett. **51**(2), 177–179 (2015)
11. Wang, Z., Mei, L., Sha, X., Leung, V.C.M.: BER analysis of WFRFT precoded OFDM and GFDM waveforms with an integer time offset. IEEE Trans. Veh. Technol. **67**(10), 9097–9111 (2018)
12. Mei, L., Sha, X., Zhang, N.: Covert communication based on waveform overlay with weighted fractional Fourier transform signals. In: 2010 IEEE International Conference on Wireless Communications, Networking and Information Security, pp. 472–475. Beijing, China (2010)
13. Yuan, L., Xinyu, D., Jialiang, W., Ruiyang, X., Zhe, Z., Hujun, L.: WFRFT modulation recognition based on HOC and optimal order searching algorithm. J. Syst. Eng. Electron. **29**(3), 462–470 (2018)
14. Cheng, Q., Fusco, V., Zhu, J., Wang, S., Wang, F.: WFRFT-aided power-efficient multi-beam directional modulation schemes based on frequency diverse array. IEEE Trans. Wirel. Commun. **18**(11), 5211–5226 (2019)
15. Da, X., et al.: Embedding WFRFT signals Into TDCS for secure communications. IEEE Access **6**, 54938–54951 (2018)
16. Fang, X., Zhang, N., Zhang, S., Chen, D., Sha, X., Shen, X.: On physical layer security: weighted fractional Fourier transform based user cooperation. IEEE Trans. Wirel. Commun. **16**(8), 5498–5510 (2017)
17. Fang, X., Sha, X., Li, Y.: MP-WFRFT and constellation scrambling based physical layer security system. China Commun. **13**(2), 138–145 (2016)
18. Ma, C., Sha, X., Mei, L., Fang, X.: An equal component power-based generalized hybrid carrier system. IEEE Commun. Lett. **23**(2), 378–381 (2019)
19. Shih, C.C.: Fractionalization of Fourier transform. Opt. Commun. **118**, 495–498 (1995)

Space-Borne Passive Location Based on a Virtual Synthetic Aperture Mechanism

Tong Zhang, Qiang Yang, Xin Zhang, and Xiaochuan Wu(✉)

Harbin Institute of Technology, Harbin, China
wxc@hit.edu.cn

Abstract. The passive location method based on space-borne phase interferometer has to design more complex structures of antenna and system to solve the conflict between phase ambiguity and location accuracy. To address the problem, a space-borne passive location method based on virtual aperture is proposed in this paper. This method utilizes the moving characteristics of the space-borne platform to apply the principles of the synthetic aperture into the virtual aperture passive location system to form a new strategy of direction finding and location. We use the equivalent squint range model (ESRM) as the range model and the parameters of the model are estimated by Doppler parameter inversion and geometric approximation. Based on the range model, the received signal is equivalent to a linear frequency modulated signal in the azimuth direction, thus the azimuth location can be completed by matched filtering, while the range location can be completed by range searching. The 2-D coordinates of the radiation source can be obtained by range searching and azimuth focusing.

Keywords: Passive location · Virtual aperture · Range-azimuth location · Space-borne single platform

1 Introduction

Passive location technology has been widely concerned in the decades due to its advantages of low power consumption, long detection distance, good concealment, and strong anti-interference ability [1].

Space-borne passive location systems can be classified as single-satellite passive location systems and multi-satellite passive location systems. Multi-satellite passive location systems have high location accuracy, whose most widely used location methods are time difference of arrival (TDOA), frequency difference of arrival (FDOA), and the joint location method, which have high location accuracy [2–4]. However, it needs to consider issues such as time synchronization and so on, which result in high system complexity and cost. In contrast, although single-satellite passive location systems are not accurate enough, it is widely used because of its low cost and simple system setup.

Traditional single-satellite location systems typically use the direction-finding location method based on the phase interferometer [5] and the location method based on the Doppler frequency [6]. The main problem of the former is phase ambiguity and low location accuracy which have a Coupling Relationship. To improve positioning

Q. Guo et al. (Eds.): WiSATS 2021, LNICST 410, pp. 583–592, 2022.
https://doi.org/10.1007/978-3-030-93398-2_52

accuracy and solve the phase ambiguity problem at the same time, it is often necessary to design more complex systems and antenna structures [7, 8]; The latter mainly uses the change of Doppler frequency generated by the relative movement between the satellite and the target. This location method has lower complexity, but the measurement of Doppler and its change rate is based on the data of different time independently, without using the coherent integration of long-term data [9].

Synthetic aperture radar (SAR) compresses the Doppler signal which is a wideband chirp signal to improve the azimuth resolution. Utilizing the synthetic aperture to passive location, the article [10] proposed a passive location algorithm for radiation sources based on synthetic aperture. However, the feasibility of the method under space-borne conditions and the influence factors of the location error lacked analysis in the article.

On this basis, a space-borne passive location method based on the virtual aperture is proposed in this paper, which uses the equivalent squint range model and obtains the location of radiation source by range searching and azimuth focusing. The parameters of the range model are estimated by the geometric approximation and its Doppler parameters. Besides, the factors which affect the location error such as the length of the virtual aperture and the altitude are analyzed in this paper.

This paper is organized as follows. The signal model under the Equivalent Squint Range Model is presented in Sect. 2. The passive location method and the range model parameters estimation method are presented in Sect. 3. Azimuth resolution, range resolution, and the error caused by the range model are investigated in Sect. 4. Simulation results and the error curves are described in Sect. 5. Finally, the conclusion that the method is practical and the location error is decreasing with the longer synthetic aperture time, the lower orbital altitude and the larger SNR is drawn in Sect. 6.

2 Signal Model

The geometric model of the passive location issues based on the virtual aperture is shown in Fig. 1. While satellite is moving, Equivalent Squint Range Model (ESRM) is a range model for low-orbit space-borne synthetic aperture imaging radar, which is shown in Fig. 2:

$$R(t_a) = \sqrt{R_c^2 + V_r^2 t_a^2 - 2R_c V_r t_a \sin\theta_r} \tag{1}$$

where R_c denotes the range to the target at the center of the beam, V_r denotes the equivalent velocity of the satellite, θ_r denotes the equivalent squint angle, t_a denotes the azimuth time. X_0 in Fig. 2 denotes the azimuth location of the radiation source.

In the case of passive location, assuming that the transmitted signal is a single frequency pulse signal $s(t)$:

$$s(t) = rect(\frac{t}{T_p}) \exp(j2\pi f_c t + \varphi) \tag{2}$$

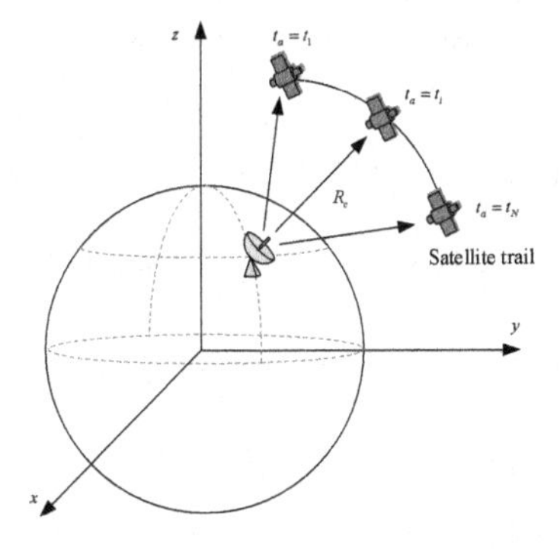

Fig. 1. The geometric model of the passive location

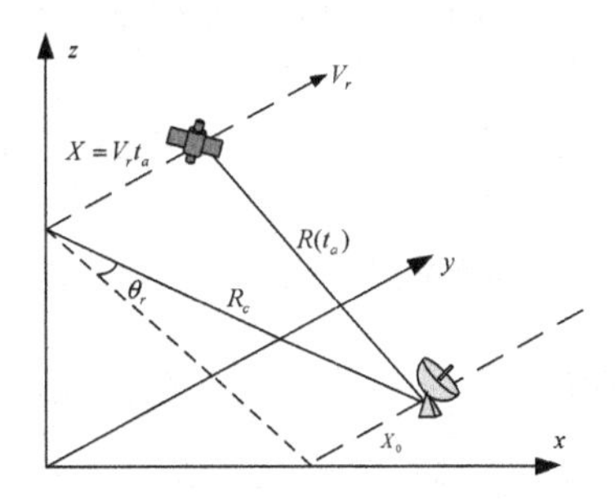

Fig. 2. Equivalent Squint Range Model (ESRM)

where $\text{rect}(u) = \begin{cases} 1 & |u| \leq 1/2 \\ 0 & |u| > 1/2 \end{cases}$, f_c is the carrier frequency, φ is the initial phase.

According to the "go-stop-go" model, the received signal can be divided into small pulses of which the pulse width is T_p and the period of each pulse is T. The data utilization mode is shown in Fig. 3. Then rearrange the data into two domains. t_r denotes the range time, $t_a = mT$ denotes the azimuth time, so the time $t = t_r + t_a$.

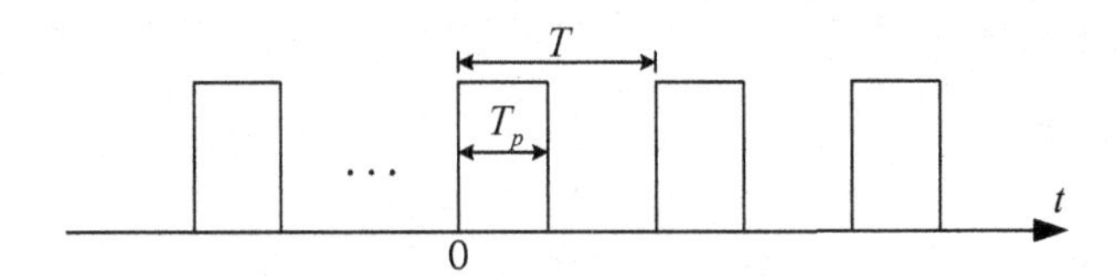

Fig. 3. The data utilization mode

The radiation signal received by at (t_r, t_a) can be represented as:

$$\begin{aligned} s_r(t_r, t_a) &= \text{rect}\left(\frac{t_r}{T_P}\right)\exp\left\{j\left[2\pi f_c\left(t_r + t_a - \frac{R(t_a)}{c}\right) + \varphi\right]\right\} \\ &= \text{rect}\left(\frac{t_r}{T_P}\right)\exp\{\text{j}(2\pi f_c t_r + \varphi)\}\exp\left\{j2\pi f_c\left(t_a - \frac{R(t_a)}{c}\right)\right\} \end{aligned} \tag{3}$$

where the second index term includes the azimuth and range information, so it can be used to locate.

3 Space-Borne Virtual Aperture Location Method

The Fourier expansion omitting cubic and higher-order terms of ESRM is:

$$R(t_a) \approx R_c - V_r \sin\theta_r t_a + \frac{V_r^2 \cos^2\theta_r}{2R_c} t_a^2 \tag{4}$$

Rewrite the received radiation signal (3) as:

$$s_r(t_r, t_a) = K\exp\{j2\pi f_c t_a\}\exp\left\{-j2\pi\frac{f_c}{c}\left(R_c - V_r \sin\theta_r t_a + \frac{V_r^2\cos^2\theta_r}{2R_c}t_a^2\right)\right\} \tag{5}$$

where $K = \text{rect}\left(\frac{t_r}{T_P}\right)\exp\{\text{j}(2\pi f_c t_r + \varphi)\}$.

The processed signal is a linear frequency modulated (LFM) signal in the slow time domain, the Doppler centroid f_{dc} and Doppler rate f_{1r} are:

$$f_{dc} = \frac{V_r \sin\theta_r}{\lambda} \tag{6}$$

$$f_{1r} = -\frac{V_r^2\cos^2\theta_r}{\lambda R_c} \tag{7}$$

According to the theory of SAR, the azimuth information can be obtained by matched filtering. The matched filter is:

$$h(t_a) = \exp\left(-j\pi\gamma t_a^2\right) \tag{8}$$

The chirp rate $\gamma = f_{1r}$, where the equivalent velocity of the satellite V_r and the equivalent squint angle θ_r need to be estimated. The traditional parameter estimation method is:

$$V_r \approx \sqrt{V_g V_s} \tag{9}$$

$$\theta_r \approx \frac{V_s}{V_g}\theta_{sq} \tag{10}$$

where V_g is the velocity at which a radar beam moves over the surface, V_s is the velocity of the satellite, θ_{sq} is the oblique view of the radar beam.

However, the above estimation method uses more geometric approximations, which will lead to larger phase errors. To make the model more accurate, we can use the Doppler parameters to estimate V_r and θ_r.

According to (6) and (7), the estimation of V_r and θ_r can be obtained as follows:

$$V_r = \sqrt{(\lambda f_{dc})^2 - \lambda R_c f_{1r}} \tag{11}$$

$$\theta_r = \arcsin(\frac{\lambda f_{dc}}{V_r}) \tag{12}$$

However, because of the passivity, the passive location can't get the range information by matched filtering in the range domain as the SAR, so R_c in (11) is unknown. So the estimation can be obtained via (9) and (12).

As for the location in the range domain, to get the range location R_c, we can use a set of azimuth function with different distances R_k:

$$h(t_a, R_k) = \exp(-j\pi\gamma_k t_a^2) \tag{13}$$

where $\gamma_k = -\frac{V_r^2\cos^2\theta_r}{\lambda R_k}$.

The corresponding matched filters can also be constructed directly by using the Principle of Stationary (PSP) in the Doppler domain, f_a denotes the Doppler frequency:

$$H(f_a, R_k) = \exp(j\pi\frac{f_a^2}{\gamma_k}) \tag{14}$$

The received radiation signal is as (5), multiply by $\exp\{-j2\pi f_c t_a\}$ and take the azimuthal Fourier transform of the signal, the Doppler spectrum is:

$$\begin{aligned} W_r(t_r, f_a) &= K' \exp\left\{j\pi\frac{f_a^2}{\gamma}\right\} \exp\left\{-j2\pi f_a \frac{R_c \sin\theta_r}{V_r\cos^2\theta_r}\right\} \\ &= K' \exp\left\{-j2\pi f_a \frac{R_c \sin\theta_r}{V_r\cos^2\theta_r}\right\} H(f_a, R_c) \end{aligned} \tag{15}$$

where $K' = K\exp\{-j2\pi\frac{f_c}{c}R_c\}\exp\left\{j\pi\frac{\sin^2\theta_r R_c}{\lambda\cos^2\theta_r}\right\}$ is a constant with respect to t_a.

Thus, through the set of matched filters as (16), the output is:

$$\begin{aligned} w(R_k, t_a) &= K'IFFT_{t_a}\{H(f_a, R_c)H(f_a, R_k)\} \otimes \delta\left(t_a - \frac{R_c \sin\theta_r}{V_r \cos^2\theta_r}\right) \\ &\approx K'IFFT_{t_a}\{H(f_a, R_c)H(f_a, R_k)\} \otimes \delta\left(t_a - \frac{X_0}{V_r}\right) \end{aligned} \quad (16)$$

When $R_k = R_c$, the processed signal will be in full focus:

$$w(R_k, t_a) \approx K'IFFT_{t_a}\left\{|H(f_a, R_k)|^2\right\} \otimes \delta\left(t_a - \frac{X_0}{V_r}\right) \quad (17)$$

The range location R_c can be obtained as (18). Thus the two-dimensional location information of the radiation source can be obtained.

$$R_c = R_k|_{\max(w(R_k, t_a)), t_a = \frac{X_0}{V_r}} \quad (18)$$

The flow of the virtual aperture passive location algorithm is shown in Fig. 4. First, Fourier transform is applied to the acquired signal in the slow time domain. Then estimate the parameters V_r and θ_r of ESRM by geometric approximation and Doppler spectrum. Construct matched filters at different distances, and pass the signal through the matched filters. Reorganize the results by distance and the range location R_c can be obtained by focusing. The radiation source is located at the highest energy.

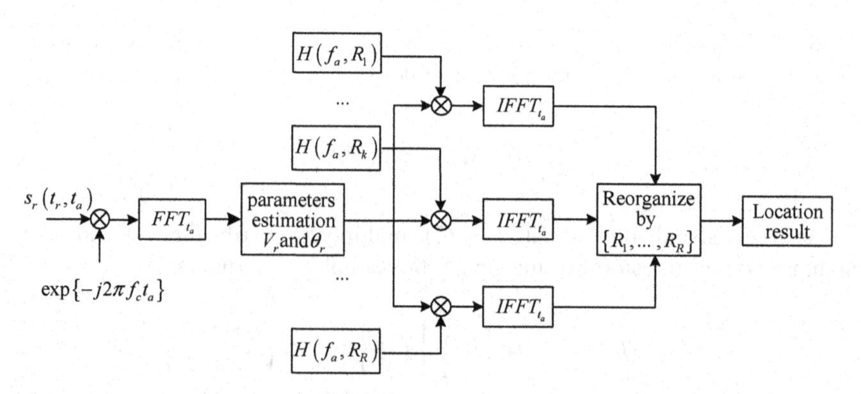

Fig. 4. The flow of the virtual aperture passive location algorithm

4 Resolution and Phase Error of the Range Model

4.1 Azimuth Resolution

The point dispersion function (PSF) determines the azimuth resolution:

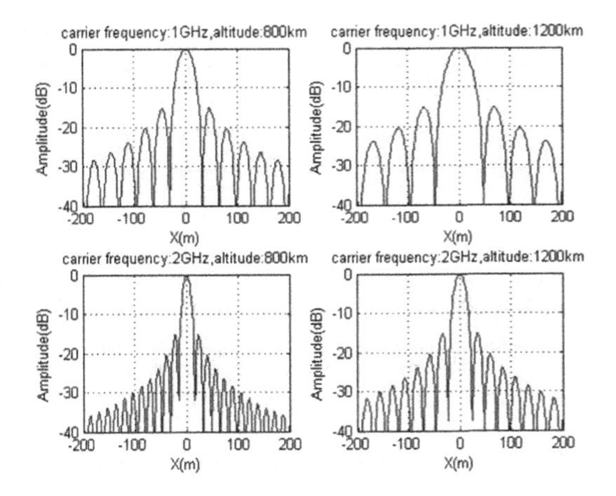

Fig. 5. The PSF in different ranges and frequency

$$psf(t_a, R_k) = IFFT_{t_a}\left\{|H(f_a, R_k)|^2\right\} \tag{19}$$

According to (19), in the case of the same number of samples in the azimuth domain, PSF depends on frequency and range. The graphs of PSF in different ranges and frequency are shown in Fig. 5. As the Fig. 5, it's easy to see that the higher frequency and the closer distance can make higher resolution.

4.2 Range Resolution

The location accuracy in the range domain depends on the distance interval of the set of matched filters. Therefore, to improve the positioning accuracy in the range domain, a step-by-step method can be adopted. First, a larger interval can be used for coarse location, and then a smaller interval around the approximate range can be used for precise location of the radiation source.

4.3 Phase Error of the Range Model

The difference between the range model and the true slant range will cause phase error. The phase error is:

$$\Delta\varphi = -\frac{2\pi}{\lambda}[R(t_a) - R_{true}(t_a)] \tag{20}$$

Analyze the phase error under different orbital altitudes and latitudes, the result is shown in Fig. 6. As Fig. 6, the longer the synthetic aperture time is, the greater the phase error is. When the synthetic aperture time is less than 1 s, the phase error is less than 0.25π.

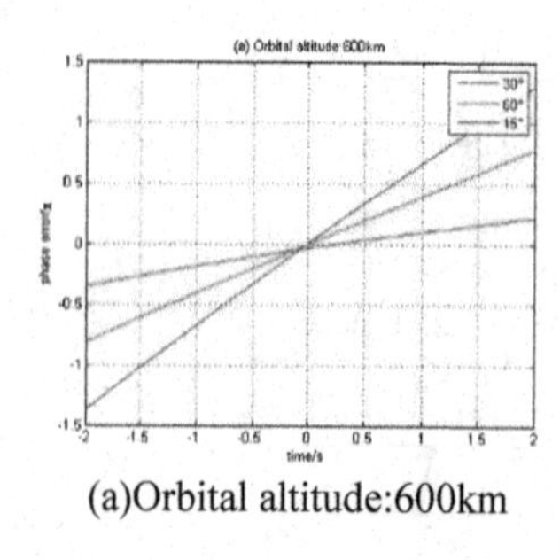

(a)Orbital altitude:600km

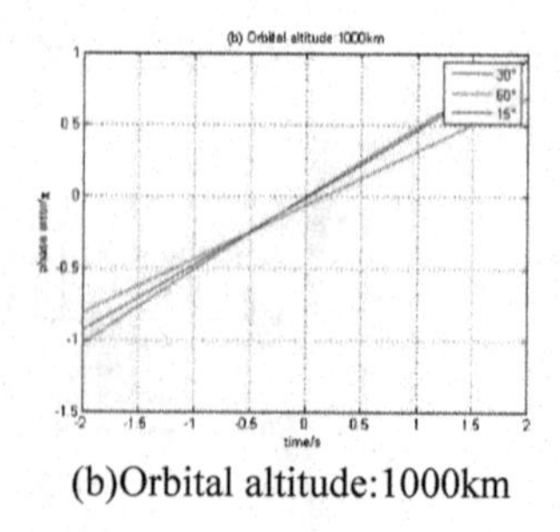

(b)Orbital altitude:1000km

Fig. 6. The phase error curves

5 Simulation

To prove the effectiveness of the algorithm, experiments are conducted. The signal parameters are shown in Table 1. The orbit parameters are shown in Table 2.

Table 1. The signal parameters

Parameters	Values
Carrier frequency	1 GHz
Sampling frequency	30 MHz
Pulse width	10 μs
Pulse repetition frequency (PRF)	5 kHz
Synthetic aperture time	1 s
SNR	0 dB

Table 2. The orbit parameters

Parameters	Values
Orbital altitude	1000 km
Eccentricity	0.0001
Inclination	98°
Argument of perigee	0°
Right ascension of ascending node	105°

The radiation source is located at $(114°\text{W}, 60°\text{N})$. The results are shown in Fig. 7. The received signal is a linear frequency modulation signal in the azimuth, so after azimuth Fourier transform the Doppler spectrum is with a certain width, which is in Fig. 7(a). The location result is shown in Fig. 7(b). By calculation, the error of the method is 1.5 km, which is more accurate than the direction-finding location method based on the phase interferometer.

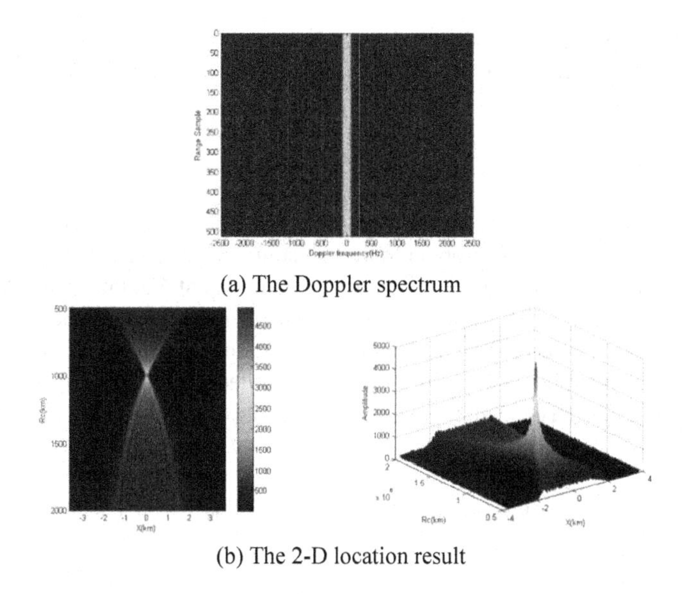

(a) The Doppler spectrum

(b) The 2-D location result

Fig. 7. The passive location result

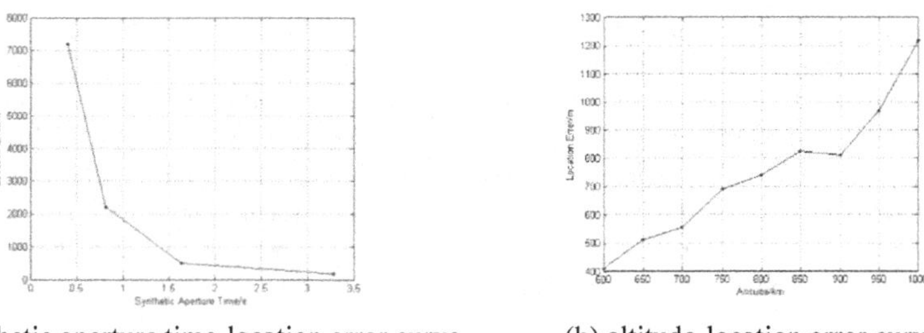

(a) synthetic aperture time-location error curve

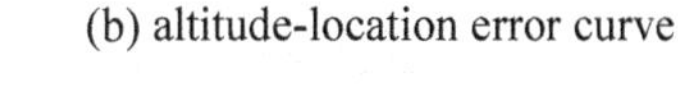

(b) altitude-location error curve

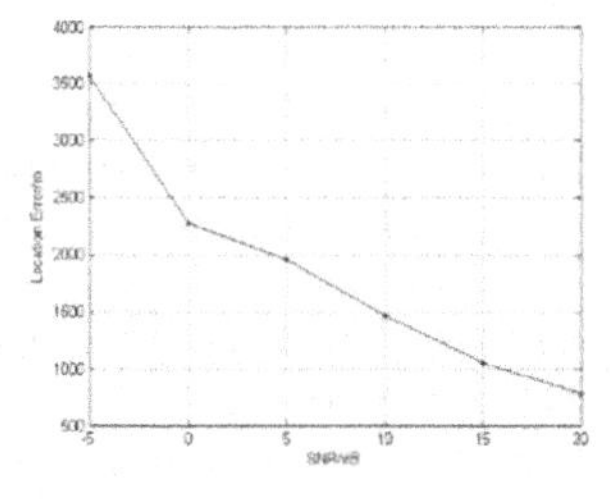

(c) SNR-location error curve

Fig. 8. The location error curves

Then simulate to discuss the factors influencing the accuracy of the location method. Considering the synthetic aperture time, orbital altitude and SNR, the error curves of location are shown in Fig. 8.

It can be seen from Fig. 8 that the location error is decreasing with the longer synthetic aperture time, the lower orbital altitude and the larger SNR.

6 Conclusion

This paper applies the virtual aperture to the space-borne passive location. The proposed method in this paper can obtain radiation source location by range searching and azimuth focusing. The signal model under the Equivalent Squint Range Model was deduced in this paper. The passive location method and the range model parameters estimation method were proposed. Azimuth resolution, range resolution and the error caused by the range model of the method were also deduced in this paper. Simulations were conducted to verify the effectiveness of the method and the factors influencing the location accuracy.

Acknowledgment. This work was supported by the Nation Scient Foundation under grant 62031014.

References

1. Huang, J.H., Barr, M.N., Garry, J.L., Smith, G.E.: Subarray processing for passive radar localization. In: 2017 IEEE Radar Conference, Seattle, WA, pp. 0248–0252 (2017)
2. CongFeng, L., Jinwei, Y., Juan, S.: Direct solution for fixed source location using well-posed TDOA and FDOA measurements. J. Syst. Eng. Electron. **31**, 666–673 (2020)
3. Liu, Z., Wang, R., Zhao, Y.: Computationally efficient TDOA and FDOA estimation algorithm in passive emitter localization. IET Radar Sonar Navig. **13**, 1731–1740 (2019)
4. Shu, F., Yang, S., Lu, J., Li, J.: On impact of Earth constraint on TDOA-based localization performance in passive multi-satellite localization systems. IEEE Syst. J. **12**, 3861–3864 (2018)
5. Dempster, A.G., Cetin, E.: Interference localization for satellite navigation systems. Proc. IEEE **104**, 1318–1326 (2016)
6. Zhu, Y., Zhang, S.: Passive location based on an accurate Doppler measurement by single satellite. In: 2017 IEEE Radar Conference, Seattle, WA, pp. 1424–1427 (2017)
7. Li, T., Guo, F., Jiang, W.: Multiple hypothesis NLS location algorithm based on ambiguous phase difference measured by a rotating interferometer. J. Electron. Inf. Technol. **34**, 956–962 (2012)
8. Kawase, S.: Radio interferometer for geosynchronous-satellite direction finding. IEEE Trans. Aerosp. Electron. Syst. **43**, 443–449 (2007)
9. Yuqi, W.: Passive localization algorithm for radiation source based on long synthetic aperture. J. Radars **9**, 185–194 (2020)
10. Wang, Y., Sun, G., Xiang, J., Xing, M., Guo, L., Yang, J.: A imaging passive localization method for wideband signal based on SAR. In: 2019 6th Asia-Pacific Conference on Synthetic Aperture Radar (APSAR), Xiamen, China, pp. 1–4 (2019)

A Cooperative Dictionary Learning and Semi-supervised Learning Framework for Sea Clutter Suppression of HFSWR

Xiaowei Ji, Qiang Yang, Xiaochuan Wu, and Xin Zhang(✉)

Harbin Institute of Technology, Nangang District, Harbin 150001, China
zhangxinhit@hit.edu.cn

Abstract. High-frequency surface-wave radar (HFSWR) has been applied in searching targets and maritime surveillance systems. However, the sea clutter is usually strong and harmful for detecting the targets. In this paper, we explore the sea clutter suppression problem for HFSWR and propose a novel sea clutter suppression method named a cooperative dictionary learning and semi-supervised learning sea clutter suppression framework (CDLSL). The semi-supervised learning can obtain abundant needed sea clutter data for the subsequent dictionary learning. The dictionary learning has ability to capture the features of sea echo and provides a desired clutter estimation. We have applied the proposed framework in the actual HFSWR data. Significant improvements in sea clutter suppression performance are achieved by the proposed method with respect to the state-of-the-art method.

Keywords: Dictionary learning · High-frequency surface-wave radar · Sea clutter suppression · Semi-supervised learning

1 Introduction

Because of the curvature of the earth, traditional microwave radar has the problem of line-of-sight and the detection blind spots. The targets over the horizon are hard to be detected by ground-based line-of-sight radar. High-frequency surface-wave radar (HFSWR) utilizes the surface-wave in the lower half of the high-frequency (HF) band to explore targets and then receives vessel and low-flying aircraft echoes over the horizon. With its own superior performance, the maximum surveillance range of HFSWR can cover Exclusive Economic Zone (EEZ) and also can monitor the ocean dynamic parameters over the horizon, such as the direction of winds and the height of the waves. In the meantime, HFSWR has the advantages of all-weather operability, and thus it is widely used in coastal warning, maritime rescue, marine resource development, and continuous monitoring for military [1–3].

Q. Guo et al. (Eds.): WiSATS 2021, LNICST 410, pp. 593–606, 2022.
https://doi.org/10.1007/978-3-030-93398-2_53

However, the strong clutter and interference severely limit the detection ability of HFSWR. As the most common clutter, sea clutter becomes a crucial problem for the accurate detection of marine targets [4, 5]. Sea clutter is produced from resonance effect between electromagnetic waves and ocean waves. Hence, it is of great importance to design suitable detectors or propose the effective sea clutter suppression methods for improving the detection ability in HFSWR.

2 Research Status Review

From 1970s, a surge of research into clutter suppression field with the development of the HFSWR all over the world. The sea clutter spectrum of shore-based HFSWR is mainly composed of first order and second order components. The first order spectrum is derived from Bragg resonance scattering and its spectrum can be represented as two discrete spectral lines. In particular, the first order Bragg peaks of sea clutter are generated by waves with specific wavelength. In target detection experiment, second-order sea clutter signal and atmospheric noise also can impact the detection effect [6].

At present, the suppression methods of sea clutter mainly include the sea clutter modeling method, cyclic iterative cancellation method [7], subspace estimation method [8, 9], and neural network method [10, 11]. The sea clutter modeling method can be roughly divided into three categories. The first one is the statistical model for describing the amplitude variation of sea clutter in time domain. The second one is the power spectrum distribution model for describing the distribution of sea clutter in the frequency domain. The last one is the sea clutter model based on the radar cross-sectional area scattering principle. It applies the sea surface electromagnetic scattering mechanism to establish a suitable interaction model between the electromagnetic wave and the scattering medium. Finally, building the radar sea surface scattering cross-sectional area equation to achieve the modeling of sea clutter [5].

In the sea clutter modeling method, Ward et al. first proposed a K-distribution model and utilized synthetic modulation to model the sea clutter of high-resolution radar [12]. In 1983, Sekine et al. proposed the Weibull distribution, and the probability density function (PDF) of Weibull distribution lies between Rayleigh and log-normal distributions [13]. In 1997, Farina et al. applied the Weibull distribution for modeling sea clutter [14]. In 2006, based on ice multi-parameter imaging X-band radar clutter data, Greco et al. proposed a generalized K-distribution model to estimate the clutter amplitude distribution. Immediately following this, some models are introduced to explicitly model sea spikes, such as the KA- and KK-distributions. In addition, Rosenberg et al. proposed the Pareto-distribution to model the longer tails in the presence of spikes [15].

In the cyclic iterative cancellation method, the sinusoidal signal is constituted by estimating parameters, and then the sinusoidal signal will be subtracted from the echo to achieve sea clutter suppression. The subspace estimation method can suppress sea clutter by clustering the property of subspace. However, the existing literatures based on subspace estimation suppression may cause the problem of target spectrum peak shift, which affects the target signal detection.

Nowadays, the machine learning has attracted much attention in machine vision and engineering applications. The machine learning approach employs abundant knowledge and information to analyze the characteristic of data and then investigates the techniques to provide a generalized treatment of the input samples and the discrimination algorithms to be applied. As a special kind of machine learning, the deep learning methods are conventionally designed for investigating the big data. In most cases, deep learning methods usually need a lot of training data and a large calculation cost, while the specific sea clutter data and interference data are limited in the experiment.

By analyzing the features of the target, clutter and the interference in Range Doppler (RD) spectrum images, Zhang et al. proposed a lightweight deep convolutional learning network based on a faster region-based convolutional neural networks (Faster R-CNN). By combining a classifier, this network first applies effective feature extraction to detect the clutter and the interference [16].

In actual environment experiment, the low signal-to-clutter ratios on the sea surface is a challenge for the researchers. Li et al. proposed a novel convolutional neural network based dual-activated clutter suppression algorithm. This framework firstly multiplies the activation weights of the last dense layer with the activation feature mapping of the upper sample layer. Then, the class activation maps (CAMs) obtained by last step are correspond to the sea clutter distribution region. By mapping the CAM inversely to the sea clutter spectrum, the framework can obtain the corresponding suppression coefficients [17].

As stated above, it is difficult to make accurate sea clutter estimation in the traditional clutter models, and the error of model estimation depends on the integrating degree of the current clutter and standard model. In the deep learning methods, initially, we still should consider the distribution of the data set, such as the subdivision of labels, multiple features and the distribution of the training and testing samples. Then, beginning the next step to extract the feature needed for suppressing the clutter.

3 Proposed Methodology

Usually, HFSWR echo data contains a large number of targets, sea clutter, ionospheric clutter and interference. Hence, the first task is asking us to accurately select the needed data, which contributes to the distinguish the clutter and the target in the experiment.

Inspired by machine learning, we think about sea clutter estimation and suppression from the perspective of semi-supervised learning. Specifically, we first select some sea clutter data (almost have no target) and the data of target coexists with sea clutter. Then, we label these two different data as labeled training samples set for subsequent classification algorithm. Of course, the training samples set also contains unlabeled samples. Next, we feed the raw radar echoes into the semi-supervised learning classification algorithm (enhanced M-training) and obtain the initial classification results. The sea clutter data and the data of target coexists with sea clutter can be automatically classified into two classes. Hence, we will obtain a large number of the sea clutter data and we only select the sea clutter data for the dictionary learning. With the help of dictionary learning and sparse representation, we will extract sea clutter components in radar echoes. In this section, a clutter suppression method that combines dictionary learning and semi-supervised learning is proposed. Through the cooperation of offline training and real-time processing, this method realizes the improvement of clutter suppression performance and accelerates the calculation speed.

3.1 STEP 1: Semi-supervised Classification – Enhanced M-training

As the classical semi-supervised classification method, Tri-training algorithm has been applied in many fields [18]. As the improvement of Tri-training algorithm, M-training algorithm has validated its effectiveness in semi-supervised classification, such as electronic nose learning technique and hyperspectral image classification. But the original M-training algorithm ignores the diversity of classifier types, which cannot achieve excellent classification results when the number of labeled samples is limited at the initial stage. In order to improve the performance of original M-training algorithm, the improved M-training algorithm with enhanced classifier diversity was proposed and obtained more promising results. Because complementing different kinds of classifiers for each other can avoid the classifier performance deterioration [19].

We set the enhanced M-training algorithm with four classifiers in two different classes, which contributes to increase the diversity of classifiers. In the experiment, we select two support vector machine (SVM) classifiers, two random forest classifiers (RF). SVM is very suitable for dealing with large input spaces and produce sparse solutions. SVM is a supervised nonparametric statistical learning technique, which doesn't rely on prior assumptions about the input samples distribution [20]. In the classification process, SVM is based on linear modeling of the classification boundary utilizing a least squares method.

Otherwise, RF has the advantages in dealing with large scale data set. When the scale of data set enlarges, the performance of RF does not appear over-fitting. Moreover, when the size of the data set is small, RF also has a strong generalization ability. When the data or some characteristic values are partially lost, RF has desired anti-noise ability and better tolerance [21].

Algorithm 2
Enhanced M-training algorithm

Input:

Initial training set: $L = \{(x_i, y_i)_{i=1}^{l}\}$

Initial unlabeled data set: $U = (x_j)_{j=1}^{u}$

Initial iteration times: $t = 0$

the number of iterations: T

While $t \leq T$:

Repeat:

1. Train classifiers C1, C2, C3, C4 by using initial training set L.
2. Choose a classifier as main classifier and others are assistant classifiers.
3. Use U as the test set. When an unlabled sample recieves the same classifiaction labels from one main classifier and three assistant classifiers, this unlabeled sample will be labled by classifiers and put it into new labeled data set $L_1(t)$, and $L_1(t) = \{x \mid x \in U, C_1(x) = C_2(x) = C_3(x) = C_4(x)\}$ (Similarly, when C2, C3, C4 as main classifier, respectively, we denote new labeled data set as $L_2(t)$, $L_3(t)$, $L_4(t)$, respectively.
4. If C1 as main classifier and $e_1(t)|L_1(t)| < e_1(t-1)|L_1(t-1)|$, update the labeled data set and unlabeled data set. $L_1(t) = L_1(t-1) \cup L_1(t)$
5. If C1 as main classifier and $e_1(t)|L_1(t)| \geq e_1(t-1)|L_1(t-1)|$, we randomly select samples as $L_1(t)$ to ensure $e_1(t)|L_1(t)| < e_1(t-1)|L_1(t-1)|$.
6. Update iteration times $t = t+1$.
7. Until $t > T$

Output: Trained classifiers

Input test samples into the trained classifiers, using majority vote strategy to obtain classification results.

In the experiment, the four classifiers (denoted as C1, C2, C3, and C4) are initially trained by the labeled samples. Each classifier has the equal probability to be the main classifier, in the meantime, the rest of classifiers as assistant classifiers. Then we will calculate the classification error rate of both the unlabeled samples and the labeled samples, respectively. The $e_i(U)$ means the error rate of unlabeled samples. The $e_i(t)$ represents the limitation of the classification error rate for assistant classifiers at t th iteration. Specifically, $e_i(U)$ and $e_i(t)$ can be estimated as:

$$e_i(U) = \frac{(n_i(u) - k_i)}{n_i(u)} \tag{1}$$

$$e_i(t) = \lambda * e_i(L) + (1 - \lambda) * e_i(U) \tag{2}$$

Where $n_i(u)$ represents the number of all the unlabeled samples, k_i represents the number of unlabeled samples with the same labels by the three assistant classifiers. λ is a weighting coefficient that tunes the tradeoff between $e_i(t)$ and $e_i(U)$. $(\lambda = 0, 0.1, 0.2 \ldots 0.9, 1)$ $L(t)$ represents the new labeled data set at t th iteration. When an unlabeled sample obtains the same class labels from all the classifiers at the same time, it can be assigned with temporary labels and begin to train classifiers from the next iteration. When the C1 as main classifier, and $e_1(t)|L_1(t)| < e_1(t-1)|L_1(t-1)|$, the original labeled data set is enlarged as $L_1(t) = L_1(t-1) \cup L_1(t)$. When $e_1(t)|L_1(t)| \geq e_1(t-1)|L_1(t-1)|$, the algorithm will randomly select samples from $L(t)$ and these selected samples will compose into $S(S = L_1(t))$. Then, this algorithm applies $L_1(t) = L_1(t-1) \cup L_1(t)$ to retrain classifier C1. Eventually, we apply the majority vote strategy to predict the class of a given unlabeled sample.

3.2 STEP 2: Dictionary Learning

In recent years, dictionary learning is an effective way to acquire more knowledge and information with less resources, which is applied in various practical cases, including visual tracking [23], face recognition [24], and classification [25]. dictionary learning (ODL) algorithm. In the current research, there has been a large amount of research on the dictionary learning. As one of the representatives for classical DL algorithms, at each iteration, K-means singular value decomposition (K-SVD) deals with the entire training set in batches. The classical DL algorithms can achieve promising performance but they require the excellent calculation ability. The essential point of dictionary learning algorithm depends on sparse representation, in which a given signal is represented as a combination of several atoms from an overcomplete dictionary [26]. The goal of the dictionary learning is to learn abundant knowledge and formulate a dictionary with less content from a mass of signal training data set. In this way, the time consumption and the cost of data collection can be greatly decreased. In the training data set, each column can contain either an original signal or features extracted from that signal.

When most elements are zero, the vector is sparse. Sparse coding is a kind of representation learning. In the sparse coding, the input data is represented by sparse vectors. By the sparse coding, the input data is broken up into a few basic elements named atoms, and a group of atoms constitutes a dictionary. The input data can be recreated with a linear combination of atoms. The sparse representation of the signal decides the approach of atoms combination for creating the signal. For each type of signal, the relevant dictionary should be established and deterministic and adaptive, such as DCT [22].

We donate $Y \in \mathbb{R}_{+}^{M\times P}$ as sea clutter data set, and each column of sea clutter data set is represented as $y_i \in \mathbb{R}_{+}^{M\times 1}, i = 1, ..., P$ $\boldsymbol{D} \triangleq [\boldsymbol{d}_1, ..., \boldsymbol{d}_k], \boldsymbol{D} \in \mathbb{R}_{+}^{M\times K}$ is donated as clutter dictionary matrix. $\boldsymbol{G} \triangleq [\boldsymbol{g}_1, ..., \boldsymbol{g}_K], \boldsymbol{G} \in \mathbb{R}_{+}^{K\times P}$ represents the sparse representation coefficient. As stated above, the dictionary learning based on sparse representation can be written as:

$$[\boldsymbol{D}, \boldsymbol{G}] = \min_{\boldsymbol{D},\boldsymbol{G} \geq 0} \frac{1}{2} \|\boldsymbol{Y} - \boldsymbol{D}\boldsymbol{G}\|_F^2 + \lambda \|\boldsymbol{G}\|_1 s.t. \|\boldsymbol{d}_i\|_F^2 \leq \alpha_i, \forall i = 1, \ldots, K \tag{3}$$

Where Frobenius norm $\|\cdot\|_{\mathrm{F}}$ is applied to measure the distance of matrix, L_1 norm is applied to represent the sparseness of $\boldsymbol{G}$. It should be emphasized that the sea clutter data set $\boldsymbol{Y}$ is denoised, but $\boldsymbol{Y}$ may contain noise or interference. Otherwise, restraining the sparseness of $\boldsymbol{G}$ and the norm of each column in dictionary also can enhance the stability of algorithm. λ represents the penalty factor and α_i is used to restrain the norm of each column in dictionary. The value of λ and α_i are greater than zero.

We apply the Block Successive Upper Bound Minimization (BSUM) to explore the optimal solution of $\mathcal{D} = \left\{\boldsymbol{D} | \boldsymbol{D} \geq 0, \|\boldsymbol{d}_i\|_F^2 \leq \alpha_i, \forall i\right\}$ and $\mathcal{G} = \{\boldsymbol{G} | \boldsymbol{G} \geq 0\}$. We can obtain the optimal value of $\boldsymbol{G}$ and $\boldsymbol{D}$ by iterative optimization. According to BSUM algorithm, the iteration of $\boldsymbol{G}$ and $\boldsymbol{D}$ is formulated as:

$$\boldsymbol{D}^{(q+1)} = \arg\min_{\boldsymbol{D} \in \mathcal{D}} \left\langle \nabla_{\boldsymbol{D}} (\|\boldsymbol{Y} - \boldsymbol{D}^{(q)} \boldsymbol{G}^{(q)}\|_F^2, \boldsymbol{D} \right\rangle + \frac{\epsilon_d^{(q)}}{2} \|\boldsymbol{D} - \boldsymbol{D}^{(q)}\|_F^2 \tag{4}$$

$$\boldsymbol{G}^{(q+1)} = \arg\min_{\boldsymbol{G} \in \mathcal{G}} \left\langle \nabla_{\boldsymbol{G}} (\|\boldsymbol{Y} - \boldsymbol{D}^{(q+1)} \boldsymbol{G}^{(q)}\|_F^2, \boldsymbol{G} \right\rangle + \frac{\epsilon_g^{(q)}}{2} \|\boldsymbol{G} - \boldsymbol{G}^{(q)}\|_F^2 + \lambda \|\boldsymbol{G}\|_1 \tag{5}$$

Where ∇_{D} represents the gradient along $\boldsymbol{D}$. $\boldsymbol{D}^{(q)}$ represents the solution of the $\boldsymbol{D}$ in q th iteration and $\boldsymbol{G}^{(q)}$ represents the solution of the $\boldsymbol{G}$ in q th iteration. $\langle \cdot, \cdot \rangle$ represents the inner product. $\mathcal{P}_{\mathcal{D}}(\cdot)$ is the projection operator to the convex set $\mathcal{D}$. Moreover, the ϵ_d and ϵ_g should be iterated as following:

$$\epsilon_d^{(q+1)} = \rho_{\max}^2 \left(\boldsymbol{G}^{(q+1)}\right) \tag{6}$$

$$\epsilon_g^{(q+1)} = \rho_{\max}^2 \left(\boldsymbol{D}^{(q+1)}\right) \tag{7}$$

3.3 Proposed Framework

In this section, we propose a cooperative dictionary learning and semi-supervised learning sea clutter suppression framework (CDLSL) to solve the sea clutter problem. Some earlier works have adequately demonstrated the effectiveness of dictionary learning approach. In essence, sparse representation applies an equipped dictionary, and dictionary learning further exploits the crucial information in dictionary. Another approach, which is named semi-supervised learning, can improve the performance by applying the unlabeled samples to increase the quantity of the training samples. To complement dictionary learning and semi-supervised learning for each other. The proposed framework applies enhanced M-training algorithm with different classifiers, sparse representation and dictionary learning to effectually detect the targets and sea clutter in HFSWR.

In our experiment, we will gather huge volumes of data in six days, and then pore over it. From the Fig. 1, HFSWR echo data usually contains a large number of targets, sea clutter, ionospheric clutter and other interference. Hence, the first task for us is that to accurately select the needed data, which contributes to the identify and distinguish the clutter and the target in the experiment. In this paper, we select a semi-supervised learning framework (enhanced M-training algorithm) to select the needed data in our experiment. In enhanced M-training algorithm, we select the sea clutter data (almost have no target) and the data of target coexists with sea clutter to train the four classifiers from A-R-D (Angle-Range-Doppler) data.

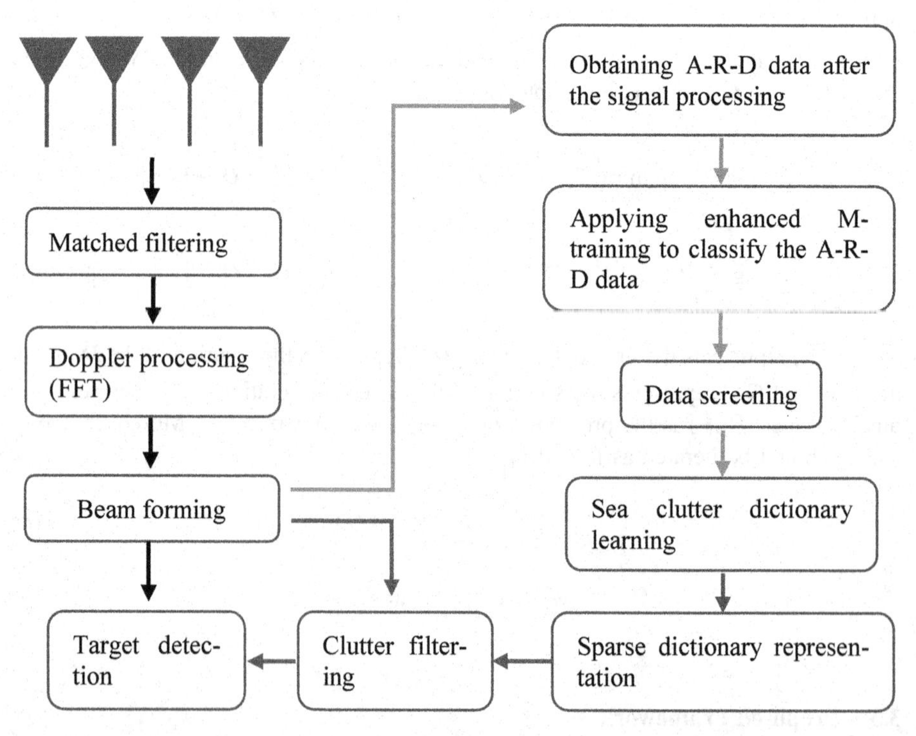

Fig. 1. Flowchart of the cooperative dictionary learning and semi-supervised learning sea clutter suppression framework (CDLSL)

We randomly divide the entire available data into two parts for each class: 50% data are used for training and 50% data are used for testing. And in the training data, we randomly select 60% samples in each class as the initial labeled data, and the remaining samples are used as the unlabeled data. Hence, after training the classifiers, we put the original radar data into the enhanced M-training algorithm. And we obtain the sea clutter data set (almost have no target). We begin to train the sea clutter frequency spectrum dictionary. Then, we will obtain the sea clutter frequency spectrum dictionary and conduct sea clutter dictionary. The above-mentioned steps are conducted in quasi real time mode and we represent the quasi real time steps in green arrows. In the traditional radar signal processing, the echo signals are received by receiving array, then echo signals are matched filtering and conducted by Fast Fourier Transform. After beam forming, the echo signals are directly detected by target detectors. The black arrows describe the above-mentioned procedures. Different from the traditional radar signal procedure, after the beam forming, CDLSL algorithm makes use of dictionary to perform sparse representation of echo and clutter suppression, and then carries out target detection on the echo signal with clutter filtering. The above-mentioned process is represented by the blue line in the Fig. 1.

4 Experimental Data Processing and Experiment Results

4.1 Experimental Data Processing

In Weihai, Shandong province, China, we have collected a large number of the radar echoes from HFSWR system. The receiver of HFSWR constantly collected echo data from August 26, 2020 to August 31, 2020. In the meanwhile, the carrier frequency (f_c) is shifty and f_c = 4–8 MHz. For just a week, the weather is always cloudy, and sometimes be sunny or light rainy. Moreover, aiming at verifying the effectiveness of the proposed framework, we choose the Sparse Dictionary Represented Optimal Filter (SDROF) algorithm as compared methods. By combining the dictionary learning and clutter filter, SDROF effectively suppresses the sea clutter for High-frequency (HF) sky-wave over-the-horizon radar (OTHR). Hence, we applied this outstanding approach in the HFSWR system as compared method.

4.2 Experiment Results

In this experiment, we select 640 samples to train the enhanced M-training and obtain the overall classification accuracy between the sea clutter data (almost have no target) and the data of target coexists with sea clutter is 77.57%. The classification results are shown in Fig. 2.

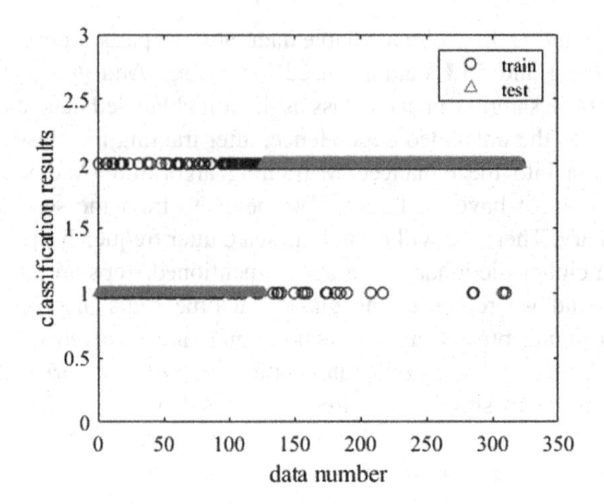

Fig. 2. The classification results of enhanced M-training.

We should emphasize that the sea clutter data is affected by the experimental weather. Hence, when we apply CDLSL framework to suppress the sea clutter, we should utilize a little earlier sea clutter data to train the dictionary. Because the ever-changing weather impacts on the physical character of sea clutter and ionospheric clutter. If we utilize the sea clutter data from a long time ago to train the dictionary, we cannot adequately mine the representative and discriminative information of current the sea clutter. Capturing the underlying patterns of near-term sea clutter is a crucial step for obtaining the adequate detection results.

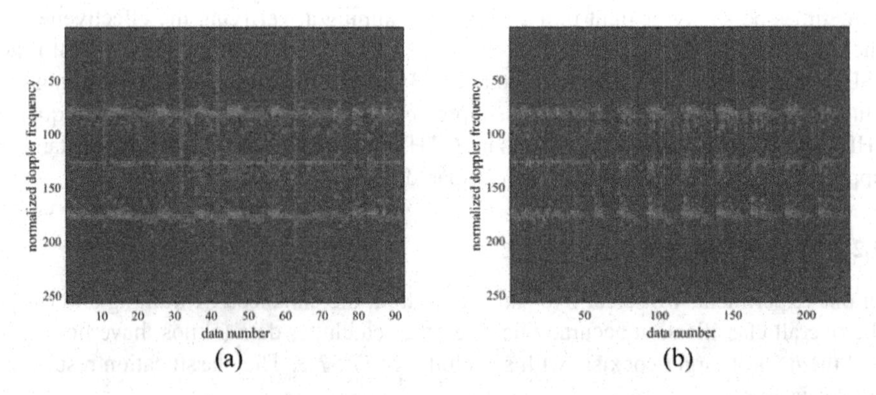

Fig. 3. The classification results of enhanced M-training algorithm (a) the sea clutter data (almost have no target); (b) sea clutter and target data

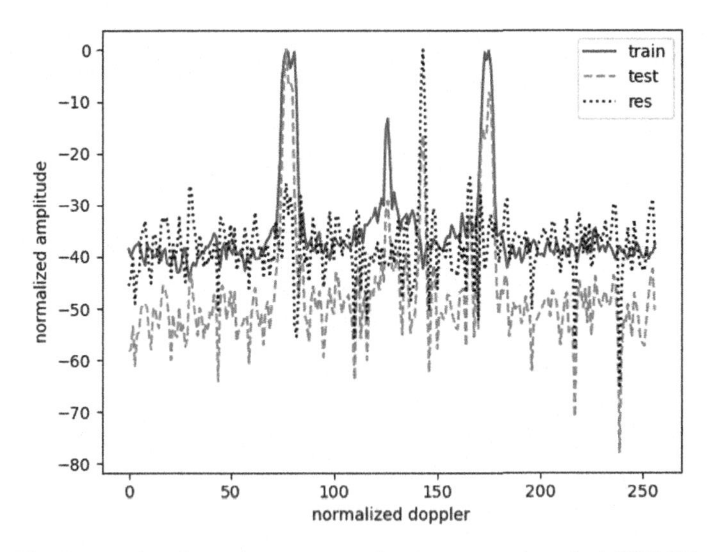

Fig. 4. Results of sea clutter suppression by proposed method (CDLSL)

Compared with dictionary learning, the enhanced M-training algorithm does not need the real-time data. And the classification results of the enhanced M-training algorithm are shown in Fig. 3. We can reserve the sea clutter data (almost have no target) for the subsequent dictionary learning. From the results map of CDLSL (Fig. 4), we can observe that the proposed framework accurately expresses the characteristics of sea clutter. Specifically, the blue curve represents the sea clutter obtained by dictionary learning. The orange curve represents current received sea clutter and will be learned the features. The black curve represents the detection target. As you can observe from the Fig. 4, the blue curve almost overlaps the orange curve which confirms that CDLSL can precisely capture the characteristics of sea clutter and construct optimal sea clutter that are estimated from the dictionary learning. The sea clutter obtained by dictionary learning is more similar to the current real sea clutter echo, the more effective the proposed method is.

We apply the same sea clutter data set to compare the performance between CDLSL and Dictionary Represented Optimal Filter (SDROF) in the HFSWR. Obviously, our proposed framework has better performance by comparing the experiment results in Fig. 4 and Fig. 5. The result obtained by SDROF is shown in Fig. 5 and it indicates that the SDROF doesn't accurately estimate the peculiarity of sea clutter in HFSWR system. The main reason may be that the sky-wave OTHR system can receive a lot of clutter echo when it is started up for daily operation, but most clutter data contains targets and interference. Irrelevant training samples will seriously affect the performance of the model, thus influences the suppression effect of algorithm. And SDROF applies fewer relevant samples, which cannot mine rich hidden sea clutter information. It also demonstrates that the real-time training data set is very important to

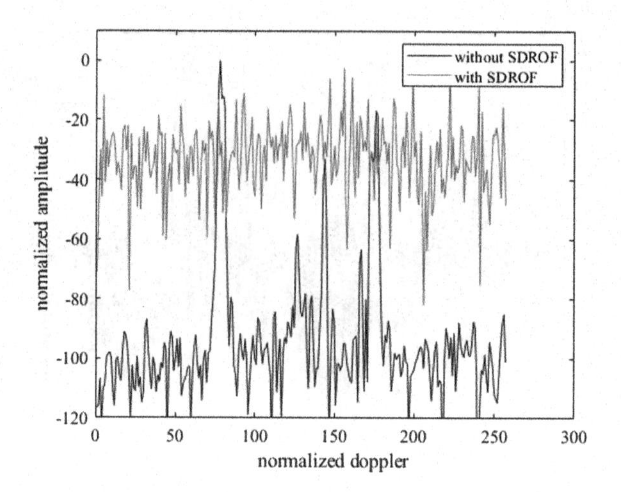

Fig. 5. Result of sea clutter suppression by SDROF

improve the suppression performance. Because our proposed framework can guarantee the accurate and real-time of the training data set for dictionary learning. Different from SDROF, by enhanced M-training algorithm, we can obtain the adequate and required sea clutter train samples for learning the physical characteristics of near-term sea clutter data. Therefore, the dictionary learning can adequately capture different sea clutter in different weather.

5 Conclusion

In this paper, we present a sea clutter suppression method for HFSWR. The proposed algorithm named a cooperative dictionary learning and semi-supervised learning sea clutter suppression framework (CDLSL). Applying enhanced M-training algorithm with different classifiers to obtain vast amounts of sea clutter data that almost contains no target. These pure sea clutter data is applied for the sparse representation and dictionary learning to capture the characteristics of sea clutter. Moreover, with the cooperation of online real time and quasi real time, the proposed framework can accurately detect the targets among the sea clutter. To further evaluate the performance of the proposed framework, we also conduct the previous research named Sparse Dictionary Represented Optimal Filter (SDROF) on the HFSWR in the same experiment setting. The experiment results effectually prove the superiority of our proposed framework (CDLSL) in HFSWR.

Acknowledgments. This work was supported in part by Hainan Province Key Research and Development Project under Grant ZDYF2019195, in part by the National Natural Science Foundation of China under Grant 62031014.

References

1. Zhang, J., Zhang, X., Deng, W., et al.: A geometric barycenter-based clutter suppression method for ship detection in HF mixed-mode surface wave radar. Remote Sens. **11**(9), 1–16 (2019)
2. Li, C., Wu, X., Yue, X., et al.: Extraction of wind direction spreading factor from broad-beam high-frequency surface wave radar data. IEEE Trans. Geosci. Remote Sens. **55**(9), 5123–5133 (2017)
3. Fujii, S., Heron, M.L., Kim, K., et al.: An overview of developments and applications of oceanographic radar networks in Asia and Oceania countries. Ocean Sci. J. **48**(1), 69–97 (2013)
4. Zhao, J., Wen, B., Tian, Y., Tian, Z., Wang, S.: Sea clutter suppression for shipborne HF radar using cross-loop/monopole array. IEEE Geosci. Remote Sens. Lett. **16**(6), 879–883 (2019)
5. Yang, Y., Xiao, S.P., Wang, X.S.: Radar detection of small target in sea clutter using orthogonal projection. IEEE Geosci. Remote Sens. Lett. **16**(3), 328–368 (2018)
6. Ji, Z., Yi, C., Xie, J., Li, Y.: Identification of the scale-free interval in HF radar sea clutter correlation dimension calculation. In: 2012 17th International Symposium on Antennas and Propagation, Nagoya, Japan, pp. 596–599 (2012)
7. Wen, B.Y., Lei, Z.Y., Cheng, F., Ma, Z.G.: Preliminary research on target detection in first-order peaks with adaptive cancellation. In: Proceedings of the 2004 3rd International Conference on Computational Electromagnetics and its Applications, Beijing, China, pp. 512–515 (2004)
8. Chen, Z., He, C., Zhao, C., Xie, F.: Using SVD-FRFT filtering to suppress first-order sea clutter in HFSWR. IEEE Geosci. Remote Sens. Lett. **14**(7), 1076–1080 (2017)
9. Xue, W.H., Mao, L., Liu, T.: A sea clutter suppression algorithm by statistics estimation and subspace analysis. In: Proceedings of the 2015 International Industrial Informatics and Computer Engineering Conference, Xian, China, pp. 1469–1474 (2015)
10. Gao, Z.Q., Chen, L.: Sea clutter sequences regression prediction based on PSO-GRNN method. In: Proceedings of the 2015 8th IEEE International Symposium on Computational Intelligence and Design (ISCID), Hangzhou, China, pp. 72–75 (2015)
11. Vicen-Bueno, R., Rosa-Zurera, M., Jarabo-Amores, M.P., de la Mata-Moya, D.: Coherent detection of swerling 0 targets in sea-ice weibull-distributed clutter using neural networks. IEEE Trans. Instr. Meas. **59**(12), 3139–3151 (2010). https://doi.org/10.1109/TIM.2010.2047579
12. Ward, K.D.: Compound representation of high resolution sea clutter. Electr lett **17**(16), 561–563 (1981). https://doi.org/10.1049/el:19810394
13. Sekine M, Musha T, Tomita Y, et al.: Weibull-distributed sea clutter. IEE Proc. F-Commun. Radar Signal Process. **130**(5), 476 (1983). https://doi.org/10.10491/ip-f-1.1983.0076
14. Farina, A., Gini, F., Greco, M.V., et al.: High resolution sea clutter data; statistical analysis of recorded live data. IEE Proc Radar Sonar Navig. **144**(3), 121–130 (1997). https://doi.org/10.1049/ip-rsn:19971107
15. Rosenberg, L., Watts, S., Greco, M.S.: Modeling the statistics of microwave radar sea clutter. IEEE Aerosp. Electron. Syst. Mag. **34**(10), 44–75 (2019). https://doi.org/10.1109/maes.2019.2901562
16. Zhang, L., You, W., Wu, Q.M.J., Qi, S.B., Ji, Y.G.: Deep learning-based automatic clutter/interference detection for HFSWR. Remote Sens. **10**(10), 1517 (2018). https://doi.org/10.3390/rs10101517

17. Li, G., Song, Z., Fu, Q.: A convolutional neural network based approach to sea clutter suppression for small boat detection. Front. Inform. Technol. Electron. Eng. **21**, 1504–1520 (2020). https://doi.org/10.1631/FITEE.1900523
18. Cui, Y., Song, G., Wang, X., et al.: Semisupervised classification of hyperspectral images based on tri-training algorithm with enhanced diversity. J. Appl. Remote Sens. **11**(4), 1 (2017)
19. Ying, C., Xueting, W., Zhongjun, L., Liguo, W.: Hyperspectral image classification based on improved M-training algorithm. J. Harbin Eng. Univ. **39**(10), 104–110 (2018). (in Chinese)
20. Zhao, D., Kang, W., Liu, G.: Long-term object tracking method based on dimensionality reduction. In: Jia, M., Guo, Q., Meng, W. (eds.) WiSATS 2019. LNICSSITE, vol. 280, pp. 529–536. Springer, Cham (2019). https://doi.org/10.1007/978-3-030-19153-5_54
21. Yao, M.H.: Random forests and its application to the classification of remote sensing image. MA thesis, Huaqiao University (2014)
22. Qin, Y., Zou, J., Tang, B., Wang, Y., Chen, H.: Transient feature extraction by the improved orthogonal matching pursuit and K-SVD algorithm with adaptive transient dictionary. IEEE Trans. Ind. Inform. **16**(1), 215–227 (2020)
23. Jia, M., Gao, Z., Guo, Q., Lin, Y., Gu, X.: Sparse feature learning for correlation filter tracking toward 5g-enabled tactile internet. Trans. Ind. Inform. **16**(3), 1904–1913 (2020)
24. Chen, Z., Wu, X., Yin, H., Kittler, J.: Noise-robust dictionary learning with slack block-diagonal structure for face recognition. Pattern Recognit. **100** (2020). https://doi.org/10.1016/j.patcog.2019.107118. Early Access
25. Yang, B.-Q., Guan, X.-P., Zhu, J.-W., Gu, C.-C., Wu, K.-J., Xu, J.-J.: SVMs multi-class loss feedback based discriminative dictionary learning for image classification. Pattern Recogn. **112**, 107690 (2021). https://doi.org/10.1016/j.patcog.2020.107690
26. Wang, Z., Shi, S., He, Z., Sun, G., Cao, J.: An ocean clutter suppression method for OTHR by combining optimal filter and dictionary learning. In: 2018 IEEE Radar Conference (RadarConf18), Oklahoma City, USA, pp. 1499–1503 (2018)

Improving Positioning Accuracy Using WLAN Optimization for Location Based Services and Cognitive Radio Networks

Sohaib Bin Altaf Khattak[1], Min Jia[1,2](✉), Qing Guo[1], and Xuemai Gu[1]

[1] Communication Research Center, School of Electronics and Information Engineering, Harbin Institute of Technology, Harbin, China
{sohaib,jiamin,qguo,guxuemaii}@hit.edu.cn

[2] CETC Key Laboratory of Aerospace Information Applications, Shijiazhuang, China

Abstract. Positioning information not only benefits the localization systems but also improve the performance of geo-location based cognitive radio (CR) networks. Most researchers focus on other aspects of CR and databases but rarely discuss the fact of how the positioning information can influence the performance of CR systems in indoor environments. WLAN is the most common technology used for indoor positioning. Optimization of WLAN access points (APs) can enhance accuracy of the localization systems. In this paper, we present an optimization algorithm for WLAN localization system. The proposed scheme estimates the optimal density of the APs required to meet the coverage demands and optimize their deployment to enhance the localization accuracy. One of our main contributions is the APs hearability-based reference points (RPs) clustering technique. Its uniqueness lies in the fact that not all installed APs participate in the localization process for all RPs. Finally, we analyze the variables governing the optimization process and the trade-off between cost, computation, and accuracy. Extensive simulations are conducted to validate the effectiveness of our algorithm. Our approach reduces the mean positioning error by 25% and the maximum error by 44% compared to the previous algorithm's performance.

Keywords: Access point · Cognitive radios · Geo-location database · Indoor localization · Location based services · WLAN optimization

1 Introduction

Location-Based Services (LBS) are one of the most facilitative and beneficial applications. These applications demand location information of the connected devices for their execution [1]. Received data without location information is of no or limited use. These connected smart devices continuously send and receive data, which is utilized for a dynamic range of applications [2]. This massive amount of data generated poses a significant challenge on the available spectrum

Q. Guo et al. (Eds.): WiSATS 2021, LNICST 410, pp. 607–621, 2022.
https://doi.org/10.1007/978-3-030-93398-2_54

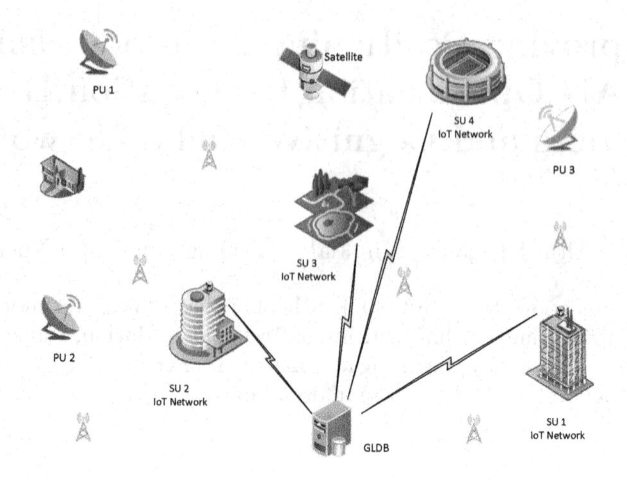

Fig. 1. Cognitive radio urban scenario.

and leads towards spectrum shortage. The ISM band is the primary source of communication in indoor environments. Heterogeneous devices use the same ISM band, which makes it overcrowded. Effective sharing techniques are required to fulfil this demand [3,4]. Instead of using the ISM band, the concept of Cognitive Radios (CR) can be applied, where the CR enabled devices to interact with each other, use a licensed spectrum when it is vacant (Fig. 1).

To find the available spectrum the Geo-location database can be used [5]. In this concept, the secondary user (SU) network is informed about the available spectrum by a spectrum database. The Primary User (PU) provide their availability information to the database location wise. On the other hand, when the SU requires to access the vacant spectrum, they also need to provide their location information. The GPS can provide reliable location information of the SUs, but when it comes to indoor environments, the GPS is ineffective. If a user or a device moves from outdoor to indoor, it must switch between GPS to an indoor positioning system (IPS). The inaccurate location information delivered to the database can cause interference between PU and SU, and also affect the utilization of the spectrum. Thus, a proper IPS needs to be connected with the CR network. Traditionally, IPS is used to locate the objects for LBS only. This location information can be used for many other services also. We aim here is to develop an efficient spectrum sharing scheme, utilizing the location information obtained from IPS.

The WLAN fingerprinting based IPS has emerged as the most popular technology [6,7]. WLAN installation requires careful design and planning. The planning and optimization of wireless networks is an important research issue, [9,10]. Much work has been done recently to guarantee the quality of service and network performance, [8,11], however, the optimal design of WLAN for IPS is still an open issue. Interestingly, just the APs optimization can lead to performance enhancement in multiple ways. Therefore, we consider AP optimization as an

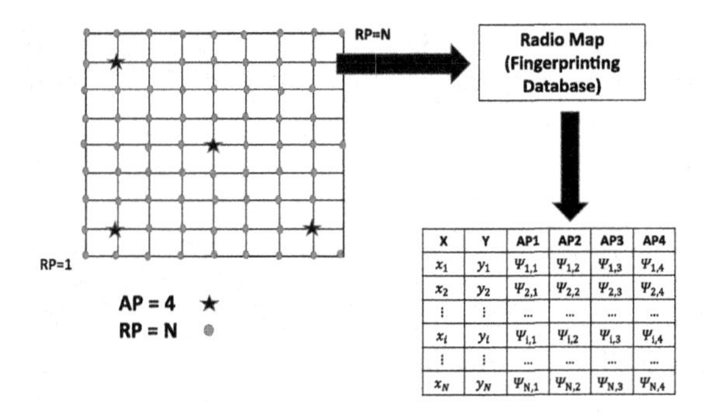

Fig. 2. Radio map for fingerprinting based indoor positioning systems.

essential topic to investigate. In our earlier work [12], we have already considered this problem for IPS. In this article, we enhance our previous work by determining the optimal APs density and improve AP deployment scheme. We test our algorithm in a more realistic and complex indoor environment for fingerprinting based positioning systems and compare it with our previous work. For ease, we call our previous algorithm as A and the enhanced algorithm as B.

2 System Model

We consider a 2D indoor environment as our area of interest (AoI) and divide it into equal spacing grids or RPs. The readings from APs are taken at these RPs and further used in optimization and positioning processes. We assume our AoI is big enough that a single AP can not provide coverage to all RPs. Multiple APs are installed to provide complete coverage. Thus, the adequate number of APs and their appropriate locations are vital factors to be investigated. Our algorithm will address this issue, ensuring positioning accuracy and signal coverage.

WLAN fingerprinting has two phases, the offline phase, and an online phase. In the offline phase a radiomap is constructed by collecting the RSS readings at all RPs. Each RP has a unique signature, based on the collected RSS values from the APs. A database is created using these unique signatures or fingerprints (FPs) corresponding to their RPs. A simple illustration can be seen in Fig. 2. In the online phase, a device sends a query to the database containing the unique FP it has recorded. The database matches this query with the available data and the best match is given as the estimated position of the device. Mathematically it can be represented as:

$$RP_j = \{(x_j, y_j)|j = (1, ...N)\} \tag{1}$$

$$\lambda = \begin{bmatrix} x_i, y_i & (\psi_{i,1}, .., \psi_{i,l}) \\ : & : \\ x_N, y_N & (\psi_{N,1}, .., \psi_{N,l}) \end{bmatrix} \tag{2}$$

here (λ) is the FP database, ψ = RSS samples from APs, (x, y) are the coordinate points of the RPs, N is the total number of RPs, and l represents the total number of APs. To simulate the indoor signal propagation, we use log-normal path loss model.

$$P(d) = P(d_0) + 10.n.log(d/d_0) + \zeta + \sum_{i=1}^{j}(\gamma_i) + \sum_{k=1}^{l}(\rho_k) \tag{3}$$

here, $P(d)$ is the RSS at a point d meters away from AP, $P(d_0)$ is the RSS at reference distance (1 m), n is the path loss coefficient (n = 3); noise (ζ), Wall attenuation factor ($\gamma = 4$), and people attenuation factor ($\rho = 3$) are the other attenuating factors.

3 Proposed Technique

Installing a large number of APs without proper planning can lead towards performance degradation, and an increase in both cost and computation. We develop a scheme to provide the optimal number of APs, satisfying the need for coverage and accuracy.

The RPs are used for FP collection and also serve as candidate positions for APs. The first AP (AP1) is placed at its desired position by the user. RSS values at each RP are calculated, and RPs with RSS greater than −80 dBm are checked. RPs satisfying the RSS condition are the average number of RPs an AP can cover. This simple calculation can give us an idea about the total number of APs required to cover all RPs. We can call this number as min APs, it is tentative and only from the viewpoint of coverage. Now, we initialize two variables, Ck_A and Ck_B. Ck_A must reach the count equal to the total number of RPs. Ck_B is the minimum number of APs that must provide coverage to each RP can be adjusted according to the requirement.

If the min APs are even in number, these APs are deployed in the AoI using symmetrical deployment, as shown in Fig. 3(a), if odd, symmetry is shown in Fig. 3(b). The basic idea of this symmetrical deployment is to divide the area into equal regions and installing each AP at the center points. After deployment, the RSS for these APs is calculated again, and if Ck_A equals N, we stop and display the number of required APs by adding one more AP. If the Ck_A condition is not satisfied, the process is repeated by increasing the min APs by one and continue the process until Ck_A condition is fulfilled. This optimal density of APs will provide coverage to every RP in AoI, with each RP having at least Ck_B APs with sufficient signal coverage. This methodology not only ensures enough number of APs for good signal coverage but also provide adequate overlapping regions for positioning purpose. As all RPs will have at least Ck_B APs to record RSS. The pseudo-code of the process can be seen in Algorithm 1.

AP deployment algorithm also starts by placing AP1 at the desired position, RSS is measured at all RPs and their average value is compared with a threshold Th, i.e. RSS between −55 dBm and −60 dBm. The reason for using this specific

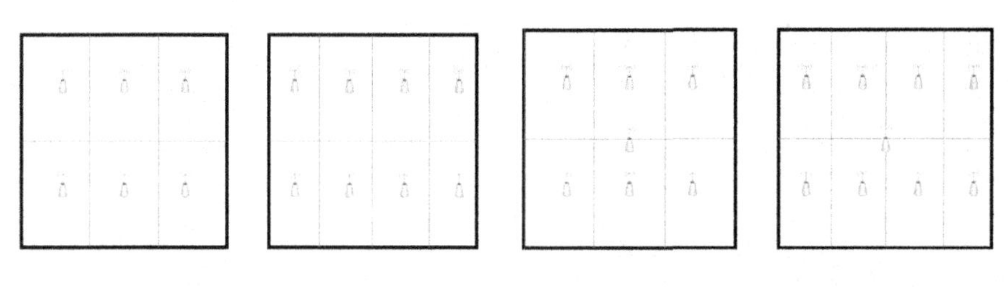

(a) Even number of APs. (b) Odd number of APs.

Fig. 3. Symmetrical deployment of APs to analyze signal coverage of installed number of APs.

threshold window has been explained in detail in our previous paper [12]. It helps create appropriate overlapping regions among APs; without a proper overlap, signal coverage can be ensured but not the positioning accuracy. Mathematically we represent the resultant RSS readings from AP1 as $J = [\psi_1, ...\psi_N]$, where N is the total number of RPs. By applying the threshold condition, J is converted into a matrix of 0s and 1s:

$$\psi_i = \begin{cases} 1, & \text{if } -55 \geq \psi_i \geq -60 \\ 0, & \text{otherwise} \end{cases} \tag{4}$$

Those RPs satisfying Eq. (4) are taken as candidate positions for next AP, and saved in $D(1)$.

$$D(1) = \begin{bmatrix} x_1 & y_1 \\ : & : \\ x_l & y_l \end{bmatrix} \tag{5}$$

l is the total number of new candidate locations, and (x, y) are the coordinates. The same process is applied for all points in $D(1)$, and the new points are saved in $D(2)$ and so on. Let us suppose the matrix $D(1)$ had four points, so $D(2)$ will have four arrays, each corresponding to the point in $D(1)$.

$$D(2) = \begin{matrix} \begin{bmatrix} x_{11} & y_{11} \\ : & : \\ x_{1l_1} & y_{1l_1} \end{bmatrix} & \begin{bmatrix} x_{21} & y_{21} \\ : & : \\ x_{2l_2} & y_{2l_2} \end{bmatrix} \\ \begin{bmatrix} x_{31} & y_{31} \\ : & : \\ x_{3l_3} & y_{3l_3} \end{bmatrix} & \begin{bmatrix} x_{41} & y_{41} \\ : & : \\ x_{4l_4} & y_{4l_4} \end{bmatrix} \end{matrix} \tag{6}$$

Here, l_1 to l_4 are the lengths of the four arrays independent of each other. After making a database of candidate positions, the next step is to make all possible APs configurations; the configurations are extracted from this database as a tree structure, explained in our previous work [12]. The updated algorithm is shown in Algorithm 2.

Algorithm 1. Optimal Density of Access points.

Input: AP1 = (x, y)
Output: Total APs = T_{AP}
1: Calculate RSS at RPs from $AP1$
2: Match RSS with Th of −80 dBm
3: No. of RPs satisfying condition (2) = avg. pts.
4: $minAPs$ = total RPs/avg. pts
5: T_{AP}=$minAPs$
6: Ck_A=0 ($Ck_A < lc$) lc= total no. of RPs
7: Check $minAPs$
8: **if** $minAPs$ is even **then**
9: Plot APs in symmetry 1
10: **else** {N is odd}
11: Plot APs in symmetry 2
12: **end if**
LOOP Process
13: **for** $i = 1$ to T_{AP} **do**
14: Calculate RSS
15: Match RSS, Ck_B APs must provide coverage above $Th2 = -100$
16: **if** Ck_A=lc **then**
17: T_{AP}=$T_{AP} + 1$
STOP: T_{AP} is optimal
18: **else**
19: T_{AP}=$T_{AP} + 1$
Go back to Step 7
20: **end if**
21: **end for**
End LOOP Process
End While
22: **return** T_{AP}

The used hybrid technique is a combination of $MaxFD$ and $MinGDOP$, i.e., maximizing fingerprint difference (FD) and minimizing Geometric Dilution of Precision (GDOP). $MaxFD$ and $MinGDOP$ have values in different ranges. We normalize it on the scale of 0 to 1.

$$MaxFD = argmax \sum_{i=1}^{N} \sum_{j=1}^{N} d(\delta_i, \delta_j) \tag{7}$$

$$MinGDOP = argmin \sqrt{Trace(H)} \tag{8}$$

$$Hybrid = argmax(MaxFD + \frac{1}{MinGDOP}) \tag{9}$$

We use a cluster-based approach for selection strategy and calculation of objective function. The Hamming distance is used to divide the data into clusters. Before applying Hamming distance, the data is transformed into a binary

Algorithm 2. Access Points Deployment strategy.

Input: AP1 = (x, y) & Total APs = T_{AP}
Output: APs Configurations

LOOP Process
1: **for** $i = 1$ to T_{AP} **do**
2: Calculate RSS at RPs from $AP(i)$
3: Match RSS with Th from Eq. (4)
4: Save the candidate positions in D{i}
5: AP{i} = D{i}
6: **end for**
End LOOP Process
7: l=1
LOOP Process
8: **for** $i = 1$ to $length(D)$ **do**
9: **for** $j = 1$ to $length(D\{i\})$ **do**
10: **for** $k = 1$ to $size(D\{i\}\{j\}, 1)$ **do**
11: Configuration(l)= [AP1 D{i}{j} D{i}{j}{k}]
12: l=l+1
13: **end for**
14: **end for**
15: **end for**
End LOOP Process
16: Configurations
17: DELETE configurations having repetition
18: Display (Configurations)

representation. Hamming distance can be defined as the number of bits between two bit sequences in which it differs. Having two bit sequences Ai and Bi, mathematically it can be represented as: $HD(Ai, Bi) = q + r$. Here, q represents the number of bits with value 1 in Ai and 0 in Bi, and r is vice versa. And also $HD(Ai, Bi) = HD(Bi, Ai)$.

Different APs have different positioning abilities; each AP contributes differently to the positioning process. Similarly, in an area where a single AP cannot provide full coverage, different sets of APs are visible at different RPs. Some sets can contribute more to the positioning process than others, based on the number of RPs covered. As an example we consider T_{APs} as 7 in AoI X. The RSS data can be represented as:

$$\Theta = \begin{bmatrix} (\psi_{1,1}, \psi_{1,2}, \psi_{1,3}, \psi_{1,4}, \psi_{1,5}, \psi_{1,6}, \psi_{1,7}) \\ \vdots \\ (\psi_{i,1}, \psi_{i,2}, \psi_{i,3}, \psi_{i,4}, \psi_{i,5}, \psi_{i,6}, \psi_{i,7}) \\ \vdots \\ (\psi_{N,1}, \psi_{N,2}, \psi_{N,3}, \psi_{N,4}, \psi_{N,5}, \psi_{N,6}, \psi_{N,7}) \end{bmatrix} \tag{10}$$

For simplicity we rewrite the matrix in row form and replace $(\psi_{i,1}, ..., \psi_{N,7})$ for RP_i by δ_i.

$$\Theta = \left[(\delta_1, \delta_2, \delta_3, ..., \delta_N)\right] \tag{11}$$

Those RPs where the RSS of an AP is below -100, is considered as a dead spot for that AP. This is considered as sensitivity threshold and indicates that a particular AP is not hearable. For mathematical simplification, we replace all the RSS values less than -100 dBm as -100 dBm. The values in this range are practically so weak that it cannot be considered for any use, whether data transmission or positioning purposes.

At an RP_i, four APs are hearable, and three are not hearable among T_{APs}. So it can be transformed into binary by representing the hearable APs as 1 and others as 0.

$$\Theta = \begin{bmatrix} (1,1,1,0,1,1,0) \\ \vdots \\ (1,1,0,0,1,1,0) \\ \vdots \\ (1,1,0,0,0,0,0) \end{bmatrix} \tag{12}$$

Suppose the dimensionality of the binary matrix is $N \times T_{AP}$, and is partitioned into K clusters, $C = (C_i, ..., C_K)$. RP_a, RP_b, RP_c, RP_d form a cluster C1, as they share the same APs set $B1 = [AP1, AP2, AP3]$, covering them, while RP_e, RP_f, RP_g, RP_h form another cluster C2, as they form FPs from the set $B2 = [AP4, AP5, AP6, AP7]$ of APs and so on. A laptop and a mobile phone can be seen in Fig. 4(a), both of these devices belong to different clusters with the visibility of different sets of APs.

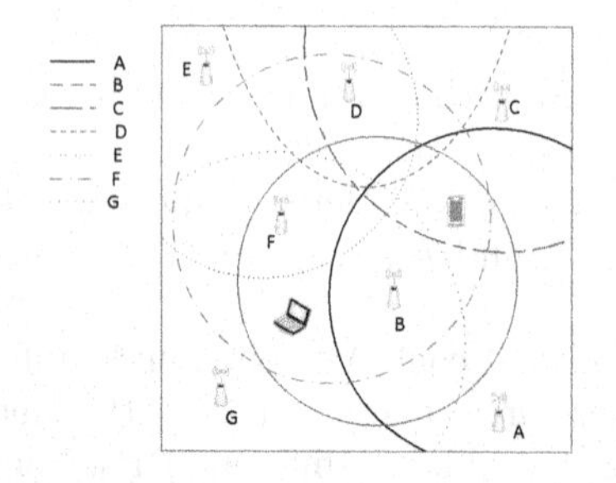

(a) Deployment of APs forming clusters.

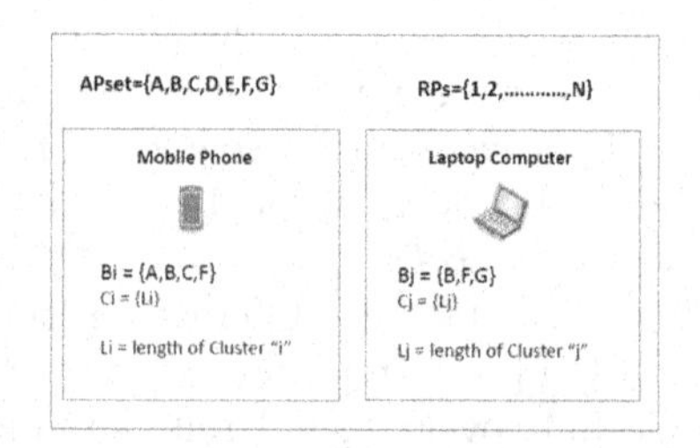

(b) Two devices with coverage from different sets of APs.

Fig. 4. APs hearability based clustering technique.

Each cluster obtained by the algorithm must be unique and sensitive to the order of APs. The criteria of clustering must satisfy certain conditions. The set of APs and RPs is $AP = [1, 2, ..., T_{AP}]$ and $RPs = [1, 2, ..., N]$ respectively, where $B_i \subset AP$. We can also say that $B_1 \cup B_2 \cup B_3 \cup .... \cup B_K = AP$ and $C_i = \{RP_i | i \in (1......N) \& B_i \subset AP\}$. An RP is assigned to one and only one cluster. All clusters lengths together $C_1 \cup C_2 .... \cup C_n =$ RPs. The size of two clusters C_i and C_j can or cannot be equal, depending upon the B_i and B_j covering the number of RPs. If the difference between B_i and B_j is zero, then C_i and C_j are same. If the set B_i of APs covers the majority of the RPs, C_i will have the most influence on the overall system. As shown in Fig. 4 and Fig. 5, few APs are not hearable at C_i so they do not participate in the positioning process.

Let us assume, in $C1$ the set of APs is $B1$, and it contains 4 number of APs, and total RPs covered in $C1$ are L. For each cluster the FPs are calculated and stored in the database. To take the FD, we take the RSS values from Eq. (2) and rewrite it as shown in Eq. 10, and Eq. 11. The difference is measured using Euclidean distance:

$$FD(i,j) = \sqrt{(\psi_{1,i} - \psi_{1,j})^2 + ... + (\psi_{4,i} - \psi_{4,j})^2} \tag{13}$$

$$\Delta = \begin{bmatrix} (\delta_i - \delta_i & \delta_i - \delta_{i+1} & .... & \delta_i - \delta_L) \\ (\delta_{i+1} - \delta_i & \delta_{i+1} - \delta_{i+1} & .... & \delta_{i+1} - \delta_L) \\ \vdots & \vdots & \vdots & \vdots \\ (\delta_L - \delta_i & \delta_L - \delta_{i+1} & .... & \delta_L - \delta_L) \end{bmatrix} \tag{14}$$

$$\Delta = \begin{bmatrix} \varphi_{1,1} & \varphi_{1,2} & .... & \varphi_{1,L} \\ \varphi_{2,1} & \varphi_{2,2} & .... & \varphi_{2,L} \\ \vdots & \vdots & \vdots & \vdots \\ \varphi_{L,1} & \varphi_{L,2} & .... & \varphi_{L,L} \end{bmatrix} \tag{15}$$

$$\Omega_i = \frac{\sum_{j=1}^{L} \varphi_{i,j}}{L} \tag{16}$$

$$O = \left[(\Omega_1, \Omega_2, \Omega_3, ..., \Omega_L)\right] \tag{17}$$

$$MaxFD_{Ci} = \frac{\sum_{i=1}^{L} \Omega_i}{L} \tag{18}$$

We now have FD Ω for each RP; to calculate the cluster-wide FD we retake the mean value. After calculating the FD for all clusters, the next step is to assign appropriate weights and calculate the final network-wide FD for this specific configuration. A cluster covering more RPs will have more influence on overall accuracy and needs to be assigned more weight as compared to a cluster covering a limited number of RPs. This weightage is assigned to the clusters, as shown below:

$$\sum_{i=1}^{K} (MaxFD_{Ci} \times (\frac{L_i}{N})) \tag{19}$$

To calculate the GDOP, we adopt the same process of clusters. Let (x, y) be the coordinates of RP and (x_i, y_i) are coordinate points of the i_{th} AP. For cluster $C1$, the GDOP will be calculated as shown below:

$$R_i = \sqrt{(x_i - x)^2 + (y_i - y)^2} \tag{20}$$

Algorithm 3. Selection strategy for optimal AP configuration.

Input: All Configurations
Output: Optimal Configuration
1: **for** $S = 1$ to length($Configurations$) **do**
2: Calculate RSS at all RPs from all APs
3: Make Fingerprints
4: Replace RSS < -100 dBm to -100 (Sensitivity Th)
5: Form Clusters on the Basis of APs hearability

6: **for** $i = 1$ to length($Cluster$) **do**
7: Calculate FD b/w all sets of two RPs within Cluster C_i
8: Cal. FD for each RP and then cluster wide FD for i
9: **end for**
10: Now calculate Network Wide FD, with assigning weights

11: **for** $i = 1$ to length($Cluster$) **do**
12: Calculate GDOP for each RP then cluster wide GDOP for i
13: **end for**
14: Calculate Network Wide GDOP
15: **end for**
16: Normalize both calculated FD and GDOP on the scale of O to 1.
17: The Configuration with the best Score will be the optimal configuration
18: Validate using both Deterministic and Probabilistic Matching Algorithms

$$A = \begin{bmatrix} \frac{x_1-x}{R_1} & \frac{y_1-y}{R_1} \\ \frac{x_2-x}{R_2} & \frac{y_1-y}{R_1} \\ \frac{x_3-x}{R_3} & \frac{y_1-y}{R_1} \\ \frac{x_4-x}{R_4} & \frac{y_1-y}{R_1} \end{bmatrix} \tag{21}$$

The matrix A is multiplied by its transpose and then its inverse is calculated. If the obtained matris is H, we calculate the GDOP by

$$MinGDOP = \sqrt{Trace(H)} \tag{22}$$

$$MinGDOP_{Ci} = \frac{\sum_{j=1}^{L} GDOP(i)}{L} \tag{23}$$

$$\sum_{i=1}^{K} (MinGDOP_{Ci} \times (\frac{L_i}{N})) \tag{24}$$

We take the normalized values of the calculated FD and GDOP. This selection scheme is summarized in Algorithm 3.

4 Simulation Results

We consider a 2-D area X with dimensions of 50 × 50 m, as shown in Fig. 5. RPs with 2 m and 3 m spacing will be used in two different scenarios. We divide the positioning area into different rooms, and each room has people's presence. H has 40, $A(1-4)$ has 20 each, $B(1,2)$ has 10 each and $D(1-4)$ has 1 each, total crowd is 144 people. This presence of people is included along with walls, obstacles, and noise in the optimization process.

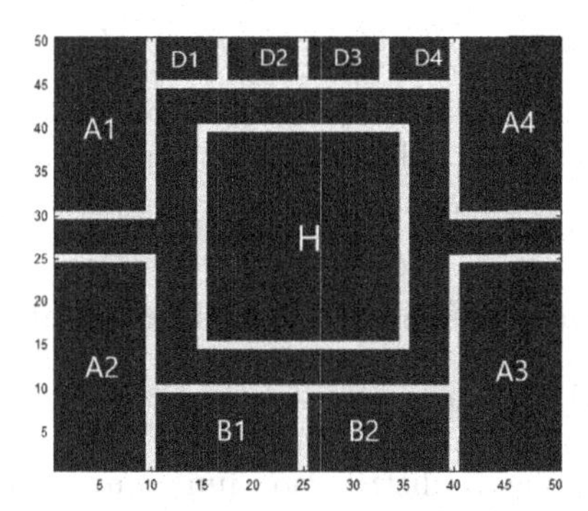

Fig. 5. Illustration of the floor plan for our simulations in MATLAB.

To evaluate the performance of obtained AP configuration, 100 random test points are generated within the AoI. The 100 queries can be presented as S in the positioning model. KNN deterministic algorithm and Bayesian probabilistic algorithm are used to validate our proposed technique.

The KNN uses Euclidean distance as the similarity measure between the online query and the FP database. Let S be the online query from the test point, and δ_i is the RSS data from the FP database λ.

$$\xi(i) = \sqrt{(S - \delta_i)^2} \tag{25}$$

$$P_{Est} = \frac{\sum_{i=1}^{K}(x_i, y_i)}{K} \tag{26}$$

In the probabilistic method, the RSS values obtained are treated as a random variable. For simplicity we represent the RPs here by $X = (x_1, x_2, ...x_N)$ The working of the probabilistic technique can be stated as we have to find the location in the radiomap that maximizes the conditional probability $P(x_i|s)$.

$$P(x \mid s) = \frac{(P(s \mid x)P(x))}{P(s)} \tag{27}$$

Without prior information about the position, we can assume that the probability of wireless device located at different places are equally likely.

$$P(x \mid s) = c \cdot P(s \mid x) \quad where \quad c = \frac{P(x)}{P(s)} \tag{28}$$

since $P(s)$ is constant for all x and if the user profile information is not known, or not used, then we can assume that all the locations are equally likely and the term $P(x)$ can be factored out from the maximization process, and the equation becomes

$$argmax_x[P(x \mid s)] = argmax_x[P(s \mid x)] \tag{29}$$

The location with the maximum probability of occurrence of the RSS vector will be the estimated location. The remaining term $P(s \mid x)$ is calculated using the radiomap, where the RSS values are stored as a joint probability distribution.

4.1 Positioning Comparison of Algorithm A and Algorithm B

AP1 is placed at [50, 1], CK_B is kept 2, T_{APs} is 6, and grid size equal to 3 m and 2 m. After applying both algorithms A and B We analyze the positioning performance using cumulative distribution function. Figure 6 shows the performance of both the algorithms with 3 m grid and KNN, where as in Fig. 7, the grid size is changed to 2 m. Probabilistic matching technique is also used for performance analysis and the results can be seen in Fig. 8. Figure 9 shows the results when the coordinates of AP1 are changed to [25, 4]. The detailed results of these graphs can be seen in Table 1. From these results; we observe enhanced algorithm performs better than our previous positioning algorithm.

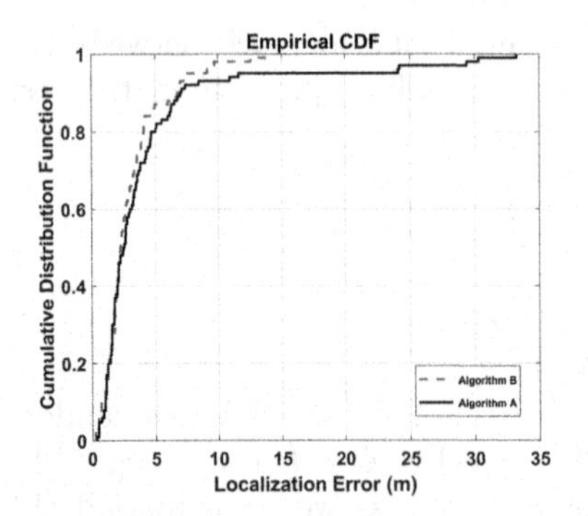

Fig. 6. Deterministic algorithm used with grid size of 3 * 3.

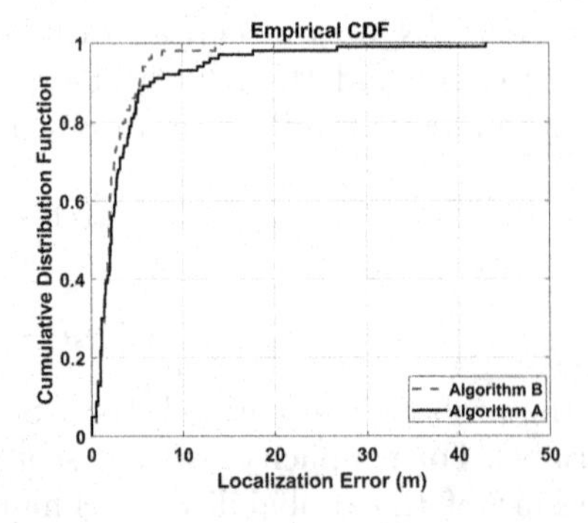

Fig. 7. Deterministic algorithm used with grid size of 2 * 2.

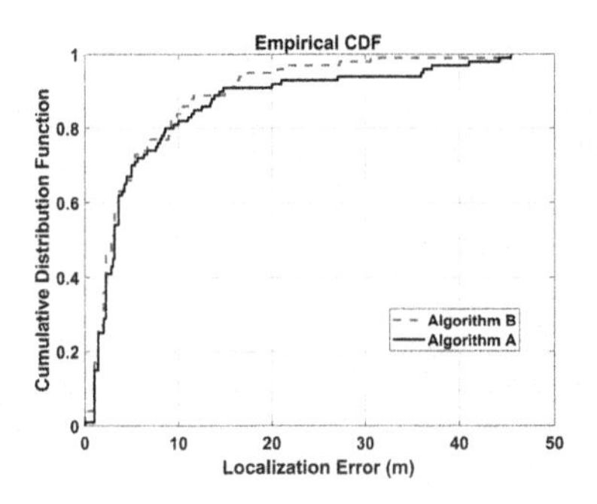

Fig. 8. Probabilistic algorithm used with grid size of 3 * 3.

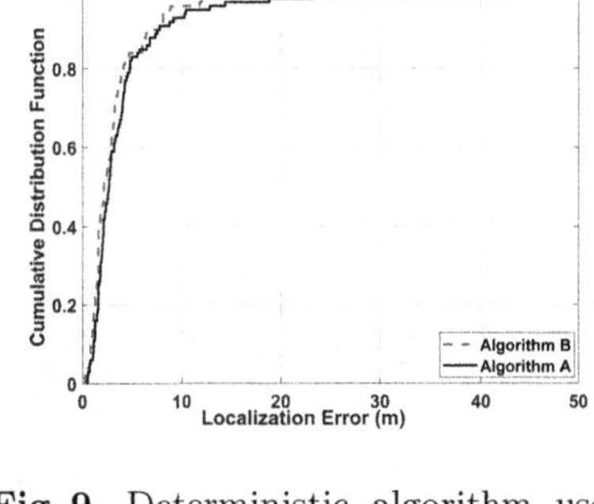

Fig. 9. Deterministic algorithm used with grid size of 3 * 3, AP1 = [25, 4].

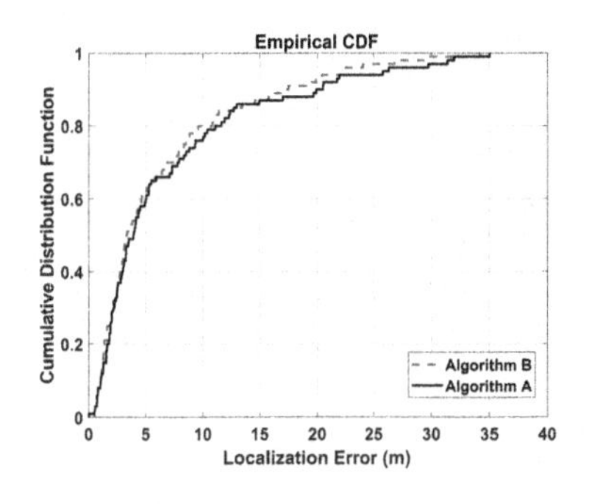

Fig. 10. $CK_B = 1, APs = 4$

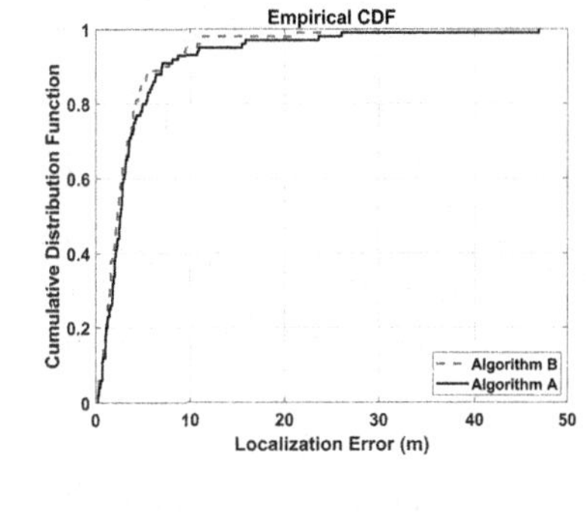

Fig. 11. $CK_B = 2, APs = 6$

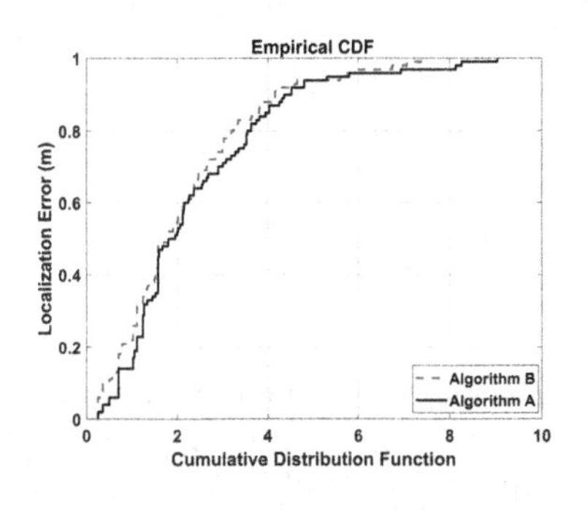

Fig. 12. $CK_B = 3, APs = 10$

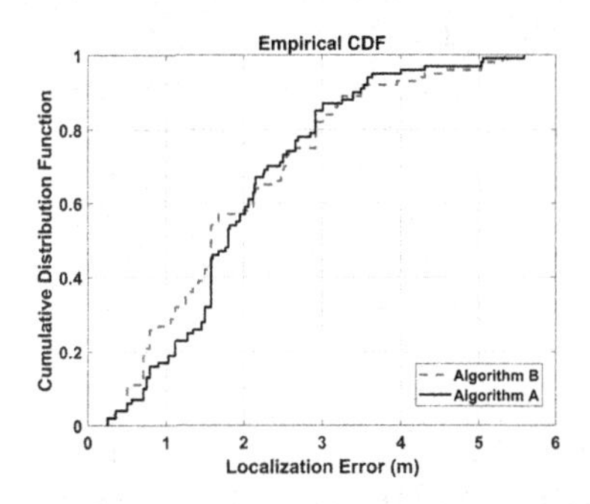

Fig. 13. $CK_B = 4, APs = 12$

4.2 Impact of Variable CK_B on the Results

The impact of CK_B on the density of APs and positioning performance can be seen in Fig. 10, 11, 12 and 13 and the details in Table 2. It is a trade-off between positioning accuracy and cost of the system, which can be adjusted according to the requirement and budget. Increasing the value of CK_B ensures

Table 1. Positioning performance (Fig. 7, 8, 9 and 10)

Figure	Algorithm	Mean error (m)	Maximum error (m)
7	A	4.2718	33.2650
7	B	3.0580	13.9754
8	A	3.6564	43.0726
8	B	2.5426	14.3962
9	A	6.8562	45.3982
9	B	5.5804	45.3982
10	A	4.1065	44.1163
10	B	3.1996	20.1556

Table 2. Impact of variable CK_B

CK_B	No. of APs	Mean error (m)	Maximum error (m)
1	4	6.4	34.21
2	6	3.3104	24.1350
3	10	2.1	9
4	12	1.93	5.3

better coverage as a larger number of APs cover a single RP, but too many APs covering every point will also lead to more interference.

The optimization algorithm deploys the APs according to the environmental dynamics and keeping in view the crowd presence. As the crowd profoundly influence the overall optimization process, which is considered in our optimization process as people attenuation factor.

5 Conclusion

In this paper, we proposed an AP optimization scheme for indoor environments within the framework of fingerprinting localization systems. This optimization improves the positioning accuracy which can be used for both LBS and CR. The scheme aims to maximize coverage and localization accuracy. A variable is initialized to decide the optimal density of APs, depending upon the requirement. The proposed optimization scheme uses the idea of clusters, based on overlapping regions of APs to select the optimal configuration. This cluster-based technique calculates the objective function cluster-wise. We also analyze the impact of the variable used for optimal AP density on localization accuracy and APs density. To validate the performance of our enhanced optimization algorithm, deterministic and probabilistic localization algorithms are used. The simulation results prove the effectiveness of our newly proposed optimization scheme.

Acknowledgment. This work was supported by the National Natural Science Foundation of China under Grants 61771163, the Natural Science Foundation for Outstanding Young Scholars of Heilongjiang Province under Grant YQ2020F001, the Science and Technology on Communication Networks Laboratory under Grants SXX19641X072 and SXX18641X028.

References

1. Khelifi, F., Bradai, A., Benslimane, A., Rawat, P., Atri, M.: A survey of localization systems in internet of things. Mob. Netw. Appl. **24**(3), 761–785 (2018). https://doi.org/10.1007/s11036-018-1090-3
2. Zanella, A., Bui, N., Castellani, A., Vangelista, L., Zorzi, M.: Internet of things for smart cities. IEEE Internet Things J. **1**(1), 22–32 (2014). https://doi.org/10.1109/JIOT.2014.2306328
3. Khan, A.A., Rehmani, M.H., Rachedi, A.: When cognitive radio meets the internet of things? In: 2016 International Wireless Communications and Mobile Computing Conference (IWCMC), Paphos, pp. 469-474 (2016). https://doi.org/10.1109/IWCMC.2016.7577103
4. Li, F., Lam, K., Li, X., Sheng, Z., Hua, J., Wang, L.: Advances and emerging challenges in cognitive internet-of-things. IEEE Trans. Industr. Inf. **16**(8), 5489–5496 (2020). https://doi.org/10.1109/TII.2019.2953246
5. Höyhtyä, M., et al.: Database-assisted spectrum sharing in satellite communications: a survey. IEEE Access **5**, 25322–25341 (2017). https://doi.org/10.1109/ACCESS.2017.2771300
6. Xue, W., Qiu, W., Hua, X., Yu, K.: Improved Wi-Fi RSSI measurement for indoor localization. IEEE Sens. J. **17**(7), 2224–2230 (2017). https://doi.org/10.1109/JSEN.2017.2660522
7. Yoo, J.: Change detection of RSSI fingerprint pattern for indoor positioning system. IEEE Sens. J. **20**(5), 2608–2615 (2020). https://doi.org/10.1109/JSEN.2019.2951712
8. Raschellà, A., Bouhafs, F., Seyedebrahimi, M., Mackay, M., Shi, Q.: Quality of service oriented access point selection framework for large Wi-Fi networks. IEEE Trans. Netw. Serv. Manag. **14**(2), 441–455 (2017). https://doi.org/10.1109/TNSM.2017.2678021
9. Jaiyeola, M.O., Young, M., Xiao, J., Medal, H., Grimes, G., Schweitzer, D.: Towards scalable planning of wireless networks. In: 2019 IFIP/IEEE Symposium on Integrated Network and Service Management (IM), Arlington, VA, USA, pp. 629-633 (2019)
10. Raschellà, A., et al.: A dynamic access point allocation algorithm for dense wireless LANs using potential game. Comput. Netw. **167**, 106991 (2020)
11. Liu, P., Meng, X., Wu, J., Yao, M., Tang, Z.: AP deployment optimization for WLAN: a fruit fly optimization approach. In: 2019 IEEE/CIC International Conference on Communications in China (ICCC), Changchun, China, pp. 478–483 (2019). https://doi.org/10.1109/ICCChina.2019.8855912
12. Jia, M., Khattak, S.B.A., Guo, Q., Gu, X., Lin, Y.: Access point optimization for reliable indoor localization systems. IEEE Trans. Reliab. **69**(4), 1424–1436 (2020). https://doi.org/10.1109/TR.2019.2955748

Meshtastic Infrastructure-less Networks for Reliable Data Transmission to Augment Internet of Things Applications

N. K. Suryadevara(✉) and Arijit Dutta

School of Computer and Information Sciences, University of Hyderabad, Hyderabad, India
nks@uohyd.ac.in

Abstract. This paper has reported a practical implementation of meshtastic networks data transmission via web interface for an IoT theme of the application. The integrated network architecture and the interconnecting mechanisms for reliable, low-cost data transmission of the infrastructure-less network are depicted. Features of meshtastic communication such as addressing and tunneling that can swiftly assimilate with co-systems of IoT were presented. The system setup, transmission, and data gathered in the real-world campus environment are intricacies. The assessment of quality parameters related to data transmission using Heltec Lora devices is described. Results are encouraging as the prototype was tested to generate real-time information exchange using mesh networks rather than a testbed scenario.

Keywords: Wireless mesh networks · Infrastructure-less networks · LoRa · Internet of Things (IoT)

1 Introduction

The technological advancements in media communications is pretty much connected with the Internet of Things (IoT) [1]. However, Telecommunications and the Internet events gave humanity limitless possibilities of information and data trade, which completely revolutionized the world. In the last decade, the growth of the data transfer between people using IoT ecosystems went huge, so did the data transmission technologies [2]. Energy utilization is a significant consideration in the event of communication between sensors for specific applications [3]. Machine to Machine (M2M) communication must ensure that capturing the sense events, nearby data processing of detected events, and packet transmission to the end nodes are efficient [3].

Public Safety and Disaster Recovery (PSDR) communications are entirely dependent on Internet communication infrastructures such as traditional landline, cellular telephony, and infrastructure-based Land Mobile Radio (LMR).

Q. Guo et al. (Eds.): WiSATS 2021, LNICST 410, pp. 622–640, 2022.
https://doi.org/10.1007/978-3-030-93398-2_55

Whereas, in environmental monitoring, the option to transmit data does not depend on existing Internet infrastructures for the applications such as biological and ecological studies, animal tracking, air quality measurements, agriculture, and natural hazards or disasters (e.g., tsunamis, earthquakes, nuclear meltdowns). Recent events have revealed significant shortcomings in PSDR communications [4]. Several aspects need to be considered in these situations when a technical solution is to be considered for effective data transmission through Infrastructure-less communications, such as lack of power supply, harsh weather, low maintenance possibilities, weight restrictions for animal-attached sensors, availability of scenario-specific sensors, and cost factors [5]. The communication technologies that evolved as trendsetters and benchmark in terms of energy efficiency and scalability without Internet infrastructures, such as ZigBee, 6LowPAN, BLE, LoRa, LoRaWAN, Thread [6,7]. The Long Range (LoRa) technology came out as a better alternative for Infrastructure-less low power long-range communication than WiFi and other Low Power Wide Area Network (LPWAN) [8].

When it comes to infrastructure-less communication, off-grid decentralized communication systems are more reliable on a global scale. Among the available Infrastructure-less networks, wireless mesh networks can give coverage to a wide geographical area without depending on a dedicated access point (AP) [9]. Presently, "Disaster.radio" [10] and "Meshtastic" [11] are open source projects that work on disaster-resilient communication networks. Meshtastic allows one to utilize economical GPS radios as extensible, long battery life, secure, network GPS communicator. These radios are extraordinary for those geographical areas where no communication tower is required to install and do not require web access. Every individual in the mesh network can generally see the area and distance of any remaining individuals, and at any instant, they can communicate through text messaging. Meshtastic radios use LoRa technology in their Physical layer component. These radios auto-create a mesh to forward data through packets and follow the publish and subscribe model [11]. The open-source community developed the corresponding firmware and python APIs to be used by the application developers. The message passing between the Meshtastic device and its corresponding application takes place through Google protobuf [12].

The primary objective of this work is to understand how LoRa technology is functioning in the Meshtastic devices for the exchange of little telemetric data that can be considered for IoT data transmission when there is an Internet/telecommunication connection failure. So that, Meshtastic device communication systems can be considered for planning and setup of frameworks that can be utilized in the arrangement of visual control systems in crisis circumstances.

Contribution: This paper aims to describe data communication architecture using meshtastic devices to transmit data in indoor and outdoor environmental scenarios. Furthermore, the network data transmission investigations and their corresponding quality parameters based on the experiments performed using the Meshtastic devices were illustrated. The experimental study analyzes the transfer of packets and received signal level indicator on different channel settings with different radio configurations of 433 MHz Meshtastic devices at various Spreading Factor (SF), Bandwidth (BW), Coding Rate (CR), Signal to Noise ratio (SNR). Accordingly, the Quality of Service (QoS) parameters such as reliability, throughput, and jitters for the Meshtastic data transmission are evaluated and presented.

The rest of the paper is organized as follows; Sect. 2 describes various Wireless Mesh Networks (WMN) and their related communication technologies. Section 3 presents the details for the setup of Meshtastic network architecture and Python APIs considered for the investigations. Furthermore, Sect. 4 describes the intricacies in the implementation, and Sect. 5 presents the results obtained for various experimental settings and their analysis. Section 6 concludes with a discussion of future work.

2 Related Works

Mudathir et al. has presented various wireless communication technologies for Infrastructure-less data communication. Figure 1 depicts the following arrangement followed by corresponding description.

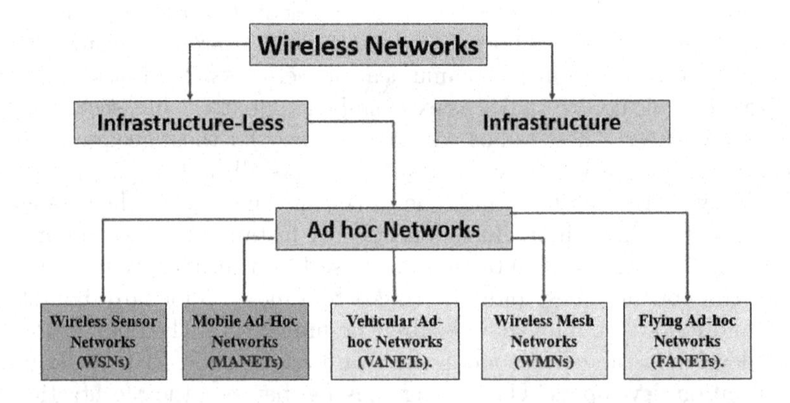

Fig. 1. Classification of Wireless Infrastructure-less Networks

An ad-hoc network comprises remote hubs that can dynamically self-arrange into a flexible and transient topology without essentially utilizing any prior configuration settings. In ad-hoc networks, every hub may convey straightforwardly to one another. Hubs that are not directly associated convey by sending the traffic through halfway hubs through multi-hop communication.

Wireless Sensor Networks (WSNs): The WSNs are interconnected sensor hubs that communicate remotely to gather information about the adjacent environment. Hubs are, for the most part, low power and dispersed in an ad-hoc, and decentralized design. It is a significant innovation empowering the majority of IoT theme of use cases.

Mobile Ad-Hoc Networks (MANET): The MANET organization is a self-governing momentary relationship of wireless devices (hubs) that communicate over remote connections. Hubs that exist within the same range can convey straightforwardly and are responsible for discovering one another. Figure 2 is an illustration of the MANET topology of various devices [13].

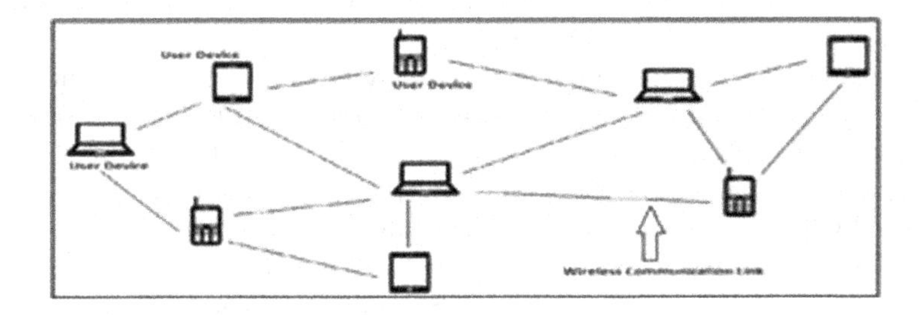

Fig. 2. Topology of MANET

Vehicular Ad-Hoc Networks (VANET): This type of wireless organization uses vehicles as versatile hubs, a subclass of MANETs to give interchanges among close-by vehicles and side of the road hardware. The hubs (vehicles) in VANETs are restricted to street geography while moving, so if the street data is accessible, we can anticipate the future situation. This is how street safety and traffic management turn out to be easy.

Wireless Mesh Network (WMN): A WMN comprises two or more wireless radios working together to share routing protocols to create an interconnected Radio Frequency (RF) pathway. No matter how many radios it includes, a wireless mesh network creates only a single name identifier or Single Set Identifier (SSID). Therefore, it is the most reliable network as every node in its range are connected with all the other nodes. Figure 3 depicts a typical WMN.

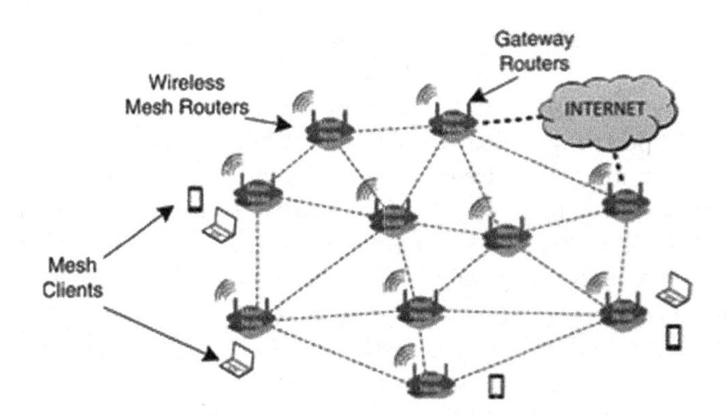

Fig. 3. Wireless Mesh Network (WMN)

Flying Ad-Hoc Networks (FANET): It is majorly used in Unmanned Aerial Vehicle (UAV) network where its importance is in several militaries, commercial and civilian applications like disaster management, crisis management, hostile environment, destroy and search operations, border surveillance, wildfire management, relaying networks, estimation of wind, civil security, agricultural remote sensing, and traffic monitoring [14].

The main advantages of ad hoc networks are flexibility, low cost, and robustness. On the other hand, a wireless mesh network design is exciting and at the same time it is challenging due to noisy channels, limited reach, and insecure wireless transmissions added to mobility and energy constraint.

The **A**LOHAnet protocol used in the computer networking system provides the wireless packet data network used for collision detection in the Medium Access Control (MAC) sublayer for wireless communication technologies [15]. A standard MAC protocol must deal with channel constraints, attenuation, and noise.

In contrast, for efficient wireless transmission in IoT theme of applications, the MAC sublayer should consider requirements, such as quality of service (QoS), low energy consumption, fairness, and scalability, as these critical parameters significantly impact the IoT system performance. Therefore, the popular protocols used in wireless mesh networks consider the QoS, low energy consumption, and scalability parameters for efficient data transmission. The most popular wireless communication technologies used in indoor and outdoor environmental monitoring conditions under the themes of IoT are Zigbee and LoRa.

ZigBee protocol uses the IEEE 802.15.4 as a base standard. One of the most used communication protocols for indoor IoT applications where low energy consumption, low data rate, low cost, and high message throughput performances are required. It can also provide high reliability, security with both encryption and authentication. It can connect up to 65000 nodes to form a WMN. Majorly, it works with the direct-sequence spread spectrum (DSSS) method, capable of transferring data at 250 kbps at 2.4 GHz. The range is around 10 to 100 m, but setting up the network is complex and costlier.

LoRaA bidirectional communication protocol uses the LoRa physical layer to provide low-power long-range communications. To achieve this, LoRa uses Chirp Spread Spectrum (CSS) modulation. With a single base station, it is possible to cover up to hundreds of square kilometers. To guarantee that all communications are completed, the LoRa MAC layer takes responsibility for joining and accepting the end-point and the gateway, scheduling the receiving slots for end-points, and confirming the reception of the received packets. LoRa uses non-licensed 433 MHz to 923 MHz industrial, scientific, and medical (ISM) radio bands according to different world countries and its telecommunication guidelines [16]. The data communication range is up to 2 to 5 km between end nodes and, for Long Range Wide Area Network (LoRaWAN) it goes up to 10 to 15 km. It supports up to 1000 nodes. Table 1 provides the primary differences between the two low-power wireless communication technologies.

Table 1. Comparison between Zigbee and LoRa

Feature	ZigBee	LoRa
Based Data Rate [Mbps]	0.25	0.11
Frequency [GHz]	2.4	433
Range [m]	10 to 100	2000 to 5000
Nodes/Masters Power	65540	1000
Complexity	Medium	Low
Security	AES 128 bit	AES 128 bit

LoRa is better suited for Low Power long-range communication. The open-source community drives the applications in low-poll rate sensing. The technology and accessible gateways for nodes to connect to community-driven networks (such as The Things Network) are suitable for developers and communities.

Considering the advantages mentioned above, the "Meshtastic project" has developed an extremely versatile mesh communicator ecosystem for various applications such as paragliding, hiking, skiing in the field of aviation, and GPS communication using available ISM sub-gigahertz frequencies of LoRa. Furthermore, its applications can be extended to disaster management and agriculture, where environmental conditions of large geographical regions are better monitored.

Islam, B. et al. [17] tested and came out with the examination of the practicality of utilizing LoRa, in indoor (house) climate. Assessments related to the dependability, coverage, strength of signals, responsiveness, and power consumption are presented. It was inferred that LoRa is more competent to use in the indoor industrial warehouse and home monitoring systems using sensor networking.

I. Bobkov, A. et al. [18] considered both 433 MHz and 868 MHz ISM band frequencies and calculated that SNR (signal to noise ratio), RSSI (received signal strength indicator), and PDR (packet delivery ratio). Their work concludes that 433 MHz frequency gives a more stable LoRa signal due to more noteworthy SNR and RSSI values at each spreading factor.

Apparently, no investigation has contemplated and talked about the presentation of LoRa in meshtastic teletype data transmission. In particular, its usage in IoT application domains, in the event of no Internet/telecommunication, the performance of teletype data transmissions over the meshtastic networks.

This paper presents the effective inter-networking mechanisms for transmitting teletype (text) data over infrastructure-less (Meshtastic) networks to support the data transmissions for IoT applications that do not require Internet communication. Reliability and delay in data packet transmission over the Meshtastic networks are presented in the results section.

3 System Description

One of the objectives of this work is to test 'hop' dynamics with a system of three Meshtastic nodes. Initially, a 'base station sender' and a 'target receiver' on opposite sides of a building are positioned to block direct LoRa transmission; and then a 'relay' node on the 'corner' of the building for messages to 'hop' around it is placed. The data transmission is initiated by the 'sender' that was located in room 1 of the building. The device is managed and controlled by the software Meshtastic-python script that would 'reply' to receive messages with the SNR of the incoming message. The idea was that from the field, it could send messages from the 'end node' (controlled via the mobile app), and if the 'relay' node was able to relay the message via a 'hop' to the 'base station sender', then eventually gets a reply relayed back to the sender at the end node.

Figure 4 shows the layout depicting key elements of the integrated system. It consists of i) wireless devices[ ] (currently support devices that use the ESP32 and the nRF52 microcontrollers), ii) Laptop running the software of the Meshtatic device, and iii) android mobile phone as hardware, and iv) meshtastic GUI desktop application v) meshtastic android mobile application as software vi) Any power bank over 500 mAh battery.

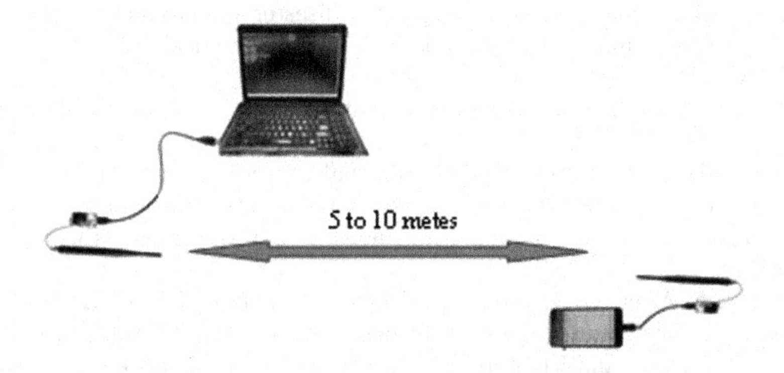

Fig. 4. Peer to Peer meshtastic system setup

There were two different settings to measure the delay, throughput, packet loss, and jitter in the Meshtastic network. For the zero hop (peer-to-peer connection) as depicted in Fig. 4, one ESP32 micro-controller is connected with Laptop where the developed meshtasticGUI software program is running. The other wireless device is at the 5 to 10-m line of sight distance. For a 'single hop' connection (one hop between the two end nodes) as depicted in Fig. 5, the wireless end nodes are kept at two different rooms partitioned by a wall, and another device is placed at a third room on the same floor which has a line of sight communication with the other two devices. The remote devices (that are not connected with the workstation) are attached to the power bank for the power supply. Distance between the hop and the end nodes are 5 to 10 m each. In order to use the Meshtastic network, the ESP32 devices are embedded with meshtastic firmware (Version 1.1.32). The 'Meshtastic' android mobile app (version 1.1.32) is also installed on the mobile.

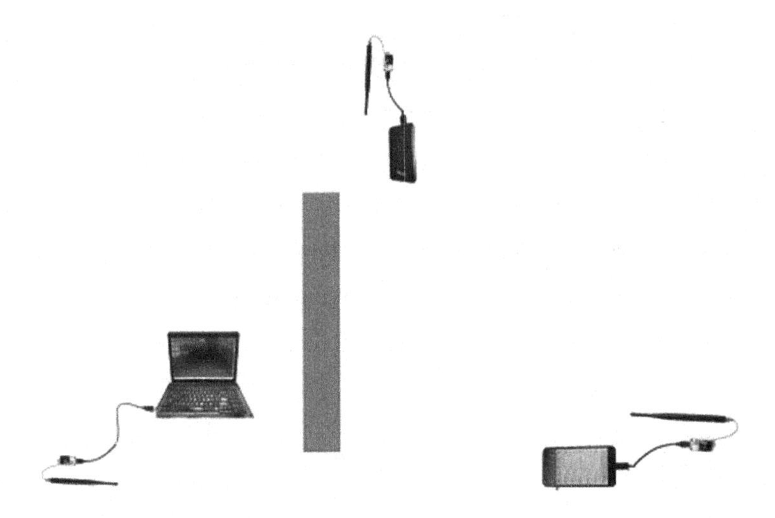

Fig. 5. Single hop meshtastic system setup

Before running the program, the following command as shown in Fig. 6 is used to sync both the devices in the same timestamp:

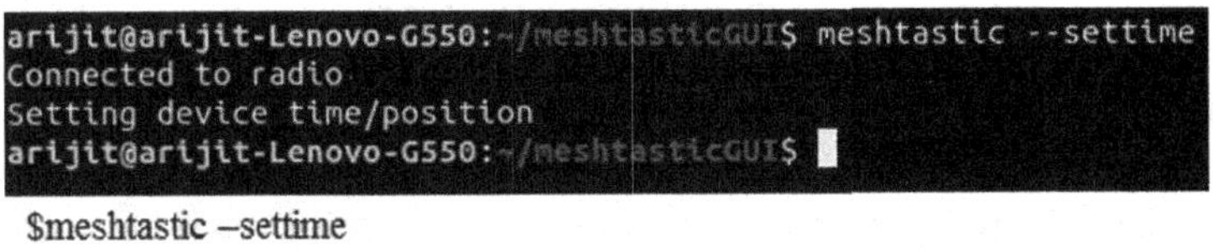

```
arijit@arijit-Lenovo-G550:~/meshtasticGUI$ meshtastic --settime
Connected to radio
Setting device time/position
arijit@arijit-Lenovo-G550:~/meshtasticGUI$
```

$meshtastic –settime

Fig. 6. Setting RTC time to zero

The command sets the RTC inside the device to default zero and also synchronize the other device's RTC with the same clock. The meshtastic python API and other GUI libraries is used to build meshtasticGUI desktop application.

After collecting the data of the transactions, the records of the success and failure of packets are merged together with the file where the sent packet record is stored. By evaluating the difference between the message send time and acknowledgment receive time, the entire round trip time of the packet transmission was calculated.

The communication process is governed by the following steps:

MeshtasticGUI Algorithm::

```
Input: Text File with the collected data
Output: Three CSV file and one JSON file
1. Opening the ports and establishing the connection between the device and the workstation.
2. send some dummy messages by typing on the Textbox of the GUI.
3. Wait till the responses from the remote devices.
4. If responses are coming as success or failure then
4.1 while end of file is true do
4.1.1 Read 223 characters from the file and send through interface.sendText(),
      record it in <filename>.csv file with timestamp;
4.1.2 Receive the incoming acknowledgment packet;
4.1.3 if acknowledgment is success then
4.1.3.1 Record it in the <filename>_success.csv with timestamp;
else if acknowledgment is failure then
4.1.3.2 Record it in the <filename>_failure.csv with timestamp;
end of If
4.1.4 wait till predefined time interval completes;
4.1.5 Record the transaction in the <filename>.JSON file as JSON array;
4.1.6 Close the connection;
end of while
5. else
5.1 Terminate the connection and rerun the program
end of If
```

A. Wireless Devices: LoRa base station consists of Heltec LoRa module connected to a Laptop with USB Type C cable. Each LoRa node for signal transmission consists of a LoRa module based on ESP32 microcontroller and SX1276 transceiver. These modules are provided by Heltec: Wi-Fi LoRa 32 module for 433 MHz and Wi-Fi LoRa 32 (v2) module for 868 MHz experiments respectively. Figure 7 depicts the corresponding Meshtastic device.

During each experiment, LoRa transmitter continuously sends and receives packets containing their identification number, text message, device IDs and acknowledgment (from the receiving node), and the base station. For this experiment, 125, 250 and 500 KHz bandwidth channels with spreading factors (SF) range between 7 to 12 conventional units and a coding rate between 5 to 8. The channels are altered using Command Line Interface. Each time before running the program, in order to gather information regarding a newly embedded device status the following command is used (device should be connected with the terminal). The Fig. 8 depicts the command used to display the Meshtastic information.

The following command is used to change the channel radio settings.

```
$meshtastic --port <port number> --setchan spread_factor <value>
  --setchan coding_rate <value> --setchan bandwidth <value>
```

Fig. 7. Heltec Wi-Fi LoRa 32 (v2) module

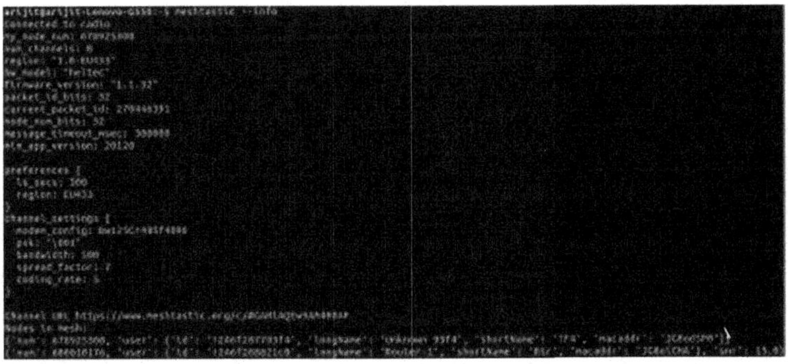

Fig. 8. $meshtastic-info and node information of devices which are using the same radio setting

There are 32 different radio settings applied on Meshtastic to observe each channel's behavior and performance. Table 2 shows the maximum throughput in LoRa for different spreading factors and bandwidth.

Table 2. Max throughput (in bps)

SF/BW	500 KHz	250 KHz	125 KHz
6	37500 bps	18750 bps	9375 bps
7	21875 bps	10937.5 bps	5468.75 bps
8	12500 bps	6250 bps	3125 bps
10	3906.25 bps	1953.125 bps	976.5625 bps
12	1171.875 bps	585.9375 bps	292.9687 bps

B. Adhoc Meshtastic Network: The Adhoc Meshtastic Network can be formed by adding any number of devices on a particular channel. Different channel settings are used in different environments depending upon the distance, speed, power consumption, and reliability. Table 3 explains the most popular channels.

Table 3. Address transformation from device to device Packet.

Channel setting	Alt channel	DR	SF/Sym	CR	BW	Link budget
Short range (but fast)	Short Fast	21.875 kbps	7/128	4/5	500	134 dB
Medium range (but fast)	Medium	5.469 kbps	7/128	4/5	125	140 dB
Long range (but slower)	Long Alt	0.275 kbps	9/512	4/8	31	153 dB
Very long range (but slow)	Long Slow	0.183 kbps	12/4096	4/8	125	154 dB

C. meshtasticGUI Interface: Meshtastic project provides a python API for the developers to built application programs for real-life usage. It can be downloaded and installed as python library and used in the application program. The meshtasticGUI interface provides a meshtastic chat environment which is also used for sending an entire file data. It is made for the computers having Linux operating system. Tkinter library is used to build the GUI interface, where several buttons are provided for some purposes. Clicking on those do the following:

SEND: To send the typed messages.
STOP: To close the connection.
READ: To start reading from a file and send the message.
EXIT: To abort the connection and stop the program.

It uses publish and subscribe model at the application layer and the serial communication between the workstation and device take place at 921600 bps through the USB cable. The field values of Tx power, radio settings, destination address, response, rx_time is set either through API or Meshtastic admin commands through terminal. The transmission packet records are stored CSV and JSON format (Fig. 9).

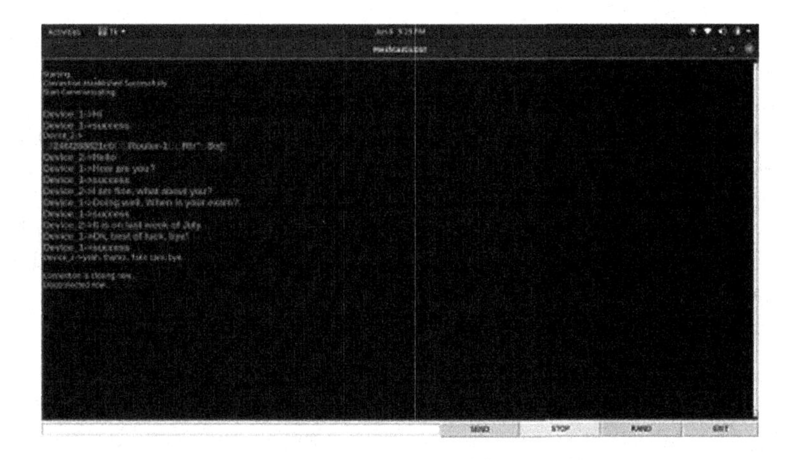

Fig. 9. Developed Meshtastic GUI interface for command execution

4 Implementation Details

This section covers the practical implementation details related to the configuration settings of the network setup.

A. Address Transformation. In Meshtastic network on a particular channel a message can either be unicast or broadcast. The addresses which are 48 byte physical (MAC) address of the data-link layer are used for the recipient identification. For, broadcast the broadcast number 0/xffffffff is used, otherwise the device name can be provided at the time of sending the message. The devices work across various vendors and implementations, using Protocol Buffers pervasively with respect to it's API clients. Before the starting of every packet delivery the channel settings are altered to different spreading factor, bandwidth and coding rate. Initially by using the "settime" option of the command line, the two devices are synchronized to a same timestamp, which is useful for the experiment. Figure 8 describes the nodes information which are on the same channel.

B. Packet Translation. MeshPacket (Meshtastic packet) in peer to peer messaging comes with a 32 bit LoRa preamble. A minimum (8 bit) preamble is used to maximize the amount of time the LORA receivers can stay asleep, which dramatically lowers power consumption.

After the preamble the 16 byte packet header is transmitted. This header is described directly by the PacketHeader class in the C++ source code. But indirectly it matches the first portion of the "MeshPacket" protobuf definition. But notably: this portion of the packet is sent directly as the following 16 bytes (rather than using the protobuf encoding). We do this to both save airtime and to allow receiving radio hardware the option of filtering packets before even waking the main CPU.

- **to (4 bytes):** the unique NodeId of the destination (or 0xffffffff for NodeNum_BROADCAST)
- **from (4 bytes):** the unique NodeId of the sender
- **id (4 bytes):** the unique (with respect to the sending node only) packet ID number for this packet. We use a large (32 bit) packet ID to ensure there is enough unique state to protect any encrypted payload from attack.
- **flags (4 bytes):** Only a few bits are currently used - 3 bits for the "HopLimit" (see below) and 1 bit for "WantAck" After the packet header the actual packet is placed onto the wire. The maximum total size of the LoRa packet is 256 bytes.

The Fig. 10 represents the Meshpacket structure

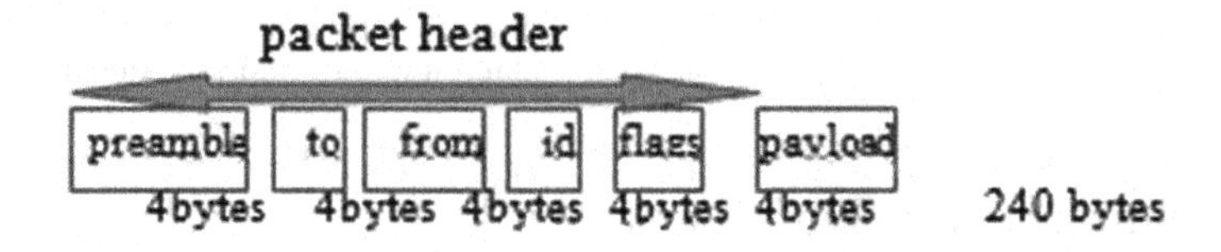

Fig. 10. Meshtastic LoRa packet structure

C. Storage of Data. The fields of the every packet can be accessed through Python program as it is in dictionary structure. While transmitting the data, the information regarding the transmitted packet and it's acknowledgment is received at the base workstation. They are stored as a JSON array which is eventually stored in a JSON type file with the field values of Text message, to, from, message ID, transmission timestamp, Acknowledgment (success or failure) and it's receiving timestamp. The following two figures are examples of a successful and failure message transmission. Then it's extracted into CSV file to do the further analysis (Figs. 11 and 12).

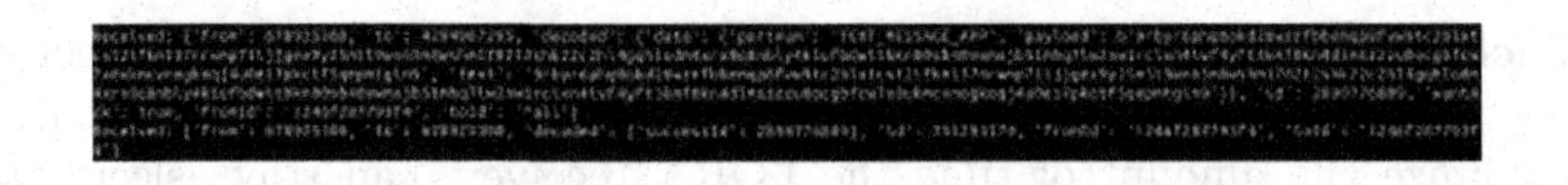

Fig. 11. Meshtastic packet success in JSON format.

Fig. 12. Meshtastic packet failure in JSON format.

5 Experimental Results

Developed system is tested in the school of Computer and Information Science building, University of Hyderabad. LoRa base station connected with the transmitter node was permanently placed on the computer lab in ground floor, while LoRa receiver was being on 10 m distance. The transmitter and receiver was placed on the same room. The experiments were conducted with total of 30 different radio settings and Table 4 shall display the results of the analysis of the data which is captured while transmission of a text file size of 58.4 kB with variable packet size between 33 bytes to 256 bytes. The result contains the total propagation delay, throughput and loss of packet during the transmission.

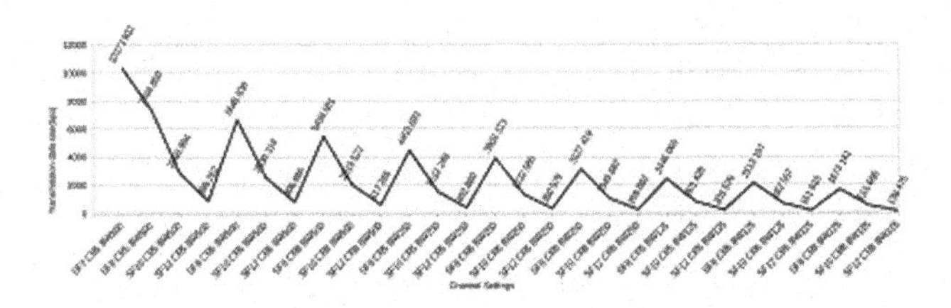

Fig. 13. Real-time graphical representation of transmission delay on different channels

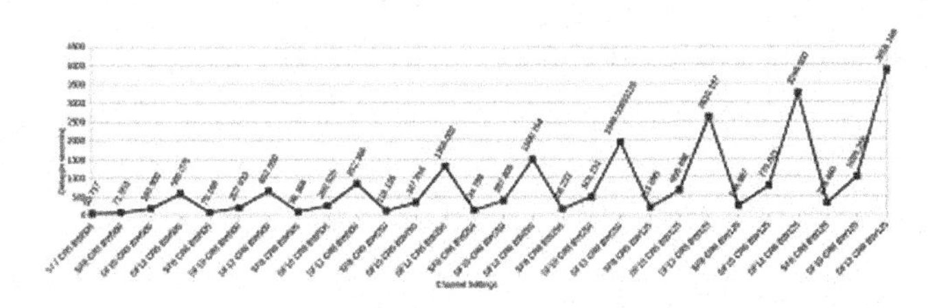

Fig. 14. Real-time graphical representation of throughput on different channels

Figure 13 and Fig. 14 represents the graphical representation of the Table 4 where the X-axis is the different radio settings and Y-axis are delay and throughput respectively. The results are also compared with a different file size to compare and contrast the reliability of the channel.

Table 5 shows the delay and throughput contrast of two different size of files in 4 different channels followed by the graphical representation in Fig. 15, 16, 17 and Fig. 18.

Table 4. Delay, throughput and packet loss of text data transmission in Meshtastic.

SNo	SF	CR	BW	Delay (sec)	Throughput (bps)	Packet loss
1	7	5	500	50.72	10373.92	4
2	8	5	500	71.05	7404.85	1
3	10	5	500	183.9	2860.99	8
4	12	5	500	585.08	899.26	6
5	8	6	500	79.17	6645.84	0
6	10	6	500	207.93	2530.32	1
7	12	6	500	652.05	806.89	3
8	8	8	500	96.37	5459.68	1
9	10	8	500	260.52	2019.52	0
10	12	8	500	852.37	617.27	28
11	8	5	250	118.13	4453.69	3
12	10	5	250	347.92	1512.25	4
13	12	5	250	1306	402.86	3
14	8	6	250	134.8	3903.12	14
15	10	6	250	397.81	1322.59	1
16	12	6	250	1500.76	350.58	22
17	8	8	250	168.23	3127.42	4
18	10	8	250	501.23	1049.69	23
19	12	8	250	1949.51	269.88	1
20	8	5	125	215.1	2446.06	2
21	10	5	125	656.5	801.43	6
22	12	5	125	2610.11	201.58	5
23	8	6	125	248.99	2113.11	23
24	10	6	125	770.83	682.56	0
25	12	6	125	3249.55	161.91	10
26	8	8	125	314.46	1673.14	0
27	10	8	125	1020.27	515.69	3
28	12	8	125	3855.17	136.48	8
29	7	5	125	135.76	4386.65	6
30	9	8	31	2999.75	198.52	2

Table 5. Comparison chart of the data transmission for 58.4 kB and 105.4 kB

SNo	SF	CR	BW	Delay (sec) 58.4 kB	Delay (sec) 105.4 kB	Throughput (bps) 58.4 kB	Throughput (bps) 105.4 kB	Packet loss 58.4 kB	Packet loss 105.4 kB
1	7	5	500	50.72	89.56	10373.92	10762.8	4	4
2	8	5	500	71.05	233.37	7404.85	4130.65	1	0
3	10	5	500	183.9	4783.4	2860.99	201.52	8	9
4	12	5	500	585.08	6917.07	899.26	139.36	6	3

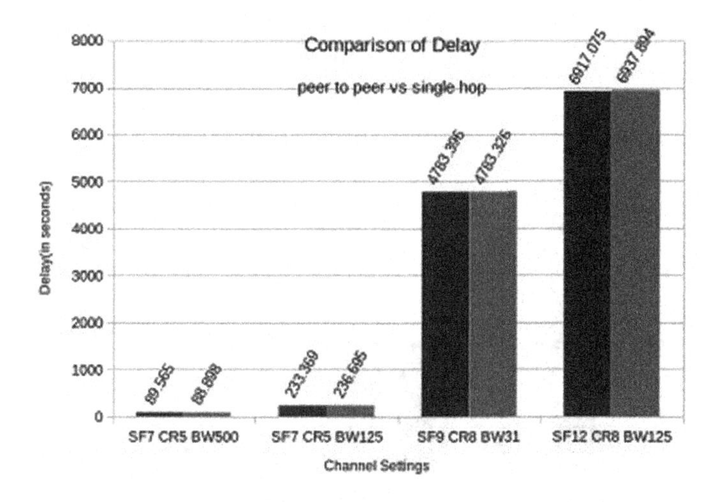

Fig. 15. Comparison graph of the delays of 105.4 kB file (zero hop vs single hop)

Throughput: Throughput is a measure of how many units of information a system can process in a given amount of time.

Jitter: Jitter is a delay variation that puts stress on the receiving endpoint, as it is trying to figure out the right sequence of data packets. There is no network free of Jitter. It affects latency-sensitive applications and hurts the user experience. Jitter introduces inconsistencies which influence the quality of communication and data transfer speeds. The following Figs. 19, 20, 21 and 22 are comparisons of Jitter in the Meshtastic network between zero hop and single hop network.

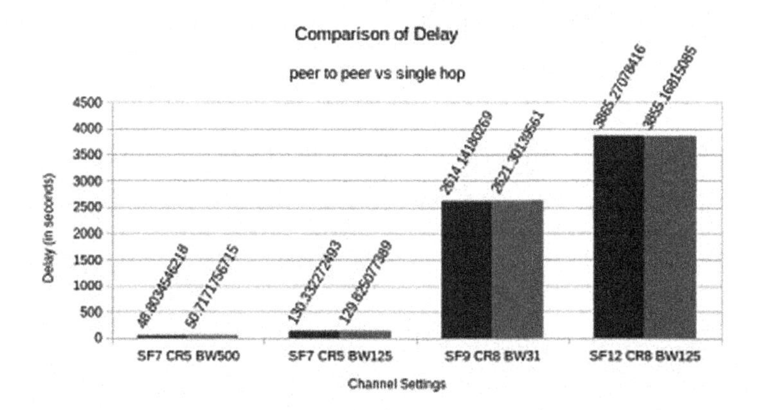

Fig. 16. Comparison graph of the delays of 58.4 kB file (zero hop vs single hop)

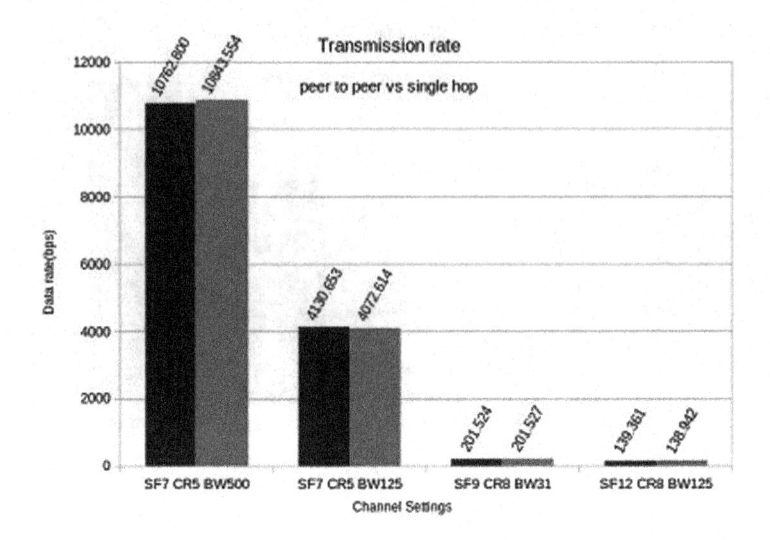

Fig. 17. Comparison graph of the throughput of 58.4 kB file (zero hop vs single hop).

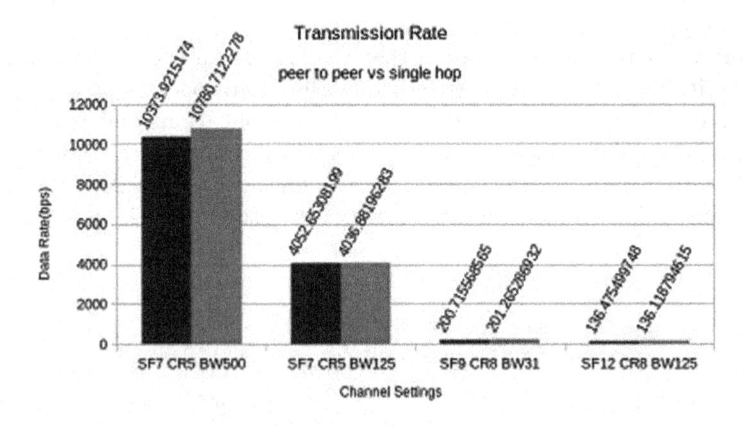

Fig. 18. Comparison graph of the throughput of 108.4 kB file (zero hop vs single hop).

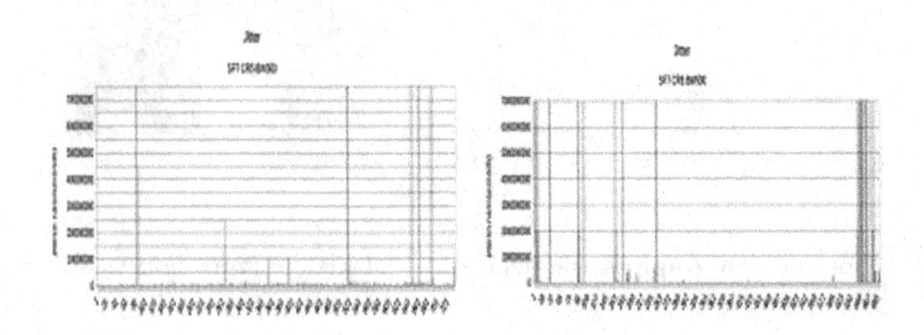

Fig. 19. Comparison of Jitter in Short range (but fast) channel

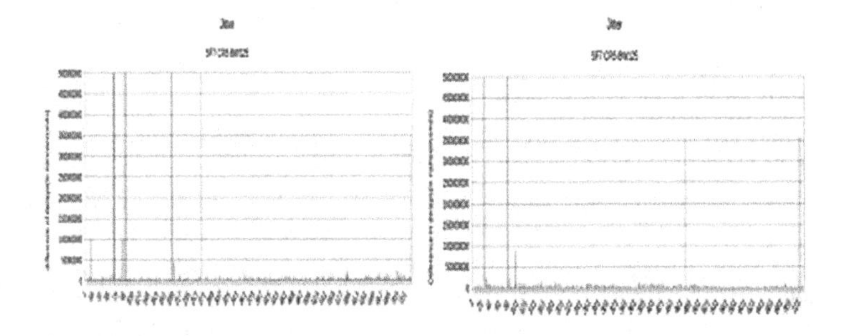

Fig. 20. Comparison of Jitter in Medium range (but fast) channel

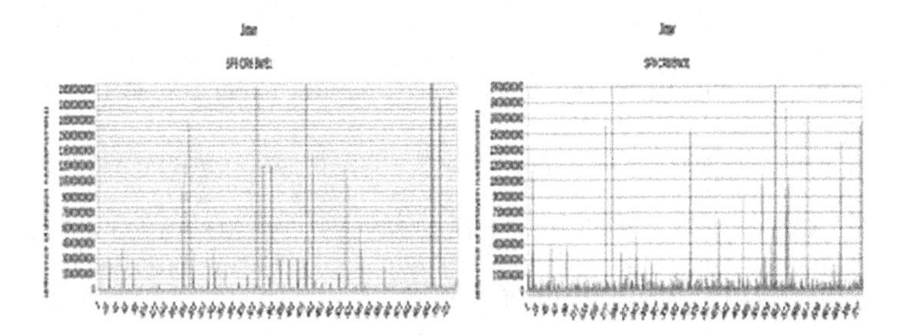

Fig. 21. Comparison of Jitter in Long range (but slow) channel

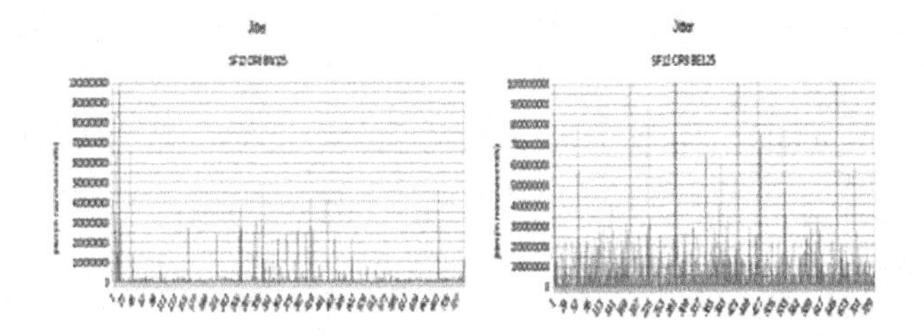

Fig. 22. Comparison of Jitter in Very Long range (but slow) channel

6 Discussion and Future Work

This article provides the experimental study results of teletype data transmission on the basis of LoRa technology using Meshtastic devices. The developed technology can be used in the situations where there is no telecommunication/Internet system for data transfer across long geographical regions. Meshtastic is an ongoing open source community project. There are future improvements that can be implemented on sending and receiving multimedia data like image, audio and video. Multi-hop unicast is a controlled broadcasting through several nodes in a channel. It will be very interesting to explore how the data transmission can be done.

Acknowledgements. The authors would like to thank the IoE PDF grant for the financial support provided in registration of the conference.

References

1. ITU-T Recommendation Y. 2060. Overview of Internet of Things, Geneva, February 2012. https://www.itu.int/en/ITU-T/gsi/iot/Pages/default.aspx
2. Kirichek, R., Kulik, V.: Long-range data transmission on flying ubiquitous sensor networks (FUSN) by using LPWAN protocols. In: Vishnevskiy, V.M., Samouylov, K.E., Kozyrev, D.V. (eds.) DCCN 2016. CCIS, vol. 678, pp. 442–453. Springer, Cham (2016). https://doi.org/10.1007/978-3-319-51917-3_39
3. Bouguera, T., Diouris, J.-F., Chaillout, J.-J., Jaouadi, R., Andrieux, G.: Energy consumption model for sensor nodes based on LoRa and LoRaWAN. Sensors **18**(7), 2104 (2018). https://doi.org/10.3390/s18072104
4. Lopresti, D., Shekhar, S.: A National Research Agenda for Intelligent Infrastructure (2021). https://cra.org/ccc/resources/ccc-led-whitepapers/. Accessed 21 Mar 2021
5. Portmann, M., Pirzada, A.A.: Wireless Mesh networks for public safety and crisis management applications. IEEE Internet Comput. **12**(1), 18–25 (2008). https://doi.org/10.1109/MIC.2008.25
6. https://www.silabs.com/whitepapers/selecting-the-appropriate-wireless-mesh-network-technology
7. Cilfone, A., Davoli, L., Belli, L., Ferrari, G.: Wireless Mesh networking: an IoT-oriented perspective survey on relevant technologies. Future Internet **11**, 99 (2019). https://doi.org/10.3390/fi11040099
8. Mekki, K., Bajic, E., Chaxel, F., Meyer, F.: A comparative study of LPWAN technologies for large-scale IoT deployment. ICT Exp. **5**(1), 1–7 (2019). https://doi.org/10.1016/j.icte.2017.12.005. ISSN 2405-9595
9. Jiang, S.: Introduction and overview. In: Wireless Networking Principles: From Terrestrial to Underwater Acoustic, pp. 1–31. Springer, Singapore (2018). https://doi.org/10.1007/978-981-10-7775-3_1
10. https://communitynetworks.group/t/long-range-low-bandwidth-mesh-projects-for-good/306
11. https://meshtastic.org/docs/about. Accessed 18 Aug 2020
12. https://github.com/meshtastic/Meshtastic-protobufs
13. Tripathy, B.K., Jena, S.K., Reddy, V., Das, S., Panda, S.K.: A novel communication framework between MANET and WSN in IoT based smart environment. Int. J. Inf. Technol. **13**(3), 921–931 (2020). https://doi.org/10.1007/s41870-020-00520-x
14. Khan, M.A., Safi, A., Qureshi, I.M., Khan, I.U.: Flying ad-hoc networks (FANETs): a review of communication architectures, and routing protocols. In: First International Conference on Latest trends in Electrical Engineering and Computing Technologies (INTELLECT) 2017, pp. 1–9 (2017)
15. Abramson, N.: The ALOHAnet - surfing for wireless data. IEEE Commun. Mag. **47**(12) (2009)
16. https://www.thethingsnetwork.org/docs/lorawan/frequencies-by-country/ . Accessed 20 Dec 2020
17. Islam, B., Islam, M.T., Kaur, J., Nirjon, S.: LoRaIn: making a case for LoRa in indoor localization. In: IEEE International Conference on Pervasive Computing and Communications Workshops (PerCom Workshops) 2019, pp. 423–426 (2019). https://doi.org/10.3390/ijgi7110440
18. Bobkov, I., Rolich, A., Denisova, M., Voskov, L.: Study of LoRa performance at 433 MHz and 868 MHz bands inside a multistory building. In: Moscow Workshop on Electronic and Networking Technologies (MWENT) 2020, pp. 1–6 (2020). https://doi.org/10.1109/MWENT47943.2020.9067427

High-Precision Monitoring System for Strong Earthquakes in Buildings Based on Distributed Intelligent Ad-Hoc Network

Xu Bai(✉), PengFei Feng, MingJie Ji, Yang Zhang, and ShiZeng Guo

School of Electronics and Information Engineering, Harbin Institute of Technology, Harbin, Heilongjiang, China
x_bai@hit.edu.cn

Abstract. Since the 20th century, the death toll caused by earthquakes in China has accounted for about half of all natural disasters in the country where the building collapse is the main cause of casualties. Therefore, it is an important issue for human society to recognize the mechanism of earthquake waves destroying buildings and design more earthquake-resistant building structures. We have developed and implemented a low-power, high-reliability and high precision earthquake monitoring system that can be deployed in buildings. The monitoring system uses high-precision sensors to collect the vibration information of each node of the building in real time, and save it in the local memory in real time, and at the same time make real-time judgments on the sampled information. When an earthquake occurs, the terminal acquisition node of the monitoring system will use the Lora ad hoc network to upload the vibration data before and after the earthquake to the cloud platform. After testing, the results show that the monitoring system operates stably in a strong earthquake environment and can provide data support for the seismic design of buildings in the future.

Keywords: Earthquake monitoring · Data acquisition · Ad-Hoc

1 Introduction

China is located between the Pacific Rim Earthquake and the Eurasian Seismic Belt, with high seismic activity frequency and high intensity [1]. This poses a serious threat to the safety of people's lives and property. Statistics show that since the 20th century, the number of people who died in earthquakes in China has reached 550,000, accounting for half of the total number of deaths due to natural disasters in the country. The main cause of casualties in the earthquake was the collapse of houses. In the 2010 Yushu earthquake in Qinghai, more than 85% of the houses collapsed. Many people were buried under the damaged houses, causing a large number of casualties and economic losses [2]. In recent years, high-rise buildings and super high-rise buildings have continuously appeared in China. They have higher sensitivity and vulnerability to earthquake disasters. Once they collapse, they will cause more serious consequences. Therefore, there is an urgent need to develop an earthquake monitoring system to measure the relative movement of buildings when an earthquake occurs, to establish an

Q. Guo et al. (Eds.): WiSATS 2021, LNICST 410, pp. 641–653, 2022.
https://doi.org/10.1007/978-3-030-93398-2_56

ideal mathematical analysis model, and to enhance the understanding of the mechanism of earthquake wave damage to the building, so as to design a more reasonable earthquake-resistant building structure. Minimize the damage caused by the earthquake to the lowest possible extent [3]. For example, the United States has established an earthquake disaster mitigation network, ANSS system and NEES system [4]. The system is equipped with about 7,000 seismic stations and sensors to measure the response of the ground and buildings to earthquake vibrations, build a seismic information database, and provide the data to the architectural design engineer in order to build stronger earthquake-resistant houses [5].

When monitoring the degree of damage to a building structure by an earthquake, the structural dynamic response information is usually obtained by measuring structural vibration information, and various data characteristics contained in the structural dynamic response information are analyzed. Therefore, the seismic monitoring system is mainly to collect the vibration information of the building. Dense distributed sensors are critical to the efficiency of vibration-based damage identification. With the emergence of high-rise buildings and super high-rise buildings, it is necessary to set up more monitoring points for these buildings, and the distance between different monitoring points is also increasing, which causes the actual wiring to be time-consuming and laborious. At the same time, in a complex building environment, compared to wireless solutions such as Wi-Fi and Bluetooth, the wireless solution based on the Lora (Long Range) protocol has more advantages in anti-interference and transmission distance, so the earthquake monitoring system uses Lora wireless Ad Hoc Networks which makes the layout of the monitoring system terminal more convenient and faster [6–8].

2 Function Realization

2.1 System Structure Model

We aim to design a low-power, high-reliability earthquake monitoring system that can be deployed in buildings. The system is mainly composed of three parts, namely the terminal acquisition node, the cloud server and the user platform. As shown in Fig. 1. In order to save resources and reduce costs, the acquisition nodes arranged in close positions use the Lora module to perform ad-hoc networking to transmit data. There are one or more terminal acquisition nodes installed with 4G modules in the same networking. The data of each node in the network is aggregated and uploaded to the cloud server. Researchers can download the vibration information of each node of the building by accessing the cloud server and analyze the data to design a more earthquake-resistant building.

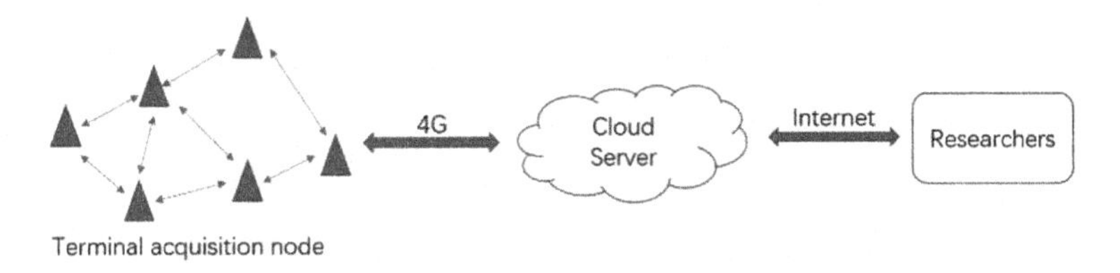

Fig. 1. Overall system structure

The specific structure diagram of the acquisition node is shown in Fig. 2, which includes a three-degree-of-freedom acceleration sensor, low-pass filter, A/D sampling module, single-chip microcomputer main control chip, storage Flash, Lora module and 4G module.

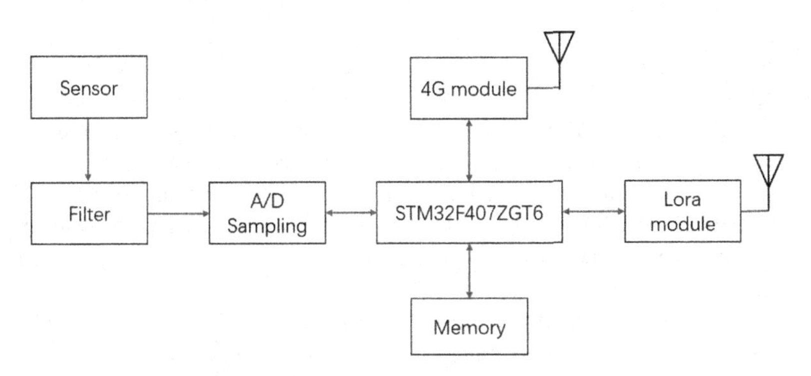

Fig. 2. The structure diagram of the acquisition node

The system uses a dual-power adaptive switching power supply module, daily use of city electricity to supply power for each terminal acquisition system and charge the backup battery, so when a high-magnitude earthquake occurs, the urban power system is destroyed, and the power supply system automatically switches to the backup battery for power supply.

The software implementation of the system is written by C language. In order to facilitate software writing, debugging and management, software programming adopts a modular structure, which mainly includes an initialization module, an acquisition module, a data storage module, and a data transmission module. The flow chart is shown in Fig. 3.

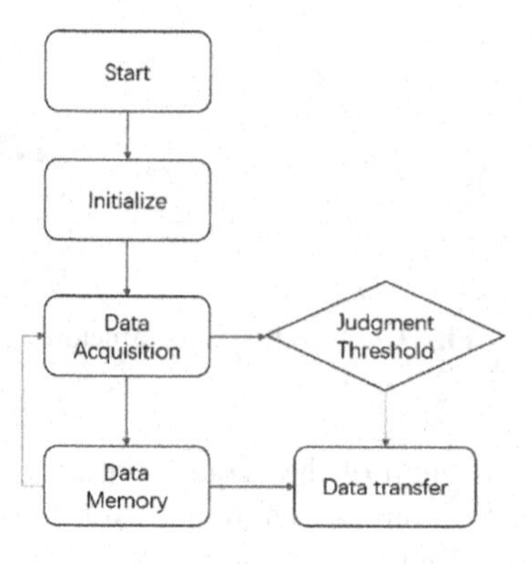

Fig. 3. Program flow diagram

2.2 Data Acquisition

The sensor converts vibration information in the three directions of X, Y, and Z into three analog voltage signals. The output voltage signal is usually a mixed signal of effective signal and interference noise. The response of buildings to earthquake vibration is often a low-frequency signal, which has the characteristics of large amplitude, concentrated energy, and slow signal change over time. Generally, it will not exceed 100 Hz. So it needs to go through a 100 Hz low-pass filter first to filter out high-frequency noise. Then the voltage signal is sent to the A/D sampling chip for sampling. The A/D converter selects the AD4111 chip, which is a low-power, low-noise, 24-bit Σ-Δ analog-to-digital converter. The chip can support 4 differential or 8 single-ended sampling channels. Support the maximum channel scan rate of 6.2 kSPS. When the A/D is sampled at a sampling rate of 200 Hz per channel, after testing, the root mean square noise is 27 μV, and the effective resolution calculation formula is:

$$6.02N + 1.76 = 20\log(noise/FSI) \tag{1}$$

It can be calculated that the effective resolution can reach 19.5 bit. Satisfy the speed and accuracy requirements for collecting building vibration data.

When the acquisition node stores and transmits the sampled data, it uses binary text for storage to save storage addresses. The frame format of each frame is shown in Table 1, including the address of the current acquisition node, the time of acquisition, the data in the three directions of X, Y, and Z, and the data synchronization bit.

Table 1. Frame format

Address	Time	X channel	Y channel	Z channel	Sync
2 bytes	4 bytes	3 bytes	3 bytes	3 bytes	1 byte

The acquisition time is set to 4 bytes, including 1 byte each for hour, minute, second, and sub-second. After sampling the three channels, by accessing the MCU's internal real-time clock register, the current time of the sampled data can be obtained, and the data can be time-labeled to facilitate the researcher to analyze the data. The data of the X, Y, and Z channels are 24-bit offset binary (offset binary). The code generated by full-scale negative voltage is 000...000. The code generated by zero differential input voltage is 100...000. The code generated by the full-scale positive voltage is 111...111. The output code of any analog input voltage can be expressed as:

$$Code = 2^{N-1} \times ((V_{IN} \times 0.1/V_{REF}) + 1) \quad (2)$$

where N = 24, V_{IN} is the analog input voltage, V_{REF} is the reference voltage, 2.5 V is selected as the reference voltage in this system and set 0x0A as the data synchronization bit at the end of each frame of data, which is convenient for researchers to process the sampled data.

2.3 Data Transmission

Since the time without an earthquake takes up a larger proportion, the information sampled at these times is useless. If the terminal monitoring node continuously sends out the sampled data, it will not only cause greater power consumption, but also occupy more channel resources, which limits the number of nodes in the local network. Therefore, when there is no earthquake, only a small amount of data is sent to the cloud to mark normal working nodes, and to check and repair abnormal working nodes in time. At the same time, real-time threshold judgment is performed on the data sampled by the terminal acquisition node. When the acceleration amplitude exceeding the threshold is sampled, start counting and observe the data sampled for the next 100 times. If 30 of the sampled data also exceed the threshold, it is considered that the node is indeed shaken, and a packet indicating the occurrence of an earthquake is sent to the sink node. If the sink node receives data packets indicating the occurrence of an earthquake by multiple nodes in the network for a period of time, it is considered that an earthquake has occurred, and the nodes in the network are ordered to send the data sampled 30 s before and after the earthquake. If only a data packet indicating the occurrence of an earthquake is received from a node.it is considered that no earthquake has occurred, This vibration caused by other reasons, such as house decoration.

When an earthquake occurs, it is inevitable that some terminal nodes will be disconnected, especially when the acquisition node with 4G function is damaged, it will affect all the acquisition nodes under the node, resulting in the loss of a large amount of sampled data. In order to minimize the loss caused by node damage, the system adopts the AODV protocol dynamic routing networking method [9]. When the route is broken, the RREQ (Route Request) message is sent out in the form of multicast to find a new path, so as to send the sampled data (Fig. 4).

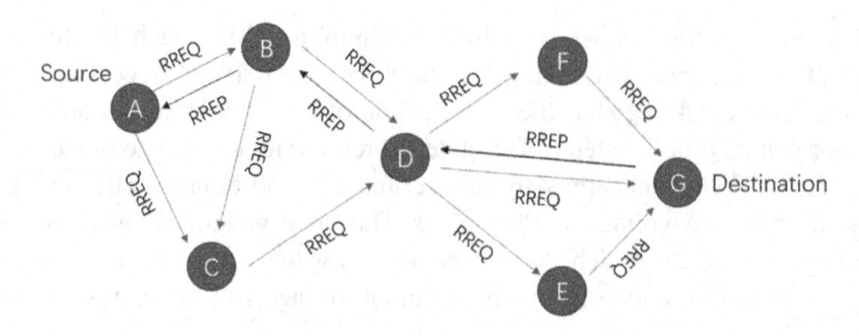

Fig. 4. Route discovery process

When an earthquake occurs, when the power of the terminal acquisition node is cut off, only a backup battery can be used for power supply. When some nodes whose communication is cut off may need to work longer, and wait for the communication to resume to transmit data. Reducing energy consumption is an effective way to extend the survival time of nodes. However, in actual engineering, in order to facilitate deployment and management, the transmit power of each node is the same. Although this setting reduces the complexity of the routing protocol. if the two nodes are relatively close, the data packet is still sent with a fixed maximum power. This caused a certain amount of energy waste. The path selection in the traditional AODV protocol is based on the minimum number of hops and does not consider the issue of network power consumption. But the least number of hops does not mean the shortest path and the least energy consumption, so the traditional AODV protocol is not suitable for this system.

Therefore, we have improved the traditional AODV protocol, and changed the route selection based on the least number of hops to the route selection based on the least energy consumption.

If you want to get the least energy consumption route, firstly, you should modify the RREQ data packet, and change the hop count in the original RREQ data packet to the sum of energy consumption. And the newly added Sending Energy table item, Sending Energy represents the power forwarded by the node, the newly modified RREQ data packet structure is shown in Table 2.

Table 2. Packet format of RREQ

Type	Reserved	Prefix Sz	Sum of energy
Sending energy			
RREQ ID			
Destination IP address			
Destination sequence number			
Originator IP address			
Originator sequence number			

RREP (Route Response) is to inform the route from the source node to the destination node that sends the RREQ message. There is also a function in the improved AODV protocol, which is to tell the destination node the total power consumed on the path, and at the same time tell the previous node through the reverse route The minimum transmit power. The newly modified RREP packet structure is shown in Table 3.

Table 3. Packet format of RREP

Type	Reserved	Prefix Sz	Sum of energy
Sending energy			
Destination IP address			
Destination sequence number			
Originator IP address			
Originator sequence number			
Lifetime			

Each node in the network must maintain a routing table. If a route is to be selected according to the minimum energy consumption, the original routing table must be modified. Change the number of hops to the destination node to the sum of the energy consumption to reach the destination node. Sending Energy is a copy of the Sending Energy table entry from RREQ, which indicates the transmitting power of the node, and the routing table structure is shown in Table 4.

Table 4. Routing table entry fields

Destination IP address
Destination sequence number
Interface
Routing flags
Sending energy
Sum of energy
Next hop
List of precursors
Lifetime

When the source node cannot find the legal route of the destination node, the route establishment process is performed. Send RREQ data packets with increasing transmit power (21 dBm, 24 dBm, 27 dBm and 30 dBm) respectively, and fill in the corresponding sending power value in the Sending Energy bit of RREQ, and calculate the value of Sum of Energy.

When a node receives an RREQ packet, it will determine whether it has received an RREQ message with the same source IP address and RREQ ID in the past period of time. If not, search the local routing table to see if there is a reverse route to the source IP address. If not, establish a reverse route. The process of establishing and updating the reverse route is as follows:

1. Compare the sequence number of the source node in the RREQ with the entry in the routing table whose destination IP is the source node, and update the sequence number in the routing table to the maximum value of the two.
2. The next hop field in the corresponding entry of the routing table is set to the IP address of the previous node that received the RREQ message.
3. The Sum of Energy value in the corresponding entry of the routing table is modified to the Sum of Energy value in the RREQ packet, and the Sending Energy value of the RREQ packet is copied into the Sending Energy entry of the routing table.

If it is not the first time to receive the RREQ packet, compare the value of Sun of Energy in the RREQ packet with the value of Sum of Energy in the corresponding entry in the routing table. If the value of Sun of Energy in the RREQ packet is smaller than the value of Sum of Energy, the reverse routing table is updated. However, updating the reverse routing table this time does not modify the sequence number in the routing table. Then the RREQ data packet is forwarded with the transmission power from small to large, and the Sending Energy value of the RREQ data packet is filled with the corresponding transmission power value to calculate the Sum of Energy value. If the Sum of Energy value in the RREQ data packet is greater than the Sum of Energy value in the reverse routing table, the RREQ data packet is discarded to avoid repeated processing.

Nodes will generate RREP packets in two situations. The first is that the node is the destination node. At this time, RREP packets will be generated and unicast back to the sending node through the reverse route. Another situation is that the current node is an intermediate node and there is a route to the destination node, then fill in the RREP according to the destination node serial number, Sum of Energy, Sending Energy in the routing table, and unicast back to the sending node through the reverse route.

When the link is cut, all adjacent nodes using the link will generate a RERR message to inform other nodes of the link failure. At this time, the affected node needs to initiate a new route request to find a new path. Before finding a new path, the data needs to be saved locally.

After experimental testing, in the same environment, using the improved AODV protocol to transmit data can save 40% of the power consumption and greatly extend the life span of each node.

When using the AODV protocol dynamic routing networking mode, the MAC layer usually selects CSMA/CA (Carrier Sense Multiple Access/Collison Avoidance) technology to avoid data packet collisions on a single shared channel as much as possible [10]. However, the CSMA/CA algorithm only uses a simple binary exponential backoff algorithm, and does not consider the network conditions. However, when an earthquake occurs, all nodes in the network will start uploading a large amount of data at the same time, which greatly increases the probability of collision, which leads to data loss, and reduces the throughput of the network and increases the delay. So we need to improve the MAC layer protocol, the improvement plan is as follows:

When the earthquake does not occur, the amount of data is small, and only a small amount of data is transmitted at intervals. At this time, the MAC layer adopts the CSMA/CA protocol to avoid data packet conflicts under the same channel. When an earthquake occurs, the nodes in the network must send data to the nodes with 4G function, and the amount of data is relatively large. At this time, the MAC layer adopts the token passing protocol [11]. Only nodes with tokens in the network can send data.

In order to evaluate the program, we use 10 sets of devices to network through the AODV protocol, of which nine sets of devices simultaneously send 100 data packets to the tenth set of devices, each with a size of 1024 bytes. In the same environment, the CSMA/CA protocol and the token protocol were used for testing. The experimental results are shown in Table 5.

Table 5. Collision test of CSMA/CA and token passing

Device	CSMA/CA	Token passing
Device 1	83/100	100/100
Device 2	87/100	100/100
Device 3	80/100	100/100
Device 4	91/100	100/100
Device 5	94/100	100/100
Device 6	78/100	100/100
Device 7	86/100	100/100
Device 8	90/100	100/100
Device 9	85/100	100/100
Total time	19 min	14 min

The test results show that: when an earthquake occurs, the MAC layer uses the token passing protocol instead of the CSMA/CA protocol, which not only avoids data packet collisions, but also has a higher channel utilization rate and can upload data to the cloud server with shorter data.

3 Test and Result Analysis

In order to verify the effectiveness of the monitoring system, we carried out a working condition verification in the model building shown in Fig. 5. The ground vibration table applied different magnitudes of acceleration in the X and Y directions of the building model. To simulate the seismic excitation that causes the building to vibrate. We arrange the acquisition equipment on the top of the first floor. The acquisition equipment will only send out the data of 30 s before and after the shaking when the seismic excitation is applied by the shaking table. After the later software reads the vibration data uploaded by the acquisition node, the acquisition waveform is shown in Fig. 6.

Fig. 5. Vibration data acquisition of building model

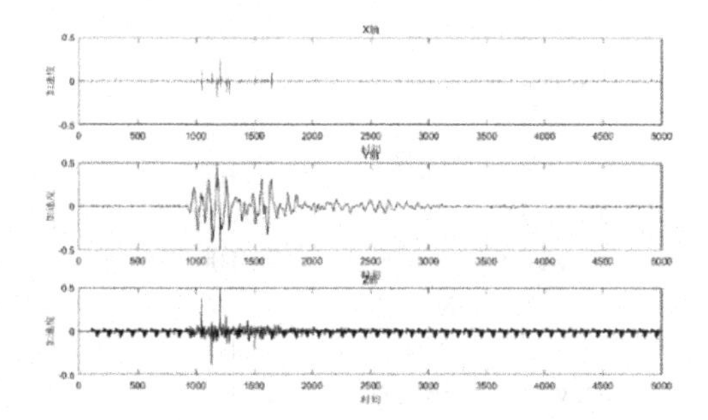

Fig. 6. The acquisition waveform of acceleration excitation with a peak value of 400 cm/s^2 is applied to the Y-axis of the building

This is the case of no interference, and it can be seen that the waveform is clearer and the noise is small. Then we add interference noise to the equipment, and the acquisition waveform is shown in Fig. 7. It can be seen that there is a lot of noise in the acquired waveform. Since noise is generally steady-state noise, it does not change with time in a short period of time. Therefore, we perform time-domain windowing and intercept the second half of the waveform without excitation for spectrum analysis. The spectrum analysis is shown in Fig. 8. It can be distinguished that the useful signal is at a low frequency, while the frequency of the noise signal is mainly concentrated around 80 Hz. So we can filter out noise by designing a filter, and the filtered signal waveform is shown in Fig. 9. It can be seen that the noise of each channel is obviously suppressed, and a better acquisition waveform can be obtained.

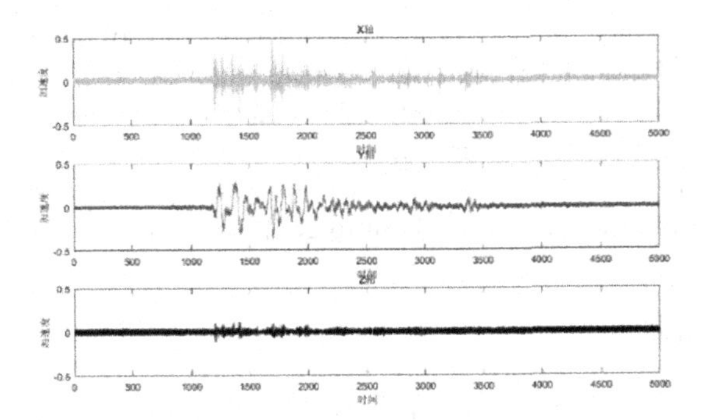

Fig. 7. In the presence of noise, the acquisition waveform of acceleration excitation with a peak value of 400 $\mathrm{cm/s^2}$ is applied to the Y-axis of the building

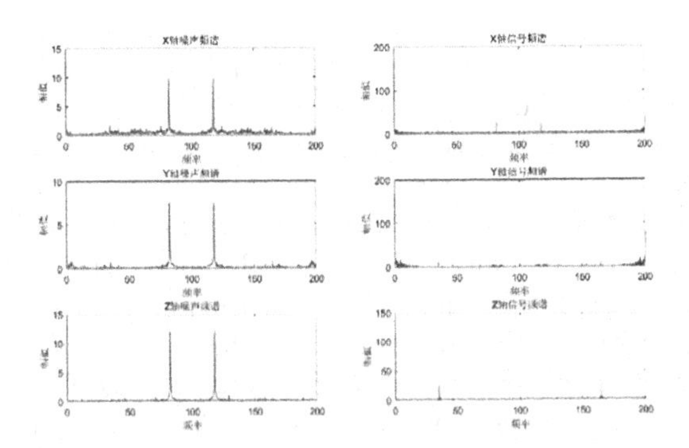

Fig. 8. Spectrum analysis

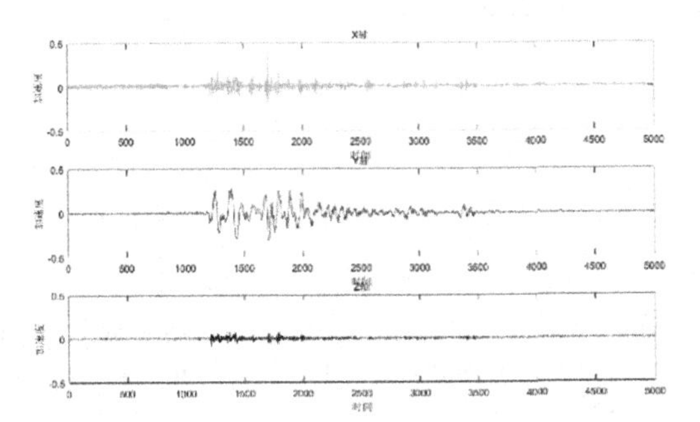

Fig. 9. Waveform after filtering the acquired data

The test results show that the system can judge the time of the earthquake through the threshold, and completely transmit the data sampled 30 s before and after the earthquake. It can be seen from the software to output the waveform that the waveform sampled is clearer without interference, and the sampled data conforms to the actual vibration situation of the building. Even if the acquisition system is disturbed due to some reasons, it can be better filtered out to get the original waveform.

4 Conclusion

The building vibration acquisition system designed in this paper has the characteristics of low cost, low power consumption and high reliability. Instead of uploading all the sampled data to the cloud server, through multiple threshold judgments on the sampled data, only the data of 30 s before and after the earthquake is uploaded. It not only reduces the power consumption of the system, but also avoids occupying channel resources to transmit a large amount of useless information, and improves the node capacity of the network. This makes it possible to densely arrange acquisition nodes in the monitored building, and use the multi-node vibration information of the building to identify damage to the building, so as to realize all-weather real-time dynamic monitoring of the building. At the same time, in view of the high destructiveness of earthquakes, AODV dynamic routing networking is adopted. Even if some nodes are damaged, data can still be transmitted in a multi-hop manner, which reduces the possibility of data loss and communication interruption, and greatly improves the robustness of the system. The test results show that the system can accurately and stably record the vibration information of the building when the earthquake occurs. Adopting the Ad-Hoc network method greatly reduces the data loss rate during transmission. The monitoring system collects the vibration information of each node of the building when an earthquake occurs, thereby constructing an earthquake information database, which is of great significance to the design and construction of buildings with stronger earthquake resistance in the future.

Acknowledgment. Supported by the National Key R&D Program of China (No. 2017YFC1500601.)

References

1. Xu, W., Liu, J., Xu, G., Wang, Y., Liu, L., Shi, P.: Earthquake disasters in China. In: Shi, P. (ed.) Natural Disasters in China. IERGPS, pp. 37–72. Springer, Heidelberg (2016). https://doi.org/10.1007/978-3-662-50270-9_2
2. Zheng, J., Liao, Y., Du, L., Niu, J.: Investigation and analysis of building damage in Yushu city caused by earthquake. Build. Struct. **A1**, 1002–1054 (2013)
3. Goulet, J., Michel, C., Kiureghian, A.D.: Data-driven post-earthquake rapid structural safety assessment. Earthquake Eng. Struct. Dynam. **44**(4), 549–562 (2015)
4. U.S. Geological Survey. Advanced national seismic system-current status, development opportunities, and priorities for 2017–2027. Translated World Seismolog. **49**(5), 397–424 (2018)

5. Elnashai, A., Di Sarno, L.: Fundamental Concepts of Earthquake Engineering. Wiley, Hoboken (2008)
6. Raza, U., Kulkarni, P., Sooriyabandara, M.: Low power widearea networks: an overview. IEEE Commun. Surv. Tutorials **19**(2), 855–873 (2017)
7. Augustin, A., Yi, J., Clausen, T., Townsley, W.M.: A study of LoRa: long range and low power networks for the Internet of Things. Sensors **16**(9), 1466 (2016)
8. Sinha, R.S., Wei, Y., Hwang, S.-H.: A survey on LPWA technology: LoRa and NB-IoT. ICT Express **3**(1), 14–21 (2017)
9. Perkins, C., Belding-Royer, E., Das, S.: RFC3561: ad hoc on-demand distance vector (AODV) routing (2003)
10. Trinh, L.H., Bui, V.X., Ferrero, F., Nguyen, T.Q.K., Le, M.H.: Signal propagation of LoRa technology using for smart building applications. In: 2017 IEEE Conference on Antenna Measurements and Applications (CAMA), pp.381–384. IEEE, December 2017
11. Moraes, R., Vasques, F., Portugal, P., Fonseca, J.A.: VTP-CSMA\CA virtual token passing approach for real-time communication in IEEE802.11 wireless networks. IEEE Trans. Ind. Inform. **3**(3), 215–224 (2007)

Resource Sharing Between Non-direct D2D Users and Cellular Users

Jiaqi Su[1], Wenbin Zhang[1,2](✉), Liyan Zhang[1], Jinwei Huang[1], and Shaochuan Wu[1]

[1] Harbin Institute of Technology, Harbin 150001, People's Republic of China
zwbgxy1973@hit.edu.cn
[2] Science and Technology on Communication Networks Laboratory, Shijiazhuang 050050, Hebei, People's Republic of China

Abstract. In recent years, mobile communication technology has developed rapidly and people have higher requirements for the communication quality. Addressing the problem of low communication quality in areas such as the edge of LTE-A cells, we introduce relay technology to expand the coverage area and improve the communication quality in hotspots. D2D technology can use the increasingly scarce spectrum resources efficiently and improve the utilization efficiency of communication system throughput and frequency band. Using layer 3 relay, we coexist with D2D direct users, D2D non-direct users, and cellular users to construct a communication network, and study the influence of different mode configurations on the throughput.

This paper considers the combination of relay technology and D2D technology in the LTE-A cell to study and analyze the communication performance of the system. Based on the LTE-A relay cellular network system architecture and link model and the communication mode of D2D technology, we combine the D2D undirected model with the LTE-A relay cellular network to establish a D2D system model, analyze the communication process, then construct a MINLP optimization problem with system throughput as the objective function and use the method of Lagrange dual to solve it. Through theoretical derivation and simulation analysis, the throughput of the LTE-A cell constructed by the combination of D2D non-direct mode and relay technology is obtained.

Keywords: LTE-Advanced cellular networks · Relay technology · D2D technology · Convex optimization

1 Introduction

1.1 LTE-A Relay Cellular Network

The introduction of relay technology in LTE-A allows relay nodes to expand the coverage area of the cell, and provide service signal coverage for areas with

Supported by National Nature Science Foundation of China (NSFC) under Grant 62071148.

Q. Guo et al. (Eds.): WiSATS 2021, LNICST 410, pp. 654–672, 2022.
https://doi.org/10.1007/978-3-030-93398-2_57

severe shadow fading in the cell, areas that cannot be covered by the signal due to the occlusion of buildings, and indoor and hotspot areas [1]. The specific implementation of the relay technology is to add some relay nodes or relay stations on the basis of the existing sites to make the distribution of the sites or antennas more dense. The user terminals use the intermediate access point, that is relay to access the network so as to obtain bandwidth services [2,3]. The use of relay nodes can reduce the loss of wireless links in free space, increase the signal-to-noise ratio, and achieve the purpose of increasing the information rate of users in edge areas. The newly added relay nodes and the existing parent base stations are connected wirelessly. When the downlink data is to be transmitted, the data to be transmitted firstly arrives at the parent base station, then it is transmitted from the parent base station to the relay node, and finally transmitted to the terminal, and vice versa is the uplink data transmission process. The data transmission method via the relay node is equivalent to narrowing the distance between the antenna and the terminal, which effectively improves the link quality in the terminal user information transmission process, and therefore can improve the system information rate and spectrum efficiency. In terms of network construction complexity and cost, the complexity of building a relay device is much lower than that of building a base station. The use of relay technology to achieve coverage in a cell can greatly reduce costs [4–7].

The LTE-A system was originally a traditional single-hop cellular network. After the introduction of relays, it becomes a relay cellular network, which is two-hop or multi-hop. This paper only studies the two-hop relay cellular network. Figure 1 shows the network architecture of the LTE-A relay cellular system under R10. Compared with the original LTE-A system architecture, it adds other interfaces such as RS. The user UE is connected to the RS through the interface Uu, and the RN is connected to the corresponding main base station DeNB through the interface Un. In the relay node protocol architecture, the DeNB is responsible for all interfaces X2 and interfaces S1 in the relay cellular network. After the introduction of relay nodes in the network, more communication links will inevitably be produced. According to the different service objects, the links in the cell where the relay exists are divided into three types, namely, the access link, the relay link, and the direct link [8–10]. In Fig. 2, there are a base station eNB, a relay node RN, and a user UE in the cell. The relay communicates with its covered users through the access link (L4), the relay communicates with the base station through the relay link (L3), and the base station communicates directly with the user through the direct link (L1, L2) [11–13].

1.2 D2D Communication Technology

How D2D Communication Technology Works. D2D communication refers to end-to-end direct communication technology under the control of cell base stations. D2D technology can use frequency band resources more efficiently, to a certain extent can alleviate the problem of shortage of communication spectrum resources, and can improve the throughput and spectrum utilization efficiency of the communication system. Figure 3 shows the D2D communication link model.

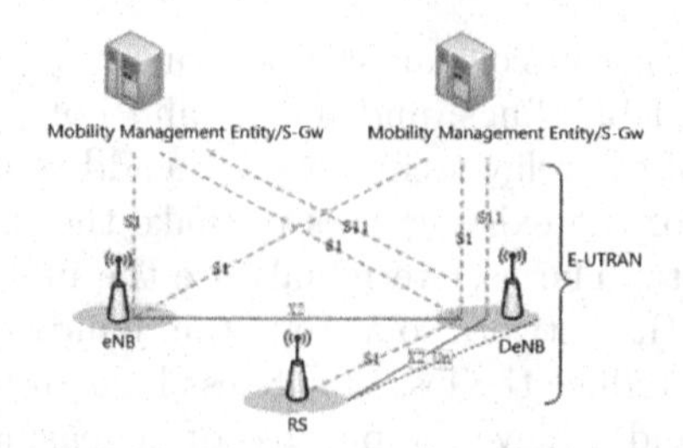

Fig. 1. Network architecture of relay cellular network

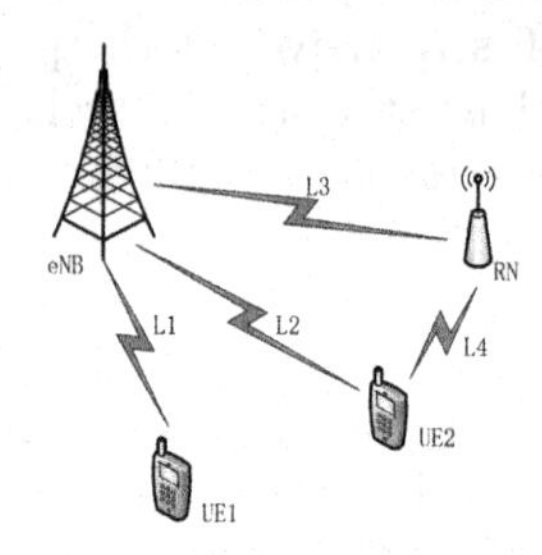

Fig. 2. Link model of relay cellular network

In the figure, the dotted line represents the control link, and the solid line represents the data link. The communication process is as follows: firstly, the user sends a request to establish D2D communication to the base station, and then the base station analyzes the request information to allocate a certain channel resource to the D2D user, and the user can use it directly to start the data communication process. Compared with all communication processes that have to be forwarded by the base station, the D2D communication method can effectively share the base station traffic and improve the system throughput under the condition of limited spectrum resources.

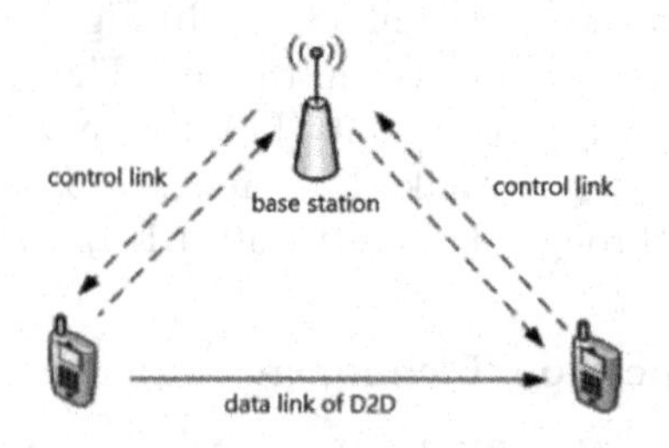

Fig. 3. D2D communication link model

D2D Communication User Working Mode. The introduction of D2D technology in a cell can increase the information rate at the edge of the cell or at the blind spot in areas that cannot be covered. D2D user clusters which contain several D2D users are mainly distributed in the edge area of a cell. Assuming that two users need to transmit information, then these two users form a D2D user pair. There are three ways for D2D user pairs to communicate: dedicated mode, multiplexing mode, and cellular mode.

In the dedicated mode, the system will allocate dedicated channel resources for D2D users to use, and the communication process of D2D users will not cause interference to the communication process of cellular users. Considering that the dedicated mode has relatively high requirements for frequency resources and will result in lower frequency utilization of the system, this mode is rarely used in actual situations.

D2D users in multiplexing mode need to share frequency (or time slot) resources with other cellular users in the cell. In this mode, there will be mutual interference between the communication process of D2D users and the communication process of cellular users, but the advantage is that it can improve spectrum utilization efficiency and increase system capacity [14,15]. Therefore, when a D2D user pair shares cellular user frequency resources, it is necessary to weigh the interference cost and communication performance, and to ensure both the communication quality of the cellular users and the D2D users, and can't save channel resources or increase the utilization rate at the expense of affecting communication performance. Subsequent chapters will analyze the interference situation in multiplexing mode in detail.

The dedicated mode and the multiplexing mode can be interchanged in some cases: when the cell has high requirements for communication reliability and the traffic volume is small, D2D users will be allowed to use the dedicated mode to communicate [16,17]. When the cell traffic is large, the system will allow D2D users to preferentially use the multiplexing mode for communication in order to save spectrum resources.

Cellular mode means that D2D users need to be relayed through the base station based on the multiplexing mode. At this time, D2D communication requires both uplink and downlink communication link resources.

Interference Analysis of D2D Communication in Multiplexing Mode. The following analyzes the interference situation when D2D users share frequency resources with other cellular users in the cell, which means the D2D users are in the multiplexing mode. Cellular user communication is divided into two communication links, namely Uplink (UL) and Downlink (DL). Uplink (UL) refers to the communication from the user terminal to the base station, while downlink (DL) refers to the communication from the base station to the user terminal. Therefore, the ways in which D2D users multiplex cellular user frequency resources can be divided into two types: multiplexing cellular user uplink and multiplexing cellular user downlink. The following will be detailed analysis of these two multiplexing situations.

Figure 4 shows the link diagram when multiplexing uplink resources, where UE1 and UE2 are a D2D user pair. UE1 is the sender, UE2 is the receiver, and UE3 is a cellular user. The interference situation is that D2D users will cause interference to the base station when communicating from the sender to the receiving end, and communications from cellular users to the base station will also cause interference to the receiving end of the D2D user. Figure 5 is the link diagram when multiplexing downlink resources. The user distribution is the same as Fig. 4. The interference situation is that D2D users cause interference to cellular users when communicating from the sender to the receiver. Also, communication from base station to cellular user will cause interference to the receiving end of D2D users. Comparing the two cases, the interference caused to the receiving end of D2D users both exists. In addition, the multiplexing of uplink frequency resources causes interference to the base station, and the multiplexing of downlink frequency resources causes interference to many cellular users. But it is more difficult to control the interference items when multiplexing the downlink. Moreover, it is mentioned in the literature that the resource utilization efficiency of the uplink communication link will be lower than the resource utilization efficiency of the downlink communication link. Therefore, through the above analysis, this paper will consider the case of D2D users multiplexing the uplink communication links of cellular users.

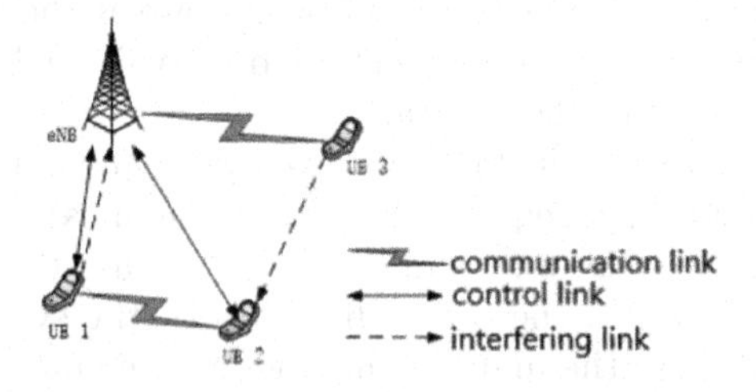

Fig. 4. D2D multiplexing cellular uplink

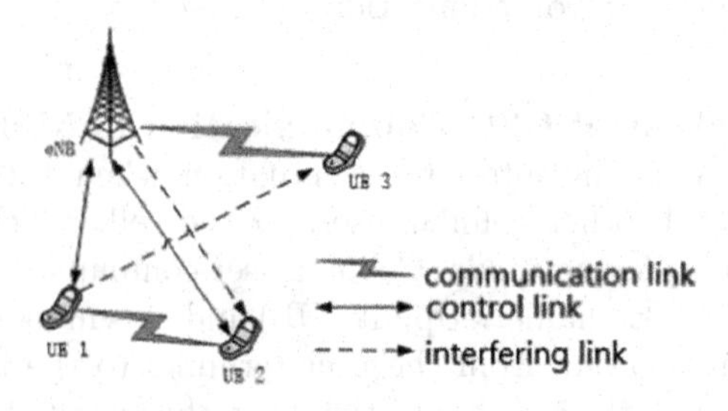

Fig. 5. D2D multiplexing cellular downlink

The rest of the paper is organized as follows. Section 2 briefly introduces the system model of D2D non-directed system. Section 3 analyses the formulation of the problem. The simulation results is shown in Sect. 4. Finally, conclusions are drawn in Sect. 5.

Notation: boldface uppercase, boldface lowercase, and lowercase letter $\mathbf{A}$, $\mathbf{a}$, a denote a matrix, vector, and scalar variable, respectively. $\mathbf{A}^T$, $\mathbf{A}^*$and $\mathbf{A}^{-1}$ represent the transpose, the conjugate transpose and the inverse of a matrix, respectively. $||$ denotes absolute value, *tr()* denotes trace of a matrix, *diag* denotes diagnoal.

2 System Model

2.1 D2D Non-directed System Model

As shown in Fig. 6, the system model diagram contains both D2D non-direct users and cellular users. The same as in Sect. 1.2, the communication process is divided into two time slots, and all relays communicate synchronously in these two time slots.

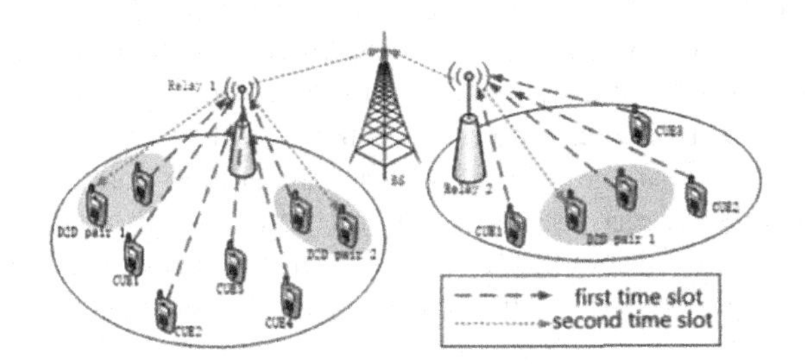

Fig. 6. A system model in which D2D non-direct users and cellular users coexist

The following analyzes the communication process of each user in turn. There are two communication modes for users under the coverage of each relay, namely:

(1) The transmitter of the D2D non-direct user pair communicates with the receiver through relay and forwarding, including the communication between the transmitter of the D2D non-direct user and the relay, and the communication between the relay and the receiver of the D2D non-direct user.
(2) Cellular users communication, including the communication between cellular users and relays, and the communication between relays and base stations.

It is assumed that the communication between the relays and base stations uses orthogonal channels, so there will be no interference between them.

The communication process in time slot one is from terminals to relays, and the communication process in time slot two includes from relays to terminals or relays to base stations. For cellular users, only the uplink is analyzed which includes two time slots. The first time slot is from the terminal to the relay, and the second time slot is from the relay to the base station. The communication process of each time slot will only be interfered by other communication processes in the own time slot.

Assume that set M represents all cellular users, and set D^c represents all non-direct D2D user pairs. The system bandwidth is divided into N resource blocks (RB), whose set is $N = \{1, 2, \cdots, |N|\}$. The usable resource block set in each relay is N, and the bandwidth of each resource block is represented by B_{RB}. The relay set is represented by $L = \{1, 2, \cdots |L|\}$, and $U_l, \forall l \in L$ is the set of all users who carry out the communication process assisted by relay l. The set of cellular users under the coverage of relay l is $M_l = M \cap U_l$, and the set of D2D non-direct users under the coverage of relay l is $D_l^c = D^c \cap U_l$. Therefore, the following formulas are established: $U_l = D_l^c \cup M_l, \forall l \in L$ $\cup_l U_l = \{D^c \cup M\}$ $\cap_l U_l = \varphi, \forall l \in L$.

According to the Shannon channel capacity formula: $R = B\log_2(1 + S/N)$. The information rate expression of each link in the communication process of the system model can be listed separately.

(1) The communication between the transmitter of D2D non-direct users or cellular users and relay: the communication process from the cellular users under the coverage of the relay and the D2D non-direct users to the relay is completed in time slot 1, and both communication processes will be affected by interference from D2D non-direct users and cellular users under other relays. According to the characteristics that similar communication process has similar interference received, the unit power SINR of the communication process is expressed as follows, that is, for the transmitter $u_l \in D_l^c$ of the D2D non-direct user, the cellular user $u_l \in M_l$, there is:

$$\gamma_{u_l,l,1}^{(n)} = \frac{h_{u_l,l}^{(n)}}{\sum\limits_{\substack{u_j \in D_j^c \cup M_j \\ j \neq l, j \in L}} Q_{u_j,j}^{(n)} \cdot g_{u_j,l}^{(n)} + \sigma^2} \tag{1}$$

s.t.

$$\sum_{u_l \in U_l} y_{u_l}^{(n)} \leq 1, \quad \forall n \in N \tag{2}$$

Where $Q_{a,b}^{(n)}$, $h_{a,b}^{(n)}$ and $g_{a,b}^{(n)}$ are the transmission power, the channel coefficient of the communication link, and the interference link gain in the communication process from the transmitting end a to the receiving end b on the resource block n respectively. $\sigma^2 = N_0 B_{RB}$, $B_{RB}$is the resource block bandwidth, and N_0 is the power spectral density of the thermal noise.

(2) The communication between the receiver of D2D non-direct users pair and relay l: the communication between the receiver of D2D non-direct users pair and relay l is carried out in time slot 2. Consider that all communication processes in the second time slot will be affected by interference in the communication process between other relay j and the receiving end of

the D2D non-direct user in its own coverage area and interference in the communication process between other relay j and the base station. Assuming that within a certain time interval, only one relay-base station link is in the working state in the second time slot for transmitting cellular user data messages to the base station, while other relay-base station links are in the dormant state. Therefore, the unit power SINR of the communication between the receiver of D2D non-direct users pair and relay l on the resource block n is expressed as the following formula, that is, for $u_l \in D_l^c$, there is:

$$\gamma_{l,u_l,2}^{(n)} = \frac{h_{l,u_l}^{(n)}}{\sum\limits_{\substack{\forall u_j \in D_j^c \cup M_j \\ j \neq l, j \in L}} Q_{j,u_j}^{(n)} \cdot g_{j,u_l}^{(n)} + \sigma^2} \tag{3}$$

(3) The communication between relay l and base station: the communication between relay l and base station means that relay l sends information from cellular users in its coverage area to the base station through resource block n and the base station forwards it in the second time slot, so as to achieve communication with users in other cells. Assuming that within a certain time interval, only one relay-base station link is in the working state in the second time slot for transmitting cellular user data messages to the base station, and other relay-base station links are in the dormant state. Therefore, in this process, the base station will also be interfered by the communication process between other relay j and the D2D non-direct user receiving end under its own coverage on resource block n. Therefore, for $u_l \in M_l$, the SINR per unit power is:

$$\gamma_{l,u_l,2}^{(n)} = \frac{h_{l,eNB}^{(n)}}{\sum\limits_{\substack{\forall u_j \in D_j^c \\ j \neq l, j \in L}} Q_{j,u_j}^{(n)} \cdot g_{j,eNB}^{(n)} + \sigma^2} \tag{4}$$

In summary, for the user u_l under the coverage of relay l, the total information rate is composed of two parts: the information rate in the first time slot is $R_{u_l,l}^{(n)} = B_{RB}\log_2\left(1 + Q_{u_l,l}^{(n)}\gamma_{u_l,l,1}^{(n)}\right), u_l \in D_l^c \cap M_l$, and the information rate in the second time slot is $R_{l,u_l}^{(n)} = B_{RB}\log_2\left(1 + Q_{l,u_l}^{(n)}\gamma_{l,u_l,2}^{(n)}\right), u_l \in D_l^c \cap M_l$. Therefore, for user u_l under the coverage of relay l, the total information rate on resource block n is denoted as $R_{u_l}^{(n)} = \frac{1}{2}\min\left\{R_{u_l,l}^{(n)}, R_{l,u_l}^{(n)}\right\}$.

3 Problem Formulation

3.1 Problem Modeling

This section discusses that under the premise that each transmitting node of the D2D link satisfies the transmission power constraint and the interference to the cellular network is less than the interference threshold, and combined with the relay and forwarding technology, how to allocate resources reasonably for cellular users and D2D non-direct users to maximize the throughput of the system.

The system capacity is maximized through resource block allocation and power allocation. For a communication process completed through two time slots, the total communication rate is determined by the smaller of the two time slot communication rates. Set the maximum value of user transmission power to $Q_{u_l}^{\max}$, and the maximum value of relay transmission power to $Q_l^{\max}$. In order to express that each resource block RB can only be used by one user under the relay, a resource block allocation factor of $y_{u_l}^{(n)}$ is introduced, and $y_{u_l}^{(n)} \in \{0,1\}$ is a binary integer variable, which means when $y_{u_l}^{(n)} = 1$, it means that resource block n is allocated to user u_l for use; otherwise, $y_{u_l}^{(n)} = 0$. For all users u_l under relay l, the total information rate is $R_{u_l} = \sum_{u_l \in U_l} \sum_{n=1}^{N} y_{u_l}^{(n)} R_{u_l}^{(n)}$. The user's QoS requirement is represented by R_{QoS}. Considering that the same resource block will be occupied by the relay in two time slots, this optimization problem can be described as:

$$\max_{y_{u_l}^{(n)}, Q_{u_l,l}^{(n)}, Q_{l,u_l}^{(n)}} \sum_{l \in L} \sum_{u_l \in U_l} \sum_{n=1}^{N} y_{u_l}^{(n)} R_{u_l}^{(n)} \tag{5}$$

s.t.

$$\sum_{u_l \in U_l} y_{u_l}^{(n)} \leq 1, \quad \forall n \in N \tag{6}$$

$$\sum_{n=1}^{N} y_{u_l}^{(n)} Q_{u_l,l}^{(n)} \leq Q_{u_l}^{\max}, \quad \forall u_l \in U_l \tag{7}$$

$$\sum_{u_l \in U_l} \sum_{n=1}^{N} y_{u_l}^{(n)} Q_{l,u_l}^{(n)} \leq Q_l^{\max}, \ \forall l \in L \tag{8}$$

$$\sum_{u_l \in U_l} y_{u_l}^{(n)} Q_{u_l,l}^{(n)} g_{u_l,l^*}^{(n)} \leq I_{th,1}^{(n)}, \ \forall n \in N \tag{9}$$

$$\sum_{u_l \in D_l^c} y_{u_l}^{(n)} Q_{l,u_l}^{(n)} g_{l,u_l^*}^{(n)} \leq I_{th,2}^{(n)}, \ \forall n \in N \tag{10}$$

$$R_{u_l} \geq R_{QoS}, \quad \forall u_l \in U_l \tag{11}$$

$$Q_{u_l,l}^{(n)} \geq 0, \quad Q_{l,u_l}^{(n)} \geq 0, \quad \forall n \in N, \ u_l \in U_l \tag{12}$$

Where constraint (6) is the condition that each indicator coefficient needs to meet, and each resource block RB can only be allocated to one user under each relay. Constraint (7) is that the transmission power of each user cannot exceed the maximum power limit. Constraint (8) means that the transmit power of each relay station cannot exceed the maximum power limit. Constraint (9) means that the interference from users under other relays needs to meet the interference threshold. Constraint (10) means that the interference from other relay stations need to meet the threshold. Constraint (11) ensures that the system meets the minimum QoS requirements. Constraint (12) means that each transmission power is non-negative.

For all users u_l on resource block n, the information rate is:

$$\begin{aligned} y_{u_l}^{(n)} R_{u_l}^{(n)} &= \tfrac{1}{2} y_{u_l}^{(n)} \min \left\{ R_{u_l,l}^{(n)}, R_{l,u_l}^{(n)} \right\} \\ &= \tfrac{1}{2} y_{u_l}^{(n)} \min \left\{ B_{RB} \log_2 \left(1 + Q_{u_l,l}^{(n)} \gamma_{u_l,l,1}^{(n)}\right), B_{RB} \log_2 \left(1 + Q_{l,u_l}^{(n)} \gamma_{l,u_l,2}^{(n)}\right) \right\} \end{aligned} \tag{13}$$

For D2D non-direct users or cellular users, the unit power SINR in the communication process in time slot 1 is:

$$\begin{aligned} \gamma_{u_l,l,1}^{(n)} &= \frac{h_{u_l,l}^{(n)}}{I_{u_l,l,1}^{(n)} + \sigma^2} \\ I_{u_l,l,1}^{(n)} &= \sum_{\substack{u_j \in D_j^c \cup M_j \\ j \neq l, j \in L}} y_{u_j}^{(n)} Q_{u_j,j}^{(n)} \cdot g_{u_j,l}^{(n)} \end{aligned} \tag{14}$$

Where $I_{u_l,l,1}^{(n)}$ is the interference item suffered by user u_l on resource block n in the first time slot. For D2D non-direct users and cellular users, the unit power SINR in the communication process in time slot 2 is:

$$\gamma_{l,u_l,2}^{(n)} = \begin{cases} \frac{h_{l,u_l}^{(n)}}{I_{l,u_l,2}^{(n)} + \sigma^2}, & u_l \in D_l^c \\ \frac{h_{l,eNB}^{(n)}}{I_{l,u_l,2}^{(n)} + \sigma^2}, & u_l \in M_l \end{cases} \tag{15}$$

Where $I_{l,u_l,2}^{(n)}$ is the interference item suffered by user u_l on resource block n in the second time slot.

$$I_{l,u_l,2}^{(n)} = \begin{cases} \sum\limits_{\substack{\forall u_j \in D_j^c \cup M_j \\ j \neq l, j \in L}} y_{u_j}^{(n)} Q_{j,u_j}^{(n)} \cdot g_{j,u_l}^{(n)}, & u_l \in D_l^c \\ \sum\limits_{\substack{\forall u_j \in D_j^c \\ j \neq l, j \in L}} y_{u_j}^{(n)} Q_{j,u_j}^{(n)} \cdot g_{j,eNB}^{(n)}, & u_l \in M_l \end{cases} \tag{16}$$

When $Q_{u_l,l}^{(n)} \gamma_{u_l,l,1}^{(n)} = Q_{l,u_l}^{(n)} \gamma_{l,u_l,2}^{(n)}$ is established, the information rate $R_{u_l}^{(n)}$ of all users u_l communicating on resource block n can reach the maximum value. At this time, $Q_{l,u_l}^{(n)}$ in the second time slot can be represented by the power in the first time slot, that is, $Q_{l,u_l}^{(n)} = \frac{\gamma_{u_l,l,1}^{(n)}}{\gamma_{l,u_l,2}^{(n)}} Q_{u_l,l}^{(n)}$. Therefore, the information rate $R_{u_l}^{(n)}$ of user u_l on resource block n can be rewritten as:

$$R_{u_l}^{(n)} = \frac{1}{2} B_{RB} \log_2 \left(1 + Q_{u_l,l}^{(n)} \gamma_{u_l,l,1}^{(n)}\right) \tag{17}$$

The optimization problem (5) contains both continuous variables and binary integer variables, and the objective function is also nonlinear, which is a MINLP problem. In order to simplify the problem, first we need to relax the resource block allocation factor $y_{u_l}^{(n)}$ to a continuous variable, that is, $y_{u_l}^{(n)} \in (0,1]$. $y_{u_l}^{(n)}$ represents the proportion of time that is allocated to the user u_l by the resource block n, and the restriction condition $\sum_{u_l \in U_l} y_{u_l}^{(n)} \leq 1, \forall n \in N$ mentioned above is still met. We also introduce power allocation variable $T_{u_l,l}^{(n)} = y_{u_l}^{(n)} Q_{u_l,l}^{(n)}$ to represent the actual transmit power of user u_l when communicating on resource block n, then the optimization problem after relaxation and adjustment can be expressed as:

$$\max_{y_{u_l}^{(n)}, T_{u_l,l}^{(n)}, \mu_{u_l}^{(n)}} \sum_{l \in L} \sum_{u_l \in U_l} \sum_{n=1}^{N} \frac{1}{2} y_{u_l}^{(n)} B_{RB} \log_2 \left(1 + \frac{T_{u_l,l}^{(n)} h_{u_l,l}^{(n)}}{y_{u_l}^{(n)} \mu_{u_l}^{(n)}} \right) \tag{18}$$

s.t.

$$\sum_{u_l \in U_l} y_{u_l}^{(n)} \leq 1, \quad \forall n \in N \tag{19}$$

$$\sum_{n=1}^{N} T_{u_l,l}^{(n)} \leq Q_{u_l}^{\max}, \quad \forall u_l \in U_l \tag{20}$$

$$\sum_{u_l \in U_l} \sum_{n=1}^{N} \frac{\gamma_{u_l,l,1}^{(n)}}{\gamma_{l,u_l,2}^{(n)}} T_{u_l,l}^{(n)} \leq Q_l^{\max} \tag{21}$$

$$\sum_{u_l \in U_l} T_{u_l,l}^{(n)} g_{u_l,l^*}^{(n)} \leq I_{th,1}^{(n)}, \quad \forall n \in N \tag{22}$$

$$\sum_{u_l \in D_l^c} \frac{\gamma_{u_l,l,1}^{(n)}}{\gamma_{l,u_l,2}^{(n)}} T_{u_l,l}^{(n)} g_{l,u_l^*}^{(n)} \leq I_{th,2}^{(n)}, \ \forall n \in N \tag{23}$$

$$\sum_{n=1}^{N} \frac{1}{2} y_{u_l}^{(n)} B_{RB} \log_2 \left(1 + \frac{T_{u_l,l}^{(n)} h_{u_l,l}^{(n)}}{y_{u_l}^{(n)} \mu_{u_l}^{(n)}} \right) \geq R_{QoS}, \ \forall u_l \in U_l \tag{24}$$

$$T_{u_l,l}^{(n)} \geq 0, \quad \forall n \in N, \ u_l \in U_l \tag{25}$$

$$I_{u_l,l}^{(n)} + \sigma^2 \leq \mu_{u_l}^{(n)}, \ \forall n \in N, \ u_l \in U_l \tag{26}$$

Where $\mu_{u_l}^{(n)}$ is an auxiliary variable, and there is $I_{u_l,l}^{(n)} = \max\left\{ I_{u_l,l,1}^{(n)}, I_{l,u_l,2}^{(n)} \right\}$. As the number of resource blocks increases, when the number is extremely large, the dual spacing of the optimization problem that satisfies the $y_{u_l}^{(n)}$ time allocation condition can be ignored. The optimization problem studied in this paper satisfies the time allocation condition, so the relaxed optimization problem has an asymptotic optimal solution. It can be seen from the above formula that the constraint (24) is convex, and other constraints are linear. If the objective function is a concave function, then the optimization problem (18) is a convex problem and there is an optimal solution. The following first proves that the objective optimization function is a concave function.

Define function $\Re\left(T_{u_l,l}^{(n)}\right) = -y_{u_l}^{(n)} B_{RB}\log_2\left(1+\frac{T_{u_l,l}^{(n)}h_{u_l,l}^{(n)}}{y_{u_l}^{(n)}\mu_{u_l}^{(n)}}\right)$, and find the Hessian matrix H of power distribution variable $T_{u_l,l}^{(n)}$ for the function is:

$$H = \left|\frac{\partial^2\Re\left(T_{u_l,l}^{(n)}\right)}{\partial T_{u_l,l}^{(n)\,2}}\right| = \left|\frac{y_{u_l}^{(n)} B_{RB} h_{u_l,l}^{(n)\,2}}{\ln 2\left(y_{u_l}^{(n)}\mu_{u_l}^{(n)} + T_{u_l,l}^{(n)}h_{u_l,l}^{(n)}\right)^2}\right| \tag{27}$$

H is a first-order matrix, so the eigenvalue is $\tilde{\lambda} = \frac{y_{u_l}^{(n)} B_{RB} h_{u_l,u_l}^{(n)\,2}}{\ln 2\left(y_{u_l}^{(n)}\mu_{u_l}^{(n)} + T_{u_l,l}^{(n)}h_{u_l,l}^{(n)}\right)^2}$. Because of $\tilde{\lambda} > 0$, $\Re\left(T_{u_l,l}^{(n)}\right)$ is convex, and the objective function (18) is concave. Therefore, the optimization problem is a convex problem, and there is an optimal solution. The KKT condition in the convex optimization theory can be used to solve this problem.

3.2 Power Distribution Method Based on Lagrange Multiplier Method

In Sect. 3.1, it has been proved that the optimization problem (18) is a convex problem. The Lagrange multiplier method is used to solve the problem, and the Lagrange multiplier of the constraints (19) to (26) of formula (18) are $\delta_n, \xi_{u_l}, \upsilon_l, \psi_n, \varepsilon_n, \lambda_{u_l}, \rho_{u_l}^{(n)}$ respectively, then the Lagrange function is:

$$\begin{aligned} L = &-\sum_{l\in L}\sum_{u_l\in U_l}\sum_{n=1}^{N}\tfrac{1}{2}y_{u_l}^{(n)} B_{RB}\log_2\left(1+\frac{T_{u_l,l}^{(n)}h_{u_l,l}^{(n)}}{y_{u_l}^{(n)}\mu_{u_l}^{(n)}}\right) \\ &+\sum_{n=1}^{N}\delta_n\left(\sum_{u_l\in U_l}y_{u_l}^{(n)}-1\right)+\sum_{u_l\in U_l}\xi_{u_l}\left(\sum_{n=1}^{N}T_{u_l,l}^{(n)}-Q_{u_l}^{\max}\right) \\ &+\upsilon_l\left(\sum_{u_l\in U_l}\sum_{n=1}^{N}\frac{\gamma_{u_l,l,1}^{(n)}}{\gamma_{l,u_l,2}^{(n)}}T_{u_l,l}^{(n)}-Q_l^{\max}\right)+\sum_{n=1}^{N}\psi_n\left(\sum_{u_l\in U_l}T_{u_l,l}^{(n)}g_{u_l,l^*}^{(n)}-I_{th,1}^{(n)}\right) \\ &+\sum_{n=1}^{N}\varepsilon_n\left(\sum_{u_l\in D_l^c}\frac{\gamma_{u_l,l,1}^{(n)}}{\gamma_{l,u_l,2}^{(n)}}T_{u_l,l}^{(n)}g_{l,u_l^*}^{(n)}-I_{th,2}^{(n)}\right)+\sum_{u_l\in U_l}\sum_{n=1}^{N}\rho_{u_l}^{(n)}\left(I_{u_l,l}^{(n)}+\sigma^2-\mu_{u_l}^{(n)}\right) \\ &+\sum_{u_l\in U_l}\lambda_{u_l}\left(R_{QoS}-\sum_{n=1}^{N}\tfrac{1}{2}y_{u_l}^{(n)} B_{RB}\log_2\left(1+\frac{T_{u_l,l}^{(n)}h_{u_l,l}^{(n)}}{y_{u_l}^{(n)}\mu_{u_l}^{(n)}}\right)\right) \end{aligned} \tag{28}$$

First, the derivative of the power distribution variable $T_{u_l,l}^{(n)}$ is obtained, and the following formula is obtained:

$$\begin{aligned} \frac{\partial L}{\partial T_{u_l,l}^{(n)}} = &-\frac{1}{2\ln 2}y_{u_l}^{(n)} B_{RB}\frac{1}{1+\frac{T_{u_l,l}^{(n)}h_{u_l,l}^{(n)}}{y_{u_l}^{(n)}\mu_{u_l}^{(n)}}}\frac{h_{u_l,l}^{(n)}}{y_{u_l}^{(n)}\mu_{u_l}^{(n)}}+\xi_{u_l}+\upsilon_l\frac{\gamma_{u_l,l,1}^{(n)}}{\gamma_{l,u_l,2}^{(n)}}+\varepsilon_n\frac{\gamma_{u_l,l,1}^{(n)}}{\gamma_{l,u_l,2}^{(n)}}g_{l,u_l^*}^{(n)} \\ &+\psi_n g_{u_l,l^*}^{(n)}-\frac{1}{2\ln 2}\lambda_{u_l}y_{u_l}^{(n)} B_{RB}\frac{1}{1+\frac{T_{u_l,l}^{(n)}h_{u_l,l}^{(n)}}{y_{u_l}^{(n)}\mu_{u_l}^{(n)}}}\frac{h_{u_l,l}^{(n)}}{y_{u_l}^{(n)}\mu_{u_l}^{(n)}} \end{aligned} \tag{29}$$

According to the KKT condition in the convex optimization theory, let $\frac{\partial L}{\partial T_{u_l,l}^{(n)}} = 0$, wee can get the optimal solution of the transmission power $Q_{u_l,l}^{(n)}$ from user u_l to relay l as follows:

$$Q_{u_l,l}^{(n)^*} = \frac{T_{u_l,l}^{(n)^*}}{y_{u_l}^{(n)^*}} = \left[\Delta_{u_l,l}^{(n)} - \frac{\mu_{u_l}^{(n)}}{h_{u_l,l}^{(n)}} \right]^*$$
$$\Delta_{u_l,l}^{(n)} = \frac{(\lambda_{u_l} + 1) B_{RB}}{2 \ln 2 \left(\xi_{u_l} + \upsilon_l \frac{\gamma_{u_l,l,1}^{(n)}}{\gamma_{l,u_l,2}^{(n)}} + \psi_n g_{u_l,l^*}^{(n)} + \varepsilon_n \frac{\gamma_{u_l,l,1}^{(n)}}{\gamma_{l,u_l,2}^{(n)}} g_{l,u_l^*}^{(n)} \right)} \tag{30}$$

Where $[\xi]^+ = \max\{\xi, 0\}$. Derivation of the resource block allocation factor $y_{u_l}^{(n)}$ after relaxation, we get:

$$\begin{aligned}
\frac{\partial L}{\partial y_{u_l}^{(n)}} &= -\tfrac{1}{2} B_{RB} \log_2 \left(1 + \frac{T_{u_l,l}^{(n)} h_{u_l,l}^{(n)}}{y_{u_l}^{(n)} \mu_{u_l}^{(n)}} \right) - \tfrac{1}{2} \lambda_{u_l} B_{RB} \log_2 \left(1 + \frac{T_{u_l,l}^{(n)} h_{u_l,l}^{(n)}}{y_{u_l}^{(n)} \mu_{u_l}^{(n)}} \right) + \delta_n \\
&\quad - \frac{1}{2 \ln 2} y_{u_l}^{(n)} B_{RB} \frac{1}{1 + \frac{T_{u_l,l}^{(n)} h_{u_l,l}^{(n)}}{y_{u_l}^{(n)} \mu_{u_l}^{(n)}}} \frac{T_{u_l,l}^{(n)} h_{u_l,l}^{(n)}}{\mu_{u_l}^{(n)}} \left(- \frac{1}{y_{u_l}^{(n)2}} \right) \\
&\quad - \frac{1}{2 \ln 2} y_{u_l}^{(n)} \lambda_{u_l} B_{RB} \frac{1}{1 + \frac{T_{u_l,l}^{(n)} h_{u_l,l}^{(n)}}{y_{u_l}^{(n)} \mu_{u_l}^{(n)}}} \frac{T_{u_l,l}^{(n)} h_{u_l,l}^{(n)}}{\mu_{u_l}^{(n)}} \left(- \frac{1}{y_{u_l}^{(n)2}} \right)
\end{aligned} \tag{31}$$

According to the KKT condition in the convex optimization theory, let $\frac{\partial L}{\partial y_{u_l}^{(n)}} = 0$, we get:

$$\delta_n = \frac{1}{2} (1 + \lambda_{u_l}) B_{RB} \left(\log_2 \left(1 + \frac{T_{u_l,l}^{(n)} h_{u_l,l}^{(n)}}{y_{u_l}^{(n)} \mu_{u_l}^{(n)}} \right) - \theta_{u_l,l}^{(n)} \right) \tag{32}$$

Where $\theta_{u_l,l}^{(n)} = \frac{T_{u_l,l}^{(n)} h_{u_l,l}^{(n)}}{\ln 2 \left(y_{u_l}^{(n)} \mu_{u_l}^{(n)} + T_{u_l,l}^{(n)} h_{u_l,l}^{(n)} \right)}$. The actual value of δ_n is not calculated by formula (32), but updated by sub-gradient iteration method. In order to obtain the integer value of resource block allocation factor $y_{u_l}^{(n)}$, the threshold variable $\chi_{u_l}^{(n)} = \frac{1}{2} (1 + \lambda_{u_l}) B_{RB} \left(\log_2 \left(1 + \frac{T_{u_l,l}^{(n)} h_{u_l,l}^{(n)}}{y_{u_l}^{(n)} \mu_{u_l}^{(n)}} \right) - \theta_{u_l,l}^{(n)} \right)$ needs to be set to obtain the discriminant of resource block allocation factor $y_{u_l}^{(n)}$ is as follows:

$$y_{u_l}^{(n)^*} = \begin{cases} 1, & \delta_n \le \chi_{u_l}^{(n)} \\ 0, & \delta_n > \chi_{u_l}^{(n)} \end{cases} \tag{33}$$

After obtaining the optimal solution of transmission power $Q_{u_l,l}^{(n)^*}$ and resource block allocation factor $y_{u_l}^{(n)^*}$ for communication from user u_l to relay l , the sub-gradient iteration method is used to update each Lagrange multiplier, and the Lagrange multiplier of the (t + 1)-th iteration is updated according to the following formulas, and the variable $\Lambda_{\alpha}^{(t)}$ is the step length in the first iteration, where $\Lambda_{\alpha}^{(t)} = a/\sqrt{t}$, and a is constant.

$$\delta_n (t+1) = \left[\delta_n (t) + \Lambda_{\delta_n}^{(t)} \left(\sum_{u_l \in U_l} y_{u_l}^{(n)} - 1\right)\right]^+ \tag{34}$$

$$\xi_{u_l} (t+1) = \left[\xi_{u_l} (t) + \Lambda_{\xi_{u_l}}^{(t)} \left(\sum_{n=1}^{N} T_{u_l,l}^{(n)} - Q_{u_l}^{\max}\right)\right]^+ \tag{35}$$

$$\upsilon_l (t+1) = \left[\upsilon_l (t) + \Lambda_{\upsilon_l}^{(t)} \left(\sum_{u_l \in U_l} \sum_{n=1}^{N} \frac{\gamma_{u_l,l,1}^{(n)}}{\gamma_{l,u_l,2}^{(n)}} T_{u_l,l}^{(n)} - Q_l^{\max}\right)\right]^+ \tag{36}$$

$$\psi_n (t+1) = \left[\psi_n (t) + \Lambda_{\psi_n}^{(t)} \left(\sum_{u_l \in U_l} T_{u_l,l}^{(n)} g_{u_l,l^*}^{(n)} - I_{th,1}^{(n)}\right)\right]^+ \tag{37}$$

$$\varepsilon_n (t+1) = \left[\varepsilon_n (t) + \Lambda_{\varepsilon_n}^{(t)} \left(\sum_{u_l \in D_l^c} \frac{\gamma_{u_l,l,1}^{(n)}}{\gamma_{l,u_l,2}^{(n)}} T_{u_l,l}^{(n)} g_{l,u_l^*}^{(n)} - I_{th,2}^{(n)}\right)\right]^+ \tag{38}$$

$$\lambda_{u_l} (t+1) = \left[\lambda_{u_l} (t) + \Lambda_{\lambda_{u_l}}^{(t)} \left(R_{QoS} - \sum_{n=1}^{N} \frac{1}{2} y_{u_l}^{(n)} B_{RB} \log_2 \left(1 + \frac{T_{u_l,l}^{(n)} h_{u_l,l}^{(n)}}{y_{u_l}^{(n)} \mu_{u_l}^{(n)}}\right)\right)\right]^+ \tag{39}$$

$$\rho_{u_l}^{(n)} (t+1) = \left[\rho_{u_l}^{(n)} (t) + \Lambda_{\rho_{u_l}^{(n)}}^{(t)} \left(I_{u_l,l}^{(n)} + \sigma^2 - \mu_{u_l}^{(n)}\right)\right]^+ \tag{40}$$

4 Simulation Results and Performance Analysis

Since each channel coefficient and interference link channel coefficient in the simulation contain random variables, for each resource block, the information rate distributed on it also has a certain randomness. In order to reduce the final randomness caused by randomness, the method of calculating the average value by multiple simulations is used to eliminate the influence of the randomness of the channel coefficient. After this process, we can obtain the fairness index curve of the information rate distribution on each resource block. This paper uses the RajJain fairness index to determine the fairness of the information rate on each resource block, and defines the fairness index: $F = \left(\sum_{n=1}^{N} R_n\right)^2 \Big/ N \sum_{n=1}^{N} R_n^2$, where N is the total number of resource blocks in the system, and is the information rate on the resource block. The simulation parameters are shown in Table 1.

Table 1. Simulation parameters and values

Parameters	Values
System bandwidth	2.5 MHz
Total number of resource blocks	13
Path loss between D2D users	102.9 + 18.7log[d(km)]
User-to-relay path loss	103.8 + 20.9log[d(km)]
Path loss from relay to base station	100.7 + 23.5log[d(km)]
Standard deviation of shadow fading among D2D users	3 dB
Standard deviation of shadow fading from user to relay	10 dB
Standard deviation of shadow fading from relay to base station	6 dB
Relay transmission power range	20–30 dBm
User transmit power range	13–23 dBm
Maximum distance between D2D direct users	20 m
Relay coverage radius	200 m
Distance between base station and relay	125 m
Noise power spectral density	−174 dBm/Hz
Interference threshold	−70 dBm

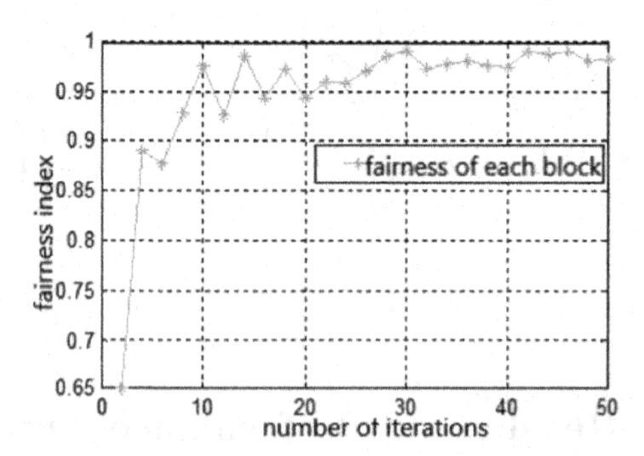

Fig. 7. Fairness index of each resource block

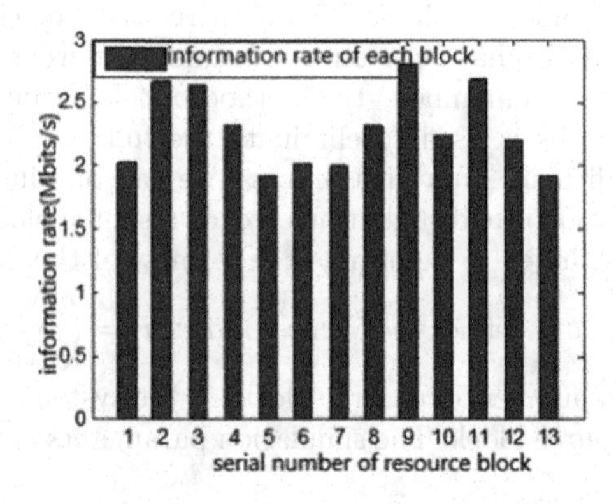

Fig. 8. Information rate distribution on each resource block

In order to avoid the inaccuracy of the final result caused by the randomness of the channel coefficient and the interference link coefficient, multiple simulations are used to average the result data to obtain the fairness index curve of the information rate distribution on each resource block. When the total system bandwidth and the total number of resource blocks are fixed, only the case of two relays is considered and the number of D2D user pairs and the number of cellular users under each relay are also the same. Still taking the number of D2D non-direcct user pairs and the number of cellular users under Relay 1 and Relay 2 is equal to 4 as an example, as shown in Fig. 7. The number of iterations gradually increases, and the fairness index of the information rate distribution of each resource block is gradually close to 1, which means the fairness becomes better. In the 50th iteration, the information rate distribution of each resource block is shown in Fig. 8, and the information rate on each resource block is distributed between 2 and 3 Mbit/s.

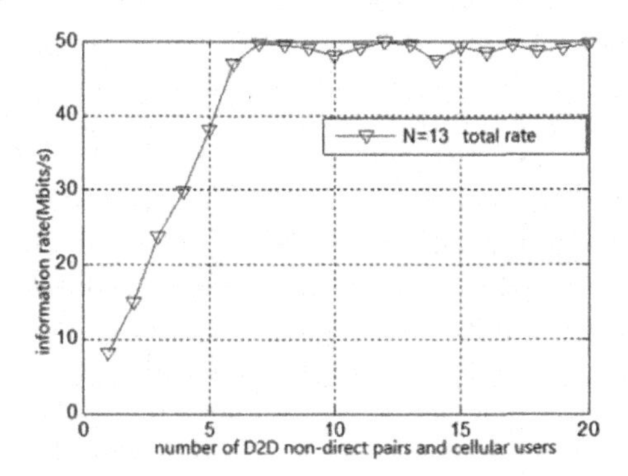

Fig. 9. Total system information rate

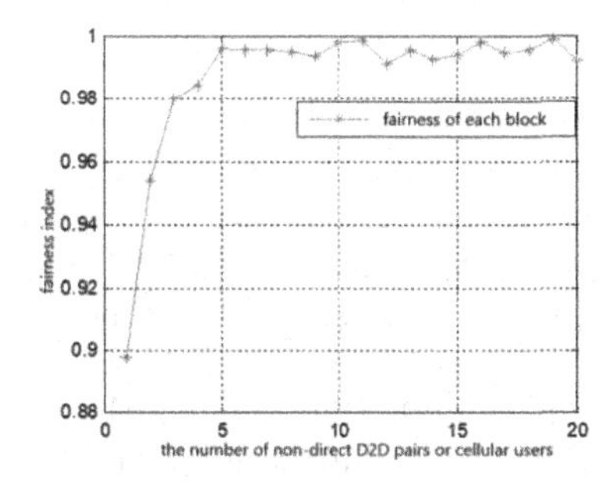

Fig. 10. Fairness of the information rate of each resource block

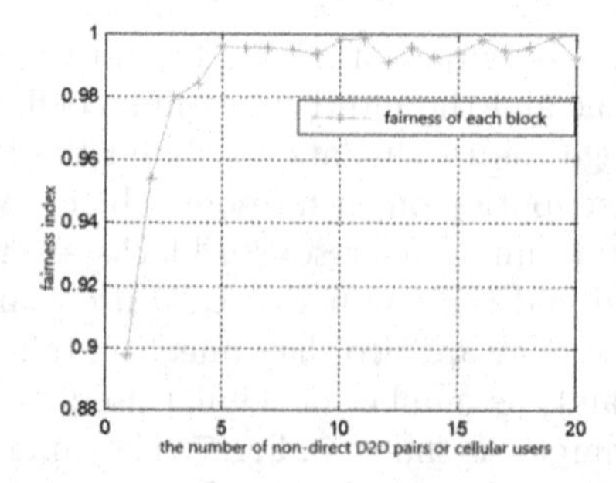

Fig. 11. Fairness of information rate of different types of users

As shown in Fig. 7 and Fig. 8, the number of iterations is 50, and the total number of resource blocks is 13. The simulation results show that as the number of D2D non-direct users and the number of cellular users under each relay increase, the total throughput of the system also first undergoes an approximately linear growth stage, and finally stabilizes. When the number of D2D non-direct users and the number of cellular users are greater than 7, the information rate reaches about 50 Mbit/s.

Next, observe the fairness of the information rate on each resource block under the simulation conditions in Fig. 9, as shown in Fig. 10 and Fig. 11, it can be seen that as the D2D pass-through user pairs under each relay As the number of users and the number of cellular users increase, the fairness index of the information rate on each resource block also increases, and gradually approaches 1. With the increase in the number of D2D direct users and the number of cellular users under each relay, the fairness of the information rate of the two shows a decreasing trend, and the greater the number of users, the worse the fairness.

5 Conclusion

This paper mainly studies the LTE-A cell throughput optimization problem that combines relay technology and D2D technology. It mainly includes two parts. The first part focuses on the actual cell system communication model, LTE-A relay cellular network architecture, link model and usage scenarios, as well as D2D communication principles and user working modes to construct a mathematical model of non-direct D2D user types. The second part uses the Lagrange multiplier method and KKT conditions to solve the constructed optimization problem with throughput as the optimization objective function.

This subject mainly completed the following work content:

(1) Under the LTE-A cellular network, combining relay technology, D2D technology and practical application scenarios together, a single cell model is constructed according to the working mode of non-direct D2D communication users. Besides, we analyze each communication process and interference

source to establish a MINLP optimization problem with power, interference, and user QoS requirements as constraints, system throughput as an objective function, user and relay transmission power, and user distribution factors on different resource blocks as optimization variables.

(2) From a mathematical point of view, the concavity and convexity of the optimization problem is proved. The result shows that the optimization problem is convex. Using the Lagrange multiplier method and the KKT conditions in the convex optimization theory, the optimal solution expression is derived. Through simulation analysis, it is found that the maximum throughput of the cellular system where D2D users are located in different working modes is very different. D2D direct users can bring greater throughput to the system than D2D non-direct users, but it reduces fairness of information rate distribution on resource blocks.

There are still many shortcomings in this article. For example, in the cell built on the basis of LTE-A cellular network, combined with relay technology and D2D technology, only power, user service quality QoS, and throughput limitations are considered. In the follow-up research process, performance indicators such as spectrum efficiency and energy efficiency can be added to continue to improve this model.

References

1. Khosrowpour, M.: An overview of 3GPP long term evolution (LTE). 13–19 (2015)
2. Phunchongharn, P., Hossain, E., Kim, D.I.: Resource allocation for device-to-device communications underlaying LTE-advanced networks. IEEE Wirel. Commun. **20**(4), 91–100 (2013)
3. Periyalwar, S., Zhang, H., Senarath, N., et al.: System and method for peer-to-peer communication in cellular systems. US, US 8260337 B2, pp. 51–60 (2012)
4. Park, K.J., Cho, H.G., Kwon, Y.H., et al.: Method for transmitting an uplink signal and feedback information, and relay apparatus using the method. US, US 20140286265 A1, pp. 56–58 (2014)
5. Zhu, K., Hossain, E.: Joint mode selection and spectrum partitioning for device-to-device communication: a dynamic Stackelberg game. IEEE Trans. Wirel. Commun. **14**(3), 1406–1420 (2015)
6. Ghaboosi, K., Vemulapalli, M.G., Oates, N.J.: Duplexing in long term evolution cellular networks. US, US20140286205, pp. 21–31 (2014)
7. Cha, W.S., Song, J.K., Kim, S.T., et al.: Method of selecting relay mode in mobile ad-hoc network. US9060386, pp. 54–68 (2015)
8. Lee, K., Lee, J., Yi, Y., et al.: mobile data offloading: how much can WiFi deliver. IEEE/ACM Trans. Netw. **21**(2), 536–550 (2013)
9. Ristanovic, N., Boudec, J.Y.L., Chaintreau, A., et al.: Energy efficient offloading of 3G networks, pp. 202–211 (2011)
10. Joung, J., Sun, S.: Power efficient resource allocation for downlink OFDMA relay cellular networks. IEEE Trans. Signal Process. **60**(5), 2447–2459 (2012)
11. Alam, M.S., Mark, J.W., Shen, X.: Relay selection and resource allocation for multi-user cooperative LTE-A uplink. In: IEEE International Conference on Communications, pp. 5092–5096. IEEE (2012)

12. Choi, Y., Ji, H.W., Park, J.Y., et al.: A 3W network strategy for mobile data traffic offloading. IEEE Commun. Mag. **49**(10), 118–123 (2011)
13. Korhonen, J., Savolainen, T., Ding, A.Y., et al.: Toward network controlled IP traffic offloading. IEEE Commun. Mag. **51**(51), 96–102 (2013)
14. Doppler, C., Ribeiro, B.: Mode selection for device-to-device communication underlaying an LTE-advanced network. IEEE Xplore **29**(16), 1–6 (2010)
15. Yu, C.H., Doppler, K., Ribeiro, C.B., et al.: Resource sharing optimization for device-to-device communication underlaying cellular networks. IEEE Trans. Wirel. Commun. **10**(8), 2752–2763 (2011)
16. Kaufman, B., Lilleberg, J., Aazhang, B.: Spectrum sharing scheme between cellular users and ad-hoc device-to-device users. IEEE Trans. Wirel. Commun. **12**(3), 1038–1049 (2013)
17. Min, H., Seo, W., Lee, J., et al.: Reliability improvement using receive mode selection in the device-to-device uplink period underlaying cellular networks. IEEE Trans. Wirel. Commun. **10**(2), 413–418 (2011)
18. Dong, H.L., Choi, K.W., Jeon, W.S., et al.: Two-stage semi-distributed resource management for device-to-device communication in cellular networks. IEEE Trans. Wirel. Commun. **13**(4), 1908–1920 (2014)
19. Tehrani, M.N., Uysal, M., Yanikomeroglu, H.: Device-to-device communication in 5G cellular networks: challenges, solutions, and future directions. IEEE Commun. Mag. **52**(5), 86–92 (2014)
20. Li, Y., Wang, Z., Jin, D., et al.: Optimal mobile content downloading in device-to-device communication underlaying cellular networks. IEEE Trans. Wirel. Commun. **13**(7), 3596–3608 (2014)

Wireless Communications and Wireless Networks

Multi-user UAV Relayed Half-Duplex Uplink Cellular Networks with Direct Links and Control and Back-Haul Links

Tong Zhang(✉), Gang Wang, and Yikun Zou

Harbin Institute of Technology, Harbin 150001, China
18B905025@stu.hit.edu.cn

Abstract. In this paper, we study a half-duplex multi-user unmanned aerial vehicle (UAV) aided communication system, with constraints of control and back-haul data rates. The UAV is connected to the cellular networks, and because of its line of sight channel, it also works as a relay to receive and transmit the information from the ground users to the base station. The users, categorized in non-orthogonal multiple access (NOMA) groups, have direct links to base stations. In the meantime, to economize spectrum resources, the UAV's control and back-haul link is also multiplexed with users. In this paper, we design the communication scheme and resource allocation strategy to achieve higher spectrum efficiency. The numerical results show the convergence of the proposed scheme and show that the proposed scheme has better performance than OMA schemes.

Keywords: Spectrum efficiency · NOMA · UAV cellular networks · Multi-user

1 Introduction

In recent years, the application of unmanned aerial vehicles (UAV) on communication networks has attracted widespread attention. The potential of UAV relays has already been mentioned in release 15. At present, relevant researches focus on UAV-assisted emergency communication and temporary communication. In release 17, the requirement for cellular networks to enable UAVs' connection has been mentioned. Benefit from their high mobility, UAVs connected to cellular networks have broad prospects as cooperative relays to enhance communication performance.

Recent studies of UAV systems are usually about the design of trajectory and the power allocation [2,4,6,9], where the UAVs are dedicated relays or base-stations [11]. The UAV's control and backhaul relay is usually not considered. With the tension of spectrum resource, UAVs in cellular networks usually have limited time and frequency resources, while the control and back-haul links of UAVs should have guaranteed data rates so as to keep the safety of UAVs'

Q. Guo et al. (Eds.): WiSATS 2021, LNICST 410, pp. 675–684, 2022.
https://doi.org/10.1007/978-3-030-93398-2_58

operation. However, how to multiplex the UAV's control and backhaul relay when the UAV joins a cellular network is still not widely studied.

There are some researches about NOMA techniques for UAV's connection to cellular networks. In [5], the authors studied the interference of base stations (BSs) to UAVs, and designed a cooperative NOMA scheme for BSs and the UAV. But the research did not consider other ground users, which is usually not ignorable. In [7], a cooperative NOMA scheme for a multi-user UAV enabled communication system is studied. The UAV works as a macro cell BS, but the UAV's control link is not considered. In [10], the authors studied a downlink full-duplex UAV relaying system, the UAV works as a relay between the base station and the users. Differently, we study an uplink communication system consisting of direct links between the base station and the UAV, and the base station and the users. The control and back haul link between the UAV and the base station is also considered to guarantee the UAV's control and relaying mission's.

2 System Illustration

2.1 System Model

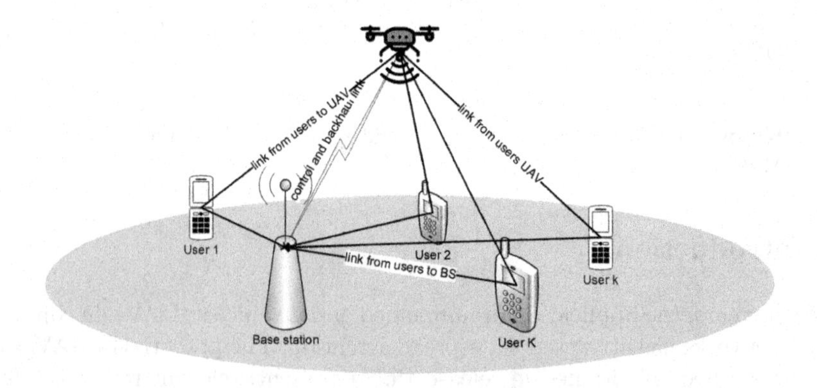

Fig. 1. The system model.

We consider a half-duplex uplink communication system, consisting of a basestation, a UAV and multiple ground nodes. The UAV flies at a given altitude and works as a decoding and forward relay. Since the UAV is connected to cellular networks as a special user, its position is given because of requirements of mission. Suppose there are K users sharing the same spectrum resource. The power transmitted from the ground nodes at time slot t is denoted as $p_{u,k,t}$. The power from the UAV to the BS is denoted as p_U. The control and backhaul power from the BS for the UAV is denoted as p_B. The noise power spectral density is N_0, and the bandwidth is B.

Denote the channel between the UAV and user k as $h_{u2U,k}$, the channel between the BS and the user k as $h_{u2B,k}$, and the channel between the BS and the UAV as h_{U2B}. Since there are line of sight links in the air-to-ground (A2G) channel, we consider $h_{u2U,k}$ and h_{U2B} as Rician channel, and the direct link between the BS and users $h_{u2B,k}$ as Rayleigh channel. Then we have

$$h_{u2U,k} = h_{u2U,k}^{large} h_{u2U,k}^{small}, \tag{1}$$

$$h_{U2B} = h_{U2B}^{large} h_{U2B}^{small}, \tag{2}$$

and

$$h_{u2B,k} = h_{u2B,k}^{large} h_{u2B,k}^{small}. \tag{3}$$

where h^{large} denotes large scale fading, which is influence by the distance and frequency. h^{small} is the small scale fading. $h_{u2U,k}^{small}$ and h_{U2B}^{small} are with Rician distribution, and $h_{u2B,k}^{small}$ is with Rayleigh distribution.

2.2 Proposed NOMA Scheme

The half duplex communication procedure is divided into two time slots. The proposed NOMA scheme is:

- In the first time slot, the UAV receives information from the users in a NOMA group, which is decided in advance according to NOMA grouping schemes [1, 3]. The BS transmits control information to the UAV. All the NOMA users and the BS share the same spectrum resource.
- In the second time slot, the user transmits information to the BS directly, since the channels between users and BS are usually not blocked completely. The users share the same spectrum resource. In the meantime, the UAV transmits back-haul information to the BS, using the same spectrum.

Since the control and backhaul channel and the users' channel share the same spectrum resource, the decoding procedure at the UAV node is designed according to the channel with the users and the BS. Firstly, we sort the channel gain by descending order as

$$|h_{u2U,a_1}|^2 \geq \cdots \geq |h_{U2B,a_j}|^2 \geq \cdots \geq |h_{u2U,a_{K+1}}|^2. \tag{4}$$

Similarly, as for the second time slot, the BS should first sort the channel gain as

$$|h_{u2B,b_1}|^2 \geq \cdots \geq |h_{U2B,b_j}|^2 \geq \cdots \geq |h_{u2B,b_{K+1}}|^2, \tag{5}$$

where a_j and b_j denotes the number of users, i.e., $a_j = k$. And we have $a_j = 0$ and $b_j = 0$ for the channel between the UAV and the BS. The indexes for the user's order is necessary, because the order of a user's channel to UAV and the order of the same user's channel to BS is usually different. But the UAV and the BS both needs to decode the information according the decrease order of the channel gains. After the decoding procedure at the second time slot, the BS will need to pair the received information of a user from the relay and the information of the user from the direct link. In fact, in the simulation procedure, we found that this sorting procedure can be finished by multiplying a matrix, which will be not increase much complexity. Note that the channel gains $|h_{U2B,a_k}|^2$ and $|h_{U2B,b_k}|^2$ are possibly listed at the head of the list because the UAV-to-BS links usually have better channels when the UAV is closer to the BS.

Table 1. Tables of parameters in the proposed NOMA scheme.

Time slot Num/item		BS	UAV	Users
Time slot 1	Power	p_B	–	$p_{u2U,k,1}$
	Channel	$h_{U2B,1}$	$h_{u2B,1}$	$h_{u2U,k,1}$
Time slot 2	Power	–	p_U	$p_{u2B,k,2}$
	Channel	$h_{U2B,2}$	$h_{u2B,2}$	$h_{u2B,k,2}$

The parameters for the proposed NOMA scheme is given in Table 1. At time slot 1, the UAV receives information from the ground users and the BS. The ground users and the BS shares the same spectrum resource. According Shannon theory, the spectrum efficiency between the user k and the UAV at time slot t are denoted as $R_{U,k,t}$, and is given by

$$R_{u2U,k,1} = \log_2\left(1 + \frac{p_{u,k,1}|h_{u2U,k,1}|^2}{I_{u2U,k,1} + N_0 B}\right). \tag{6}$$

The spectrum efficiency from the BS to the UAV $R_{U2B,t}$, and from the UAV to the BS $R_{U2B,t}$ are both

$$R_{U2B,t} = \log_2\left(1 + \frac{p_U|h_B|^2}{I_{U2B,k,t} + N_0 B}\right). \tag{7}$$

Note that as for (7), the interference at time slot 1 and time slot 2 are different, because at time slot 1, $R_{U2B,1}$ is the control data rate to the UAV, the interference consists of the users' information to the UAV; as for time slot 2, $R_{U2B,2}$ is the back-haul link from the UAV to BS, the interference of which consists of the data of the direct link from the users to the BS. The spectrum efficiency from the users to the BS is given by

$$R_{u2B,k,2} = \log_2\left(1 + \frac{p_{u,k,2}|h_{u2B,k,2}|^2}{I_{u2B,k,2} + N_0 B}\right). \tag{8}$$

Then at time slot 1, at the UAV relaying node, the interference is the received information from the users with lower channel gain than the user k in (4).

$$I_{u2U,k,1} = \sum_{j=a_{k+1}}^{a_{K+1}} \widetilde{p}_{j,1}|\widetilde{h}_{j,1}|^2, \tag{9}$$

similarly, at time slot 2, at the BS, the interference is

$$I_{u2B,k,2} \text{ or } I_{U2B,k,2} = \sum_{j=b_{k+1}}^{b_{K+1}} \widetilde{p}_{j,2} |\widetilde{h}_{j,2}|^2. \tag{10}$$

Note that in (9), $\widetilde{p}_{j,1}$ and $\widetilde{h}_{j,1}$ represent p_B or $p_{u2U,j,1}$ and $h_{u2U,j}$ or $h_{U2B,j}$ in (4). And in (10), $\widetilde{p}_{j,2}$ and $\widetilde{h}_{j,2}$ represent p_U or $p_{u2B,j,2}$ and $h_{u2B,j}$ or $h_{U2B,j}$ in (5).

3 Mathematical Problem Formulation

Considering the communication quality for each users, we maximize the minimum achievable rate per Hz of the users, under the constraints of the available power of users, the UAV and the BS. In addition, we also guarantee the quality of the UAV's control and back-haul data rates. The problem is formulated as

$$\max_{p_{u,k,t}, p_U, p_B} \quad \min \frac{R_{u2U,k,1} + R_{u2B,k,2}}{2}, k = 1, \cdots, K \tag{P1}$$

$$\text{s.t. } R_{U2B,1} \geq R_{Backhaul}, \tag{11}$$

$$R_{U2B,2} \geq \sum_{k=1}^{K} R_{u2U} + R_{Backhaul}, \tag{12}$$

$$\sum_{t=1}^{2} p_{u,k,t} \leq P_{u,k}, k = 1, \cdots, K, \tag{13}$$

$$p_U \leq P_U, \tag{14}$$

$$p_B \leq P_B, \tag{15}$$

where (11) is the channel capacity constraint of the control link from the BS to the UAV. (12) is to guarantee the users' data all transmitted to the BS through the UAV relay, as well as ensuring the back-haul data of the UAV transmitted to the BS with required data rates. (13), (14) and (15) are the constraints of power consumption. The problem (P1) is a non-convex problem because of the objective function, (11) and (12), which is hard to be solved. In the next section, we propose the suboptimal solution to design the power allocation scheme of the users, the UAV and BS iteratively.

4 Solution Approach

Since the problem (P1) is a non-convex problem, we iteratively solve it referring to difference of convex functions (DC) programming [8]. According to DC programming, the minimization of the difference of two convex functions

$\mathbf{q}(x) = \mathbf{f}(x) - \mathbf{g}(x)$ can be approached iteratively by solving its convex upper bound $\mathbf{q}(x) = \mathbf{f}(x) - \mathbf{g}^{(l)}(x) - \nabla \mathbf{g}(x^{(l)})^T(x - x^{(l)})$. After using slake parameter $R_{u,min}$, the problem (P1) is given by

$$\max_{p_{u,k,t}, p_U, p_B} \quad R_{u,min} \tag{P2}$$

$$\text{s.t. } (13), (14), (15) \tag{16}$$

$$R_{u,min} + \frac{1}{2}(-R_{u2U,k,1})^{ub} + \frac{1}{2}(-R_{u2B,k,2})^{ub} \le 0, k = 1, \cdots, K, \tag{17}$$

$$R_{backhaul} + (-R_{U2B,1})^{ub} \le 0, \tag{18}$$

$$R_{backhaul} + \sum_{k=1}^{K} R_{u2U,k,1}^{ub} + (-R_{U2B,2})^{ub} \le 0, \tag{19}$$

where $[\cdot]^{ub}$ is the upper bound according to DC programming. For example,let

$$\mathbf{f}(p_{u,k,1}) = -\log_2 \left(\sum_{j=a_k}^{a_K+1} \widetilde{p}_{j,1} |\widetilde{h}_{j,1}|^2 + N_0 B \right), \tag{20}$$

$$\mathbf{g}(p_{u,k,1}) = \left[-\log_2 \left(\sum_{j=a_{k+1}}^{a_K+1} \widetilde{p}_{j,1} |\widetilde{h}_{j,1}|^2 + N_0 B \right) \right], \tag{21}$$

then the derivative of $\mathbf{g}(p_{u,k,1})$ is

$$\nabla \mathbf{g}(p_{u,k,1}^{(l)}) = -\frac{[|\widetilde{h}_{a_{k+1},1}|^2, \cdots, |\widetilde{h}_{a_K,1}|^2]^T}{\ln 2 \left(\sum_{j=a_{k+1}}^{a_K+1} \widetilde{p}_{j,1}^{(l)} |\widetilde{h}_{j,1}|^2 + N_0 B \right)}, \tag{22}$$

where l is the iterative index. Then the upper bound of $-R_{u2U,k,1}$ is

$$(-R_{u2U,k,1})^{up} = \mathbf{f}(p_{u,k,1}) - \mathbf{g}(p_{u,k,1}^{(l)}) - \nabla \mathbf{g}(p_{u,k,1}^{(l)})(p_{u,k,1} - p_{u,k,1}^{(l)}) \tag{23}$$

Similarly, as for the upper bound of $R_{u2B,k,2}$, let

$$\mathbf{f}(p_{u,k,2}) = -\log_2 \left(\sum_{j=b_k}^{b_K+1} \widetilde{p}_{j,2} |\widetilde{h}_{j,2}|^2 + N_0 B \right), \tag{24}$$

$$\mathbf{g}(p_{u,k,2}) = \left[-\log_2 \left(\sum_{j=b_{k+1}}^{b_K+1} \widetilde{p}_{j,2} |\widetilde{h}_{j,2}|^2 + N_0 B \right) \right], \tag{25}$$

then the derivative of $\mathbf{g}(p_{u,k,2})$ is

$$\nabla \mathbf{g}(p_{u,k,2}^{(l)}) = -\frac{[|\widetilde{h}_{b_{k+1},2}|^2, \cdots, |\widetilde{h}_{b_K,2}|^2]^T}{\ln 2 \left(\sum_{j=b_{k+1}}^{b_K+1} \widetilde{p}_{j,2} |\widetilde{h}_{j,2}|^2 + N_0 B \right)}. \tag{26}$$

Then the upper bound of $-R_{u2B,k,2}$ is

$$(-R_{u2B,k,2})^{up} = \mathbf{f}(p_{u,k,2}) - \mathbf{g}(p_{u,k,2}^{(l)}) - \nabla \mathbf{g}(p_{u,k,2}^{(l)})(p_{u,k,2} - p_{u,k,2}^{(l)}). \quad (27)$$

Since the function of spectrum efficiency form can be expressed as the difference between two log functions, other upper bounds can be derived similarly. Note that the upper bounds of the functions are all convex. Accordingly, the problem (P2) is a convex optimization problem, which can be solved efficiently using CVX tools or interior methods in Matlab. Then the power allocation algorithm is proposed is Algorithm 1

Algorithm 1. Design of power allocation for UAV NOMA communication system

1: Initialize the iteration number $l = 0$, and the initial values of $p_{u,k,t}, p_U, p_B$.
2: **repeat**
3: Solve the convex optimization problem (P2), the solutions of the variables are used to update $p_{u,k,t}^{(l+1)}, p_U^{(l+1)}$, and $p_B^{(l+1)}$;
4: Update the iteration number $l = l + 1$.
5: **until** The value of the objective function reaches a convergence

5 Numerical Results and Discussion

In this section, we show the numerical results of the proposed algorithms. In the system, we consider a square place with length 1000 m and width 1000 m. At the studied time slots, the UAV's position is given as (400 m, 200 m, 100 m). The base-station is fixed at (0 m, 500 m, 0 m). The number of users is 4. Their positions are given randomly in the area of (500 m, 1000 m) for x, (0 m, 1000 m) for y, and $z = 0$ m. The communication system works at 5 GHz with the bandwidth of 20 MHz. Thus the parameter of large-scale path-loss factor at reference distance is -46 dB, the path-loss exponent is set to be 2. The noise spectrum density is -150 dBm/Hz.

We consider OMA scheme as a comparison. As for the OMA scheme, in the first time slot, the UAV also receives information from the ground users and receives control data from the base station. But all the users requires separate spectrum resources. As for the second time slot, the UAV transmits received information as well as control and back-haul data to the BS. And the users transmit data with their separate spectrum resources. For both of the two schemes, we optimize the power of users, BS and UAV to maximize the minimum spectrum efficiency of users. The channels of users to UAV and UAV to BS are Rician channel, with the Rician factor of 5. The channel between users and BS at the second time slot is Rayleigh channel.

Figure 2 shows the convergence of the proposed algorithms.

Figure 3 shows the influence of the available power of the ground nodes. The minimum back-haul data rate per Hz of the base-station is set to be 0.02 bps/Hz, the maximum transmitted power from the base-station for the control and back-haul information to the UAV is 0.02 Watt. The maximum communication power consumption for the UAV is 0.02 Watt. As can be seen from the Fig. 3, with more power available, the minimum spectrum efficiency of the users increases. Also, the results show that the proposed NOMA scheme outperforms the OMA shceme.

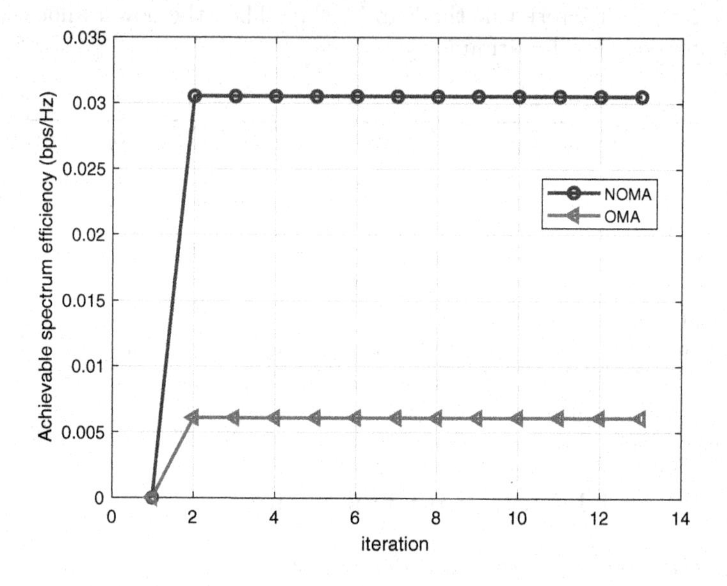

Fig. 2. The convergence of the proposed algorithm.

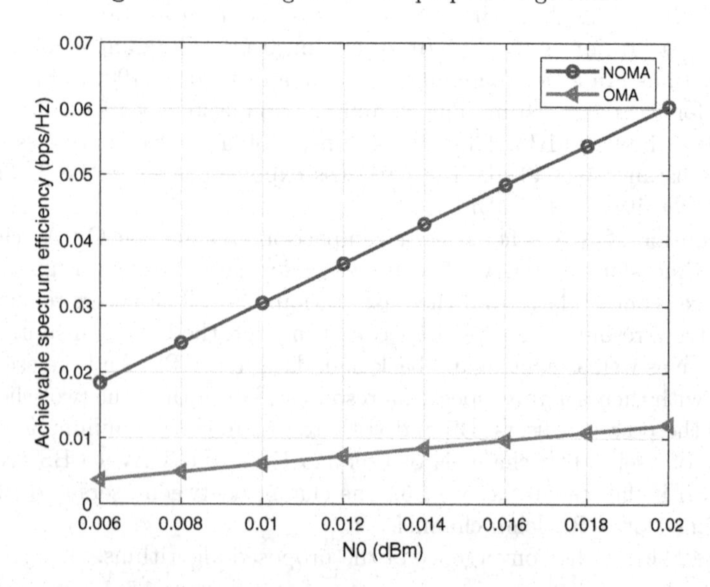

Fig. 3. The influence of the available power for ground users.

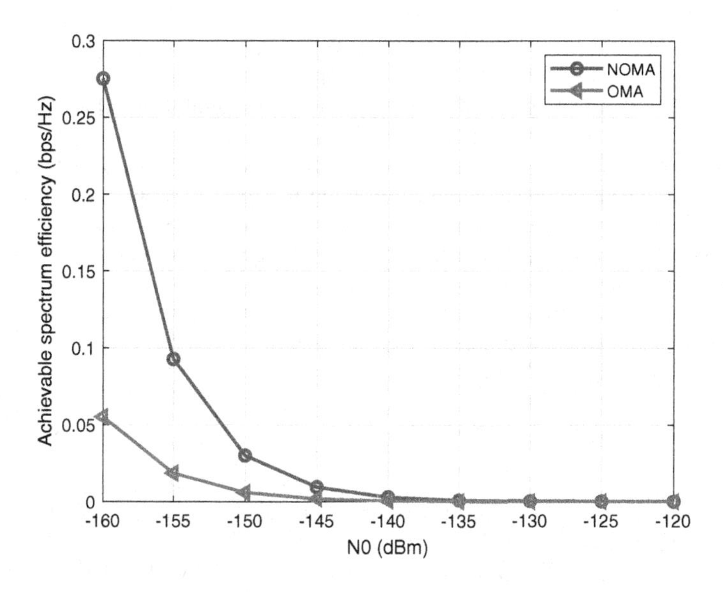

Fig. 4. The influence of noise power spectral density.

Figure 4 shows the influence of noise power. The maximum power from each user is set to be 0.01 Watt. The spectrum efficiency decreases with the increasing of noise power spectral density. It is observed that with the channel condition getting worse, the achievable spectrum efficiencies of NOMA and OMA schemes are both getting worse.

6 Conclusion

In this paper, we studied a cellular network with a UAV and multiple ground users. The UAV communicates with the BS to receive and transmit control and back-haul information as well as works as a relay to enhance the communication performance of ground users. To maximize the spectrum efficiency, we designed the power allocation scheme by solving the non-convex problem using DC programming. Numerical Results show the convergence of the proposed algorithm and prove that the cooperative NOMA scheme can achieve better spectrum efficiency than the OMA scheme. The future work will be extended to user grouping. Also, the influence of the UAV's dynamic trajectory will be studied in following work.

References

1. Chen, W., Zhao, S., Zhang, R., Chen, Y., Yang, L.: Machine learning-based generalized user grouping in NOMA. In: GLOBECOM 2020–2020 IEEE Global Communications Conference, pp. 1–6 (2020)

2. Feng, W., et al.: NOMA-based UAV-aided networks for emergency communications. China Commun. **17**(11), 54–66 (2020)
3. Kang, J.M., Kim, I.M.: Optimal user grouping for downlink NOMA. IEEE Wireless Commun. Lett. **7**(5), 724–727 (2018)
4. Masaracchia, A., Nguyen, L.D., Duong, T.Q., Yin, C., Dobre, O.A., Garcia-Palacios, E.: Energy-efficient and throughput fair resource allocation for TS-NOMA UAV-assisted communications. IEEE Trans. Commun. **68**(11), 7156–7169 (2020)
5. Mei, W., Zhang, R.: Uplink cooperative NOMA for cellular-connected UAV. IEEE J. Sel. Top. Signal Process. **13**(3), 644–656 (2019)
6. Na, Z., Liu, Y., Shi, J., Liu, C., Gao, Z.: UAV-supported clustered NOMA for 6g-enabled Internet of Things: trajectory planning and resource allocation. IEEE Internet Things J. 1 (2020)
7. Nguyen, T.M., Ajib, W., Assi, C.: A novel cooperative NOMA for designing UAV-assisted wireless backhaul networks. IEEE J. Sel. Areas Commun. **36**(11), 2497–2507 (2018)
8. Parida, P., Das, S.S.: Power allocation in OFDM based NOMA systems: a DC programming approach. In: 2014 IEEE Globecom Workshops (GC Wkshps), pp. 1026–1031 (2014)
9. Wang, B., et al.: Graph-based file dispatching protocol with D2D-enhanced UAV-NOMA communications in large-scale networks. IEEE Internet Things J. **7**(9), 8615–8630 (2020)
10. Youssef, M.J., Farah, J., Nour, C.A., Douillard, C.: Full-duplex and backhaul-constrained UAV-enabled networks using NOMA. IEEE Trans. Veh. Technol. **69**(9), 9667–9681 (2020)
11. Zhang, T., Liu, G., Zhang, H., Kang, W., Karagiannidis, G.K., Nallanathan, A.: Energy-efficient resource allocation and trajectory design for UAV relaying systems. IEEE Trans. Commun. **68**(10), 6483–6498 (2020)

Energy-Efficient Joint Offloading and Resource Allocation Strategy in Vehicular Networks

Wei Wu(✉), Ning Wang, Xuanli Wu, and Lin Ma

Communication Research Center, Harbin Institute of Technology, Harbin 150080, China
{kevinking,xlwu2002,malin}@hit.edu.cn, 19S105158@stu.hit.edu.cn

Abstract. In the vehicular networks integrated with mobile edge computing (MEC), vehicle users are permitted to offload latency-sensitive and computation-intensive tasks to nearby MEC servers, which can extend battery life of the vehicle while improving the experience of users. In this paper, we consider a multi-user computation offloading scenario in vehicular networks with MEC server, in which tasks are executed at vehicle and MEC server parallelly through partial offloading. However, the finite communication and computation resource limit the flexibility of offloading. We propose a joint offloading and resource allocation algorithm based on improved hybrid particle swarm and simulated annealing to reduce the system energy consumption as much as possible. The simulation results demonstrate that our algorithm performs well in convergence and energy consumption under strict time constraint.

Keywords: Vehicular networks · Mobile edge computing · Partial offloading · Communication and computation resource allocation

1 Introduction

With the rapid development of emerging mobile applications, the communication and computation requirements of latency-sensitive and computation-intensive applications place higher demands on vehicle networks, which requires powerful computation capability to meet such applications with low latency requirements. In addition, with the introduction of Green 5G, energy consumption has also become the focus of many studies. MEC makes up for the limited computation capability and battery capacity of the vehicle by deploying computation resource and storage resource on the network edge. Moreover, compared with cloud computing, lower latency can be obtained through executing tasks at the MEC server closer to users, which is in line with the communication and computation requirements of the Internet of Vehicles [1].

In practice, the vehicle may need to run streaming data processing applications (such as video analysis). For such applications, the data partition can be used to divide the task into sub-tasks, which can be calculated parallelly at the vehicles and MEC servers [2].

There have been many studies on MEC, most of which focused on offloading decision and resource allocation strategies. On partial offloading, the interaction

Q. Guo et al. (Eds.): WiSATS 2021, LNICST 410, pp. 685–696, 2022.
https://doi.org/10.1007/978-3-030-93398-2_59

between the edge cloud and users is modeled with a Stackelberg game, and uniform and differentiated pricing schemes to find optimal offloading proportion for maximizing MEC server's computation revenue and minimizing users' cost are proposed in [3]. Some works researched the strategy of offloading decision and resource allocation jointly. [4] studied the offloading of VR applications via downlink in vehicular networks, and proposed a three-stage heuristic algorithm to minimize the maximal task completion time by jointly determining offloading proportion and resource allocation. In [5], the scenario where users offload tasks to macro base station through backhaul was considered, and the system computation overhead was minimized through joint optimization of task offloading, wireless backhaul bandwidth resource and computation resource allocation. Based on convex optimization and Hungarian algorithm, [6] proposed a bi-level programming algorithm to optimize binary offloading, power and subcarriers allocation to minimize the system energy consumption jointly. [7] proposed a collaborative scheme of optimizing binary offloading and computation resource allocation to maximize the utility fuction in Cloud-MEC based vehicular networks. [8] applied particle swarm optimization and Nash equilibrium to solve binary offloading and resource allocation respectively, and alternately iterated the two sub-algorithms to maximize the total utility of all vehicles under limited delay, wireless and computation resource. [9] studied the MEC system with physical layer security, and a scheme based on a convex algorithm was proposed to optimize the allocation of local computing tasks and CPU frequency, offloading power and timeslots allocation collaboratively.

The aforementioned works have rarely studied the energy-efficient multi-user MEC system with strict time constraint, limited communication and computation resource for streaming data processing applications. In this case, offloading proportion and resource allocation need to be considered jointly.

In this paper, aim to minimize the system energy consumption while meeting the time constraint in the multi-user vehicular networks, we propose a joint optimization of the offloading proportion, communication and computation resource allocation. The main contributions can be summarized as:

1) We establish a joint optimization model for streaming data processing applications in the vehicular networks with limited resource, in which the system energy consumption is minimized while task completion deadline is met by jointly optimizing offloading proportion and allocation of communication and computation resource.
2) We propose a joint task offloading and resource allocation algorithm based on improved hybrid particle swarm and simulated annealing (IHPS-JORA) to solve the NP-hard problem. The simulation results show that the proposed algorithm can converge fast with strong global optimization capability, and can significantly reduce system energy consumption with time constraint by determining partial offloading decision and resource allocation reasonably.

The rest of this paper is organized as follows. In Sect. 2, system model is illustrated in detail. In Sect. 3, the problem of system energy consumption minimization is formulated. In Sect. 4, we propose a joint task offloading and resource allocation algorithm based on improved hybrid particle swarm and simulated annealing, and the execution steps are described in detail. The simulation results are presented in Sect. 5 to indicate the performance of our proposed algorithm. In Sect. 6, the conclusions are given.

2 System Model

In this part, we describe the task offloading scenario in vehicular networks in detail, including models of the network, communication and computation.

2.1 Network Model

As shown in Fig. 1, we considered a vehicular network covering a section of urban road, which consists of a micro base station (MBS) equipped with a MEC server and several vehicles. We assume that all vehicles are running streaming data processing applications. Since the MBS is equipped with a MEC server, it can provide communication resource and computation resource for vehicles in the coverage of MBS. Vehicles also have limited computation resource. In this network, tasks can be partitioned into two parts, which are called sub-tasks, at any proportion. One is transmitted to the MBS and executed by MEC server through partial offloading, the other can only be executed by vehicles locally. Here, we ignored the transmission time for return of the computation results the computation results from MBS [10, 11], since the output data is usually much less than input data. For simplicity, we assume that all sub-tasks offloaded to MEC server can be executed in parallel without queuing.

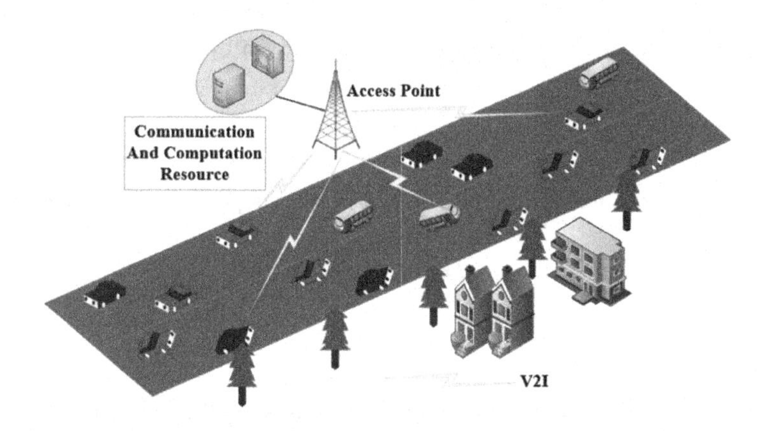

Fig. 1. Multi-user vehicular networks with MEC.

The set of vehicles is expressed as by $\mathbf{N} = \{1,2 ,\ldots, n,\ldots, N\}$. Task of vehicle n is described by a four-field variable $T_n = \{D_n, \beta_n, \tau_n, a_n\}$. This variable contains the size of the task input-data D_n (in Mb), the computation intensity β_n (in CPU cycles per bit), the requested completion deadline τ_n (in msec) and offloading proportion a_n, a_n [0,1]. Then sub-task a_nD_n will be calculated by MEC server and sub-task $(1 - a_n)D_n$ will be executed by vehicle n. The offloading proportion vector is expressed as $\mathbf{A} = [a_1, a_2,\ldots, a_n,\ldots, a_n]$.

2.2 Communication Model

K available sub-channels can be allocated for wireless transmission of vehicles. The bandwidth of each sub-channel is W. The number of sub-channels allocated to vehicle n is denoted by b_n. Then the achievable data transmission rate for vehicle n is expressed as

$$R_n = b_n W \log_2\left(1 + \frac{p_n^t h_n \gamma_n d_n^{-\theta}}{N_0}\right) \tag{1}$$

Here, the path loss is modeled as $d_n^{-\theta}$, where d_n is the distance between vehicle n and the MEC server, θ denotes the path loss exponent. h_n and γ_n represent Rayleigh fading and shadow fading, respectively. p_n^t is the transmission power of vehicle n, and N_0 is the power of white noise.

Since the number of sub-channels of MBS is limited, the allocation of sub-channels is limited by

$$\sum_{n=1}^{N} b_n \leq K \tag{2}$$

The communication resource allocation vector is defined as $\mathbf{B} = [b_1, b_2, \ldots, b_n, \ldots, b_N]$. For vehicle n, the uplink transmission time is given by

$$t_n^{trans} = \frac{a_n D_n}{R_n} \tag{3}$$

Then the corresponding transmission energy consumption can be given by

$$e_n^{trans} = p_n^t t_n^{trans} \tag{4}$$

2.3 Computation Model

In this paper, the task n can be divided into 2 sub-tasks. a_n is the proportion of subtask with is computed remotely. Therefore, $a_n D_n$ and $(1 - a_n) D_n$, is the sub-task size of remote execution and local execution respectively. Vehicle uses these two execution modes parallelly to complete the execution of task.

With respect to different execution modes, time and energy consumption are presented separately as below.

Local Execution. For vehicle n, the frequency of local CPU is $f_{n,loc}$, thus the local execution time of sub-task at vehicle is expressed by

$$t_n^{loc} = \frac{(1 - a_n) D_n \beta_n}{f_{n,loc}} \tag{5}$$

The energy consumption for each CPU cycle of the vehicle n is $k_u f_{n,loc}^2$ according to [12], where k_u is a constant. Then, the energy consumption of sub-task at vehicle for local execution can be given by

$$e_n^{loc} = k_u(1 - a_n)D_n\beta_n f_{n,loc}^2 \tag{6}$$

Remote Execution. We assume that MEC server's CPU frequency is F_{MEC}, which is much higher than that of any vehicle. The computation resource allocated by the MEC server to vehicle n is denoted by f_n, thus the remote execution time is given by

$$t_n^{exe} = \frac{a_n D_n \beta_n}{f_n} \tag{7}$$

The computation resource allocation vector is denoted as $\mathbf{F} = [f_1, f_2, \ldots, f_n, \ldots, f_N]$. Since the the MEC server's computation capability of is limited, the allocation of computation resource is limited by

$$\sum_{n=1}^{N} f_n \leq F_{MEC} \tag{8}$$

The remote execution time includes transmission time and actual computation time, thus it is given by

$$t_n^{off} = t_n^{trans} + t_n^{exe} \tag{9}$$

Similarly, the energy consumption of sub-task at MEC server for remote execution can be given by

$$e_n^{exe} = k_s a_n D_n \beta_n f_n^2 \tag{10}$$

Where k_s is a constant, as a rule, $k_s < k_u$. It can be seen that computation time is shorter, and the corresponding energy consumption is higher, when the computation resource MEC server allocates to vehicle is more. Then, the total energy consumption for vehicle n is given by

$$e_n = e_n^{loc} + e_n^{trans} + e_n^{exe} \tag{11}$$

3 Problem Formulation

In this paper, it is considered that time and energy consumption for streaming data processing applications, aiming to reduce the system energy consumption as much as possible while the task completion deadline is met. The system energy consumption is the total energy consumption of all vehicles. The optimal problem can be formulated as

$$\min_{\mathbf{A},\mathbf{F},\mathbf{B}} E(\mathbf{A},\mathbf{B},\mathbf{F}) = \sum_{n=1}^{N} e_n \tag{12}$$

$$s.t. \quad \text{C1: } \max\left\{t_n^{off}, t_n^{loc}\right\} \le \tau_n, \forall n \in \text{N}$$

$$\text{C2: } \sum_{n=1}^{N} b_n \le K$$

$$\text{C3: } \sum_{n=1}^{N} f_n \le F_{MEC}$$

$$\text{C4: } 0 \le a_n \le 1, \forall n \in \text{N}$$

$$\text{C5: } b_n > 0, \forall n \in \text{N}$$

$$\text{C6: } f_n > 0, \forall n \in \text{N}$$

Where $\mathbf{A} = [a_1, a_2,\ldots, a_n,\ldots, a_N]$, $\mathbf{B} = [b_1, b_2,\ldots, b_n,\ldots, b_N]$, $\mathbf{F} = [f_1, f_2,\ldots, f_n,\ldots, f_N]$. Constraint C1 limits the completion time of all tasks. C2 is the constraint on the number of available sub-channels of MBS. C3 constrains the allocation of computation capability of MEC server. C4 indicates the offloading proportion's range. C5 and C6 represents that the resource allocation of communication and computation is always positive.

Aim to minimize the system energy consumption while all the constraints are satisfied, the solution of the optimal offloading proportion solution $\mathbf{A}^*$, the solution communication resource allocation $\mathbf{B}^*$ and the solution of computation resource allocation $\mathbf{F}^*$ for all vehicles need to be find.

4 Joint Offloading and Resource Allocation Algorithm

This optimization problem is a mixed integer nonlinear programming (MINLP) problem. It can be proved to be a NP-hard problem. Next, the intelligent optimization algorithms will be applied to design a joint task offloading and resource allocation algorithm.

To solve the problem, global optimization algorithm is needed. Particle Swarm Optimization (PSO) is one of them. It is a swarm intelligence algorithm and can converges fast. PSO is usually used to solve unconstrained optimization problems, then we introduce the exterior penalty function method to enable it to handle problems with constraints. PSO tends to be trapped in the local optimal solution, while simulated annealing (SA) algorithm has a strong ability in breaking away from the local optimal solution. In view of this, we consider combining PSO with SA.

The proposed joint task offloading and resource allocation algorithm based on improved hybrid particle swarm and simulated annealing (IHPS-JORA) is illustrated in **Algorithm 1**. The corresponding specific steps are detailed as:

Step 1: Initialization. Inertia weights $\omega_{\max}$ and $\omega_{\min}$, acceleration factors c_1 and c_2, number of particles *M*, maximum iterations *S* of PSO, initial temperature T_0, freezing rate R_T, maximum iterations *C* of SA. Then, the position matrix P_m and velocity matrix V_m of each particle are generated randomly under constraints, that is, the initial offloading proportion and resource allocation of all vehicles are specified. And calculate fitness $f(P_m)$, i.e. the system energy consumption.

Step 2: SA Iterations. Each particle is based on SA independently. For the particle *m*, its position of the *s*-th iteration is given as

$$P_m(s) = \left(P_m^1, P_m^2, \ldots, P_m^n, \ldots, P_m^N\right) \tag{13}$$

Where $P_m^n = (a_n, b_n, f_n)$ represents the offloading proportion, communication and computation resource allocation for vehicle *n*.

At present temperature $T(s)$ of the *s*-th iteration, a new local optimal position $P_{best}^{m,trial}$ of particle *m* by adding perturbation on the previous P_{best}^m is generated, then boundary processing and exterior penalty. Exterior penalty can force particles that do not meet the constraints of (12) to return to feasible region.

If $f(P_{best}^{m,trial}) \leq f(P_{best}^m)$ then $P_{best}^m = P_{best}^{m,trial}$. Otherwise, whether to accept $P_{best}^{m,trial}$ according to Metropolis Criterion. This step is repeated until iterations *C* of SA is reached.

Step 3: Update the position of particles and velocity. For the particle *m*, its velocity is given as

$$V_m(s) = \left(V_m^1, V_m^2, \ldots, V_m^n, \ldots, V_m^N\right) \tag{14}$$

For each iteration of PSO, updating of velocity and position in the *d*-dimension of particle *m* is given by

$$\omega = \omega_{\max} - \left(\frac{\omega_{\max} - \omega_{\min}}{S}\right)s \tag{15}$$

$$V_{md}(s+1) = \omega V_{md}(s) + c_1 \mathrm{rand}_1\left(P_{best}^m(s) - P_{md}(s)\right) + c_2 \mathrm{rand}_2\left(G_{best}(s) - P_{md}(s)\right) \tag{16}$$

$$P_{md}(s+1) = P_{md}(s) + V_{md}(s+1) \tag{17}$$

$$P_{md}(s+1) = round(P_{md}(s) + V_{md}(s+1)) \tag{18}$$

Where rand_1 and rand_2 are random variables in [0,1], P_{best}^m is the local optimal position of particle *m*, and G_{best} represents the global optimal position of swarm. Note that the offloading proportion and computation resource allocation are both continuous variables, and positions are updated according to (17); the wireless resource allocation is a discrete variable, which needs to be updated according to (18).

Step 4: Check for improvement in the local and global optimal position. Boundary processing and exterior penalty on the new position of each particle, then update P^m_{best} and G_{best}.

Step 5: Convergence. If iterations S of PSO is reached, output G_{best} as the best solution, i.e.$G_{best} = [\mathrm{A}^*, \mathbf{B}^*, \mathbf{F}^*]$. Otherwise, update temperature as (19) and go to step 2.

$$T(s) = R_T T(s-1) \tag{19}$$

Algorithm 1: IHPS-JORA Algorithm

Input: $\omega_{\max}$, $\omega_{\min}$, c_1, c_2, M, S, T_0, R_T, C

Output: $G_{best} = [\mathbf{A}^*, \mathbf{B}^*, \mathbf{F}^*]$: optimal offloading proportion policy and resource allocation

1: Initialization
2: while $s \leq S$ do
3: $s = s + 1$;
4: for $m = 1 : M$ do
5: while $c \leq C$ do
6: $c = c + 1$;
7: Generate new $P^{m,trial}_{best}$ by adding perturbation, then boundary processing, exterior penalty and calculate $f(P^{m,trial}_{best})$;
8: if $f(P^{m,trial}_{best}) \leq f(P^m_{best})$, $P^m_{best} = P^{m,trial}_{best}$ then
9: else Whether to accept $P^{m,trial}_{best}$ according to Metropolis Criterion;
10: end if
11: Update G_{best};
12: end while
13: Update velocity and position of each particle according to (15), (16) and (17), (18) respectively, then boundary processing;
14: Exterior penalty and update P^m_{best} and G_{best};
15: end for
16: Update temperature according to (19);
17: end while
18: return G_{best}

5 Performance Evaluations

We verify our proposed algorithm's performance through simulations.

5.1 Parameters

In the simulations, a vehicular network with multiple vehicles and one base station with MEC server is taken into account. Note that in practice, the computation capability, transmission power, and computation intensity of each vehicle are different, but the heterogeneity of users has no effect on the performance of our algorithm. The basic simulation parameters are shown in the Table 1.

Table 1. Simulation parameters.

System parameters	Values
Shadow fading standard deviation	8 dB
Sub-channel bandwidth W	0.4 MHz
Number of sub-channels K	150
Transmission power of vehicle $p_n^t, n \in \mathbf{N}$	30 dBm
Computation capacity of MEC server F_{MEC}	50 GHz
Computation capacity of vehicle $f_{n,loc}, n \in \mathbf{N}$	0.8 GHz
Capacitance coefficient of MEC server k_s	10^{-25}
Capacitance coefficient of vehicle k_u	10^{-24}
Input data size $D_n, n \in \mathbf{N}$	1.8–2 Mb
Computation intensity $\beta_n, n \in \mathbf{N}$	60 cycles/bit
Task completion deadline $\tau_n, n \in \mathbf{N}$	80 ms

5.2 Simulations

Since the proposed algorithm is an iterative algorithm, we first verify its convergence. As shown in Fig. 2, as the iterative number increasing, the system energy consumption gradually reduces, and eventually tends to stabilize and converge. In addition, as the number of vehicles increases, the convergence of the our algorithm will not deteriorate.

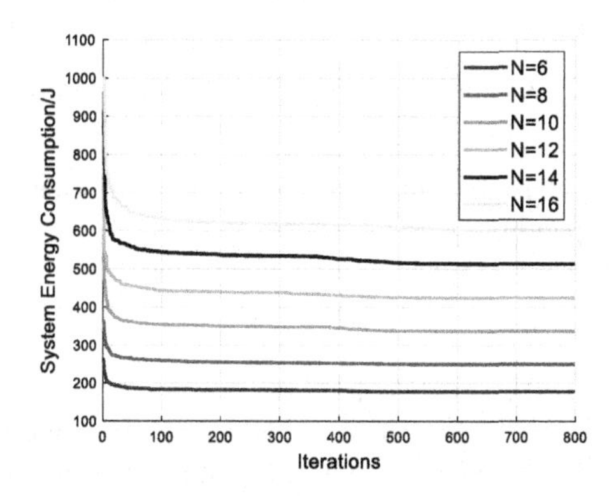

Fig. 2. Convergence for different numbers of vehicles.

Furthermore, we compared proposed algorithm with other joint offloading and resource allocation algorithms, such as particle swarm algorithm, another hybrid particle swarm algorithm and simulated annealing algorithm [13], named as P-JORA and HPS-JORA respectively, as shown in Fig. 3. It shows that, for the same number of vehicles, the proposed IHPS-JORA converges faster and can obtain a better approximate global optimal solution. Besides, when there are fewer vehicles, the convergence of the three algorithms is very close; when there are more vehicles, the convergence of the three algorithms differs greatly. It is because that with the number of vehicles

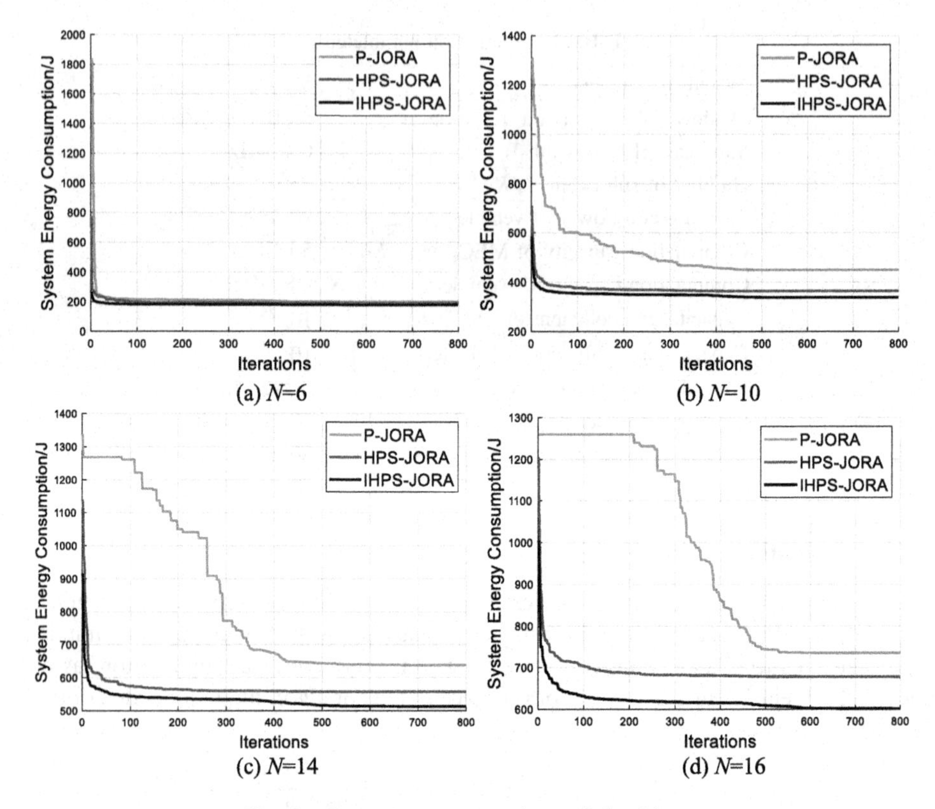

Fig. 3. Convergence comparison of algorithms.

increases, resource gradually becomes insufficient, which further tests the algorithm's global optimization capability.

We introduce two baseline schemes as comparisons. In the All-Local, all tasks are computed at vehicles; in the All-Edge, all tasks are offloaded to the MEC server for compution.

In Fig. 4, as the number of vehicles increases, the system energy consumption continues to increase. Then, when there are fewer vehicles, the performance of All-Edge, P-JORA, HPS-JORA and IHPS-JORA is similar. This is because the resource is sufficient and can easily meet the requirements of all vehicles. With the number of vehicles increases, resource is gradually becoming insufficient. Compared with other algorithms, the proposed IHPS-JORA algorithm always achieves lower system energy consumption. Note that for the simulation parameters of this paper, the execution time of All-Local scheme is at least 135 ms, which cannot meet the limit of task completion time.

Figure 5 presents the relationship between the usage of computation resource of edge server and the number of vehicles. When the number of vehicles increases, the computation resource usage of MEC server for all algorithms continues to increase. Then, for the same number of vehicles, the computation resource usage of server of IHPS-JORA is the least. The less the MEC server's computation resource is used, the

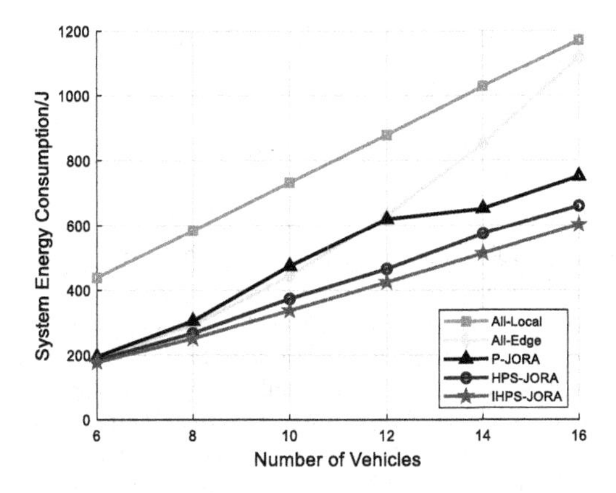

Fig. 4. System energy consumption vs. the number of vehicles.

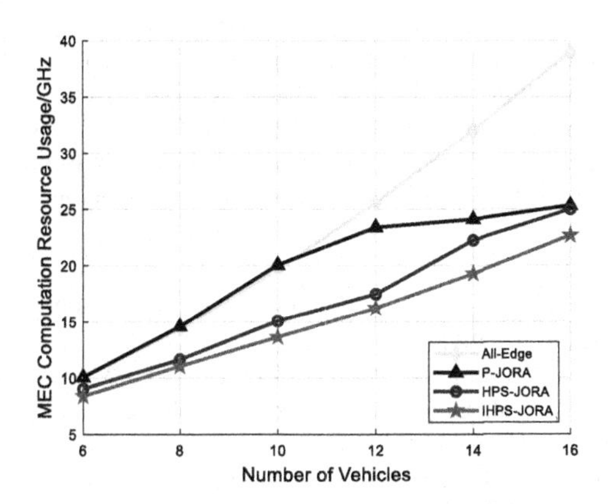

Fig. 5. The computation resource usage of MEC server for different number of vehicles.

more computation resource can be reserved to carry more new offloading tasks. Therefore, under the premise of meeting the predetermined goals, it is hoped that the usage of MEC server's computation resource is as small as possible.

6 Conclusion

In this paper, we studied the multi-user vehicular networks integrated with MEC. The aim is to minimize the system energy consumption for task under strict time constraint, and we formulated a joint optimization model for offloading proportion and resource allocation. We proposed an IHPS-JORA algorithm to achieve the approximate optimal solution to solve this problem. The simulation results reveal that our algorithm converges fast while having strong global optimization capability. In addition, our algorithm can reduce system energy consumption significantly while taking up less computation resource of MEC server, so it has better task offloading and resource allocation performance.

References

1. Abbas, N., Zhang, Y., Taherkordi, A., Skeie, T.: Mobile edge computing: a survey. IEEE Internet Things J. **5**(1), 450–465 (2018)
2. Jiang, C., Cheng, X., Gao, H., Zhou, X., Wan, J.: Toward computation offloading in edge computing: a survey. IEEE Access **7**, 131543–131558 (2019)
3. Liu, M., Liu, Y.: Price-based distributed offloading for mobile-edge computing with computation capacity constraints. IEEE Wirel. Commun. Lett. **7**(3), 420–423 (2018)
4. Zhou, J., Wu, F., Zhang, K., Mao, Y., Leng, S.: Joint optimization of offloading and resource allocation in vehicular networks with mobile edge computing. In: 2018 10th International Conference on Wireless Communications and Signal Processing (WCSP), Hangzhou, China (2018)
5. Pham, Q., Le, L.B., Chung, S., Hwang, W.: Mobile edge computing with wireless backhaul: joint task offloading and resource allocation. IEEE Access **7**, 16444–16459 (2019)
6. Cheng, K., Teng, Y., Sun, W., Liu, A., Wang, X.: Energy-efficient joint offloading and wireless resource allocation strategy in Multi-MEC server systems. In: 2018 IEEE International Conference on Communications (ICC), Kansas City, pp. 1–6 (2018)
7. Zhao, J., Li, Q., Gong, Y., Zhang, K.: Computation offloading and resource allocation for cloud assisted mobile edge computing in vehicular networks. IEEE Trans. Veh. Technol. **68** (8), 7944–7956 (2019)
8. Wu, W., Wang, Q., Wu, X., Zhang, N.: Joint offloading and resource allocation for scalable vehicular edge computing. In: VTC2020-Fall, pp. 1–5 (2020)
9. Wang, J.-B., Yang, H., Cheng, M., Wang, J.-Y., Lin, M., Wang, J.: Joint optimization of offloading and resources allocation in secure mobile edge computing systems. IEEE Trans. Veh. Technol. **69**(8), 8843–8854 (2020)
10. You, C., Huang, K., Chae, H., Kim, B.: Energy-efficient resource allocation for mobile-edge computation offloading. IEEE Trans. Wirel. Commun. **16**(3), 1397–1411 (2017)
11. Mao, S., Leng, S., Yang, K., Huang, X., Zhao, Q.: Fair energy-efficient scheduling in wireless powered full-duplex mobile-edge computing systems. In: GLOBECOM 2017 - 2017 IEEE Global Communications Conference, Singapore, pp. 1–6 (2017)
12. Mao, Y., You, C., Zhang, J., Huang, K., Letaief, K.B.: A survey on mobile edge computing: the communication perspective. IEEE Commun. Surv. Tutor. **19**(4), 2322–2358 (2017)
13. Javidrad, F., Nazari, M.: A new hybrid particle swarm and simulated annealing stochastic optimization method. Appl. Soft Comput. **60**, 634–654 (2017)

PAPR Reduction Scheme for Localized SC-FDMA Based on Deep Learning

Hao Lu[1], Yu Zhou[2], Yue Liu[1], Rui Li[1(✉)], and Ning Cao[1]

[1] Hohai University, Nanjing 210098, China
luhao@hhu.edu.cn
[2] Marketing Service Center, State Grid Jiangsu Electric Power Co. Ltd., Nanjing, China

Abstract. Large peak-to-average power ratio (PAPR) hinders the development of the localized single carrier frequency division multiple access (SC-LFDMA). In this paper, autoencoder (AE) is introduced in SC-LFDMA to reduce PAPR, known as AE-SC-LFDMA. In AE-SC-LFDMA, the Encoder and Decoder of AE are used to encode and decode the modulated symbols of conventional SC-LFDMA based on deep neural network (DNN). This process aims to make AE-SC-LFDMA achieve lower PAPR as well as be more robust to the nonlinear distortion (NLD) of high power amplifier (HPA). Simulation results show that the proposed scheme outperforms conventional schemes both in bit error rate (BER) and PAPR.

Keywords: SC-LFDMA · AE · DNN · HPA

1 Introduction

Single carrier frequency division multiple access (SC-FDMA) has been adopted in the long term evolution (LTE) uplink [1]. It can be described as a version of orthogonal frequency division multiplexing (OFDMA) in which pre-coding and inverse pre-coding stages are added at the transmitter and receiver ends respectively. SC-FDMA has similar throughput and complexity as OFDMA [2]. With lower peak-to-average power ratio (PAPR), SC-FDMA has been seen as a good replacement of OFDMA in some power-efficient scenarioes.

Nevertheless, to acquire long-range detection capabilities and improve power efficiency, the high power amplifier (HPA) in the radar transmitter always operates in saturation. PAPR of the localized SC-FDMA (SC-LFDMA) is not negligible under the nonlinear (NL) HPA. It requires a large input back off (IBO) of the transmit hard power amplifier (HPA) from its output saturation point, which leads to very low power efficiency [3].

Some research has focused on the PAPR reduction of the SC-LFDMA. Selective mapping (SLM) [4] is a notable technique for PAPR reduction. However, the

Q. Guo et al. (Eds.): WiSATS 2021, LNICST 410, pp. 697–708, 2022.
https://doi.org/10.1007/978-3-030-93398-2_60

receiver needs to know the phase factor correctly. Otherwise, the BER performance is greatly degraded. Secondly, pulse shaping method is also proposed. [5] analytically derives the time and frequency domain parametric linear pulse for PAPR reduction. [6] considers an envelope-constrained filter design to optimize the impulse response of a hybrid filter in terms of PAPR reduction. Although this technique is widely used to reduce PAPR, the spectral efficiency and computation complexity are still needed to be improved.

Recently, deep learning (DL), an important branch of machine learning (ML) has shown great potentials in optimizing wireless communication system. As a purely data-driven method, we can learn the properties and the parameters of a DL model directly from the data, without handcraft or ad-hoc designs [7,8]. Autoencoder (AE), a special case of DL, consists of auto-encoder and auto-decoder. Both of them are represented by neural networks (NNs) and trained in an end-to-end manner. Due to the similarity between auto-encoder and transmitter as well as auto-decoder and receiver, AE has been used in some publications to solve communication problems [9]. A novel PAPR reducing network (PRNet) based on AE is proposed in [10]. Convolutional autoencoder (CAE) for PAPR reduction under NL HPA is introduced in [11]. NNs trained on the active constellation extension (ACE) signals is used to reduce the PAPR of OFDM signals in [12]. [13] extends the end-to-end learning to OFDM with cyclic prefix (CP) and compares with conventional OFDM over frequency-selective fading channels. [14] handles the joint transmitter and noncoherent receiver optimization for multiuser single-input multiple-output communications through unsupervised deep learning. Tone reservation network is proposed in [15] to improve the performance of the tone reservation technique. The successful applications indicate that the AE is capable of optimizing the communication system.

To the best of our knowledge, there is no previous work solving the PAPR problem of SC-LFDMA through DL. In this paper, we introduce DL not only to reduce PAPR of SC-LFDMA but also to make it more robust to the nonlinear distortion (NLD) from HPA. The main contribution of this paper is listed as follows,

- We propose a novel PAPR reduction system for SC-LFDMA named AE-SC-LFDMA. In this system, coding and decoding of the modulated symbols in each subcarrier are optimized jointly based on DNN.
- The performance of the proposed system is evaluated using computer simulations. We find that AE-SC-LFDMA has lower PAPR than SC-LFDMA and other PAPR reduction schemes. Moreover, AE-SC-LFDMA shows more robust to the nonlinear effects.

This paper is organized as follows. System model is introduced in the Sect. 2. Then, proposed scheme is given in Sect. 3. Performance is evaluated in Sect. 4. Finally, Sect. 5 concludes this paper.

2 System Model

2.1 System Model

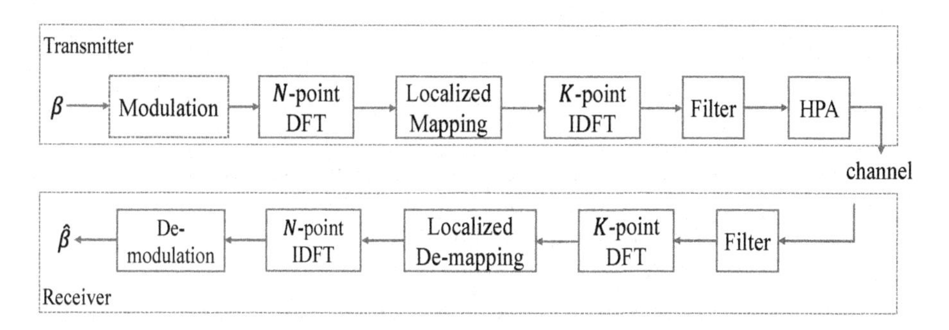

Fig. 1. System Model.

Transmitter. Considering a single user SC-LFDMA in one time block, the system model is shown in Fig. 1. Firstly, the modulated signal is given,

$$s(t, \beta) = \exp(j\phi(t; \beta)), \tag{1}$$

where β is the binary information and $\phi(t; \beta)$ is the signal phase.

Then, $\boldsymbol{s} = [s_0, s_1, \cdots, s_{Q-1}]$, sampled $s(t, \beta)$, is first linearly precoded by an Q by Q DFT precoding matrix. The resulting frequency-domain data vector $\boldsymbol{S} = [S_0, S_1, \cdots, S_{Q-1}]^T$ is,

$$S_i = \sum_{q=0}^{Q-1} s_n e^{-j\frac{2\pi iq}{Q}}, \quad 0 \le i \le Q-1. \tag{2}$$

There are total K subcarriers in SC-LFDMA and K is an integer multiple of Q. The frequency domain data $\boldsymbol{S}$ are mapped across the frequency band by localized method [1]. For LFDMA, chunks of adjacent subcarriers are allocated to each user,

$$\boldsymbol{S}^{'} = \overbrace{[0, 0, \cdots, 0, S_1, S_2, \cdots, S_Q, 0, 0, \cdots, 0]}^{K}, \tag{3}$$

Then, the frequency domain elements $\boldsymbol{S}^{'}$ are processed with the K-point IDFT and converted to analog domain. Assuming there are N_s symbols and symbol duration is T_s, the formed SC-LFDMA signal is,

$$\begin{aligned} x(t) &= e^{j2\pi f_c t} \sum_{\mu=0}^{N_s-1} \sum_{k=0}^{K-1} S^{'}_{\mu,k} e^{j2\pi f_k t} \\ &\cdot g(t - \mu T_s), \end{aligned} \tag{4}$$

where $S'_{\mu,k}$ is the symbol in the kth subcarrier of μth symbol; f_c is the carrier frequency; N_s is the number of SC-FDMA symbols contained in each pulse; $f_k = k\Delta f$, $\Delta f = \frac{1}{T_s}$ is the frequency interval between subcarrier; $g(t)$ is root raised cosine filter (RRC) pulse shaping filter.

HPA. $x(t)$ is fed to an HPA to produce an amplified time domain signal $x'(t)$,

$$x'(t) = x(t) + \Phi_{NL}(t), \tag{5}$$

where $\Phi_{NL}(t)$ denotes the NL term caused from HPA and distributed following the complex circular Gaussian random variable [16].

Based on (4), $x'(t)$ can be expressed as,

$$\begin{aligned} x'(t) &= e^{j2\pi f_c t} \sum_{\mu=0}^{N_s-1} \sum_{k=0}^{K-1} S^T_{\mu,k} e^{j2\pi f_k t} \cdot g(t-\mu T_s) \\ &= e^{j2\pi f_c t} \sum_{\mu=0}^{N_s-1} \sum_{k=0}^{K-1} (S'_{\mu,k} + S^{NL}_{\mu,k}) e^{j2\pi f_k t} \\ &\cdot g(t-\mu T_s), \end{aligned} \tag{6}$$

where $S^T_{\mu,k}$ is the final transmitted symbol in the kth subcarrier of μth symbol and $S^{NL}_{\mu,k}$ is the corrupted term due to the band-limited RRC filter and HPA.

Receiver. Considering communication link, Vehicle B receives the digital domain RCI signal,

$$r_n = x'_n + w_n, \tag{7}$$

where w_n is additive white Gaussian noise (AWGN) with zero mean and one-sided PSD. We can see that the nonlinear part $\Phi_{NL}(t)$ (5) decreases the signal to interference plus noise ratio (SINR). Signal processing in the communication receiver is shown in Fig. 1. The received signal r_n is first filtered by RRC and is transformed into the frequency domain via DFT, $r_n \xrightarrow{\text{DFT}} R_{\mu,k}$. Next, collecting these coefficients of $R_{\mu,k}$ that correspond to each user by inverse localized mapping. We obtain the recovered modulated symbol $R_{\mu,k} \xrightarrow{\text{IDFT}} \tilde{\mathbf{s}}$, which includes NLD and Gaussian noise. Noted that for simplicity, we denote the steps, DFT, mapping, IDFT and RRC as SC-LFDMA MOD and the inverse process as SC-LFDMA DEMOD.

2.2 Design Insights

Recovery of transmitted communication information is paramount for communication system. From (6) and (7), band-limited filter, HPA and AWGN affect the communication quality. Signal's PAPR and robustness determine the degree of these effects on them. Therefor, we can improve SC-LFDMA system in two ways. First, the transmitted symbol are modulated and encoded to have lower PAPR and be robust to these effects. Second, the receiver is able to identify the correct information from the corrupted signal. Hence, we not only need to search the efficient modulation and code of the $S^{'}_{\mu,k}$ but also require an efficient receiver. Traditional model-based methods tend to optimize the transmitter and receiver separately and rarely consider the joint transceiver design. To achieve an efficient SC-LFDMA system, we introduce AE to optimize the transmitter and receiver jointly.

3 Proposed Scheme

In this section, we first briefly discuss the AE general concept. Then, we provide our proposed scheme.

3.1 Introduction of Autoencoder

As one specific type of feedforward NNs, AE consists of Encoder and Decoder. We denote x as the input, $f(x)$ as the Encoder, and $g(x)$ as the Decoder, respectively. The final output of AE is $g(f(x))$. To achieve one or more specific objects, the loss function is set as $L(x, g(f(x)))$. With the goal of minimizing $L(x, g(f(x)))$, AE can be trained through some optimization methods such as adaptive moment estimation (Adam), Stochastic Gradient Descent (SGD) and so on. For example, in [8], to improve the block error rate (BLER) performance, a conventional communication system is represented as an AE. AE is trained to recover the input at the output under various channel conditions. The loss function is set as $L = -\log(b_s)$, where b_s is the sth element of the message probability vector $\boldsymbol{b}$. To minimize the loss function, SGD is used as the optimization method to update the parameters of AE. Finally, the estimation of original input can be obtained from the largest element in $\boldsymbol{b}$.

3.2 Overall Structure of AE-SC-LFDMA

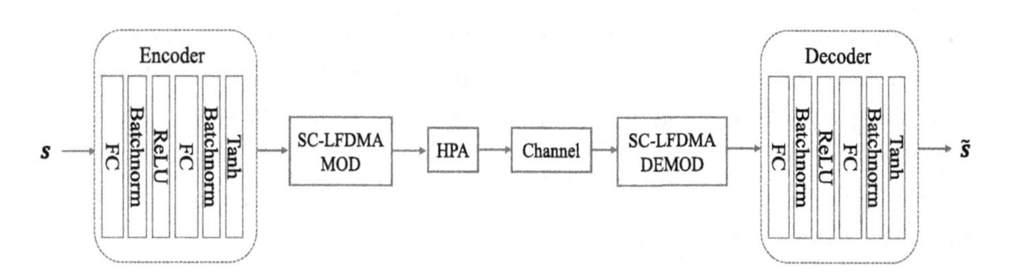

Fig. 2. Proposed AE-SC-LFDMA scheme.

The overall structure of the AE-SC-LFDMA is shown in Fig. 2. Similar to a conventional communication system, the AE-SC-LFDMA system contains three modules, namely, the integrated transmitter/radar receiver, the communication receiver, and the channel. The integrated transmitter takes the modulated symbol $\boldsymbol{s}$ to generate transmitted signal. Noted that modulated symbol $\boldsymbol{s}$ is complex values vector, we assume each element in this vector is $s_q = a_q + b_q i$. Since pytorch cannot operate with complex values, we should convert the input vector $\boldsymbol{s}$ to the real value vector $\{a_0, b_0, a_1, b_1, \cdots, a_Q, b_Q\}$. The Encoder outputs $f(\boldsymbol{s})$ are fed through the SC-FDMA MOD block. Then, the transmitted RCI signal can be expressed as

$$x(t) = \text{MOD}(f(\boldsymbol{s})). \tag{8}$$

The received signal passes through SC-LFDMA DEMOD block. Similar to the Encoder, the symbol should be divided into real and imaginary parts. The obtained vector is fed into Decoder. Finally, the recovered symbol is,

$$\tilde{\boldsymbol{s}} = g(\text{DEMOD}(H(x(t)))), \tag{9}$$

where $H(\cdot)$ means the channel effect.

The used AE block is shown in Fig. 2. The Encoder and Decoder contain the sub-blocks with the same structure. Each sub-block contains fully connected (FC) layers, batch normalization (BN), and activation function. In the FC, $\boldsymbol{x}_{FC}$ is the input of the mth layer, and the FC output $\boldsymbol{y}_{FC}$ can be expressed as $\boldsymbol{y}_{FC} = \boldsymbol{w}_m \boldsymbol{x}_{FC} + \boldsymbol{b}_m$, where $\boldsymbol{w}_m$ and $\boldsymbol{b}_m$ are weights and bias of the mth layer. The output of the FC enters the Batchnorm unit and is used to normalize the input of the activation function. The mathematical representation of Batchnorm is $\text{BN}(\boldsymbol{y}_{FC}) = \gamma \frac{\boldsymbol{y}_{FC} - \mathbb{E}\{\boldsymbol{y}_{FC}\}}{\sqrt{Var\{\boldsymbol{y}_{FC}\} + v}} + \beta$, where γ and β denote the scaling and shift factors, respectively. $\mathbb{E}\{\boldsymbol{y}_{FC}\}$ and $Var\{\boldsymbol{y}_{FC}\}$ denote the mean and variance of $\boldsymbol{y}_{FC}$, respectively. In addition, $v = 0.001$ prevents division by zero. γ and β can be obtained through training and learning. The normalized value then enters the activation function $\phi(\cdot)$. In this solution, ReLU and Tanh serve as the activation function. The output of the ReLU on the mth layer is $\max(\text{BN}(\boldsymbol{y}_{FC}), 0)$, while the output of the Tanh on the mth layer is $\frac{e^{\boldsymbol{y}_{FC}} - e^{-\boldsymbol{y}_{FC}}}{e^{\boldsymbol{y}_{FC}} + e^{-\boldsymbol{y}_{FC}}}$. The used Encoder is composed of $L_f = 2$ sub-blocks. The final output can be expressed as $f(\boldsymbol{r}) = \phi_{L_f}(\text{BN}(\boldsymbol{w}^f_{L_f} \phi_{L_f - 1}(\cdots \phi_1(\text{BN}(\boldsymbol{w}^f_1 \boldsymbol{r} + \boldsymbol{b}^f_1)) \cdots)))$, where $\boldsymbol{w}^f_{L_f}$ and $\boldsymbol{b}^f_{lf}$ are the weights and biases for the l_f-th FC of the Encoder. Similar to the Encoder, the Decoder can be expressed as $g(\boldsymbol{r}) = \phi_{L_f}(\text{BN}(\boldsymbol{w}^f_{L_f} \phi_{L_f - 1}(\cdots \phi_1(\text{BN}(\boldsymbol{w}^f_1 \boldsymbol{r} + \boldsymbol{b}^f_1)) \cdots)))$, where $\boldsymbol{w}^f_{L_f}$ and $\boldsymbol{b}^f_{lf}$ are the weights and biases for the l_f-th FC of the Decoder.

3.3 Training of Network

To optimize both the transmitter and receiver performance under the NLD effect, we jointly train the Encoder and Decoder to adjust the parameters $\theta = \{\theta_f, \theta_g\}$, where θ_f and θ_g denotes the parameters in the Encoder block and Decoder block, respectively. In the following, two objective functions required for network training are provided.

Firstly, to make signal be robust to the NLD, the first objective function is set to minimize the distance between recovered symbols and original symbols $\mathcal{L}_1(\boldsymbol{s}, \tilde{\boldsymbol{s}})$,

$$\mathcal{L}_1(\boldsymbol{s}, \tilde{\boldsymbol{s}}) = \|\boldsymbol{s} - \tilde{\boldsymbol{s}}\|_2, \tag{10}$$

where $\boldsymbol{s}$ denotes the original symbols and $\tilde{\boldsymbol{s}}$ is the recovered symbols. As the first objective function $\mathcal{L}_1(\boldsymbol{s}, \tilde{\boldsymbol{s}})$ is minimized, communication receiver are more likely to obtain the correct communication information.

Secondly, the transmitted signal is expected to have lower PAPR. We propose to minimize PAPR as the goal. Then, the second objective function is to minimize $\mathcal{L}_2(\boldsymbol{s})$,

$$\mathcal{L}_2(\boldsymbol{s}) = \text{PAPR}\{x[n]\} = \text{PAPR}\{\text{MOD}(f(\boldsymbol{s}))\} \tag{11}$$

The PAPR of the transmitted signal vector $\boldsymbol{x} = [x_0, x_1, \cdots, x_{N-1}]^T$ is defined as,

$$\text{PAPR}\{\boldsymbol{x}\} = \frac{\max_{0 \leq n \leq N-1} |x_n|^2}{\mathbb{E}\{|x_n|^2\}},$$

where $\mathbb{E}\{|x_n|^2\}$ is the average power of $x[n]$ over $0 \leq n \leq N-1$.

The objective funtion $\mathcal{L}_1(\boldsymbol{s}, \tilde{\boldsymbol{s}})$ is not only used to train the demodulator to reconstruct the transmitted symbol correctly but also train the transmitted symbol more robust to the bandlimited filter, NLD and AWGN. The objective funtion $\mathcal{L}_2(\boldsymbol{s})$ constrains the PAPR of the transmitted signal. To achieve both the recovery of transmitted symbol and the reduction of PAPR, the Encoder and Decoder modules need to be jointly trained. Hence, the joint loss function $\mathcal{L}(\boldsymbol{s}, \tilde{\boldsymbol{s}})$ is defined as,

$$\mathcal{L}(\boldsymbol{s}, \tilde{\boldsymbol{s}}) = \mathcal{L}_1(\boldsymbol{s}, \tilde{\boldsymbol{s}}) + \omega * \mathcal{L}_2(\boldsymbol{s}) \tag{12}$$

where ω represents weighting factor of the objective function $\mathcal{L}_2(\boldsymbol{s})$.

The training progress is listed as follows:

- Data preparation: We generate the binary information randomly and modulate them by quadrature phase shift keying (QPSK). The Data set is divided into three parts. 70% of the data is used for training; 20% is used for validation; 10% is used for testing.
- Model and loss function: AE-SC-LFDMA is used for training. We set the loss function as $\mathcal{L}(\boldsymbol{s}, \tilde{\boldsymbol{s}})$ (12) and deploy Adam to minimize $\mathcal{L}(\boldsymbol{s}, \tilde{\boldsymbol{s}})$.
- Training: The training data is trained iteratively. In each iteration, Adam updates the parameters of AE-SC-LFDMA to gradually approach minimum of $\mathcal{L}(\boldsymbol{s}, \tilde{\boldsymbol{s}})$. The model is evaluated on validation set after every five iterations. Finally, we can obtain an AE-SC-LFDMA with optimal parameters after training.
- Testing: Ultimately, the communication and radar performance are presented through deploying the final AE-SC-LFDMA model on the testing data set.

4 Simulation Results

In this section, the performance of our proposed AE-RCI system is compared to a conventional SC-FDMA scheme by considering important attributes such as complementary cumulative distribution function (CCDF) of the PAPR and BER. We compare the proposed scheme with the common traditional scheme, SLM [4]. Besides, the proposed scheme is also compared with the convolutional neural network (CNN) shown in [11].

4.1 Parameter Setting

We consider an SC-LFDMA scheme with parameter setting listed in Table 1. QPSK is selected as modulation and AWGN channel is assumed. Parameters used for the AE setup is provided in Table 2. Both Encoder and Decoder are made of three layers and 128 neurons for hidden layer. For training the autoencoder, we use the batch size of 1000. Adam [17] optimizer and learning rate of 0.05 are used. Moreover, the SNR for training is set to be 10 dB. The training SNR is determined by training AE with 2 dB SNR increment per step starting from 0 dB to 20 dB until the SNR with the balanced performance with the BER and PAPR.

In this paper, HPA uses the TWTA model [18]. The amplified output, i.e., the final transmitted signal is,

$$x'(t) = A(\rho(t))e^{j(\phi(t)+\Phi(\rho(t)))} \tag{13}$$

where $\rho(t)$ and $\phi(t)$ represent the envelope and phase of $x(t)$, respectively. And $A(\cdot)$ and $\Phi(\cdot)$ represent amplitude to amplitude modulation (AM/AM) and amplitude to phase modulation (AM/PM) conversions, respectively. The expressions for $A(\cdot)$ and $\Phi(\cdot)$ are given,

$$A(r) = \frac{\alpha_a r}{1+\beta_a r^2}, \qquad \Phi(r) = \frac{\alpha_\phi r^2}{1+\beta_\phi r^2}.$$

A plausible choice of the above parameters is $\alpha_a = 2.1587$, $\beta_a = 1.1517$, $\alpha_\phi = 4.0033$, $\beta_\phi = 9.1040$ [19].

Input back off (IBO) is an important parameter which describes the amplifier operating point by relating the saturation power of the HPA to the average power of the input signal. It is defined as

$$\text{IBO} = \frac{P_{in}^{sat}}{P_{in}}, \tag{14}$$

where P_{in}^{sat} is the input saturation power and $P_{in} = \mathbb{E}[|x[n]|^2$.

4.2 PAPR Comparison

The PAPR measure is commonly used as an indicator for the required amount of IBO for HPA operation. Higher PAPR means that larger IBO is needed in

Table 1. Parameter setting of RCI system

Parameter	Symbol	Value
Carrier frequency	f_c	24 GHz
Frequency interval	Δf	400 kHz
Symbols of each user	Q	40
Symbol number	N_s	5
Subcarrier number	K	80

Table 2. Parameters used for the AE setup

Encoder module		Decoder module	
Parameter	Value	Parameter	Value
Size of input layers	16	Size of input layers	16
Size of hidden layers	128	Size of hidden layers	128
Size of output layers	16	Size of output layers	16
Hidden layer activation	'ReLU'	Hidden layer activation	'ReLU'
Optimizer algorithm		Adam	
Learning rate		0.05	
Training SNR		10 dB	
Batch size		1000	
Weighting		$\omega = 2$	

HPA. In other words, HPA can not operate in the saturation (IBO = 0dB). This can lead to the power loss of the system. Hence, low PAPR is conducive to the improvement of system power efficiency. The numerically calculated CCDF of the PAPR of AE-SC-LFDMA, original SC-LFDMA, SC-LFDMA with SLM and CAE-SC-LFDMA are given in Fig. 3. We can see that, compared with SC-LFDMA, the corresponding gain is about 3 dB for CCDF = 10^{-3} and 2–3 dB for high CCDF percentiles. For SC-LFDMA with SLM, the gain is about 1dB for CCDF = 10^{-3}. Although SC-LFDMA with SLM outperforms in the high CCDF percentiles. The side information needed in the SLM detection lowers the system efficiency. Moreover, for CAE-SC-FDMA, AE-SC-LFDMA outperforms about 1–2 dB in the overall CCDF. Thus, the NL distortion is expected to be smaller and the power efficiency can be improved in the proposed AE-SC-LFDMA scheme.

4.3 BER Comparison

BER performance comparison of these schemes in the AWGN channel under IBO = 0dB is shown in Fig. 4.

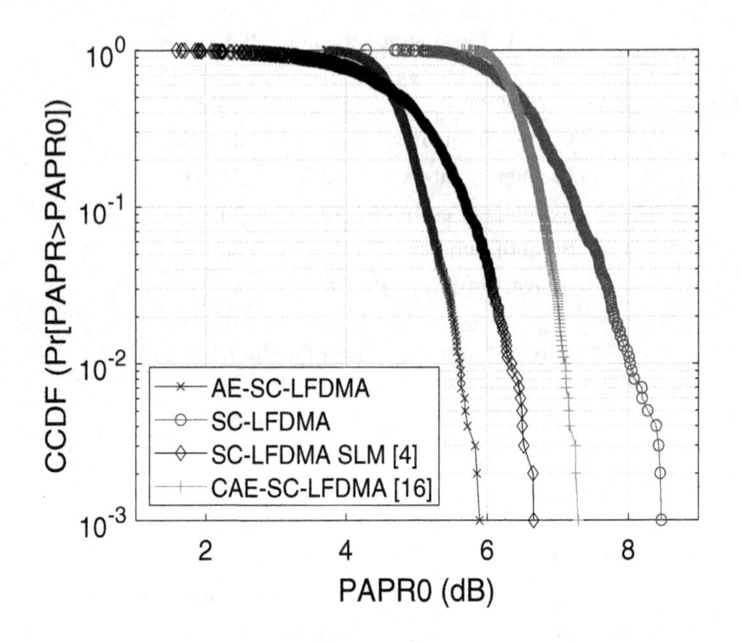

Fig. 3. PAPR performance comparison.

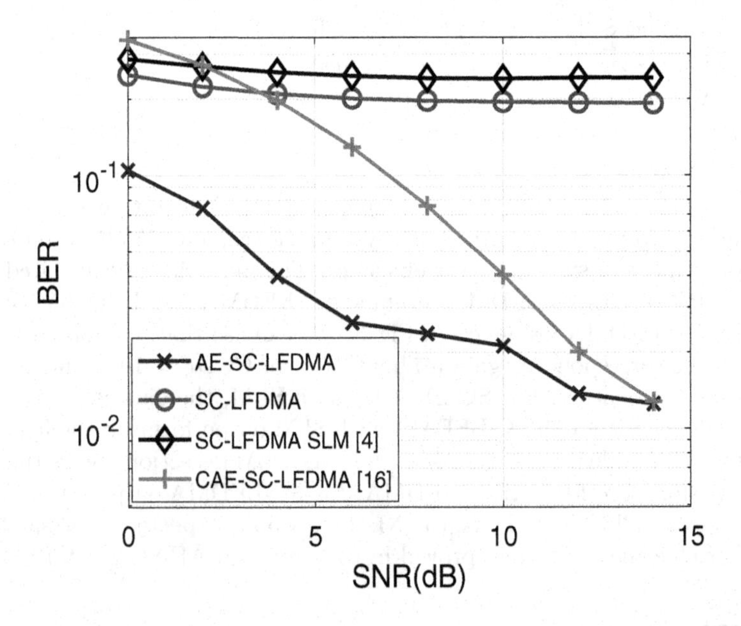

Fig. 4. BER performance comparison (IBO = 0 dB).

The BER performance of the AE-SC-LFDMA is obviously better than that of other SC-LFDMA schemes in the entire SNR range. Compared with original SC-LFDMA and SC-LFDMA with SLM, proposed AE-SC-LFDMA outperforms significantly in the entire range. Besides, AE-SC-LFDM achieves about 5dB performance over CAE-SC-LFDMA in the low SNR range. In a word, the proposed AE-SC-FDMA can achieve satisfied BER performance even under the IBO = 0dB. This shows that AE-SC-FDMA is more robust to the NLD.

5 Conclusion

This paper proposes a novel autoencoder (AE)-based PAPR reduction scheme for localized SC-FDMA, namely AE-SC-LFDMA. Through optimizing transceiver jointly based on deep neural network (DNN), AE-SC-LFDMA is capable of reduing PAPR while improving robustness to the nonlinear effects of high power amplifier. Simulation shows that the proposed AE-SC-LFDM can achieve lower PAPR and BER than conventional SC-LFDMA. Futher work focusing on usage of sparse autoencoder is needed.

Acknowledgements. This work was supported by Research on Performance Evaluation and Optimization Technology of Local IOT for Client-side Metering Equipment under grant No. 5700-202118203A-0-0-00.

References

1. Myung, H.G., Goodman, D.J.: Single Carrier FDMA: A New Air Interface for Long Term Evolution, vol. 8. Wiley, Hoboken (2008)
2. Chen, G., Song, S., Letaief, K.B.: A low-complexity precoding scheme for PAPR reduction in SC-FDMA systems. In: 2011 IEEE Wireless Communications and Networking Conference, pp. 1358–1362. IEEE (2011)
3. Ji, J., Ren, G., Zhang, H.: PAPR reduction of SC-FDMA signals via probabilistic pulse shaping. IEEE Trans. Veh. Technol. **64**(9), 3999–4008 (2014)
4. Mohammad, A., Zekry, A., Newagy, F.: A time domain SLM for PAPR reduction in SC-FDMA systems. In: 2012 IEEE Global High Tech Congress on Electronics, pp. 143–147. IEEE (2012)
5. Meza, C.A., Lee, K., Lee, K.: PAPR reduction in single carrier FDMA uplink system using parametric linear pulses. In: ICTC 2011, pp. 424–429. IEEE (2011)
6. Kamal, S., Meza, C.A.A., Tran, N.H., Lee, K.: Low-PAPR hybrid filter for SC-FDMA. IEEE Commun. Lett. **21**(4), 905–908 (2016)
7. Ye, H., Liang, L., Li, G.Y., Juang, B.: Deep learning-based end-to-end wireless communication systems with conditional GANs as unknown channels. IEEE Trans. Wirel. Commun. **19**(5), 3133–3143 (2020)
8. Dörner, S., Cammerer, S., Hoydis, J., Ten Brink, S.: Deep learning based communication over the air. IEEE J. Sel. Top. Sig. Process. **12**(1), 132–143 (2017)
9. O'shea, T., Hoydis, J.: An introduction to deep learning for the physical layer. IEEE Trans. Cogn. Commun. Networking **3**(4), 563–575 (2017)
10. Kim, M., Lee, W., Cho, D.H.: A novel PAPR reduction scheme for OFDM system based on deep learning. IEEE Commun. Lett. **22**(3), 510–513 (2017)
11. Huleihel, Y., Ben-Dror, E., Permuter, H.H.: Low PAPR waveform design for OFDM system based on convolutional auto-encoder. arXiv preprint arXiv:2011.06349 (2020)

12. Sohn, I.: A low complexity PAPR reduction scheme for OFDM systems via neural networks. IEEE Commun. Lett. **18**(2), 225–228 (2014)
13. Felix, A., Cammerer, S., Dörner, S., Hoydis, J., Ten Brink, S.: OFDM-autoencoder for end-to-end learning of communications systems. In: 2018 IEEE 19th International Workshop on Signal Processing Advances in Wireless Communications (SPAWC), pp. 1–5. IEEE (2018)
14. Xue, S., Ma, Y., Yi, N., Tafazolli, R.: Unsupervised deep learning for MU-SIMO joint transmitter and noncoherent receiver design. IEEE Wirel. Commun. Lett. **8**(1), 177–180 (2018)
15. Wang, B., Si, Q., Jin, M.: A novel tone reservation scheme based on deep learning for PAPR reduction in OFDM systems. IEEE Commun. Lett. **24**(6), 1271–1274 (2020)
16. Balti, E., Guizani, M.: Impact of non-linear high-power amplifiers on cooperative relaying systems. IEEE Trans. Commun. **65**(10), 4163–4175 (2017)
17. Kingma, D.P., Ba, J.: Adam: a method for stochastic optimization. arXiv preprint arXiv:1412.6980 (2014)
18. Candreva, E.A., Tarchi, D., Vanelli-Coralli, A., Corazza, G.E.: Robust SC-FDMA subcarrier mapping for non-linear channels. In: 2014 7th Advanced Satellite Multimedia Systems Conference and the 13th Signal Processing for Space Communications Workshop (ASMS/SPSC), pp. 360–365. IEEE (2014)
19. Saleh, A.A.: Frequency-independent and frequency-dependent nonlinear models of TWT amplifiers. IEEE Trans. Commun. **29**(11), 1715–1720 (1981)

A Power Allocation Method for Downlink MUSA

Haoran Zhang(✉), Shaochuan Wu, Qiuyi Sui, and Rundong Zuo

Harbin Institute of Technology, Harbin, China
scwu@hit.edu.cn

Abstract. Multi-user shared access (MUSA) is a kind of Non-Orthogonal Multiple Access (NOMA), which is suitable for multi-user transmission and has better channel capacity. This paper analyzes the influencing factors of power allocation on the performance of downlink MUSA from the perspective of sum rate of users and bit error ratio (BER) performance. This paper considers a two-user scenario. A power allocation method based on sum rate of users is proposed and a closed-form expression of the sum rate and power allocation coefficients of users is given in the MUSA downlink scenario. In order to analyze and discuss the impact factors of power allocation on MUSA system performance, extensive simulation results are provided to evaluate the performance of the proposed method under different user weight-ratio scenarios. Through these simulation results, the proposed method is proved to be realistic. Based on this result, a power allocation algorithm for a multi-user scenario is presented in this paper.

Keywords: MUSA · Power allocation · BER · Sum rate

1 Introduction

In order to meet the unprecedented massive data demand for wireless services in the future, such as high spectrum utilization, large system throughput and energy efficiency. New wireless access technology is needed. Traditional technology Orthogonal Multiple Access (OMA) is not enough for 5G scenarios. NOMA is proposed to overcome these limitations. Users can share the same time, code and frequency in NOMA [1]. Multi-User Shared Access (MUSA) is a kind of NOMA technology in the code domain. In this technology, the modulation symbols of each user are extended by a complex spreading sequence to realize superimposed transmission of multiple user data. In addition, this technology uses Serial Interference Cancellation (SIC) technology to complete multi-user detection [2]. The principle of SIC is to treat other user data as the interference of user data to be detected. MUSA is very suitable for the Internet of Things (IoT) business due to the diversity of its plural sequence and low cross-correlation [3]. It has been clarified that the channel capacity of MUSA is better thanks to non-zero code spreading, and that the advantages of MUSA are most strongly evidenced in the scenarios of high SNR and multi-user transmission. Thus, MUSA is suitable for 5G massive Machine Type Communications (mMTC). However, because SIC detection has error propagation phenomenon, that is, the accuracy of the user data

Q. Guo et al. (Eds.): WiSATS 2021, LNICST 410, pp. 709–718, 2022.
https://doi.org/10.1007/978-3-030-93398-2_61

detected earlier will affect the detection of user data later [4]. The power allocation directly affects the SINR of each user, and then affects the detection performance of the MUSA system. Therefore, the impact of power allocation on MUSA system performance needs to be analyzed and discussed.

Current research focuses on the power allocation strategy of NOMA in the power domain. Energy efficient power allocation strategies are studied in [5, 6] and outage based power allocation is investigated in [7, 8]. Although the above research is in the field of NOMA in power domain, there are still some reference values. In [9], it is proved that downlink MUSA can achieve better BER performance than NOMA over Rayleigh fading channel by simulation. Besides, its simulation results show that a reasonable power allocation is the key to improve BER performance of MUSA. However, the simulation results lack of theoretical basis and there is no general rule.

In this paper, we give a closed-form expression of the sum rate and power allocation coefficients of users in downlink MUSA scenario, which is only studied in NOMA of power domain, and combine the simulation results to design a user power allocation method based on channel gain. Besides, we analyze the impact of power allocation on MUSA performance from the perspective of actual Bit Error Rate (BER) detection, which analyzes the actual detection BER of the system and the influence of the power distribution coefficient, and also verifies the correctness of the channel capacity analysis. This paper is based on the analysis two users' case and the multi-user conditions are inferred to reach general conclusions.

2 System Model

We consider a downlink MUSA transmission system that includes a single-antenna base station and K single-antenna users. In this system, the base station allocates the transmission power of K users. These users share the same time and frequency channel resources. The received signals of K users contain the data of all of the users, so each user needs to use MMSE-SIC to detect the data they need. The channel gain from the Base Staion(BS) to the k-th user is h_k, $k \in K$. Users could be assumed to be sorted such that $|h_1| \leq |h_2| \leq \cdots \leq |h_k| \leq \cdots \leq |h_K|$. The transmitted signal of the base station can be expressed as

$$S = \sum_{k=1}^{K} \sqrt{P_k} w_k x_k \tag{1}$$

where $P_k = c_k P$. P is the total transmit power. c_k is the power allocation coefficient of the base station to user k, $\sum_{k=1}^{K} c_k = 1$.w_k is the extended sequence of user k, x_k is the modulated signal of user k. Then, the received signals y_k at the user k from the BS are given by

$$y_k = h_k S + n_k \tag{2}$$

where n_k represents the additive white Gaussian noise (AWGN) at the user k (Fig. 1).

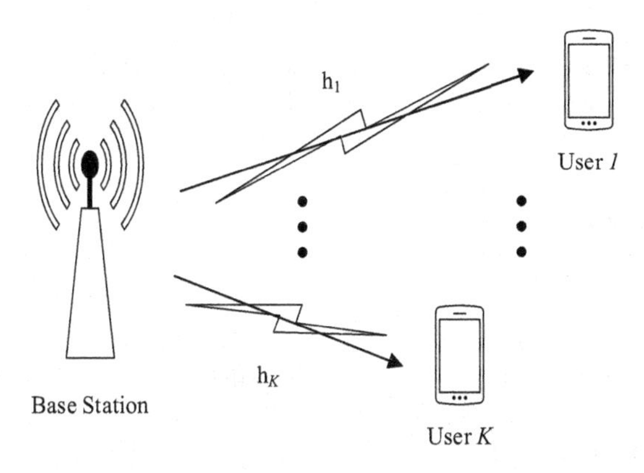

Fig. 1. Downlink MUSA system with K users

According to the detection principle of MMSE-SIC, users with larger SINR are first detected at the receiver, and then the remaining users are detected after reconstruction and elimination. In order to achieve a better sum rate of users, the power distribution coefficient c_k should be allocated reasonably. Assuming ideal power distribution, the SINR of users is

$$\mathrm{SINR}_k = \frac{c_k P|h_k|^2 \|w_k\|^2}{P|h_k|^2 \sum_{i=k+1}^{K} c_i \|w_i\|^2 + N} \tag{3}$$

Since the extended sequence $\|w_k\|^2$ has been normalized by power, Eq. (3) is further simplified as

$$\mathrm{SINR}_k = \frac{c_k P|h_k|^2}{P|h_k|^2 \sum_{i=k+1}^{K} c_i + N} \tag{4}$$

Further, the channel capacity of each user can be obtained, that is, the achievable rate of each user is

$$\mathrm{R}_k = \log_2(1 + \mathrm{SINR}_k) = \log_2\left(1 + \frac{c_k P|h_k|^2}{P|h_k|^2 \sum_{i=k+1}^{K} c_i + N}\right) \tag{5}$$

However, solution to the weighted sum rate maximization problem is not convex and finding a solution is not straight forward [10]. To find the most efficient way of power allocation, we consider the scenario of two users.

3 Channel Capacity Analysis of Power Allocation

In the two-user scenario, c_1 and c_2 are the power distribution coefficients of user *1* and user *2* respectively. In order to draw a general conclusion, we classify and discuss the value of *c*.

3.1 Classify and Discuss

Case (i): $c_1 < c_2$
In this case, the power allocated to user *1* by the BS is less than the power allocated to user *2*. At this time, according to the detection principle of MMSE-SIC, at user 1, the data of user 2 is first detected, and then the detected data of user *2* is reconstructed and then subtracted from the received signal, then the new received signal is used to detect the data of user *1*; on the user 2 side, the data of user 2 is also detected first, but because user *2* does not need the data of user *1*, there is no need to perform reconstruction and elimination operations, and the detected data of user *2* is directly obtained.

First, on the side of user *1*. The date of user *2* is first detected, the SINR of user *2* is

$$\mathrm{SINR}_2^{(1)} = \frac{c_2 P|h_1|^2}{c_1 P|h_1|^2 + N} \tag{6}$$

After the data of user *2* is detected, the data of user *2* is reconstructed and eliminated, and the new received signal is

$$y_1^{new} = h_1 \sqrt{c_1 P} w_1 x_1 + n \tag{7}$$

Using the new received signal to detect user *1*'s data, the SINR of user *1* can be obtained as

$$\mathrm{SINR}_1^{(1)} = \frac{c_1 P|h_1|^2 \|w_1\|^2}{N} = \frac{c_1 P|h_1|^2}{N} \tag{8}$$

The achievable rate of each user on the side of user *2* is

$$\mathrm{R}_1^{(1)} = \log_2\left(1 + \frac{c_1 P|h_1|^2}{N}\right) \tag{9}$$

$$\mathrm{R}_2^{(1)} = \log_2\left(1 + \frac{c_2 P|h_1|^2}{c_1 P|h_1|^2 + N}\right) \tag{10}$$

Because user *2* no longer performs reconstruction and elimination operations, and only needs to obtain the user *2*'s rate as shown in Eq. (11).

$$\mathrm{R}_2^{(2)} = \log_2(1 + \frac{c_2 P|h_2|^2}{c_1 P|h_2|^2 + N}) \tag{11}$$

Hence the sum rate of two users in case(i) is

$$\begin{aligned} \mathrm{R}_{\mathrm{sum}}^{c_1 < c_2} &= \mathrm{R}_1^{(1)} + \mathrm{R}_2^{(2)} \\ &= \log_2(\frac{c_1 P|h_1|^2 + N}{N} \bullet \frac{P|h_2|^2 + N}{c_1 P|h_2|^2 + N}) \end{aligned} \tag{12}$$

Considering $\frac{P}{N} \to \infty$, that is, the signal power is larger than the noise power, and the noise power can be ignored. At this time, the user sum rate is

$$\mathrm{R}_{\mathrm{sum}}^{c_1 < c_2} = \log_2(\frac{P|h_1|^2}{N}) \tag{13}$$

Case (ii): $c_1 > c_2$
In this case, the power allocated to user *1* by the BS is larger than the power allocated to user *2*. At this time, according to the principle of MMSE-SIC detection, on the user *1* side, the data of user *1* is first detected. Because user *1* does not need the data of user *2*, the detection can be stopped after the data of user *1* is obtained. On the user *2* side, the data of user *1* is still first detected. Because user *2* needs the data of user *2*, then the data of user *1* detected will be reconstructed, and then removed from the received signal, and then used. The received signal detects user *2*'s data. Similar to case (i), we get the user sum rate.

$$\begin{aligned} \mathrm{R}_{\mathrm{sum}}^{c_1 > c_2} &= \log_2(\frac{P|h_1|^2 + N}{(1 - c_1)P|h_1|^2 + N} \bullet \frac{(1 - c_1)P|h_2|^2 + N}{N}) \\ &\overset{\frac{P}{N}\to\infty}{=} \log_2(\frac{P|h_2|^2}{N}) \end{aligned} \tag{14}$$

Case (iii): $c_1 = c_2$
In this case, the power allocated to user *1* by the BS is equal to the power allocated to user *2*. The receiver can either choose to detect the data of user *1* first, or choose to detect the data of user *2* first.

If we choose to detect the data of user 1 first, the channel capacity is

$$\mathrm{R}_{\mathrm{sum}}^{c_1 = c_2} = \mathrm{R}_{\mathrm{sum}}^{c_1 > c_2} = \log_2(\frac{P|h_1|^2 + N}{(1 - c_1)P|h_1|^2 + N} \bullet \frac{(1 - c_1)P|h_2|^2 + N}{N}) \tag{15}$$

On the contrary, if we choose to detect the data of user *2* first, the channel capacity is

$$\mathrm{R}_{\mathrm{sum}}^{c_1=c_2} = \mathrm{R}_{\mathrm{sum}}^{c_1<c_2} = \log_2\left(\frac{c_1P|h_1|^2+N}{N} \bullet \frac{P|h_2|^2+N}{c_1P|h_2|^2+N}\right) \tag{16}$$

3.2 Simulation and Analysis

Case(iii) is a special case of case(i) and case(ii). From case(i) and case(ii), we can conclude that sum rate of users is only related to the channel coefficient $|h|^2$ and has nothing to do with the power distribution coefficient c_1 and c_2 when $\frac{P}{N} \rightarrow \infty$. Under the condition of $|h_1|^2 < |h_2|^2$, the relationship curve between sum rate and the power allocation coefficient of user *1* under the two detection strategies is obtained. We assume that $|h_1|^2 = 0.15$ and $|h_2|^2 = 0.2$. The simulation result is below (Fig. 2).

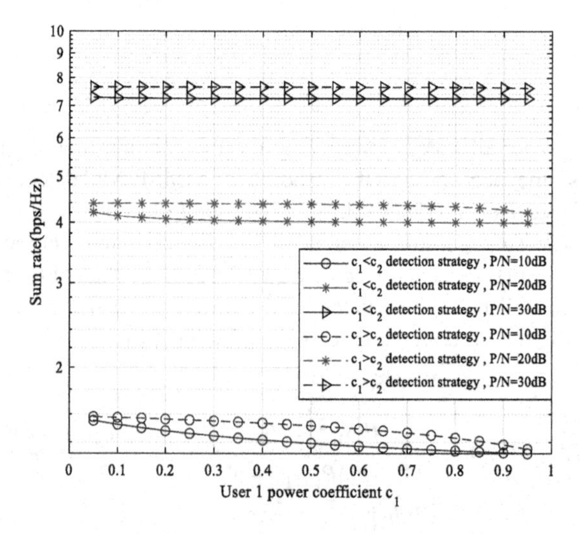

Fig. 2. Sum rate simulation

The above figure compares and analyzes the situation of sum rates under different P/N conditions. According to the curve in the figure, with the increase of P/N, the relationship curve between sum rate and power distribution coefficient of user *1* becomes smoother, that is, the larger the signal power is compared to the noise power, the smaller the relevance between sum rate and power distribution coefficient. At the same time, it is verified that the user and rate are only related to the channel gain, not to the power allocation when the immediate noise power can be ignored compared to the signal power.

It can also be seen that under the condition of $|h_1|^2 < |h_2|^2$, the sum rate of the $c_1 > c_2$ detection strategy are optimal, that is, the strategy of first detecting the data of user *1* and then detecting the data of user *2* can achieve the optimal sum rate. However, considering that in the actual detection, if in the case of $c_1 > c_2$, the data of user *1* is still detected first, the SINR of user *1* is small, which will cause the detection accuracy to be low, and due to the existence of the SIC error propagation phenomenon, the detection accuracy of user *2* will also be low, resulting in poor system detection performance. Aiming at the actual detection situation, this article will conduct a more in-depth analysis from the perspective of detecting BER later.

4 BER Performance

Considering that in the multi-user situation, there are many user power allocation parameters and the simulation parameter setting is more complicated. Therefore, this section mainly simulates and analyzes the situation of two users. The simulation parameter configuration table is shown in Table 1.

Table 1. Simulation parameter configuration

Simulation parameter	Configuration
Coding scheme	Turbo coding with code rate 1/2
Modulation scheme	QPSK
Spreading sequence category	Complex ternary sequence
Spreading sequence length	4
Amount of users	2
Antenna configuration	1Tx,1Rx
Channel	AWGN
Channel estimation	Ideal estimate
Receiver algorithm	MMSE-SIC

Set the channel gain between user 1 and the base station is $|h_1|^2 = 0.15$, and the channel gain between user 2 and the base station is $|h_2|^2 = 0.2$. Figure 3 is the simulation curve.

It can be seen from the simulation results that under the conditions of $|h_1|^2 < |h_2|^2$, the detection BER of the MUSA system in the case of $c_1 > c_2$ is lower and the detection performance is better, which is consistent with the previous sum rate analysis results.

According to the simulation results, the BER of user *1* first decreases and then increases within the range of $c_1 < 0.5$, and the BER continues to decrease within the range of $c_1 \geq 0.5$, while the BER of user *2* continues to increase within the range of $c_1 < 0.5$, and first decreases and then increases within the range of $c_1 \geq 0.5$; the average BER of user *1* and user *2* first decreases and then increases in the range of $c_1 < 0.5$, and also decreases first and then increases in the range of $c_1 \geq 0.5$, and the average BER is obtained the minimum value in $c_1 = 0.65$. The reasons for this trend are analyzed as follows.

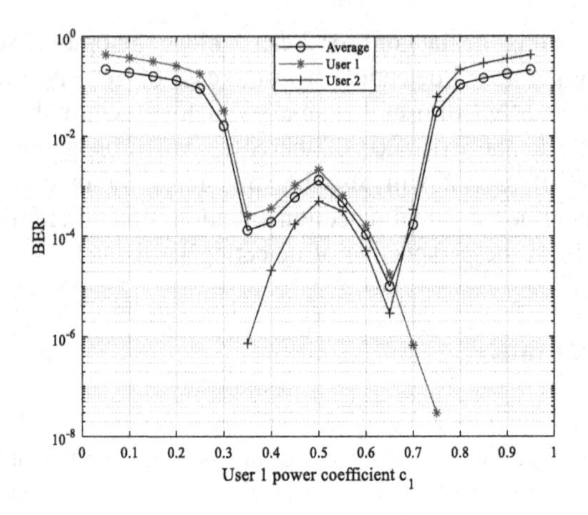

Fig. 3. BER simulation ($|h_1|^2 = 0.15$, $|h_2|^2 = 0.2$)

In the case of $c_1 < c_2$, user *2* directly detects its data. As user *1*'s power allocation coefficient increases, user *2*'s SINR gradually decreases. When the data of user *2* is detected, the MAI of user *1* increases, sot the detection performance deteriorates, and the detection BER gradually increases. On the user *1* side, the data of user 2 is detected first, and then the data of user *2* is reconstructed and eliminated, and then the data of user *1* is detected. Considering the influence of error propagation, the detection accuracy of user *2* data will affect the data of user *1* detection. In the case of $c_1 < 0.35$, User *2* data detection is affected by user *1*'s MAI smaller, and the detection accuracy of user *2* data was higher. Therefore, the detection accuracy of user *1* data gradually improved with the increase of c_1, and the detection BER of user *1* gradually decreased; In the case of $0.35 < c_1 < 0.5$, User *2* data detection is subject to user *1*'s MAI. As c_1 increases, the detection accuracy of user *2* gradually deteriorates. Due to the effect of error propagation, the detection accuracy of user *1* also gradually deteriorates. The BER of user *1* gradually increases. Therefore, in the case of $c_1 < c_2$, the average BER of user *1* and user 2 also shows a trend of first decreasing and then increasing.

In the case of $c_1 \geq c_2$, user *1* directly detects the data of user *1*, so as the power allocation coefficient of user *1* increases, the detection performance of user *1* gradually becomes better, and the detection BER of user *1* gradually decreases. On the other hand, user 2 first detects user *1*'s data, then reconstructs and eliminates user *1*'s data, and then detects user *2*'s data. However, considering the error propagation phenomenon, the accuracy of user 2 data detection is affected by user *1* data detection. In the case of $0.5 \leq c_1 \leq 0.65$, with the increase of c_1, the detection accuracy of user *1* gradually improved, so the detection accuracy of user 2 gradually improved, and the detection BER of user 2 gradually became smaller; In the case of $c_1 > 0.65$, with the increase of c_1, the detection accuracy of user *1* is still gradually getting better, but c_1 becomes very large at this time, and therefore c_2 becomes very small. User 2 cannot obtain good detection performance when detecting the user 2 data, so as the increase of c_1, the detection performance of user 2 gradually deteriorates, and the detection BER gradually increases. Therefore, in the case of $c_1 \geq c_2$, the average BER of user *1* and user *2* also shows a trend of first decreasing and then increasing.

To get a general conclusion, we also set the channel gain between user 1 and the base station is $|h_1|^2 = 0.05$, the channel gain between user 2 and the base station is $|h_2|^2 = 0.2$. The trends of the three curves in the Fig. 4 are basically the same as the simulation curves in Fig. 3.

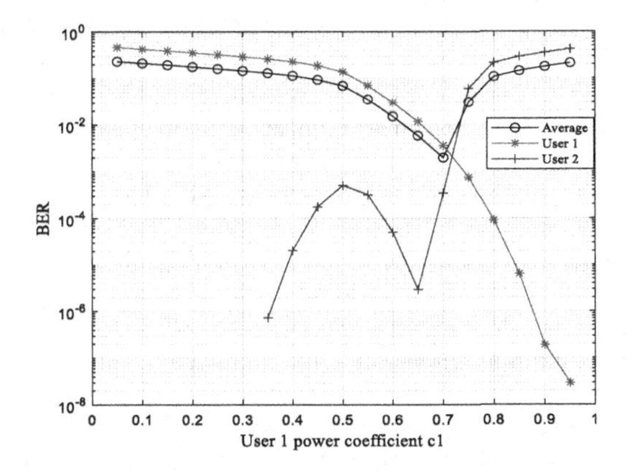

Fig. 4. BER simulation ($|h_1|^2 = 0.05$, $|h_2|^2 = 0.2$)

In summary, through the analysis of channel capacity and detection of BER, it can be concluded that the user's power allocation will affect the detection performance of MUSA.

5 Conclusion

This paper analyzes the impact of power allocation on the performance of the MUSA system from the perspective of channel capacity and detection BER, and gives a closed-form expression of the sum rate of users and user power allocation coefficient in the MUSA downlink scenario, and combines the simulation results to design user power allocation method based on the channel gain. That is, in the case of $|h_1|^2 \leq |h_2|^2 \leq \cdots \leq |h_k|^2 \leq \cdots \leq |h_K|^2$, when the power allocation coefficient satisfies $c_1 \geq c_2 \geq \cdots \geq c_k \geq \cdots \geq c_K$, that is, users with lower channel gains are allocated more transmit power, which enables the MUSA system to effectively improve the BER performance on the premise of ensuring high user sum rate.

Acknowledgements. This research is supported by the National Key R&D Program of China (Under Grant: 2018YFC0806803) and the National Science Foundation of China (Under Grant: 61671173)".

References

1. Dai, L., Wang, B., Yuan, Y., Han, S., Chih-Lin, I., Wang, Z.: Non-orthogonal multiple access for 5G: solutions, challenges, opportunities, and future research trends. IEEE Commun. Mag. **53**, 74–81 (2015)
2. Wang, B., Wang, K., Lu, Z., Xie, T., Quan, J.: Comparison study of non-orthogonal multiple access schemes for 5G. In: 2015 IEEE International Symposium on Broadband Multimedia Systems and Broadcasting, Ghent, Belgium, pp. 1–5 (2015)
3. Yuan, Z., Yu, G., Li, W., et al.: Multi-user shared access for Internet of Things. In: IEEE Vehicular Technology Conference, Nanjing, China, pp. 26–31 (2016)
4. Patel, P., Holtzman, J.: Analysis of a simple successive interference cancellation scheme in a DS/CDMA system. IEEE J. Sel. Areas Commun. 796–807 (2002)
5. Fang, F., Zhang, H., Cheng, J., Leung, V.C.: Energy-efficient resource allocation for downlink non-orthogonal multiple access network. IEEE Trans. Commun. **64**, 3722–3732 (2016)
6. Yi, Z., et al.: Energy-efficient transmission design in non-orthogonal multiple access. IEEE Trans. Veh. Technol. **66**, 2852–2857 (2017)
7. Cui, J., Ding, Z., Fan, P.: A novel power allocation scheme under outage constraints in NOMA systems. IEEE Sig. Process. Lett. **23**, 1226–1230 (2016)
8. He, B., Liu, A., Yang, N., Lau, V.K.N.: On the design of secure non-orthogonal multiple access systems. IEEE J. Sel. Areas Commun. **35**, 2196–2206 (2017)
9. Xu, Y., Wang, G., Zheng, L., Liu, R., Zhao, D.: BER performance evaluation of downlink MUSA over Rayleigh fading channel. In: Gu, X., Liu, G., Li, B. (eds.) MLICOM 2017. LNICSSITE, vol. 226, pp. 85–94. Springer, Cham (2018). https://doi.org/10.1007/978-3-319-73564-1_9
10. Sindhu, P., Deepak, K.S., KM, A.H.: A novel low complexity power allocation algorithm for downlink NOMA networks. In: 2018 IEEE Recent Advances in Intelligent Computational Systems (RAICS), Thiruvananthapuram, pp. 36–40 (2018)

Downlink Power Allocation Strategy in Multi-antenna Ultra-dense Networks Based on Non-cooperative Game

Donglai Zhao[1(✉)], Gang Wang[1], Haoyang Liu[1], and Shaobo Jia[2]

[1] Communication Research Center, Harbin Institute of Technology, Harbin, China
gwang51@hit.edu.cn

[2] School of Information Engineering, Zhengzhou University, Zhengzhou, China
ieshaobojia@zzu.edu.cn

Abstract. This paper investigates the downlink power allocation strategy for a multi-antenna spectrum sharing ultra-dense small cell network in order to suppress the inter-cell interference and improve the system spectral efficiency (SE). The non-cooperative game is adopted to transform the system SE maximization problem into several convex subproblems which maximize the utility function of each user. By designing a dynamic pricing, each Nash equilibrium (NE) of the game is a stationary point of the original optimization problem. In addition, an interference power constraint is applied to guarantee the quality-of-service (QoS) of the key user. Under the game theory framework, an iterative dynamic pricing power allocation (DPPA) algorithm is designed, which is proved to be convergent to the NE of the game model. Furthermore, in order to reduce the signaling overhead and improve the resource utilization, an approximate dynamic pricing power allocation (ADPPA) algorithm is also proposed. Simulation results show that the proposed DPPA algorithm achieves a better performance than benchmark methods and the proposed ADPPA algorithm effectively reduces the signaling overhead with a little performance loss.

Keywords: Power allocation · Ultra-dense network · Non-cooperative game · Spectral efficiency (SE) · Quality-of-service (QoS)

1 Introduction

In the era of data explosion, data traffic in mobile wireless networks has grown explosively. Ultra-dense networks (UDNs) technology is emerging as one of the most promising solutions to meet the requirements of users' wireless traffic volume [1]. UDNs can efficiently improve the network coverage and boost the network capacity by densely deploying the various small cell base stations (SBSs)

This work is supported in part by National Natural Science Foundation of China (No. 61671184).

Q. Guo et al. (Eds.): WiSATS 2021, LNICST 410, pp. 719–730, 2022.
https://doi.org/10.1007/978-3-030-93398-2_62

in the service area of macro cells. Actually, the SBS is a general term, it includes micro, pico, femto, relay and remote radio head (RRH).

Due to the intensive deployment of various types of SBSs, the inter-cell interference in the network is quite serious. Moreover, the interference scenarios are fairly complex with the flexible deployment of small cell base stations. Therefore, interference management for ultra-dense small cell networks becomes an important issue to be resolved. The authors in [2] summarized the advanced interference management techniques for UDNs, including coordinated multi-point, muliple access methods, parallel interference cancellation and power allocation. In [3], the authors designed a multi-domain interference management scheme which combined power control, multiple access scheduling and interference alignment (IA) technology in ultra-dense small-cell networks. Many existing works attempted to manage interference in the power domain. In [4], the authors proposed a contract-based traffic offloading and resource allocation mechanism for the software-defined heterogeneous UDN. Since the mechanism needs a centralized controller to manage the entire network globally, the information exchanged is very huge. To reduce the signaling overhead and computational complexity of the central management mechanism, the authors in [5] presented a local information-based iterative distributed power allocation algorithm by constructing the interfering domains. In [6], a novel energy efficient dynamic power allocation strategy applying a new receiver puncturing technique was proposed for the fifth generation systems. In [7], the authors proposed a cluster-based energy-effcient resource allocation scheme to mitigate the interference and boost energy effciency.

It has been investigated that efficient power allocation strategy can greatly mitigate interference and improve system throughput. As we know, power optimization problems are usually non-convex, so it is a challenging task to obtain the global optimal solution. Fortunately, we can use the game theory to transform non-convex problems into various convex problems. In addition, compared with the centralized power allocation scheme, the distributed algorithms based on game theory require less channel state information (CSI). The authors in [8] investigated a non-cooperative game with penalty factor for UDN, and proved the Nash equilibrium (NE) of the game is unique and Pareto-efficient. Under the game framework, a power allocation algorithm with virtual cell local information was poposed. In [9], a stackelberg game was established to study the joint utility maximum problem subject to a interference power constraint, then the power allocation strategy with non-uniform price was presented for densely deployed scenario. In [10], a distributed interference-aware power control scheme was designed to mitigate interference. In order to reduce the huge interference-related information exchange between the players, the authors formulated a robust mean field game. However, there are still some shortcomings in the above-mentioned research work. First of all, it remains unclear about the gap between the NE points of aforementioned game models and the globally optimal solutions of original optimization problems. Secondly, no interference power constraint has been adopted to guarantee the QoS of key users. Besides, the convergence of proposed iterative algorithms are not proved theoretically. Last but not least, the system models are too simple, it is assumed that the base stations are equipped with a single antenna.

In this paper, we investigate the downlink power allocation problem for a spectrum sharing multi-antenna UDN. The main contributions of our work are: (i) We use the non-cooperative game theory to design a iterative dynamic pricing power allocation algorithm, which converges to the NE of the game as well as the stationary point of the original SE maximization problem. (ii) An interference power constraint is applied to guarantee the QoS of the key user. (iii) An approximate power allocation algorithm is also proposed, which aims to reduce the signaling overhead and improve the resource utilization.

The remainder of this paper is organized as follows. The system model is introduced in Sect. 2. Section 3 formulates the non-cooperative game. In Sect. 4, the power allocation algorithms are proposed. Numerical results and discussions are presented in Sect. 5. Finally, Sect. 6 concludes this paper.

2 System Model

Consider a spectrum sharing ultra-dense small cell network, where the SBSs are equipped with M antennas. The distribution of BSs with density λ is modeled as a homogeneous Poisson point processes (PPPs) $\Phi = \{1, 2, \cdots, N\}$, where N is the number of SBS in region $\mathcal{S}$. The users with single antennas are also modeled by an independent homogeneous PPP of density λ^{u}. Each user is associated with the closest SBS. We assume $\lambda^{\mathrm{u}} \gg \lambda$, so all SBS in the system are active. Each SBS serves one user at each time slot, as the time division multiple access (TDMA) is adopted. For the sake of brevity, we use user n to represent the user connected to SBS n. Although there is no intra-cell interference, the severe inter-cell interference limits the system capacity. The interference links are described in Fig. 1.

The power allocation strategy is divided into two levels. At the system level, the total transmit power p_n of SBS n is allocated under the limitation of the maximum transmit power $p_n^{\max}$; At the cell level, the SBS n allocates the transmit power for each antenna. In this paper, we focus on the system level. As for the cell level, the maximum ratio transmission is adopted.

The channel power gain between SBS i and user j is $h_{i,j}$, modeled as $d_{i,j}^{-\alpha}\beta_{i,j}$, where $d_{i,j}$ is the distance between the transmitter and the receiver, α is the path loss exponent, and $\beta_{i,j}$ is independent random fading coefficient. It should be noted that $\beta_{i,j} \sim \mathrm{Gamma}(1,1)$ when $i \neq j$, otherwise, $\beta_{i,j} \sim \mathrm{Gamma}(M,1)$. Under the above framework, the signal-to-interference-plus-noise ratio (SINR) of user of small cell n can be denoted by

$$\mathrm{SINR}_n = \frac{p_n h_{n,n}}{\sum\limits_{i \neq n} p_i h_{i,n} + w_n} \tag{1}$$

where w_n denotes the additive white Gaussian noise power. Then, the SE of small cell n is given by

$$R_n = \log_2(1 + \mathrm{SINR_n}) \tag{2}$$

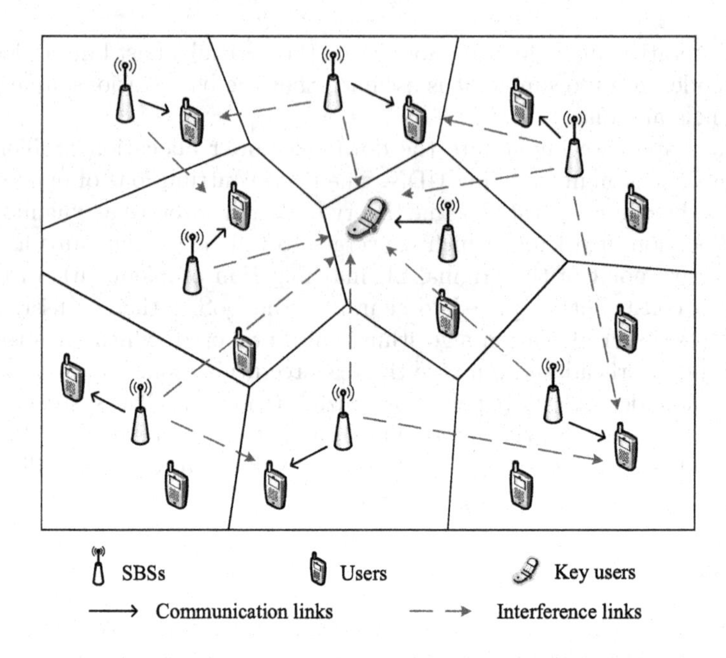

Fig. 1. System model. For clarity, only part of the interference links are plotted.

Without loss of generality, we assume that the key user is user x_0. The interference power constraint Q is used to limit the aggregate interference received by the key user, which can be written as

$$\sum_{i \neq x_0} p_i h_{i,x_0} \leq Q \tag{3}$$

Then, the optimization problem of maximizing system SE can be expressed as

$$\text{Problem 1}: \max_{\mathbf{p}} R_{\text{sum}}(\mathbf{p}) = \sum_{n=1}^{N} R_n(\mathbf{p}) \tag{4a}$$

$$s.t. \quad 0 \leq p_n \leq p_n^{\max} \tag{4b}$$

$$\sum_{i \neq x_0} p_i h_{i,x_0} \leq Q \tag{4c}$$

where $\mathbf{p} = (p_1, p_2, \cdots, p_N)$ is the SBS transmit power vector. Obviously, problem 1 is non-convex, so it is a challenging task to obtain its global optimal solution. The model of non-cooperative game will be adopted to decouple the problem into several convex subproblems in next section.

3 Problem Formulation

In this section, the non-cooperative game model based on dynamic pricing is formulated.

3.1 Game Model

In order to obtain the maximum transmission rate, each SBS will continuously increase its own transmit power. This process can be regarded as a non-cooperative game. However, as each SBS constantly pursues its own utility maximization, it will also causes severe interference to other users. In order to limit this interference, it is necessary to introduce a pricing mechanism in the non-cooperative game model.

We model each SBS as a self-interested game player, the non-cooperative game can be denoted as $G = [C, \{\rho_n | n \in \mathrm{C}\}, \{U_n | n \in C\}]$. $C = \{1, 2, \cdots, N\}$ is the set of game players. $\rho_n = \{p_n | 0 < p_n < p_n^{\max},\ p_n h_{n,x_0} + \sum\limits_{i \neq n} p_i h_{i,x_0} \leq Q\}$ is the set of available power strategy for player n. U_n is the utility function of player n, it is denoted by

$$U_n(p_n | \mathbf{p}_{-n}) = R_n(p_n | \mathbf{p}_{-n}) + \theta_n p_n \tag{5}$$

where $\mathbf{p}_{-n} = (p_1, \cdots, p_{n-1}, p_{n+1}, \cdots, p_N)$ is the transmit power of all the players except player n. θ_n can be regarded as a kind of dynamic price for the transmit power, it is given by

$$\theta_n = \sum_{i \neq n} \frac{\partial}{\partial p_n} R_i(\mathbf{p}) \tag{6}$$

Obviously, U_n is a convex function of p_n, so we can obtain a convex approximation of the system SE $R_{\text{sum}}(\mathbf{p})$ in (4a), it can be written as

$$\tilde{R}_{\text{sum}}(\mathbf{p}) \triangleq \sum_{n=1}^{N} U_n(p_n | \mathbf{p}_{-n}) \tag{7}$$

Under the above non-cooperative game model, problem 1 can be decomposed into N convex sub-problems as follows:

$$\text{Problem 2}: \max_{p_n} U_n(p_n | \mathbf{p}_{-n}),\ \ \forall n \in C \tag{8a}$$

$$s.t.\quad 0 \leq p_n \leq p_n^{\max} \tag{8b}$$

$$p_n h_{n,x_0} + \sum_{i \neq n} p_i h_{i,x_0} \leq Q \tag{8c}$$

So the system SE maximization problem can be equivalent to maximizing the SE of each small cell. We can use the convex optimization method to find the globally optimal solution of problem 2, which can be expressed as

$$b_n = \arg \max_{p_n \in \rho_n} U_n(p_n | \mathbf{p}_{-n}) \tag{9}$$

3.2 Analysis of Nash Equilibrium

Definition 1. *The transmit power vector* $\mathbf{p}^*$ *is the NE point of game G if and only if the transmit power of each player is the best strategy corresponding to the transmit power of others, i.e.,* $p_n^* = b_n, \forall n \in C$.

Once the NE is reached, no player will attempt to change its strategy, due to

$$U_n(p_n^*|\mathbf{p}_{-n}^*) \geq U_n(p_n|\mathbf{p}_{-n}^*), \ \forall n \in C \tag{10}$$

Each NE of game G (i.e., problem 2) is a stationary point of the non-convex problem 1 and vice versa [11]. As we know, the globally and locally optimal solutions must be stationary points. Therefore, by finding the NE of game G, we may obtain the locally optimal solutions of the original problem, even the globally optimal solutions if we are lucky enough.

4 Distributed Iterative Power Allocation Algorithms

In this section, the dynamic pricing power allocation (DPPA) algorithm is proposed, and the convergence analysis shows that it converges to the NE of game G. In order to reduce the signaling overhead, the approximate dynamic pricing power allocation (ADPPA) algorithm is also presented.

4.1 The Dynamic Pricing Power Allocation Algorithm

Since the individual utility function U_n and the constraint set ρ_n are convex, problem 2 is a typical convex optimization problem. The Lagrange function of problem 2 is given by

$$\begin{aligned} L_n(p_n, \mu_1, \mu_2, \mu_3) = & -R_n(p_n|\mathbf{p}_{-n}) - \theta_n p_n + \mu_1(p_n - p_n^{\max}) \\ & -\mu_2 + \mu_3(p_n h_{n,x_0} + \sum_{i \neq n} p_i h_{i,x_0} - Q) \end{aligned} \tag{11}$$

Then the Lagrangian dual function is denoted by

$$g_n(\mu_1, \mu_2, \mu_3) = \inf_{p_n} L_n(p_n, \mu_1, \mu_2, \mu_3) \tag{12}$$

It can be seen that the strong dual condition is satisfied, the duality gap is zero. By solving the Karush-Kuhn-Tucher (KKT) conditions, we obtain the optimal solution of the original problem, that is given by

$$b_n = \max\{0, \min\{p_n^Q, p_n^{\text{op}}, p_n^{\max}\}\} \tag{13}$$

where

$$p_n^Q = \frac{1}{h_{n,x_0}}(Q - \sum_{i \neq n} p_i h_{i,x_0}) \tag{14}$$

$$p_n^{\text{op}} = \frac{1}{\sum_{i \neq n} \frac{h_{i,i} p_i h_{n,i}}{\sigma_i(\sigma_i + h_{i,i} p_i)}} - \frac{\sigma_n}{h_{n,n}} \tag{15}$$

with $\sigma_j, \forall j \in C$ representing the interference-plus-noise experienced by user j, it is denoted by

$$\sigma_j = \sum_{i \neq j} p_i h_{i,j} + p^{\text{m}} h_j^{\text{ms}} + w_j, \ \forall j \in C \tag{16}$$

The DPPA algorithm is designed based on above analytical solutions, as shown in Algorithm 1. In each iteration, given the current power vector $\mathbf{p}^t$, each SBS computes its optimal transmit power by (13). Then update the transmit power vector and step-size. The NE of game G is obtained until the termination condition is satisfied, i.e., $\left|\tilde{R}_{\text{sum}}(\mathbf{p}^{t+1}) - \tilde{R}_{\text{sum}}(\mathbf{p}^t)\right| < \delta$, where δ is an arbitrary small positive constant.

Algorithm 1. Dynamic pricing power allocation algorithm

Input: The transmit power vector $\mathbf{p}$, the step-size r, constant δ and ε.
Output: The optimal transmit power vector $\mathbf{p}^*$.
1: **Initialization:** Set the initial value $\mathbf{p}^0$, $r^0 \in (0, 1]$, δ and $\varepsilon \in (0, 1)$.
2: **while** $\left|\tilde{R}_{\text{sum}}(\mathbf{p}^{t+1}) - \tilde{R}_{\text{sum}}(\mathbf{p}^t)\right| \geq \delta$. **do**
3: As termination condition is not satisfied, set $t \leftarrow t + 1$.
4: Given the transmit power vector $\mathbf{p}^t$, each SBS independently computes its optimal transmit power $b_n^t, n \in C$ by (13).
5: Update the transmit power vector by $\mathbf{p}^{t+1} = \mathbf{p}^t + r^t(\mathbf{b}^t - \mathbf{p}^t)$, where $\mathbf{b}^t = (b_0^t, b_1^t, \cdots, b_N^t)$.
6: Update the step-size by $r^{t+1} = r^t(1 - \varepsilon r^t)$.
7: Compute the value of $\tilde{R}_{\text{sum}}(\mathbf{p}^{t+1})$.
8: **end while**
9: **return** $\mathbf{p}^{t+1}$

4.2 Convergence Analysis

Theorem 1. *The transmit power vector* $\mathbf{p}^t$ *converges to the NE of game* G *and* $\tilde{R}_{\text{sum}}(\mathbf{p}^t)$ *converges to a finite value via Algorithm 1.*

Proof. The approximate system SE function $\tilde{R}_{\text{sum}}(\mathbf{p})$ is Lipschitz continuous on feasible domain, so there exists a positive constant η such that

$$\left\|\tilde{R}_{\text{sum}}(\mathbf{p}^{t+1}) - \tilde{R}_{\text{sum}}(\mathbf{p}^t)\right\| \leq \eta \left\|r^t(\mathbf{b}^t - \mathbf{p}^t)\right\| \tag{17}$$

Based on Descent Lemma [12], we obtain the following inequality

$$\tilde{R}_{\text{sum}}(\mathbf{p}^{t+1}) \leq \tilde{R}_{\text{sum}}(\mathbf{p}^t) + r^t \tilde{R}_{\text{sum}}(\mathbf{p}^t)^{\text{T}}(\mathbf{b}^t - \mathbf{p}^t) + \frac{\eta (r^t)^2}{2}\left\|\mathbf{b}^t - \mathbf{p}^t\right\|^2 \tag{18}$$

$\mathbf{b}^t - \mathbf{p}^t$ is a descent direction of function $\tilde{R}_{\text{sum}}(\mathbf{p})$ at $\mathbf{p}^t$ such that

$$\tilde{R}_{\text{sum}}(\mathbf{p}^t)^{\text{T}}(\mathbf{b}^t - \mathbf{p}^t) \leq -\tau\left\|\mathbf{b}^t - \mathbf{p}^t\right\|^2 \tag{19}$$

where τ is a positive constant. Resorting to (18) and (19), we obtain

$$\tilde{R}_{\text{sum}}(\mathbf{p}^{t+1}) \leq \tilde{R}_{\text{sum}}(\mathbf{p}^t) - r^t(\tau - \frac{\eta r^t}{2})\left\|\mathbf{b}^t - \mathbf{p}^t\right\|^2 \tag{20}$$

Since $r^t \to 0$, we can find a positive constant ψ when t is sufficiently large, i.e., $t \geq \bar{t}$, such that

$$\tilde{R}_{\text{sum}}(\mathbf{p}^{t+1}) \leq \tilde{R}_{\text{sum}}(\mathbf{p}^t) - r^t \psi \left\| \mathbf{b}^t - \mathbf{p}^t \right\|^2 \quad (21)$$

Since $\tilde{R}_{\text{sum}}(\mathbf{p})$ is coercive, using the Robbins-Siegmund Theorem [13], we derive that $\tilde{R}_{\text{sum}}(\mathbf{p}^t)$ converges to a finite value and the following inequation is hold.

$$\sum_{t \geq \bar{t}} r^t \left\| \mathbf{b}^t - \mathbf{p}^t \right\|^2 < \infty \quad (22)$$

As $\sum_{t \geq \bar{t}} r^t \to \infty$, then $\lim_{t \to \infty} \|\mathbf{b}^t - \mathbf{p}^t\| = 0$, i.e., $\lim_{t \to \infty} \mathbf{b}^t = \mathbf{p}^t$. The proof is completed.

4.3 An Approximate Form

Using the DPPA algorithm, each SBS needs the transmit power of others and the channel power gains of signal and interference links to compute its optimal power strategy. The number of signaling required for each SBS is $\frac{1}{2}N^2 + \frac{3}{2}N - 1$ in each iteration, so the signaling overhead is enormous when the number of small cells increases.

In order to reduce the signaling overhead, the ADPPA algorithm is proposed. The iterative process of ADPPA is the same as the DPPA algorithm, but the calculation method for solving b_n^t in step 4 is slightly different. $h_{ij}(i \neq j)$ in (14)–(16) is replaced with $\bar{h}_{ij}(i \neq j)$, which is approximate channel power gain model, where the influence of Rayleigh fading is ignored and the distance between SBS i and user j is replaced with the distance between SBSs.

As the distance between SBSs is constant, it is unnecessary to repeatedly transmit $\bar{h}_{ij}(i \neq j)$ during the iterative process. Using the proposed ADDPA algorithm, each SBS only requires $2N - 1$ signaling including: (i) The transmit power of other SBSs; (ii) The channel power gain between each SBS and its own user.

5 Numerical Results

In this section, the performance of proposed algorithms are compared with state-of-art algorithms. For simplicity, it is assumed that $w_n = w, \forall n \in C$, $p_n^{\max} = p^{\max}$, $\forall n \in C$, and the key user is located at the origin of the Euclidean plane. The main simulation parameters are shown in Table 1.

The performance of proposed algorithms are compared with the penalty factor based power allocation (PFPA) algorithm in [8] and the non-uniform pricing based power allocation (NPPA) algorithm in [9]. We set the average SE of each small cell as the indicator.

The results in Fig. 2 are obtained by the Monte Carlo method, and the number of simulations is 10,000. We can see that the average SE of each small cell achieved by the proposed DPPA algorithm is the highest, and the performance

Table 1. Parameters setting

Parameter	Value
Pass loss exponent α	3.7
Thermal noise density	−174 dBm/Hz
Bandwidth	10 MHz
Maximum transmit power of each SBS $p^{\max}$	150 mW
Region $\mathcal{S}$	1 Km^2
Interference power constraint Q	−35 dBm

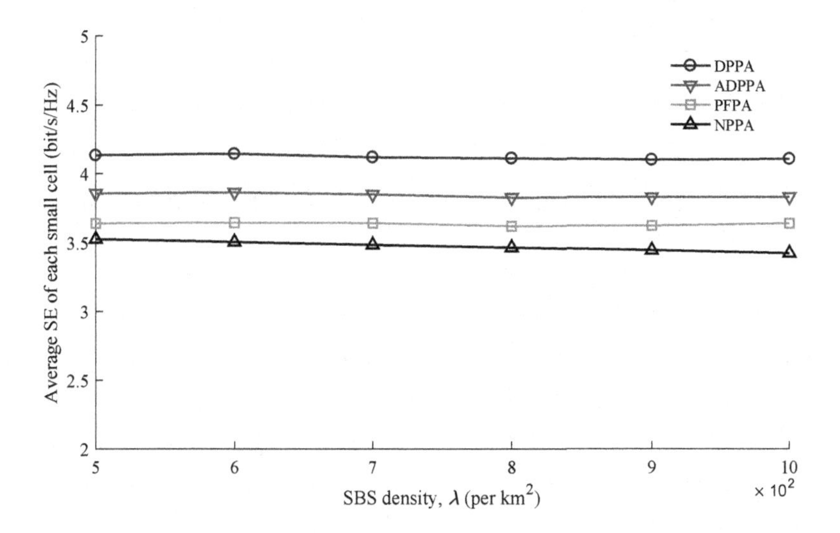

Fig. 2. Average SE of each small cell versus the SBS density. ($M = 4$.)

of the proposed ADDPA algorithm is better than the benchmark. Moreover, the performance of algorithms hardly change with the density of SBSs, which is counter-intuitive. The dense deployment of SBSs makes the average coverage area of each SBS smaller, and the distance between the SBS and the user is closer. Therefore, the negative effect caused by the severe interference is cancelled by the stronger signal.

Figure 3 shows the convergence performance of different algorithms. Although the convergence speed of the DPPA algorithm is slightly slower than the PFPA algorithm due to dynamic pricing, the average SE of each small cell obtained at the convergence time is higher. The ADPPA algorithm has similar convergence performance with the DPPA algorithm. Since the solving process of the NPPA algorithm does not require iteration, it is a horizontal line in the figure.

Figure 4 shows the effects of the number of antennas M on the performance of power allocation algorithms. As it shows, the average SE of each small cell achieved by algorithms is improved with the increase of M. The reason is obvious,

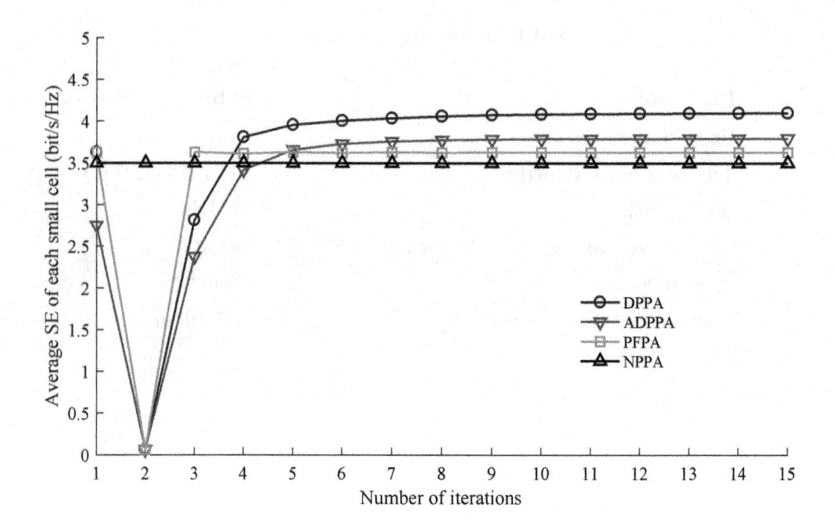

Fig. 3. Convergence performance of the proposed algorithms with benchmark algorithms. ($\lambda = 5 \times 10^2$ per km^2, $M = 4$.)

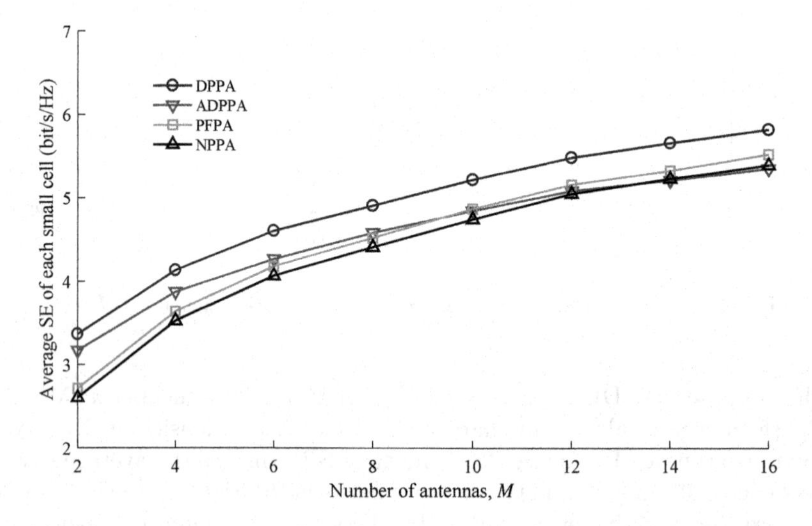

Fig. 4. Average SE of each small cell versus the number of antennas. ($\lambda = 5 \times 10^2$ per km^2.)

because a larger antenna number increases the diversity gain. In addition, it is shown that the proposed DPPA algorithm has the best performance, and the ADPPA algorithm outperforms the benchmark algorithms when M is less than 10, which means the proposed approximate algorithm is more suitable for the case with a small number of SBS antennas.

Figure 5 shows the average signaling overhead required for each SBS per iteration of the proposed algorithms. It is assumed that transmitting a floating

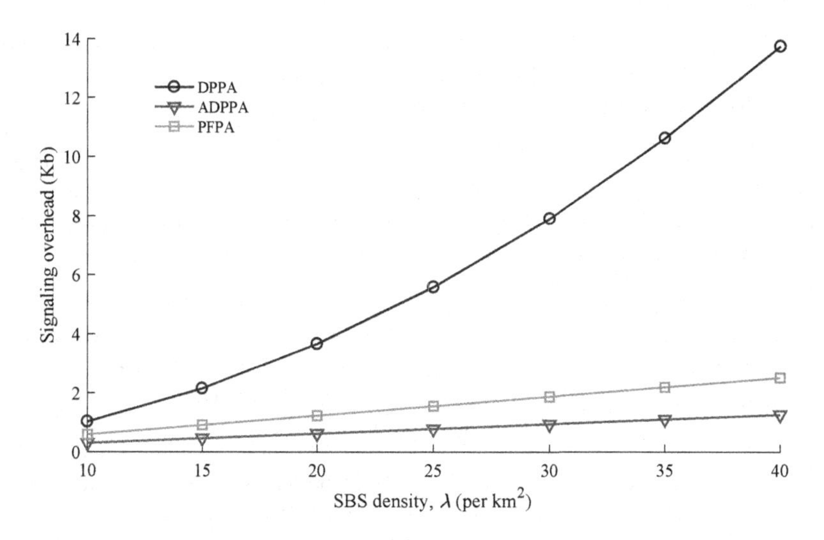

Fig. 5. Signaling overhead analysis.

number needs 16 bits. It can be seen that with the increase of SBS density, the signaling overhead of the proposed DPPA algorithm increases significantly, while it increases slowly for the ADPPA algorithm.

6 Conclusion

In this paper, distributed iterative downlink power allocation strategies were investigated for a spectrum sharing UDN with multi antennas resorting the game theory. The non-cooperative game model was adopted to decouple the non-convex system SE maximization problem into several convex subproblems. The dynamic pricing designed guarantees that each NE of the game is a stationary point of the original optimization problem. Moreover, the interference power constraint was adopted for the key user. Under the game theory framework, we proposed the DPPA algorithm required global information and the ADPPA algorithm which needs limited information. Simulation results show that the DPPA algorithm outperforms the benchmark, and the ADPPA algorithm with a little performance loss can effectively reduce the signaling overhead compared with the DPPA algorithm.

References

1. Ge, X., Tu, S., Mao, G., Wang, C., Han, T.: 5G ultra-dense cellular networks. IEEE Wirel. Commun. **23**(1), 72–79 (2016)
2. Liu, J., Sheng, M., Liu, L., Li, J.: Interference management in ultra-dense networks: challenges and approaches. IEEE Network **31**(6), 70–77 (2017)

3. Xiao, J., Yang, C., Anpalagan, A., Ni, Q., Guizani, M.: Joint interference management in ultra-dense small-cell networks: a multi-domain coordination perspective. IEEE Trans. Commun. **66**(11), 5470–5481 (2018)
4. Du, J., Gelenbe, E., Jiang, C., Zhang, H., Ren, Y.: Contract design for traffic offloading and resource allocation in heterogeneous ultra-dense networks. IEEE J. Sel. Areas Commun. **35**(11), 2457–2467 (2017)
5. Zheng, J., Wu, Y., Zhang, N., Zhou, H., Cai, Y., Shen, X.: Optimal power control in ultra-dense small cell networks: a game-theoretic approach. IEEE Trans. Wirel. Commun. **16**(7), 4139–4150 (2017)
6. Kim, H., Villardi, G.P., Ma, J.: Energy efficient radio resource allocation scheme using receiver puncturing technique for 5G networks. In: 2017 IEEE 86th Vehicular Technology Conference (VTC-Fall), pp. 1–7. IEEE, Toronto (2017)
7. Liang, L., Wang, W., Jia, Y., Fu, S.: A cluster-based energy-efficient resource management scheme for ultra-dense networks. IEEE Access **4**, 6823–6832 (2016)
8. Wang, X., Liu, B., Su, X.: A power allocation scheme using non-cooperative game theory in ultra-dense networks. In: 2018 27th Wireless and Optical Communication Conference (WOCC), pp. 1–5. IEEE, Hualien (2018)
9. Kang, X., Zhang, R., Motani, M.: Price-based resource allocation for spectrum-sharing femtocell networks: a Stackelberg game approach. IEEE J. Sel. Areas Commun. **30**(3), 538–549 (2012)
10. Yang, C., Dai, H., Li, J., Zhang, Y.: Distributed interference-aware power control in ultra-dense small cell networks: a robust mean field game. IEEE Access **6**, 12608–12619 (2018)
11. Scutari, G., Facchinei, F., Song, P., Palomar, D.P., Pang, J.: Decomposition by partial linearization: parallel optimization of multi-agent systems. IEEE Trans. Sign. Process. **62**(3), 641–656 (2014)
12. Bertsekas, D.: Nonlinear Programming, 2nd edn. Athena Scientific, Belmont (1999)
13. Polyak, B.T.: Introduction to Optimization. Optimization Software, NewYork (1987)

Reliable Transport Mechanism Based on Multi-queue Scheduling

Chenyang Ding, Hongyan Li(✉), Peng Wang, Qin Liu, Hongyuan Zhang, and Fan Qi

State Key Laboratory of Integrated Service Networks, Xidian University, Xi'an 710071, China
hyli@xidian.edu.cn

Abstract. In the Integrated Space and Onboard Network, transmission demands arrive randomly. In addition, the routing algorithm calculate the end-to-end paths the average occupied bandwidth over a period of time, causing microburst of data at the forwarding nodes. The existing queue scheduling methods usually queue the burst data and discard the amount of data which exceed the length of queues, which will cause more severe congestion with the re-transmission of discarded data. In order to solve this problem, a multi-queue scheduling method is proposed to guarantee reliable transmission without loss. Specifically, a token-based mechanism is designed to process the irregular data which has not been sent on time or has been over time in the queue. This mechanism will arrange new transmission opportunities for the data which miss them, avoiding data loss and retransmission. The simulation results show that the proposed reliable transmission mechanism can significantly decrease the network packet loss rate by about 80%, compared to existing queue scheduling methods.

Keywords: Queue scheduling algorithm · CBS (Commit Burst Size) · Multi-Queue scheduling · Space-Ground integrated network

1 Introduction

With the rapid development Integrated Space and Onboard Network (ISON), the data volume needed to be transmitted becomes more and more massive, causing huge transmission pressure for the current ISON. Specially, network congestion is inevitable due to the randomness of data arrival and inaccurate computation of route. How to deal with it will significantly affect the end-to-end delay packet loss through put, and network resource utilization.

In fact, when congestion at forwarding nodes occurs, the queue scheduling mechanism will significantly effect the performance of the network, since it can solve the shortcomings of insufficient link bandwidth and ensure that each service stream can receive works. According to their designing principles, these scheduling algorithms can be roughly divided into First-Come-First-Served, Weighted Round-Robin, and Weighted Fair scheduling [3].

Q. Guo et al. (Eds.): WiSATS 2021, LNICST 410, pp. 731–741, 2022.
https://doi.org/10.1007/978-3-030-93398-2_63

- The First-Come-First-Served (FCFS) algorithm is characterized by low cost, simple structure and easy implementation. However, FCFS cannot provide isolation technology. When a burst packet occupies a large portion of the buffer space, other packets suffer the risk of being discarded.
- The Weighted Fair Scheduling algorithm (WFQ) could serve multiple queues fairly at the same time. However, when some real-time services appear in the network, the weights cannot be changed in the WFQ algorithm, causing that the order of forwarded packets cannot be changed, which affects the quality of the real-time services.
- The weighted Round-Robin scheduling algorithm (WRR) assigns a weight to each queue. There is a counter in each queue to calculate the weight. The weight of the counter is checked before the round-robin scheduling. And if the counter weight is not 0, the corresponding packet in this queue can be forwarded automatically and then the weight of the counter will be set as zero. However, the WRR algorithm is very sensitive to packet length changes. The fairness cannot be guaranteed if the packet length changes.

As a result, the above queue scheduling algorithms do not deal with the packets that cannot be forwarded in the current time slot and have their own disadvantage. In fact, the delayed data will conflict with the current data needed to be forwarded, which affects forwarding the current data. To overcome the problem [4], a Credit-Based Flow-Control mechanism is proposed to adjust the transmission rate of delayed data in the next time slot. Next, to decrease the effects of delayed data on the data of current time slot, a Credit-Based Multi-Queue scheduling is proposed, which delays the forwarding data packets that cannot be completed in the current time slot, and adjusts the forwarding rate, it will reduce the impact on normal data. This mechanism can reduce network congestion caused by random services and reduce the PLP of delayed data packets. Finally, the performance of the scheme was verified by simulation.

2 System Model

The purpose of the time deterministic network is to realize the boundedness of end-to-end delay jitter, the determinism of forwarding delay, and the low packet loss rate and high reliability of network transmission. The forwarding scheduling strategy plans the forwarding method at each node, which is the key to realizing time deterministic network [5]. This chapter introduces the Credit-Based Flow-Control mechanism and it's system model.

2.1 A Credit-Based Flow-Control Mechanism

The Credit-Based Flow-Control mechanism was proposed by the AVB working group firstly [5] in order to solve the problem of excessive business bursts occupying low-priority business bandwidth. Through priority classification, the transmission of high-priority services is guaranteed first, and the next best service is assigned a credit value to describe its accumulated transmission opportunity. The credit value will continue increases when it is not transmitted, and the credit value decreases when it is transmitted. Then, change the slope of the credit value to achieve the purpose of controlling the service transmission rate.

2.1.1 Types of Business in CBS

TSN uses the 3-bit Priority Code Point (PCP) in the virtual local area network (VLAN) tag defined in 802.1Q to distinguish different levels of data streams, which are divided into control data traffic (Control Data traffic, CDT), Class A, Class B and BE business.

2.1.2 CBS Queue Model

CDT is the highest priority type of service in the TSN network. Class A and Class B are generally audio and video services, those services has higher priority. Without CDT service, the Credit-Based Flow-Control mechanism will be the first choice. BE is a best-effort service, it does not has credit, and transmits in the gap of high-priority transmission. The queue model of CBS is shown in the figure below (Fig. 1):

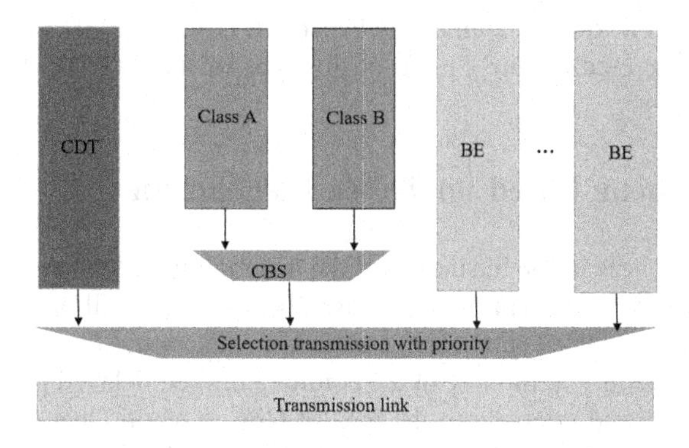

Fig.1. CBS queue scheduling

In order to control the transmission rate of Class A and Class B, the CBS algorithm gives the corresponding credit value for these two types of services to decide which service should be selected for transmission. For each level, there are two parameters used to control the increase and decrease speed of the credit value: Send slope and Idle slope. The Idle slope determines the speed of accumulating transmission opportunities when the transmission is idle. The Send slope determines the opportunities for other services to accumulate bandwidth during transmission, its size is related to the transmission rate of Class A and Class B. The calculation method is:

$$\text{idleslope} = \frac{\text{RB}}{\text{CMI}} = \text{R} \tag{1}$$

R represents the bandwidth resource allocated for this type of service, and RB is the maximum frame length of the data frame. Define the Send slope and Idle slope as S_i and I_i.CBS should meet the following conditions:

1. $S_i < 0$, $I_i > 0$;
2. $\sum_{i=0}^{N} I_i < C$,Where N represents the number of data streams, and C represents the total bandwidth of the link;
3. $S_i = I_i - C$;

Under the control of S_i and I_i, the specific sending criteria of the CBS algorithm are as follows:

1. When a certain type of data is not sent and meanwhile the queue is not empty, the credit value $Credit_i$ of this type of data rises at a slope of I_i. When the service is being transmitted, the credit value $Credit_i$ of this type of data decreases at the slope of S_i;
2. Each transmission preferentially selects the business data frame with higher priority for transmission (Class A priority is greater than Class B);
3. Data frames with $Credit_i$ less than 0 will lose transmission opportunities and cannot be transmitted;
4. When the current queue is empty, when there is no data of a certain type to be transmitted, the credit value $Credit_i$ of this type of data will be 0.

3 A Multi-queue Based on Token Mechanism

Due to the randomness of the business and the uncertainty of the network environment, which will lead to the problem of packet loss [6]. This part will propose a forwarding strategy that delayed packets that cannot be send in the current time slot will be send in the next two time slots [7]. In view of the conflict between delayed packets and normal forwarded packets, and referring to the idea of CBS, a Multi-Queue Based On Token mechanism is proposed to adjust the transmission rate of delayed services.

3.1 Analyze the Problems in the Time-Deterministic Network Forwarding Mechanism

Due to the business arrival randomly. The traditional Internet uses a Best-Effort transmission method [8], and the node queue length is limited. When the burst service exceeds the queue length, the packets will be lost, which will affect the reliability of the network. The IEEE TSN working group proposed the queue management mechanism of CQF and CSQF, which uses a single queue to send and other queues to receive,this method will achieve time determinism [8]. In this method, the service that can be processed by a single time slot is still limited by the length of the receiving queue. When the burst service exceeds the length of the receiving queue, it will loss packet yet, which does not meet the requirements of low packet loss rate. As shown in Fig. 2.

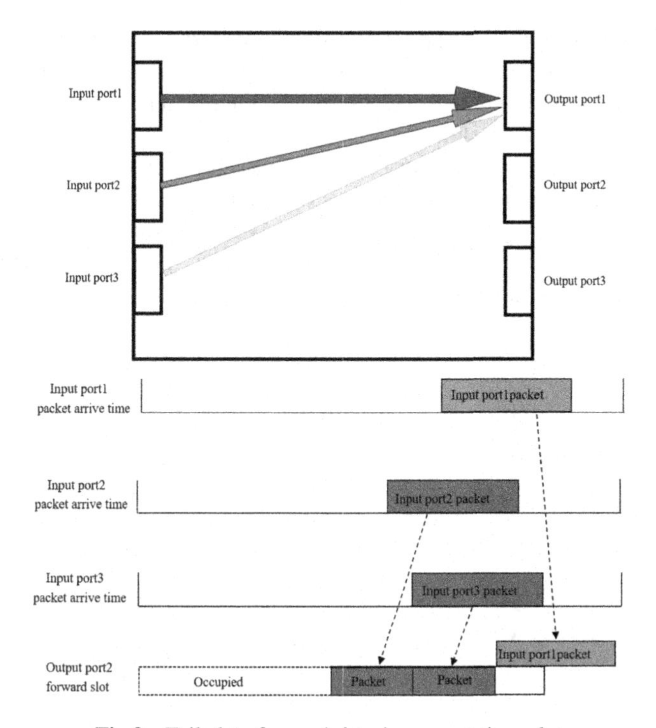

Fig.2. Failed to forward data in current time slot

3.2 A Credit-Based Multi-queue Scheduling

In order to solve the problems summarized above, a method is to delay the data packets that cannot be send in the current time slot to the next time slot, and then perform Multi-Queue joint scheduling, store the data exceeding the time slot capacity in the sending queue and participate in joint scheduling in the next time slot [9]. When the delayed transmission data packet conflicts with the normal transmission data packet, a Credit-Based Multi-Queue scheduling mechanism is proposed to adjust the transmission rate of delayed service.

3.2.1 Formulate Packet Forwarding Queue Rules

Firstly, for the data that cannot be sent in the time slot, use a reasonable queue mechanism to store the data [10]. The forwarding rule adopts the method of time slot division, and uses a multi-queue cycle method to ensure time determinism. The queue rules are as follows:

(1) Take the cyclic forwarding structure of three queues as an example. The three queues are named SQ, TQ1 and TQ2. At the same time, only the queue SQ can be opened to forward data packets, and all three queues can receive data packets. The data that exceeds the time slot resources and cannot be sent in the current time slot will be retained in the SQ, waiting for the next time slot to be scheduled.

(2) The three queues are opened periodically in turn, and the length of the open time is T, representing the current time slot and the next two time slots. The size of each queue is the maximum data that can be transmitted by the link corresponding to the outgoing port in a time slot.
(3) The data packet carries forwarding time information, according to the information carried in the header of the packet, it is determined in which time slot to forward, and enters the corresponding queue [11]. Assuming that the resources of each queue will be reserved in advance by some services, we can know the percentage of queue resources has been reserved.

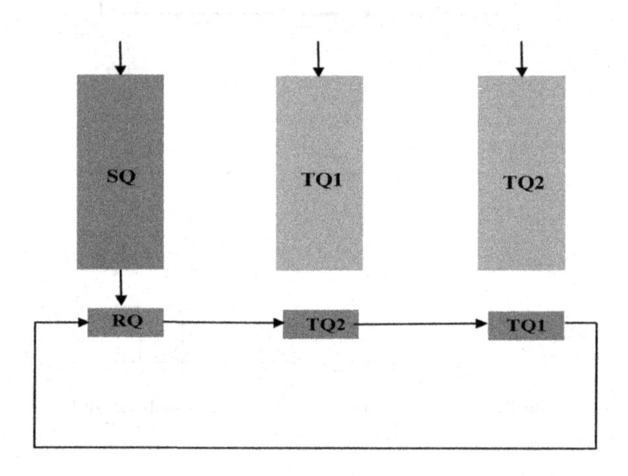

Fig. 3. Design forwarding queue

The queue diagram is shown in Fig. 3. The queue opened in the current time slot is called SQ, TQ1 is the queue opened in the next time slot, TQ2 is the queue opened with a delay of two time slots, and TQ1 and TQ2 are used to enqueue data packets that arrive early. With reference to CSQF, the three-queue structure can balance some of the effects caused by processing delay, and the number of queues can be expanded according to the size of the delay jitter. Packets arriving within 3T can enter the queue, and packets arriving early will wait in the queue until the queue is opened before completing the forwarding. SQ can also be received during transmission, so that the queue can accommodate data that exceeds its capacity, and services that exceed the current time slot resources are stored in the queue, facilitating joint scheduling in the next time slot. For the business that SQ saved when it was closed and failed to complete the forwarding, it is necessary to formulate a strategy for processing [12].

3.2.2 A Scheme of Credit-Based Multi-queue Scheduling

When transmitting delayed data, it is necessary to formulate an appropriate mechanism to adjust its transmission rate and reduce the impact on normal data. Refer to CBS for AVB service, which provides a multi-queue scheduling mechanism that can regulate the output traffic of different queues. In order to characterize the transmission opportunities accumulated by the business, credits are given to different types of business queues, adjust the transmission rate of the service by adjusting idle slop. The larger the idle slop is, the faster the service accumulates credit value will be, and more transmission opportunities can be accumulated at the same time. The smaller the idle slop is, the slower the accumulating credit value will be, and the faster the credit value will decrease to a negative value, leaving more transmission opportunities for other businesses. In accordance with this idea, this section formulates a credit-based multi-queue scheduling scheme to control the transmission rate of delayed data services.

When dealing with the problem of delayed data transmission, the data that needs to be transmitted in the current time slot can be divided into three types: normal data; delayed by one time slot; delayed by two time slots. In order to facilitate problem analysis, the total set of the three types of data is recorded as $P_{present}$, P_{1delay} and P_{2delay}.

In order to adjust the transmission rate, assign credit values to different types of data. The credit value is not set for $P_{present}$ data services, and the priority is the highest. Once the current queue has business accumulation, it will transmitted the packets directly,and the delay type data is transmitted in the gap of $P_{present}$ data service transmission. Define different credit values for type P_{1delay} and P_{2delay} data, control their sending slope by adjusting their idle slope, and make them transmit according to the bandwidth. The transmission selection of these three types of data in a time slot can be transformed into the queue model of Fig. 4.

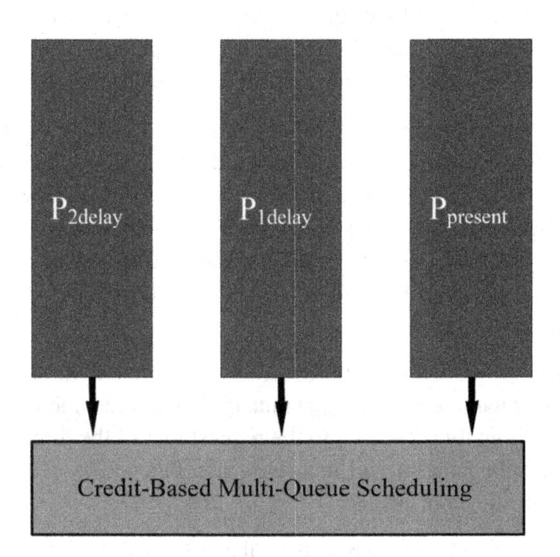

Fig. 4. Delayed data and normal data in Credit-Based Multi-Queue Scheduling

For P_{1delay} and P_{2delay} data queues, the credit values are defined as $Credit_1$ and $Credit_2$, representing their accumulated transmission opportunities. The corresponding transmission slopes of each queue are S_1 and S_2, and the idle slopes are I_1 and I_2.

The three types of data are planned to be forwarded in the current time slot. The respective transmission resources are related to the total size of the data that needs to be transmitted. The request rate is used to characterize the bandwidth resources they need to complete the forwarding in the current time slot. The total size of the data packets of $P_{present}$, P_{1delay} and P_{2delay} are respectively defined as B_0, B_1 and B_2, Then their request rates are R_0, R_1 and R_2, which are:

$$R_0 = \frac{B_0}{T}, R_1 = \frac{B_1}{T}, R_2 = \frac{B_2}{T} \tag{2}$$

Then the total required bandwidth C_{sum} for all three types of services in the current time slot to complete the transmission is:

$$C_{sum} = R_0 + R_1 + R_2 \tag{3}$$

If the current bandwidth is C, all three types of services need to follow $C_{sum} > C$ to complete the transmission in the current time slot; if $C_{sum} \leq C$, the bandwidth resource is less than the total bandwidth required by the service, and the data in the three types of queues will inevitably be transmitted in the current time slot.

CBS stipulates that I_i and S_i satisfy $S_i = I_i - C$, and I_i is related to the reserved bandwidth of data i, so S_i can be regarded as a bandwidth resource reserved for other data. In order to limit the impact of delayed data on normal data, in the case of $C_{sum} > C$, P_{1delay} and P_{2delay} only allocate the remaining bandwidth excluding $P_{present}$ data reservation. At the same time, the transmission of $P_{present}$ data should not accumulate the credit value of delayed data. It is stipulated that I_i and S_i satisfy $S_i = I_i - (C - R_0)$. The specific queue scheduling rules are as follows:

(1) When $P_{present}$ data is being transmitted, $Credit_1$ and $Credit_2$ remain unchanged;
(2) When P_{1delay} data is transmitted, $Credit_1$ decreases with the slope of S_1, and $Credit_2$ increases with the slope of I_2; when the P_{2delay} service is transmitted, $Credit_2$ decreases with the slope of S_2, and $Credit_1$ increases with the slope of I_1;
(3) When $Credit_1$ or $Credit_2$ is less than zero, the corresponding queue is closed;
(4) When the credit values of the two types of delayed services are both 0, the P_{1delay} queue is sent first;

In the case of $C_{sum} \leq C$, bandwidth resources are sufficient. In the best case, the data in all queues can be guaranteed to be transmitted. At this time, the bandwidth resources are allocated proportionally according to the request rate of the three types of data. The credit value of the delayed data will increase when blocked by $P_{present}$ data. define as $S_i = I_i - C$. The specific transmission rules are as follows:

(1) When $P_{present}$ data is being transmitted, the credit values of $Credit_1$ and $Credit_2$ increase at the slope of I_1 and I_2;
(2) When P_{1delay} data is transmitted, $Credit_1$ decreases with the slope of S_1, and $Credit_2$ increases with the slope of I_2; when the P_{2delay} service is transmitted, $Credit_2$ decreases with the slope of S_2, and $Credit_1$ increases with the slope of I_1;
(3) When $Credit_1$ or $Credit_2$ is less than zero, the corresponding queue is closed;
(4) When the credit values of the two types of delayed services are both 0, the P_{1delay} queue is sent first;

4 Simulation

This chapter uses different time slot lengths to simulate the performance of the Credit-Based Multi-Queue scheduling and the dropped delay data scheme, and analyze the simulation results.

The design simulation scenario is shown in Fig. 5. Sent several time-determined service data packets randomly from the source node A to Router1. The data is designated to be forwarded in a certain time slot. The total amount of data sent is the ratio of the reserved bandwidth multiplied by the length of the time slot. The data packet arrives at Router1 randomly, and forwarded to Router2 through Router1. The size of the data packet is a random value in the range of 50Byte–1450Byte, the bandwidth C is 10 Gbits, and a certain amount of burst services exceeding the time slot capacity is set.

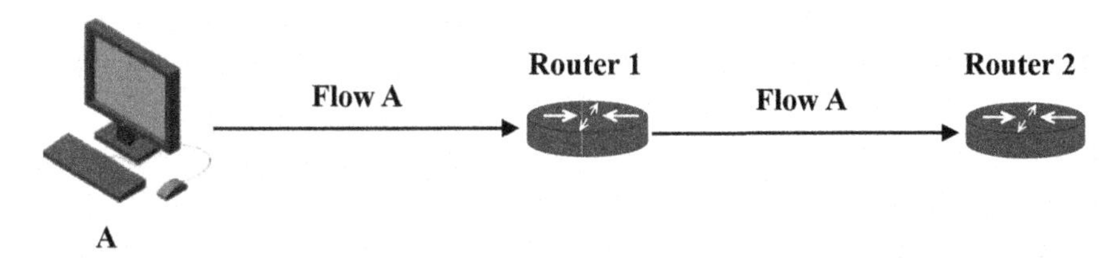

Fig. 5. Simulation scenario

The processing method for failing to send is to delay until the next time slot. The multi-queue scheduling scheme will affect the forwarding delay of service packets. In order to describe the impact of this mechanism, the definition of credit-based packet forwarding mechanism adds Average delay, the size is the difference between the average forwarding delay of all data packets under the two schemes in each experiment, and then average the difference. Change the ratio of the forwarding time slot length to the service bandwidth reservation, and then generate a certain amount of burst services that exceed the time slot resources. Perform 100 sets of experiments each time to calculate the average packet loss rate and the average increase time of the comparison experiment strategy and the multi-queue strategy Extension.

Comparative test of changing the length of the time slot:

It can be seen that the multi-queue joint scheduling mechanism can significantly reduce the packet loss rate of data packets from Fig. 6. The smaller the time slot division, the greater the packet loss rate, and the more obvious the ability of the optimized solution to reduce the packet loss rate.

It can be seen that the greater the time slot division, the greater the average delay increase caused by the optimization plan from Fig. 7. The optimization plan needs to delay forwarding the data that cannot be forwarded in the time slot, because the priority of the delayed data is lower, it is generally sent at the end of the time slot, so the larger the time slot division, the greater the increase in the average delay. When division a

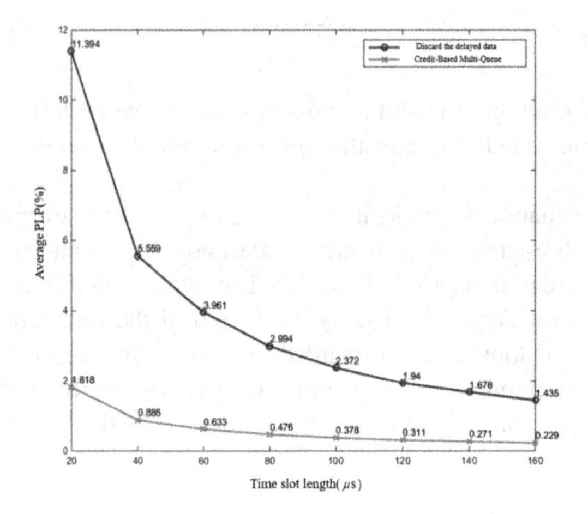

Fig. 6. Compare the packet loss rate of the two schemes under different time slot lengths

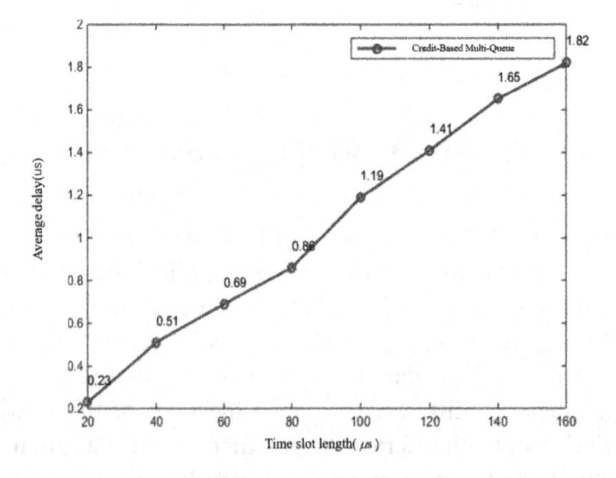

Fig. 7. Average increase delay of Multi-Queue scheduling mechanism under different time slot lengths

large time slot, the probability of packet loss will also decrease. At this time, you can consider adjusting the optimization target, it will increasing the output rate of the delayed data queue and reduce the impact of the average delay caused by the delayed data packet.

Acknowledgments. This work is supported by the National Natural Science Foundation of China 61931017).

References

1. Liyi: Study on dynamic scheduling strategy of intelligent RGV based on DP-FCFS algorithm. J. Chengdu Technol. Univ. 48–53 (2020)
2. Zhang, K.: Research and application of communication queue scheduling algorithm. Wuhan Huazhong Univ. Sci. Technol. 1–68 (2019)
3. Qiang, L., Liu, B., Delei, Y.U., et al.: Large-scale deterministic network forwarding technology. Telecommun. Sci. (2019)
4. Jiang, X., et al.: Low-latency networking: Where latency Lurksand how to tame it. Proc. IEEE **107**, 280–306 (2019)
5. Tucker, C., Zhang, J.: Long tail or steep tail? a field investigation into how online popularity information affects the distribution of customer choices. Working Pap. **6**(1), 6 (2007)
6. Al-Maqri, M.A., Alrshah, M.A., Othman, M.: Review on QoSprovisioning approaches for supporting video traffic in IEEE802.11e: Challenges and issues. IEEE Access, **6**, 55202–55219 (2018)
7. Garner, G.M., Ryu, H.: Synchronization of audio/video bridging networks using IEEE 802.1 AS. IEEE Commun. Mag. **49**(2), 140–147 (2011)
8. Craciunas, S.S., Oliver, R.S., Chmelík, M., et al.: Scheduling real-time communication in IEEE 802.1Qbv time sensitive networks. In: International Conference on Real-time Networks and Systems. ACM (2016)
9. Specht, J.: IEEE draft standard for local and metropolitan area networks–Media Access Control (MAC) Bridges and Virtual Bridged Local AreaNetworks Amendment: Asynchronous Traffic Shaping, Standard IEEE P802.1Qcr/D0.5, January 2018
10. IEEE standard for local and metropolitan area networks–bridges and bridged networks corrigendum 1: technical and editorial corrections. IEEE Std. 8021q-/cor. IEEE (2016)
11. Pop, P., Raagaard, M.L., Gutierrez, M., et al.: Enabling fog computing for industrial automation through time-sensitive networking (TSN). IEEE Commun. Stan. Mag. **2**(2), 55–61 (2018)
12. Geng, X., Chen, M., Li, Z., et al.: DetNet configuration YANG model. Internet Eng. Task Force, Fremont, CA, USA, Internet-Draft draft-geng-detnet-conf-yang-00 (2017)

Ergodic Lower Bound and Optimal Power Allocation for Secrecy-Capacity-Optimization-Artificial-Noise in MIMO Wireless System

Yebo Gu(✉), Zhilu Wu, and Zhendong Yin

School of Electronics and Information Engineering,
Harbin Institute of Technology, Harbin, China
16B305002@hit.edu.cn

Abstract. The transmission security problem is considered in this paper. A new artificial noise is added into transmitted signal to improve the secrecy performance of the system. A expression for the ergodic lower bound of SCO-AN system is derived. Based on the expression, optimal power distribution for SCO-AN system is studied. Moreover, optimal ratio of power distribution can be calculated. Moreover, the influence of channel estimation error is considered. As a result, it is necessary to generate more SCO-AN if the channel information error is considered.

Keywords: Artificial noise · Transmission security · Secrecy capacity · SCO-AN · Power distribution

1 Introduction

The transmission security of information is fundamental problems in wireless communication [1]. Wireless communication is useless unless a secure transmission of information is guaranteed. The foundation of the physical layer security (PLS) was proposed in [3]. Wyner introduced wiretap channel model which contains an information transmitter, a receiver, and an eavesdropper. Later in 2003, Csiszár and Körner considered common conditions of the model and studied the transmission of broadcasting information [4].

[5] introduces the Artificial noise (AN). AN is orthogonal to receiver's channel, so AN will not reduce ability of the receiver's channel to obtain information. Meanwhile, AN reduces the capable performance of eavesdropper's channel.

Based on AN, SCO-AN is introduced in [6]. [6] studies the issues in the secure communication of fading channels. A expression for secrecy capacity in fading channels which adds SCO-AN is derived.

Q. Guo et al. (Eds.): WiSATS 2021, LNICST 410, pp. 742–748, 2022.
https://doi.org/10.1007/978-3-030-93398-2_64

The main contribution is summarized as follows: This paper creatively uses statistical knowledge to find the objective function for SCO-AN under ideal conditions, and finds the optimal power distribution coefficient under ideal conditions. We reach crucial conclusion: the secrecy capacity of the legitimate channel depends entirely on the partition of power and the number of receiving antennas. On the contrary, the allocation objective function is not related to power at all.

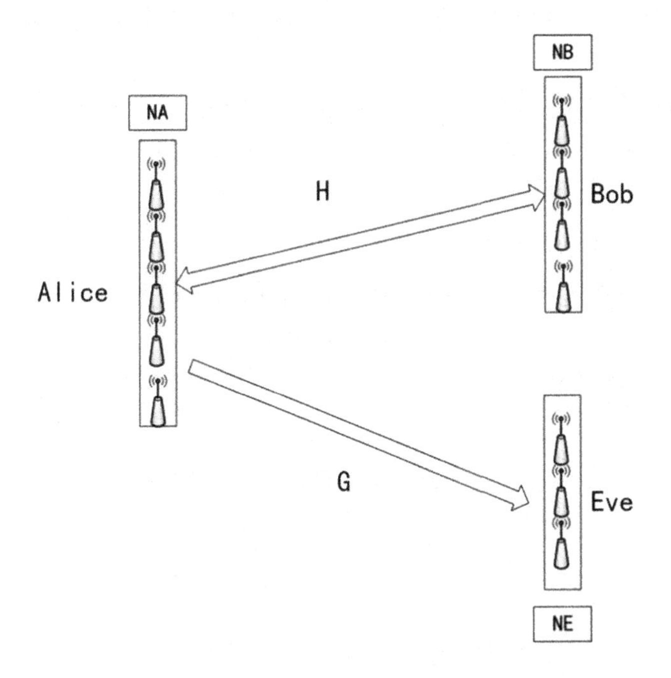

Fig. 1. Wiretap communication system model

According to the previously obtained expression for ergodic secrecy capacity, the optimal power distribution ratio of SCO-AN communication system is investigated. We consider the imperfect channel information as well. The expression for ergodic secrecy capacity with channel information error is derived. We conclude that once the channel estimation error increases, SCO-AN should be generate to more power.

2 System Model

Secure communication system over a multiple access wiretap channel is shown in Fig. 1. Alice is the signal transmitter. Bob is the receiver who equipped N_B antennas. Eve is the eavesdropper who equipped with N_E antennas. To facilitate discussion, we assumed that $N_A > N_E$, which means that the transmitter will have more freedom to assign the transmitted signal to SCO-AN. In the next section, we assume that N_E is a large number to apply the strong law of large numbers. If N_E is a large number and $N_A < N_E$, the study in Sect. 3 will be invalid. Eve is a passive eavesdropper. So:

$$y_{B1} = \mathbf{Hx} + n, \tag{1}$$

$$y_{E1} = \mathbf{Gx} + e, \tag{2}$$

Bob receives the signal y_{B1}, Eve receives the signal y_{E1}. $\mathbf{H}$ denotes legitimate channel, $\mathbf{G}$ is eavesdropping channel; $\mathbf{x}$ is the transmitted signal. n and e are AWGN. In this paper,

The SCO-AN designs as follows: $W = [\mathbf{p}_1 \ \mathbf{P}_2]$ is a $N_A \times N_B$ matrix, where $\mathbf{p}_1 = \mathbf{H}^{\dagger} / \|\mathbf{H}\|$. x is a $N_A \times 1$ vector, and $\mathbf{x} = \mathbf{p}_1\mathbf{u} + \mathbf{P}_2\mathbf{v}$. $\mathbf{u}$ is transmitting signal. The power of $\mathbf{u}$ is σ_u^2. So:

$$y_{B1} = \mathbf{Hp}_1\mathbf{u} + \mathbf{HP}_2\mathbf{v} + n = \|\mathbf{H}\|\,\mathbf{u} + \mathbf{H}_1\mathbf{v} + n, \tag{3}$$

$$y_{E1} = \mathbf{Gp}_1\mathbf{u} + \mathbf{GP}_2\mathbf{v} + e = \mathbf{G}_1\mathbf{u} + \mathbf{G}_2\mathbf{v} + e, \tag{4}$$

where $\mathbf{H}_1 = \mathbf{HW}_2$, $\mathbf{G}_1 = \mathbf{Gw}_1$ and $\mathbf{G}_2 = \mathbf{Gw}_2$.

The transmission power P, so $P = \sigma_u^2 + (N_A - 1)\sigma_v^2$. The percentage of signal is ϕ. Therefore,

$$\sigma_u^2 = \phi P, \tag{5}$$

$$\sigma_v^2 = (1 - \phi)P/(N_A - 1), \tag{6}$$

according to (5) and (6), Alice could change the power allocation strategy by changing ϕ.

3 SCO-AN: Optimal Power Allocation

3.1 Ergodic Lower Bound of Secrecy Capacity

The boundaries of SCO-AN are determined by $\mathbf{H}$ and $\mathbf{G}$. The legitimate channel capacity is:

$$\begin{aligned} C_A &= E_1\{\log_2(1 + \frac{\sigma_u^2\|\mathbf{H}\|^2}{\sigma_v^2\|\mathbf{H}\|^2})\} \\ &= E_1\{\log_2(1 + \frac{\sigma_u^2}{\sigma_v^2})\} \\ &= E_1\{\log_2(1 + \frac{\phi P}{(1-\phi)P/(N_A - 1)})\} \\ &= E_1\{\log_2(1 + \frac{\phi(N_A - 1)}{(1-\phi)})\}, \end{aligned} \tag{7}$$

From (7), we see the receiver's channel capacity is entirely unrelated to the transmitted power. It is an exciting discovery that the receiver's channel capacity only relates to the power partition coefficient of ϕ. The receiver's channel capacity is easy to adjust.

Next, the eavesdropping channel capacity will be discussed.

The upper bound of the eavesdropping channel capacity is:

$$\begin{aligned} C_B &= E_{h,\mathbf{G}_1,\mathbf{G}_2}\{\log_2 \left| I + \sigma_u^2 \mathbf{G}_1 \mathbf{G}_1^\dagger (\sigma_v^2 \mathbf{G}_2 \mathbf{G}_2^\dagger)^{-1} \right| \} \\ &= E_{h,\mathbf{G}_1,\mathbf{G}_2}\{\log_2 (1 + \frac{N_A - 1}{r - 1} \mathbf{G}_1^\dagger (\mathbf{G}_2 \mathbf{G}_2^\dagger)^{-1} \mathbf{G}_1 \}, \end{aligned} \tag{8}$$

r is defined as ϕ^{-1}. In (8), $\mathbf{G}_2\mathbf{G}_2^\dagger$ is invertible so the assumption of $N_A > N_E$ is guaranteed.

As we know, the secrecy capacity is $C = [C_1 - C_2]^+$ and $[a]^+$ means $\max\{0, \alpha\}$. The ergodic secrecy capacity is:

$$\begin{aligned} C_B = \frac{1}{\ln 2}[E_h\{\log_2(1 + \frac{r * N_A - 1}{(r-1)}) - \sum_{n=0}^{r*N_A-1} \binom{r * N_A - 1}{k} \frac{r * N_A - 1}{r+1} \\ \times D(k+1, r * N_A - 1 - k) \times F_1(1, k+1; N_A; \frac{r - r * N_A}{r+1}))\}], \end{aligned} \tag{9}$$

in the following sections, the above expressions leads us to get the optimal power allocation.

3.2 Infinite N_A Analysis

This subsection discusses the ergodic secrecy capacity lower bound when N_A approaches infinity asymptotically. Here, $\lim_{N_C \to \infty} \mathbf{G}_2\mathbf{G}_2^\dagger/(N_C - 1) = I$. So we can rewrite (9) as:

$$\lim_{N_A \to \infty} C_B = \lim_{N_A \to \infty} E_{\mathbf{G}_1,\mathbf{G}_2}\{\log_2 \left(1 + \frac{1}{r-1}\mathbf{G}_1^\dagger (\frac{\mathbf{G}_2\mathbf{G}_2^\dagger}{N_A - 1})^{-1} \mathbf{G}_1\right)\} = e_k(r+1)\sum_{n=1}^{N_E} E_k(r+1), \tag{10}$$

the gain of Rayleigh fading channel among N_E non-colluding eavesdroppers obey an exponential distribution. So $\|\mathbf{G}_1\|^2$ obeys a Gamma distribution with parameters $(N_E, 1)$.

When N_A approaches infinity asymptotically, so

$$C = E_h\{\log_2(1 + \frac{\phi(N_A - 1)}{(1-\phi)})\} - e_k(r+1)\sum_{n=1}^{N_E} E_k(r+1) \tag{11}$$

3.3 The Optimal Allocation

We study the case where N_A approaches infinity asymptotically. It then follows that $N_E = 1$. (13) gives $dC_B \backslash dz = 0$. Substituting $N_E = 1$ in the expression of C_B in (11) and taking its derivative with respect to z gives

$$\begin{aligned} dC_B/dz &= (e_k(r+1)E_1(r-1) - \exp(r-1)E_x(r+1))/\ln 2 \\ &= \left(e_k(r+1)e_k(r+1) - (r+1)^{-1}\right)\Big/\ln 2. \end{aligned} \tag{12}$$

The solution to (12) is $r = 1.8$. Therefore when N_A approaches infinity asymptotically.

Hence, for sufficiently large N_A, the optimal $\phi = 0.55$. We can see that there is no significance difference in the optimal power distribution between the case where N_A takes its smallest possible value and the case where N_A approaches infinity asymptotically.

4 Results and Discussion

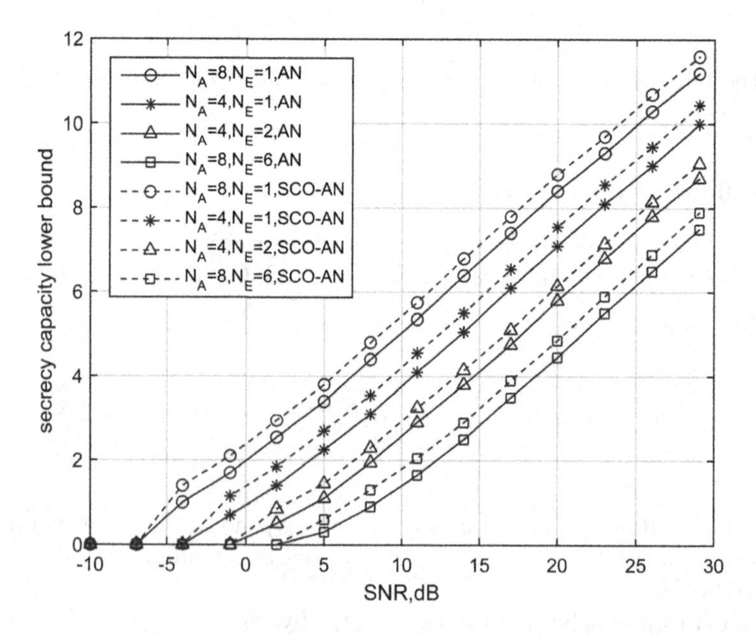

Fig. 2. The lower bound of secrecy capacity versus SNR with different numbers of antennas

Figure 2 shows the variation of C calculated using (9) with the optimal ϕ. The solid lines show the system's secrecy capacity (SC), which uses SCO-AN. The dash lines show the SC, which uses AN. The variation of SCO-AN and AN have

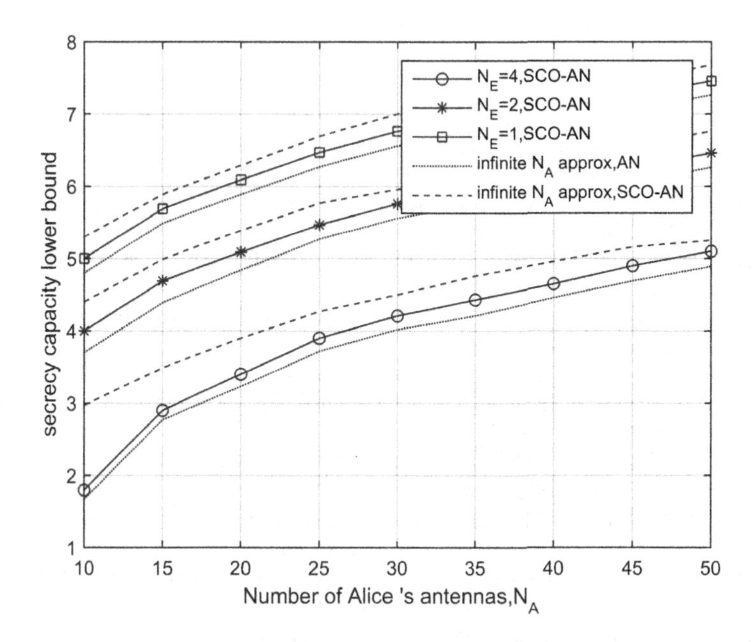

Fig. 3. The SC versus SNR with different numbers of eavesdropper's antennas and infinite N_A

similar behaviour. Furthermore, C increases with N_A when SNR is from -10 dB to 30 dB. C decreases with N_E as well. C reduced to closer to zero when the SNR is small.

The dashed line in Fig. 3 denotes the approximations of C in (11) as N_A approaches infinity. The dotted line denotes the approximations of SC of AN as N_A approaches infinity. Figure 3 shows that SC is correlated with N_A and N_E. The SC increases with an increasing of N_A. The infinite N_A lower bound of secrecy capacity for SCO-AN is always greater than that of secrecy capacity for AN. When the N_A approaches infinite asymptotically, the secrecy capacity of AN is the smallest. The secrecy capacity decreases with a increase of N_E.

5 Conclusion

This paper has explored and discussed the power distribution coefficients between the transmitted signal and SCO-AN under ideal conditions. The expression of secrecy capacity is obtained. We then prove the secrecy capacity for SCO-AN is always greater than that of secrecy capacity for AN. The optimal power allocation ratio is computed. In comparison with AN, SCO-AN has better secrecy performance when presented with less power for extra artificial noise. It then follows that SCO-AN is a better option for artificial noise than AN. This paper also considers the channel estimation error on power allocation. The result of this paper provides a much-valued help to the design of a secure wireless communication system.

References

1. Hashem, T., Hasan, R.: Method and system for secure transmission of information. US (2004)
2. Barros, J., Rodrigues, M.R.D.: Secrecy capacity of wireless channels. In: 2006 IEEE International Symposium on Information Theory, pp. 356–360. IEEE (2006)
3. Wyner, A.D.: The wire-tap channel. Bell Syst. Tech. J. **54**(8), 1355–1387 (1975)
4. Csiszar, I., Korner, J.: Broadcast channels with confidential messages. IEEE Trans. Inf. Theory **24**(3), 339–348 (2003)
5. Negi, R., Goel, S.: Secret communication using artificial noise. In: IEEE Vehicular Technology Conference (2005)
6. Yebo, G., Zhilu, W., Yin, Z., et al.: The secrecy capacity optimization artificial noise: a new type of artificial noise for secure communication in MIMO system. IEEE Access **7**, 58353–58360 (2019)
7. Oggier, F., Hassibi, B.: The secrecy capacity of the MIMO wiretap channel. IEEE Trans. Inf. Theory **57**(8), 4961–4972 (2011)
8. Gopala, P.K., Lai, L., El Gamal, H.: On the secrecy capacity of fading channels. IEEE Trans. Inf. Theory **54**(10), 4687–4698 (2008)
9. Mahdavifar, H., Vardy, A.: Achieving the secrecy capacity of wiretap channels using polar codes. IEEE Trans. Inf. Theory **57**(10), 6428–6443 (2011). 1628–1631
10. Khisti, A., Wornell, G.W.: Secure transmission with multiple antennas: the MIMOME channel. IEEE Trans. Inf. Theory (to be published). http://allegro.mit.edu/pubs/posted/journal/2008-khisti-wornell-it.pdf2547-2553
11. Ekrem, E., Ulukus, S.: The secrecy capacity region of the Gaussian MIMO multi-receiver wiretap channel. IEEE Trans. Inf. Theory **57**(4), 2083–2114 (2011)
12. Gao, H., Smith, P.J., Clark, M.V.: Theoretical reliability of MMSE linear diversity combining in Rayleigh-fading additive interference channels. IEEE Trans. Commun. **46**(5), 666–672 (1998)

Radar Systems

Slow-Time MIMO in High Frequency Surface Wave Radar

Wang Linwei[1], Li Bo[1], Yu Changjun[1(✉)], and Ji Xiaowei[2]

[1] Harbin Institute of Technology at Weihai, Weihai, China
yuchangjun@hit.edu.cn
[2] Harbin Institute of Technology, Harbin, China

Abstract. High Frequency Surface Wave Radar (HFSWR) has been widely used for remote sensing and maritime target detection. However, the HFSWR system contains a huge antenna array for accurate angle measurement. Miniaturization of antenna array has been one of the most urgent problems in this field. Applying the Multiple-Input Multiple-Output (MIMO) technology to radar can effectively improve the angle estimation accuracy in the case of small aperture through virtual channels. Hence, the superiority of MIMO contributes to improve the performance of HFSWR system. Slow-time MIMO is a special form of MIMO radar through Doppler diversity, and it can operate at same time and same frequency. This paper introduces the experiment of slow-time MIMO in HFSWR system in recent days. The results validated on real data can prove the benefit of MIMO technology in angle estimating, and we analyze the specific impact of sea clutter, ionospheric clutter and radio interference for Slow-time MIMO HFSWR.

Keywords: High frequency surface wave radar · Slow-time MIMO · Aperture expanding · Beamforming · Clutter suppression

1 Introduction

The High Frequency Surface Wave Radar (HFSWR), which is applied on surface wave propagation, provides an excellent capability to detect ocean surface dynamical environment over-the-horizon [1]. And HFSWR operates in the High Frequency (HF) frequency band of 3 to 30 MHz. Nowadays, HFSWR systems has been widely applied in many fields, such as detecting the sea surface conditions, surveilling the large Exclusive Economic Zone(EEZ), as well as military defense [2]. Due to the large surveillance area, many countries apply the HFSWR to continuously supervise the maritime of activity within a nation's EEZ, which is a crucial question in protection of national sovereignty. At present, hundreds of HFSWRs have been deployed all over the world, and most of the typical HFSWR systems are lied close to offshore and apply the large array structures with aperture sizes of hundreds or even thousands of meters, which can occupy a large area of coastal. However, with the development of national economy, the coastal resources and environment become more precious than before. Obviously, the occupation of scarce coastal resources in a large area has become an important reason that limits the development of HFSWR system. Therefore, how to reduce the

Q. Guo et al. (Eds.): WiSATS 2021, LNICST 410, pp. 751–761, 2022.
https://doi.org/10.1007/978-3-030-93398-2_65

high land rental and construction cost is a hot research topic in the future development. With the excellent performance and distinctive design, Multiple-Input Multiple-Output (MIMO) antenna systems have the ability to markedly improve the performance of communication systems over single antenna systems. Multipath effect will adverse impact the signal quality in the transmission. Hence, the traditional antenna systems pay attention on eliminating multipath effects. However, the novelty of MIMO system is that it takes the opposite view. Specifically, MIMO explores the independence between signals at the array elements and applies the multipath effect to improve communication quality. In a MIMO system, the transmitter and receiver utilize multiple antennas to communicate simultaneously. MIMO systems often employ sophisticated signal processing techniques to significantly enhance reliability of signal, transmission range, and throughput. The transmitter uses these techniques to simultaneously send multiple radio frequency signals and the receiver recovers data from the radio frequency signals. As stated above, by transmitting different streams of information, MIMO communication systems solve the problems caused by fading from several irrelevant transmitters. Inspired by MIMO communication systems, Fishler et al. [3] proposed the MIMO radar, which also has the same advantages in multiple signal transmission. Specifically, in order to explore the radio cross section of target scintillations, the MIMO radar system transmits the different signals from some decorrelated transmitters. The receiver will receive a superposed signal that formed by independently faded signals, so the average signal to noise ratio (SNR) of the received signal is more or less constant. In this case, the MIMO radar system avoid the problem that received SNR contrasted sharply.

The HFSWR system and MIMO radar have their own superiority, so complementing them for each other becomes a natural choice. Many of the existing works have demonstrated that this combination is a promising approach for target detection in long range. In 2014, Dzvonkovskaya [4] proposed a new approach that applies MIMO technique to compact HF radar system WERA deployments and obtain positive economic benefits without degrading system detection performance. The WERA system adopts modular design, which can easily meet the requirements of practical application. Most of the signal processing steps are done by software, so this system can be easily adapted to different application. Initially, in the standard configuration of WERA receiving system has 12 receiving antennas and these receiving antennas are located in a linear array and their spacing is about one half of a radar wavelength. The test MIMO geometry of WERA system has 6 real antennas. These 6 real antennas are taken as real elements and other antenna elements are skipped [5]. The separation distance between the transmitters is L, which roughly 3 times of the radar wavelength. Therefore, the positions of virtual receiving antenna elements are placed after the real elements. The carrier frequency shift of the first and second transmitters is equal to one third of the distance cell frequency, which contributes to separate the echo between sea surface and targets in every real antenna. Windowed discrete Fourier transform is used to estimate Doppler frequency and windowed beamforming is used to estimate Angle. Because of the transmitted signals, the phase difference caused by transmitting antennas along with phase differences in receiving antennas can form a larger virtual array by a small number of antennas to proceed with the beamforming technique. In 2015, Jangal et al. [6] proposed a new HFSWR deployed for the European I2C project (i.e. Integrated

System for Interoperable sensors & Information sources for Common abnormal vessel behaviour detection & Collaborative identification of threat). The crucial improvement of ONERA's HFSWR is deploying key technologies as combination. These technologies are: employing proven hardware, slow rate equipment, real time processing, MIMO architecture, full digital system and possibility of located in rough terrain. The ONERA's HFSWR is deployed on Levant Island in the South East of France and applies Multi-carrier frequency technology (surface wave radar is operated at 5 and 9 MHz). Like other HFSWR systems, the important requirement of the Levant Island system is to deal with the Target to Clutter Mode(TCM) and reduce the radar cell means. For obtaining the aforementioned coverage on small and slow targets, The ONERA's HFSWR system increases the diversity and the array length. Simultaneously, ONERA's HFSWR system will increase the transmitting power and use FMCW mode instead of pulse mode to realize the detection of large target or fast target. In North America, A.M. Ponsford et al. [2] presented Canada's Third Generation High Frequency Surface Wave Radar System, which is a monostatic pulse Doppler radar. The radar is simultaneously operated by two independent frequencies in an interleaved pulse mode, and makes full use of time diversity and frequency diversity. For each frequency, the data collected on the thread is processed in multiple parallel paths optimized for different classes of ships.

However, none of the aforementioned methods is applicable to the large number of transmitting antennas. In 2006, Mecca et al. [7] proposed a MIMO method that in conventional radar waveforms are phase-coded to be orthogonal after Doppler processing at the receiver, i.e. in "slow-time". Slow-time MIMO has the advantages of high bandwidth efficiency, easy implementation, and doesn't need to modify the receiver before range pulse compression. Thanks to the advantage of Slow-time MIMO, many improvement approaches are proposed and applied in over horizon radar system. In the following study, Mecca et al. [8] presented beamspace multiple-input multiple-output space-time adaptive processing(STAP) to suppress radar clutter subject to multipath propagation between transmitter and receivers. In 2008, Frazer et al. [9] designed the Australian HILOW experiment to examine the MIMO radar concepts based on OTHR. Yu et al. [10] presented a technique for generating slow-time MIMO waveforms by using a low-cost passive frequency mixer and an existing (non-MIMO) radar structure. In this study, the concept of double-sideband orthogonal MIMO slow time radar and the experimental results of DUKE S-band radar test stand are introduced. In 2011, Rossum et al. [11] introduced article a novel waveform for MIMO radar, named Random Slow-Time Code Division Multiple Access (ST-CDMA). Compared with the conventional CDMA, the feature of ST-CDMA is that the waveforms of the different transmitters are orthogonal per burst. In fact, the typical MIMO-OTHR waveforms with existing coded modulation are not strictly orthogonal. Zhang et al. [12] presented slow time random phase coding waveform design based on Walsh matrix, which can improve the orthogonal performance of the coded waveform and the target detection capability of the radar system. In the meanwhile, in order to solve the problem that the autocorrelation performance of STRPC waveform encoded by Walsh matrix is unsatisfying, the MIMO-OTHR mismatch filtering can further improve the principal and sidelobe ratio of pulse pressure with a certain loss of mismatch filtering. Recently, the slow-time MIMO concept is employed into automotive radars, In the framework of Generalized Likelihood Ration Test (GLRT), Wang et al. [13] proposed an explicit signal model considering waveform separation residuals and presented a target detector based on Kronecker subspace. The precise theoretical analysis has verified the proposed target detection scheme.

Slow-time MIMO can operate at same time and frequency, which can save spectrum resources and do not have many extra hardware requirements. Due to the above advantages of Slow-time MIMO, it is suitable for HFSWR. In order to apply Slow-time MIMO to HFSWR, we have designed an experimental system, and conducted related experiment. This paper will introduce the experiment and results of slow-time MIMO in high frequency surface wave radar system in recent days. The rest of the paper is structured as follows. Section 2 introduces the signal model of slow-time MIMIO. Section 3 shows the experimental system and its configuration. Section 4 analyzed the results of the experiment, and a simple summary and outlook are given at the end.

2 Slow-Time MIMO Signal Model

Consider a MIMO radar system with M transmitters and N receivers. The transmitting signals are modulated by additional Doppler frequency in each pulse based on conventional pulse compression waveform, such as Linear Frequency Modulation (LFM). The q-th transmitting pulse of the m-th transmitter is given by

$$s_{mq}(t) = u(t)e^{j2\pi w_m q T_{\mathrm{pp}}} \tag{1}$$

where $u(t)$ is a conventional pulse compression waveform, $w_m, m = 1, 2, \ldots, M$ is the additional Doppler frequency of m-th transmitting signal, T_{pp} is the pulse repetition period.

Without loss of generality, LFM signal is taken as an example of $u(t)$ in this paper,

$$u(t) = \begin{cases} e^{j\pi K t^2} & 0 \leq t < T_{\mathrm{pw}} \\ 0 & t \geq T_{\mathrm{pw}} \end{cases} \tag{2}$$

where $K = B/T_{\mathrm{pw}}$ is the modulation slope, B is the signal bandwidth, and T_{pw} is the pulse width. To divide the whole Doppler bandwidth into M orthogonal sub-channels whose bandwidth is f_{a}/M, w_m is defined as

$$w_m = \frac{f_{\mathrm{a}}}{2}\left(\frac{2m-1}{M} - 1\right) \tag{3}$$

where $f_{\mathrm{a}} = 1/T_{\mathrm{pp}}$ is the Pulse Repetition Frequency (PRF).

Consider a M elements uniform linear transmitting antenna array and a N elements uniform linear receiving antenna array. The transmitting steering vector and the receiving steering vector can be written as

$$\begin{aligned} \mathbf{a}(\theta) &= \begin{bmatrix} 1 & e^{-j2\pi d\sin(\theta)/\lambda} & \cdots & e^{-j2\pi(M-1)d\sin(\theta)/\lambda} \end{bmatrix} \\ \mathbf{b}(\varphi) &= \begin{bmatrix} 1 & e^{-j2\pi d\sin(\varphi)/\lambda} & \cdots & e^{-j2\pi(N-1)d\sin(\varphi)/\lambda} \end{bmatrix} \end{aligned} \tag{4}$$

where θ is the Direction Of Departure (DOD), φ is the Direction Of Arrival (DOA), d is the spacing between array elements, λ is the wave length of the operating frequency. Assume a far field target with a Doppler frequency f_{d}, and the DOD and DOA of the target is θ_{t} and φ_{t}. The echo of the target from n-th receiving antenna element in q-th pulse is

$$r_{nq}(t) = \sum_{m=1}^{M} s_{mq}(t) a_m(\theta_{\mathrm{t}}) b_n(\varphi_{\mathrm{t}}) e^{j2\pi f_{\mathrm{d}} q T_{\mathrm{pp}}} \tag{5}$$

where $a_m(\theta) = e^{-j2\pi(m-1)d\sin(\theta)/\lambda}$ is the phase shift of the m-th element in transmitting antenna array, and $b_n(\varphi) = e^{-j2\pi(n-1)d\sin(\varphi)/\lambda}$ is the phase shift of the n-th element in receiving antenna array. After pulse compression, we get

$$x_n(q) = \sum_{m=1}^{M} a_m(\theta_{\mathrm{t}}) b_n(\varphi_{\mathrm{t}}) e^{j2\pi w_m q T_{\mathrm{pp}}} e^{j2\pi f_{\mathrm{d}} q T_{\mathrm{pp}}} \tag{6}$$

When the frequency of targets is less than the frequency interval between w_m, we can separate echoes from different transmitters in Doppler domain.

3 Radar System Configuration

The MIMO radar experimental system used in this paper is site in Weihai, China. The system has a uniform linear array consisting of 8 transmitting-receiving antennas with 11.5 m spacing. Because our experimental MIMO System has 8 transmitters and 8 receivers, we can obtain 64 virtual channels and this design is shown in Fig. 1.

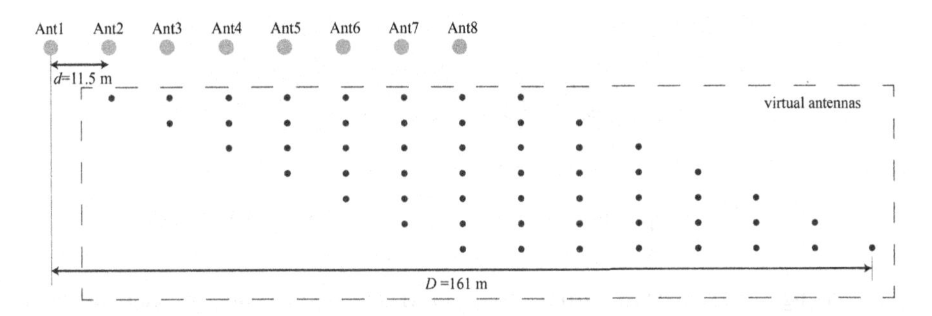

Fig. 1. The position of real antennas and virtual antennas

However, many virtual channels have the same phase shift, because the spacing between transmitting antenna elements is equal to the spacing of receiving antenna elements. So, the virtual array aperture is 161 m, which equivalents to a 15 elements linear antenna array.

The system can generate independent transmitting signals for 8 transmitters relying on a high-speed DAC module, so it can conveniently switch the working mode to single transmitter mode, phased array mode and MIMO mode. We obtained the transmit signals for slow-time mode whose parameters is shown in Table 1. The other working modes use similar parameters in this experiment.

Table 1. Transmitting signals parameters

Parameters	Value
Number of transmitters	8
Carrier frequency	4.7 MHz
Bandwidth	30 kHz
Transmit power	100 W
Pulse repetition period	55 ms
Doppler interval	2.27 Hz

4 Results and Analysis

4.1 Working in MIMO Mode

For further processing, the detection data is recorded after pulse compression. RD spectrum is obtained by Fourier transform in the slow-time dimension, as shown in Fig. 2. The 8 transmit signals are completely separated.

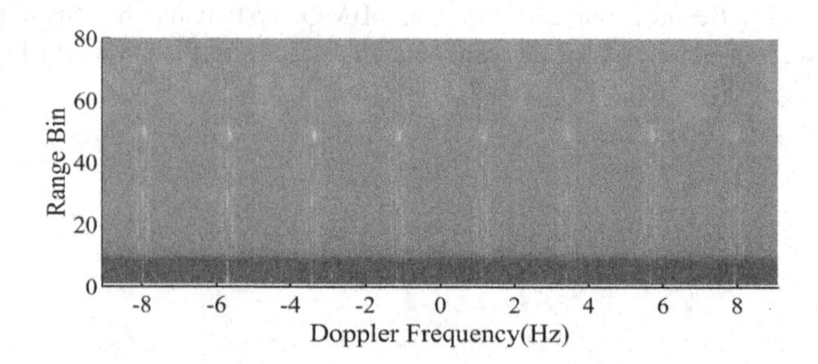

Fig. 2. RD spectrum with 8 transmitters in slow-time MIMO radar

Dividing the RD spectrum from each receiver into 8 equal parts, we can obtain the whole 64 virtual channels' data. Figure 3 shows the RD spectrum from one of the virtual channels.

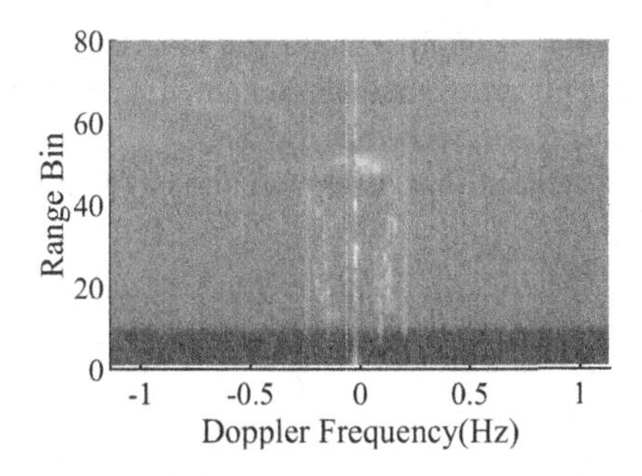

Fig. 3. RD spectrum in one of the virtual channels

In the experiment, we set the working mode of radar system as slow-time MIMO mode, phased array mode and single transmitter mode in turn. The time interval for switching the working mode is about 10 min, and the output powers of the transmitters remain unchanged during this process. Figure 4 shows the amplitudes of slow-time MIMO, single transmitter and phased array at a same range bin. The amplitude of Phased array mode is about 10 dB higher than that of slow-time MIMO and single transmitter. Meanwhile, slow-time MIMO and single transmitter have similar echo amplitudes.

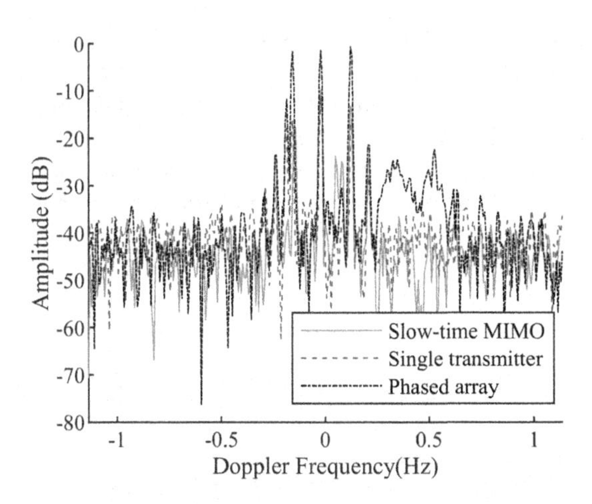

Fig. 4. Comparison of echo amplitudes of different radar working modes

4.2 Angle Estimation Performance

MIMO technology can improve the accuracy of angle measurement and the ability of angle resolution by expanding the aperture of the array. Figure 5 shows the beam-forming results of a target echo when the radar system is working in slow-time MIMO mode. The blue line is the result of the scene that Ant1 transmits signal. The red line is the result of the scene that Ant1 and Ant8 transmit signals. The black line is the result of the scene that all antennas transmit signals. The main lobe width of the scene with 2 transmitting antenna elements and 8 transmitting antenna elements are almost the same, which is about half of the scene with one transmitting antenna element. Besides, the side lobe in scene with 8 transmitting antenna elements is lower.

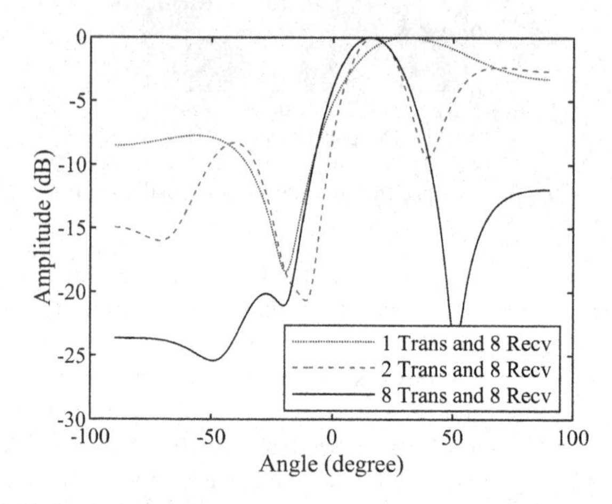

Fig. 5. Beamforming results of a target (Color figure online)

The range resolution capability of HFSWR is weak, and there may be multiple targets within the same range cell. When the speeds of the targets are similar, it is difficult to separate the targets with the small-aperture antenna array. Benefit from the aperture expanding by the MIMO technology, the ability to distinguish targets from different angles is enhanced. Figure 6 shows the results of two targets in a same range cell and Doppler cell, which can be separated in MIMO mode, but can't be separated with single transmitter.

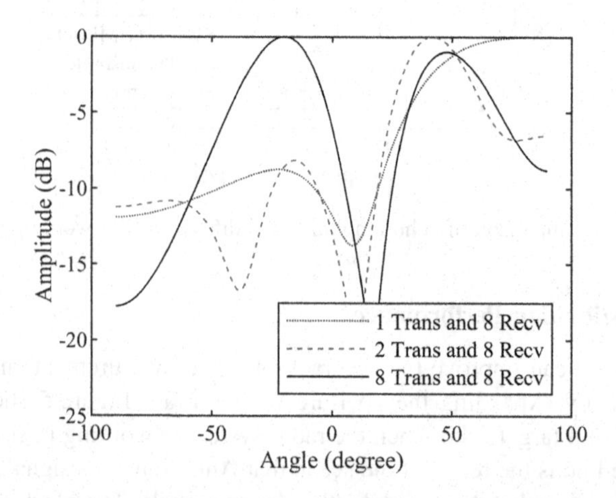

Fig. 6. Beamforming results of two targets with the same range and Doppler

4.3 Clutter and Interference

HFSWR has many clutters an interference such as sea clutter, ionospheric clutter, radio interference, etc. The sea clutter, mainly the first-order spectrum, has a low Doppler shift, located near the ship target echoes. So, the structure and characteristics of sea clutter in slow-time MIMO mode is similar to that in single transmitter mode. However, because of the slow-time MIMO mode has 64 independent virtual channels, the freedom degree and the amount of information that the slow-time MIMO mode can provide are square times that of the single transmitter mode, which has great significance for sea clutter detection and suppression.

Unlike sea clutter, ionospheric clutter may have a higher Doppler shift. Large-scale ionospheric clutter will repeat and overlap between Doppler sub-channels, shown in Fig. 7. The ionospheric clutter overlap will causes more severely affect in slow-time MIMO mode than in others, so the suppression of ionospheric is important in slow-time MIMO.

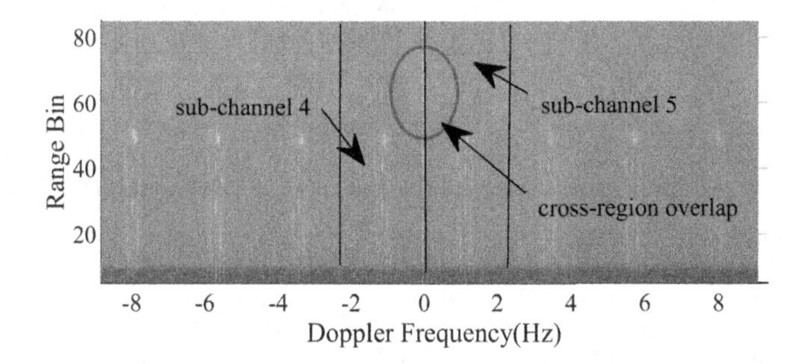

Fig. 7. Ionospheric clutter in cross-region

Contrary to ionospheric clutter, radio interference usually affects only a narrow bandwidth in Doppler domain. Figure 8 shows the RD spectrum with a simulated radio signal adding in. The radio interference affects only one transmit channel, rather than all channels.

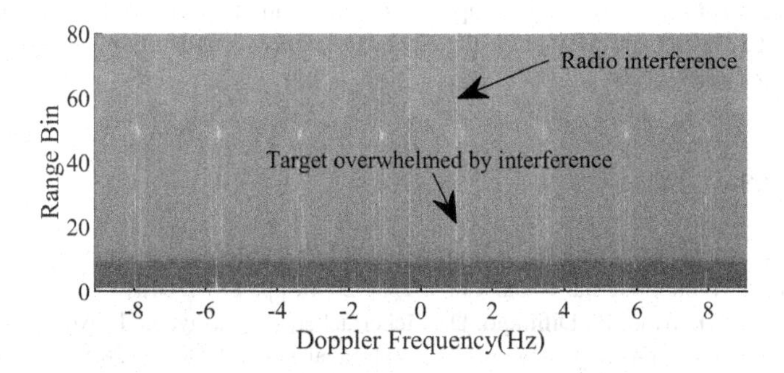

Fig. 8. Simulation radio interference add-in RD spectrum

Many virtual channels of slow-time MIMO are redundant, so deleting a part of them will not affect the detection performance seriously. Supposing a radio signal indenting from −10°, the beamforming result is affected by the interference, and the main lobe direction is shifted from 13.2° to −5.3°. After data deletion, the angle is restored to 13.2° with some side lobe elevation (Fig. 9).

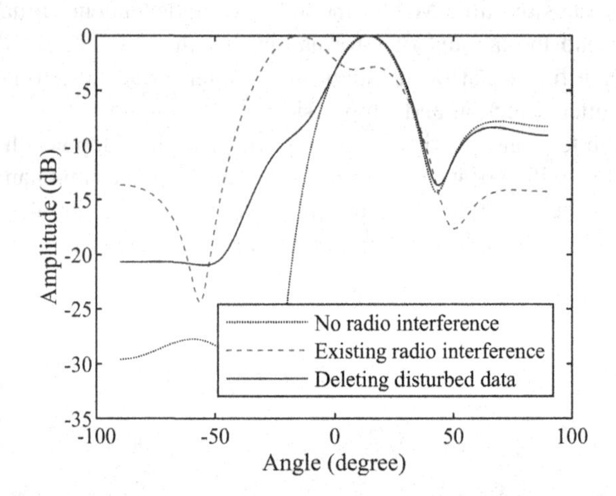

Fig. 9. Result of radio interference suppression by deleting disturbed data

5 Conclusion

This paper introduces the experiment of HFSWR based on slow-time MIMO. The experimental results have proved the practicability of slow-time MIMO in HFSWR. The MIMO technology can improve the performance of angle estimating, and the MIMO system (consists of 8 transmitting-receiving antennas) has more desired detection accuracy compared with the conventional system (consists of 1 transmitting antenna and 8 receiving antennas). In addition, the influence of interference and clutter on slow-time MIMO HFSWR is analyzed in this paper. And how to effectively suppress the clutter in MIMO HFSWR system is the following study in the future.

Acknowledgements. This work is supported by the National Science Foundation of Chain under Grant 62031015 and 61971159, the National Science Foundation of Shandong Province under Grant ZR2020MF007.

References

1. Dang, B., Yang, Q., Deng, W.: Angular resolution performance analysis for HF-MIMO radar. In: IET International Radar Conference 2015, pp. 1–4 (2015)
2. Ponsford, A.M., Moo, P., Difilippo, D., Mckerracher, R., Kashyap, N., Allard, Y.: Canada's third generation high frequency surface wave radar system, **10**, 21–28 (2015)

3. Fishler, E., Haimovich, A., Blum, R., Chizhik, D., Cimini, L., Valenzuela, R.: MIMO radar: an idea whose time has come. In: Proceedings of the 2004 IEEE Radar Conference (IEEE Cat. No.04CH37509), pp. 71–78 (2004)
4. Dzvonkovskaya, A., Helzel, T., Petersen, L., Merz, C.R., Liu, Y., Weisberg, R.H.: Initial results of ship detection and tracking using WERA HF ocean radar with MIMO configuration. In: 2014 15th International Radar Symposium (IRS), pp. 1–3 (2014)
5. Dzvonkovskaya, A., Merz, C.R., Liu, Y., Weisberg, R.H., Helzel, T., Petersen, L.: Initial surface current measurements on the West Florida shelf using WERA HF ocean radar with multiple input multiple output (MIMO) synthetic aperture. In: 2014 Oceans - St. John's, pp. 1–4 (2014)
6. Jangal, F., Menelle, M.: French HFSWR contribution to the European integrated maritime surveillance system I2C. In: IET International Radar Conference 2015, pp. 1–5 (2015)
7. Mecca, V.F., Ramakrishnan, D., Krolik, J.L.: MIMO radar space-time adaptive processing for multipath clutter mitigation. In: Fourth IEEE Workshop on Sensor Array and Multichannel Processing, 2006, pp. 249–253 (2006)
8. Mecca, V.F., Krolik, J.L., Robey, F.C.: Beamspace slow-time MIMO radar for multipath clutter mitigation. In: 2008 IEEE International Conference on Acoustics, Speech and Signal Processing, pp. 2313–2316 (2008)
9. Frazer, G.J., Abramovich, Y.I., Johnson, B.A., Robey, F.C.: Recent results in MIMO over-the-horizon radar. In: 2008 IEEE Radar Conference, pp. 1–6 (2008)
10. Yu, J., Krolik, J.: Quadrature slow-time MIMO radar with experimental results. In: 2010 Conference Record of the Forty Fourth Asilomar Conference on Signals, Systems and Computers, pp. 2134–2137 (2010)
11. van Rossum, W., Anitori, L.: Doppler ambiguity resolution using random slow-time code division multiple access MIMO radar with sparse signal processing. In: 2018 IEEE Radar Conference (RadarConf18), pp. 0441–0446 (2018)
12. Zhang, K., Chen, K.W., Bao, Z.: Waveform design of slow time random phase coding based on Walsh matrix. J. Air Force Early Warning Acad. **32**, 21–24+30 (2018)
13. Wang, P., Boufounos, P., Mansour, H., Orlik, P.V.: Slow-time MIMO-FMCW automotive radar detection with imperfect waveform separation. In: ICASSP 2020 - 2020 IEEE International Conference on Acoustics, Speech and Signal Processing (ICASSP), pp. 8634–8638 (2020)

A Radar Target Detection Method Based on RBF Neural Network

Hu Jurong, Wu Tong(✉), Lu Long, and Li Xujie

School of Computer and Information, Hohai University, Nanjing 210098, China

Abstract. In recent years radar target detection environment is more and more complex. Traditional CFAR (Constant False Alarm Rate) is a technology in which the radar system discriminates the output signal and noise of the receiver to determine whether the target signal exists under the condition that the False Alarm probability is kept Constant. In order to improve the radar target detection performance, a radar target detection method based on NN (Neural Network) is proposed. In this paper, the radar signal received by a single RBFNN is used for network training, and the probability of detection target is studied by combining the binary detection theory. Simulation results show that the proposed algorithm can effectively improve the radar target detection probability.

Keywords: Radar · Constant false alarm rate · Neural network · Target detection

1 Introduction

An important task performed by the radar system is target detection [1]. There are various interferences in the target detection environment of radar signals. These interferences include thermal noise inside the receiver antenna and clutter interferences caused by obstacles, wind or rain, sea waves and so on, sometimes there are active interference and passive interference from the enemies. The intensity of clutter and enemies' interference is often much higher than the internal noise level of the radar receiver. Therefore, to extract signals from strong interference, not only a certain signal-to-noise ratio is required, but also constant false-alarm rate (CFAR) [2] on the signal.

Neural network [3, 4] has a distributed structure, which is similar to the human brain, and it has strong robustness and fault tolerance. When we use the neural network to process the disturbed signal, it will guarantee the authenticity of the signal. Neural network is a non-liner model [5, 6], so we can simulate many engineering nonlinear problems through neural network. Using the chaotic characteristics of echo [7], artificial neural network [8] can be used as a clutter simulator to train the single radar echoes which in a targetless state, then use the trained network as a predictor to predict

This work was supported in part by the Provincial Key Research and Development Program of Jiangsu under Grant BE2019017, in part by the Provincial Water Science and Technology Program of Jiangsu under Grant 2020028.

Q. Guo et al. (Eds.): WiSATS 2021, LNICST 410, pp. 762–769, 2022.
https://doi.org/10.1007/978-3-030-93398-2_66

the echo received by the radar at the next moment. According to the comparison result of the prediction error of the radar echo signal and the decision threshold [9–11], we can classify the two kinds of interference to complete the target detection. Simulation result shows that the detection probability of this method id better than the traditional cell average CFAR (CA-CFAR) detection method.

2 CFAR Target Detection

Define that $r(t)$ is the signal received by the radar receiver, and $r(t)$ is composed of target signal $\mathrm{x_T}(t)$, interference signal $j(t)$ and noise $\mathrm{g}(t)$, $\mathrm{g}(t)$ is a random signal with zero-mean. They are independent of each other, then the signal $r(t)$ can be expressed as:

$$r(t) = \mathrm{x_T}(t) + j(t) + g(t) \tag{1}$$

The simulation model of the target signal $\mathrm{x}(t)$ with radar single scan is:

$$x_T(t) = \frac{\mathrm{K_T}}{\mathrm{R}^2(\mathrm{t})}\cos[\omega_d(t - \frac{2\mathrm{R_0}}{\mathrm{c}} - kT_r) + (\omega_k + \pi b)(t - \frac{2\mathrm{R(t)}}{\mathrm{c}} - kT_r)] \tag{2}$$

K_T is a constant and is determined by factors such as radar transmit power, antenna gain, transmission loss or target cross-sectional area; $\mathrm{R}(t)$ is the instantaneous distance between the radar and the target; $\mathrm{R_0}$ is the distance between the target and the radar when the kth pulse meets the target; ω_d is Doppler frequency; ω_k is frequency increment; c is the speed of light; T_r is pulse repetition period; b is linear FM scan rate.

We take for example that the interference signal $j(t)$ is an FM signal:

$$j(t) = U_i\cos[w_it + 2\pi K\int_0^t u_f(\tau)d\tau + \varphi] \tag{3}$$

Where U_i represent the amplitude of FM signal; Time-varying functions $u_\mathrm{f}(t)$ represent a stationary random process with zero-mean; And φ is a uniform distribution on $[0, 2\pi]$ and independent of $u_\mathrm{f}(t)$; ω_i is the central frequency of the noise FM signal; K_f is FM slope.

Divide the intercepted signal represented by formula (1) into multiple time series u (k), then construct a binary hypothesis testing:

$$\begin{cases} H_0 : \mathrm{r}_{n,k} = \mathrm{j}_{n,k} + \mathrm{g}_{n,k} \\ H_1 : \mathrm{r}_{n,k} = \mathrm{x}_{n,k} + \mathrm{j}_{n,k} + \mathrm{g}_{n,k} \end{cases} \tag{4}$$

Where $\mathrm{g}_{n,k}$ and $\mathrm{r}_{n,k}$ represent the value of the nth sampling point in the kth time series; $\mathrm{x}_{n,k}$ and $j_{n,k}$ represent signal term and error term respectively. H_0 means no target, while H_1 means target exists.

By the N-P criterion, we can define the false alarm probability P_f(k) and detection probability P_d(k) in the time series u(k):

$$\begin{cases} P_f(\mathrm{k}) = \mathrm{P}\{\mathrm{r}_{n,k} > V|H_0\} \\ P_d(\mathrm{k}) = \mathrm{P}\{\mathrm{r}_{n,k} > V|H_1\} \end{cases} \tag{5}$$

In the observation time, the average false alarm probability is $\overline{\mathrm{P}}_f$ and the average detection probability is $\overline{\mathrm{P}}_d$:

$$\begin{cases} \overline{P_f} = \frac{1}{M}\sum_{\mathrm{i}=1}^{\mathrm{M}} P_f(i) \\ \overline{P_d} = \frac{1}{M}\sum_{\mathrm{i}=1}^{\mathrm{M}} P_d(i) \end{cases} \tag{6}$$

Assume that the noise in the intercepted signal conforms to the zero-mean Gaussian distribution, and its power is δ^2:

$$g(\mathrm{n}) \sim N\left(0, \delta^2\right) \tag{7}$$

Perform n-point discrete-time Fourier transform on formula (3) to get G(*f*), the variance can be expressed as:

$$N'\delta^2 = \delta^2 \sum_{n=1}^{N} h_n^2 \tag{8}$$

$|\mathrm{G}(f)|$ is the magnitude of G(f), and $|\mathrm{G}(f)|$ obeys the Rayleigh distribution with parameter $\lambda = \sqrt{N'/2\delta}$, we can get that:

$$P_f = e^{-V_1^2/N'\delta^2}, \quad \eta = \sqrt{-N'\delta^2 \ln P_f} \tag{9}$$

Where P_f is false alarm probability and η is detection threshold.

According to the false alarm probability formula:

$$P_{fa} = \int_{\eta}^{\infty} p_n(x)dx = 1 - \int_0^{\eta} p_n(x)dx \tag{10}$$

We can get:

$$\int_0^{\eta} p_n(x)dx = 1 - P_{fa} \tag{11}$$

The threshold η can be obtained by formula (11), then we can conclude that the thresholds corresponding to different probability density functions can get a constant false alarm probability. That is, as the probability density function transforms, there is a corresponding adaptive threshold.

3 Target Detection Based on Neural Network

Assume that the echo signal is described by the following model.

RBF neural network has three layers, its input layer and hidden layer have M and K neurons respectively, and it has only one output neuron. The input and output relationship of the neural network is shown in Fig. 1.

The activation function of the RBF neural network can be represented by the Gaussian function as:

$$R(r_p - c_i) = \exp(-\frac{1}{2\delta_k^2}\left\|r_p - c_i\right\|^2) \tag{12}$$

Where r_p is the pth input sample, c_i is the ith center point. $R(r_p - c_i)$ is a nonlinear mapping that maps the input vector to the hidden layer.

The output of RBF neural network can be represented as:

$$y_j = \sum_{i=1}^{K} w_{ij} \exp(-\frac{1}{2\delta_k^2}\left\|r_p - c_i\right\|^2) \quad j = 1, 2, \ldots, n \tag{13}$$

Where ω_{ij} is adaptive update through RLS algorithm, c_i is randomly selected from the trained array, K is obtained by the least absolute deviation which found through multiple trials. Besides, $\delta_k^2 = d^2/K, \delta_k$ is standard deviation, d is the maximum distance between selected centers.

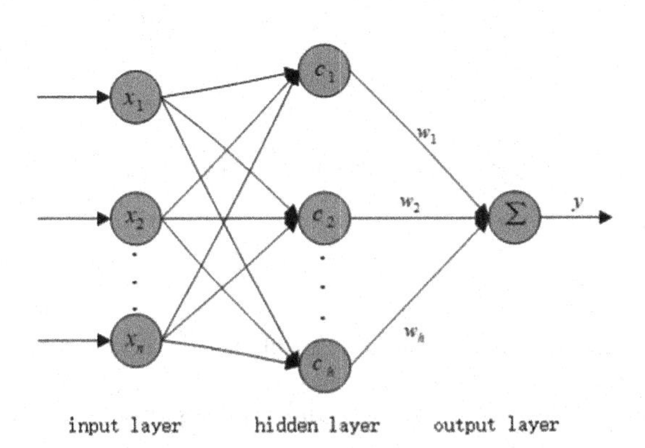

Fig. 1. The schematic diagram of neural network

The prediction absolute error can be expressed as:

$$\varepsilon(t) = \left|r(t) - \hat{r}(t)\right| \tag{14}$$

Binary detection:

$$r(t) \in \begin{cases} H_0 & \varepsilon(t) \geq \eta \\ H_1 & \varepsilon(t) < \eta \end{cases} \tag{15}$$

Where H_0 means there is no target, and H_1 is the hypothesis of the target exists.
The flowchart of the method of predicting and detecting targets based on neural network is shown in Fig. 2:

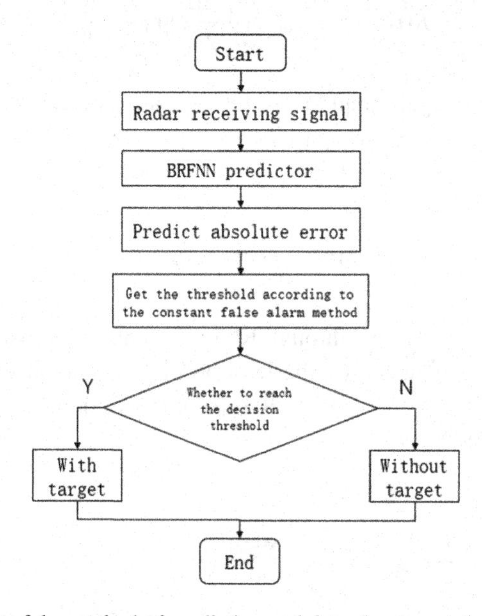

Fig. 2. The flowchart of the method of predicting and detecting targets based on neural network

The specific three steps of the detection method based on neural network target detection:

The first step: Determine the training sample set, and train the parameters of the neural network;
The second step: Use the single-step prediction method to predict the clutter state of the next moment;
The third step: Obtain the detection threshold according to the constant false alarm rate method, then compare the prediction absolute error with the detection threshold to get the target detection result.

4 Simulation

This experiment uses FM signal as the interference signal, collect two sets of FM signal data, with 2000 training samples and 2000 test samples. The Doppler frequency is 80 Hz, and the length of the time-series is 64 and 128 respectively. Perform network training to get the RBF predictor, then pass the signal $r(t)$ which obtained by the radar receiver through the RBF predictor. Use the simulation to compare the effects of different signal-to-noise ratios on the ability of neural network target detection.

4.1 The Influence of Sampling Frequency on the Mean Square Error of Neural Network Training

In order to study the influence of sampling frequency on the mean square error of neural network training, we will try to find the best sampling frequency through a large number of experiments. Assuming that the interference signal is an FM signal, the neural network parameter c_i is randomly selected from the trained array, parameter K is obtained through multiple experiments to find the smallest prediction absolute error, δ_k^2 is calculated by the formula $\delta_k^2 = d^2/K$. Then selected a set of FM interference data with a sequence length of 64 to train the RBF network, and the time length is 30 s. When the sampling frequency is set to 2 Hz, 3 Hz, 4 Hz, the mean square error that varies with the number of samples can be obtained, as shown in Fig. 3.

According to Fig. 3, we can get that when the sampling frequency is 2 Hz and 3 Hz respectively, as the number of samples increases, the mean square error gradually decreases, gradually approaching a certain order of magnitude 10^{-16} and 10^{-17} respectively. When the sampling frequency is 4 Hz, at the beginning, as the number of samples increase, the mean square error gradually decreases but still large. When the number of samples reaches a certain value, the mean square error tends to increase for a short time. This is because the sample data has great randomness. If the sample is too small, the training requirements cannot be met, and the prediction model cannot fully reflect the distribution characteristics of the interference signal. Similarly, if there are too many samples, the mean square error of the sampling frequency of 4 Hz in Fig. 3

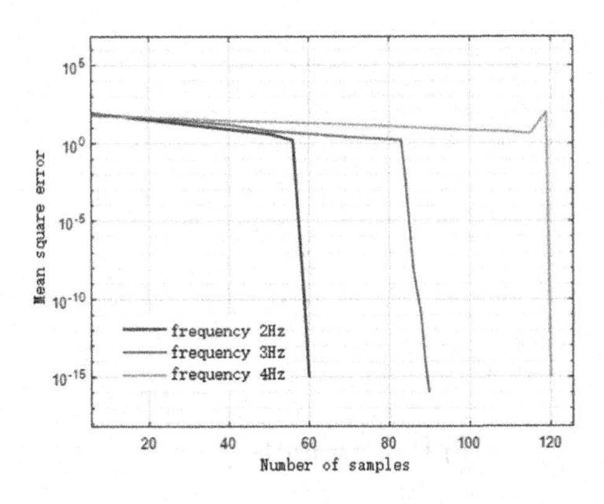

Fig. 3. Mean square error curve under different sampling frequencies

will become larger in a short time, and cannot achieve better prediction. Therefore, a large number of simulation experiments are needed to find that the sampling frequency is 3 Hz, the number of training samples is moderate, and the mean square error is also approaching 0.

4.2 Target Detection Based on RBF Neural Network

Use FM signal as interference signal to study the two cases where the length of the data sequence is 128 and 256 in a single scan, under the condition that the false alarm probability $P_f = 10^{-13}$, f_d = 120 Hz, and the sampling frequency is 3 Hz. Pass the signal through the RBF network predictor for simulation experiment, then compare the prediction result with the traditional CFAR detection performance. The target detection probability of the signal-to-noise ratio (SNR) of different sequence lengths is shown in the Fig. 4.

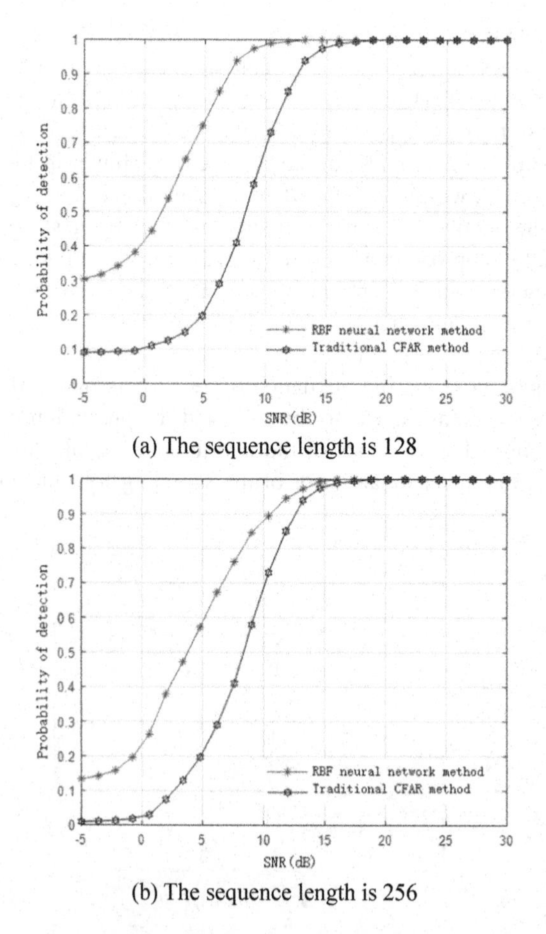

Fig. 4. Target detection probability curves of different sequence lengths

It can be seen from Fig. 4(a)–(b) that when the SNR is low, the detection performance of the method based on neural network is better than the traditional CFAR method in the FM interference environment. As the SNR increases, the detection probability continues to increase. When the SNR increases to 6, the detection probability can reach more than 90%. When the signal-to-noise ratio is low, the detection probability of the longer sequence length is low. As the SNR increases, the detection probability of the two methods continues to increase, and eventually approach 1.

5 Conclusion

We studied the target detection method based on neural network and the constant false alarm target detection method through theory and simulation. And we take the FM signal as the interference signal, and study to use the prediction method based on neural network to predict the modulated interference signal. The simulation result shows that this method can effectively improve the radar's detection probability of the target under the modulation interference environment, and the detection performance is better than the traditional CFAR method.

References

1. Li, J., Stoica, P.: MIMO Radar Signal Processing. Wiley, Hoboken (2009)
2. Nagle, D.T., Saniie, J.: Performance analysis of linearly combined order statistic CFAR detectors. IEEE Trans. Aerosp. Electron. Syst. **31**, 522–533 (1995)
3. Xing, H., Xu, W.: The neural networks method for detecting weak signals under chaotic background. Acta Phys. Sin. **56**(7), 3771–3776 (2007)
4. Si, W., Tong, N., Wang, Q.: Weak targets detection research under sea clutter background. J. Signal Process. **01**, 106–111 (2014)
5. Harpham, C., Dawson, C.W.: The effect of different basis function on a radial basis function network for time series prediction: a comparative study. Neurocomputing **69**, 2161–2170 (2006)
6. Hennessey, G., Leung, H., Yip, P.C.: Sea-clutter modeling using a radial-basis-function neural network. IEEE J. Oceanic Eng. **26**(3), 358–372 (2001)
7. Xie, N., Leung, H., Chan, H.: A multiple-model prediction approach for sea clutter modeling. IEEE Trans. Geosci. Remote Sens. **41**(6), 1491–1502 (2003)
8. Haykin, S., Thomson, D.J.: Signal detection in a non-stationary environment reformulated as an adaptive pattern classification problem. Proc. IEEE **86**(11), 2325–2344 (1998)
9. Zheng, H., Xing, H., Xu, W.: Detection of weak signal embedded in chaotic background using echo state network. J. Signal Process. **03**, 336–345 (2015)
10. Zhao, F.: Detection and Extraction of Weak Target in Sea Clutter Based on RBF. Jilin University, Changchun (2013)
11. Zheng, K., Zhu, X., Wang, H.: Suppression effects analysis of DRFM jamming on radar CFAR detection. Electron. Inf. Warfare Technol. (01), (2016)
12. Wang, J., Hou, R.: CFAR detection algorithm of electronic reconnaissance system in frequency domain. Shipboard Electron. Countermeas. **40**(2), 60–63 (2017)

Ground Clutter Suppression Method for Three-Coordinate Air Search Radar Based on Adaptive Processing in Beam Domain

Lei Yu(✉), Yueyu Guo, Jiahao Liu, and Yinsheng Wei

Harbin Institute of Technology, Harbin, China
yu.lei@hit.edu.cn

Abstract. A novel ground clutter suppression method is proposed for the three-coordinate air search radar. According to the theory of statistical optimal beamforming, the optimal solution is given in the background of clutter and noise. The minimum variance distortionless response (MVDR) algorithm is used to process the received signals. When estimating the covariance matrix, in addition to the traditional selection of samples in the range domain, the samples in different scans can be selected based on cognition. In the case that the received data is the beam domain data and there is no information about the geometric configuration of the array, the beam domain response of the obvious target can be used as the target steering vector. Then the optimal weighted vector can be calculated to filter the received data. The processing results of the real data of the three-coordinate radar show the effectiveness of the proposed method in ground clutter suppression.

Keywords: Adaptive processing · MVDR · Clutter covariance matrix · Target steering vector

1 Introduction

The algorithm of clutter suppression in beam domain based on adaptive processing is adopted for various radar systems [1]. As the backbone radar in modern air defense system, three-coordinate radar can detect and track multiple targets in designated air-space, and measure the range, azimuth and altitude of targets at the same time [2, 3]. The ground clutter masks the low-altitude aircraft in the received signal of three-coordinate radar [4]. Therefore, by suppressing the clutter, the target detection capability of three-coordinate radar can be greatly improved [5].

Compared with the ground clutter, the target velocity is generally much larger, their Doppler difference can be used for clutter suppression [6]. Moving target indication (MTI) and moving target detection (MTD) are commonly used for ground clutter suppression. The MTI technology is equivalent to using a high-pass filter. By weighting and summing data of successive pulses, the ground clutter is mitigated. Thus the signal-to-clutter ratio and the probability of target detection can be improved [7, 8]. In the MTD technology, by covering the whole repetition frequency range through a set

Q. Guo et al. (Eds.): WiSATS 2021, LNICST 410, pp. 770–779, 2022.
https://doi.org/10.1007/978-3-030-93398-2_67

of band-pass filters, the target energy can be accumulated, while the clutter is suppressed [9, 10].

In the three-coordinate radar, the MTI and MTD technologies are generally used for clutter suppression, but spatial clutter suppression algorithms are rarely used. In this paper, we proposed a novel clutter suppression method in the spatial domain. Spatial adaptive processing is a practical signal processing technology, which has wide applications in interference suppression, clutter suppression and other fields [11, 12]. In this proposed method, the spatial covariance matrix of the clutter are first estimated. Then by using the minimum variance distortionless response (MVDR) algorithm, the optimal weighting vector is obtained. Since the array steering vector is unavailable in the beam domain data, the target steering vector in the beam domain is used to replace the conventional array steering vector [13, 14]. At last, by weighting the received signal in the beam domain using the obtained optimal weighting factor, the signal-to-clutter-plus-noise ratio (SCNR) of the received signal can be maximized. Thus the clutter is suppressed effectively.

2 Adaptive Processing Theory

2.1 Adaptive Processing in Array Element Domain

The received signal of a uniform linear array of M elements can be expressed as

$$\mathbf{x} = \mathbf{a}_0 s_0 + \mathbf{x}_{c+n} \tag{1}$$

where $\mathbf{a}_0 \in \mathbb{C}^{M \times 1}$ is the spatial steering vector of the target signal, $s_0 \in \mathbb{C}^{1 \times 1}$ is the received target signal, and $\mathbf{x}_{c+n} \in \mathbb{C}^{M \times 1}$ is the sum vector of clutter and noise.

The target steering vector $\mathbf{a}_0 \in \mathbb{C}^{M \times 1}$ can be expressed as

$$\mathbf{a}_0(\theta) = \left[1, \exp\left(j2\pi \frac{d\sin(\theta)}{\lambda}\right), \cdots, \exp\left(j2\pi(M-1)\frac{d\sin(\theta)}{\lambda}\right)\right]^T \tag{2}$$

where d is the distance between array elements, λ is the wavelength, θ is the desired elevation angle of the target.

The received signal is filtered by a spatial weighted vector $\mathbf{w} = [w_1, w_2, \cdots, w_M]^T$, and the output $y \in \mathbb{C}^{1 \times 1}$ is

$$y = \mathbf{w}^H \mathbf{x} \tag{3}$$

By using the well known MVDR algorithm, the optimal output signal to clutter and noise ratio (SCNR) can be achieved and the optimal solution $\mathbf{w}_{opt}$ is given as

$$\mathbf{w}_{opt} = \frac{\mathbf{R}_{c+n}^{-1}\mathbf{a}_0}{\mathbf{a}_0^H \mathbf{R}_{c+n}^{-1}\mathbf{a}_0} \tag{4}$$

The clutter covariance matrix $\mathbf{R}_{c+n} = E\{\mathbf{x}_{c+n}\mathbf{x}_{c+n}^H\} \in \mathbb{C}^{M \times M}$ has to be estimated from the clutter samples. The maximum likelihood estimation under the independent identically distributed Gaussian samples is

$$\hat{\mathbf{R}}_{c+n} = \frac{1}{K_\Gamma}\sum_{i\in\Gamma}\mathbf{x}_i\mathbf{x}_i^H \quad (5)$$

where $\mathbf{x}_i \in \mathbb{C}^{M\times 1}$ is the clutter sample, i is the index of range cell, Γ is the set of sample index, K_Γ is the number of samples Γ. The samples are selected in the range domain. The selected samples should have similar spatial distribution. In order to prevent the target from being eliminated, the clutter samples are selected by setting protection cells on the target.

2.2 Adaptive Processing in Beam Domain

The adaptive processing can also be processed in beam domain for dimension reduction. The array element domain signal is transformed to the beam domain by a linear transformation such as

$$\mathbf{z} = \mathbf{T}\mathbf{x} \quad (6)$$

where $\mathbf{z} \in \mathbb{C}^{N\times 1}$ is the beam domain signal, and $\mathbf{T} \in \mathbb{C}^{N\times M}$ is the transforming matrix used for conventional beamforming.

$$\mathbf{T} = [\mathbf{a}_0(\theta_1), \cdots, \mathbf{a}_0(\theta_N)]^T \quad (7)$$

where $\mathbf{a}_0(\theta_1), \cdots, \mathbf{a}_0(\theta_N)$ are the N beamforming vectors of interest, and $N < M$ is generally used for dimension reduction.

Then the filter output $\bar{y} \in \mathbb{C}^{1\times 1}$ is

$$\bar{y} = \overline{\mathbf{w}}^H\mathbf{z} \quad (8)$$

The optimal solution $\overline{\mathbf{w}}_{opt} = [w_1, w_2, \cdots, w_N]^T$ of the weighting vector can be obtained by

$$\overline{\mathbf{w}}_{opt} = \frac{\overline{\mathbf{R}}_{c+n}^{-1}\bar{\mathbf{a}}_0}{\bar{\mathbf{a}}_0^H\overline{\mathbf{R}}_{c+n}^{-1}\bar{\mathbf{a}}_0} \quad (9)$$

Accordingly, the estimated clutter covariance matrix $\overline{\mathbf{R}}_{c+n} = E\{\mathbf{z}_{c+n}\mathbf{z}_{c+n}^H\} \in \mathbb{C}^{N\times N}$ in the beam domain is

$$\hat{\overline{\mathbf{R}}}_{c+n} = \frac{1}{K_\Gamma}\sum_{i\in\Gamma}\mathbf{z}_i\mathbf{z}_i^H \quad (10)$$

where $\mathbf{z}_i \in \mathbb{C}^{M \times 1}$ is the clutter sample and i is the index of range cell. The method of selecting clutter samples is similar to the process of the array element domain, and then the clutter covariance matrix of the beam domain is calculated.

After the array element domain signal is linearly transformed into the beam domain, the steering vector $\overline{\mathbf{a}}_0 \in \mathbb{C}^{N \times 1}$ will also change accordingly.

$$\overline{\mathbf{a}}_0 = \mathbf{T}\mathbf{a}_0 \tag{11}$$

According to the array geometry and the beam direction information in the array element domain, the transformation matrix $\mathbf{T}$ for conventional beamforming is calculated. By substituting the above expressions, the clutter covariance matrix, target steering vector and spatial weighted vector in beam domain can be obtained, and then the beam domain signal can be filtered in linear spatial domain.

3 Implementation Methods of Adaptive Processing in Beam Domain

3.1 Adaptive Processing in the Same Scan

When the received data are beam domain data and lack of array geometry information, the transform matrix $\mathbf{T}$ is unknown. The clutter suppression can still be carried out directly in the beam domain data. In this case, the estimation of the clutter covariance matrix $\overline{\mathbf{R}}_{c+n}$ and the steering vector $\overline{\mathbf{a}}_0$ in the beam domain are necessary.

In this paper, the data of real target in the beam domain under the background of noise is considered as the steering vector in the beam domain. In a pulse, we first find the range cell where the maximum value of the target is located in the range-beam domain. Then the data of the beam domain of this range cell is extracted to form a vector. The target steering vector in the beam domain is obtained by normalizing the maximum value of the vector, as shown in Fig. 1. The coordinates of the maximum value are (1423, 6). The beam domain vector corresponding to the 1423 range cell is selected as the spatial target steering vector.

In order to suppress clutter, it is necessary to obtain the statistical characteristics of clutter spatial distribution. A certain amount of clutter samples with the same or similar spatial distribution are selected for covariance matrix estimation. In addition to the homogeneity of samples, it is also required that the samples should not contain the target signal component. The schematic diagram of clutter sample selection in the range domain is shown in Fig. 2. Several protection cells are set near the cell under test, and then enough cells are selected on both sides of the protection cells as clutter samples. Then the clutter covariance matrix in the beam domain can be calculated by using the selected samples.

Next, the clutter covariance matrix and the target steering vector are used to calculate the weighting vector of the beam domain, and the beam domain signal can be filtered in linear spatial domain. The flowchart of the proposed clutter suppression method in the one scan is shown in Fig. 3.

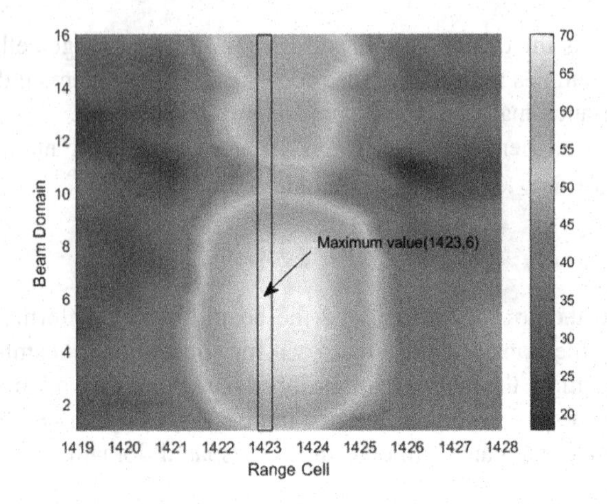

Fig. 1. Range-beam pattern of real target under noisy background

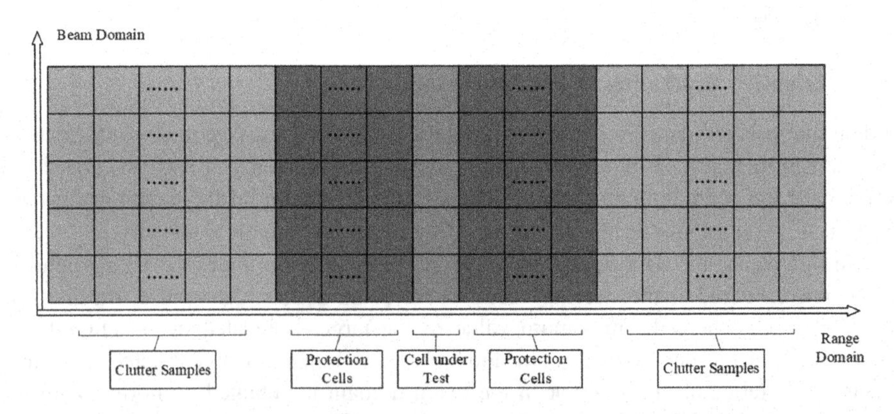

Fig. 2. Schematic diagram of clutter sample selection in the range domain

3.2 Adaptive Processing Based on Cognition

Generally speaking, modern radar systems mainly use the received data for processing, while ignoring external targets and environmental information. Therefore, the detection performance improvement faces a major bottleneck in the complex geographical and electromagnetic environment. To solve this problem, the concept of cognitive radar has been proposed in recent years. Cognitive radar has the ability of online perception and memory of environment and target information. Combined with the prior knowledge, it can optimize the transmitting and receiving processing mode in real time. By achieving the optimal match with the target and environment, target detection performance can be improved.

In the application of three-coordinate radar, we assume that the radar's location is stationary, and the received clutter is mainly ground clutter. The clutter information can be recorded and learned by multiple scans after the radar has been turned on. Such

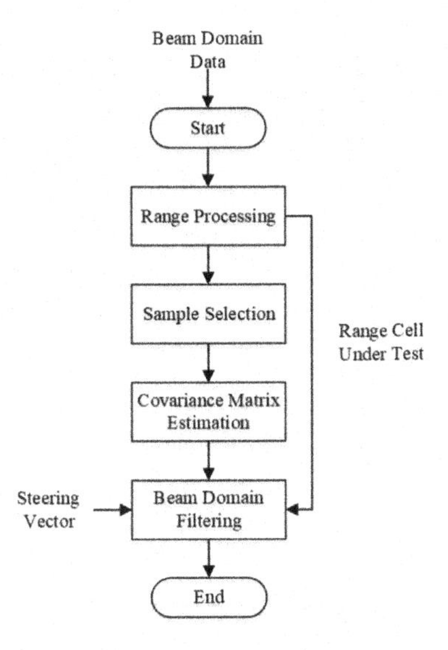

Fig. 3. Flowchart of proposed clutter suppression method in the one scan

information can then be used to assist in clutter suppression in the subsequent target detection. In the process of target detection, the information in clutter database can be updated periodically to adapt to the slow changing environment.

Specifically, for the beam domain adaptive filtering, we can use the multi-scans data of the same range cell under test as samples to estimate the clutter covariance matrix. When the adaptive filtering performance of different range cells is degraded due to the change of terrain, we can choose the historical data as samples for clutter suppression. The selection of clutter samples is shown in Fig. 4. According to the position of the cell under test, the data of the cell under test and its adjacent range cells in the previous scans are selected as the clutter samples. The target steering vector is the same as before, and the overall processing flow is shown in Fig. 5.

4 Experiments and Results

In this section, the real data are used to illustrate the performance of the proposed methods. The key parameters are as follows: the radar transmitted signal is LFM pulse signal, which is scanned mechanically at 360% in azimuth and electrically in elevation domain, and 16 beams are formed at 0–25%. After beamforming, the data format is beam-range-pulse, and then the beam domain adaptive processing is carried out by the proposed method.

Starting with the received beam domain signal, clutter samples are selected in the range domain of the same scan for covariance estimation. According to the beam domain information of the real target in the noise background, the steering vector is referenced, and then the adaptive filtering processing in the beam domain is completed. The results before and after clutter suppressing are shown in Fig. 6. The covariance

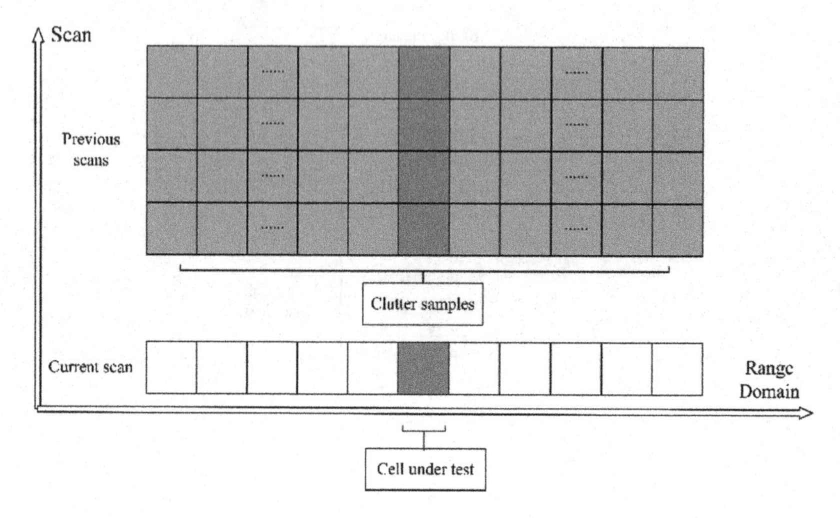

Fig. 4. Schematic diagram of clutter sample selection in previous scans

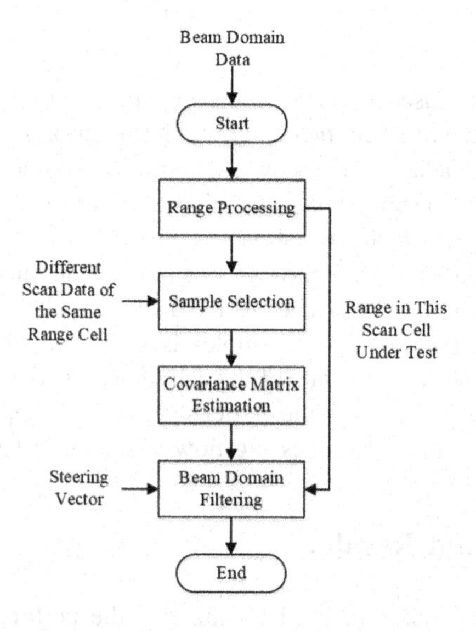

Fig. 5. Flowchart of proposed clutter suppression method based on cognition

matrix estimation is completed by range domain sampling based on cognition, and the results before and after clutter suppression are shown in Fig. 7.

It can be seen from Fig. 6 that the adaptive processing in beam domain by sampling in the range domain of the same scan can effectively suppress the clutter, and the SCNR improvement reaches 25.80 dB. It can also be seen from Fig. 7 that the adaptive processing in beam domain based on cognition has a limited suppression degree to clutter, and the improvement factor reaches 6.96 dB. The clutter suppression performance is obviously not as good as the former.

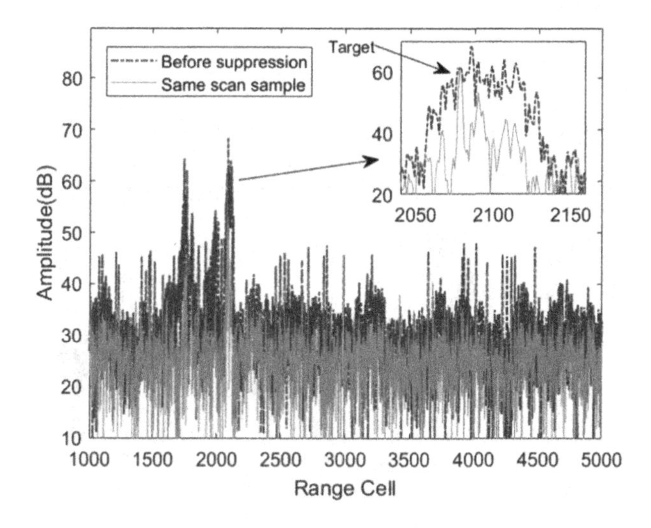

Fig. 6. Clutter suppression based on samples in the same scan

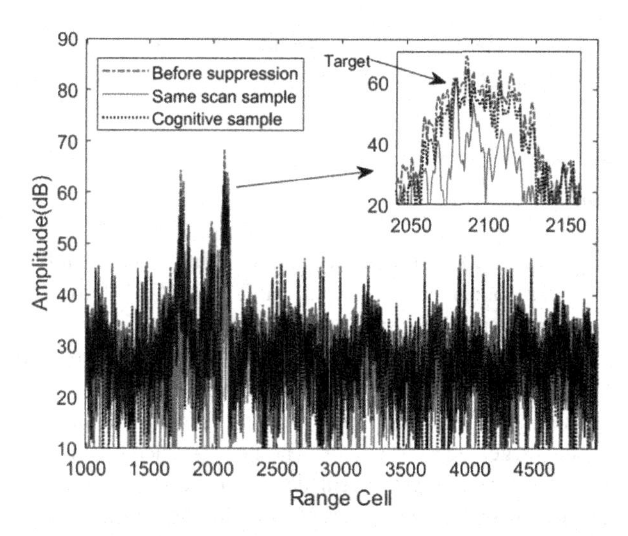

Fig. 7. Clutter suppression based on cognitive samples

Since the radar system may have certain errors when scanning in different directions, the azimuth of two adjacent scans is not completely aligned. This will increase the difference of data in different scans. In the ideal case of the system, the multiple measurement results of different static clutter in the same resolution cell should have good independent and identically distributed sample properties. In such ideal case, the suppression results should be much better. According to the simulated ideal cognitive data, the processing result is shown in Fig. 8.

It can be seen from Fig. 8 that the clutter suppression performance of adaptive processing in beam domain based on ideal cognitive data near target range cell is better

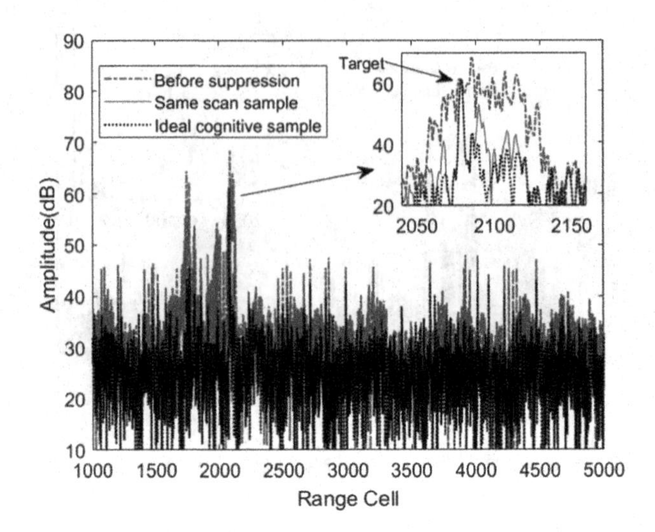

Fig. 8. Clutter suppression is based on ideal cognitive samples

than that of the sampling method in the range domain of the same scan. he suppression performance of other range cells is the same, and the calculated improvement factor is 27.08 dB.

5 Conclusion

In this paper, a spatial clutter suppression method based on adaptive processing is proposed for the three-coordinate radar system. The MVDR algorithm framework is adopted to clutter suppression in beam domain. When estimating the covariance matrix, in addition to the selection of samples in the range domain, we also proposed a method of selecting samples in different scans based on cognition. In the case that the received data is beam domain data, we proposed to use the beam domain response of the obvious target as the target steering vector. The performance of the methods on clutter suppression are verified by using the real data and the improvement factors are calculated. In addition, this method can be cascaded with the traditional slow-time domain clutter suppression algorithms to further improve the performance.

References

1. Li, Z.Z., Yang, G., Jiang, B., Li, H.T., Xu, J.F.: Performance simulation and experimental verification of fast broadband MVDR algorithm. In: 2019 IEEE 11th International Conference on Communication Software and Networks (ICCSN), China, pp. 616–620 (2019)
2. Shu, T., Liu, X.Z., Yu, W.X.: Target height finding in narrowband ground-based 3D surveillance radar using beamspace approach. In: 2009 IEEE Radar Conference, USA, pp. 1–6 (2009)

3. Seong-Taek, P., Jang, G.L.: Improved Kalman filter design for three-dimensional radar tracking. IEEE Trans. Aerosp. Electron. Syst. **37**(2), 727–739 (2001)
4. Amer, M., et al.: Statistical analysis of measured high resolution land clutter at X-band and clutter simulation. In: 2015 European Radar Conference (EuRAD), France, pp. 105–108 (2015)
5. Chen, J.W., Xu, D.H., Liu, B.Q.: Performance analysis of meter band radar height-finding approach for low-angle tracking. In: 2006 IEEE Antennas and Propagation Society International Symposium, USA, pp. 1157–1160 (2006)
6. Liu, Z., Dominic, K.C.H., Xu, X.Q., Yang, J.Y.: Moving target indication using deep convolutional neural network. **6**, 65651–65660 (2018)
7. Chang, C.H., Wong, C.M., Liu, W.X., Fu, J.S.: Radar MTI/MTD implementation and performance. In: ICMMT 2000. 2000 2nd International Conference on Microwave and Millimeter Wave Technology Proceedings (Cat. No.00EX364), China, pp. 674–678 (2000)
8. Sun, Q., Zhang, Q.L., Sun, Y.: Evaluation to the performance of MTI radar anti-passive jamming. In: IET 3rd International Conference on Wireless, Mobile and Multimedia Networks (ICWMNN 2010), China, pp. 282–284 (2010)
9. Dai, Q.N., Tian, Y.H.: Optimal design of MTD filter based on FIR. In: 2019 IEEE International Conference on Signal, Information and Data Processing (ICSIDP), China, pp. 1–4 (2019)
10. Rytel-Andrianik, R.: Design of MTD filters with maximal average improvement factor. In: 2014 15th International Radar Symposium (IRS), Poland, pp. 1–5 (2014)
11. Yang, L.S., Matthew, R.M., Romain, C.: High-dimensional MVDR beamforming: optimized solutions based on spiked random matrix models. **66**(7), 1933–1947 (2018)
12. Ibrahim, E.N., Khalil, E.: Improve the robustness of MVDR beamforming method based on steering vector estimation and sparse constraint. In: 2019 International Symposium on Advanced Electrical and Communication Technologies (ISAECT), Italy, pp. 1–5 (2019)
13. Zhang, L., Liu, W., Yu, L.: Statistical analysis of the finite sample forward-backward MVDR beamformer. In: 2011 IEEE Statistical Signal Processing Workshop (SSP), France, pp. 421–424 (2011)
14. Qian, R.R., Sellathurai, M., Wilcox, D.: A study on MVDR beamforming applied to an ESPAR Antenna. **22**(1), 67–70 (2015)

Integrated Radar and Communication System Design Based on Constant Envelope Waveform

Hao Lu[1], Yu Zhou[2], Yue Liu[1(✉)], Rui Li[1], and Ning Cao[1]

[1] Hohai University, Nanjing 210098, China
luhao@hhu.edu.cn
[2] State Grid Jiangsu Electric Power Co., Ltd.
Marketing Service Center, Nanjing, China

Abstract. To perform both radar and communication functions for multiuser and improve the effectiveness of power resources, a novel integrated system based on a constant envelope waveform is proposed. In this system, first, the integrated waveform combining discrete Fourier transform spread orthogonal frequency-division multiplexing (DFT-s OFDM) and continuous phase modulation (CPM), referred as DFT-s OFDM-CPM, is generated to achieve 0 dB peak to average power ratio (PAPR). Second, the detailed communication receiver and radar processing is provided. Finally, the experiment verifies the effectiveness of the proposed integrated system and its superiority to the widely used constant envelope OFDM (CE-OFDM) integrated waveform.

Keywords: Radar-communication integration · Waveform design · Continuous phase modulation · Orthogonal frequency division multiplexing · Ant lion optimizer

1 Introduction

The rapid increment of user's quantity and quality put forward a demanding requirement for the radio frequency spectrum. The concept of radar-communication integration has been seen as a promising solution to this radio frequency congestion by allowing radar and communication to share the same resources and operate in a single platform [1,2]. Among various radar and communication co-existence schemes, incorporating information bearing symbols into radar emission to form an integrated waveform is an effective approach. It can perform radar and communication functions at the same time without interference [3,4].

Currently, with more degrees of freedom in waveform synthesis and flexibility inherent to orthogonal multiple access, multicarrier waveform design represented

Q. Guo et al. (Eds.): WiSATS 2021, LNICST 410, pp. 780–789, 2022.
https://doi.org/10.1007/978-3-030-93398-2_68

by orthogonal frequency division multiplexing (OFDM) is gaining interest. Multiobjective optimal waveform design for OFDM integrated radar and communication systems is presented in [5]. The optimization of radar performance within the structure imposed by a coded OFDM format required to achieve an acceptable communication link is considered in [6]. Power minimization-based robust OFDM radar waveform design for radar and communication systems in coexistence is considered in [7]. An adaptive orthogonal frequency division multiplexing integrated radar and communications waveform design method is proposed in [8].

The above research has greatly promoted the development of joint radar-communication systems based on OFDM, but further study is still needed. Since OFDM signal is the sum of many subcarriers component via inverse discrete Fourier transform (IDFT), it has a high peak to average power ratio (PAPR) problem. This may result in the signal distortion and spectrum expansion. Besides, high power amplifier in radar transmitter may not be driven into saturation and then the power efficiency will be reduced [9]. Therefore, constant envelope OFDM (CE-OFDM) is proposed in [10] using OFDM signal to modulate phase. Nevertheless, this can not maintain the usage of subcarriers' orthogonality which means that some advantages of OFDM such as low complexity frequency equalization, frequency multiplexing and so on cannot be used.

In this letter, to address the high PAPR problem, a novel radar-communication integrated system is proposed. Advantages of this system include three points. First, discrete Fourier transform spread orthogonal frequency-division multiplexing (DFT-s OFDM) and continuous phase modulation (CPM) are combined to generate an integrated waveform, DFT-s OFDM-CPM, with constant envelope. This operation allows the power amplifier in radar transmitter working in a saturated region, gaining farther effective working distance and better performance. Second, the spread spectrum and interleaved frequency division multiple access (I-FDMA) subcarrier mapping methods enable the proposed system to allow multiple users to work simultaneously and due to the orthogonality between the subcarriers, there will be no interference between users. Third, to eliminate the dependence of radar detection performance on the transmitted symbol, radar processor in frequency domain is carried out. Simulation results show that the proposed system can realize the superior performance both in radar and communication.

This paper is organized as follows. System framework and workflow are introduced in Sect. 2. Simulation results are given in Sect. 3 and Sect. 4 concludes this letter.

2 System Model

2.1 Basic Framework of Integrated System

Considering a traffic controlling scenario (Fig. 1), there is a control center containing U controllers and each controller manages one or more ship targets.

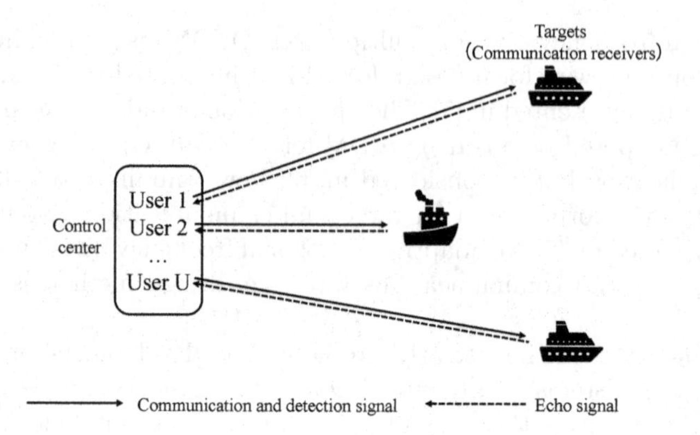

Fig. 1. Typical operational scenario.

U managers need to send commands to the targets and keep track of the targets' states including their position and speed. Each target, i.e., communication receiver is asked to decode commands and preform related operations after receiving the signal.

Firstly, the controller u sends integrated signal $x_u(t)$ to realize information transmission and targets tracking,

$$x_u(t) = \exp(j2\pi f_c t) \sum_{\mu=0}^{N_s-1} \sum_{n=0}^{K-1} S_{Tu}(\mu, n) \exp(j2\pi f_n t) \cdot \mathrm{rect}(\frac{t-\mu T_s}{T_s}) \tag{1}$$

where f_c is the carrier frequency; N_s is the number of OFDM symbols contained in each pulse; K is the subcarriers number contained in one time block; T_s is the OFDM symbol duration; $f_n = n\Delta f$, $\Delta f = \frac{1}{T_s}$ is the frequency interval between subcarrier. $S_{Tu}(\mu, n)$ denotes the code in the nth subcarrier of the μth symbol. We consider that a constant envelope waveform may be achieved through replacing these conventional data symbols with samples from a CPM waveform in time domain. The specific operations is given in the next subsection.

2.2 Generation of Data Symbols

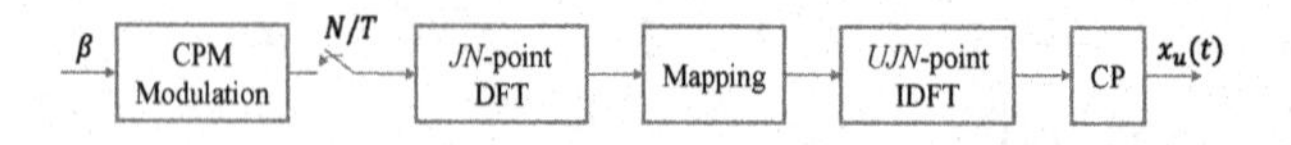

Fig. 2. Signal model.

For convenience, we discuss the generation of data symbols of user u in one time block. The signal model is shown in Fig. 2. Firstly, we give the CPM signal definition [11],

$$s(t, \beta) = \exp(j\phi(t; \beta)). \tag{2}$$

$\phi(t; \beta)$ is the signal phase,

$$\phi(t; \beta) = 2\pi h \sum_i \beta_i q(t - iT), \tag{3}$$

where h is the modulation index, information symbols $\beta_i \in \{\pm 1, \cdots, \pm(M-1)\}$ and M is modulation order. $q(t)$ is the phase shaping function. The derivative of $q(t)$ is called the frequency shaping function and is denoted by $g(t)$. The function $g(t)$ has a support of L symbol intervals and an underlying area of 1/2. This paper considers the full response CPM with $L = 1$.

We observe $s(t; \beta)$ over JT seconds (J is an integer) and sample at rate $f_{sa} = N/T$ to generate the vector of signal samples $\boldsymbol{s} = [s_0 \cdots s_{JN-1}]^{\mathrm{T}} = [s(t = \frac{0N}{T}; \beta) \cdots s(t = \frac{(JN-1)N}{T}; \beta)]^{\mathrm{T}}$. Then, we pass the sampled output into JN-point DFT,

$$S_i = \sum_{n=0}^{JN-1} s_n \exp(\frac{-j2\pi in}{NJ}), \tag{4}$$

where $i = 1, \cdots, JN - 1$ denotes the discrete frequency index. The DFT outputs will be treated as the data symbols in each subcarrier of OFDM. The key observation is this, if the time domain symbols at DFT input are constant modulus, then the time-domain symbols at the IDFT output will also be constant modulus.

Each user obtains the data symbols according to the above steps and maps them into subcarriers of IDFT using I-FDMA approach [12], where the subcarriers are equally spaced over the entire system bandwidth,

$$S_{Tu}(1, n) = \begin{cases} S_i, & n = u + iU \\ 0, & \text{otherwise} \end{cases} \tag{5}$$

where $n = 1, \cdots, K$, $K = UJN$. I-FDMA mapping maintains the input time symbols in each sample and has lower PAPR [13]. Besides, due to the orthogonality between subcarriers, the interference between different users can be avoided. Therefor, the transmitted waveform (1) is expected to own constant envelope in each time block and be friendly toward the use of nonlinear radar transmitter. We refer this waveform as DFT-s OFDM-CPM.

2.3 Communication Receiver

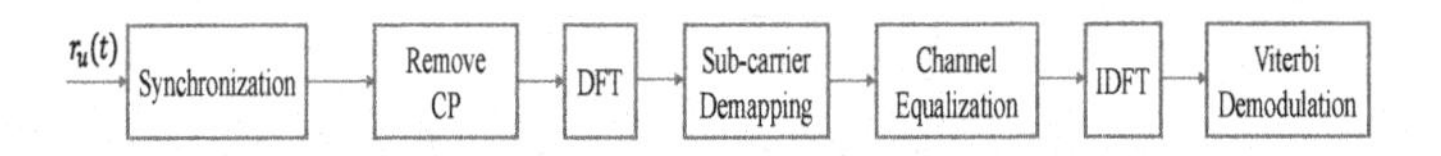

Fig. 3. Communication receiver.

Each target is seen as a communication receiver to receive and decode the controller's commands. As shown in Fig. 3, they receive the integrated signal

$$r_u(t) = h(t) * x_u(t)\exp(j2\pi\epsilon t) + w(t), \tag{6}$$

where ρ and ϵ represent time and frequency offset, respectively. $h(t)$ represents the channel impulse response and $w(t)$ is awgn. The receiving end first performs symbol positioning and frequency compensation through the CP-based ML maximum likelihood synchronization algorithm. Then remove the CP and perform DFT and inverse mapping to obtain the symbol of user u in the frequency domain. If the frequency selection channel is used, the receiver needs to equalize the received signal. Here, the receiver can use MMSE frequency domain equalization with lower complexity. The equalized signal is transformed into the time domain by IDFT, and finally the famous VA algorithm [11] is used to obtain the transmitted CPM symbol.

2.4 Radar Processing in Frequency Domain

Assuming a moving target managed by user u is at the range R with relative velocity v, the user u receives the target echo is,

$$\begin{aligned} y_u(t) &= \sum_{\mu=0}^{N_s-1}\sum_{n=0}^{K-1} S_{Ru}(\mu,n)\exp(j2\pi f_n t)\mathrm{rect}(\frac{t-\mu T_s}{T_s}) \\ &= \sum_{\mu=0}^{N_s-1}\sum_{n=0}^{K-1} S_{Tu}(\mu,n)\exp(j2\pi\mu f_d T_s) \\ &\cdot \exp(-j2\pi f_n\frac{2R}{c})\exp(j2\pi f_n t)\mathrm{rect}(\frac{t-\mu T-\frac{2R}{c}}{T_s}), \end{aligned} \tag{7}$$

where c is the speed of light and f_d denotes the Doppler frequency shift caused by the targets movement.

Since the information code on some subcarriers in each user's transmitted waveform is 0, the traditional radar detection method that relies on symbol correlation is no longer applicable. To eliminate the dependence of radar processing on the correlation of the transmitted information, we can compare the transmitted information $S_{Tu}(\mu,n)$ and the received information $S_{Ru}(\mu,n)$ at the output of the $y(t)$ de-multiplexer to obtain targets' states. The frequency domain channel transfer function is given by calculating an element-wise division,

$$\begin{aligned} D(\mu,n) &= \frac{S_{Ru}(\mu,n)}{S_{Tu}(\mu,n)} \\ &= \exp(j2\pi\mu f_d T_s)\exp(-j2\pi f_n\frac{2R}{c}). \end{aligned} \tag{8}$$

It is evident that the range and Doppler influences are orthogonal. We denote $k_R(n) = \exp(j2\pi f_n \frac{2R}{c})$ and $k_D(\mu) = \exp(j2\pi\mu f_d T_s)$. Note that we only operate on DFT-s OFDM-CPM subcarriers that contain data information. Assuming $U = 2$, range of target can be calculated through IDFT of $k_R(n)$,

$$r(p) = \mathrm{IDFT}(k_R(n)) = \frac{1}{JN}\sum_{n=0}^{JN-1} k_R(2n)\exp(j2\pi\frac{n}{JN}p), \tag{9}$$

with $p = 0, \cdots, JN-1$. A peak of $r(p)$ will occur at $p = \lfloor\frac{2R\Delta fK}{c}\rfloor$. The relative velocity can be solved in a similar method to $k_v(\mu)$ and a peak will occur in the $\lfloor\frac{2vf_cN_s}{\delta fc}\rfloor$.

3 Simulation Results

In this section, the performance of the integrated system in communication and detection is illustrated with the spectral efficiency (SE), bit error rate (BER) as well as radar range and velocity profile. Without loss of generality, we consider an integrated system operating with the parameter setting: carrier frequency f_c = 24 GHz; frequency interval Δf = 400 kHz; user number U = 2; symbols of each user J = 40; symbol number N_s = 5; sample number N_{sa} = 2; CP length N_{cp} = 20; frequency offset ϵ = 0.25; the modulation order M = 4; modulation index $h = 1/4$. We consider AWGN channel. For reference, performance of conventional OFDM modulated by 8PSK and the well known constant envelope OFDM (CE-OFDM) [10] by phase modulating OFDM are also assessed.

Firstly, We give the signal normalized amplitude comparison in Fig. 4. We can see that the DFT-s OFDM-CPM maintains the constant envelope as that of CE-OFDM while OFDM has a large amplitude variation.

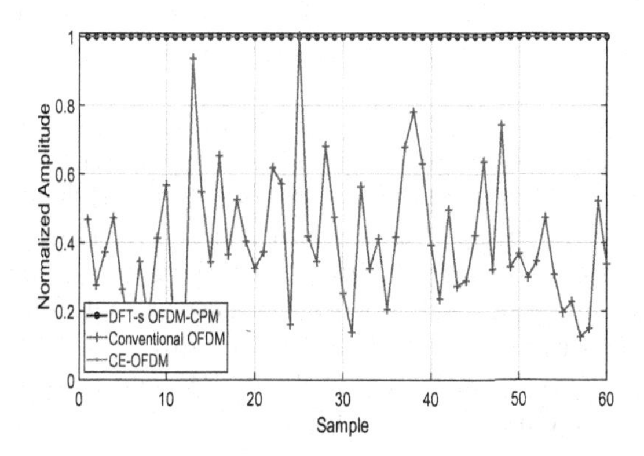

Fig. 4. Signal normalized amplitude comparison.

3.1 Spectral Efficiency

SE is defined as $SE = \frac{R_b}{W}$, where R_b is the bit rate and W is the signal occupied bandwidth. For one thing, we consider the bit rate of the three modulations. The bit rate of OFDM is $R_b = N_c \log_2 M/T(b/s)$. Since DFT-s OFDM-CPM needs to sample symbol of the original CPM signal with two points, the signal rate is $R_b/2$. In addition, in order to ensure that the input of the phase modulator is a real number, the input to the IDFT is a conjugate symmetric, zero-padded data vector, so its bit rate is less than $R_b/2$. For another, the bandwidth occupied by DFT-s OFDM-CPM and traditional OFDM is $W = N_c * \Delta f$. Because CE-OFDM requires phase modulation of the OFDM signal. Its bandwidth is $B = \max(2\pi h, W)$Hz. Therefore, when $h = 1/4$, the highest spectral efficiency is OFDM, DFT-s OFDM -CPM is higher than CE-OFDM. According to the parameter settings, the spectral efficiency of the three modulation methods is shown in Table 1.

Table 1. Spectral efficiency of the three modulations

Modulations	SE (bits/s/Hz)
DFT-s OFDM-CPM	1
Conventional OFDM	2
CE-OFDM	0.637

3.2 BER Performance Comparison

Here, BER performance comparison between the three modulation methods is provided in Fig. 5. First, under the premise of perfect frequency synchronization, we can see that the DFT-s OFDM-CPM has the best BER performance. The demodulation process depends on the demodulation of symbol in each subcarrier after a series of operations opposite to the transmitter. Thus, due to the symbol memory inherent in CPM, the BER performance of DFT-s OFDM-CPM is outstanding. Then, considering frequency shift synchronization. To maintain the constant envelope of DFT-s OFDM-CPM, CP-based ML synchronization [14] without additional training sequences is introduced here. The Gaussian weighted average of the estimated frequency shift value of each symbol is used as the final frequency shift estimation. Among the three methods, CE-OFDM is least sensitive to imperfect frequency synchronization due to its phase modulation and unwrapping operations. Both DFT-s OFDM-CPM and traditional OFDM are sensitive to carrier frequency offset (CFO).

3.3 Radar Detection Performance

Considering one controller manages two targets with distance 28 m, velocity 650 m/s and distance 33 m, velocity 350 m/s, respectively. The radar range profile

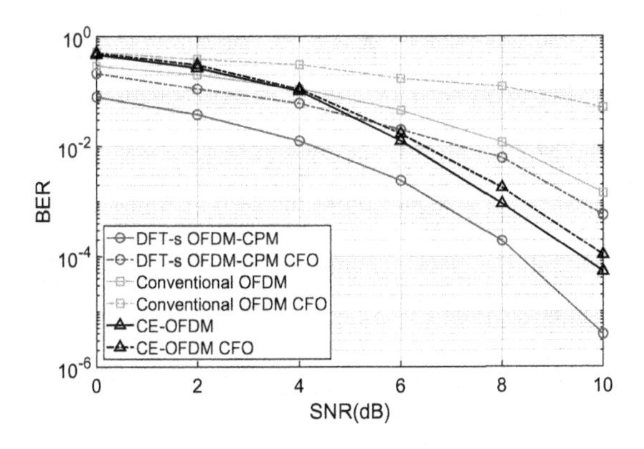

Fig. 5. BER comparison under AWGN channel.

comparison is given in Fig. 6. Assuming the same amount of transmitted data symbols, we can see that only the two main peaks in 28m and 33m of the DFT-s OFDM-CPM are prominent, the middle depression is about 20dB, and the side lobe is the lowest. It can well distinguish targets 5m apart. The range resolution of the radar is inversely related to the number of subcarriers. Considering the number of transmitted data symbols is J, the number of subcarriers of DFT-s OFDM-CPM is UJN, CE-OFDM is $2J + 2$, and OFDM is J. Therefore, the radar range resolution of DFT-s OFDM-CPM is optimal.

Finally, the velocity profile comparison is given in Fig. 7. The proposed system exhibits the similar velocity resolution performance to that of conventional OFDM with about 13 dB drop between two targets. This paper focuses on distance resolution. Different application scenarios can adjust parameters according to their own needs to obtain the required resolution.

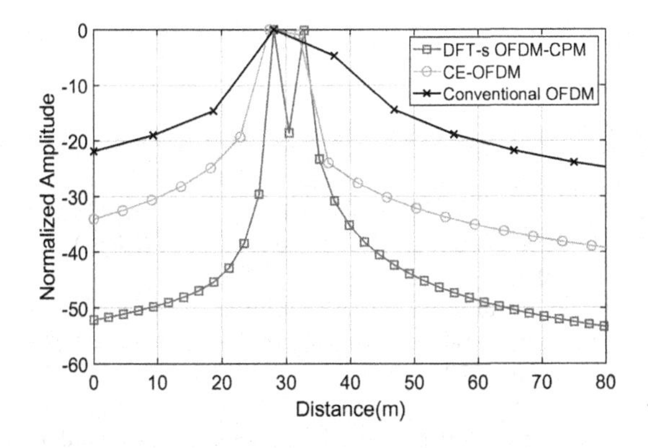

Fig. 6. Radar range profile.

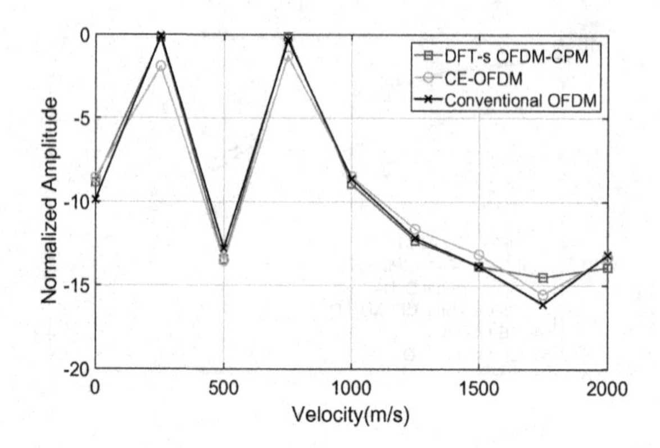

Fig. 7. Radar velocity profile.

4 Conclusion

In this letter, a DFT-s OFDM-CPM scheme for an joint radar-communication system is studied to solve the high PAPR problem. The detailed communication and radar processing are provided. The proposed system exhibits competitive communication and radar performance. We believe that our design may contribute to improving radar and communication integration performance. Future work will entail refining our model by considering different kinds of channel and enhancing the robustness to the frequency shift.

Acknowledgements. This work was supported by Research on Performance Evaluation and Optimization Technology of Local IOT for Client-side Metering Equipment under grant No. 5700-202118203A-0-0-00.

References

1. Chiriyath, A.R., Paul, B., Bliss, D.W.: Radar-communications convergence: coexistence, cooperation, and co-design. IEEE Trans. Cogn. Commun. Netw. **3**(1), 1–12 (2017)
2. Sturm, C., Wiesbeck, W.: Waveform design and signal processing aspects for fusion of wireless communications and radar sensing. Proc. IEEE **99**(7), 1236–1259 (2011)
3. Li, Q., Dai, K., Zhang, Y., Zhang, H.: Integrated waveform for a joint radar-communication system with high-speed transmission. IEEE Wirel. Commun. Lett. **8**(4), 1208–1211 (2019)
4. Zhang, Q., Zhou, Y., Zhang, L., Gu, Y., Zhang, J.: Waveform design for a dual-function radar-communication system based on CE-OFDM-PM signal. IET Radar Sonar Navig. **13**(4), 566–572 (2018)
5. Liu, Y., Liao, G., Yang, Z., Xu, J.: Multiobjective optimal waveform design for OFDM integrated radar and communication systems. Signal Process. **141**, 331–342 (2017)
6. Ellinger, J., Zhang, Z., Wu, Z., Wicks, M.C.: Dual-use multicarrier waveform for radar detection and communication. IEEE Trans. Aerosp. Electron. Syst. **54**(3), 1265–1278 (2017)

7. Shi, C., Wang, F., Sellathurai, M., Zhou, J., Salous, S.: Power minimization-based robust OFDM radar waveform design for radar and communication systems in coexistence. IEEE Trans. Signal Process. **66**(5), 1316–1330 (2017)
8. Liu, Y., Liao, G., Xu, J., Yang, Z., Zhang, Y.: Adaptive OFDM integrated radar and communications waveform design based on information theory. IEEE Commun. Lett. **21**(10), 2174–2177 (2017)
9. Wylie-Green, M.P., Perrins, E., Svensson, T.: Introduction to CPM-SC-FDMA: a novel multiple-access power-efficient transmission scheme. IEEE Trans. Commun. **59**(7), 1904–1915 (2011)
10. Huang, Y., et al.: Constant envelope OFDM RadCom fusion system. EURASIP J. Wirel. Commun. Netw. **2018**(1), 104 (2018)
11. Anderson, J.B., Aulin, T., Sundberg, C.E.: Digital Phase Modulation. Springer, Heidelberg (2013)
12. Sorger, U., De Broeck, I., Schnell, M.: Interleaved FDMA-a new spread-spectrum multiple-access scheme. In: 1998 IEEE International Conference on Communications, Conference Record, Affiliated with SUPERCOMM 1998, ICC 1998 (Cat. No. 98CH36220), vol. 2, pp. 1013–1017, June 1998
13. Myung, H.G., Goodman, D.J.: Single Carrier FDMA: A New Air Interface for Long Term Evolution, vol. 8. Wiley, Hoboken (2008)
14. Doğan, H., Odabaşıoğlu, N., Karakaya, B.: Time and frequency synchronization with channel estimation for SC-FDMA systems over time-varying channels. Wirel. Pers. Commun. **96**(1), 163–181 (2017)

[illegible] Bahrami M. Zhang [illegible] for minimizing [illegible] OFDM [illegible] communication systems [illegible] IEEE Trans. Signal Process. [illegible]
[illegible] Liao [illegible] Yang [illegible] Zhang [illegible] OTFS [illegible] communications [illegible] based [illegible] theory. IEEE Commun. [illegible] (10) [illegible] (2018)
[illegible] M.P.R. [illegible] CPM-SC-FDE [illegible] efficient [illegible] IEEE Trans. Commun. [illegible]
[illegible] OFDM [illegible] IEEE Trans. Commun. [illegible] 2018 [illegible] (2018)
[illegible] Pulse [illegible]

Access Technology

Deep Leaning Aided NOMA Combining Different NOMA Schemes

Qiuyi Sui, Shaochuan Wu(✉), and Haoran Zhang

Harbin Institute of Technology, Harbin, China
scwu@hit.edu.cn

Abstract. Non-orthogonal multiple access (NOMA) is a promising technique for future wireless communication. Compared with orthogonal multiple access (OMA), it provides high spectral efficiency and the ability to support a large number of users. A novel DNN-NOMA scheme is proposed in this paper. Both the encoder and decoder of it are composed of deep neural networks (DNN). This DNN-NOMA scheme combines power-domain NOMA, sparse code multiple access (SCMA) and multi-user shared access (MUSA), which means all of these NOMA schemes can be regarded as a special case of it. DNN-based encoders and decoders are able to generate appropriate codebooks and decode received signals automatically. The orthogonal resources (OR) used by each user are automatically determined in the process of optimizing network parameters. The simulation results prove that the BER of this novel DNN-NOMA scheme is lower than other NOMA schemes. Moreover, because there is no need to design the codebook manually, it can easily adapt to different numbers of users and ORs.

Keywords: NOMA · Deep learning · SCMA

1 Introduction

As one of the most important and key technologies, multiple access has always been a focus of mobile communication. With the number of users increasing rapidly, non-orthogonal multiple access NOMA [1] becomes an inevitable choice. Although it has higher computational complexity compared with OMA, it improves the spectrum efficiency greatly [2].

SCMA evolved from low density signature (LDS) [3]. This scheme uses codebooks to complete spreading and constellation mapping. Signals from multiple users are overlaid on the same time-frequency resources sparsely. At the receiver, message passing algorithm (MPA) is the most commonly used decoding algorithm. Codebook design is a key problem and has a strong influence on the performance of SCMA. In [4], a basic codebook design method using lattice rotation technique is proposed. According to the lattice theory, SCMA receiver achieves better performance in higher dimensions [5]. Thus, in [5] and [6], new SCMA codebook design methods which could construct large size codebooks have been proposed.

Power domain NOMA distinguishes data of different users by allocating different transmitting power. Compared to OMA, the spectral efficiency of power domain

Q. Guo et al. (Eds.): WiSATS 2021, LNICST 410, pp. 793–801, 2022.
https://doi.org/10.1007/978-3-030-93398-2_69

NOMA can be improved by 30% [7]. In [8], the optimal power allocation under different conditions and constraints has been studied.

MUSA uses complex sequences as spreading sequences [9]. Each user chooses a spreading sequence and carry out spreading. The choice of spreading sequences and the algorithm used at the receiver have great impact on the performance of MUSA scheme [10]. In [11], a multi-stage partial parallel interference cancellation detection algorithm which does not require repeated ordering and repeated matrix inversion has been proposed. It can reduce the computational complexity with a good error rate when the overload rate is low. In [12], a blind multi-user detection without reference signal has been proposed.

In recent years, deep learning has developed rapidly, and its application in the physical layer has become more and more widespread. In [13], a belief propagation algorithm based on deep learning aiming to decode high-density parity check (HDPC) code was proposed. In [14], a DNN based encoder and decoder was used in SCMA. This encoder is able to generate codebooks automatically.

The key to all NOMA schemes is serving multiple users in the same time frequency resource. Thus, in this paper, a novel DNN-NOMA model is proposed, which is a generalized NOMA model. Multiple NOMA schemes (power-domain NOMA, SCMA, MUSA) can be regarded as special cases of it. They can be implemented by adjusting the parameters in the DNN or simply changing the network structure. The simulation shows that the proposed DNN-NOMA model performs better than other NOMA schemes.

The rest of this paper is organized as follows. The system model of DNN-NOMA is described in the second section. In the third section, we discussed the structure and training method of it in detail. The simulation results are shown in the fourth section. Finally, the conclusion is given in the fifth section.

2 System Model and DNN-NOMA Structure

2.1 System Model

Assume that the signals of J users is transmitted to the receiver through K ORs, e.g. OFDMA tones. In NOMA schemes, set $K < J$ so that non-orthogonal resource allocation is inevitable. The user overload rate (OR) is defined as $\mathrm{OR} = \frac{K}{J}$. More than one user's data is transmitted simultaneously on each resource. The received signal can be expressed as below.

$$\mathbf{y} = \sum_{j=1}^{J} \sqrt{p_j}\, diag(\mathbf{h}_j)\mathbf{x}_j + \mathbf{n} \tag{1}$$

where $\mathbf{x}_j \in \mathbb{C}^{K \times 1}$ indicates the normalized signal of the jth user, p_j is the transmit power of the jth user, $\mathbf{h}_j = (h_{1j}, h_{2j}, \ldots, h_{Kj})$ is the channel vector between the jth user and the receiver, and $\mathbf{n}$ is the additive white Gaussian noise (AWGN). The total power of the transmit signals for all users is $P = \sum_{j=1}^{J} p_j$ and the number of multiplexed layers is J.

In power-domain NOMA, the vector $\mathbf{x}_j$ usually contains only one nonzero element, which means each user only transmits signals on one resource. In SCMA, $\mathbf{x}_j$ is a sparse vector containing zero elements. $x_{kj} = 0$ means the *j*th user does not transmit signals on the *k*th resource. In MUSA, each user transmits signals on all resources, so there is no zero element in $\mathbf{x}_j$. Moreover, p_j is transmit power allocated to the *j*th user. p_j is 1 in SCMA and MUSA, because these two schemes are usually implemented without power allocation.

2.2 DNN-NOMA Model

We built DNN encoder and decoder at the transmitter and receiver respectively. The input of the encoder is the data from users, which is binary. The encoder consists of a few DNN units and each DNN unit serves one user. The output of DNN encoder is the modulated signals which can be used to transmit on the channel. Then, the signals reach the receiver through the channel and binary data reconstructed in the decoder.

$f(\cdot)$ and $g(\cdot)$ are used to denote the encoder and decoder respectively. Suppose that $\mathbf{r}$ represents users' data and $\hat{\mathbf{r}}$ represents the reconstructed data after encoding and decoding. Then, the relationship between $\mathbf{r}$ and $\hat{\mathbf{r}}$ can be expressed as $\hat{\mathbf{r}} = g(f(\mathbf{r}))$. By minimizing the Euclidean distance between $\mathbf{r}$ and $\hat{\mathbf{r}}$, the network parameters can be optimized.

3 Proposed Scheme

In this part, the DNN-NOMA network structure, the physical meaning of each part in DNN and training procedure is explained in detail. The structure of DNN-NOMA is shown in Fig. 1. Data from users are encoded in the encoder and transmit on the orthogonal resources, and the decoder reconstruct them according to the received signals.

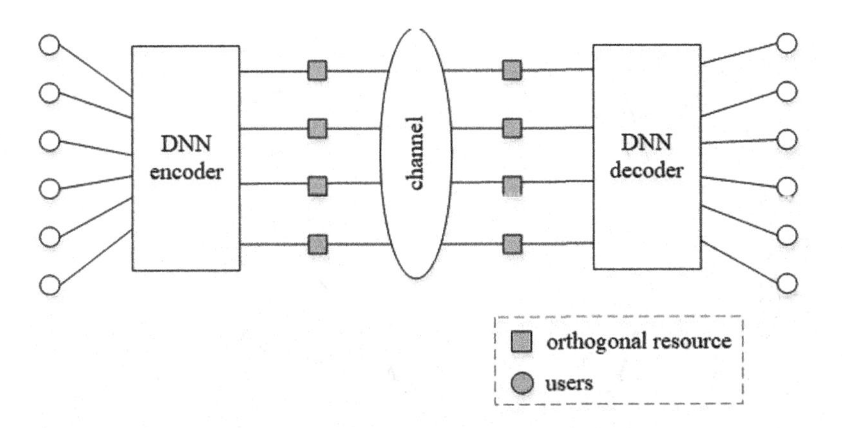

Fig. 1. DNN-NOMA structure.

3.1 DNN Encoder

The structure of DNN encoder is shown in Fig. 2. The encoder consists of multiple DNN units and each DNN unit is connected to a user. User data is encoded by the DNN unit connected to it. Hence, the number of DNN units equals to the number of users. Unlike [14], each DNN unit is connected to all ORs in this scheme, so the users can choose the ORs they use and find the best scheme during training.

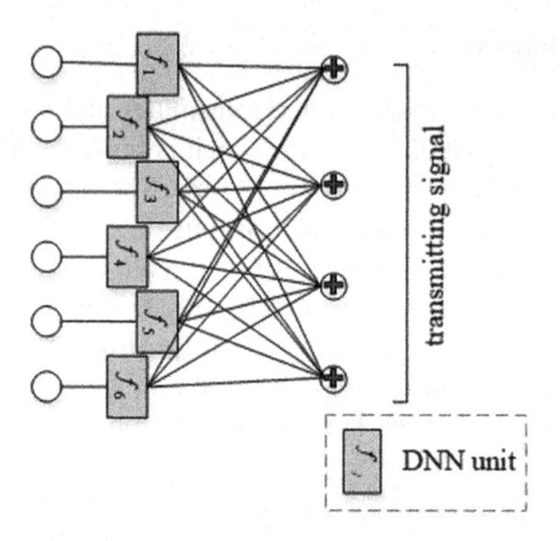

Fig. 2. The structure of DNN encoder.

The structure of each DNN unit is depicted in Fig. 3. Each DNN unit is composed of many fully connected layers. There are M input nodes because the encoder is M-ary. The number of hidden layers is 4 and each hidden layers consists of 32 nodes. Since DNN cannot calculate complex numbers, we split the real and imaginary parts of the signal. Thus, there are $2K$ output nodes in the output layer. Each part corresponds to one node in the output layer of the unit.

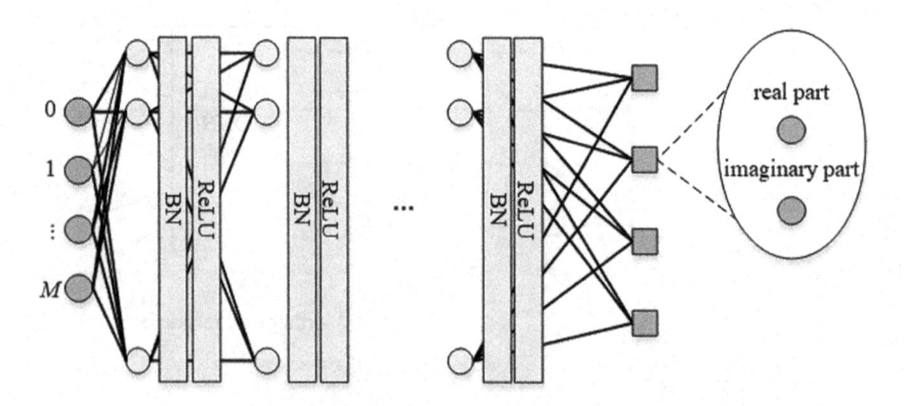

Fig. 3. The structure of DNN unit.

Then, the received signal can be written as below:

$$\mathbf{y}_{\mathrm{DNN}} = \sum_{j=1}^{J} \mathbf{h}_j f_j(r_j) + \mathbf{n}_j \tag{2}$$

3.2 DNN Decoder

In some NOMA schemes, each user may transmit signals on more than one OR, and the signals of different users are multiplexed over orthogonal resources. Hence, when decoding the signal of a certain user, the signals of the users which use the same ORs will affect the decoding result.

In order to combine the information scattered on all resources, we use a fully connected (Fully Connected, FC) network as decoder. The structure is shown in Fig. 4. The input layer is connected to all ORs. The input of the decoder is the received signals and the output represents the decoding results. The result of each user is represented by the value of M output nodes. There are 4 hidden layers and each layers has 64 neural nodes in the decoder.

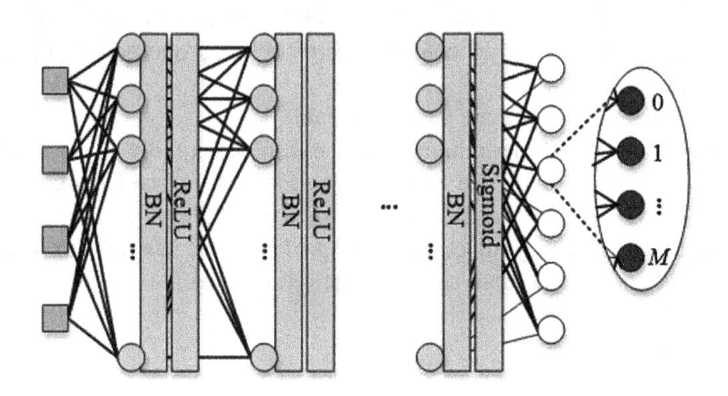

Fig. 4. The structure of decoder.

3.3 Training

Obviously, it is difficult to express the principle of optimal codebook design, so we train the network by minimizing the BER. The encoder and decoder are regarded as one network when training. We use randomly generated binary data stream as training set and test set. The training process of the network can be regarded as an optimization process according to the least square.

$$\min_{\lambda} \|\mathbf{r} - g(\mathbf{y}_{\mathrm{DNN}})\| = \min_{\lambda} \left\| \mathbf{r} - \sum_{j=1}^{J} \mathbf{h}_j f_j(r_j) + \mathbf{n}_j \right\|_2 \tag{3}$$

where λ represents the parameters of encoder and decoder, $\mathbf{r}$ is the original data, $g(\mathbf{y}_{\mathrm{DNN}})$ is the reconstructed data, and $\|\cdot\|_2$ represents the Euclidean distance.

The loss function can be defined as

$$L(\mathbf{r}, \hat{\mathbf{r}}; \lambda_f, \lambda_g) = L(\mathbf{r}, g(\mathbf{h}f(\mathbf{r}; \lambda_f) + \mathbf{n}; \lambda_g) = \left\|\mathbf{r} - g(\mathbf{h}f(\mathbf{r}; \lambda_f) + \mathbf{n}; \lambda_g)\right\|_2 \tag{4}$$

where λ_f and λ_g represent the parameters in DNN encoder and DNN decoder respectively.

According to the loss function, we use mini-batch gradient descent (MBGD) algorithm to update the parameters. The update process can be expressed as

$$(\lambda_f, \lambda_g)^+ := (\lambda_f, \lambda_g) - \alpha \nabla L(\mathbf{r}, \hat{\mathbf{r}}; \lambda_f, \lambda_g) \tag{5}$$

where α is the learning rate, determining the step size of updating in each iteration. Here we set $\alpha = 0.0001$.

4 Simulation Results

The simulation results are shown in this section. Among the three non-orthogonal multiple access technologies (power-domain NOMA, SCMA, MUSA), SCMA has the best BER performance. Thus, we compare the results of proposed DNN-NOMA with conventional SCMA.

The simulation parameters are shown in Table 1.

It is noted that SNR of the training data needs to be carefully selected. On the one hand, if the SNR is too high, the anti-noise performance of codebooks designed by DNN-encoder will be bad. On the other hand, if the SNR is too low, noise in the received signal will seriously affect the training of decoder. We refer to literature [14] and choose SNR of training data after exhaustive searches.

Table 1. The simulation parameters for proposed DNN-NOMA scheme

Simulation parameters	Parameters configuration
Training algorithm	MBGD
Modulation	QAM
Loss function	mean-square error (MSE)
Learning rate	0.0001
Number of training samples	1000000
SNR of training data	4 dB, 6 dB, 8 dB
Channel	AWGN
Number of orthogonal resources	4, 8, 6
Number of users	6, 12, 24
Overloading rate	1.5

The results of proposed DNN-NOMA and conventional SCMA are shown in Fig. 5. The number of users and number of orthogonal resources are 6 and 4 in the simulation respectively. According to Fig. 5, the proposed DNN-NOMA scheme significantly outperforms the conventional SCMA.

The main reasons for the better performance of DNN-NOMA are summarized as follows.

1) Unlike SCMA, in DNN-NOMA scheme, the orthogonal resources used by each user is not fixed. They are automatically determined in the process of training.
2) In SCMA, the performance heavily depends on the codebook. In MUSA, it depends on the spreading code. In the proposed DNN-NOMA scheme, if the number of hidden layers and nodes of the neural network is sufficient, the best codebook and transmitted signal can be obtained theoretically.

The results of DNN-NOMA with different numbers of users and ORs are shown in Fig. 6. According to the Fig. 6, the performance of DNN-NOMA improves with the increase of the number of users and codebook dimensions. This phenomenon is similar to SCMA [5, 6].

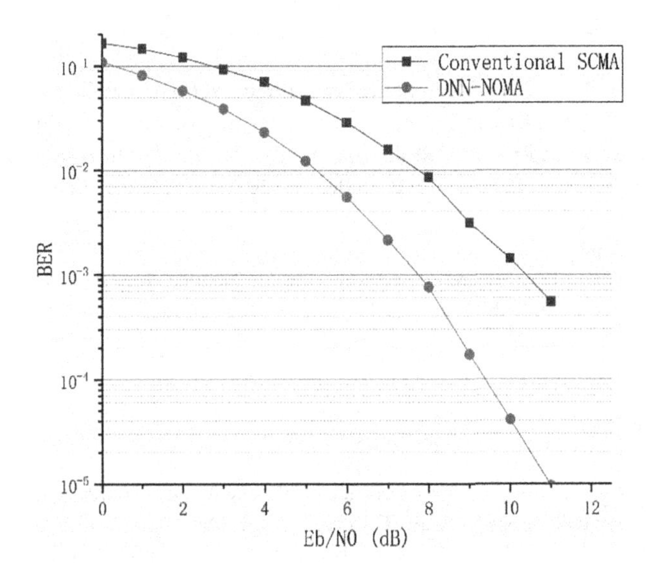

Fig. 5. BER of conventional SCMA and DNN-NOMA.

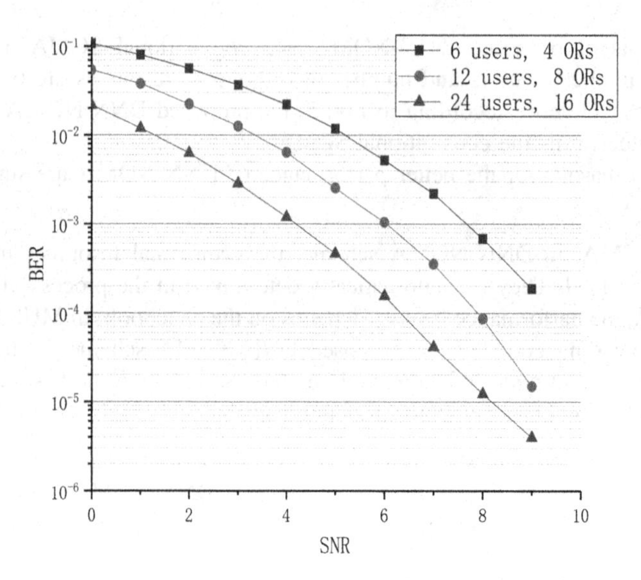

Fig. 6. BER of DNN-NOMA with different number of users.

5 Conclusion

We proposed a DNN-NOMA scheme and three NOMA schemes (power-domain NOMA, SCMA, MUSA) can be regarded as special cases of it. In this scheme, user data can be automatically encoded and decoded. The simulation results show that the DNN-NOMA scheme can achieve better BER performance than other NOMA schemes.

Acknowledgements. This research is supported by the National Key R&D Program of China (Under Grant: 2018YFC0806803) and the National Science Foundation of China (Under Grant: 61671173).

References

1. Wang, P., Xiao, J., Ping, L.: Comparison of orthogonal and non-orthogonal approaches to future wireless cellular systems. IEEE Veh. Technol. Mag. **1**(3), 4–11 (2006)
2. Islam, S.M.R., Avazov, N., Dobre, O.A., Kwak, K.: Power-domain non-orthogonal multiple access (NOMA) in 5G systems: potentials and challenges. IEEE Commun. Surv. Tutor. **19** (2), 721–742 (2017)
3. Nikopour, H., Baligh, H.: Sparse code multiple access. In: International Symposium on Personal Indoor and Mobile Radio Communications, pp. 332–336 (2013)
4. Taherzadeh, M., Nikopour, H., Bayesteh, A., Baligh, H.: SCMA codebook design. In: 80th Vehicular Technology Conference (VTC2014-Fall), pp. 1–5 (2014)
5. Li, L., Ma, Z., Fan, P.Z., Hanzo, L.: High-dimensional codebook design for the SCMA down link. IEEE Trans. Veh. Technol. **67**(10), 10118–10122 (2018)

6. Fontana da Silva, B., Silva, D., Uchôa-Filho, B.F., Le Ruyet, D.: A multistage method for SCMA codebook design based on MDS codes. IEEE Wirel. Commun. Lett. **8**(6), 1524–1527 (2019)
7. Benjebbour, A., Saito, Y., Kishiyama, Y., Li, A., Harada, A., Nakamura, T.: Concept and practical considerations of non-orthogonal multiple access (NOMA) for future radio access. In: International Symposium on Intelligent Signal Processing and Communication Systems, pp. 770–774 (2013)
8. Zhu, J., Wang, J., Huang, Y., He, S., You, X., Yang, L.: On optimal power allocation for downlink non-orthogonal multiple access systems. IEEE J. Sel. Areas Commun. **35**(12), 2744–2757 (2017)
9. Yuan, Z., Yu, G., Li, W., Yuan, Y., Wang, X., Xu, J.: Multi-user shared access for internet of things. In: 83rd Vehicular Technology Conference (VTC Spring), pp. 1–5 (2016)
10. Wang, B., Wang, K., Lu, Z., Xie, T., Quan, J.: Comparison study of non-orthogonal multiple access schemes for 5G. In: IEEE International Symposium on Broadband Multimedia Systems and Broadcasting, pp. 1–5 (2015)
11. Liang, Y., Wu, H., Wang, G.: Multi-stage partial parallel interference cancellation algorithm for MUSA system. Telecommun. Comput. Electron. Control **14**(4), 1390–1396 (2016)
12. Yuan, Z., Hu, Y., Li, W., Dai, J.: Blind multi-user detection for autonomous grant-free high-overloading multiple-access without reference signal. In: 87th Vehicular Technology Conference (VTC Spring), pp. 1–7 (2018)
13. Nachmani, E., Be'ery, Y., Burshtein, D.: Learning to decode linear codes using deep learning. In: 54th Annual Allerton Conference on Communication, Control, and Computing (Allerton), pp. 341–346 (2016)
14. Kim, M., Kim, N.I., Lee, W., Cho, D.H.: Deep learning-aided SCMA. IEEE Commun. Lett. **22**(4), 720–723 (2018)

A Novel Combining Method for NOMA and OMA in Cell-Free Massive MIMO System

Qiuyi Sui, Shaochuan Wu(✉), and Haoran Zhang

Harbin Institute of Technology, Harbin, China
scwu@hit.edu.cn

Abstract. Cell-Free massive MIMO (cf-mMIMO) is an incarnation of distributed massive MIMO, which provides high spectral efficieny (SE). However, when the number of users increases, the SE of cf-mMIMO will decrease. In order to increase the SE when the number of users is large, many researchers investigated cf-mMIMO with non-orthogonal multiple access (NOMA). In the existing research, cf-mMIMO with NOMA requires the number of users in each cluster to be equal, which is inflexible. This paper investigates the multiple access technology for cf-mMIMO. A more flexible scheme which uses the orthogonal multiple access (OMA) and NOMA at the same time is proposed in this paper. This scheme does not require the number of users in each cluster to be equal and it combines the advantages of both NOMA and OMA. An achievable SE of this scheme is derived. It is proved that under certain number of users and the length of the channel's coherence time conditions, using OMA and NOMA at the same time in the cf-mMIMO is able to obtain higher SE than using OMA or NOMA alone.

Keywords: Cell free massive MIMO · NOMA · OMA · Successive interference cancellation

1 Introduction

Massive MIMO is considered to be one of the key technologies in the fifth generation (5G). However, one of the limitations of it is that when the location of the users are at the edge of the cell, it will lead to low throughput, poor reliability, and increased communication delay [1, 2]. Therefore, a distributed, network-based massive MIMO, Cell-Free massive MIMO has been proposed and raised lots of concern. In cf-mMIMO, all access points (AP) serve all users simultaneously, so the problems mentioned above can be solved and higher SE can be provided [3, 4].

However, when the number of users is large, the length of the pilots increases, which will decrease the SE of cf-mMIMO. NOMA is a solution to this problem. As one of the most important key technologies in wireless communication, multiple access technology has always been a focus of research. As the number of users growing, the performance of NOMA is getting better [5, 6].

There are many researchers investigating cf-mMIMO with NOMA. The simulation results prove that when the number of users is large, the sum rate of cf-mMIMO with

Q. Guo et al. (Eds.): WiSATS 2021, LNICST 410, pp. 802–812, 2022.
https://doi.org/10.1007/978-3-030-93398-2_70

NOMA is higher than OMA. However, when the number of users is small, the performance of NOMA is worse than that of OMA due to intra-cluster interference and error propagation of imperfect SIC [7]. The number of users determine which technology provides higher sum rate. Hence, OMA should be adopted when the number of users is small, and switch to NOMA when the number of users is large. In [8] and [9], the authors discussed how to select the appropriate switching point and the specific switching method according to the actual situation and the number of users. Conjugate beamforming (CB) and normalized conjugate beamforming (NCB) is used to achieve spatial multiplexing. That is because they do not require APs to share channel state information (CSI) [3], which is very suitable for cf-mMIMO.

Although some researchers have considered the switching technique of OMA and NOMA, this switching technique cannot achieve ideal results near the switching point. Therefore, we combine NOMA and OMA, which means combining the advantages of them. Simulation results prove that this scheme provides higher SE under certain number of users and the length of the channel's coherence time conditions.

2 System, Channel and Signal Model

2.1 System and Channel Model

We let g_{mlk} denote the channel coefficient between the kth user in the lth cluster and mth AP. The channel coefficient g_{mlk} is modelled as follows:

$$g_{mlk} = \beta_{mlk}^{1/2} h_{mlk} \quad (1)$$

where h_{mlk} represents the small-scale fading, and $\beta_{mlk}^{1/2}$ represents the large-scale fading. Because each AP and user is distributed in a large area discretely, $h_{mlk}, m = 1, 2, \ldots M, k = 1, 2, \ldots, K$ is independent and identically distributed (i.i.d.).

In cf-mMIMO, the entire data transmission process is divided into three stages: uplink channel estimation, uplink transmission and downlink transmission. In uplink channel estimation stage, the users send pilots to APs, and APs obtain CSI using received pilot signals. The users do not know the accurate channel state information, but only know the statistical information of channels. As the number of APs grows sufficiently large, the underlying channels harden, which means the statistical CSI is fixed in several coherence intervals. Therefore, it is sufficient for users to decode.

Let τ be the length of the channel's coherence time, which is equal to the product of the channel's coherence time and coherence bandwidth. In this paper, τ is defined as the number of samples for each coherence interval. The coherence time depends on the users' moving speed. τ_p is the length of pilot for each user. τ_u and τ_d are the uplink and downlink transmission time respectively. Obviously, $\tau = \tau_p + \tau_u + \tau_d$. In this paper, we assume that the uplink and downlink transmission time is equal, that is $\tau = \tau_p + 2\tau_d$.

The model of cf-mMIMO with NOMA and OMA is shown in the Fig. 1. The model of cf-mMIMO with NOMA and OMA.

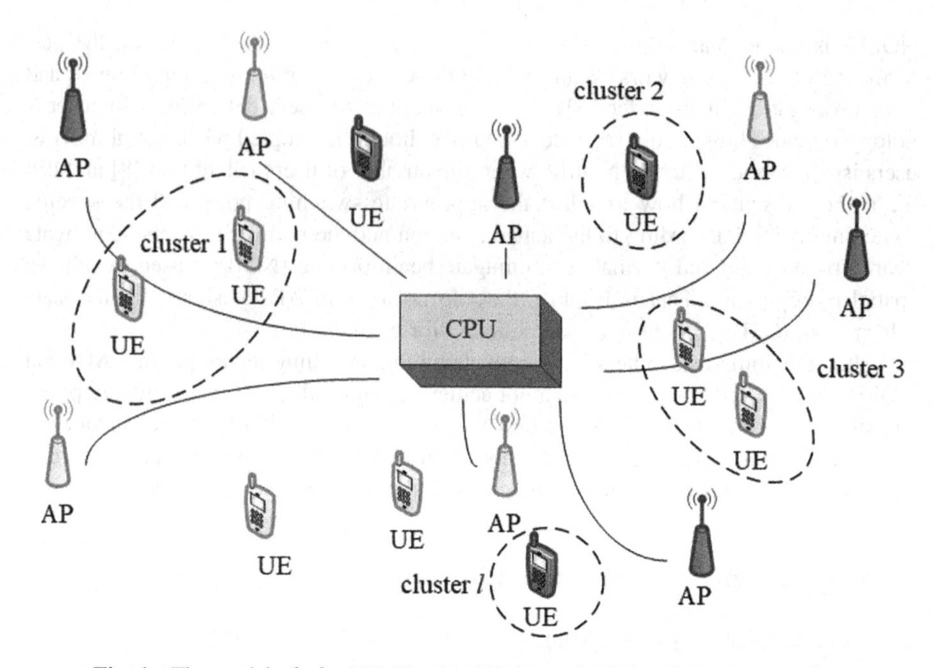

Fig. 1. The model of cf-mMIMO with NOMA and OMA. (Color figure online)

As Fig. 1 shows, all APs and user equipments (UEs) are randomly distributed. Each AP has N antennas, and each UE has 1 antenna. All APs and users are divided into 2 parts. The first part of users uses NOMA (yellow UEs). There are more than one user in each cluster, that is $k_l > 1$. All k users in one cluster use the same pilot. The set of these clusters is represented by Ω_{N}. Some APs serve them and the set of these APs is represented by Φ_{N}. The other part of users uses OMA (red UEs). There is only one user in each cluster, that is $k_l = 1$. The set of these clusters is represented by Ω_{O}. The other APs serve them and the set of these APs is represented by Φ_{O}.

Suppose that users in the system are divided into L_{total} clusters. The number of NOMA clusters is L_1, and the number of users in each cluster is K_1 ($K_1 > 1$). The number of OMA clusters is L_2, and the number of users in each cluster is K_2 ($K_2 = 1$). Then, the total number of users in the system is $K_{total} = K_1L_1 + K_2L_2$. There are MAPs. The numbers of APs serving NOMA users and OMA users are M_1 and M_2 respectively.

In the following paper, L is used to represent the number of NOMA or OMA clusters, and K is used to represent the number of users in the cluster. When $m \in \Phi_{\mathrm{N}}$ or $l \in \Omega_{\mathrm{N}}$, there are $L = L_1$ and $K = K_1$. When $m \in \Phi_{\mathrm{O}}$ or $l \in \Omega_{\mathrm{O}}$, there are $L = L_2$ and $K = K_2$.

2.2 Signal Model

In the channel estimation stage, all users send pilots to the APs. The users in one cluster use the same pilot and the users in different clusters use orthogonal pilots. The pilot sequence allocated for lth cluster is denoted as $\boldsymbol{\varphi}_l \in \mathbb{C}^{\tau_p \times 1}$, satisfying $\|\boldsymbol{\varphi}_l\|^2 = 1$.

The received signal of mth AP can be written as

$$\mathbf{y}_m = \sqrt{\tau_p \rho_p} \sum_{l=1}^{L_{total}} \sum_{k=1}^{K} g_{mlk} \boldsymbol{\varphi}_l + \mathbf{n}_m \tag{2}$$

where ρ_p is the pilot transmit power and $\mathbf{n}_m$ is additive noise at the mth AP.

Although in [3], the uplink pilot is used to estimate the channel gain of each user, estimating a linear combination of the users' channel in the same cluster provides a better performance according to [10]. Hence, we estimate the linear combination of the users' channel.

$$f_{ml} = \sum_{k=1}^{K} g_{mlk}, \forall m, \forall l \tag{3}$$

The received signal at mth AP is projected onto $\boldsymbol{\varphi}_l$ as

$$\breve{\mathbf{y}}_{ml} = \boldsymbol{\varphi}_l^H \mathbf{y}_m = \sqrt{\tau_p \rho_p} f_{ml} + \boldsymbol{\varphi}_l^H \mathbf{n_m} \tag{4}$$

The MMSE estimate of f_{ml} given $\breve{\mathbf{y}}_{ml}$ is written as

$$\hat{f}_{ml} = c_{ml} \breve{\mathbf{y}}_{ml} \tag{5}$$

where $c_{ml} = \frac{\sqrt{\tau_p \rho_p} \sum_{k'=1}^{K} \beta_{mlk'}}{\tau_p \rho_p \sum_{k'=1}^{K} \beta_{mlk'} + 1}$.

The downlink transmission uses conjugate beamforming (CB), so the transmit signal at the mth AP can be expressed as

$$x_m = \sqrt{\rho_d} \sum_{l=1}^{L} \sum_{k=1}^{K} \sqrt{\eta_{mlk}} \hat{f}_{ml}^* s_{lk} \tag{6}$$

where η_{mlk} is the power allocated to the kth UE in the lth cluster at the mth AP, s_{lk} ($\mathrm{E}\left\{|s_{lk}|^2\right\} = 1$) is the transmit symbol, and ρ_d is the downlink transmit power.

Then the normalized transmit power of the mth AP can be derived as

$$\mathrm{E}\{|x_m|^2\} = \rho_d \sum_{l=1}^{L} \sum_{k=1}^{K} \eta_{mlk} \gamma_{ml} \tag{7}$$

where $\gamma_{ml} = \mathrm{E}\left\{|f_{ml}|^2\right\} = \frac{\tau_p \rho_p \left(\sum_{k'=1}^{K} \beta_{mlk'}\right)}{\tau_p \rho_p \sum_{k'=1}^{K} \beta_{mlk'} + 1}$.

Each AP needs to satisfy the power constraint:

$$\sum_{l=1}^{L} \sum_{k=1}^{K} \eta_{mlk} \gamma_{ml} \leq \frac{1}{N}, \forall m \tag{8}$$

Suppose that in the lth cluster, user UE-$l1$ is the user with the best channel, and user UE-lK is the user with the worst channel.

The received signal of NOMA and OMA at the kth user in the lth cluster is given by (9) and (10) respectively.

$$\begin{aligned} y_{lk.\mathrm{N}} &= \sum_{m=1}^{M} g_{mlk} x_m + n_{lk} \\ &= \sqrt{\rho_d} \sum_{m \in \Phi_\mathrm{N}} g_{mlk} \sqrt{\eta_{mlk}} \hat{f}_{ml}^{*} s_{lk} + \sqrt{\rho_d} \sum_{m \in \Phi_\mathrm{N}} \sum_{k'=1, k' \neq k}^{K_1} g_{mlk} \sqrt{\eta_{mlk'}} \hat{f}_{ml}^{*} s_{lk'} \\ &+ \sqrt{\rho_d} \sum_{m \in \Phi_\mathrm{N}} \sum_{l' \neq l}^{L_1} \sum_{k'=1}^{K_1} g_{mlk} \sqrt{\eta_{ml'k'}} \hat{f}_{ml'}^{*} s_{l'k'} + \sum_{m \notin \Phi_\mathrm{N}} \sum_{l' \neq l}^{L_2} \sum_{k'=1}^{K_2} g_{mlk} \sqrt{\eta_{ml'k'}} \hat{f}_{ml'}^{*} s_{l'k'} + n_{lk} \end{aligned} \tag{9}$$

$$\begin{aligned} y_{lk.\mathrm{O}} &= \sum_{m=1}^{M} g_{mlk} x_m + n_{lk} \\ &= \sqrt{\rho_d} \sum_{m \in \Phi_\mathrm{O}} g_{mlk} \sqrt{\eta_{mlk}} \hat{f}_{ml}^{*} s_{lk} + \sqrt{\rho_d} \sum_{m \in \Phi_\mathrm{O}} \sum_{k'=1, k' \neq k}^{K_2} g_{mlk} \sqrt{\eta_{mlk'}} \hat{f}_{ml}^{*} s_{lk'} \\ &+ \sqrt{\rho_d} \sum_{m \in \Phi_\mathrm{O}} \sum_{l' \neq l}^{L_2} \sum_{k'=1}^{K_2} g_{mlk} \sqrt{\eta_{ml'k'}} \hat{f}_{ml'}^{*} s_{l'k'} + \sum_{m \notin \Phi_\mathrm{O}} \sum_{l' \neq l}^{L_1} \sum_{k'=1}^{K_1} g_{mlk} \sqrt{\eta_{ml'k'}} \hat{f}_{ml'}^{*} s_{l'k'} + n_{lk} \end{aligned} \tag{10}$$

The first term in the above expression is the desired signal, the second term is the interference caused by other users in the cluster, the third term is the interference caused by other clusters, the fourth term is the interference caused by OMA/NOMA users, and the fifth term is noise.

The users are assumed to be ordered to satisfy [11]:

$$\mathrm{E}\{\log_2(1 + \mathrm{SINR}_{lj}^{lk})\} \geq \mathrm{E}\{\log_2(1 + \mathrm{SINR}_{lk}^{lk})\}, \forall j < k, \forall l \tag{11}$$

where SINR_{lj}^{lk} refers to the effective SINR of the jth user in the lth cluster when the jth user in lth cluster is decoding the signal intended for the kth user in the same cluster.

Finally, the achievable rate of the kth user in the lth cluster can be expressed as

$$R_{lk}^{lk,final} = \min(\mathrm{E}\{log_2(1 + \mathrm{SINR}_{lj}^{lk})\}, E\{log_2(1 + \mathrm{SINR}_{lk}^{lk})\}), \forall l, k \tag{12}$$

3 Achievable Spectral Efficieny

The signal actually used for decoding after SIC for NOMA and OMA can be written as (13) and (14) respectively.

$$\begin{aligned} r_{lk,\mathrm{N}}^{lk} &= r_{lk} - \sqrt{\rho_d}\sum_{k''=k+1}^{K_1}\mathrm{E}\{\sum_{m\in\Phi_\mathrm{N}}\sqrt{\eta_{mlk''}}g_{mlk}^T\hat{f}_{ml}^*\}s_{lk''} \\ &= \sqrt{\rho_d}\sum_{m=1}^{M}\sum_{l'=1}^{L}\sum_{k'=1}^{K}\sqrt{\eta_{ml'k'}}g_{mlk}\hat{f}_{ml'}^*s_{l'k'} + n_{lk} - \sqrt{\rho_d}\sum_{k''=k+1}^{K_1}\mathrm{E}\{\sum_{m\in\Phi_\mathrm{N}}\sqrt{\eta_{mlk''}}g_{mlk}\hat{f}_{ml}^*\}s_{lk''} \\ &= \underbrace{\sqrt{\rho_d}\mathrm{E}\{\sum_{m\in\Phi_\mathrm{N}}\sqrt{\eta_{mlk}}g_{mlk}\hat{f}_{ml}^*\}s_{lk}}_{\mathrm{DS}_{lk}} + \underbrace{\sqrt{\rho_d}(\sum_{m\in\Phi_\mathrm{N}}\sqrt{\eta_{mlk}}g_{mlk}\hat{f}_{ml}^* - \mathrm{E}\{\sum_{m\in\Phi_\mathrm{N}}\sqrt{\eta_{mlk}}g_{mlk}\hat{f}_{ml}^*\})s_{lk}}_{\mathrm{BU}_{lk}} \\ &+ \underbrace{\sum_{k'\neq k}^{k-1}\sqrt{\rho_d}\sum_{m\in\Phi_\mathrm{N}}\sqrt{\eta_{mlk'}}g_{mlk}\hat{f}_{ml}^*s_{lk'}}_{\mathrm{IUI}_{lk'}} + \underbrace{\sum_{k''=k+1}^{K_1}\sqrt{\rho_d}(\sum_{m\in\Phi_\mathrm{N}}\sqrt{\eta_{mlk''}}g_{mlk}\hat{f}_{ml}^* - \mathrm{E}\{\sum_{m\in\Phi_\mathrm{N}}\sqrt{\eta_{mlk''}}g_{mlk}\hat{f}_{ml}^*\})s_{lk''}}_{\mathrm{ISIC}_{lk''}} \\ &+ \underbrace{\sum_{l'\neq l}^{L_1}\sum_{k'=1}^{K_1}\sqrt{\rho_d}\sum_{m\in\Phi_\mathrm{N}}\sqrt{\eta_{ml'k'}}g_{mlk}\hat{f}_{ml'}^*s_{l'k'}}_{\mathrm{ICI}_{lk''}} + \underbrace{\sum_{m\notin\Phi_\mathrm{N}}\sum_{l'=1}^{L_1}\sum_{k'=1}^{K_1}\sqrt{\rho_d}\sqrt{\eta_{ml'k'}}g_{mlk}\hat{f}_{ml'}^*s_{l'k'}}_{\mathrm{OUI}_{l'k'}} + n_{lk} \end{aligned} \tag{13}$$

$$\begin{aligned} r_{lk,\mathrm{O}}^{lk} &= r_{lk} - \sqrt{\rho_d}\sum_{k''=k+1}^{K_2}\mathrm{E}\{\sum_{m\in\Phi_\mathrm{O}}\sqrt{\eta_{mlk''}}g_{mlk}\hat{f}_{ml}^*\}s_{lk''} \\ &= \sqrt{\rho_d}\sum_{m=1}^{M}\sum_{l'=1}^{L}\sum_{k'=1}^{K}\sqrt{\eta_{ml'k'}}g_{mlk}\hat{f}_{ml}^*s_{l'k'} + n_{lk} - \sqrt{\rho_d}\sum_{k''=k+1}^{K_2}\mathrm{E}\{\sum_{m\in\Phi_\mathrm{O}}\sqrt{\eta_{mlk''}}g_{mlk}\hat{f}_{ml}^*\}s_{lk''} \\ &= \underbrace{\sqrt{\rho_d}\mathrm{E}\{\sum_{m\in\Phi_\mathrm{O}}\sqrt{\eta_{mlk}}g_{mlk}\hat{f}_{ml}^*\}s_{lk}}_{\mathrm{DS}_{lk}} + \underbrace{\sqrt{\rho_d}(\sum_{m\in\Phi_\mathrm{O}}\sqrt{\eta_{mlk}}g_{mlk}\hat{f}_{ml}^* - \mathrm{E}\{\sum_{m\in\Phi_\mathrm{O}}\sqrt{\eta_{mlk}}g_{mlk}\hat{f}_{ml}^*\})s_{lk}}_{\mathrm{BU}_{lk}} \\ &+ \underbrace{\sum_{k'\neq k}^{k-1}\sqrt{\rho_d}\sum_{m\in\Phi_\mathrm{O}}\sqrt{\eta_{mlk'}}g_{mlk}\hat{f}_{ml}^*s_{lk'}}_{\mathrm{IUI}_{lk'}} + \underbrace{\sum_{k''=k+1}^{K_2}\sqrt{\rho_d}(\sum_{m\in\Phi_\mathrm{O}}\sqrt{\eta_{mlk''}}g_{mlk}\hat{f}_{ml}^* - \mathrm{E}\{\sum_{m\in\Phi_\mathrm{O}}\sqrt{\eta_{mlk''}}g_{mlk}\hat{f}_{ml}^*\})s_{lk''}}_{\mathrm{ISIC}_{lk''}} \\ &+ \underbrace{\sum_{l'\neq l}^{L_2}\sum_{k'=1}^{K_2}\sqrt{\rho_d}\sum_{m\in\Phi_\mathrm{O}}\sqrt{\eta_{ml'k'}}g_{mlk}\hat{f}_{ml'}^*s_{l'k'}}_{\mathrm{ICI}_{lk''}} + \underbrace{\sum_{m\notin\Phi_\mathrm{O}}\sum_{l'=1}^{L_1}\sum_{k'=1}^{K_1}\sqrt{\rho_d}\sqrt{\eta_{ml'k'}}g_{mlk}\hat{f}_{ml'}^*s_{l'k'}}_{\mathrm{NUI}_{l'k'}} + n_{lk} \end{aligned} \tag{14}$$

In (13) and (14), DS_{lk} and BU_{lk} represent the desired signal and beamforming uncertainty for the kth UE in the lth cluster, and $\mathrm{IUI}_{lk'}$ represents the inter-user-interference imposed by the k' th UE in the lth cluster, $\mathrm{ISIC}_{lk''}$ represents the interference caused by imperfect successive interference cancellation, $\mathrm{ICI}_{l'k'}$ is the inter-cluster-interference, and $\mathrm{OUI}_{l'k'}$ and $\mathrm{NUI}_{l'k'}$ represent the OMA-user-interference and NOMA-user-interference respectively.

The SINR can be derived. The specific derivation process is the same as appendix A in [9].

$$\mathrm{SINR}_{lk,\mathrm{N}}^{lk} = \frac{N^2\left(\sum\limits_{m\in\Phi_\mathrm{N}}\sqrt{\eta_{mlk}}\frac{\gamma_{ml}\beta_{mlk}}{\sum\limits_{i=1}^{K_1}\beta_{mli}}\right)^2}{N^2\sum\limits_{k'=1}^{k-1}\left(\sum\limits_{m\in\Phi_\mathrm{N}}\sqrt{\eta_{mlk'}}\frac{\gamma_{ml}\beta_{mlk}}{\sum\limits_{i=1}^{K_1}\beta_{mli}}\right)^2 + N\sum\limits_{l'=1}^{L_1}\sum\limits_{k'=1}^{K_1}\sum\limits_{m\in\Phi_\mathrm{N}}\eta_{ml'k'}\beta_{mlk}\gamma_{ml'} + \frac{1}{\rho_d}} \tag{15}$$

$$\mathrm{SINR}_{lj,\mathrm{N}}^{lk} = \frac{N^2\left(\sum\limits_{m\in\Phi_\mathrm{N}}\sqrt{\eta_{mlk}}\frac{\gamma_{ml}\beta_{mlj}}{\sum\limits_{i=1}^{K_1}\beta_{mli}}\right)^2}{N^2\sum\limits_{k'=1}^{k-1}\left(\sum\limits_{m\in\Phi_\mathrm{N}}\sqrt{\eta_{mlk'}}\frac{\gamma_{ml}\beta_{mlj}}{\sum\limits_{i=1}^{K_1}\beta_{mli}}\right)^2 + N\sum\limits_{l'=1}^{L_1}\sum\limits_{k'=1}^{K_1}\sum\limits_{m\in\Phi_\mathrm{N}}\eta_{ml'k'}\beta_{mlj}\gamma_{ml'} + \frac{1}{\rho_d}} \tag{16}$$

$$\mathrm{SINR}_{lk,\mathrm{O}}^{lk} = \frac{N^2\left(\sum\limits_{m\in\Phi_\mathrm{O}}\sqrt{\eta_{mlk}}\frac{\gamma_{ml}\beta_{mlk}}{\sum\limits_{i=1}^{K_2}\beta_{mli}}\right)^2}{N^2\sum\limits_{k'=1}^{k-1}\left(\sum\limits_{m\in\Phi_\mathrm{O}}\sqrt{\eta_{mlk'}}\frac{\gamma_{ml}\beta_{mlk}}{\sum\limits_{i=1}^{K_2}\beta_{mli}}\right)^2 + N\sum\limits_{l'=1}^{L_2}\sum\limits_{k'=1}^{K_2}\sum\limits_{m\in\Phi_\mathrm{O}}\eta_{ml'k'}\beta_{mlk}\gamma_{ml'} + \frac{1}{\rho_d}} \tag{17}$$

$$\mathrm{SINR}_{lk,\mathrm{O}}^{lk} = \frac{N^2\left(\sum\limits_{m\in\Phi_\mathrm{O}}\sqrt{\eta_{mlk}}\frac{\gamma_{ml}\beta_{mlj}}{\sum\limits_{i=1}^{K_2}\beta_{mli}}\right)^2}{N^2\sum\limits_{k'=1}^{k-1}\left(\sum\limits_{m\in\Phi_\mathrm{O}}\sqrt{\eta_{mlk'}}\frac{\gamma_{ml}\beta_{mlj}}{\sum\limits_{i=1}^{K_2}\beta_{mli}}\right)^2 + N\sum\limits_{l'=1}^{L_2}\sum\limits_{k'=1}^{K_2}\sum\limits_{m\in\Phi_\mathrm{O}}\eta_{ml'k'}\beta_{mlj}\gamma_{ml'} + \frac{1}{\rho_d}} \tag{18}$$

According to the expressions (15) and (16), we use $\tilde{\mathbf{g}}_{lk} = [\gamma_{1l}w_{1lk}, \gamma_{2l}w_{2lk}, \ldots, \gamma_{Ml}w_{Mlk}]^T, \forall l, k$ to sort users, where $w_{mlk} = \frac{\beta_{mlk}}{\sum_{i=1}^{K} \beta_{1li}}$. After sorting, we derive $|\tilde{\mathbf{g}}_{l1}|^2 \geq |\tilde{\mathbf{g}}_{l2}|^2 \geq \ldots \geq |\tilde{\mathbf{g}}_{lK}|^2$.

The final downlink SE of single user for NOMA and OMA is derived as

$$S_{lk}^{lk,final} = \frac{1}{2}\left(1 - \frac{\tau_p}{\tau}\right)\log_2\left(1 + \mathrm{SINR}_{lk}^{lk,final}\right) \tag{19}$$

where $\mathrm{SINR}_{lk}^{lk,final} = \min\left(\mathrm{SINR}_{lj}^{lk}, \mathrm{SINR}_{lk}^{lk}\right), \forall l, k$.

4 Power Allocation

We separately allocate power to NOMA and OMA users by using max-min SINR algorithm [9].

$$\mathrm{P1}: \max_{\eta_{mlk}} \min_{k=1\ldots K, l=1\ldots L} SINR_{lk}^{lk,final}$$

$$\text{s.t.} \sum_{l=1}^{L}\sum_{k=1}^{K} \eta_{mlk}\gamma_{mlk} \leq \frac{1}{N}, \forall m, \ \eta_{mlk} \geq 0, \forall m, \forall l, \forall k \tag{20}$$

Define $\varsigma_{mlk} = \sqrt{\eta_{mlk}}$ and introduce slack variables $\upsilon_m, \tilde{\upsilon}_m, \lambda_{lk'j}, \tilde{\lambda}_{lk'k}$. We reformulate (20) as follows:

$$\mathrm{P2}: \min_{\{\varsigma_{mlk}, \lambda_{lk'j}, \upsilon_m\}} \sum_{m=1}^{M}\sum_{l'=1}^{L}\sum_{k'=1}^{K} \gamma_{ml'}\varsigma_{ml'k'}^2$$

$$\text{s.t.} \left\|\mathbf{z}_{lj}\right\| \leq \frac{N\sum_{m=1}^{M} \varsigma_{mlk} \frac{\gamma_{ml}\beta_{mlj}}{\sum_{i=1}^{K}\beta_{mli}}}{\sqrt{t}}, \forall j < k, \ \left\|\tilde{\mathbf{z}}_{lk}\right\| \leq \frac{N\sum_{m=1}^{M} \varsigma_{mlk} \frac{\gamma_{ml}\beta_{mlk}}{\sum_{i=1}^{K}\beta_{mli}}}{\sqrt{t}}$$

$$\sum_{l' \neq l}^{L}\sum_{k'=1}^{K} \gamma_{ml}\varsigma_{ml'k'}^2 + \sum_{k'=1}^{K} \gamma_{ml}\varsigma_{mlk'}^2 \leq \upsilon_m^2, \forall m, \forall j < k, 0 \leq \upsilon_m \leq \frac{1}{\sqrt{N}}, \forall m$$

$$\sum_{m=1}^{M} \varsigma_{mlk'} \frac{\gamma_{ml}\beta_{mlj}}{\sum_{i=1}^{K}\beta_{mli}} \leq \lambda_{lk'j}, 1 \leq k' \leq k-1, \forall j < k$$

$$\left(\sum_{l'\neq l}^{L}\sum_{k'=1}^{K}\gamma_{ml'}\varsigma_{ml'k'}^{2}+\sum_{k'=1}^{K}\gamma_{ml}\varsigma_{mlk'}^{2}\right)\leq\tilde{\upsilon}_{m}^{2},\forall m,0\leq\upsilon_{m}\leq\frac{1}{\sqrt{N}},\forall m$$

$$\sum_{m=1}^{M}\varsigma_{mlk'}\frac{\gamma_{ml}\beta_{mlk}}{\sum_{i=1}^{K}\beta_{mli}}\leq\tilde{\lambda}_{lk'k},1\leq k'\leq k-1$$

$$\varsigma_{mlk}\geq 0,\forall m,\forall l,\forall k,\sum_{l=1}^{L}\sum_{k=1}^{K}\varsigma_{mlk}^{2}\gamma_{ml}\leq\frac{1}{N},\forall m \tag{21}$$

where $\|\mathbf{z}_{lj}\|\triangleq\left[N\mathbf{v}_{lj,1}^{T}\quad\sqrt{N}\mathbf{v}_{lj,2}^{T}\quad\frac{1}{\sqrt{\rho_d}}\right]^{T}$, $\|\tilde{\mathbf{z}}_{lk}\|\triangleq\left[N\tilde{\mathbf{v}}_{lk,1}^{T}\quad\sqrt{N}\tilde{\mathbf{v}}_{lk,2}^{T}\quad\frac{1}{\sqrt{\rho_d}}\right]^{T}$, $\mathbf{v}_{lj,1}=\left[\lambda_{l1j}\quad\cdots\quad\lambda_{l(k-1)j}\right]^{T}$, $\mathbf{v}_{lj,2}=\left[\sqrt{\beta_{1lj}}v_{1}\quad\cdots\quad\sqrt{\beta_{Mlj}}v_{M}\right]^{T}$, $\tilde{\mathbf{v}}_{lk,1}=\left[\tilde{\lambda}_{l1k}\quad\cdots\quad\tilde{\lambda}_{l(k-1)k}\right]^{T}$, and $\tilde{\mathbf{v}}_{lk,2}=\left[\sqrt{\beta_{1lk}}\tilde{v}_{1}\quad\ldots\quad\sqrt{\beta_{Mlk}}\tilde{v}_{M}\right]^{T}$. Using bisection search algorithm shown in Table 1 can solve P2.

Table 1. Bisection search algorithm

Initialize $t_{\max}$, $t_{\min}$ and ε
Do while
Set $t=\frac{t_{\max}+t_{\min}}{2}$ and solve P2
If P2 is feasible, set $t_{\min}=t$
else, set $t_{\max}=t$
until $(t_{\max}-t_{\min})<\varepsilon$

5 Numerical Results

The simulation parameters are shown in Table 2.

Table 2. The simulation parameters for cf-mMIMO with OMA and NOMA

Simulation parameters	Parameters configuration
Carrier frequency	1.9 GHz
Bandwidth	20 MHz
Noise figure	9 dB
Downlink transmit power	200 mW
Pilot transmit power	100 mW
The number of antenna on each AP	10
The standard deviation of shadow fading	8 dB

There are 40 APs. 20 APs are randomly selected to serve NOMA users and the other 20 APs serve OMA users. There are 20 users in total. 10 users are randomly selected to use NOMA and the other 10 users use OMA. The result is shown in Fig. 2.

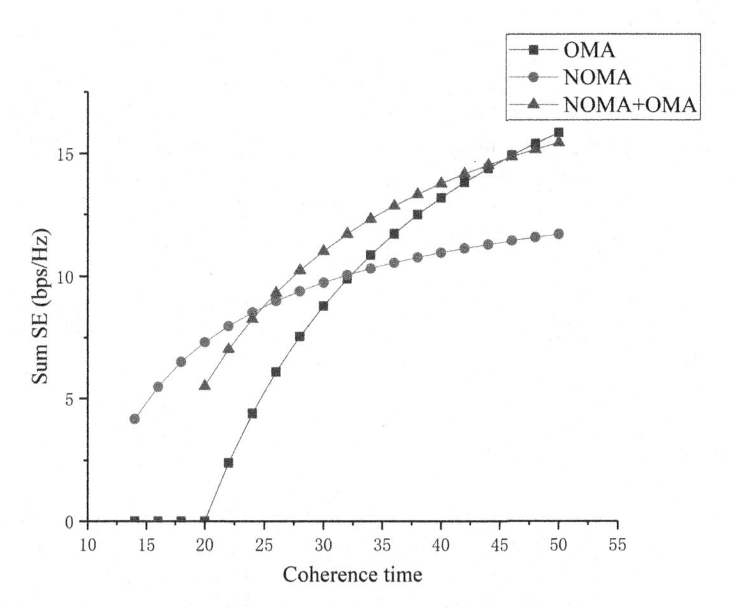

Fig. 2. Comparison of SE among NOMA, OMA and NOMA combined with OMA.

As shown in Fig. 2, using OMA and NOMA at the same time in the cf-mMIMO is able to obtain higher sum SE than using OMA or NOMA alone under certain number of users and the length of the channel's coherence time conditions.

The advantage of OMA is that there is no interference caused by the users in the same cluster and imperfect SIC. Therefore, the sum rate of OMA is higher than NOMA. However, OMA requires a large number of orthogonal pilots, which means the length of the pilot sequences is large. It takes a long time to send the pilots, resulting in a shorter time that can be used for uplink and downlink data transmission, which will seriously affect SE when the coherence time is short. The advantage of NOMA is that fewer orthogonal pilots are required, which means the length of pilot sequences can be shorter. However, there is interference caused by the users in the same cluster and imperfect SIC. When the coherence time is long, the SE of NOMA is lower than OMA. Additionally, in cf-mMIMO with NOMA scheme, the number of users in each cluster must be equal (usually 2), which is inflexible.

Using OMA and NOMA at the same time is a compromise. When the coherence time is not too long or too short, cf-mMIMO with OMA and NOMA combine the advantages of OMA and NOMA. Some users use NOMA to shorten the length of pilot sequences, and the other users use OMA to get a higher sum rate. In addition, it does not require the number of users in each cluster to be equal. However, when the coherence time is too short, the length of pilots is greater than that of NOMA, make the SE of it lower than NOMA. And when the coherence time is too long, the length of pilots has little effect, but the interference caused by NOMA users decreases the sum rate and SE. Therefore, cf-mMIMO with OMA achieves the best performance.

6 Conclusion

This paper studies the multiple access technology in cf-mMIMO, and proposes a cf-mMIMO with NOMA and OMA scheme. Unlike cf-mMIMO with NOMA, this scheme does not require the number of users in each cluster to be equal. The simulation results prove that under certain number of users and the length of the channel's coherence time conditions, using OMA and NOMA at the same time in the cf-mMIMO is able to obtain higher sum SE than using OMA or NOMA alone.

Acknowledgements. This research is supported by the National Key R&D Program of China (Under Grant: 2018YFC0806803) and the National Science Foundation of China (Under Grant: 61671173).

References

1. Xu, Y., Yue, G., Mao, S.: User grouping for massive MIMO in FDD systems: new design methods and analysis. IEEE Trans. Wirel. Commun. **14**(12), 6827–6842 (2014)
2. Bashar, M., Ngo, H.Q., Burr, A.G., Maryopi, D., Cumanan, K., Larsson, E.G.: On the performance of backhaul constrained cell-free massive MIMO with linear receivers. In: 52nd Asilomar Conference on Signals, Systems, and Computers, pp. 624–628 (2018)
3. Ngo, H.Q., Ashikhmin, A., Yang, H., Larsson, E.G., Marzetta, T.L.: Cell-free massive MIMO versus small cells. IEEE Trans. Wirel. Commun. **16**(3), 1834–1850 (2017)
4. Nguyen, L.D., Duong, T.Q., Ngo, H.Q., Tourki, K.: Energy efficiency in cell-free massive MIMO with zero-forcing precoding design. IEEE Commun. Lett. **21**(8), 1871–1874 (2017)
5. Xiang, L., Ng, D.W.K., Ge, X., Ding, Z., Wong, V.W.S., Schober, R.: Cache-aided non-orthogonal multiple access: the two-user case. IEEE J. Sel. Top. Signal Process. **13**(3), 436–451 (2019)
6. Saito, Y., Kishiyama, Y., Benjebbour, A., Nakamura, T., Li, A., Higuchi, K.: Non-orthogonal multiple access (NOMA) for cellular future radio access. In: IEEE 77th Vehicular Technology Conference (VTC Spring), pp. 1–5 (2013)
7. Li, Y., Aruma Baduge, G.A.: NOMA-aided cell-free massive MIMO systems. IEEE Commun. Lett. **7**(6), 950–953 (2018)
8. Bashar, M., Cumanan, K., Burr, A.G., Ngo, H.Q., Hanzo, L., Xiao, P.: NOMA/OMA mode selection-based cell-free massive MIMO. In: International Conference on Communications (ICC), pp. 1–6 (2019)
9. Bashar, M., Cumanan, K., Burr, A.G., Ngo, H.Q., Hanzo, L., Xiao, P.: On the performance of Cell-Free massive MIMO relying on adaptive NOMA/OMA mode-switching. IEEE Trans. Commun. **2**(68), 792–810 (2020)
10. Cheng, H.V., Björnson, E., Larsson, E.G.: Performance analysis of NOMA in training-based multiuser MIMO systems. IEEE Trans. Wireless Commun. **17**(1), 372–385 (2018)
11. Hanif, M.F., Ding, Z., Ratnarajah, T., Karagiannidis, G.K.: A minorization-maximization method for optimizing sum rate in the downlink of non-orthogonal multiple access systems. IEEE Trans. Signal Process. **64**(1), 76–88 (2016)

Accurate SER Quantization for Uplink Multi-user NOMA System

Yu Xu, Zhenyong Wang(✉), Chen Cui, and Qing Guo

School of Electronics and Information Engineering, Harbin Institute of Technology, Harbin 150001, China
{xu_yu,ZYWang,cuichen,QGuo}@hit.edu.cn

Abstract. Non-orthogonal multiple access (NOMA) is presented as a promising solution to support a massive number of communication devices in wireless networks. Due to the error propagation problem of successive interference cancellation (SIC) decoding, it is very difficult to quantify error rate performance accurately for the NOMA system. In this paper, accurate quantization expression of symbol error rate (SER) for the uplink NOMA system over additive white Gaussian noise (AWGN) channel is given. The quantified expression is effective regardless of the number of access users. Finally, we validate the provided expression and confirm the effectiveness through Monte Carlo experiments.

Keywords: Non-orthogonal multiple access (NOMA) · Multi-user · Symbol error rate (SER) · Error propagation

1 Introduction

NOMA, where multiple users are multiplexed in the power domain, is identified as a fundamental technique that can enable trillions of Machine-Type Communication (MTC) devices to communicate with the base station (BS) in cellular IoT use cases [1]. Due to its potential, NOMA is studied thoroughly in recent years [2–4]. Among these contributions, many problems are addressed from the perspective of system performance, such as system capacity and outage probability based on the Shannon formula. Channel codes currently used are designed to approach the channel capacity for long data packets. However, the transmission of small data packets associated with IoT poses a challenge. That is, due to the significant performance degradation, we cannot use the channel codes to approach the channel capacity of the NOMA system. In this regard, evaluating the error rate performance helps to develop effective algorithms, which can be utilized to improve the performance of the entire NOMA system [5,6]. Unfortunately, this has not been studied in depth so far.

Due to the error propagation problem of SIC [7], it is very intractable to analyze error rate performance for the NOMA system. Only a few works addressed this issue. Some papers quantified the error rate performance accurately [8–10],

Q. Guo et al. (Eds.): WiSATS 2021, LNICST 410, pp. 813–823, 2022.
https://doi.org/10.1007/978-3-030-93398-2_71

but the authors limited their research to two-user cases. It is worth mentioning that NOMA aims to serve as many users as possible with good performance, i.e., not suggested to serve only two users. Meanwhile, in [11–13], the error rate performance is analyzed and focused on multi-user cases. However, this research can only provide approximate results when considering the error propagation problem of SIC. Given the above situation, an accurate SER quantization expression for the uplink NOMA system over the AWGN channel is given in this paper. This expression is effective under multi-user scenarios. As current cellular IoT technologies like NB-IoT use low-rate modulations, i.e., BPSK and QPSK. We obtain the results in the BPSK case, then we extend our results to the QPSK case.

The rest of the paper is organized as follows. In Sect. 2, an uplink NOMA system with L users is presented. The SIC decoder of the NOMA system is introduced and the error propagation problem is analyzed. In Sect. 3, the SER performance of the multi-user NOMA system is quantified. The expression is verified through Monte Carlo experiments in Sect. 4. The paper concludes in Sect. 5.

2 Preliminary

2.1 System Model

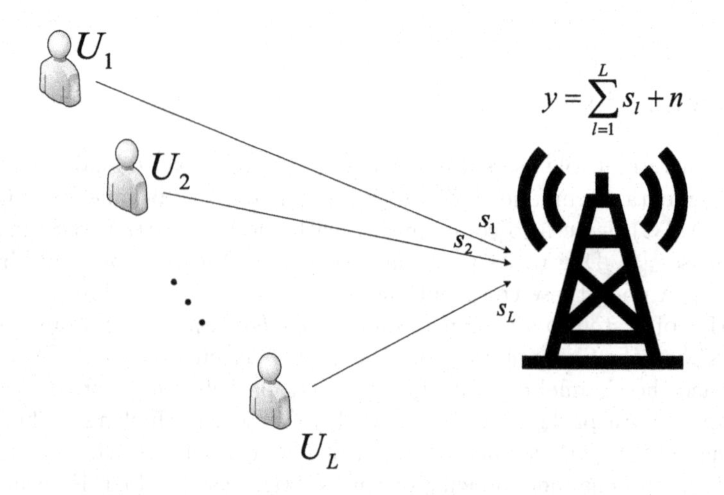

Fig. 1. Uplink NOMA systems with L users.

An uplink NOMA system with L users, denoted by $U_1, U_2, ..., U_L$, and one BS is illustrated in Fig. 1. For simplicity, we assume that both the BS and MTC devices are equipped with a single antenna. All users adopt the same modulation, that is BPSK or QPSK. Besides, the carrier is recovered perfectly and the symbol is synchronized at the BS. We suppose that different users are multiplexed in the power domain by transmitting equal maximum power. Because of their different channel conditions, their received power at the BS is different. Therefore, the superimposed received signal at the BS is given by

$$y = \sum_{l=1}^{L} s_l + n, \tag{1}$$

where s_l is the corresponding signal transmitted by U_l with received power ε_l at the BS; n is the zero mean AWGN with variance $N_0/2$. Without loss of generality, we assume that users are sorted in ascending order according to their received power, i.e. $\varepsilon_1 < \varepsilon_2 < ... < \varepsilon_L$; ε_L and ε_1 correspond to the maximum and minimum received power, respectively.

2.2 Successive Interference Cancellation Decoding

At the BS, the receiver adopts SIC to decode all users' data. The user's signal, which has the strongest power, is detected first while treating others' signals as interference. The first user's signal is detected using maximum likelihood (ML) detection according to the following rule

$$\hat{s}_L = \arg \min_{\hat{s}_L \in D_L} |y - s_L|, \tag{2}$$

where D is the set of signal constellation. If U_l is BPSK modulated, D_l can be denoted by $D_l = \{\sqrt{\varepsilon_l}, -\sqrt{\varepsilon_l}\}$. Once the receiver decodes s_L correctly, it reconstructs s_L and subtracts s_L from the received signal y. Then s_{L-1} is decoded. Repeating this operation in descending order according to their received power until all users' information has been decoded.

Since the received signal of the U_l is interfered by the rest $l-1$ users' signals, the BS should select users whose channel conditions are significantly different and multiplex them in the power domain. To fulfill this condition, here we assume that

$$\sqrt{\varepsilon_l} > \sum_{j=1}^{l-1} \sqrt{\varepsilon_j}, 1 < l \leq L. \tag{3}$$

2.3 Error Propagation Problem of SIC

We use k_j, $l < j < L$, to indicate whether the former U_j's information is correctly decoded. $k_j = 1$ denotes U_j's information is decoded correctly and $k_j = -1$ denotes decoded incorrectly. For U_l, $1 \leq l < L$, the process of demodulating is given as

$$\hat{s}_l = \arg \min_{\hat{s}_l \in D_l} \left| y - \sum_{j=l+1}^{L} k_j s_j - s_l \right|. \tag{4}$$

If the messages of previous users are not correctly decoded at the receiver, the wrong reconstructed signal is subtracted from the composite multiuser signal, thus resulting in interference to the remaining users. It leads to the accumulation of decoding errors. The error propagation problem of SIC makes the SER quantization for the NOMA system intractable.

3 SER Quantization for Multi-user NOMA System

In this section, we derive the expression of the SER performance for the NOMA system. First, we quantify SER performance focus on two-user cases. Then we extend to the multi-user cases. For simplicity, we assume that all users adopt BPSK modulated signals. The results for the QPSK modulation case are given at the end of this section.

Considering the system model of $L = 2$, the BS receives the superimposed signal transmitted by U_1 and U_2 with power ε_1 and ε_2, respectively. We can model the received signal as a Gaussian mixture model. A mixture distribution is made up of several component distributions and the corresponding probability distribution function is expressed as

$$p_2 = \sum_{s_1 \in D_1, s_2 \in D_2} \alpha\phi(x|\theta), \tag{5}$$

where p_i denotes the Gaussian mixture probability distribution consist of i users. The components $\phi(x|\theta)$ are Gaussian distribution. Each component has a separate parametrized mean $\theta = s_1 + s_2$ and uniformly covariance $N_0/2$. The parameters of a Gaussian mixture specify the prior probability α given to each component. In our case, the signal s_l transmitted by U_l, $l \in \{1, 2\}$, is chosen from the set $D_l = \{\sqrt{\varepsilon_l}, -\sqrt{\varepsilon_l}\}$ with equal probability of 1/2 at the receiver. There are four components therefore α equals 1/4. The following is our main result that SER performance of two-user uplink NOMA system:

Theorem 1. *We use P_{i-j} to indicate the SER of U_j while i users are multiplexed in the power domain. Therefore, the SER of U_2 can be expressed as*

$$P_{2-2} = \frac{1}{2}[Q(\sqrt{\frac{2\varepsilon_2}{N_0}} + \sqrt{\frac{2\varepsilon_1}{N_0}}) + Q(\sqrt{\frac{2\varepsilon_2}{N_0}} - \sqrt{\frac{2\varepsilon_1}{N_0}})], \tag{6}$$

and the SER of U_1, is expressed as

$$\begin{aligned} P_{2-1} = \frac{1}{2}[2Q(\sqrt{\frac{2\varepsilon_1}{N_0}}) - Q(\sqrt{\frac{2\varepsilon_1}{N_0}} + \sqrt{\frac{2\varepsilon_2}{N_0}}) - Q(\sqrt{\frac{2\varepsilon_1}{N_0}} - \sqrt{\frac{2\varepsilon_2}{N_0}}) \\ + Q(\sqrt{\frac{2\varepsilon_1}{N_0}} + 2\sqrt{\frac{2\varepsilon_2}{N_0}}) + Q(\sqrt{\frac{2\varepsilon_1}{N_0}} - 2\sqrt{\frac{2\varepsilon_2}{N_0}})], \end{aligned} \tag{7}$$

where Q function is defined as $Q(x) = \frac{1}{\sqrt{2\pi}}\int_x^{\infty} \mathrm{e}^{-\frac{t^2}{2}} \mathrm{d}t$.

Proof. The receiver decodes s_2 while treating s_1 as interference.

$$\begin{aligned} P_{2-2} &= 2\int_{-\infty}^{0} p_2(s_2=\sqrt{\varepsilon_2})\mathrm{d}x \\ &= \frac{1}{2}\int_{-\infty}^{0} \mathcal{N}(\sqrt{\varepsilon_2}+\sqrt{\varepsilon_1},\frac{N_0}{2})+\mathcal{N}(\sqrt{\varepsilon_2}-\sqrt{\varepsilon_1},\frac{N_0}{2})\mathrm{d}x \\ &= \frac{1}{2}[Q(\sqrt{\frac{2\varepsilon_2}{N_0}}+\sqrt{\frac{2\varepsilon_1}{N_0}})+Q(\sqrt{\frac{2\varepsilon_2}{N_0}}-\sqrt{\frac{2\varepsilon_1}{N_0}})], \end{aligned} \tag{8}$$

where $\mathcal{N}(m,\sigma)$ denotes the probability density functions of Gaussian random variables with mean m and covariance σ. The (6) is obtained.

When U_2 's signal is correctly decoded, the SER of U_1 can be expressed as

$$\begin{aligned} P_{2-1}{}' &= 2[\int_{-\infty}^{0} p_2(s_1=\sqrt{\varepsilon_1},s_2=\sqrt{\varepsilon_2})\mathrm{d}x+\int_{-\infty}^{0} p_2(s_1=\sqrt{\varepsilon_1},s_2=-\sqrt{\varepsilon_2})\mathrm{d}x] \\ &= \frac{1}{2}[\int_{-\sqrt{\varepsilon_2}}^{0} \mathcal{N}(\sqrt{\varepsilon_1},\frac{N_0}{2})\mathrm{d}x+\int_{-\infty}^{0} \mathcal{N}(\sqrt{\varepsilon_1},\frac{N_0}{2})\mathrm{d}x] \\ &= \frac{1}{2}[2Q(\sqrt{\frac{2\varepsilon_1}{N_0}})-Q(\sqrt{\frac{2\varepsilon_1}{N_0}}+\sqrt{\frac{2\varepsilon_2}{N_0}})]. \end{aligned} \tag{9}$$

The receiver adopts SIC to decode s_2, reconstructs, and subtracts s_2 from the received signal as shown in (9). It's worth noting that when $s_1=\sqrt{\varepsilon_1}, s_2=\sqrt{\varepsilon_2}$, the subtraction process of SIC affects the lower bound of the integration interval.

Similarly, when U_2's signal is incorrectly decoded, the SER of U_1 can be expressed as

$$\begin{aligned} P_{2-1}{}'' &= \frac{1}{2}[\int_{-\infty}^{0} \mathcal{N}(\sqrt{\varepsilon_1}+2\sqrt{\varepsilon_2},\frac{N_0}{2})\mathrm{d}x+\int_{-\sqrt{\varepsilon_2}}^{0} \mathcal{N}(\sqrt{\varepsilon_1}-2\sqrt{\varepsilon_2},\frac{N_0}{2})\mathrm{d}x] \\ &= \frac{1}{2}[Q(\sqrt{\frac{2\varepsilon_1}{N_0}}+2\sqrt{\frac{2\varepsilon_2}{N_0}})+Q(\sqrt{\frac{2\varepsilon_1}{N_0}}-2\sqrt{\frac{2\varepsilon_2}{N_0}})-Q(\sqrt{\frac{2\varepsilon_1}{N_0}}-\sqrt{\frac{2\varepsilon_2}{N_0}})]. \end{aligned} \tag{10}$$

The total SER of U_1 can be expressed as

$$P_{2-1}=P_{2-1}{}'+P_{2-1}{}''. \tag{11}$$

Thus, the (7) is obtained. ■

In multi-user case, we rewrite the received signal as $s_l=i_l\sqrt{\varepsilon_l}, 1\leq l\leq L$, where i_l is chosen from the set $\{1,-1\}$ with equal probability.

Theorem 2. *Considering the system with L users, the SER for U_L can be expressed as*

$$P_{L-L}=\frac{1}{2^{L-1}}\sum_{i_1,i_2,\ldots,i_{L-1}\in\{-1,1\}} Q(\sqrt{\frac{2\varepsilon_L}{N_0}}+\sum_{j=1}^{L-1} i_j\sqrt{\frac{2\varepsilon_j}{N_0}}). \tag{12}$$

For $1 < l < L$, the SER of U_l, can be calculated as

$$
\begin{aligned}
P_{L-l} = & \frac{1}{2^{L-1}} \sum_{i_1,\ldots,i_{l-1}\in\{-1,1\}} \{ \sum_{i_{l+1},\ldots,i_L\in\{-1,1\}} \sum_{k_{l+1},\ldots,k_L\in\{-1,1\}} \\
& 2^{\frac{1}{2}\sum\limits_{n=l+1}^{L}(k_n+1)} Q[\sqrt{\frac{2\varepsilon_l}{N_0}} + \sum_{j=1}^{l-1} i_j \sqrt{\frac{2\varepsilon_j}{N_0}} \\
& + \sum_{n=l+1}^{L} (1-k_n) i_n \sqrt{\frac{2\varepsilon_n}{N_0}}] - \sum_{m=l+1}^{L} \sum_{i_{l+1},\ldots,i_L\in\{-1,1\}} \sum_{k_{m+1},\ldots,k_L\in\{-1,1\}} \\
& 2^{\frac{1}{2}\sum\limits_{n=m+1}^{L}(k_n+1)} Q[\sqrt{\frac{2\varepsilon_l}{N_0}} + \sum_{j=1}^{l-1} i_j \sqrt{\frac{2\varepsilon_j}{N_0}} + \sum_{n=l+1}^{m} i_n \sqrt{\frac{2\varepsilon_n}{N_0}} \\
& + \sum_{m\neq L, n=m+1}^{L} (1-k_n) i_n \sqrt{\frac{2\varepsilon_n}{N_0}}]\}.
\end{aligned} \tag{13}
$$

Specially, the SER for U_1 can be expressed as

$$
\begin{aligned}
P_{L-1} = & \frac{1}{2^{L-1}} \{ \sum_{i_2,\ldots,i_L\in\{-1,1\}} \sum_{k_2,\ldots,k_L\in\{-1,1\}} 2^{\frac{1}{2}\sum\limits_{n=2}^{L}(k_n+1)} \\
& Q[\sqrt{\frac{2\varepsilon_1}{N_0}} + \sum_{n=2}^{L} (1-k_n) i_n \sqrt{\frac{2\varepsilon_n}{N_0}}] \\
& - \sum_{m=2}^{L} \sum_{i_2,\ldots,i_L\in\{-1,1\}} \sum_{k_{m+1},\ldots,k_L\in\{-1,1\}} 2^{\frac{1}{2}\sum\limits_{n=m+1}^{L}(k_n+1)} \\
& Q[\sqrt{\frac{2\varepsilon_1}{N_0}} + \sum_{n=2}^{m} i_n \sqrt{\frac{2\varepsilon_n}{N_0}} + \sum_{m\neq L, n=m+1}^{L} (1-k_n) i_n \sqrt{\frac{2\varepsilon_n}{N_0}}]\}.
\end{aligned} \tag{14}
$$

Proof. The derivation process of (12) is similar to (8). For $1 < l < L$, the SER can be expressed as

$$
\begin{aligned}
P_{L-l} &= 2 \int_{-\infty}^{0} p_L(s_l = \sqrt{\varepsilon_l}) \mathrm{d}x \\
&= \frac{1}{2^{L-1}} [\sum_{i_1,\ldots,i_{l-1},i_{l+1},\ldots i_L\in\{-1,1\}} \sum_{k_{l+1},\ldots,k_L\in\{-1,1\}} \\
& \int_{LB}^{0} \phi(x|s_1,\ldots,s_l=\sqrt{\varepsilon_l},\ldots,s_L) \mathrm{d}x],
\end{aligned} \tag{15}
$$

where

$$
s_j = \begin{cases} i_j\sqrt{\varepsilon_j}, \text{if } 1 \le j < l, \\ (1-k_j) i_j \sqrt{\varepsilon_j}, \text{if } l < j \le L. \end{cases} \tag{16}
$$

LB is the lower bound of the integration interval and can be calculated as follows

$$LB = \begin{cases} -\sqrt{\varepsilon_m}, \text{if } m \notin \emptyset \text{ and } m = l+1, \\ -\sqrt{\varepsilon_m} + \sum\limits_{n=l+1}^{m-1} \sqrt{\varepsilon_n}, \text{if } m \notin \emptyset \text{ and } m \neq l+1, \\ -\infty, \text{if } m \in \emptyset, \end{cases} \tag{17}$$

where m is intermediate variable and can be calculated as

$$m = \min_{i_j k_j > 0, l < j \leq L} j. \tag{18}$$

We substitute (16)–(18) into (15) and (13)–(14) are obtained. ■

When all users use QPSK modulated signals, it is in fact two binary phase-modulation signals in phase quadrature. The SER can be calculated as

$$P'_{i-j} = 1 - (1 - P_{i-j})^2, \tag{19}$$

where P_{i-j} and P'_{i-j} correspond to SER when all users adopt BPSK and QPSK modulation, respectively.

4 Numerical Results

In this section, several sets of experiments using Monte Carlo simulation are designed to verify the expression obtained beforehand. We first illustrate the limitations of the approximate expression used in [12], which we call the approximate results, through the case of a two-user NOMA system. Then we show that our quantization expression is effective under the multi-user case.

In Fig. 2 and Fig. 3, we illustrate the SER performance comparisons between our method and the approximate quantization method through a two-user NOMA system. In our experiment, we change the received power ε_1 of U_1 while keeping the received power ε_2 of U_2 unchanged. The x-axis represents the ε_1/N_0 of U_1. Both users adopt BPSK modulation and we set $\varepsilon_2/N_0 = 12\,\text{dB}$ and $\varepsilon_2/N_0 = 10\,\text{dB}$ respectively. Monte Carlo simulation results, our quantified expression and the approximate quantified expression shows consistency in Fig. 2. However, in Fig. 3, the approximate quantization method cannot fit the SER performance curve of user 1. It is only suitable for occasions where the error propagation phenomenon is not obvious, and cannot accurately quantify the performance.

Figure 4 illustrates the SER performance curve of four-user NOMA system. All users adopt BPSK modulation. It shows perfect consistency between our expression and Monte Carlo simulation. In our experiment, we change the received power ε_1 of U_1 while keeping the received power of other users unchanged, to explore the relationship between the SER performance of each user and the corresponding ε_1/N_0. We set the received power of other users as follows. $\varepsilon_4/N_0 = 23.2\,\text{dB}$, $\varepsilon_3/N_0 = 17.2\,\text{dB}$, $\varepsilon_2/N_0 = 11.1\,\text{dB}$. It can be observed

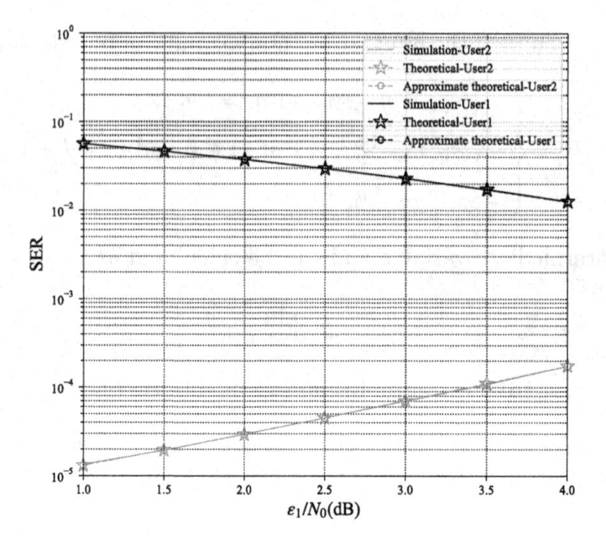

Fig. 2. SER performance comparisons of two-user NOMA system in AWGN channel ($\varepsilon_2/N_0 = 12\,\text{dB}$).

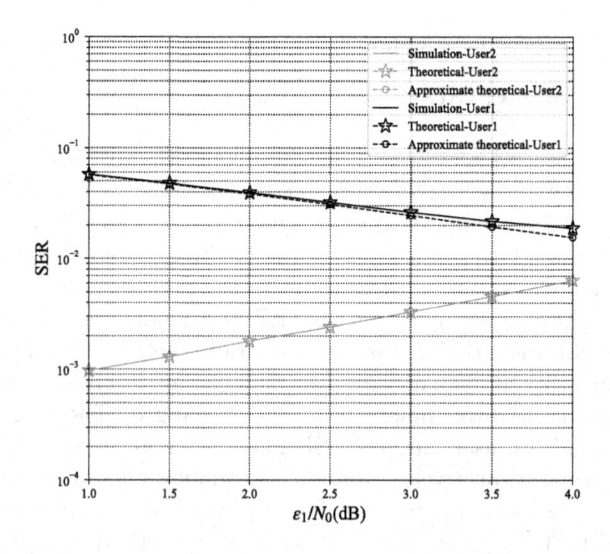

Fig. 3. SER performance comparisons of two-user NOMA system in AWGN channel ($\varepsilon_2/N_0 = 10\,\text{dB}$).

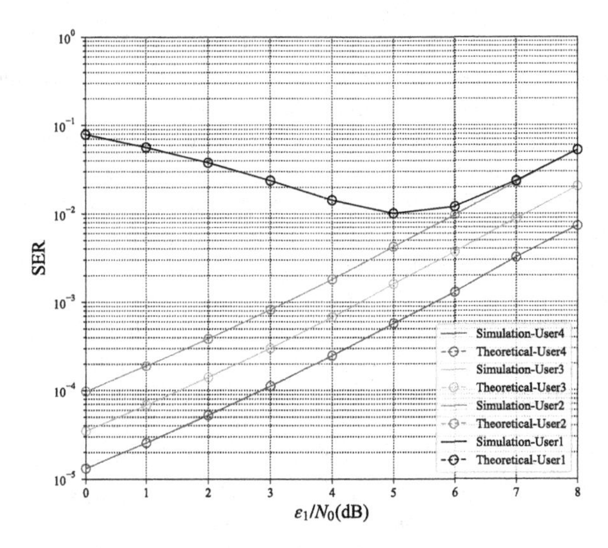

Fig. 4. SER performance of 4-user NOMA system in AWGN channel.

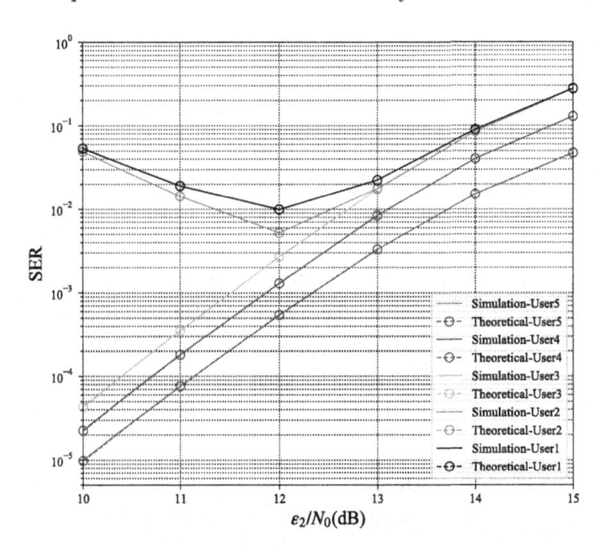

Fig. 5. SER performance of 5-user NOMA system in AWGN channel.

from the figure that the SER performance of other users decreases as ε_1/N_0 increases. This is because when other users are being decoded, U_1's signal is regarded as interference. The stronger the power of U_1, the stronger the interference, which causes performance degradation. Meanwhile, the SER of U_1 increases and then decreases as ε_1/N_0 increases. The main reason for this phenomenon is that when ε_1/N_0 is lower than 5 dB (5 dB is the approximate minimum point. The exact point can be calculated by the obtained expression), the SER performance improves as the signal power increases. When ε_1/N_0 exceeds 5 dB, the extra power causes previous users to suffer more interference during decoding, which leads to error propagation. In the decoding process of U_1, the performance degradation caused by error propagation exceeds the performance improvement brought about by the increase of its power, and the overall performance is degraded. We use the same experimental method to explore the performance of the QPSK case. In Fig. 5, the SER performance of the five-user NOMA system is tested. In this experiment, we change the received power ε_2 of U_2 while keeping the received power of other users constant. The received power settings of other users are as follows. $\varepsilon_5/N_0 = 30.0$ dB, $\varepsilon_4/N_0 = 24.0$ dB, $\varepsilon_3/N_0 = 18.0$ dB, $\varepsilon_1/N_0 = 6.0$ dB. From the figure we can draw the same conclusion as in Fig. 4. It is worth noting that when ε_2/N_0 is higher (greater than 14 dB), the SER curves of U_1, U_2 and U_3 coincide. This is because in this case, the error propagation phenomenon is severe, and the decoding failure of the previous user almost certainly causes the decoding failure of the subsequent users. As can be seen from Fig. 4 and Fig. 5, our quantizied expression can fit the performance curve accurately even when the error propagation is severe. That confirms the effectiveness.

5 Conclusion

NOMA is identified as a promising technology for massive IoT applications. This paper quantified the SER performance accurately for the uplink multi-user NOMA system over the AWGN channel. The corresponding simulation results verified the correctness of the derived expression. Based on the expression proposed in this paper, we can solve many problems effectively which cannot be solved by the traditional Shannon formula, thus improve the performance of the entire system.

References

1. Shirvanimoghaddam, M., Dohler, M., Johnson, S.J.: Massive non-orthogonal multiple access for cellular IoT: potentials and limitations. IEEE Commun. Mag. **55**(9), 55–61 (2017)
2. Shahini, A., Ansari, N.: NOMA aided narrowband IoT for machine type communications with user clustering. IEEE Internet Things J. **6**(4), 7183–7191 (2019)
3. Mostafa, A.E., Zhou, Y., Wong, V.W.S.: Connection density maximization of narrowband IoT systems with NOMA. IEEE Trans. Wirel. Commun. **18**(10), 4708–4722 (2019)
4. Shirvanimoghaddam, M., Condoluci, M., Dohler, M., Johnson, S.J.: On the fundamental limits of random non-orthogonal multiple access in cellular massive IoT. IEEE J. Sel. Areas Commun. **35**(10), 2238–2252 (2017)
5. Ng, B.K., Lam, C.: Joint power and modulation optimization in two-user non-orthogonal multiple access channels: a minimum error probability approach. IEEE Trans. Veh. Technol. **67**(11), 10693–10703 (2018)

6. Dutta, A.K.: MBER criterion assisted power NOMA design and performance analysis with estimated channel. IEEE Trans. Veh. Technol. **68**(12), 11816–11826 (2019)
7. Patel, P., Holtzman, J.: Analysis of a simple successive interference cancellation scheme in a DS/CDMA system. IEEE J. Sel. Areas Commun. **12**(5), 796–807 (1994)
8. Wang, X., Labeau, F., Mei, L.: Closed-form BER expressions of QPSK constellation for uplink non-orthogonal multiple access. IEEE Commun. Lett. **21**(10), 2242–2245 (2017)
9. He, Q., Hu, Y., Schmeink, A.: Closed-form symbol error rate expressions for non-orthogonal multiple access systems. IEEE Trans. Veh. Technol. **68**(7), 6775–6789 (2019)
10. Lee, I., Kim, J.: Average symbol error rate analysis for non-orthogonal multiple access with M-ary QAM signals in Rayleigh fading Channels. IEEE Commun. Lett. **23**(8), 1328–1331 (2019)
11. Bariah, L., Muhaidat, S., Al-Dweik, A.: Error probability analysis of non-orthogonal multiple access over Nakagami-m fading channels. IEEE Trans. Commun. **67**(2), 1586–1599 (2019)
12. Zheng, J., Zhang, Q., Qin, J.: Average block error rate of downlink NOMA short-packet communication systems in Nakagami-m fading channels. IEEE Commun. Lett. **23**(10), 1712–1716 (2019)
13. Aldababsa, M., Göztepe, C., Kurt, G.K., Kucur, O.: Bit error rate for NOMA network. IEEE Commun. Lett. **24**(6), 1188–1191 (2020)

Multi-user Shared Access Research in Cell-Free Massive MIMO

Haoran Zhang and Shaochuan Wu(✉)

Harbin Institute of Technology, Harbin, China
scwu@hit.edu.cn

Abstract. With the advent of the 5G era, massive device access and explosive data traffic grow rapidly, it is imperative to improve network throughput. As a new network architecture, Cell-Free Massive MIMO can effectively improve network throughput. MUSA (Multi-user shared access), as a kind of NOMA (non-orthogonal multiple access technology), has a great gain in spectrum efficiency. Integrating MUSA's related technologies into the Cell-Free Massive MIMO system can effectively improve the spectrum efficiency and meet the massive access requirements.

In this paper, by using MUSA, the Cell-Free Massive MIMO system is improved. A clustering method is proposed. Each cluster corresponds to a pilot. Users in the cluster use MUSA to distinguish and detect multi-users. For the new model, theoretical analysis and simulation of spectrum efficiency are carried out. The results show that the improved Cell-Free Massive MIMO system has better spectrum efficiency.

Keywords: Cell-free massive MIMO · MUSA · Spectrum efficiency

1 Introduction

The three major business scenarios of 5G include enhanced mobile bandwidth, massive machine communication, ultra-high reliability and low-latency communication. In the application of the Internet of Vehicles, the ultra-high reliability scene is the mainstream.

From the perspective of ultra-high reliability, the current cellular-based network architecture is no longer suitable for the continued evolution and development of the Internet of Vehicles. In the scene of high-speed vehicle movement, the traditional cellular architecture inevitably has the inherent defect of inter-cell handover, which causes the complicated signaling overhead and decrease of reliability. Although technologies such as cooperative communication have been proposed internationally, they can only alleviate the problem of handover between cells to a certain extent. This way of dividing the network into cells, with the massive use of high frequency bands such as millimeter waves in the future, the problem of reliable communication will become more serious. High attenuation is a significant feature of the millimeter wave frequency band. At this time, in order to effectively cover, the layout of outdoor

Q. Guo et al. (Eds.): WiSATS 2021, LNICST 410, pp. 824–832, 2022.
https://doi.org/10.1007/978-3-030-93398-2_72

wireless network base stations will be denser, and the network coverage of the base stations will also become smaller, which will cause more serious handover problems. From the above analysis, it is not difficult to see that the traditional cellular architecture has seriously affected the application of the scene, and the new Cell-Free network architecture has gradually entered the field of vision of researchers.

In [1], Cell-Free Massive MIMO is proposed as a new type of distributed network architecture, which uses a large number of distributed antenna access points to serve a small number of users distributed in a large area. All units are connected to the central processing unit through the backhaul link, so there is no concept of cell boundaries.

The current research on Cell-Free Massive MIMO mainly focuses on the signal processing part of the Cell-Free system and the performance comparison between Cell-Free Massive MIMO and Cell-Free. The signal processing in the Cell-Free system mainly includes channel estimation and uplink signal. Precoding and power allocation schemes for detection and downlink. For channel estimation, the literature [1] adopts the time division duplex (TDD) mode and uses criterion for channel estimation in the uplink and channel disparity to obtain the channel information of the downlink at the same time. This scheme avoids downlink channel estimation and saves a lot of pilot overhead. However, since the channel hardening characteristics of the Cell-Free system are actually not perfect, this solution has the problem of inaccurate downlink channel estimation. In the literature [2–4], the downlink channel is estimated. Compared with the uplink channel estimation, the spectrum efficiency is indeed improved, but it causes a lot of pilot overhead. Regardless of the uplink channel estimation or the downlink channel estimation, pilot pollution exists and causes certain estimation errors. For uplink signal detection, literature [1] uses matched filtering on each AP for signal detection, which does not require CPU processing and reduces the burden on the backhaul link. Literature [5] uses part or all of the CPU processing, and the results show that the spectrum efficiency can be significantly improved. In the literature [6], four different levels of CPU and AP cooperative processing are compared, and the results show that the higher the degree of CPU processes, the higher the spectral efficiency is. For the precoding and power allocation of the downlink, it is currently mainly divided into CB (conjugate beamforming precoding) and ZF (zero-forcing precoding). The literature [5] compares the performance of CB and ZF. ZF can reduce the interference between users, but increases the data transmission of the backhaul link. Literature [7] proposed an improved conjugate beamforming scheme, which completely eliminates the self-interference in Cell-Free and maintains the simplicity of the original conjugate beamforming.

The above documents are rarely research on how to combine Cell-Free Massive MIMO and NOMA in the downlink, and in the literature [8] the author gives the spectral efficiency of Cell-Free Massive MIMO-NOMA in the uplink closed expression. At the same time, the problem of maximizing spectrum efficiency based on the quality of service, power limit and limitation of each user is proposed, and an iterative GP (Geometric Programming) algorithm is proposed to solve this problem. To combine Cell-Free Massive MIMO and MUSA, our research is as follows.

2 Cell-Free Massive MIMO Model

As shown in Fig. 1, this article considers a cell-free network composed of K single-antenna UEs (users) and L APs (wireless access points). Each AP is equipped with an antenna and is arbitrarily distributed in covered area. These APs are connected to the edge cloud processor in any way, called the CPU. This kind of setting can perform consistent joint transmission and reception for the UE in the entire coverage area, and when L and K are relatively large, it is called Cell-Free Massive MIMO; usually assume that L > K, that is, the number of APs is more than the number of users.

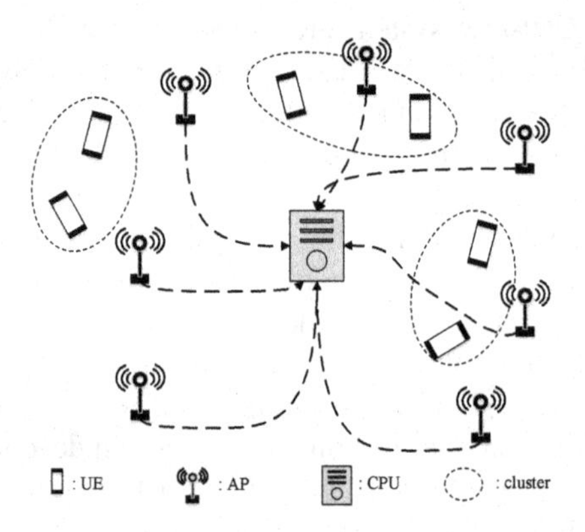

Fig. 1. Cell-free massive MIMO network

For channel estimation, it can be considered that the transmission channel between the AP and the UE is reciprocal, that is, the channel is the same in both directions. In the time division duplex mode, the same frequency resources are used for uplink and downlink transmission. The assumption of reciprocity means that the channel only needs to be characterized in one direction. The uplink channel is a better choice, because the user only needs to send one pilot signal, which is received by all APs. The complexity of channel estimation is proportional to the number of users, rather than the number of antennas in the array. This is very important, because the user may be moving, so channel estimation needs to be performed multiple times. Testing channel information based on the uplink has another great advantage, that is, all channel estimation and signal processing tasks are completed at the base station, rather than at the user end. These tasks are heavy and not suitable for the user end. In the Cell-Free Massive MIMO system, these tasks are all done in the CPU.

It is assumed that the AP and UE perform transmission according to the TDD protocol, where the TDD duplex includes an uplink transmission pilot phase and a downlink data transmission phase for channel estimation. Each coherent block is divided into channels used for uplink pilot, and used for downlink data, so that. The channel between the AP and the UE is denoted by, and the channel parameters of the UE from all APs are denoted by. In each coherent block, assume that the independent Rayleigh fading is, where is the spatial correlation coefficient (assuming the antenna is

1Tx, 1Rx). The Gaussian distribution model is used to model small-scale fading, and large-scale fading is described, including path transmission loss, shadow fading, antenna gain, and spatial channel correlation. Assuming that the channel parameters of different APs are independently distributed, for the channel parameters.

$$\mathbb{E}\left\{h_{kn}(h_{kl})^{\mathrm{H}}\right\} = 0, \quad l \neq n \tag{1}$$

2.1 Pilot Transmission and Channel Estimation

In order to use large-scale antennas more efficiently, in each coherent block, the base station needs to estimate the channel response of all user equipment. In this article, this task is performed by the CPU.

Suppose there are p mutually orthogonal pilot signals of length τ_p, where τ_p is a constant independent of K. When UEs access the network, they are allocated to UEs. When these UEs send pilots, the received signal at the AP is

$$\mathbf{y}_{tl}^{\text{pilot}} = \sum_{i \in \mathcal{S}_t} \sqrt{\tau_p p_i} \mathbf{h}_{il} + \mathbf{n}_{tl} \tag{2}$$

Using the above formula, the MMSE (Minimum Mean Square Error) of this channel is obtained as

$$\widehat{\mathbf{h}}_{kl} = \sqrt{p_k \tau_p} \mathbf{R}_{kl} \Psi_{tl}^{-1} \mathbf{y}_{tl}^{\text{pilot}} \tag{3}$$

where

$$\Psi_{tl} = \mathbb{E}\left\{\mathbf{y}_{tl}^{\text{pilot}} \left(\mathbf{y}_{tl}^{\text{pilot}}\right)^{\mathrm{H}}\right\} = \sum_{i \in \mathcal{S}_t} \tau_p p_i \mathbf{R}_{il} + \sigma^2 \mathbf{I}_N \tag{4}$$

is the correlation matrix. Similar to the situation in traditional Massive MIMO, mutual interference generated by multi-user pilot sharing UE will cause pilot pollution, thereby reducing system performance. There are two main consequences of pilot contamination. First, it will reduce the estimation quality, thereby reducing the efficiency of coherent transmission; Second, the estimated values $\widehat{\mathbf{h}}_{kl}$ are correlated with each other, which leads to additional interference. Both of these effects will affect the performance of the UE.

After using the MMSE criterion for channel estimation in the uplink, the channel reciprocity can be used to obtain the downlink channel information at the same time. This scheme avoids downlink channel estimation and can save a lot of pilot overhead.

2.2 Downlink Data Transmission

Let $\mathrm{w}_{il} \in \mathbb{C}^N$ denote the precoding allocated to the UE by the AP. During the downlink transmission, the received signal at the UE is

$$y_k^{\mathrm{dl}} = \sum_{l=1}^{L} \mathrm{h}_{kl}^{\mathrm{H}} \sum_{i=1}^{K} \mathrm{w}_{il}\zeta_i + n_k = \mathrm{h}_k^{\mathrm{H}} \sum_{i=1}^{K} \mathrm{w}_i\zeta_i + n_k \tag{5}$$

Where $\zeta_i \in \mathbb{C}$ is the unit power data signal used for the UE i. The system model is mathematically equivalent to a downlink single-cell massive MIMO system with correlated fading.

Therefore, the achievable downlink spectrum efficiency in Cell-Free Massive MIMO can easily be derived from the massive MIMO literature with related fading.

3 MUSA Improvement of Cell-Free Massive MIMO

3.1 Cell-Free Massive MIMO Cluster Model

Pilot pollution often greatly reduces the performance of Cell-Free Massive MIMO, and the cause of pilot pollution is that multiple users share the same pilot. In order to reduce pilot pollution and make full use of the limited pilot resources, this paper proposes a solution for user clustering, that is, users in a cluster share one pilot, and different clusters use different pilots. Non-orthogonal multiple access technology to distinguish users. In this way, the influence of pilot pollution is effectively avoided, and the spectrum efficiency is improved.

As shown in Fig. 1, we divide users into K clusters, tentatively there are two users in each cluster. There are three clustering schemes, namely random clustering, closest clustering and farthest clustering. In this paper, the first scheme is randomly clustered for modeling, and the other two schemes are waiting for in-depth study later.

3.2 Spectrum Efficiency Research

Because the channel hardening phenomenon is a reasonable assumption. Therefore, users use channel statistics instead of instantaneous CSI to perform serial interference cancellation (SIC), relying on the average value of the effective channel gain as an estimate of the channel gain. Suppose that in the cluster l, the user l_1 has the best received signal, so it can decode other user information, but l_k is the weakest signal. It can only decode on its own and cannot decode other user signals. In other words, MUSA is only used in each cluster, not between clusters. When users can use instantaneous CSI, users in each cluster can be sorted according to their effective channel gain. In order to successfully implement SIC at stronger users to decode weaker user signals, the following necessary conditions should be met:

$$\mathbb{E}\left\{\log_2\left(1+\mathrm{SINR}_{mj}^{mk}\right)\right\} \geq \mathbb{E}\left\{\log_2\left(1+\mathrm{SINR}_{mk}^{mk}\right)\right\}, \forall j<k, \forall m \tag{6}$$

That is, in the same cluster, user j with high SNR can decode the user k's signal.

Based on the above conditions, the achievable rate of the first user can be written as:

$$R_{lk}^{lk,\text{final}} = \min\left(\mathbb{E}\left\{\log_2\left(1 + \text{SINR}_{lj}^{lk}\right)\right\}, \mathbb{E}\left\{\log_2\left(1 + \text{SINR}_{lk}^{lk}\right)\right\}\right), \forall l, k \tag{7}$$

That is, the user with the worst channel condition in the cluster can be detected by other users. We should take the smaller rate. The reason is that when the formula is not satisfied, that is to say, when the SINR calculated by the user with poor channel gain is larger than the previous user, it is necessary to consider reducing the data transmission rate of this user to facilitate the detection on the user.

Based on the above analysis, the spectrum efficiency of the improved Cell-Free Massive MIMO system can be derived:

$$\text{SE}_{mk} = \left(1 - \frac{\tau_p}{\tau_c}\right)\log_2\left(1 + \text{SINR}_{mk}\right) \tag{8}$$

With the aid of conjugate beamforming, the achievable SINR of the user is obtained. Then we use MUSA's SIC detection scheme, in which users with higher received power first detect their signals, then we demodulate them, and subtract the corresponding signals from the composite received signal, thus leaving the lower power users without interference.

With the previous assumptions, users are sorted according to their channel quality. MUSA uses the code domain to send multiple signals on the same resource, and performs SIC at the receiver to decode the corresponding signals.

3.3 System Performance Analysis

We set up a scenario in which L APs and K UEs are independently and randomly distributed in a square area of km^2, assuming that the APs and UEs in the area have only a single antenna. Specific configuration parameters are shown in Table 1.

Table 1. Simulation parameter configuration

Simulation parameter	Configuration
Channel condition	Rayleigh
Channel estimation	MMSE
AP number	400
UE number	0–400
Antenna configuration	1Tx, 1Rx
Uplink power	1 W
Downlink power	100 mW
Resource block length	200
Precoding scheme	MR

First, compare the spectrum efficiency between the improved Cell-Free Massive MIMO system and the improved system before. It should be noted that the default clustering method in this article is that there are two users in each cluster, and random clustering is used to pair users. Figure 2 is a study of the spectrum efficiency distribution of the Cell-Free Massive MIMO system before and after the improvement in the scenario of 400 APs and 100 users. The improved system contains 50 clusters, each of which contains two users.

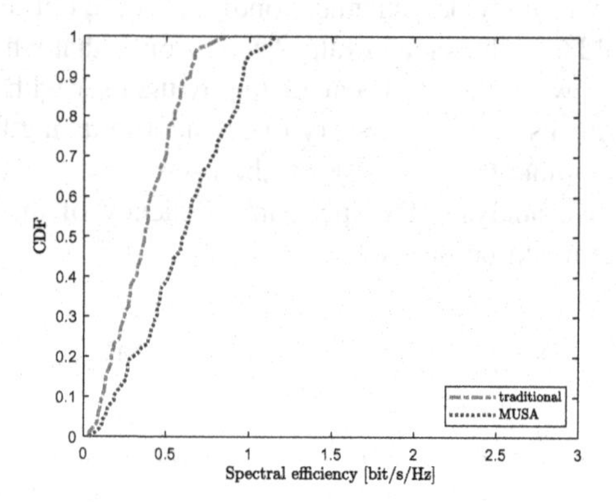

Fig. 2. CDF of spectrum efficiency

It can be seen from Fig. 2 that the improved Cell-Free Massive MIMO system has a greater improvement in spectrum efficiency compared to the previous system. The reason is that the existence of clustering improves the utilization rate of pilots, thereby increasing the resources used for downlink data transmission, thereby effectively improving the spectrum efficiency of the Cell-Free Massive MIMO system. The existence of MMSE precoding and SIC suppresses the interference between users of multiple users in the same cluster, thereby improving the overall signal-to-interference and noise ratio of the user receiving end, and also has a great effect on enhancing the spectrum efficiency.

Considering the influence of interference between users, this article also explores the user service carrying capacity of this hypothetical scenario. The specific method is: still keep the number of APs at 400, and the length of related resource blocks at 200. On this basis, the total number of users accessed is adjusted, and the number of users per cluster is still two. Count the spectrum efficiency of all users.

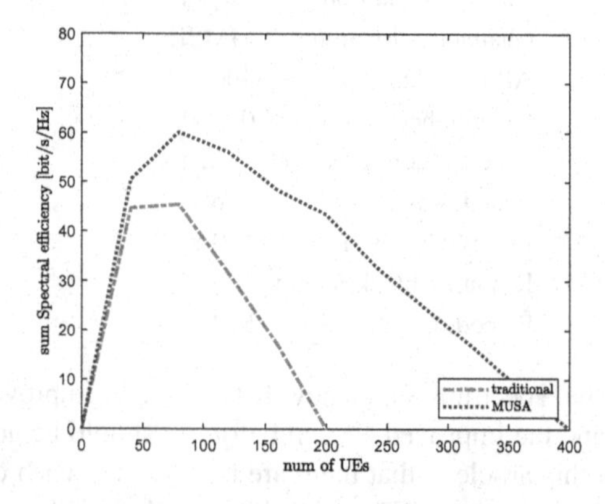

Fig. 3. Total spectrum efficiency with number of users changing

It can be seen from Fig. 3 that as the number of users access increases, the total spectrum efficiency of the Cell-Free Massive MIMO system has a tendency to increase first and then decrease to zero. This is because when the number of users is appropriate, the sum rate of the system increases as the number of users increases, and the corresponding inter-user interference is not enough to reduce the total spectrum efficiency due to the small number of users. However, when the number of connected users reaches a peak, there will be users accessing again, which will cause the spectrum efficiency of the system to decrease. This is because the spectrum efficiency is also related to the effective correlation time. An increase in the number of users will lead to an increase in the number of pilots. Thereby reducing the time of data transmission. Therefore, it can be concluded that the number of users that the Cell-Free Massive MIMO system can access has a certain limit, and blindly increasing user access will only cause the loss of system performance. In order to specifically analyze the changes in the spectrum efficiency under each user number scenario, this paper also studies the average spectrum efficiency. Keep the number of users in each cluster at two and change the total number of users connected. As shown in Fig. 4.

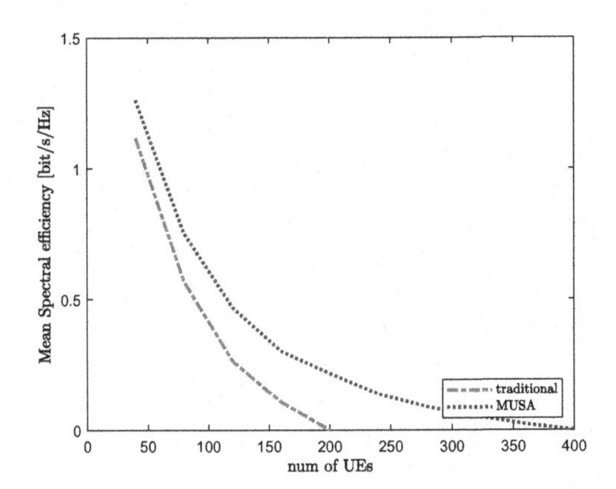

Fig. 4. Average spectrum efficiency

It can be seen from Fig. 4 that the average spectral efficiency of all users will gradually decrease as the number of users increases, until it decreases to zero. Compared with the system before the improvement, the average spectrum efficiency of the system after the improvement decreases more slowly, and the number of users reaching 0 is also more than that before the improvement. It shows that the improved system has greater user carrying capacity. Because the relevant resource blocks are limited and the number of users contained in each cluster is fixed, both of which are two. Therefore, the increase of the number of users will increase the total number of clusters in the area, thereby increasing the resources occupied by the pilot and compressing the resources for downlink data transmission. As a result, the spectrum efficiency will gradually decrease as the number of users increases. The improved Cell-Free Massive MIMO system has better user bearing capacity because the clustering algorithm improves the utilization of pilots.

4 Conclusion

In this paper, the Cell-Free Massive MIMO system is improved on the MUSA non-orthogonal multiple access technology, and a clustering method is proposed. Users in the same area are divided into clusters according to random matching. Each cluster contains two users. Users in each cluster share the same pilot and different clusters uses different pilot, thereby improving the utilization rate of the pilot. Users in the same cluster use MUSA to distinguish code domains and use MMSE-SIC to detect multi-user information. The result shows that the improved Cell-Free Massive MIMO system has better spectrum efficiency. In the future, we plan to change the number of users in each cluster, which may improve the spectrum efficiency and consider different clustering methods.

Acknowledgements. This research is supported by the National Key R&D Program of China (Under Grant: 2018YFC0806803) and the National Science Foundation of China (Under Grant: 61671173)".

References

1. Ngo, H.Q., Ashikhmin, A., Yang, H., et al.: Cell-free massive MIMO versus small cells. IEEE Trans. Wirel. Commun. **16**(3), 1834–1850 (2017)
2. Interdonato, G., Ngo, H.Q., Frenger, P., et al.: Downlink training in cell-free massive MIMO: a blessing in disguise. IEEE Trans. Wirel. Commun. **18**(11), 5153–5169 (2019)
3. Kim, S., Shim, B.: FDD-based cell-free massive MIMO systems. In: Proceedings of the 2018 IEEE 19th International Workshop on Signal Processing Advances in Wireless Communications (SPAWC), 25–28 June 2018 (2018)
4. Interdonato, G., Ngo, H.Q., Larsson, E.G., et al.: How much do downlink pilots improve cell-free massive MIMO? In: Proceedings of the 2016 IEEE Global Communications Conference (GLOBECOM), 4–8 December 2016 (2016)
5. Nayebi, E., Ashikhmin, A., Marzetta, T.L., et al.: Precoding and power optimization in cell-free massive MIMO systems. IEEE Trans. Wirel. Commun. **16**(7), 4445–4459 (2017)
6. Bjrnson, E., Sanguinetti, L.: Making cell-free massive MIMO competitive with MMSE processing and centralized implementation. IEEE Trans. Wirel. Commun. **19**, 77–90 (2019)
7. Attarifar, M., Abbasfar, A., Lozano, A.: Modified conjugate beamforming for cell-free massive MIMO. IEEE Wirel. Commun. Lett. **8**(2), 616–619 (2019)
8. Wang, B., Wang, K., Lu, Z., et al.: Comparison study of non-orthogonal multiple access schemes for 5G. In: Proceedings of the 2015 IEEE International Symposium on Broadband Multimedia Systems and Broadcasting, 17–19 June 2015 (2015)

Author Index

Printed in the United States
by Baker & Taylor Publisher Services